Handbook of
Urban Studies

Handbook of
Urban Studies

Edited by
RONAN PADDISON

SAGE Publications
London • Thousand Oaks • New Delhi

SAGE Publications Ltd
6 Bonhill Street
London EC2A 4PU

SAGE Publications Inc
2455 Teller Road
Thousand Oaks, California 91320

SAGE Publications India Pvt Ltd
32, M-Block Market
Greater Kailash - I
New Delhi 110 048

British Library Cataloguing in Publication data

A catalogue record for this book is
available from the British Library

ISBN 0 8039 7695 X

Library of Congress catalog card record available

Typeset by M Rules
Printed in Great Britain by
The Cromwell Press Ltd, Trowbridge, Wiltshire

Contents

Preface ix
Acknowledgements xii
Notes on Contributors xiii

1 Studying Cities 1
Ronan Paddison

PART I IDENTIFYING THE CITY

Introduction 11

2 Defining the City 14
William H. Frey and Zachary Zimmer

3 Urban Ecology 36
Peter Saunders

4 Impartial Maps: Reading and Writing Cities 52
Hana Wirth-Nesher

PART II THE CITY AS ENVIRONMENT

Introduction 67

5 The Physical Form of Cities: A Historico-Geographical Approach 69
J.W.R. Whitehand

6 Housing in the Twentieth Century 88
Ray Forrest and Peter Williams

7 Transport and the City 102
Tom Hart

8 Managing Sustainable Urban Environments 124
Roberto Camagni, Roberta Capello and Peter Nijkamp

PART III   THE CITY AS PEOPLE

Introduction   141

9   Urbanization, Suburbanization, Counterurbanization and
     Reurbanization   143
     Tony Champion

10   Social Segregation and Social Polarization   162
     Chris Hamnett

11   Race Relations in the City   177
     Joe T. Darden

12   Communities in the City   194
     Ronan Paddison

13   Women, Men, Cities   206
     Linda M. McDowell

14   Urban Crime in the USA and Western Europe: A Comparison   220
     Paula D. McClain

PART IV   THE CITY AS ECONOMY

Introduction   241

15   Urban Scale Economies   243
     J. Vernon Henderson

16   Cities in the Global Economy   256
     Saskia Sassen

17   The Post-fordist City   273
     W.F. Lever

18   The Post-Industrial City   284
     Douglas V. Shaw

19   The New Urban Economies   296
     Donald McNeill and Aidan While

20   The Growth of Urban Informal Economies   308
     Colin C. Williams and Jan Windebank

PART V   THE CITY AS ORGANIZED POLITY

Introduction   323

21   Urban Governance   325
     Michael Goldsmith

22 Cities and Services: a Post-Welfarist Analysis 336
Stephen J. Bailey

23 Social Policy and the City 351
Susanne MacGregor

PART VI POWER AND POLICY DISCOURSES IN POSTMODERN CITIES

Introduction 367

24 Communicative Planning, Emancipatory Politics and Postmodernism 369
Louis Albrechts and William Denayer

25 Planning, Power and Conflict 385
James Simmie

26 Power, Discourses and City Trajectories 402
Mark Boyle and Robert J. Rogerson

PART VII CITIES IN TRANSITION

Introduction 417

27 Cities in Pacific Asia 419
David W. Smith

28 Post-Socialist Cities in Flux 451
Grigoriy Kostinskiy

29 The Cities of Sub-Saharan Africa: From Dependency to Marginality 466
Richard Stren and Mohamed Halfani

Index 486

PREFACE

The arrival of the new millennium was marked by another major milestone in the history of humankind: for the first time half, and according to some censuses slightly more than a half, of the world's population live in cities. Among the advanced economies this rapid development of cities spans some 200 years, whereas in the less developed economies it is much more recent in origin. Indeed, it is in the latter that urbanization, including the development of mega-cities with populations of more than 8 million, has been the most spectacular, and in which the prospects for future urban growth are likely to continue to focus, in some cases – China, for example – to a daunting scale. Yet, if the scale of urban growth in the developing world has eclipsed that of the advanced economies, the deepening effects of globalization ensure the continued domination of the cities in the north in the world economy.

Accompanying this growth of cities has been the development of academic endeavour aimed at understanding cities. While the proper study of cities is little over 100 years old, linked closely to the development of the social sciences, urban analysis has become a multi-faceted, eclectic body of knowledge embracing a wide range of disciplines. Reflecting this breadth and eclecticism is the variety of academic journals covering the study of cities, the majority of which are devoted to specific disciplinary perspectives or facets of urban life. Journals devoted to urban analysis in any holistic sense are far outnumbered by those which cover the field from a disciplinary perspective, including geography, economics, culture, law, planning and anthropology. Other journals focus on a particular aspect of urban life, such as education or the regeneration of urban economies.

The paradox is that as much as the study of cities has become increasingly specialized, and to a degree fragmented, there is a growing need for a more holistic appreciation of the city. Recognition of the interdependent nature of cities and of urban living, as, explicitly, in the currently fashionable concept of the sustainable city, lends weight to the notion that while urban analysts tend to be of necessity specialists in a particular aspect of the city, the need to appreciate the wider implications of urban development and change has increased.

The *Handbook of Urban Studies* is an attempt to meet such a need. While the volume does not pretend to embrace in any encyclopaedic sense the field of urban studies – even were this possible – the essays span the major disciplines which have sought to unravel an understanding of the city. The Handbook represents a distillation of the knowledge of the field, synthesizing the existing literature while representing the diversity of issues which are generated by urban development.

APPROACHING THE FIELD OF URBAN STUDIES

One of the immediate problems raised – and an issue which is discussed in different ways in the early chapters of this volume – is that what might be termed the field of urban studies has fuzzy boundaries and lacks any unified consensus as to its definition. Even the definition of what constitutes a city is itself contested. Part of the problem here stems from the fact that by their nature cities can and have been studied from different perspectives. Where these perspectives relate to the different social science disciplines, it is only to be expected that disciplinary viewpoints will prioritize particular types of questions. Further, social science disciplines frequently work within different epistemologies.

These differences need to be seen as a strength. Cities are multi-faceted, and any volume attempting to reflect the breadth of issues raised by urban development will need in turn to reflect these different disciplinary approaches. Economists, geographers, political scientists, economic historians, sociologists and others need to be represented. Similarly the different paradigms through which cities have been studied – including positivist, behavioural, political economy and poststructuralist – need to be embraced, approaches which even if in some cases contradictory, collectively have had a cumulative effect on building an understanding of the city.

To understand cities, then, it is necessary to take account of the diversity of different forms of analysis, historical as well as contemporary. Cities are too diverse for them to be capable of being understood through a single perspective. It is this rationale, above all, which has underlain the editorial approach adopted in this handbook.

ORGANIZATION OF THE HANDBOOK

The organization of the Handbook is structured around seven sections. An introductory chapter examining how cities have been studied is followed by Section I, which takes up definitional issues of the city and city living. Subsequently, the volume focuses on specific aspects of the city, followed by a final section, which by focusing on cities in three different world regions aims to take account of the diversity of urban development and change.

Part I is foundational – how are cities to be defined, how are they delimited in physical, economic and social terms and how are they represented to us through literature. These are broad but fundamental issues, moving us from the 'factual' to the more discursive. We can no more expect consensus on the definition of what constitutes the city as we could expect novelists (or other artists) to agree on how the city should, and could, be represented. Such issues will always be the subject of debate.

The following Parts, II-V, forming the bulk of the Handbook, examine different specific aspects of the city. The presentation is divided into sections, examining successively the city as environment, as people, as economy, and as polity, with Part VI devoted to the analysis of power and policy discourses in the contemporary city.

Where much of the discussion in the previous sections tends to focus on the experience of urban development in the advanced economies, the final section of the Handbook is devoted to looking at the urban experience in other contrasting regions of the world.

KEY DECISIONS ON PRESENTATION

An edited volume which aims to do justice to such a diverse field as that of urban studies inevitably involves the making of decisions which mould presentation. Diversity is a source of richness, but it also creates major problems forcing an editor to include certain approaches and issues, and hence exclude others. Given the overall limitations of the size of the volume set by the publishers exclusions become inevitable.

One key editorial decision taken at the outset was to give contributors a relatively generous word length in which to operate. This has enabled authors to develop theoretical and empirical arguments in greater depth. Obviously, such a decision has had an impact on the range of topics which could be covered, excluding the separate consideration of a range of issues such as health and education. On the other hand, by extending the word length and by identifying those key issues of the city, housing, transport, sustainability, crime, social segregation and so forth, the presentations are able to achieve a degree of depth in their treatment of topics which in many cases have been the subject of (smaller!) edited volumes in their own right.

Editorial decisions need to respond to diversity in other ways. Other edited volumes on the city – though less ambitious in scope than the Handbook – have had to make similar decisions on the format of the volume, for example, as to whether the approach taken should be more geographical or more systematic in focus. Recent emphases on localization within an increasingly globalized world have re-emphasized the uniqueness of cities, highlighting how geography matters. It is not implausible to imagine a *Handbook of Urban Studies* which was organized wholly around a selection of cities which represented different 'positions' in the global economy.

Such a presentation would give maximum currency to the uniqueness of cities and of urban development. Yet, it would deny the fact that there are underlying issues, and processes, common to cities; if their manifestation is unique, their analysis and resolution is not. Further, any more systematically (or generically) focused analysis needs to draw on empirical example(s) for demonstration or verification. Focusing on the issues and processes common to cities, and particularly those which are critical to understanding how they function and change, does not, and indeed should not, exclude the discussion of specific cities.

Even so, and bearing in mind the word and overall constraints within which authors and the editor were working, key decisions on presentation became necessary. Thus, the emphasis of the Handbook draws on urban conditions in the more advanced economies. While some urban analysts have begun to talk in terms of the convergence of urban issues and problems in northern and southern economies, with some justification, fundamental differences in the processes underpinning urban structure and change persist. Of course, there is the added point that we know a lot more about cities in the more developed economies. Yet, this is not to deny that ideally a fuller treatment of cities in the developing world in a yet larger volume would be warranted. It is for this reason that the final section is devoted to looking at cities in different regions. Again, there is no attempt here to achieve blanket geographical coverage – Latin America and the Middle East, for example, are excluded. Nevertheless, by looking at three regions occupying very different 'positions' in the global economy it is possible to begin to grasp the diversity of the urban experience.

ACKNOWLEDGEMENTS

Editing a major volume such as this would not be possible without the contribution of the authors, the editorial board and the publishers. At the outset of the exercise the editorial board provided valuable comments on the overall structure and approach of the volume, suggesting in some cases how key gaps might be filled and potential authors to approach. Of course, the volume would not have been possible without the willing cooperation of the authors, responding to editorial suggestions on the contents of their chapters and to comments on drafts.

In a venture of this size, and with the number of authors involved, it was perhaps inevitable that the road to receiving completed drafts was not always smooth. Altogether, completing the volume became a more protracted exercise than either I or the publishers, not to mention the authors who had completed their chapters, would have wished. Several commissioned authors failed to meet extended deadlines, and rather than sustain major gaps in the volume, new authors were approached. In one case a new author needed to be found because of the death of the intended contributor. These problems resulted in inevitable delays, and I am grateful to all those who participated in the exercise for their patience, not least the publishers who allowed the authors to update their chapters (where necessary) at proof stage.

Of the many individuals who have helped in the production of this volume special mention must be made to those at Sage, particularly Simon Ross and Rosemary Campbell. Simon (who has since moved to a more senior position in another publishing house) initiated the proposal and provided enthusiasm and support throughout. Rosemary entered towards the end of the process, ensuring production and relieving me of a number of tasks at times when my university commitments were mounting. Thanks are also due to Bill Lever, my co-Managing Editor on the journal *Urban Studies* who was instrumental in formulating the structure of the volume. Equally, Isabel Burnside, Editorial Assistant on the journal, gave freely of her support and encouragement, ensuring that the infrastructural help necessary to ensure that authors were kept in touch was provided.

Ronan Paddison
University of Glasgow

Notes on Contributors

Louis Albrechts is Professor of Planning at the Catholic University of Leuven. He is editor of European Planning Studies and past president of the Association of European Schools of Planning. From 1992 to 1996 he was responsible for the structure plan for Flanders and from 1999 to 2000 he was coordinator of the transport plan for Flanders. He has published widely on regional issues, planning matters and on transportation questions. His current research is focused on public involvement in planning, changes in governance and integrated territorial planning and policies. He is presently co-editing a book entitled *The Changing Institutional Landscape of Planning*.

Stephen J. Bailey is Professor of Public Sector Economics at Glasgow Caledonian University. He has published widely in the field of local government economics and local government finance, including journal articles, book chapters and books dealing with intergovernmental grants, local taxation, charges, public choice theory and local government reform etc. His books include: *Local Government Economics: Principles and Practice* (1999), *Public Sector Economics: Theory, Policy and Practice* (1995) and *Local Government Charges: Policy and Practice* (1993, with P. Falconer and S. McChlery).

Mark Boyle is Lecturer in Geography at the University of Strathclyde and is Chair of the Urban Geography Research Group of the RGS-IBG. His recent publications have been in the area of the politics of local economic development, including papers in *Environment and Planning A* (1997, 1999), *Progress in Planning* (2000), *Political Geography* (2000) and *The Growth Machine Concept: twenty years on* (1999). He is book review editor for *Space and Polity*.

Roberto Camagni is Professor of Economics and Urban Economics at the Politecnico of Milan. He has been Head of the Department for Urban Affairs at the Presidency of the Council of Ministers, Rome, under the Prodi Government. He is the President of the GREMI, an international association for the study of innovative environments or 'milieux', located in Paris, Sorbonne University. He has been President of AISRe, the Italian Section of the Regional Science Association, and a Member of the Council of the ASRDLF, the French Regional Science Association. He is the author of a textbook on urban economics in Italian, published also in French by Economica, Paris and forthcoming in Spanish.

Roberta Capello is Associate Professor of Regional and Urban Economics at the University of Molise, and of Economics at the Politecnico of Milan. She has been National Secretary of AISRe, the Italian Section of the Regional Science Association. She is a member of the Organising Committee of the European Regional Science Association and is the Treasurer of the same Association. She is a member of NECTAR, an international research network on Transport, Communication and Regional Development, and of GREMI, Groupe de Recherche Européen sur les Milieux Innovateurs. She has a PhD in Economics at the Free University of Amsterdam.

Tony Champion is Professor of Population Geography at Newcastle University. His research interests include urban and regional changes in population distribution and composition, with particular reference to counterurbanization and population deconcentration in developed countries and the policy implications of changes in local population profiles. He is the author of *Population Matters: The Local Dimension* (1993) and co-author of *The Population of Britain in the 1990s* (1996). Previous publications include *Migration Processes and Patterns* (1992), *People in the Countryside* (1991), *Contemporary Britain: A*

Geographical Perspective (1990) and *Counterurbanization* (1989).

Joe T. Darden is Professor of Geography and Urban Affairs at Michigan State University. He received his PhD in urban geography at the University of Pittsburgh in 1972. From 1971 to 1972 he was a Danforth Foundation Fellow at the University of Chicago. He received Michigan State University's Distinguished Faculty Award in 1984. His research interests are racial residential segregation and neighbourhood socio-economic inequality in multi-racial, white-dominated societies. He is the author of more than 100 publications, including the co-authored *Detroit: Race and Uneven Development* (1987). From 1997 to 1998 he was a Fulbright Scholar in the Department of Geography at the University of Toronto. He is presently writing a book entitled, *Toronto: The Significance of White Supremacy in a Canadian Metropolis.*

William Denayer graduated with a PhD in political science at the State University of Leyden (his thesis being on the political theory of Hannah Arendt). His interests are now in political philosophy (Arendt, Heidegger, Latour), internationalization and the restructuring of the welfare state, poverty and cities and actor network theory. He is a research fellow at the Department of 'Space and Society' at the University of Ghent, Belgium.

Ray Forrest is Professor of Urban Studies at the University of Bristol and Visiting Professor at the City University of Hong Kong. He has published extensively on housing, urban and related issues. Books include *Selling the Welfare State: The Privatisation of Public Housing* (with Alan Murie), *Home Ownership: Differentiation and Fragmentation* (with Alan Murie and Peter Williams) and *Housing and Family Wealth: Comparative International Perspectives* (with Alan Murie). His current research interests include the role of the residential neighbourhood in contemporary society and cross-national comparative analysis of economic instability and the housing market.

William H. Frey, is Research Scientist on the Faculty of the Population Studies Center at the University of Michigan. He is also Senior Fellow of Demographic Studies at the Milken Institute in Santa Monica, CA. He has written widely on issues relating to migration, population distribution and the demography of metropolitan areas. He is the author of *Regional and Metropolitan Growth and Decline in the United States* (with Alden Speare, Jr, 1988) and *Metropolitan America: Beyond the Transition* (1990).

Michael Goldsmith is Professor of Government and Politics at the University of Salford. He is the author of several books and articles in the fields of comparative local government and urban politics, including (with H. Wolman) *Urban Politics and Policy* (1992) and with K. K. Klaussen (eds) *Europeanisation and Local Government* (1997). His current research interests and writing are concerned with the impact of globalization on local governments; metropolitan government; and the relationship between the size of governmental units and democratic performance.

Mohamed Halfani is Senior Lecturer at the Institute of Development Studies in the University of Dar es Salaam. His field of research is institutional revitalization and local governance, with a special focus on urban development. Currently, he is serving as Director of Cabinet at the Organization of African Unity (OAU) in Addis Ababa.

Chris Hamnett is Professor of Human Geography at King's College London. Previously he was Professor of Urban Geography at the Open University. He has held a number of visiting appointments including the Banneker Professorship at the Center for Washington Area Studies, George Washington University, Washington DC, Norman Chester Senior Research Fellow, Nuffield College Oxford and Visiting Research Fellow at the Netherlands Institute of Advanced Studies. He is the author of numerous papers in the areas of housing and social change, the home ownership market in Britain, wealth and inequality in global cities. He is the author or co-author of *Cities, Housing and Profits*, *Winners and Losers: the Home Ownership Market in Modern Britain* and co-author of *Shrinking the State: The Political Underpinnings of Privatisation* and *Safe as Houses: Housing Inheritance in Britain*. He is also the editor or co-editor of numerous other books including *Housing and Labour Markets*.

Tom Hart is an Honorary Research Fellow in the Department of Economic and Social History at the University of Glasgow. With degrees in history and law he has specialized in transport and regional history and in comparative urban studies. He maintains major interests in contemporary and future transport policies, fiscal issues and settlement patterns and has written extensively on sustainable development and links between transport, environmental, economic and social policies. He chairs the Scottish Transport Studies Group and was an invited participant in the OECD Vancouver Conference on Sustainable Transportation in 1996.

J. Vernon Henderson is Eastman Professor of Political Economy at Brown University. He is currently doing research on systems of cities, industrial location, urban productivity, environmental regulation, development of urban sub-centres, as well as tax and public service competition among cities, focusing on the United States, Indonesia and Korea. Professor Henderson is on the editorial boards of the *Journal of Urban Economics, Regional Science and Urban Economics* and *Journal of Economic Growth*. He is a member of the National Bureau of Economic Research and the Boston Research Data Center of the US Census Bureau and has done research work for the National Science Foundation, Guggenheim Foundation, Earhart Foundation and the World Bank. He recently served as a team member of the World Development Report 1999–2000, with its focus on political decentralization and urbanization in developing countries.

Grigoriy Kostinskiy is Senior Research Fellow at the Institute of Geography, Russian Academy of Science. His interests are in social and economic geography, urban studies and the theory of geography. He is a corresponding member of the Commission on History of Geographical Thought, International Geographical Union. He is editor of the book *Russia and the CIS: Disintegrational and Integrational Processes*. Among his practical investigations are the social assessment studies for the World Bank.

William F. Lever is Professor of Urban Studies at the University of Glasgow. With a first degree in human geography and a doctorate in demography from Oxford University, he has researched industrial change and urban policy in Britain, Europe and North America. He has authored or edited several books on urban economic regeneration and is currently working on urban competitiveness in Scotland and Europe, and on the effects of market enlargement in the EU. He has worked for the EU, OECD and government. He is currently a Director of the European Urban Institute and a Managing Editor of the journal *Urban Studies*.

Paula D. McClain is Professor of Political Science and Law at Duke University. A Howard University PhD, her primary research interests are in racial minority group politics, particularly inter-minority political and social competition, and urban politics, especially public policy and urban crime. Her most recent articles have appeared in the *Journal of Politics, American Political Science Review* and *American Politics Quarterly*. Westview Press published the second edition update of her most recent book, '*Can We All Get Along?': Racial and Ethnic Minorities in American Politics* (co-authored with Joseph Stewart, Jr) in early

1999. She is a past vice-president of the American Political Science Association, and past president of the National Conference of Black Political Scientists, served as Program Co-Chair for the 1993 annual meeting of the American Political Science Association and served as Program Chair for the 1999 annual meeting of the Midwest Political Science Association.

Linda M. McDowell is Professor of Economic Geography at University College London. Her interests are in urban and economic geography, especially in new forms of work in cities. She is currently undertaking a project funded by the Joseph Rowntree Foundation about young men's labour market entry and precarious forms of employment. Her recent publications include *Capital Culture* (1997) and *Gender, Identity and Place* (1999) and she is also joint editor (with Joanne Sharp) of *A Feminist Dictionary of Human Geography* (1999).

Susanne MacGregor is Professor of Social Policy at Middlesex University, London. She has written extensively on urban issues and urban policy. Her publications include *Tackling the Inner Cities* (edited with Ben Pimlott, 1991) and *Transforming Cities: Contested Governance and New Spatial Divisions* (edited with Nick Jewson, 1997). She has recently conducted joint research on community development and drugs prevention; neighbourhood images and regeneration experiences; policies on social security, poverty and social exclusion; and women at risk in cities. She is currently Programme Co-ordinator for the UK Department of Health's Drugs Misuse Research Initiative.

Donald McNeill is a lecturer in Human Geography at the University of Southampton. He is currently conducting research into cultural and political identity in the European city, with particular reference to Paris, Madrid and Barcelona.

Peter Nijkamp is Professor of Regional Economics and Economic Geography at the Free University, Amsterdam. He has published extensively on urban and regional issues, on transportation questions and on environmental and resource matters. At present he is vice-president of the Royal Netherlands Academy of Sciences.

Ronan Paddison is Professor of Geography at the University of Glasgow and Managing Editor of *Urban Studies*. He is currently working on an ESRC project on urban competitiveness and social cohesion in Central Scotland. He has published widely on the themes of urban participation and decentralization, on which he has completed research projects for central and local governments.

Robert J. Rogerson is Senior Lecturer and Head of Department in the Department of Geography at the University of Strathclyde, and is Director of the Quality of Life Research Group based in the University. He has published widely on issues associated with urban quality of life, including contributions to *Urban Quality of Life: critical issues and options* (1999), *Planning for a better quality of life in cities* (2000), *Progress in Planning* (1996, 2000) and *Urban Studies* (1999). His main research interests include the role of quality of life in urban regeneration and community development, and the analysis of property developers in shaping urban developments. He acts as the European representative on the International Quality of Life Network, and is journal editor of *Applied Geography*.

Saskia Sassen is Professor of Sociology at the University of Chicago and Centennial Visiting Professor at the London School of Economics. Her most recent books are *Guests and Aliens* (1999) and *Globalization and its Discontents* (1998). *The Global City* is coming out in a new updated edition in 2000. Her books have been translated into 10 languages. Her edited book, *Cities and their Cross-Border Networks*, sponsored by the United Nations University, will appear in 2000. She is completing her research project on 'Governance and Accountability in a Global Economy'.

Peter Saunders is Professor of Sociology at the University of Sussex, UK, and he has also held visiting positions at universities and research institutes in Australia, New Zealand, Germany and the USA. His book *Social Theory and the Urban Question* represented a major contribution to the development of a 'new urban sociology' in the 1980s, and he has also published influential works on home ownership, urban politics, British politics, privatization, capitalism, social stratification and social mobility.

Douglas V. Shaw is Associate Professor of Public Administration and Urban Studies at the University of Akron, Akron, OH. He has published in the areas of immigration, ethnic tensions, transportation and public culture. He is author of *Immigration and Ethnicity in New Jersey History* (1994).

James Simmie is Professor of Innovation in the School of Planning at Oxford Brookes University. He worked previously at University College London and the School of Land Management and Development at the University of Reading. His books include *Citizens in Conflict* (1974), *Power, Property and Corporatism* (1981) and *Planning at the Crossroads* (1991). He has also edited major research collections such as *Yugoslavia in Turmoil* (1991), *Planning London* (1994), and *Innovation, Networks and Learning Regions?* (1996). His current research is focused on innovation and its contributions to urban and regional endogenous local economic growth. Major projects include 'Innovation Clusters and Competitive Cities in the UK and Europe'. This forms a part of the ESRC programme on 'Cities: Competitiveness and Cohesion'.

David W. Smith obtained his BA and MA from the Department of Geography, at the University of Aberystwyth. After a brief encounter with the Department of Geography at Glasgow University (1966–68), David moved to the University of Hong Kong (1968–73) where he took up a lectureship. Here, he obtained his PhD, which was part of a large interdisciplinary study initiated by the History, Sociology and Law departments of HKU. David's study examined and analysed the accommodation provision by the then colonial administration, comparing it with that of Singapore, an independent state. While in Hong Kong, he became interested in Southeast Asia and the subject of development. A brief sojourn at Durham University (1973–75) enabled him to work on a project with Bill Fisher which centred on accommodation/needs in the gececondu areas of Ankara. In 1975, he took up a post as Research Fellow in the Department of Geography, RSPacS, ANU, Canberra and extended his geographical interests to include Pacific Asia . A joint project with Gerry Ward and Terry McGee on the processes of the industrialization of food in the Pacific Island economies also increased his interests in development processes. In Australia he researched the Aboriginal fringe camps of Alice Springs and began to focus his thoughts on marginalization and internal-colonialism. After five years in Canberra (1975–80) David and his family returned to the UK and to the Geography Department at Keele University where, via the British Council, links were established with the Departments of Geography in Zimbabwe and later, Fiji. The link with Zimbabwe helped to develop his notions of urban sustainability and urban food production. While in Keele, he was anxious to maintain some links with Southeast Asia, wishing ultimately to research in Vietnam, an emerging independent state where development and its multi-faceted forms could be observed at an early stage. Thus he began a joint research project in Singapore. At Keele he obtained his Senior Lectureship (1985) and later his Personal Chair (1988). In 1995 he became Professor of Economic Geography at Liverpool University, where he remained until his death in December 1999. In 1996 he was presented with the Sir Edward Heath Award for his contribution to Development Studies. It was unfortunate that the latter years, from 1996 onwards, were clouded by the knowledge of his illness which preyed on him both mentally and physically. Before his death, and to his credit, he managed to complete the reworking of his book, *The Third World City*, which was published posthumously in May 2000. His success as an academic, as a team player and as a teacher and researcher, has not passed unremarked. His written works stand as testament.

Richard Stren is Professor of Political Science and former director of the Centre for Urban and Community Studies at the University of Toronto. Over the past 30 years he has carried out research in many African cities, including Mombasa, Nairobi, Dar es Salaam, Abidjan and Makurdi. Currently he is working on comparative themes involving Africa, Latin America and Asia.

Aidan While is a research officer at the Centre for Urban Development and Environmental Management at Leeds Metropolitan University. His research interests have centred on the changing role of the state in urban and regional economic development.

J.W.R. Whitehand is Professor of Urban Geography and Head of the Urban Morphology Research Group in the University of Birmingham, UK. His publications include *The Changing Face of Cities* and *The Making of the Urban Landscape*. He edits the international journal *Urban Morphology* and is a member of the Council of the International Seminar on Urban Form. He was formerly Chairman of the Urban Geography Study Group of the Institute of British Geographers and Editor of *Area*.

Colin C. Williams is Senior Lecturer in Economic Geography at the Department of Geography, University of Leicester. His research interests cover the geographies of informal economic activity, the theory and practice of the 'new localism' and formulating policy initiatives to rebuild social capital. His books include *A Helping Hand: Harnessing Self-Help to Combat Social Exclusion* (1999), *Informal Employment in the Advanced Economies: Implications for Work and Welfare* (1998), both co-authored with Jan Windebank, *Consumer Services and Economic Development* (1997) and *Examining the Nature of Domestic Labour* (1988).

Peter Williams is Visiting Professor at the Centre for Urban Studies, University of Bristol, and Deputy Director General at the Council of Mortgage Lenders, London, UK. His research interests cover housing policy, housing markets, housing history and urban change. His books include *Gentrification of the City* (edited with Neil Smith), *Class and Space* (edited with Nigel Thrift), *Home Ownership Differentiation and Fragmentation* (with Ray Forrest and Alan Murie) and *Surviving or Thriving? Managing Change in Housing Organisations*. He is Policy Review Editor for *Housing Studies*.

Jan Windebank is Senior Lecturer in French Studies and Associate Fellow of the Political Economy Research Centre (PERC) at the University of Sheffield. Jan's research interests cover cross-national research methodology, gender divisions of work and Anglo-French comparisons of women's work situations. Her books include *Women's Work in France and Britain: Practice, Theory and Policy* (1999), *A Helping Hand: Harnessing Self-help to Combat Social Exclusion* (1999), *Informal Employment in the Advanced Economies: Implications for Work and Welfare* (1998) (both co-authored with Colin Williams), and *The Informal Economy in France* (1991).

Hana Wirth-Nesher is Associate Professor of English at Tel Aviv University and Coordinator of the Samuel L. and Perry Haber Chair on the Study of the Jewish Experience in the United States. She is the author of *City Codes: Reading the Modern Urban Novel*, and the editor of *What is Jewish Literature?*, *New Essays on 'Call It Sleep'*, and the forthcoming *Cambridge Companion to Jewish American Literature*. She has published numerous essays on the work of nineteenth- and twentieth-century writers, including Charles Dickens, Mark Twain, Henry James, Joseph Conrad, James Joyce, Virginia Woolf, Isaac Bashevis Singer, Philip Roth, Franz Kafka, D.M. Thomas, and Cynthia Ozick.

Zachary Zimmer is currently a Research Associate in the Policy Research Division of the Population Council in New York. He graduated with a PhD from the Population Studies program at the University of Michigan in 1998 and subsequently spent two years as an Assistant Professor in the Department of Sociology at the University of Nevada–Las Vegas. His primary areas of interest include the demography of health and ageing in Asian societies and his research has been published in a variety of demographic and gerontological journals.

1

Studying Cities

RONAN PADDISON

Befitting the nature of the urban, the complex multi-faceted make-up of the city, how cities have been studied is itself an entangled story. By their nature, the understanding of cities – how they grow and decline, are structured and function, give meaning to social life – intersects with each of the principal social science disciplines, sociology, economics, geography and political science in particular, as well as urban planning. To varying degrees, each social science discipline has developed distinctive epistemologies by which to understand the city, and the key issues of structure, process and change. Alongside this endeavour have been other social scientists, including anthropologists and (economic) historians, working in different environments and historical periods, who too have sought to unravel urban social processes, structure and change. The understanding of cities is of necessity an eclectic project; small wonder, then, that the historiography of urban studies is a complex story, cutting across, as well as at times constrained by, disciplinary lines.

An immediate problem here is that, partly because of the difficulties in defining the term urban, and partly because of the breadth of the subject area, delimiting the boundaries of the field of urban studies is by no means a straightforward task. Further, because the city has become a (if not usually the) principal locus through which economic growth and a change is conducted, cities feature prominently in the demographic, economic, social and political structures and working of virtually all of the separate states making up the world political map. It may be that it is in the advanced economies that the dominance of the urban is most pronounced – where, for example, 80 per cent and more of the national population may be classified as living in urban areas – but in less developed countries, too, the significance of the city is no less apparent, often in terms of their growth but also in acting as the conduit with the core economies.

Taken together – the multi-faceted nature of cities, the breadth of the subject area involved in understanding cities, and their centrality in economic and social life in advanced as well as less developed economies – it is hardly surprising that in each of the social science disciplines the study of cities constitutes a separately identifiable branch, more or less central to the discipline itself. Reflecting the trend are the labels those studying cities frequently give themselves. Urban scholars tend to describe themselves as urban economists or urban geographers, or urban political scientists, rather than urbanists *per se*, a term which has not gained the widespread currency it deserves. (If anything the term urbanist has been appropriated by those concerned with the planning and redesign of cities, though in North American universities separate faculties of urban studies are more commonplace than they are in the United Kingdom.) In one measure this reflects the importance disciplinary perspectives have had, and continue to have, in defining separate fields of knowledge. In another it reflects the complexity of cities, the intellectual challenge in being able to unravel the interlocking processes and facets that explain how they function and develop. While these disciplinary perspectives do not work in isolation – indeed there are (and have been) strong interconnections between them – such intellectual fragmentation runs contrary to the need to understand the city as an holistic entity.

Matching the dynamic nature of cities, how they have been studied has itself not been static, reflecting the influences of the paradigmatic shifts which have underlain the development of the social sciences, and its separate disciplines, over the last century and more. Equally the agenda of urban studies has reflected contemporary interpretations of what analysts from different disciplinary backgrounds have identified as the key questions underpinning cities, whether this is construed in economic, spatial, social, cultural or political terms, or some mix of these. In the rest of this introductory chapter we shall explore some of the implications arising from these two observations.

STUDYING CITIES

While the diversity and breadth of urban studies reflects both the complex nature of the city, this is not to suggest that there is not an identifiable core as to what should constitute its study. Such a core would centre on understanding the structure of the city, its growth and change (whether expressed in physical, economic, social or other terms), the nature of urban social processes, their interplay with class, gender, race as cleavage-forming dimensions in the making of the different geographies of the city, social organization and disorganization, and economic and political processes linked to their structure and change. Equally, where the analysis of cities needs to take account of their contextual location, emphasis needs to be given to the contemporary conditions underpinning change and to the diversity of cities characterizing economies and societies at different locations in the world economy. Finally, any study of the city needs to acknowledge that there are different ways by which it can be defined and can be (and has been) studied.

The structure of the Handbook broadly reflects this core. Beginning with different approaches to the study of the city, the book is divided into seven sections dealing successively with the city as environment, as people, as economy and as polity. The final two parts are concerned with contemporary shifts in the urban condition, and in particular to the intersection between market forces and political factors underpinning change, and to the diversity of urban experience in three different regions, transitional economies, rapidly developing economies and marginalized economies.

Not unexpectedly, the interpretation of the field of urban studies represented in the volume – wide-ranging rather than comprehensive, in any absolute sense – reflects contemporary appreciation of the nature of cities by the several social science disciplines. Predictably too, it reflects current concerns with the meanings and change of cities in the contemporary period, itself one of very rapid, but not unprecedented, socio-technological change. In other words, the volume is a reflection of its time – of the key issues facing the contemporary city, current methodological approaches and policy problems and prescriptions.

While our concern is to understand the contemporary urban condition, such an analysis is able to draw on a rich historical vein of theoretical and empirical work. Much of our understanding of cities, in theoretical and empirical terms, stems from work conducted in the twentieth century. Admittedly, there had been important studies conducted earlier, some of which, such as the work of Ferdinand Tönnies (1883/1995), and, in a very different vein, Booth (1891), writing in the latter part of the nineteenth century, was to remain durable. Nor should we overlook earlier commentators on the city. Precisely because cities create such distinctive types of settlement, earlier scholars from the Classical period through to the early modern period, such as Thomas Stowe, have sought to understand the nature of the city, or as in the case of Classical scholars, their place within the polity. These earlier contributions apart, what might be termed the main historical avenues of study, developed by the principal social science disciplines, largely over the last 100 years, continue to have, albeit to varying degrees, an influence on contemporary analyses.

That the intellectual study of the city is limited largely to the last century or so is hardly surprising given that the growth of large cities was linked in the First World nations of Western Europe and North America to the emergence of the industralized economy of the nineteenth century. By comparison with its growth in the nineteenth and twentieth centuries, urban development previously had been muted, certainly in terms of the size of cities as well as in terms of the overall importance and rate of urbanization. This is not to deny, of course, that in pre-industrial societies cities lacked significance, rather that urbanized societies had to await the development of industrialism. Historically, then, only a handful of cities had ever attained a population of one million, whereas by the end of the nineteenth century few industrial nations were not able to boast at least one, and often several, such cities. By the end of the following century there were to be more than 300 'millionaire' cities. In contrast, then, to the relatively 'slow' development of the city over the longer historical duration, the spread of industralism spawned rapid urban growth, the development of large cities and the urbanized society.

One of the truisms of the historiography of urban studies is that how cities have been studied has reflected perceptions of their contemporary condition. The traumatic nature of the changes of the nineteenth century is reflected in those issues which became the key concerns of the urban research agenda amongst early urban scholars, as well as other contemporary observers such as novelists. In the first industrial nation, Britain, the social transformation from a rural to an urbanized society is barely of more than a century in standing, and had itself taken less than 100 years. It was not just the rapidity of urbanization which demanded understanding. Urban-industrial societies created unprecedented economic wealth and progress, but they also created unprecedented social and political problems. Even if the problems posed by the city to social and political order were not new, their scale was, which made the imagined (and sometimes real) threat of 'the mob' and of disease, the more palpable to the elite as well as to the burgeoning middle classes. As much of nineteenth century observers of the city were in awe of it, they saw the need to understand it.

In the nineteenth century, then, it is not surprising to find that so much of the commentary on the urban condition was directed at those aspects of it which reflected the major transformations which had taken place, and which distinguished it from the recent historical past. Tönnies's work was to reflect the broad change from a rural to an urbanized society, while Booth's work, an analysis of social conditions in London, together with surveys conducted in a number of other British cities, charted some of the traumatic changes that had been brought by rapid urban growth. Contemporary observers of the Victorian city were in awe of its achievements, and often its spectacle, but there was no denying the squalor of the city which so many of its inhabitants had to endure, and which was so graphically described by novelists of the time.

By comparison with its early study, the agenda of urban research as it has developed over the last 100 years, has broadened and deepened, and been enriched through the development of alternative theoretical perspectives and empirical analysis. The developing agenda of urban studies reflects the development and growing significance of the social sciences within the academic curriculum, and the centrality of cities in the social, economic and political life in virtually all types of economy. Equally, it represents a growing appreciation of the complexities of the city together with its importance to economic growth.

The lesson of history here is that where the research agenda of urban analysis has altered it has tended to reflect changing interpretations of the key factors associated with the contemporary city. In the late nineteenth century cities were growing at unprecedented rates in both Western Europe and North America. Yet, their growth was not haphazard; market factors and socio-cultural forces helped impose a social ordering of the city. Theorizing the city needed to come to terms with the forces underlying its social ordering.

It is against this background that the rise of the human ecology school in the early decades of the twentieth century developed, and which became so intertwined with theorizing the socio-spatial formations of Chicago. Though the city could trace its origins to the eighteenth century, it was in the next century, and particularly after the coming of the railway in the 1850s, that Chicago was to develop rapidly, as an entrepot and as a manufacturing centre. In population terms the development of the city was phenomenal, but as remarkable as its rate of growth was the cosmopolitan make-up of its population. Small wonder, then, that the human ecologists, notably Robert Park but also Ernest Burgess and Robert Zorbaugh among others, were at pains to demonstrate how and why the resultant socio-economic and ethnic mix of the city became sorted spatially, and how the ecological map of the city altered over time. There were other aspects of the city's development which were scrutinized, not least the connections between urban growth, location and transport networks. But it was the sheer contrasts of the city at the turn of the new century, between affluence and poverty and between ethnic groups, and its segregation, against the backgrounds of immigration and urban development, which not only provided the fascination of the city for these early sociologists, but also for some of them some of the key tenets which helped distinguish the nature of the city.

By the end of the twentieth century an observer might be forgiven for thinking that so much of the agenda of urban studies had become dominated by the over-arching theme of globalization and of its multiple impresses on the restructuring of cities within what is widely, if not uncritically, perceived as an increasingly global world (see, for example, Borja and Castells, 1996; Knox and Taylor, 1995). Apart from the world-city literature itself, beginning with Friedmann and Wolff's seminal paper in 1982, which by a little over ten years later had spawned several hundred papers on the subject of world-cities alone, much of the contemporary analysis is focused on the multiple restructurings consequent upon, and contributory to, the major economic, social, political and cultural shifts linked to the contemporary condition. How 'old' industrial cities can regenerate their economies, the role of urban marketing, of public–private partnerships, the need to decentralize and empower citizens, particularly those in deprived neighbourhoods, addressing through policy the objectives and urban consequences of Structural Adjustment Programs in cities in developing economies, the shift towards deregulation, marketization and liberalization have become common concerns of contemporary *urban* restructuring. The agenda of urban research – even the concepts used to proffer policy solutions to urban problems – has to a degree itself become globalized, leading some to talk of the growing convergence of the urban condition and of urban problems between North and South (Cohen, 1996). While such issues by no means constitute the totality of what constitutes present-day urban analysis, as any casual scanning of the academic journals in the field would testify, they constitute a major part of current research concern.

That the agenda of urban analysis has shifted alongside its broadening and deepening, and that at successive points it has reflected those key aspects imagined as defining or problematic to the contemporary urban condition, is not to deny that there have been continuities in what aspects of the city it is that should be studied. Fundamental questions linked to the development of the city – including the processes linking the city with the wider economy, their spatial structure and change, their governance – have persisted as research nodes studied by urban scholars originating from different disciplines. Thus, explaining the nature of segregation within the city, and the role it plays in giving meaning to social life as well as in maintaining social order, has been an ongoing focus of interest since Engels's classic work on Manchester in the 1840s, through to the contemporary concern of postmodern analysis showing its contribution to defining identity and difference.

These trends – of continuities and changing emphases in the content matter of urban analysis – are overlain by, and related to, epistemological shifts in the way cities have been studied. How they have been studied has shifted in accordance with the dominant ideas (to some, paradigms) underpinning social science analysis or its separate disciplines. Certain schools of thought have become the dominant (in a Kuhnian sense) paradigm through which urban analysis has been conducted, subsequently to become discredited or, at least, superseded by the need to focus on a different aspect of the city, in turn requiring different theoretical and methodological tools.

Historically, then, the study of cities is identifiable with continuities and with discontinuities – continuities in terms of the basic questions cities pose, discontinuities in terms of how they have been studied and theorized. While historically such analysis is linked to different, often disciplinary-driven, perspectives, the shifts in the ways in which cities have been studied is closely connected to what urban issues have been successively prioritized. In turn this has resulted in different perspectives, sometimes radically so, being brought to bear on the same urban characteristic. The concern urban analysis had with the geographic patterning of residential segregation, particularly through the development of factorial ecology techniques in the 1960s, is in stark contrast with current studies of segregation, frequently based on ethnographic analyses, and concerned with the nuanced meanings and connections of segregation for the individual rather than the identification of spatial patterns (see, for example, Schnore and Winsborough, 1972 and Mumford, 1997).

STUDYING CITIES – INHERITED TRADITIONS AND CONTEMPORARY PERSPECTIVES

The principal ways in which cities have been studied mirror the interests of the main social science disciplines linked to their analysis. Each has left a legacy of theoretical and empirical work of more or less enduring significance for subsequent work. Taking the period 1890 to 1980 Dunleavy (1982) has suggested that there has been five main traditions: locational analysis, studies of systems of cities, and socio-cultural, institutional and political economy approaches. Locational analyses and studies of systems of cities are both concerned essentially with spatial relationships, at intra- and inter-urban scales, and are identified principally with the work of geographers and economists. Socio-cultural and institutional approaches have been developed principally by sociologists, urban anthropologists and political scientists. The development of political economy, including neo-Marxist approaches to the city, differs in two respects from the other traditions, in being, by definition, less identifiable with a specific discipline, while their development dates principally from the 1960s as opposed to having been represented throughout the period since the latter part of the nineteenth century.

Dunleavy's typology is useful, but not exhaustive. There are other important strands to urban research as it has developed from the (late) nineteenth century onwards. One important tradition has been represented throughout the period, the management of the city, focusing on the planning and reconstruction of cities and the derivation of urban policy. Further, with the (not uncontested) notion of the development of postmodern society, the study of cities has taken a new turn, providing powerful, and often novel, ideas and methodologies by which to understand the contemporary urban condition. Both issues, the management of cities and the intersections between postmodern society and the city, have furnished other important ways in which cities have been theorized and empirically investigated.

Table 1.1 summarizes these different approaches to the study of cities. By looking at each of the major approaches, the table is not intended to be comprehensive, and clearly could not hope to be, in view of the wide scope of theoretical and empirical work connected with the city; rather it aims to delimit the major ways in

Table 1.1 *Major approaches and milestones in the study of cities*

Perspective	Major periods of development	Representative authors/key issues
Spatial/Economic		
(1) Locational Analysis	19th century, intermittent	Effects of distance from urban market place, i.e. accessibility, on patterning of agricultural land use, von Thunen, (1826). Vertical and lateral patterning of social classes in the city Kohl, (1841)
	Early 20th century	Patterning of city land values, accessibility and site characteristics/potential Hurd, (1903)
	1930s–1960s	Studies of urban land use e.g. Hoyt (1939), on growth of residential neighbourhoods, on development of urban form Gottmann, (1961) Development of Urban Economics. Alonso (1960) on the urban land market, derivation of bid-rent theory and accessibility.
	1970s–present	Development of new urban economics, Richardson (1977)
(2) Spatial Networks	1930s–1960s	Spatial distribution of systems of cities, ranking, development. Christaller (1933) on central place theory, geometric patterning of cities differentiated by specialization of services. Subsequent theoretical development by Losch (1939), Isard (1956). Empirical assessment of central place theory, Berry (1950s, 1960s)
	1980s–present	Formulation of linkages between world economy, urban economic restructuring, competition. World-city (Friedmann and Wolff, 1982), global cities (Sassen, 1991), urban restructuring in regional (e.g European) arenas
Sociological		
(1) Socio-cultural	Late 19th century, early 20th century	Nature and meaning of urban social life; Tönnies (1883/1995), definition of community and society ideal-types. Weber (1905) on ideal-type city as loci of civilization and historical change.
	1900s–1940	Simmel (1903) on mental adaptations to city life and social behaviour. Chicago School – understanding of competition between land uses and groups for urban space using ecological ideas. Derivation of principles of competition, succession-invasion by which groups/uses dominate City Spaces, Park (1916, 1929). Concentric patterning of urban growth and social differentiation, Burgess (1925), progenitor of later work on socio-spatial differentiation. Formulation of 'principles' of size, density and heterogeneity as defining urbanism as a way of life, Wirth (1938).
	1950s/1960s–present	Development of social area analysis – Shefsky and Bell (1955) on impress of urban-industrialism on socio-spatial differentiation. Use of factorial ecology techniques to identify socio-spatial patterning, Timms (1971). Empirical studies of distinctive ways of urban life, e.g. Whyte's (1955) study of gangs in Boston. Late accounts of social nature and meaning of urban life Sennett (1970, 1977, 1994)

Table 1.1   *cont.*

Perspective	Major periods of development	Representative authors/key issues
(2) Socio-Political Community/Community and Urban Power Studies	Late 19th century	Empirical studies of urban social life especially of poor. Large-scale detailed surveys, e.g. Booth (1891); poverty in York and Rowntree (1902)
	1920s–1960s	Development of community studies. Detailed empirical studies of community life, Lynd and Lynd (1929) and community power Lynd and Lynd (1937).
	1950s–1960s	Focused community studies, Young and Willmott (1962). Early community/urban power studies. Alternative theories of power distribution.
	1960s–1980s	Hunter's (1952) study of local elites, Dahl's (1961) formulation of dispersed/pluralist sources of power. Development of neo-Weberian perspectives, the role of institutions and agency in influencing social outcomes: Rex and Moore's (1967) study, *Race Community and conflict* showing housing outcomes in Birmingham as not only products of occupational class, but also race and immigrant status. Formulation of urban managerialism/gatekeeping controlling access to key resources, Pahl (1970).
Neo-Marxist	1960s–present	Emphasis on social processes allied to urban development and social conditions/inequalities. Alternative approaches, e.g. application of Marxist analysis to urban issues, Harvey (1973); urban government as local state, Cockburn (1977); urban ideology and collective consumption conflicts, Castells (1977); dependency theory and the formation and role of colonial cities.
Urban Planning/ Reconstruction/ Policy-Making	Late 19th–early 20th centuries	Early planning visionaries, the conversion of social ideals to 'good' urban design, Morris (1891), Howard (1898/1945) on the development of garden cities. Early development of planning theory and methodology, Geddes (1915). Capital city (re)construction, development and spread of Haussmannization, Choay (1969).
	1920s–1950s	Modernist visions of the city, Le Corbusier (1929/1947) and (1933/1964), Frank Lloyd Wright (1935) urban reconstruction in its regional setting, New York Regional Plan (1922), Greater London Plan (1944), city and region, Dickinson (1946).
	1960s–present	Critics of urban planning, Jacobs (1967) Urban economic reconstructuring and policy formulation, the development of urban entrepreneuralism, Harvey (1989) Housing analysis.
Postmodern accounts	1980s–present	Eclectic suite of ideas confronting the assumptions of positivism, the determinacy and meta-theoretical explanation of neo-Marxism and the theory and practice of modernist planning. Accounts emphasize different aspects of city life – cities as centres of consumption, Mort (1998), of recreation, Hannigan (1998), of image, Gottdiener (1995). Representational cities – messages encoded in environment read as texts, Jacobs (1996)

Source: Based in part on Dunleavy (1982)

which cities have been studied, identifying important accounts which reflect the different approaches. The perspectives crystallize around the key questions posed by cities – spatial, economic, social, political and their planning – in which our understanding of these different dimensions has been incremental, though periodically linked with radically different theorizations of these basic questions. As different as are these perspectives, they can be identified with the key themes which have underpinned much of the analysis of cities.

One such theme has focused on the structure of the city. A great deal of the analysis of the city has been concerned with their structure and with the processes underlying their structure and change. Thus the spatial structure of the city and of networks of cities at different scales of analysis has been an enduring question. Indeed, some of the earliest theorizing of the city centred on its spatial structure, from which developed attempts to delineate the patterning of the city, as of social groups in the case of the early work on Manchester (England) by Engels (1969) in the 1840s. Nearly one hundred years later, but in terms of understanding the city as a network of cities, the development of central place theory (Christaller, 1933) was concerned with the identification of patterns within a theoretical framework. A concern with pattern was evident in some of the more frequently quoted work of the Chicago School too, although attention was focused also on the processes underlying such patterns, and their theorizing.

A second theme underpinning the study of cities has been the continuing, and increasing, attention to the theorization and empirical scrutiny of the processes underpinning structure. The trend is as laudable as it has been necessary. Yet its influence in shifting the ways in which cities should be studied has been profound. The study of urban retailing is a case in point, in which even scarcely a generation ago one of its chief concerns was the study of structure, particularly in spatial terms. Berry's text published in the mid-1960s reflected the concerns of urban retail geography of the day, its focusing around the tenets of central place theory as devised by Christaller, and modified by later workers, and the search for patterns and regularities of both retail structure and consumer behaviour. Even where the current research agenda is drawn to the spatial structure of urban retailing, its analysis is far more concerned with understanding the processes linked to decentralization (see Guy, 1994) than it is with the search for spatial regularities. Further, a much richer contemporary research vein is concerned with identifying how retailing and consumer behaviour intersects with the emergence of postmodern consumption-based identities (see, for example, Mort, 1996). Problematizing the processes linked to (spatial) structures in and of the city has become the more dominant research endeavour than the identification of structures *per se*.

A continuing theme associated with the study of the city is that there has been an ongoing tradition of normative analysis concerned particularly with how cities could, and should, be better managed and organized. Early planning visionaries were guided by the need for cities to avert the economic, transport and social problems of the rapidly developing urban-industrial cities of the nineteenth century *and* the need to design cities which would be efficient economic engines and socially equitable. More fundamentally there was the belief that as human creations, it should be possible to create cities which achieved both criteria and be liveable, a canon which reached its zenith in high modernism. If, by the end of the twentieth century, the critique of postmodernism (Dear, 1986; Sandercock, 1998) had put to the rack the aspirations of modernism, urban planners themselves tenaciously hung on to much of the inherited tradition, while urban restructuring emphasized the need for urban management to be more socially inclusive.

As in other fields of knowledge, much of our understanding of cities has been built incrementally with major new insights offered periodically by the development of radically different epistemologies. The Chicago School, the adoption of neo-Marxism and the contemporary influence of the postmodern turn are among the more obvious examples, each of which provided fertile theoretical and conceptual ground on which subsequent research could flourish. Yet, such shifts have not necessarily supplanted earlier epistomologies. Rather, there has been a gradual accretion of different perspectives able to theorize alternative facets of the urban condition set against an intellectual background in which the inheritance of previous traditions continues to influence the research agenda.

How cities are and have been studied, then, is a complex weave of new theorizations and techniques of analysis along with an inheritance from the past of those ideas which have proved durable. In some cases – ethnographic forms of analysis, for example – current theorizations have helped breathe new life into a well-tried methodology. While disciplinary perspectives remain identifiable, more than lip service is paid to the need for the study of cities to cut across disciplinary boundaries. Cities are too complex for it to be otherwise.

REFERENCES

Alonso, W. (1960) 'A Theory of the urban land market', *Papers and Proceedings of the Regional Science Association*, 6. pp. 149–57.

Berry, B.J.L. (1965) *Geography of Market Centres and Retail Distribution*. Englewood Cliffs, N.J: Prentice-Hall.

Booth, C. (1891) *Life and Labour of the People of London*, 5 vols. Fairfield N.J: Kelley.

Borja, J. and Castells, M. (1996) *Local and Global: Management of Cities in the Information Age*. London: Earthscan.

Burgess, E.W. (1925) 'The growth of the city', in R.E. Park, E.W. Burgess and R.D. McKenzie, *The City*. Chicago: Chicago University Press.

Castells, M. (1977) *The Urban Question*. London: Edward Arnold.

Choay, F. (1969) *The Modern City: Planning in the Nineteenth Century*. London: Studio Vista.

Christaller, W. (1933) *Die Zentralen Orte in Suddeutschland*. Jena. (Translated by C.W. Baskin (1966) *Central Places in Southern Germany*. Englewood Cliffs, N.J: Prentice-Hall.)

Cockburn, C. (1977) *The Local State*. London: Pluto Press.

Cohen, M.A. (1996) 'The hypothesis of urban convergence: are cities in the North and South becoming more alike in an age of globalisation', in M.A. Cohen, R.A. Ruble, J.S. Tulchin and A.M. Garland (eds), *Preparing for the Urban Future*. Washington: Woodrow Wilson Center Press. 25–38.

Dahl, R. (1961) *Who Governs: Democracy and Power in an American City*. New Haven: Yale University Press.

Dear, M. (1986) 'Postmodernism and planning', *Society and Space*, 4: 367–84.

Dear, M. (2000) *The Postmodern Urban Condition*. Oxford: Basil Blackwell.

Dickinson, G.E. (1946) *City and Region*. London: George Allen and Unwin.

Dunleavy, P.J. (1982) 'The scope of urban studies in social science', Units 3/4 *Urban Change and Conflict*. Milton Keynes: The Open University Press.

Engels, F. (1969) *The Condition of the English Working Classes in 1844*. St Albans: Panther (first published in England 1892).

Friedmann, J. and Wolff, G. (1982) 'World city formation: and agenda for research and action', *International Journal of Urban and Regional Research*, 6(3): 309–44.

Geddes, P. (1915) *Cities in Evolution*. London: Williams and Norgate.

Gottdiener, M. (1995) *Postmodern Semiotics: Material Culture and the Forms of Postmodern Life*. Oxford: Basil Blackwell.

Gottman, J. (1961) *Megalopolis*. New York: Twentieth Century Fund.

Guy, C.M. (1994) *The Retail Development Process*. London: Routledge.

Hannigan, J. (1998) *Fantasy City: Pleasure and Profit in the Postmodern Metropolis*. London: Routledge.

Harvey, D. (1973) *Social Justice and the City*. London: Edward Arnold.

Harvey, D. (1989) 'From managerialism to entrepreneurialism: the transformation of urban governance in late capitalism', *Geografiska Annaer*, 71(B): 3–17.

Howard, E. (1898/1945) *Garden Cities of Tomorrow*. London: Faber and Faber.

Hoyt, H. (1939) *The Structure and Growth of Residential Neighbourhoods in American Cities*. Washington, DC: Federal Housing Association.

Hunter, F. (1952) *Community Power Structure*. Chapel Hill: University of North Carolina Press.

Hurd, R.M. (1903) *Principles of City Land Values*. New York: Arno Reprint.

Isard, W. (1956) *Location and Space-Economy: a General Theory Relating to Industrial Location, Market Areas, Land Use, Trade and Urban Structure*. Cambridge, MA, MIT: Wiley, Chapman and Hall.

Jacobs, J. (1967) *Death and Life of Great American Cities*. Harmondsworth: Penguin.

Jacobs, J. (1993) 'The city unbound: qualitative approaches to the city', *Urban Studies*, 30: 827–48.

Jacobs, J. (1996) *Edge of Empire: Postcolonialism and the City*. London: Routledge.

Knox, P. and Taylor, P.J. (eds) (1995) *World Cities in a World System*. Cambridge: Cambridge University Press.

Kohl, J.G. (1841) *Der Verkehr und die Ansiedlung der Menschen in ihrer Abhhangikgit von Gestaltung der Erdoberflache*. Leipzig: Arnoldische Buchhandlung.

Le Corbusier (1929/1947) *The City of Tomorrow*. London: Architectural Press.

Le Corbusier (1933/1964) *The Radiant City*. London: Faber and Faber.

Lloyd Wright, F. (1935) 'Broadacre City: a new community plan', *Architectural Record*, 77.

Losch, A. (1939) *The Economics of Location*: Translated by W.H. Woglom and W.F. Stolper (1954). New Haven: Yale University Press.

Lynd, R. and Lynd, H. (1929) *Middletown*. New York: Harcourt Brace (reprinted in 1964, London: Constable).

Lynd, R. and Lynd, H. (1937) *Middletown in Transition*. New York: Harcourt Brace (reprinted in 1964 London: Constable).

Morris, W. (1891) 'News from nowhere', reprinted in A. Briggs (ed.) (1962) *William Morris: Selected Writings and Designs*. Harmondsworth: Penguin.

Mort, F. (1996) *Cultures of Consumption*. London: Routledge.

Mumford, K.J. (1997) *Interzones: Black/White Sex Districts in Chicago and New York in the Early Twentieth Century*. New York: Columbia University Press.

Pahl, R.E. (1970) *Whose City?* London: Longman.

Park, R.E. (1916) 'The city: suggestions for the investigation of human behaviour in the urban environment', *American Journal of Sociology*, XX: 577–612.

Park, R.E. (1929) *Human Communities*. New York: Free Press 1952.

Rex, J. and Moore, B. (1967) *Race, Community and Conflict*. London: Oxford University Press.

Richardson, H.W. (1977) *The New Urban Economics and Alternatives*. London: Pion.

Rowntree, B.S. (1902) *Poverty: A Study of Town Life*. London: Macmillan.

Sandercock, L. (1998) *Towards Cosmopolis: Planning for Multicultural Cities*. Chichester: John Wiley.

Sassen, S. (1991) *The Global City: New York, London and Tokyo*. Princeton, NJ: Princeton University Press.

Schnore, L.F. and Winsborough, H. (1972) 'Functional classification and the residential location of social classes', in B.J.L. Berry (ed.), *City Classification Handbook*. New York: Wiley Interscience. pp. 124–51.

Sennett, R. (1970) *The Uses of Disorder: Personal Identity and City Life*. New York: Random House.

Sennett, R. (1977) *The Fall of Public Man*. New York: Alfred A Knopf Inc.

Sennett, R. (1994) *Flesh and Stone: The Body and the City in Western Civilisation*. London: Faber and Faber.

Shefsky, E. and Bell, W. (1955) *Social Area Analysis*. Stanford: Stanford University Press.

Simmel, G. (1903) 'The metropolis and mental life', reprinted in R. Sennett, (ed.) (1969) *Classical Essays on the Culture of Cities*. New York: Appleton-Century-Crofts. pp. 47–60.

Timms, D. (1971) *The Urban Mosaic: Towards a Theory of Residential Differentiation*. Cambridge: Cambridge University Press.

Tönnies, F. (1883/1995) *Community and Society*. New York: Harper and Row.

von Thunen, J.H. (1826) *Der Isolierte Staat*. (Published in English as *The Isolated State*, P. Hall (ed.) 1968.) London: Oxford University Press.

Weber, M. (1905) *The City*. London: Heinemann, 1958.

Whyte, W.F. (1955) *Street Corner Society: Social Structure of an Italian Slum*. Chicago: Chicago University Press.

Wirth, L. (1938) 'Urbanism as a way of life', *American Journal of Sociology*, 44 (1): 1–12.

Young, M. and Willmott, P. (1962) *Family and Kinship in East London*. Harmondsworth: Penguin.

Part I

IDENTIFYING THE CITY

The question of what is a city has occupied the attention of many urban scholars. Indeed, one of the paradoxes of studying cities is that how the city is to be defined has proved as (if not more) problematic than has the question of how they should be studied. As it was argued in the introductory chapter the city is a complex, multifaceted social organization. Little wonder, then, that how they have been studied has inevitably reflected different theoretical and disciplinary perspectives. A similar conclusion can be drawn to the question as to what constitutes a city.

Cities have many different 'faces'. What, then, gives the city its significance, its individuality from other types of socio-spatial organization has been given different emphases. Consider two views of the city, one by Raymond Williams (1973) in his classic book *The Country and the City*, the other by Lewis Mumford (1938) in *The Culture of Cities*. To Mumford:

> The city, as one finds it in history, is the point of maximum concentration for the power and culture of a community. It is the place where the diffused rays of many separate beams of life fall into focus, with gains in social effectiveness and significance. The city is the form and symbol of an integrated social relationship; it is the seat of the temple, the market, the hall of justice, the academy of learning. Here in the city the goods of civilisation are multiplied and manifold; here is where human experience is transformed into viable signs, symbols of conduct, systems of order. (1938/1995: 104)

The defining features of the city are linked to its strategic functioning to the wider community, its importance as a civilizing force besides its part in facilitating the market. Such a depiction of how the city is to be understood contrasts with the description offered by Raymond Williams:

> The great buildings of civilisation, the meeting places, the libraries and the theatres and domes; and often more moving than these, the houses, the streets, the press and excitement of so many people with so many purposes. I have stood in so many cities and felt this pulse: in the physical differences of Stockholm, and Florence, Paris and Milan. (1973: 14)

Both are concerned to identify the city as a symbol of civilization and culture, but while Mumford expresses the significance of the city in functional terms, Williams's emphasis is more experiential. Cities present many different faces no one of which should be privileged as constituting *the* defining element(s) of it.

Paradoxically, in everyday language there is a sense in which, albeit somewhat negatively, there is little doubt as to what constitutes the city. Popular discourse draws sharp boundaries between the urban and the rural; in other words, the urban is definable in terms of what it is not, the rural. Of course, such a definition is hardly enlightening as to what it is that defines the urban, except that, again in popular discourse, it is through antonyms that the image of the rural, and hence by implication the urban, is interpreted. Exurban movement in the advanced economies is spurred by images of what are read as the benefits of living in rural environments within commuting distance of the city, cleaner, safer environments with better schooling. In fact, these represent imagined attributes of rural areas/smaller places (or as opposites of the large city), rather than defining features of it; closer scrutiny may identify them as more mythical than real.

This is not to dismiss a distinction which in everyday parlance does have real meaning. Rather, like other popular discourses it is the stereotypes through which the subject becomes represented, more than the subject itself, which is problematic. Urban and rural environments *are* fundamentally different precisely because of their different spatiality. It is the density of the city which marks it out as so different from rural environments, densities not only of people but also of economic, cultural and leisure activities. Nor is it just density, or put another way proximity, which marks out the urban from the rural; cities are also characterized by difference, by their mix of culture and race. Combined, it is density and difference, which contribute so much to the excitement of the city, clearly felt by Williams.

Where intuitively urban and rural are distinguishable, and where initially this can be conceptualized in terms of their different spatiality, it is not surprising that spatial definitions of the city are a frequently used method of delimitation. In pre-industrial, as well as in early industrial settlements the distinction between what was urban, and was not, was often sharp, at least in a physical sense, no more so than in the walled city. But even in the contemporary low-density city, sprawling relatively uncontrolled into the surrounding countryside (among many cases the metropolitan area of Perth in Western Australia, for example) the edge of the urban area is readily apparent, precisely because of the changes in population density.

Yet, such a boundary rarely marks the limit of the urban, particularly in the more developed economies where increased mobility has enabled urban populations to live at a distance from the city. Census definitions frequently give recognition to the fact that urban population growth occurs beyond the continuously built-up area. In the more highly urbanized advanced economies – the Netherlands and Britain, for example – the spread of the functional urban region has meant that only the most geographically remote areas lie outwith the influence of a city.

Effectively, the experience of urban development in the advanced economies has meant that decentralization has brought the urban and rural every more entwined physically as well as functionally. But it is not only in the more developed economies that urban and rural have become more entangled. In many sub-Saharan African cities recent commentators have commented how urbanization has spread into the surrounding countryside, within the so-called peri-urban zone, but also that there has been a progressive 'ruralization' of the urban (Stren, 1994). Thus where there is the potential for it, urban farming in African cities has become commonplace to meet basic household needs, and even, for some, generate a small surplus. In South-East Asia McGee (1991) has defined a distinctive pattern of urban growth, *desakota*, in which urbanization has spread successively (and often rapidly) into the surrounding countryside.

The intertwining of town and country has problematized not only the spatial definition of urban and rural, but also whether in social terms the distinction remains real. To some, and particularly in the highly urbanized society, the distinction has little meaning, at least in the sense of there being an urban society distinct from a rural one; as Pahl (1970: 202) memorably said 'In an urbanised society, urban is everywhere and nowhere: the city cannot be defined, and so neither can urban sociology'.

Yet, such a denial is confronted by popular understandings of the differences between urban and rural in those societies which Pahl was describing. At the time of writing this introduction there is an ongoing, and not politically insignificant, conflict in Britain centring on the attempt, not for the first time by any means, to end fox-hunting. Its interpretation by the media and by the advocates of the 'sport' is as a contest between 'townies' and rural dwellers. The holding of mass demonstrations in London, spearheaded by the Countryside Alliance, a loose conglomerate of pressure groups representing rural (especially pro-hunting) interests, has politicized the issue in ways which rekindled an urban–rural divide more characteristic of early, than post, industrial Britain. Urban society has been posited as having different values from rural society, and whether such differences are manufactured for political convenience or not, the conflict has helped solidify the impression of the social differences between the two types of environment.

Though much criticized Wirth's frequently quoted article of 1938 'Urbanism as a way of life' still provides important pointers to the distinctiveness of this urban environment. In essence, Wirth argued this was due to the impress of three main factors, each of which helped characterize the city, size, density and heterogeneity. The size of cities increased the likelihood of impersonal, transitory contact; their density spelt 'close physical contact yet distant social relations'; their heterogeneity was a product of the ethnic and cultural pluralities characteristic of urban populations. Wirth's analysis exemplifies a second way of defining and interpreting the city, in social terms. It is not divorced from the first, the spatial – indeed, size, and particularly density and heterogeneity marry powerfully with the spatial character of the city.

A third more recently pioneered path into the analysis of the city is based on the power of narrative in guiding social action and of informing us of the concepts and experience of the city. Story-telling, as Finnegan (1998) shows us using the example of Milton Keynes, a new city north of London, has much to tell us of the city, not only in the way in which it was planned, but also in the way it is differently experienced by its citizens, and in the multiple readings of the city as text. The power of narrative stems in part from its ubiquity. Clearly, a rich source of narrative arises from the analysis of the novel, which gives lucid, and often particularistic, accounts delimiting the nature of 'cityness'.

These three types are representative of different forms of analysis by which cities have been understood. Methodologically they are distinct. Spatially-based methods of defining the city are reliant on quantitative analysis – cities are measurable, their limits are definable, with greater or lesser precision. Other forms of analysis are more qualitative, relying on textual analysis or in the conducting of interviews to source the narratives which help define the nature, experience and interpretation of the city.

REFERENCES

Finnegan, R. (1998) *Tales of the City: a Study of Narratives and Life*. Cambridge: Cambridge University Press.

McGee, T.G. (1991) 'The emergence of *desakota* regions in Asia: expanding a hypothesis', in N. Ginsburg, B. Koppel and T.G. McGee (eds), *The Extended Metropolis: Settlement Transition in Asia*. Honolulu: University of Hawaii Press. pp. 3–25.

Mumford, L. (1938/1995) The Culture of Cities. New York: Harcourt, Brace. Reprinted in D.L. Miller (ed.) *The Lewis Mumford Reader*. Athens: University of Georgia Press.

Pahl, R. (1970) *Whose City?* London: Longman.

Stren, R. (1994) 'Urban research in Africa 1960–92', *Urban Studies*, 31 (4/5): 729–44.

Williams, R. (1973) *The Country and the City*. London: Chatto & Windus.

Wirth, L. (1938) 'Urbanism as a way of life', *American Journal of Sociology*, 44: 1–24.

2

Defining the City

WILLIAM H. FREY AND ZACHARY ZIMMER

The concept of urban and the phenomenon of urbanization are somewhat new to human populations. In fact, compared to the entire history of human evolution, it has only been fairly recently that people have begun to live in relatively dense urban agglomerations. None the less, the speed at which societies have become urbanized is striking, and the extent to which societies of today are urbanized and the size of present day agglomerations is unprecedented. Davis (1969) notes that before 1850, no society could be described as being fundamentally urban in nature. Today, all industrial nations, and many of the less developed countries, could be described as being urban societies. Moreover, the world is overall becoming more and more urban with the passage of time, as those living in less developed societies move toward the urban living patterns that have been common in more advanced societies for some while.

Despite this rapid transformation of societies from primarily rural to primarily urban, and the importance of this evolution for the study of human populations, the notion of urban remains fleeting, changing from time to time, differing across political boundaries, and being modified depending upon the purpose that the definition of urban would serve. At times, urban populations are defined in terms of administrative boundaries, at times in terms of functional boundaries, and at times they are defined in terms of ecological factors such as density and population size. Although many of today's social problems involve living in very large urban agglomerations, these divergences in defining the notion of urban has made it difficult to conduct comparative studies on urban populations across time and across borders. In a sense, then, the difficulties encountered in defining 'urban' create barriers to understanding the phenomenon completely and finding solutions to a host of social problems that involve the urban population.

In order to place the city of today into context, the following chapter is divided into two discussions. First, we examine past, present and future trends in urban growth. To do this, we begin with a brief review of the history of urban formation. Following that, we examine trends in urban growth, and in specific cities, in the recent past and into the near future using population projections. In the second part of the chapter, we elaborate on a number of concepts concerning the meaning of the term 'urban'. We begin by defining the city of today in terms of several criteria, such as function and space. We then go on to concentrate on a single example, the United States, to further clarify the evolution of the city definition. We conclude by suggesting a new definition of the city that better defines today's agglomerations, the 'Functional Community Area'.

URBAN GROWTH, PAST, PRESENT AND FUTURE

A Brief History of Urban Growth

There is evidence that cities emerged in the world as early as 5,500 years ago, the first of which were in Mesopotamia, the Nile Valley, the Indus Valley and the Hoang-ho Valley. There were several organizational factors that may have precipitated the formation of these early cities, including commercial and trade, religious and political factors. Chandler and Fox (1974) document relatively large agglomerations existing in Babylon (250,000), Patna (350,000) and Rome (650,000) between about 400 BC and AD 100. The number of cities increased during medieval times, although, according to Davis (1969), they remained small. Populations, in general, remained rural and were overwhelmingly involved in agricultural production. In fact, the proportion of people living in urban areas fluctuated between 4 per cent and 7 per cent throughout history, until about 1850 (Lowry, 1991).

The real change in population distributions and urban living patterns occurred with the industrial revolution in the nineteenth century, which made it possible for large numbers of people to live in urban centres. A number of factors are often cited as creating a favourable environment for urban growth. They include, first, the mechanization in rural areas which increased agricultural production and yields per acre, creating the surplus needed to sustain large urban populations that were not involved in agricultural production; secondly, the development of mass production in manufacturing and industry which made obsolete pre-industrial handicrafts; and thirdly, sophistication of transportation and communication systems, brought about in part by the steam engine and railway system, which liberalized trade between places, making urban locations centres of mercantilism. In turn, industries were no longer tied to locations near energy sources and could establish themselves in centralized locales (McVey and Kalbach, 1995). Centralized, diversified economies, made possible by mass production, generated jobs for those who were no longer able to find employment in agriculture. In the end, although the city was dependent upon rural areas for food production, higher production capacity and diversification of production allowed cities to contribute to the economy while improved transportation and communication created international trade routes. New forms of organization in industry allowed for the absorption of substantial numbers of individuals.

The industrial revolution was accompanied by more than just changes in industry. In fact, revolutions occurred in such diverse areas as philosophy, science, government, technology, education, administration, politics and the military. This created the need for interdependence. Urban areas began to thrive on specialization. Multiple functions needed to be conducted within close proximity, creating high population densities. Indeed, even today, it is the diversity of functions that often demarcates between urban and rural areas.

The implications of urban development were far reaching, impacting not merely on the economy, but on the social and ecological order within the city (McVey and Kalbach, 1995). Changes to the city included the emergence of a middle class, an emphasis on social reform, the development of world trade, the founding of financial institutions, the centralization of industry, the decentralization of the upper class, and the shift away from family operated handicraft industry to urban factories. The industrial transformation has also been credited with the nucleation and mobilization of the family, two changes that go hand in hand.

Most point to Europe as being the area that precipitated the modern urban centre. Europe was generally an agrarian society up until the industrial transformation. Only 1.6 per cent of the total population was living in urban areas in AD 1600, which increased to only 2.2 per cent by the beginning of the nineteenth century. Proportions living in urban areas began to double rapidly, however, from that point on. In England and Wales, for instance, the proportion living in cities doubled twice between 1800 and 1900. Davis (1969) also points out the later a country became industrialized, the faster was the pace of urbanization. The change from 10 per cent to 30 per cent living in an urban area with a population size of 100,000 or more occurred over 79 years in England and Wales, among the first countries to industrialize, but required only 36 years in Japan, one of the last of the present day modern societies.

It is important to note how the growth in city population during this period occurred. In the early phases of industrialization, urban areas were characterized by unhealthy living conditions and had excess mortality. Indeed, the adverse conditions faced by the working class in London led to a series of writings from both Friedrich Engels and Karl Marx, which have become among the must influential works in the social sciences. Birth rates in urban areas also tended to be lower than in the countryside. Hence, the growth in the urban population could only have occurred through mass rural to urban migration. Considering the adverse living conditions faced by migrants, the urban area grew by offering advantages to the rural inhabitant in the form of demand for employment and increases in wages from those found in the rural areas.

Modern urban development patterns in the more industrialized world have taken on newer forms as a consequence of improvements in the transportation, the communications revolution and the increasing connectedness of places in a world economy (Castells, 1985; Champion, 1989; Frey, 1993; Frey and Speare, 1988; Sassen, 1991 and this volume). Individual urban areas are now more polarized in their growth tendencies, juxtaposing at extremes: the corporate headquarters-centred 'world cities' as contrasted to low-level manufacturing areas specializing in less than competitive industries. New urban areas are now being developed in a 'low-density mode', following a different model than that offered by the single-core/hinterland development experiences around older cities. Since the Second World War there has been an extensive suburbanization and, more recently, 'exurbanization' which has further blurred the distinction between urban and rural areas, especially as the latter have come to rely less on farming and extractive activities and more on new production, service and recreation industries. Finally, international migration has become an increasingly important source of population growth in selected urban areas in developed countries, in many cases supplanting the more traditional rural-to-urban flows (Champion, 1994; Frey, 1996). As a result, urbanization in the developed world is occurring at a far less accelerated pace than in the past; and it is dependent much more on the establishment of new cities than the growth of older ones (see Chapter 9 below).

Thus far, the discussion has been concentrated on the development of the city in the Western industrialized context. Historically, the development of the modern city occurred much later and under differing circumstances in the less developed nations (Chandler, 1987; Gilbert and Gugler, 1982; Gugler, 1988; Harris, 1992; Kasarda and Parnell, 1993; McGee and Robinson, 1995; United Nations Center for Human Settlements, 1996). In comparison to the more developed and industrialized regions, urban growth in these regions has occurred much more rapidly. In a matter of 20 or 30 years, these areas have experienced the urban growth that took place over centuries in Europe. These rapid rates of urban growth are due to several factors, including: a lack of employment opportunity in rural areas, itself due to rapid changes to the agricultural structure of these areas; the opportunity of employment in urban areas; social networks being set up in urban areas making more accessible transitions to urban life; and ease of communication and transportation into urban areas.

A particular problem in less developed areas is that urban population, although it is a smaller proportion of the total population than in industrialized countries, is more concentrated in fewer cities. There is often just one primate city attracting all rural migrants, a phenomenon known as *macrocephaly*. Such a situation tends to breed an array of problems associated with the inability of a single city to absorb masses of rural migrants. Urban growth that outpaces employment opportunities is termed *overurbanization*. Cairo, Calcutta, Jakarta, Lagos, Seoul are often noted as cities that have become overurbanized. Migrants to these cities are often characterized as being unskilled and illiterate. They tend to live with friends and relatives in already crowded conditions. They place a strain on the water resources available. New inhabitants to these cities tend to live in slums and shantytowns on the outskirts of cities. In addition, urban growth of

this nature poses problems for government, draining the economy of resources, which are used for the increasing transportation burden, water supply and other infrastructures. Countries which face resource limitations have difficulty distributing wealth to rural areas and smaller urban centres. Some would then argue that urban migration of this sort is, therefore, an effective method of redistributing resources and contributing to economic development (Gugler, 1983). Hence, there is some debate as to whether rapid urban growth creates a situation of overurbanization. In addition, some argue that the process of rapid urban growth in less developed countries has begun to slow (Gilbert, 1993). None the less, in the pages to follow, which describe recent trends in urban growth, the role of less developed countries (LDCs), as the dominant region in that growth process, will become apparent.

Recent Trends in Urbanization

We now move on to examine recent urban growth and trends that are observed on a worldwide scale based on statistics compiled by the United Nations.[1] Urban growth has been occurring rapidly over the recent past, and will continue into the near future. These trends are highlighted in Figure 2.1, which displays the urban population in LDCs (less developed countries) and MDCs (more developed countries) starting from 1950 and projecting into the near future, and juxtaposes two measures. First, the bars indicate the percentage of the total world urban population living in either LDCs or MDCs. Second, the lines plotted indicate the percentage of the population in MDCs or in LDCs that live in urban areas.

Looking first at the share of the world urban population living in different areas, we notice that in 1950 about 60 per cent lived in MDCs and 40 per cent in LDCs. Since that time the share living in LDCs has increased rapidly, while the percentage living in MDCs has declined. By 1975, more than half of the world's population was living in LDCs. The shift will continue into the future, so that by the year 2025, it is projected that 80 per cent of the world's urban population will be living in LDCs.

Urbanization and Urban Growth

Urbanization levels, measured by the percentage of the population living in urban areas, is increasing in both LDCs and MDCs, and the increase is clearly more dramatic in the former (see plotted lines on Figure 2.1). In 1950, less than 30 per cent of the world's total population

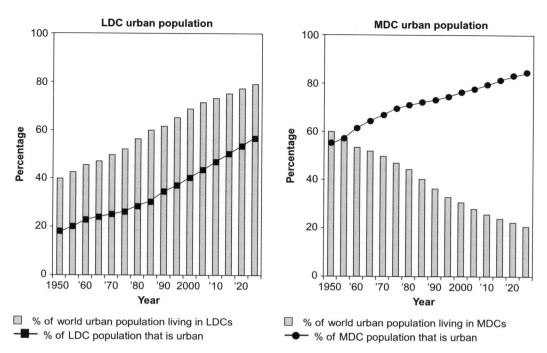

Figure 2.1 The urban population in LDCs and MDCs as a percentage of the world urban population and the percentage living in urban areas

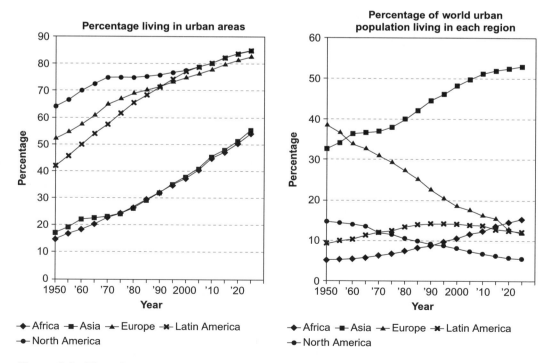

Figure 2.2 The urban population in five world regions as a percentage of the world urban population and the percentage living in urban areas

were living in urban areas. There was, however, great variation between MDCs and LDCs, with levels typically much higher in the former. In 1950, less than 20 per cent of LDC inhabitants lived in urban areas, compared to almost 55 per cent of their MDC counterparts. To give examples using some of the world's most populous countries, in 1950 about 11 per cent of the population in China, 17 per cent in India, 12 per cent in Indonesia, and 36 per cent in Brazil were living in urban areas. This compares to 64 per cent in the United States, 50 per cent in Japan, 72 per cent in Germany and 84 per cent in the United Kingdom.

But, the growth in the urban population has been occurring, and will continue to occur, much faster in LDCs than in MDCs. The result is that although a higher proportion of the MDC population will live in urban areas in comparison to the total LDC population, the gap has been closing and will continue to close into the future. By the year 2025, it is expected that about 84 per cent of those in MDCs will be living in urban areas, as will 57 per cent of those in LDCs. There will continue to be variation between nations in LDCs as it is expected that the proportion who live in urban areas will be, for example, about 55 per cent in China, 45 per cent in India, 61 per cent in Indonesia and 89 per cent in Brazil. There will be less variation in MDCs with, for example, 85 per cent in the USA and Japan, 86 per cent in

Germany and 93 per cent in the United Kingdom living in urban areas.

There are also major differences in urbanization trends between and among major geographical regions of the world, and Figure 2.2 presents this information, this time comparing major world regions (Oceania not included). The left-hand graph shows the percentage of the population, within each region, who live in urban areas. There appear to be two groups of regions. First, there is Africa and Asia. These regions were about 15 per cent urbanized in 1950, but drastic increases in the percentage living in urban areas are taking place in these two regions. By 2000, over 35 per cent will be living in urban areas in Africa and Asia, while by 2025, over half of the population in these areas will be urbanized.

Secondly, there is North America, Europe and Latin America. In 1950 there was major variation in the percentage living in urban areas among these regions: about 40 per cent of those in Latin America, 50 per cent of those in Europe and about 65 per cent of those in North America. These three regions are, however, experiencing some convergence in their urbanization trends. By 2025, between 80 and 85 per cent of the population in each of these three areas will be urbanized. This means, of course, that urbanization growth has been occurring more rapidly in Latin America than in the other two regions, but future growth will be very similar.

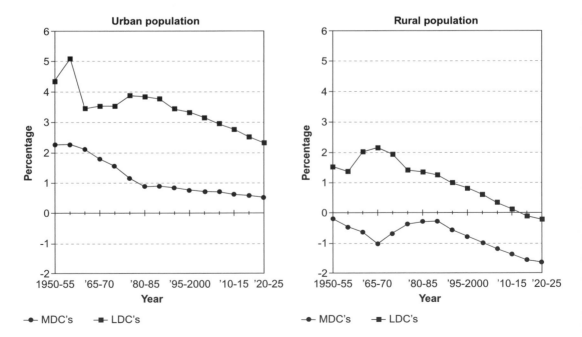

Figure 2.3   The average annual percentage rate of change of the urban and rural population in MDCs and LDCs, 1950–55 to 2020–25

The right hand side of the graph shows the percentage of the total world urban population living in these five regions, and the trends here mimic those shown in Figure 2.1. That is, although urbanization is occurring world wide, a great shift is occurring in the share of the urban population living in various regions. In 1950, the largest share of the world's urban population (about 40 per cent) lived in Europe. At that time, only about 5 per cent of the world's urban population lived in Africa, about 10 per cent lived in Latin America, about 15 per cent lived in North America, and a little over 30 per cent lived in Asia. By 2025, the proportion of the world's urban population in Europe will fall to fewer than 15 per cent. At that time, well over one-half of the world's urban population will live in Asia. About 15 per cent will live in Africa and Latin America, and only about 7 per cent will live in North America.

Urban and rural growth trends

In the remainder of this section, we will attempt to place the trends introduced above into context by presenting a series of additional figures which compare urban growth to trends in rural growth, in LDCs versus MDCs, and among world regions, for the recent past and projected into the near future.

Figure 2.3 presents the average annual rates of change, or growth rates, for urban and rural populations, for MDCs and LDCs, presented in five-year age periods from 1950 to 2025.[2] Looking at the figures for urban areas, growth rates started at between 4 per cent and 5 per cent in LDCs around 1950, and are declining slowly. Today, growth rates in LDCs for urban areas are about 3 per cent, and they are expected to decrease to about 2 per cent in the near future. In MDCs, urban growth rates have fallen from just over 2 per cent per year to just under 1 per cent, but they will remain relatively stable in the near future.

Rural growth rates have been somewhat lower in LDCs since 1950, and, in fact, are negative in MDCs. In other words, the rural population is only increasing slightly in LDCs and is declining in MDCs. Rates of change are expected to be negative in LDCs in the near future. Specifically, rural growth rates in LDCs have been between about 1 per cent and 2 per cent in the recent past and they are declining steadily to near 0 and negative figures over the next 20 years. In MDCs, rural populations have been declining since 1950, when rural growth rates were about –0.25. Yet, the pace of decline slacked off during the 1970s, following a world-wide trend (Champion, 1989; Fielding, 1992; Vining and Kontuly, 1978). By the year 2025, rural areas in MDCs are expected to be losing about 1.5 per cent of their population base per year.

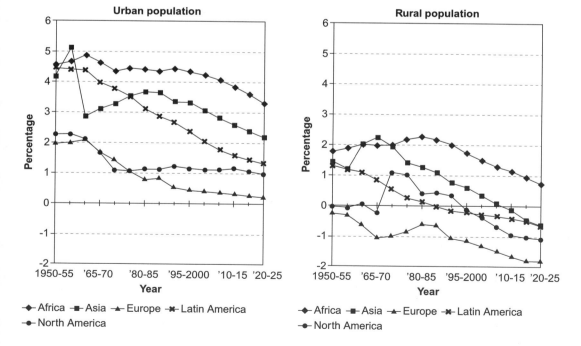

Figure 2.4   The average annual percentage rate of change of the urban and rural population in five world regions, 1950–55 to 2020–25

These rural and urban growth rates can also be viewed in the context of the total world population growth. That is, total world population growth can be partitioned into four growth shares: urban growth in MDCs, urban growth in LDCs, rural growth in MDCs, and rural growth in LDCs. The addition of these four rates is 100 per cent of the total world population growth. When world population growth is considered this way, almost all of the increase in the population is accounted for by increases in the urban population in LDCs. For instance, in 1995, urban growth in LDCs was responsible for about two-thirds of the world population growth. Rural LDC growth accounted for about 20 per cent, and MDC rural and urban growth accounted for a negligible amount. But, the growth in the urban LDCs is increasing so rapidly, that by the year 2010, urban LDC growth will be responsible for 100 per cent of the world growth. Rural areas will either be not growing or declining in population, while urban MDC growth will stabilize at less than 10 per cent of the total world growth.

Average annual rates of growth are shown for five world regions in Figure 2.4. Urban population growth rates are highest for Africa, followed by Asia. Latin American urban growth rates have been declining rapidly, from about 4.5 per cent in the 1950–55 period, to under 2 per cent today. European and North American rates have been declining from over 2 per cent in 1950–55 to between 0 and 1 per cent today and into the near future. Rural population growth is declining in all world regions quite rapidly. It is still positive in Africa and Asia, is near zero in Latin America and North America, and is well below zero in Europe. By the year 2025, only Asia will be experiencing a positive growth in its rural population.

These growth rates translate into changes in actual population sizes between and among world regions. In Figure 2.5, the population totals living in urban and rural areas are plotted, for MDCs and LDCs, beginning in the year 1950 and projected into the future. Looking first at the urban population, between 1950 and 1970 the number of people living in urban areas, world wide, was about equal in MDCs and LDCs. After that point, urban LDC population has outgrown urban MDC population several fold. By 2025, there will be four times as many urban dwellers in LDCs as there will be in MDCs. In 1950, there were about 450 million living in urban areas in MDCs and about 300 million in urban areas in LDCs. The MDC urban population will pass the 1 billion (thousand million) mark around the year 2020, and will be slightly over 1 billion in the year 2025. At the same time, the LDC urban population passed the 1 billion mark around the year 1985 and will be over 4 billion by 2025.

The rural population displays a different trend.

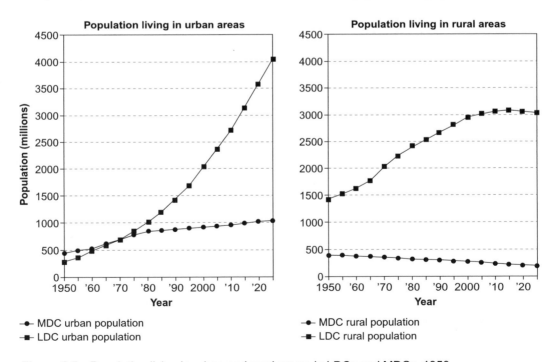

Figure 2.5   Population living in urban and rural areas, in LDCs and MDCs, 1950 to 2025

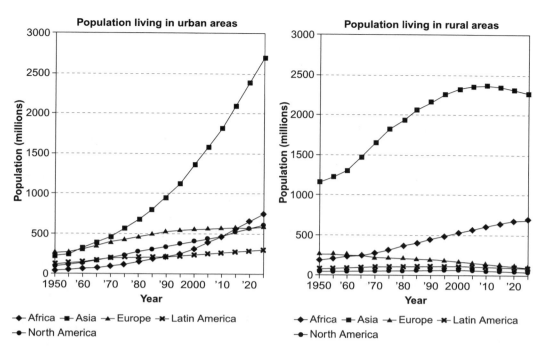

Figure 2.6 Population living in urban and rural areas, in five world regions, 1950 to 2025

LDC rural population has outnumbered MDC rural population since 1950. The differential has increased somewhat between 1950 and 1995, but the change is not as dramatic as is seen in the urban population. Specifically, the LDC rural population, less than 1.5 billion in 1950, will grow to over 3 billion by the year 2005, but will stabilize thereafter and remain at about 3 billion until the year 2025. In MDCs, the rural population was about 350 million in 1950, and has fallen steadily over the years. By 2025, it is expected that there will be less than 200 million living in rural areas in MDCs.

Urban and rural population sizes are displayed in Figure 2.6 for five world regions. The influence of Asia on the change in the world population size in both urban and rural areas, but particularly in the former, can be clearly seen. The Asian urban population was just over 200 million in 1950. By the year 2020, that number is expected to be almost 2.5 billion. At the same time, other areas are experiencing an increase in urban population, although only in Africa is it dramatic. African urban population was just over 30 million in 1950. The total is expected to be well over 600 million by 2020.

Rural population sizes have been increasing in Asia since 1950, but they are expected to stabilize, and perhaps even drop, in the future. African rural population is also increasing, although the population in the other world regions has been either dropping or is increasing at a negligible rate.

To summarize, it is clear that recent urban growth can be accounted for mainly by growth in LDC areas, although there continues to be some urban growth in MDCs. In terms of major world regions, Asia dominates the growth in the urban population in the recent past and will continue to do so into the near future. On the other hand, rural populations are declining in MDCs and are stabilizing in LDCs. In the future, nearly all of the world population growth will be due to increases in urban populations in LDCs. This, for the most part, means increases in the size of the urban population in Asia and in Africa.

Trends in Urban Agglomerations

Large Agglomerations

The shifts in the world urban population structure, discussed in the previous section, means that changes are also taking place to the structure and distribution of the world's largest urban agglomerations.[3] Table 2.1 presents data on the ten largest urban agglomerations for the recent past, the present, and projected to the year 2015. We provide both population sizes and annual average

Table 2.1 *Top ten world cities, in population size, 1955, 1975, 1995 and 2015, and annual growth rates*

1955				1975			
City	Country	Population (m)	Growth rate (%)	City	Country	Population (m)	Growth rate (%)
New York	USA	13.22	1.38	New York	USA	19.77	3.66
London	England	8.93	0.45	London	England	15.88	−0.39
Tokyo	Japan	8.82	4.86	Tokyo	Japan	11.44	0.51
Shanghai	China	6.87	5.05	Shanghai	China	11.24	4.29
Paris	France	6.27	2.84	Paris	France	9.89	4.08
Buenos Aires	Argentina	5.84	2.95	Buenos Aires	Argentina	9.84	0.95
Essen	Germany	5.82	1.90	Essen	Germany	9.13	1.64
Moscow	Russia	5.75	1.41	Moscow	Russia	8.93	1.27
Chicago	USA	5.44	1.91	Chicago	USA	8.89	0.89
Los Angeles	USA	5.16	4.85	Los Angeles	USA	8.55	1.10
1995				2015			
Tokyo	Japan	26.84	1.41	Tokyo	Japan	28.70	0.03
Sao Paulo	Brazil	16.42	2.01	Mumbai	India	27.37	2.40
New York	USA	16.33	0.34	Lagos	Nigeria	24.44	3.27
Mexico City	Mexico	15.64	0.73	Shanghai	China	23.38	1.69
Mumbai	India	15.09	4.22	Jakarta	Indonesia	21.17	1.98
Shanghai	China	15.08	2.29	Sao Paulo	Brazil	20.78	0.70
Los Angeles	USA	12.41	1.60	Karachi	Pakistan	20.62	3.22
Beijing	China	12.36	2.57	Beijing	China	19.42	1.73
Calcutta	India	11.67	1.67	Dacca	Bangladesh	18.96	3.44
Seoul	S. Korea	11.64	1.95	Mexico City	Mexico	18.79	0.68

'Cities' represent urban areas, urban agglomerations, or cities as defined in the 1994 revisions, Estimates and Projections of Urban and Rural Populations and of Urban Agglomerations (United Nations, 1995). See note [1] of text.

growth rates for the previous five-year period. That is, the 1955 data includes average annual growth rates for the years 1950 to 1955.

In 1955, the only agglomeration with a population of more than 10 million was New York, with a population of over 13 million. Of the ten largest agglomerations, only two, Buenos Aires and Shanghai, could be considered to be in less developed countries. The number of the top ten that are in these areas over the years highlights the growth of urban agglomerations in LDCs. In 1975, Shanghai, Mexico City, Sao Paulo, Buenos Aires and Beijing all become top ten urban agglomerations. By 1995, the number of MDC agglomerations in the largest ten is limited to three. By 2015, only Tokyo will remain as a top ten city in an MDC country. New York, which was the largest urban agglomeration in the world in 1955, will no longer be in the top ten, and the smallest population among the top ten in 2015 will be almost 19 million. Seven urban agglomerations in 2015 will have a population of more than 20 million, with six of these being cities in LDC countries.

Agglomeration growth rates determine how a city population will change. Large agglomerations with high growth rates (say 3 per cent or more), will accumulate population very quickly. For instance, an urban area with a population of 1 million, and a 4 per cent growth rate, will have a population of over 2.2 million in 20 years given a constant growth rate over that time period. Lagos, Nigeria, with a population of over 24 million in 2015, and a growth rate of 3.27 per cent, would be expected to have a population of 47 million by 2035 given a constant rate of growth. These dramatic figures indicate that cities with high growth rates become mega cities fairly quickly. Most of the agglomerations listed in the top 10 have had high rates of growth in the past. For instance, although the growth rate for Jakarta is 1.98 per cent in 2015, its high population is due to a 4.35 per cent growth rate in the year 1995. Mexico City's growth rate was over 5 per cent, and Sao Paulo's growth rate was close to 7 per cent at the time that these cities were growing rapidly.

These figures suggest that there may be a problem with overurbanization, or urban growth that is so large as to become problematic in being able to absorb newcomers. It would seem to make some ecological sense that urban agglomerations with extraordinarily high rates of growth will show a slowing in that growth when they become very large. The advantages of a large agglomeration, in terms of employment opportunities, might become overshadowed by the disadvantages in terms of pollution, over-crowdedness, and other such environmental concerns. However, such is not always the case. Mumbai (formerly Bombay) India, for instance, with a population of over 15 million in 1995, still maintains a growth rate of over 4 per cent. Its population will reach 27 million by the year 2015.

Table 2.2 presents the number of large urban agglomerations in various world regions with very high current rates of growth. (Appendix 2.1 provides the names of all of these cities, their location, growth rates and their population size.) The table shows that most of these very high growth areas are located in Asia. For instance, of the 127 large world cities with growth rates of 3 per cent or higher, 92 are located in Asia. Thirty-eight of these Asian cities are located in China, but ignoring these, the rest of Asia still accounts for 54 of the world's fastest growing large urban agglomerations. There are 13 cities with 5 per cent or higher growth rates. More than half of these are located in Africa and the rest are in Asia. Clearly, there are a number of African urban agglomerations that will soon rate among the most populous in the world. Europe contains none of the urban agglomerations with the fastest growth rates, while North America contains only one (Norfolk). Latin America contains 11 of these, and all have growth rates of between 3 and 4 per cent.

As noted above, the combination of a large

Table 2.2 *Number of cities with a population of one million or more in 1995 and 3 per cent or higher average annual growth between 1995 and 2000, by region and growth rate*

Region	Average annual growth			
	3–4%	4–5%	5% or higher	Total no. cities
Africa	9	7	7	23
Asia	65	21	6	92
Europe	0	0	0	0
Latin America	11	0	0	11
North America	1	0	0	1
Total	86	28	13	127

population and a high growth rate may be the recipe needed for creating overurbanization. There are 13 world cities that have a population of 2 million or more and growth rates of over 4 per cent. These not only have a high current population, but the increase over the next couple of decades will be dramatic and potentially problematic. Seven of these cities are located in Africa (Lagos, Nigeria; Abidjan, Côte d'Ivoire ; Khartoum, Sudan; Maputo, Mozambique; Addis Ababa, Ethiopia; Luanda, Angola; Nairobi, Kenya), while the other six are located in Asia (Karachi, Pakistan; Dacca, Bangladesh; Hyderabad, India; Riyadh, Saudi Arabia; Kabul, Afghanistan; Lucknow, India). Some of these already have population sizes that place them in the current list of world's largest urban agglomerations, but the others are fast growers and will be in the largest cities category in decades to come. Table 2.3 presents the population sizes for these

Table 2.3 *Population of cities of 2 million or more in 1995 and 4 per cent or higher average annual growth from 1995 to 2000, for selected years*

City	Country	Population (m)				
		1975	1985	1995	2005	2015
Lagos	Nigeria	3.30	5.83	10.29	17.04	24.44
Karachi	Pakistan	3.98	6.34	9.86	14.64	20.62
Dacca	Bangladesh	1.93	4.41	7.83	12.95	18.96
Hyderabad	India	2.09	3.19	5.34	8.04	10.66
Abidjan	Côte d'Ivoire	0.96	1.65	2.80	4.41	6.61
Riyadh	Saudi Arabia	0.71	1.40	2.58	3.87	5.12
Khartoum	Sudan	0.89	1.53	2.43	3.77	5.78
Maputo	Mozambique	0.53	1.09	2.23	4.09	5.76
Addis Ababa	Ethiopia	0.93	1.49	2.21	3.47	5.85
Luanda	Angola	0.67	1.24	2.21	3.64	5.55
Nairobi	Kenya	0.68	1.13	2.08	3.48	5.36
Kabul	Afghanistan	0.67	1.24	2.03	3.55	5.38
Lucknow	India	0.89	1.25	2.03	3.02	4.06

fast-growing urban agglomerations for 10-year intervals beginning in 1975 and projected to 2015. Several examples will demonstrate the dramatic nature of the population growth presently occurring in these places. Lagos, Nigeria had a population of 3.3 million in 1975 and has increased in size by about 7 million over the past 20 years. Hyderabad, India, had a population of about 2 million in 1975 and its population will reach 10.6 million by 2015, an increase of well over 8 million people in 40 years. Maputo, Mozambique was a city of about half a million people in 1975. Today, the city has about 2 1/4 million, and by 2015 it will have nearly 6 million inhabitants.

It is clear, from this discussion, that the most spectacular changes in population are occurring in Asian and African cities. Not only do these regions have several selected urban agglomerations that are growing rapidly, but as Table 2.4 illustrates, the actual number of cities with 1 million or more in population has grown most rapidly in Asia in comparison to other regions. In 1950, there were slightly more large cities located in Europe than in Asia. By 1965, Europe had 39 cities of 1 million or more, compared to 50 located in Asia. The number of these cities has increased somewhat in Europe over the years, while the number has virtually exploded in Asia. By 1995, the number of Asian cities of 1 million or more far outnumbered those in Europe. As can be seen in Table 2.4, Latin America, North America and Africa all have fewer numbers of these large cities, but all three appear to be catching up with Europe. The number of African cities of 1 million or more has grown from just a couple in 1950 to 31 in 1995. Latin American cities with over 1 million in population now number 44, while there are 39 North American cities of this size, mostly located in the United States.

Table 2.4 *Number of cities with a population of one million or more by year and region, 1950–1995*

Region	No. of cities			
	1950	1965	1980	1995
Africa	2	5	14	31
Asia (incl. Oceania)	27	50	84	142
Europe	29	39	56	62
Latin America	7	15	24	44
Northern America	14	23	32	39
Total	79	132	210	318

Distribution by City Size

We finally turn to the distribution of the world urban population by city size. Although we have, up to now, been examining the large and largest urban agglomerations in the world, a majority of urban dwellers live in places with less than 1 million in population. None the less, this too has been changing, as more and more of the urban population becomes accounted for in large agglomerations. Figure 2.7 shows the distribution of the urban population by agglomeration size for selected years between 1950 and 1995. In 1950, 74 per cent of the world urban population lived in cities under 1 million in population. In comparison, only 2 per cent lived in extremely large agglomerations of 10 million or more. The remainder lived in cities between 1 and 10 million in size, with 8 per cent being in cities of between 1 and 2 million, 10 per cent in cities between 2 and 5 million, and 6 per cent in cities between 5 and 10 million. Over the years, the percentage living in urban agglomerations with under 1 million in population has fallen, and was 65 per cent in 1995. The proportions living in cities of various sizes between 1 and 10 million has increased but only slightly. For instance, the percentage living in cities of 1–2 million has increased to 11 per cent of the total urban population. The largest increase has come in the percentage living in the very largest agglomerations, with populations of 10 million or more. This proportion has quadrupled since 1950, and doubled since 1980, to 8 per cent of the total world urban population.

What this means in terms of actual population is that more and more people are living in larger and larger urban centres. In 1950, about 547 million of the 737 million living in urban areas lived in cities with populations of less than one million, while 12 million people were living in the largest sized urban agglomerations, places with populations of 10 million or more. By 1995, the number living in cities of less than one million was about 1.6 billion, an increase of about three times that of 1950. However, the population living in cities between 1 and 2 million, 2 and 5 million and 5 and 10 million increased about fourfold. The population living in cities of 10 million or more increased to over 17 times what it was in 1950, to 211 million.

Urban growth in general is occurring more rapidly in LDCs than in MDCs, and in Asia and Africa more than in other world regions; moreover, the very large urban agglomerations, which have up to recent times been concentrated in more industrialized nations such as Germany, the United States, and England, have begun to characterize the urban distribution in LDCs. Although there continues to be a substantial number of cities with large populations in MDCs, large LDC agglomerations are beginning to outnumber these. In addition, looking at the distribution of the urban population by city size, we noted that more and more of this population is inhabiting the world's mega cities, which have populations of over 10 million, and more of these mega cities are becoming common in LDCs. Judging by current

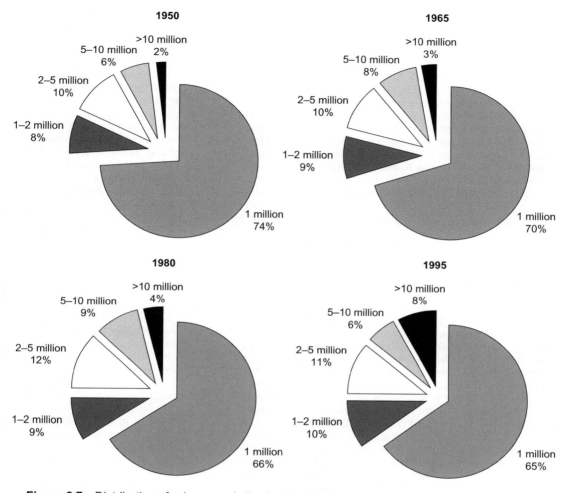

Figure 2.7 Distribution of urban population by city size for selected years

city growth rates, the future will see an even greater expansion in the number of these largest agglomerations in Asia and Africa.

DEFINING THE CITY

Thus far the discussion of urban centres and urban growth has taken place without any specific denotation of what we mean by the term 'urban' or 'urbanization', or how a city is defined. We now will move on to this topic.

Criteria for Defining an Urban Area

The city and the concept of urban are interrelated, and in order to define a city it is necessary to first establish what is meant by urban. We may consider the city, in fact, to be an administrative definition that places a boundary on a contiguous urban area. Problems in terms of analysis arise when urban is defined differently across administrative and national boundaries. Generating a specific definition is difficult. The difficulty arises from the fact that the concept of urban, from which city boundaries are determined, is an abstraction that involves a series of interrelated factors, some of which are: population size; population density; space; economic and social organization; economic function; labour supply and demand; and administration. Yet, for practical statistical collection purposes, national statistical definitions often identify urban places in terms of criteria such as administrative boundaries or in terms of size of population living in a given area. Hence, such official definitions are often simplifications of the broader set of concepts, implying that where, for instance, population size is large enough, economic functions, social organization and other factors tend to resemble what is normally thought to represent other elements of urban (Goldstein and Sly, 1975). Further, countries differ widely in their urban definitions, making comparative analyses difficult.[4] Although the definition of urban may be fixed, city boundaries tend to be fluid, often changing with alterations to a local urban environment, like the addition of population to outskirt areas which could widen the city boundaries. As such, urban growth can occur due to reclassification rather than population growth within a given space.

Three Elements of the Urban Concept

In its essence, any definition of the concept of urban needs to differentiate between it and the non-urban part of the settlement system. It is possible to denote three elements which best distinguish between a rural and urban character. First, there is an *ecological element*. Spatial considerations of urban normally revolve around factors such as population size and density. It is here that great differences exist between nations. In the United States, for instance, areas with populations of 2,500 or more are considered to be urban. For Denmark, a population of 250 or greater is considered to be urban, while in India, it is 5,000 or more. An example of a density-defined urban area is Japan's 'Densely Inhabited Districts', which considers contiguous districts having a population of at least 5,000 and of a given population density. Such definitions ignore any administrative boundaries, or economic functions taking place within the densely inhabited area.

Second, there is an *economic element*, which considers the function of the urban area and the activities that take place within the area. In comparison to a rural area, the majority of the economic activity in an urban centre is organized around non-agricultural production. A definition of urban based on non-agricultural functioning captures a different dimension than those based on density and population size. For instance, Davis (1969), examining national differences in urban definitions, noted that the simple ratio of agricultural workers to rural population in LDCs can be high, in part because a high number of those in urban areas are involved in agricultural production. Therefore, it is advisable to consider that in addition to non-agricultural activity, indicators of a diversity of such functions, as well as a measure that shows their concentration in the (urban) area. The variety of economic functions that take place in an urban centre includes various types of production, but also educational,

political, administrative and socially related economic activities which tend to employ a diversely orientated labour force.

An important related concept is that of 'agglomerative economies', which are a concentration of economic functions that operate external to a particular firm but make it advantageous for a firm to locate there. For instance, other firms, banking, credit, transportation and storage facilities tend to exist in and around urban centres. All of these are essential to the growth of a particular firm. Agglomerative economies tend to attract population, and increase the density of a given area, and economic activities, in turn, become concentrated in areas with agglomerative economies.

With increases in diversity of function, and agglomeration of economies, tends to come increased movement in and around the urban area as people commute to and from work. Commuting patterns are often used as a criterion for defining an urban space. Cities in a number of developed regions have expanded in size due to a *deconcentration* of the population, as individuals choose to live further and further away from a central core in suburban areas. This results in both an expanded urban territory in terms of size and an increased density in areas surrounding urban cores. These trends require frequent transformations to the definition of an urban boundary, often resulting in a reclassification of boundaries.

The need for flexibility has resulted in the development of the 'metropolitan area' concept. In many developed nations, the metropolitan area has taken over as the single most important notion that determines where one urban agglomeration ends and another begins. Metropolitan areas are generally considered to be contiguous areas that are under the primary influence of an urban core. Metropolitan areas then encompass both the densely settled areas and their less dense surroundings that are clearly under the dominance of the urban core in terms of economic functions. Complicating matters, however, is the fact that in some large metropolitan areas, there can be subareas which are more rural in nature, that is, small areas within a metropolitan area that are sparsely populated, have few amenities, have some agricultural production, and so on. In addition, some rural areas have urban characteristics, such as banks, hospitals or non-agricultural economic activity.

Ultimately, it may be the nature of the people that defines urban and, hence, the third element distinguishing the rural from the urban is the *social character* of the area. It is common to consider the differences, for example, in the way rural and urban people live, their behavioural characteristics, their values, the way they perceive the world, and the way they interrelate. The social element can be referred to as the degree of 'urbanism', which is a term referring to the way of life that is associated with urban areas. Related to this are environmental differences that shape the social character of the urban centre. These can be site characteristics, such as piped water, lighting, or the entertainment facilities that tend to be located in an urban area. These characteristics are often referred to in a positive sense. They can also be negative characteristics, such as crime, congestion and pollution.

There are two difficulties in distinguishing urban areas on the basis of social character. First, particularly in less developed countries, rural traditions often remain strong among those who migrate to urban areas, making such demarcations difficult to establish. Secondly, particularly in more advanced countries, many rural areas begin to take on the characteristics of an urban core, despite a lower population density. For instance, entertainment, services, styles of speech and values may tend to be relatively similar. This suggests that there is no real or single demarcation dividing an urban from a rural area, and the two exist on a continuum, with some areas being more urban than others. This would call for an index of the degree of urban character displayed by a given centre or area. Yet, for the purposes of policy, statistical tabulation and research, it is often necessary to dichotomize rural and urban areas, creating definitions that clearly differentiate one from the other.

The question that remains is how have these elements influenced the definitions of urban areas that are commonly used? Here we note that governments and demographers tend to define cities in terms of a combination of the above stated elements, that is, size, density, function and degree of urbanism. Yet, it is important to note that most of these definitions tend to serve varying purposes, such as those important for administration versus research and policy. The types of definitions required may then differ according to the applied purpose. Policy analysts dealing with the environment, for instance, may be interested in definitions that take advantage of spatial elements. Others may find it more advantageous to accept already defined administrative boundaries. Demographic and economic types of analyses may be better served with definitions that consider the function of the urban area and the degree of interaction that takes place within a given area.

For the most part, researchers tend to accept administrative boundaries due to the difficulty in obtaining adequate data for defining urban areas. It is therefore necessary to consider whether definitions of urban areas are fixed or fluid. Where they are fixed, changes in urban population must occur by demographic processes only: births, deaths and migration. Where they are fluid, reclassification becomes an important element in urban population change. Fluid definitions make comparisons over time difficult, but tend to better characterize the changes that identify urban settlement patterns. Where boundaries are more fixed, urban growth may not be fully reflected in statistics. Where they are fluid, population may be shown to increase despite declines to the core. Here, it is necessary to recognize whether the boundaries of an urban area are underbounded or overbounded. In the former case, there are additional contiguous areas built up around an administratively defined core that could be considered part of the urban centre given alternate definitions. In the latter case, the urban centre includes low-density areas that more resemble rural areas in character.

Adjustments to the urban boundary are made more frequently where fluid definitions are used. Adjustment problems occur when definitional parameters change. In fact, the flexibility in defining the city will determine, to some extent, the degree of over- or underboundedness of a centre and the speed of change. Underbounded cities tend to arise in places that utilize more rigid definitions, while changes to these boundaries tend to occur rather quickly.

In sum, we have put forward three elements that best distinguish urban areas: the ecological, the economic and the social. Yet, these three elements in themselves are difficult to measure and define. Hence, we are often left with definitions that tend either to support particular purposes or be predefined by administration. In the following section we concentrate on a specific example, that is, defining the metropolitan area in the United States. It will be seen that here it is functional features that become the essential element in the definition of metropolitan areas, and we will later suggest that this may become the most efficacious way of defining the modern city.

The USA as an Example

With 35 metropolitan areas with a population of one million or more, the USA has more urban areas of that size than any other MDC, and has the world's second highest number of large urban agglomerations next to China (which had 47 in 1995). The USA also has the largest number of people living in urban areas in MDCs (201 million in 1995) and has the third highest urban population in the world next to China (369 million) and India (251 million). The sheer number of urban areas makes the USA an interesting case study for the development of the urban notion and for suggesting policy for the future definitions of what constitutes urban. This notion has been developed since the 1940s, with changes that were intended to reflect the changes to the settlement patterns. By reviewing a case study of this nature, we hope to alert the reader to the complexity in defining the city in a highly populated nation with numerous urban centres. We will first review the historical development of the urban concept in the United States, then assess the relevance of this concept in light of changing settlement patterns and discuss the notion of the Functional Community Area, a possible substitute for the commonly used term metropolitan area.

The first distinction in regards to an urban centre was made in 1874 (Peters and Larkin, 1993), where towns with a population of greater than 8,000 were considered to be urban. In 1900, the census lowered the size criterion to 2,500, and in 1950 a density criterion was added. The formulation of the modern metropolitan concept in the United States is dependent upon the notion of a metropolitan area, originally seen as an economic unit where a cluster of activities in a core location dominated export, import and service functions that sustained the population of a surrounding hinterland that was economically and socially integrated with the core area.[5] Historically, this functional definition coincided with the physical properties that were common to most metropolitan areas in the 1940s when the metropolitan concept was first conceived. Socio-demographic, industrial and land use characteristics also patterned themselves in common ways as distance from the core increased. Because of the correspondence between functional and physical space, the metropolitan area could be operationalized by identifying core areas with population size and density criteria, and hinterland areas by measures of integration with the core.

Operationalizations of the metropolitan concept have changed over the years, as have the definitions, although the basic concept has remained the same. In 1949 the Standard Metropolitan Area (SMA) was based on a large population nucleus together with adjacent communities that have a high degree of integration with the nucleus. Integration was defined mainly by commuting trips. MSAs were defined in terms of counties or county equivalents. For all censuses between 1950 and 1990, metropolitan areas have been defined as including a densely settled urban core with a population of at least 50,000, the rest of the county in which most of this core was located, and any contiguous counties which met both the criteria of metropolitan character and the criteria of integration with the core.

Various changes to the criteria for defining SMAs have occurred: a 1950 revision stipulating two contiguous cities with a combined population of 50,000 as a nucleus of a metropolitan area, providing that the smaller had at least 15,000; a 1971 modification to allow a city of 25,000 to qualify if the total population of the city and surrounding places with a density of 1,000 or more persons per square mile was at least 50,000; a 1980 concept enlarging the urban core to include urbanized areas of 50,000 without necessarily a larger central city if the metropolitan area had a population of 100,000. Changes occurred in congruence to changes to the term SMA to Standard Metropolitan Statistical Area (SMSA) and Metropolitan Statistical Area (MSA).

Since 1949, adjacent counties were defined as adding to the metropolitan area if they met criteria of a metropolitan character and social and economic integration. In 1950 this meant that one-half of the population were living in minor civil divisions with a density of 150 or more persons per square mile and less than one-third of labour working in agriculture, 15 per cent working in the central city county of 25 per cent commuting from the central county. In 1958 this changed to 75 per cent of population of a contiguous county being employed in non-agricultural areas. In 1980, the minimum agricultural content was dropped and a sliding scale combining integration and metropolitan character was adopted. For instance, an adjacent county in which 50 per cent of the workers commute can be added with a density as low as 25 persons per square mile, while if 15 per cent of workers commute, the density needed to be 50 persons per square mile.

The Standard Consolidated Area (SCA) was added to the definition in 1960. This was meant to provide an alternative aggregate unit that included two or more SMSAs that were closely integrated. In 1975, definite criteria of size and integration were established and the name was changed to Standard Consolidated Statistical Areas (SCSA). In the 1983 revision, metropolitan areas with over one million population which contained two or more counties, were divided into two or more Primary Metropolitan Statistical Areas (PMSA) if local opinion supported such a division. The original metropolitan area was known as a Consolidated Metropolitan Statistical Area, and the components were called Primary Metropolitan Statistical Areas.

The original concept of the metropolitan area fitted the settlement patterns that existed prior to 1950, that is, functional areas could be approximated with physical attributes, areas contained a dense central city, with spreading areas declining in density. The application of the criteria for defining metropolitan areas has been criticized, however, the argument being that the nation's settlement areas evolved in ways not anticipated by the original concept. On the one side, those who feel that a metropolitan area should be a relatively autonomous economic area have pointed out that most officially defined metropolitan areas are underbounded in terms of including all of the population which depends upon the area for certain services such as public utilities, retail shopping, medicine, education and other personal services. On the other side are those who associate metropolitan character with size, density and the performance of certain functions. These critics feel that the concept has been stretched to allow more and more marginal areas to qualify for federal programmes targeted for metropolitan areas. Most notable among these critics is Calvin Beale who has pointed out that the new metropolitan areas which were designated in the 1970s lack many of the facilities which might be expected of a 'metropolitan area', such as a television station, a Sunday newspaper, local bus service, a four-year college and specialized hospital services (Beale, 1984).

It is true that since the metropolitan concept was put into use, there have been massive shifts in the patterns of settlement in the USA which have called into question the applicability of this concept for future decades. Improved transportation and communication technologies, together with massive federal subsidies, have led to a continued spread of residential, retail and manufacturing activities away from the central cities (Long, 1981; Zimmer, 1975). As a result, cities have experienced suburbanization, multinucleated suburbs and suburbs with diversified economies. Moreover, low-density development has taken place in previously undeveloped areas. This has particularly occurred over the past two decades, where the expansion of metropolitan population was the result of the continued spread of population into new territory around existing metropolitan areas, and the establishment of new metropolitan areas in less densely populated parts of the country (Frey and Speare, 1988; Long and DeAre, 1988).

This redistribution has been followed in employment shifts as metropolitan core jobs began shifting to outside suburban areas. Stanback (1991) contends that these changes brought about a new era of metropolitan development, with suburban areas taking on many of the functions of the old core, such as wholesaling and business related services. Other suburbs have achieved agglomeration economies in competition with the old core with respect to key export services. These areas house high-tech and office complexes, divisional offices, sales centres and corporation headquarters. Surrounding these are hotels, retail stores and entertainment. The result is a deconcentration that questions the original city–hinterland model and calls for a new definition that includes both newly developed and older areas as metropolitan areas.

One final aspect of the national settlement system that has changed since the mid-century is the nature of those areas that lie outside of metropolitan areas, as currently defined. In the 1940s the territory outside of metropolitan areas was more predominantly rural and less integrated into the national economy than has been the case for the past two decades. While the population and economic characteristics and territory now classed as 'non-metropolitan' still shows some distinction from that in metropolitan areas, improvements in transportation, communication and the organization of production have served to integrate economic activities in non-metropolitan areas to those in the rest of the country (Fuguitt et al., 1989). Also, around 1970, residential and employment activities began to deconcentrate around many small and moderate sized places, following a pattern that has not heretofore existed in metropolitan areas.

Towards a New Definition of the City

The US example has highlighted several limitations with the current concept of the metropolitan area. First, the definition of the metropolitan area seems to be too wedded to the central core–hinterland concept of settlement area. The need for high central densities no longer exists and there is no reason why a modern post-industrial settlement could not be developed around a set of dispersed labour market areas that could be entirely 'suburban' in character while providing employment, shopping and recreation for its inhabitants. Second, the MSA definitions are tied to county building blocks, which may be too large to adequately define functional or activity space. Third, with the present metropolitan statistical system, much of the country may be left out. The vast territory now classified as 'non-metropolitan' has become more integrated both with the metropolitan economy, and, internally, on the basis of local labour market areas. Moreover, to the extent that government agencies and private sector analysts find metropolitan areas useful in their planning, they may ignore the population living in non-metropolitan territory simply because a manageable classification scheme is not available.

It is beyond the scope of this chapter to develop a new definition of the city. However, as 'food for thought' we are able to pass along some suggestions that may be helpful in the development of such a conceptualization. Following Frey and Speare (1995), we suggest that new definitions of the city be based upon a notion of a Functional Community Area (FCA), which represents, to the extent possible, a self-contained local labour market within an area characterized by high frequencies of daily interaction.[6] In other words, the city may be best defined in terms of a functional metropolitan community. This notion agrees with Hawley's (1971) conception of an enlarged area of local life:

> The concept of the metropolitan area lends itself to various definitions . . . It may apply to an enlarged area of local life, i.e., with a radius of twenty-five to thirty miles, or it may refer to a much broader area in which the scattered activities have come under the administrative supervision of a metropolis. The former is what is usually denoted when the term *metropolitan area* or *metropolitan community* is used; *metropolitan region* is ordinarily reserved for the latter.
>
> The principle of the metropolitan community, as well as the metropolitan region, is delineated by the frequency with which outlying residents and institutions transact their affairs in the metropolis, whether through direct visitation or through indirect means of communication. These frequencies . . . decline in gradient fashion with distance from the center. Thus figuratively speaking, one might rotate a gradient on its center and sweep out a zone in which the residents routinely engage in a given frequency of communication with the center. (Hawley, 1971: 149–150)

As such, the FCA is not tied to any physical configuration, such as population size and density criteria, or location in urbanized areas. They may be specified solely on the basis of measures of interaction. Because the same kind of interaction procedures are used to designate FCAs in rural as well as urban parts of the country, they are not formally distinguished on the basis of 'metropolitan' and 'non-metropolitan' status. As such, they encompass a much larger portion of the national territory than the current system.

The measures of integration that may be used to designate FCAs are those traditionally used to define local labour markets. A labour market is an area within which a worker can commute to work. We assume that it is possible to identify spatially distinct labour markets on the basis of commuting data. The FCA concept does not presume to identify homogeneous areas on physical characteristics. Neither does it presume to identify homogeneous areas on population or housing attributes. The main criteria for identifying these areas are high levels of interaction.

One commuting-based procedure for identifying FCAs has been conducted by Killian and Tolbert (1991). Their procedure starts with a county by county matrix of place of work by place of residence. Unlike the procedure for defining current MSAs, their procedure uses flows in both directions between pairs of counties. There is no attempt to define one county as 'central'. They used a two-step procedure in which a computer algorithm was first used to group counties into commuting clusters and these clusters were then aggregated into labour market areas. Recognizing the need for a minimum size for labour market areas either to provide reliable estimates for some measures based on samples or to protect confidentiality on public use samples of individual data, they aggregated adjacent commuting clusters to provide labour market areas with at least 100,000 population. This aggregation was based primarily on commuting flows between clusters and secondarily upon pure proximity when flows were too weak to link clusters with less than 100,000. In this fashion, they divided the USA into 382 labour market areas.

How small should a local labour market be? If an area had a square shape and residences and workplaces were randomly distributed throughout the area, the average commuting distance would be about 0.6 times the length of one side of the square. Assume an average commuting distance of 6 kilometers, an area of 10 kilometers by 10 kilometers, or 100 square kilometers, would be large enough to contain a single labour market area. Yet, there should also be a minimum size for building blocks, based on population size.

The above described FCA designation has a purely functional basis. The end goal is to obtain labour market areas that closely reflect actual commuting areas. The data required for such formulations are commuting and population data, which can be provided by a national census. Because this data needs to be reliable, we would not advocate updating the system between census enumerations unless reliable commuting data can be obtained elsewhere. In the end, we see the FCA designation as more applicable to the current patterns of urban growth.

This proposal for the USA is based on a generalization of the functional concept which is already used with the current metropolitan area definitions in the United States and those employed in many other developed countries (for example, see Coombes et al., 1982). Yet, the US federal government's Office of Management and Budget, along with the Census Bureau, has begun to examine several different formulations for a new metropolitan area concept that they plan to establish in the next decade (see Dahmann and Fitzsimmons, 1995). The above proposal represents one of these, but other suggestions incorporate both ecological (i.e., morphological) *and* functional underpinnings (Adams, 1995; Berry, 1995; Morrill, 1995) and additional proposals for the new system will continue to be solicited. Similar attempts to harmonize and redefine urban areas of the countries of the European Union have been under way, where specifications are being drawn for both morphological and functional urban area concepts

(Byfuglien, 1995; Pumain et al., 1992). Clearly, the greater spread and inter-urban connectedness of settlement in most developed regions of the world, fostered by the rise of the service economy and the communications revolution, calls for a re-examination of what represents a 'city' in these regions. The new concept should be more flexible, emphasizing interactions and functional ties rather than specific ecological or morphological forms.

CONCLUSION

In order to conceptualize the present world urban situation and the current notion of urban, we have concentrated our discussion in this chapter on two issues. First, we reviewed the past, present, and future trends in urban growth. By doing this, we have highlighted the role of less developed regions – Asia and Africa, in partic-ular – as being the most influential in current world urban changes. Second, we discussed the notion of the city by defining what is meant by the term urban, by reviewing a specific example of defining a city, and by suggesting a new notion, the 'Functional Community Area'. There is a connection to be made between these two discussions. The notion of the metropolitan area allows for a more comparable definition of a city in MDCs. Yet, urban growth is occurring with more rapidity in LDCs. In these regions, the city has been more often defined strictly in terms of spatial characteristics, such as population size. With the growth of the urban area in these regions, it is clear that some of the notions of metropolitan area need to be employed in defining LDC cities.

Variations in definitions of cities result in various levels of underbounded or overbounded-ness. For instance, in more advanced countries, the term metropolitan area includes zones with urban features, like economic functions, and commuting patterns which stream into the core. But, settlement around the core may be sparse, suggesting that metropolitan areas can often be overbounded. Although less advanced city areas are more often defined in terms of population size, urban boundaries are often surrounded by squatter settlements. City boundaries that ignore such settlements may be underbounded. City boundaries that include such settlements ignore the fact that these individuals tend often to be socially quite rural. Those living in these outskirt areas are, however, often employed in the core area, though often in tertiary and fringe occupa-tions.

These issues suggest that definitions of the city may need to differ by levels of development and city function. Despite difficulties involved in data collection, it is obvious that a more universal defi-nition of the city would facilitate future studies, comparisons and development or policy in the area of urban studies. We suggested the notion of the FCA, which could be examined in terms of its validity across a variety of nations in various stages of development. It is only after a more universal determination of the city is developed that we may go on to study the city in a compar-ative way across societies.

NOTES

We are indebted to the late Professor Alden Speare, Jr of Brown University, whose ideas and collaborative work with the senior author have contributed substantially to this chapter.

1. These statistics are drawn from the United Nations Population Division's 1994 Revision of Estimates and Projections of Urban and Rural Populations and of Urban Agglomerations (United Nations, 1995). Urban areas (referred to in the text as either urban areas, urban agglomerations, or cities) represent, for the most part, urban agglomerations that include a city or town plus the suburban fringe lying outside of, but adjacent to, the city boundaries. However, in some cases, adjustments were made from the original data, and in other cases, statistics refer to the city proper or (in the case of the United States) metropolitan areas.

2. These figures are calculated by taking the differ-ence in population between two years, converting it to a percentage with the earlier year as a base, and averaging these growth percentages over a five year period.

3. We have not yet defined a city in a theoretical sense, hence, in this section we will use the general term urban agglomeration, urban area, or 'city'. These statis-tics, compiled by the United Nations, are not compa-rable across countries (see note 1). We have more to say about definition comparability below.

4. A review of existing urban definitions and data-bases has been undertaken by the Network for Urban Research in the European Community (NUREC) (Byfuglien, 1995; Pumain et al., 1992) and has concluded that the level of comparability across nations remains low, partly because the basic geographical building blocks differ greatly in area. Individual comparative studies include: Brunet (1989), Davis (1959, 1969, 1972), Dickinson (1964), Eurostat (1994), Hall and Hay (1980), International Institute of Statis-tics (1963), OECD (1988), United Nations (1995), Vandenberg et al. (1982).

5. The metropolitan area, a functional concept, is the urban spatial unit most widely used in the United States by government agencies, local planners, scholars, private sector marketing specialists, and the public at large. As such, it is defined by the Office of Management and

Budget, an agency of the federal government's executive branch. Nevertheless, the Census Bureau also defines an urban area concept, the urbanized area, on the basis of physical settlement patterns (US Bureau of the Census, 1993). An urbanized area comprises one or more central places and the adjacent densely settled surrounding territory that together include at least 50,000 people. The density criterion is foremost in this definition such that the urban fringe surrounding the central place must have a population density of at least, 1,000 square miles. Most metropolitan areas have an urbanized area lying within it.

6. An especially coordinated effort, following the 1980–81 round of censuses, was conducted in a joint project at the Department of Geography, University of Reading (UK) and at the International Institute for Applied Systems Analysis (IIASA). The project identified comparable Functional Urban Regions (FURs) for most Eastern and Western European countries as well as Japan (Hall and Hay, 1980; Kawashima and Korcelli, 1982) that followed a functional definition similar to the metropolitan area definitions utilized in the United States (Frey and Speare, 1988) and that used in Canada and Australia (Bourne and Logan, 1976). For the most part, these areas have not been used for official government policy purposes, but have been useful in a number of scholarly works requiring cross-national comparability in urban area units (see, for example, Cheshire et al., 1988; Frey, 1988).

REFERENCES

Adams, John S. (1995) 'Classifying settled areas of the United States: conceptual issues and proposals for new approaches', in Donald C. Dahmann and James D. Fitzsimmons (eds), *Metropolitan and Nonmetropolitan Areas: New Approaches to Geographical Definition*. Working Paper No. 12. Washington, DC: US Bureau of the Census, Population Division. pp. 9–83.

Beale Calvin L. (1984) '"Poughkeepsie's" complaint or defining metropolitan areas', *American Demographics*, January, pp. 28–48.

Berry, Brian J.L. (1995) 'Capturing evolving realities: statistical areas for the American future', in Donald C. Dahmann and James D. Fitzsimmons (eds), *Metropolitan and Nonmetropolitan Areas: New Approaches to Geographical Definition*. Working Paper No. 12. Washington, DC: US Bureau of the Census, Population Division. pp. 85–138.

Bourne, Larry S. and Logan, M.I. (1976) 'Changing urbanization patterns at the margin: the examples of Australia and Canada', in Brian J.L. Berry (ed.), *Urbanization and Counterurbanization*. Beverly Hills, CA: Sage. pp. 111–43.

Brunet, R. 1989 *Les Villes 'Européenes'*. Paris: La Documentation Francaise.

Byfuglien, Jan (1995) 'On the search for an urban definition at the European level', *Norsk Geograpfisk Tidsskrift*, 49: 83–5.

Castells, Manuel (1985) 'High technology, economic restructuring and the urban–regional process in the United States', in Manuel Castells (ed.), *High Technology, Space and Society*. Beverly Hills, CA: Sage Publications.

Champion, A.G. (ed.) (1989) *Counterurbanization: The Changing Pace and Nature of Population Deconcentation*. London: Edward Arnold.

Champion, A.G. (1994) 'International migration and demographic change in the developed world', *Urban Studies*, 31: 653–78.

Chandler, Tertius (1987) *Four Thousand Years of Urban Growth: An Historical Census*. Lewiston, NY: St David's University Press.

Chandler, Tertius and Fox, Gerald (1974) *3000 Years of Growth*. New York: Academic Press.

Cheshire, P., Hay, D., Carbonaro, G. and Bevan, N. (1988) *Urban Problems and Regional Policy in the European Community*. Commission des Communautés Européennes.

Coombes, M.G., Dixon, J.S., Godard, J.B., Openshaw, S. and Taylor, P.J. (1982) 'Functional regions for the population census of Great Britain', in D.T. Herbert and R.J. Johnston (eds), *Geography and the Urban Environment: Progress in Research and Applications*. Chichester: Wiley. pp. 63–112.

Dahmann, Donald C. and Fitzsimmons, James D. (eds) (1995) *Metropolitan and Nonmetropolitan Areas: New Approaches to Geographical Definition*. Working Paper No. 12. Washington, DC: US Bureau of the Census, Population Division.

Davis, Kingsley (1959) *The World's Metropolitan Areas*. Los Angeles: University of California Press.

Davis, Kingsley (1969) *World Urbanization, Volume I, Basic Data for Cities, Counties, and Regions*. Berkeley, CA: Institute of International Studies, University of California.

Davis, Kingsley (1972) *World Urbanization 1950–1970, Volume II, Analysis of Trends, Relationships, and Developments*. Berkeley, CA: Institute for International Studies, University of California.

Dickinson, Robert E. (1964) *City and Region*. London: Routledge and Kegan Paul.

Eurostat (1994) *Delimitation of European Agglomerations by Remote Sensing: Results and Conclusions*. Pilot project. Luxembourg: Eurostat.

Fielding, A.J. (1992) 'Counterurbanization in Western Europe', *Progress in Planning*, 17 (1): 1–52.

Frey, William H. (1988) 'Migration and metropolitan decline in developed countries: A comparative national study', *Population and Development Review*, 14 (4): 595–628.

Frey, William H. (1993) 'The new urban revival in the United States', *Urban Studies*, 30 (4/5): 741–74.

Frey, William H. (1996) 'Immigration, domestic migration and demographic Balkanization in America: new evidence for the 1990s', *Population and Development Review*, 22 (4): 741–63.

Frey, William H. and Speare, Alden, Jr (1988) *Regional*

and Metropolitan Growth and Decline in the United States. New York: Russell Sage.

Frey, William H. and Speare, Alden, Jr (1995) 'Metropolitan areas as functional communities', in Donald C. Dahmann and James D. Fitzsimmons (eds), *Metropolitan and Nonmetropolitan Areas: New Approaches to Geographical Definition*. Working Paper No. 12. Washington, DC: US Bureau of the Census, Population Division, pp. 139–90.

Fuguitt, Glen V., Brown, David L. and Beale, Calvin L. (1989) *Rural and Small Town America*. New York: Russell Sage Foundation.

Gilbert, Alan (1993) 'Third world cities: the changing national settlement system', *Urban Studies*, 30: 721–40.

Gilbert, Alan and Gugler, Josef (1982) *Cities, Poverty, and Development: Urbanization in the Third World*. New York: Oxford University Press.

Goldstein, Sidney and Sly, David F. (1975) *Working Paper I: Basic Data Needed for the Study of Urbanization*. Liege: International Union for the Scientific Study of Population.

Gugler, Josef (1983) 'Overurbanization reconsidered', *Economic Development and Cultural Change*, 31: 173–89.

Gugler, Josef (ed.) (1988) *The Urbanization of the Third World*. Oxford: Oxford University Press.

Hall, Peter and Hay, Dennis (1980) *Growth Centers in the European Urban System*. London: Heinemann.

Harris, Nigel (ed.) (1992) *Cities in the 1990s: The Challenges for Developing Countries*. New York: St Martin's Press.

Hawley, Amos H. (1971) *Urban Society: An Ecological Approach*. New York: The Ronald Press.

International Institute of Statistics (1963) *Statistiques Demographiques des Grandes Villes*. The Hague: IIS

Kasarda, John D. and Parnell, Allan M. (eds) (1993) *Third World Cities: Problems, Policies and Prospects*. Newbury Park, CA: Sage.

Kawashima, T. and Korcelli, Piotr (eds) (1982) *Human Settlement Systems: Spatial Patterns and Trends*. CP-82-S1. Laxenburg, Austria: International Institute for Applied Systems Analysis.

Killian, Molly S. and Tolbert, Charles M. (1991) 'A commuting-based definition of metropolitan and nonmetropolitan local labor markets in the United States'. Presented at the Annual Meeting of the American Statistical Association, Atlanta, GA.

Long, John F. (1981) *Population Deconcentration in the United States*. Special Demographic Analysis CDS-81-5. Washington, DC: US Government Printing Office.

Long, Larry and DeAre, Diana (1988) 'US population redistribution: a perspective on the nonmetropolitan turnaround', *Population and Development Review*. 14: 433–50.

Lowry, Ira S. (1991) 'World urbanization in perspective', in Kingsley Davis and Mikhal S. Bernstam (eds), *Resources, Environment, and Population*. New York: Oxford University Press. pp. 148–79.

McGee, T.G. and Robinson, Ira M. (1995) *The Mega-Urban Regions of Southeast Asia*. Vancouver: University of British Columbia Press.

McVey, Wayne William, Jr and Kalbach, Warren E. (1995) *Canadian Population*. Toronto: Nelson and Sons.

Morrill, Richard L. (1995) 'Metropolitan Concepts and Statistics Report', in Donald C. Dahmann and James Fitzsimmons (eds), *Metropolitan and Nonmetropolitan Areas: New Approaches to Geographical Definition*. Working Paper No. 12. Washington, DC: US Bureau of the Census, Population Division. pp. 191–250.

OECD (1988) *Statistiques Urbanies dans les pays de l'OCDE*. Programmes des affaires urbanies. Paris: OECD.

Peters, Gary L. and Larkin, Robert P. (1993) *Population Geography, 4th edn*. Dubuque, Iowa: Kendall/Hunt Publishing.

Pumain, D., Saint-Julien, T., Cattan, N. and Rozenblat C. (1992) *The Statistical Concept of the Town in Europe*. Luxembourg: Eurostat.

Sassen, Saskia (1991) *The Global City: New York, London, Tokyo*. Princeton, NJ: Princeton University Press.

Stanback, Thomas M., Jr (1991) *The New Suburbanization: Challenge to the Central City*. Boulder, CO: Westview Press.

United Nations (1989) *World Population Prospects*. New York: United Nations.

United Nations (1995) *World Urbanization Prospects: The 1994 Revision*. New York: United Nations.

United Nations Center for Human Settlements (1996) *An Urbanizing World: Global Report on Human Settlements, 1996*. Oxford: Oxford University Press.

US Bureau of the Census (1993) *A Guide to State and Local Census Geography*. 1990 Census of Population and Housing CPH-1-18. Washington, DC: US Bureau of the Census.

Vandenberg, L., Drewett, R., Klassen, L.H., Rossi, A. and Vijeverberg, C.H.T. (1982) *A Study of Growth and Decline*. Oxford: Pergamon Press.

Vining, Daniel R. and Kontuly, Thomas P. (1978) 'Population dispersal from major metropolitan regions: an international comparison', *International Regional Science Review*, No. 3: 49–73.

Zimmer, Basil G. (1975) 'The urban centrifugal drift', in Amos H. Hawley and Vincent P. Rock (eds), *Metropolitan American in Contemporary Perspective*. Beverly Hills, CA: Sage. pp. 23–91.

APPENDIX 1: CITIES WITH A POPULATION OF 1 MILLION OR MORE IN 1995 AND 3 PER CENT OR HIGHER AVERAGE ANNUAL GROWTH FROM 1995 TO 2000

Country	City	Growth rate (%)	Population	Country	City	Growth rate (%)	Population
Afghanistan	Kabul	6.08	2.03	Dominican Rep.	Santiago de los	3.01	1.01
Algeria	Algiers	3.61	4.43	Egypt	Shubra El-Khema	4.10	1.16
Angola	Luanda	5.26	2.21	Ethiopia	Addis Ababa	4.29	2.21
Bangladesh	Chittagong	3.79	2.41	Ghana	Accra	3.92	1.69
Bangladesh	Dacca	5.27	7.83	Guinea	Conakry	5.43	1.51
Bolivia	La Paz	3.18	1.25	Haiti	Port-au-Prince	3.86	1.27
Brazil	Belem	3.55	1.57	India	Bangalore	3.04	4.75
Brazil	Campinas	3.02	1.61	India	Bhopal	4.06	1.30
Brazil	Fortaleza	3.17	2.66	India	Bombay	3.66	15.09
Brazil	Manaus	3.51	1.19	India	Delhi	3.34	9.88
Cameroon	Douala	4.88	1.32	India	Hyderabad	4.46	5.34
Cameroon	Yaounde	5.27	1.12	India	Jaipur	3.56	1.80
China	Anshan	3.11	1.65	India	Kochi (Cochin)	4.44	1.42
China	Baotou	3.11	1.41	India	Lucknow	4.27	2.03
China	Benxi	3.15	1.06	India	Ludhiana	4.47	1.26
China	Changchun	3.15	2.52	India	Meerut	4.04	1.03
China	Changsha	3.24	1.56	India	Pune (Poona)	3.45	2.94
China	Chengdu	3.06	3.40	India	Srinagar	3.28	1.02
China	Dalian	3.75	3.13	India	Surat	4.42	1.89
China	Daqing	3.38	1.15	India	Thiruvananthapur	4.08	1.01
China	Datong	3.09	1.28	India	Ulhasnagar	4.35	1.32
China	Fushun	3.08	1.60	India	Vadodara	3.60	1.33
China	Fuzhou	3.12	1.54	India	Vijayawada	3.93	1.02
China	Guiyang	3.12	1.79	India	Visakhapatnam	4.82	1.34
China	Handan	3.26	1.32	Indonesia	Bandung	3.39	2.98
China	Hangzhou	3.07	1.58	Indonesia	Bogor	5.07	1.22
China	Hefei	3.22	1.14	Indonesia	Jakarta	4.06	11.50
China	Hohhot	3.31	1.07	Indonesia	Medan	3.71	2.22
China	Jilin	3.18	1.51	Indonesia	Palembang	4.23	1.46
China	Jinan	3.87	3.02	Indonesia	Semarang	3.91	1.49
China	Jinzhou	3.72	1.09	Indonesia	Surabaja	3.12	2.74
China	Jixi	3.19	1.03	Indonesia	Tanjung Karang	7.41	1.23
China	Kaohsiung	3.21	1.73	Indonesia	Ujung Pandang	4.28	1.24
China	Kunming	3.11	1.94	Iran	Esfahan	6.33	1.92
China	Lanzhou	3.02	1.75	Iran	Mashhad	3.26	2.01
China	Luoyang	3.28	1.41	Jordan	Amman	4.12	1.19
China	Nanchang	3.36	1.65	Kenya	Nairobi	5.45	2.08
China	Nanjing	3.06	2.97	Libya	Benghazi	3.60	1.06
China	Nanning	3.57	1.50	Libya	Tripoli	3.91	3.27
China	Qingdao	3.08	1.59	Mexico	Naucalpan	3.78	1.79
China	Qiqihar	3.09	1.64	Morocco	Rabat	3.76	1.58
China	Shijiazhuang	3.24	1.55	Mozambique	Maputo	6.85	2.23
China	Taipei	3.21	3.42	Myanmar	Yangon	3.14	3.85
China	Taiyuan	3.14	2.50	Nicaragua	Managua	3.85	1.20
China	Urumqi	3.59	1.64	Nigeria	Ibadan	3.14	1.48
China	Wuxi	3.14	1.09	Nigeria	Lagos	5.37	10.29
China	Xian	3.18	3.28	Pakistan	Faisalabad	4.09	1.88
China	Xuzhou	3.24	1.08	Pakistan	Gujranwala	4.82	1.66
China	Yichun	3.15	1.05	Pakistan	Hyderabad	3.91	1.11
China	Zhengzhou	3.15	2.00	Pakistan	Karachi	4.05	9.86
Congo	Brazzaville	4.15	1.01	Pakistan	Lahore	3.97	5.09
Côte d'Ivoire	Abidjan	4.73	2.80	Pakistan	Multan	4.16	1.26

APPENDIX 1: *cont.*

Country	City	Growth rate (%)	Population	Country	City	Growth rate (%)	Population
Pakistan	Peshawar	5.10	1.68	Turkey	Adana	3.46	1.07
Pakistan	Rawalpindi	4.06	1.29	Turkey	Bursa	4.37	1.03
Philippines	Davao	3.43	1.01	Turkey	Istanbul	3.51	7.82
Philippines	Metro Manila	3.04	9.28	Turkey	Izmir	2.18	2.03
Saudi Arabia	Jeddah	3.60	1.47	UR Tanzania	Dar-Es-Salaam	3.66	1.73
Saudi Arabia	Riyadh	4.49	2.58	United States	Norfolk	3.07	1.68
Senegal	Dakar	3.94	1.99	Venezuela	Maracaibo	3.11	1.60
S. Korea	Inchon	3.89	2.34	Venezuela	Valencia	3.79	1.26
S. Korea	Kwangchu	3.33	1.42	Zaire	Kinshasa	3.90	4.21
Sudan	Khartoum	4.36	2.43	Zambia	Lusaka	5.25	1.33
Syria	Aleppo	3.79	1.86	Zimbabwe	Harare	4.09	1.04
Syria	Damascus	3.21	2.05				

3

Urban Ecology

PETER SAUNDERS

This chapter starts out from three basic axioms. First, social life in cities is 'patterned', and cities are in this sense 'socially organized'. It is this 'patterned' and 'organized' character which enables us to talk of cities as 'urban systems'. Second, cities are part of what Anthony Giddens has called the 'created environment' (Giddens, 1984). It may sound trite and obvious, but it is important to recognize at the outset that in analysing cities we are dealing with the products of human intervention in the material world. Third, this intervention occurs over an extended period of time, and at any one time, the possibilities of intervention (the conditions of action) will depend upon the legacy bequeathed by the efforts of earlier generations. Furthermore, millions of individuals are involved in shaping the social structure of the city, both consciously and unconsciously, and the combined results of their actions may turn out to be very different from what any of them specifically intended.

Drawing these three points together, it is clear that, while cities may appear to be organized systems created through human intervention, we cannot deduce from this that their organized character is necessarily the result of deliberate human agency. My aim in this chapter is to consider the extent to which the social organization of cities may arise spontaneously as an unplanned outcome rather than the product of conscious human design and intentionality.

To explore this question, I shall first return to the insights of the pre-war Chicago School of human ecology. Although this work has long since lost its attraction for most contemporary urban sociologists, I shall identify two linked but distinct themes which arise from its focus on 'biotic competition' as the driving force shaping urban social organization. These are, first, the idea of a 'natural' process shaping social life in cities, and second, the idea of cities as 'evolved' systems. I shall conclude, albeit tentatively, that

both of these insights continue to have some validity, and that the post-war reaction against the biologism and evolutionism of the Chicago School has blinded contemporary urban theory to some crucial processes which still shape the organization of urban systems. Robert Park and his colleagues at Chicago did not simply found the new subject of 'urban sociology', but they also bequeathed it some fundamental tools which remain central to its future scientific development.

CREATED ENVIRONMENTS AND HUMAN DESIGN

According to Anthony Giddens, nature was for our ancestors a threatening and uncontrollable force, but we have learned (up to a point) how to tame nature and to transcend many of the limitations which it once imposed upon us. In the modern period, we live in a 'created environment'.

Giddens emphasizes that the created environment is found everywhere. For him, the old distinction between town and country has ceased to have much meaning or significance, for the rural landscape is no less of a human creation than the urban one. Nevertheless, modern cities are perhaps the clearest expression of this transcendence of nature. In a city like New York, where it is possible to shop for one's groceries at four o'clock in the morning if one is so inclined, the division between daytime (a period of natural light during which social activity is possible) and night time (a period of darkness in which social activity is closed down) has collapsed. The seasons, too, pass virtually without notice as we pursue our year-round rhythms of life insulated by central heating from the winter snows and by air conditioning from summer heatwaves. The original physical topography of the city has also long since been obliterated, for we daily defy the contours and constraints of 'nature' as we hurtle through the earth in subway trains or speed half a mile upwards in the elevator of some monumental skyscraper. Even distance itself has largely been overcome, not only by revolutions in transportation technology such as the automobile and the jet aeroplane, but also by advances in electronic communications including the telephone, television, and latterly, the internet.

The urban environment in which we live out our lives is, therefore, a testament to human will, ingenuity and creativity. Like Sherman McCoy, in Tom Wolfe's New York novel *The Bonfire of the Vanities* (1988), those who today participate in creating this environment – planners, developers, politicians, architects, business executives – could be excused for believing themselves 'masters of the universe', limited in their designs for the future only by the constraints of their own imaginations.

For much of the twentieth century, these putative 'masters of the universe' exercised their imaginations to the limit. The history of the past hundred years is littered with blueprints for revolutionary urban environments which sought to harness the potential of modern technology in order to realize the ideals of modern social and political Utopias. Ebenezer Howard's plans for 'garden cities' harmoniously combining the advantages of town and country; Le Corbusier's dreams of 'cities in the sky'; Albert Speer's designs for an imperial Berlin as the showpiece centre for a thousand year Reich; the miles of concrete barracks stretching out across the landscape of eastern Europe to house the new socialist men and women of the Stalinist empire; the 'New Towns' and sprawling high-rise council estates of post-war Britain; the demented (and now aborted) plans of the Rumanian dictator Ceausescu for a mega-palace in the heart of Bucharest – all are evidence of the desire of powerful men (for there are few mistresses of the universe) to sweep away the clutter of urban 'chaos' bequeathed by an earlier age and to replace it with the authoritative stamp of a rational and planned new social order.

Looking back on this history, two things are immediately apparent. The first is that even the most determined 'masters of the universe' have often encountered limits on the effective realization of their dreams. In the late 1960s, which was perhaps the highpoint of the twentieth century's flirtation with rational planning and technocratic social engineering, a British urban sociologist, Ray Pahl, proposed a framework for analysing cities which became known as 'urban managerialism' (Pahl, 1975). According to Pahl, the city could be seen as an organized system of resource allocation, a system which is recognizably

patterned and which gives rise to systematic inequalities. In order to explain this pattern and to account for these inequalities, Pahl argued that we should look to the people who determine how resources shall be allocated – the 'urban managers' – whom he deemed to be the crucial 'independent variables' in any sociological investigation.

It soon became clear, however (not least to Pahl himself), that 'urban managers' – planners, politicians, housing agencies and the like – were not operating in a vacuum. Their actions were constrained, partly by economic and political factors, but also partly by the operation of what is best described as a 'spatial logic'. Those who appeared to be running the urban 'system', the individuals who decided what should be built where and who should get access to which locations and facilities, were in fact themselves the subject of 'forces' beyond their control. It was not just that local managers were constrained by national governments (which were themselves limited by international movements of capital), nor even that planning so often encountered the brute force of a market logic which could not simply be 'wished away', but that they also ran up against a 'spatial logic'. At its simplest, this dictated that no two people could simultaneously occupy the same point in space (which means that inequalities of distance are to some extent unavoidable under any social arrangements), that the existing use of any one piece of land tends to set limits on its possible future uses (a problem which became particularly acute for socialist planners working in cities developed during an earlier, pre-socialist, era), and that the potential uses of any one area of land are limited by the existing pattern of use of the areas around it.

The tyranny of higher level politicians, the tyranny of market forces, and the tyranny of distance all served to limit what was possible, and by the mid-1970s Pahl had virtually abandoned his earlier formulation, recognizing that even the most powerful urban actors merely 'intervene in' or 'mediate' processes which, ultimately, are under nobody's control. As he vividly expressed it, holding urban managers responsible for the social consequences of their actions is 'rather like the workers stoning the house of the chief personnel manager when their industry faces widespread redundancies through the collapse of world markets' (Pahl, 1975: 284).

The second point is that, where 'masters of the universe' have been able to refashion urban environments to express some ideal of how people 'should' live, the results have generally been disappointing, if not downright disastrous. Of course, just as Marxists have long argued that Marx's blueprint for a future socialist society was faultless, but that Lenin, Stalin, Mao, Pol Pot, Honecker and the rest somehow consistently failed properly to implement it, so too defenders of urban utopias have always been able to claim that the original vision was flawless but that it was implemented badly (defenders of Le Corbusier, for example, have often claimed that the high rise housing disaster in Britain does nothing to undermine the brilliance of the original idea). The point, however, is that when blueprints are implemented from the pristine white pages of a planning document, they encounter real-world obstacles and flesh-and-blood people. If, repeatedly, the ideas do not work as intended when they are put into practice, it is because there is something about the real world which fouls them up. I am reminded in this regard of Marx's caustic critique of nineteenth-century Hegelian idealism in which he tells of the 'valiant fellow' who believed that 'men were drowned in water only because they were possessed with the idea of gravity. If they were to knock this notion out of their heads . . . they would be sublimely proof against any danger from water' (Marx and Engels, 1970: 37). The same caustic response is appropriate in the case of the twentieth century's urban visionaries who blame reality when their ideas fail.

The test of ideal blueprints is whether they work as they should, and by and large, planned urban utopias have not worked. In Britain, for example, New Towns designed to break down social divisions and to sustain a more rounded and fulfilling way of life for their inhabitants have ended up as soulless suburban deserts with gangs of youths roaming the streets on a Saturday night with nothing to do once the pubs have shut. High-rise housing blocks, intended to create 'communities in the air', have ended up with their lifts vandalized and their stairwells scarred by graffiti and urine, and in many cases, they have been dynamited within 20 or 30 years of their proud unveiling. Huge council estates designed to replace the clutter of nineteenth-century inner city back streets with planned environments 'fit for heroes to live in' have ended up as 'hard-to-let' estates avoided if at all possible by all but the most desperate of homeless families, their 'community centres' wrecked on a regular basis by the very youngsters for whom they were designed.

Planned visions for urban living are, therefore, hard to implement, and when implemented, they tend not to work. Why is this?

My argument in this chapter is that large-scale designs for urban living tend not to work because they violate certain human and social 'needs' of which we are still only dimly aware. Over many centuries, human beings have, largely unconsciously, evolved forms of living which reflect these 'needs', but these forms have wilfully and deliberately been replaced in the modern period by consciously articulated designs which have been implemented through large-scale interventions intended to embody rational plans for social improvement and refashioning.

It is true, as we have seen, that all urban environments are to a large extent humanly created. But the fact that human beings actively create their environments does not mean that we can collectively decide to recreate them in accordance with one or other blueprint for living and expect the result to work. It is also true, as Pahl recognized, that all urban environments are generally patterned, and that in this sense they have some of the properties of organized 'systems'. But it does not follow from this that the pattern has been the result of deliberate human design, nor that the 'system' has been the result of conscious planning and coordination.

There is a crucial distinction to be drawn between the idea of the 'created environment' and the idea of the 'designed environment'. For at least five thousand years, human beings have been busy shaping the landscape, taking what was bequeathed them by earlier generations and working on it further, refashioning it and modifying it as their requirements and their technology have changed. Throughout this time, the environment has perpetually been created and recreated through the purposeful activity of individuals, but there has been no overall plan to which these developments have been forced to conform, and no grand social objective to which they were expected to contribute. It is only in the modern period that we have attempted to plan the landscape in accordance with some single, grand design. Convinced by a belief in their own omniscience and omnipotence, twentieth-century masters of the universe have impatiently swept away the accumulated legacy of earlier generations and have replaced it across the board with environments – housing estates, shopping malls, leisure complexes – which have been planned from scratch on rational–technical principles. The results, very often, have been environments which do not 'work'.

SOCIAL EVOLUTION AND BIOLOGICAL REDUCTIONISM

Evolutionary change is non-teleological. If, as a result of a long-term process of evolution, we end up with a 'system' which 'works', it is not because any single agency planned it that way, but because, through trial and error, people have come to learn which arrangements best enable them to fulfil their own individual needs and requirements, and how these arrangements can best be adapted to those developed by other people around them. Evolutionary change entails a process of constant innovation at an individual level coupled with iterative mechanisms for mutual adjustment at the level of the wider community. Nobody runs such systems, nobody directs their course, and they have no final endpoint. Evolved systems 'work', not because they are consciously ordered or coordinated, but because variants which do not work so well (i.e. those which do not 'fit' the environments in which they arise) are jettisoned along the way.

In sociology, evolutionary theory has been a long time dead. The idea that, like natural organisms, societies have systemic elements which become increasingly complex and functionally interdependent through time as a result of a process of unconscious evolution was powerfully expressed in the work of the late nineteenth-century sociologists such as Herbert Spencer and Emile Durkheim, and, as late as the 1960s, Talcott Parsons applied evolutionary theory to an explanation of why and how societies 'develop' towards modernization (Parsons, 1964). From the 1960s onwards, however, evolutionary theory became discredited and fell out of fashion in orthodox sociological circles. There were at least four reasons for this.

First, growing political sensitivities about the newly independent nations of the 'Third World' and the historic role of Western imperialism in fostering what Marxists saw as their 'underdevelopment' could not easily be reconciled with evolutionary thinking, which seemed at least to imply that the West was at a 'higher' stage of evolution than its former colonies. Furthermore, the idea that the countries of the 'Third World' would have to pass along much the same road of capitalist development in order to achieve affluence was anathema to those who held out the prospect of an alternative, socialist, route to modernization. Second, a resurgence of interest in 'social action' approaches, in which individual agency figures prominently as a source of change, flew in the face of evolutionary perspectives with their focus on 'system properties' and 'evolutionary universals' implying little scope for individuals (including politicians and sociologists) to bring about desired outcomes through a mere act of will. Third, increasing agitation to effect radical social change in the Western capitalist countries, expressed within sociology by the groundswell of neo-Marxist writing from the 1960s onwards, could not easily be squared with forms of theory that emphasized the historic significance of gradualism and adjustment as against tumultuous upheaval and destruction. And fourth, there was a new commitment in many areas of sociology to forms of 'humanist' thinking which rejected any

approach, such as evolutionary theory, which seemed to suggest or imply that human action and human institutions could in any way be explained with reference to 'natural' or biological processes.

It was this last consideration which was probably the key one. Social theories of evolution have their roots in the Darwinian revolution within biology. The same fundamental principles discovered by Darwin – notably the principle of survival through competition (natural selection) and the idea that species development entails growing functional interdependence and complexity – are replicated (often by means of an 'organic analogy' such as that found in Durkheim and the later ideas of structural functionalism) within sociological theories of evolution. Of itself, this does not necessarily result in 'biological reductionism' (the explanation of social phenomena with reference to ultimate biological causes). Durkheim, for example, was strongly committed to an evolutionary explanation of social change (set out mo st clearly in his *The Division of Labour in Society,* 1984), but he was also equally strongly committed to the irreducibility of sociological explanation, arguing (in *The Rules of Sociological Method*, 1938) that social phenomena have social causes and social effects which are independent of psychological or biological factors. As we shall see later, it is quite possible to analyse social phenomena (like cities) as complex evolved systems without necessarily implying that they are in some way the result of 'natural' causation.

Nevertheless, biologism and evolutionism are related, if only through the pervasive influence of the organic analogy in sociological accounts of evolution. It is but a short step from arguing that 'societies function in a similar way to living organisms' to arguing that 'social arrangements come to reflect natural, biological needs of human organisms'. Given sociology's long-standing resistance to all forms of psychological or biological reductionism, and its commitment (until recently) to Enlightenment values which hold out the promise of using rationality to escape from nature, it is not surprising that evolutionary theory, with its pessimistic message about the possibilities of refashioning and directing social development, should have been abandoned.

HUMAN ECOLOGY: EVOLUTION THROUGH BIOTIC COMPETITION

The abandonment of evolutionary theory within sociology as a whole was reflected within urban sociology by a sustained attack on the theory of human ecology. Described by Leonard Reissman as late as 1964 as 'the closest we have come to a systematic theory of the city', human ecology was perhaps the most explicit example anywhere in sociology of an attempt to analyse patterns of social life with reference to (a) natural or biological forces operating beyond the consciousness of human agents, and (b) social organization as the product of unconscious evolution.

The theory was developed in the first 30 years of the twentieth century by Robert Park and his associates in the Department of Sociology at the University of Chicago (Park, 1952). Looking at Chicago, Park was struck by the systemic properties which the city exhibited. With little in the way of purposive planning or consciously directed coordination, the city – a seething mass of several million individuals – nevertheless appeared to be 'organized'. In particular, certain kinds of people and activities tended to locate in certain kinds of areas and, somehow, these turned out more often than not to be locations in which they could thrive. Designating these locations as 'natural areas', Park hypothesized that there was some underlying process at work through which cities spontaneously grow while simultaneously ensuring that different social functions – retailing, family housing, prostitution or whatever – come to be located in the most appropriate spatial locations.

The process which he went on to identify he termed 'biotic struggle', and he contrasted this with more conscious activity – economic trading, political decision-making, law-enforcement, moral socialization and the like – which he summarized under the heading of 'society' or 'culture'. This distinction between 'biotic' processes of unconscious competition and adaptation, and 'cultural' or social processes entailing economic, political and moral cooperation or conflict, lay at the heart of the theory of human ecology. Park was well aware that much of what happens in a city is a product of more-or-less conscious, inter-subjectively meaningful and purposeful activity pursued by human agents in social relationships with one another. This 'level' of activity corresponds to the traditional subject matter of the social sciences (for example, to what Weber defined as 'social action' – Weber, 1968: 4). But he also insisted that, behind and underpinning this social action, there was an unconscious process operating which was akin to the process of ecological competition taking place in plant and animal communities. Furthermore, it was only as a result of unconscious biotic struggle that human beings found themselves in a position to develop social–cultural institutions, for it is biotic struggle which sifts and sorts us into socially complementary spatial units where we can find a functional niche, put down roots, and begin to interact positively with those around us.

According to Park, biotic struggle entails a process of competition for space. Those social activities which are functionally best suited to a given location will gradually come to dominate that space, driving out and repelling alternative uses which will then gravitate towards other locations where they in turn can achieve dominance. As certain areas become established for certain uses, so there arise symbiotic relationships between different types of users which can thrive through close proximity with each other, and the resulting ecological system tends towards a state of equilibrium as different social functions based in different areas come to adapt to the wider environment of which they form a part. If this equilibrium is disturbed (for example, by an influx of population, or the introduction of a new technology), biotic competition is once again sparked off, and different groups begin jostling to establish a new niche in the changed ecology within which they find themselves. Old uses find themselves displaced by newer ones which are now better suited to that location, and in this way, different areas of the city undergo a process of succession from one dominant use to another. As things once again gradually bed down, so the intensity of biotic competition recedes, and social (cultural) activity reasserts itself within and across the new communities which have been formed.

Cities therefore grow and sustain themselves as functioning systems through a process of evolution which entails recurring periods of unconscious competition between different social groups, invading, defending and dominating the natural areas to which they are functionally best adapted. Overlaid upon this ecological system is a social system with its economic institutions, its political agencies and its cultural forms.

Park's theory was subjected to criticism from very early on, and the principal target was his distinction between biotic and cultural processes. Basically, critics charged that the distinction could not be demonstrated empirically, for whenever we look at urban areas, we see only social-cultural processes at work. Biotic competition, precisely because it is an *underlying* process, can never be identified distinct from the social institutions and social actions through which it is expressed. Firey (1945), for example, cited evidence from Boston to show that so-called 'biotic' and 'cultural' factors were inextricably linked and could not empirically be distinguished. Certain areas – the upper class housing on Beacon Hill, the Italian slum area, the city centre graveyards – were invested with such a degree of symbolic importance by various residents that the biotic processes of competition posited by Park had been modified, mediated or even transcended to a point where the processes of invasion and succession predicted by the theory had apparently ceased to operate. With its location so close to the expanding Central Business District, for example, Beacon Hill should, according to Park's theory, have been successfully invaded, and the traditional WASP residents should have moved out to a new area. The fact that this had not happened suggested either that biotic competition can be blocked by cultural factors (in this case, the sentimental attachment of the residents to this area), or that biotic competition does not actually exist at all.

In retrospect, we can see that Park's theory encountered much the same sort of problem as both Marx (for example, in his claim that a 'material base' shapes a social 'superstructure') and Durkheim (for example, in his claim that a 'suicidogenic current' shapes the observed 'social rate' of suicide) had encountered before him (see Saunders, 1986 for a more extended discussion). What are we to make of the claim that some underlying process exists when it cannot be directly observed independently of the phenomenon which it is said to generate?

The problem as it applied to Park's theory was summarized by Alihan (1938), who questioned the existence of an underlying process – biotic competition – which, by definition, could not be observed empirically. Park argued that this process created observable entities – 'natural communities' such as the Central Business District, the red light area, the immigrant slum area or the suburb – but (as Firey's examples showed), it was impossible to find evidence to support such a claim. It is noticeable, for example, that much of the inter-war output of the Chicago School of human ecology took the form of ethnographies which provided detailed accounts of 'ways of life' of different urban groups – Italian slum dwellers, hobos, women working in taxi-dance halls, the inhabitants of a Jewish ghetto – living in different 'natural areas' of the city (for a review, see Hannerz, 1980), yet none of these studies was able (or even tried) to demonstrate the existence of a 'biotic struggle', for 'natural areas' are, according to Park's own theory, precisely those areas where biotic competition has receded and 'social' or 'cultural' processes have come to the fore.

Alihan suggested that Park's problem could have been resolved by presenting the biotic and cultural levels of human organization as 'ideal types', analytical tools for thinking about urban processes rather than empirically observable categories. This, however, would have dramatically weakened the theory, for there is a world of difference between claiming that biotic competition exists, and claiming that it can be useful to classify observations by imagining that it exists. More recently, Peter Dickens (1990) has drawn on so-

called 'realist' epistemology to highlight a second way out of the problem. Biotic forces can, he says, be represented as 'real' yet 'unobservable' tendencies in human organization, tendencies which may or may not become manifest according to 'contingent' conditions operating at the level of social organization. This would certainly help to get around the sort of problem posed by Firey's research, for it could enable us to claim that, just as gravity is a real force which continues to operate on aeroplanes in flight even though we do not observe them plummeting towards the earth, so too biotic competition is a real force which is always operating in cities, even though its effects may fail to manifest themselves due to countervailing contingent cultural conditions (such as symbolic attachments to certain areas, as in Boston).

As I have suggested elsewhere (Saunders, 1986, Appendix), however, this sort of theorizing is still problematic for as long as it remains impossible to identify whether or not the theorized underlying force really does exist. Unlike physical scientists, who can set up experimental conditions in which contingent effects are held constant to enable us to observe the operation of the underlying forces, social scientists have rarely been able to demonstrate independent evidence for the underlying processes they have theorized. Is there a 'material base' behind the superstructure? Is there a 'suicidogenic current' behind the suicide statistics? Is there a 'biotic struggle' behind the cultural processes at work in the urban environment? In every case, we end up unable finally to answer the question, in which case we may be excused for asking why we should accept the theory as valid in the first place.

Park himself never resolved this problem. Given his commitment to an empiricist epistemology (he once famously described his methodology as 'walking the streets'), it was in principle irresolvable, for no amount of walking the streets was ever going to uncover direct evidence of biotic competition, and this proved to be his Achilles' heel. As things turned out, later social ecologists like Amos Hawley simply side-stepped the problem by dropping altogether Park's concern with observing cities and their natural areas. Post-war human ecology limited itself to developing Park's theoretical concern with processes of differentiation, dominance and functional interdependence while ignoring his specific empirical concern with observing these processes at work in different city locations. Human ecology thus survived into the post-war years, but its association with urban sociology was gradually weakened as the theory became more general and less applied. By the 1960s, ecological theory, with its emphasis on questions of adaptation and functional interdependence, had become almost indistinguishable from general sociological theories of evolution and structural functionalism – in his 1964 article, for example, Otis Dudley Duncan explicitly tied ecological analysis to evolutionary theory in general. Not surprisingly, therefore, the revolution in sociology which swept evolutionary theories out of mainstream debate from the 1960s onwards took ecological theory along with it.

WHAT CAN BE SALVAGED FROM HUMAN ECOLOGY?

What, if anything, can be salvaged from Park's theory of human ecology? His ideas are still discussed in most urban sociology courses and textbooks, but they are normally represented as little more than an intellectual museum piece, an example of where the subject started rather than a theory which has anything much to offer to contemporary thinking about the city. Like Park, urban sociologists still recognize that cities are patterned and that they have many of the properties of an organized system, but attempts to account for these patterns and to explain the genesis and functioning of the system generally avoid any reference to 'natural' or 'biotic' processes.

Today's theorizing about the city can be divided into two broad types of approaches. In the first, emphasis is placed on the activities of powerful individuals or groups who more or less consciously shape the city to their own requirements. Pahl's early version of the urban managerialist thesis is one obvious example of such an approach. So too is the tradition of 'community power' research and its legacy in later work on the city as a 'growth machine' and in so-called 'regime theories' of urban politics (see Judge et al., 1995, Part I, for a review). What all of this work shares in common is a focus on purposeful human agency as the core explanation for patterns in urban systems.

The alternative approach derives from the resurgence of Marxist theory in urban sociology from the 1970s onwards. Influenced by Castells's attempt to analyse cities through the role they play in the reproduction of labour power in advanced capitalism (for example, Castells, 1977), and by Harvey's attempt to link patterns of urban development to flows of capital between different 'circuits' of investment (for example, Harvey, 1982), Marxist geographers and political economists have gone on to analyse cities in the context of changes in the international division of labour, processes of industrial restructuring, and regimes of accumulation (see Savage and Warde, 1993 for a review). This type of theorizing seeks to explain urban systems with reference to the changing requirements of capital accumulation as mediated by interventionist states (which operate largely in the interests of one or other 'fraction' of capital) and the ebb and flow of class struggle and resistance.

It is not my purpose here to criticize either of these approaches. Adequate sociological accounts are always multi-faceted, and it would be absurd to deny that cities are influenced by the actions of powerful groups or organizations (political organization), or that they are subject to the exigencies of global capitalist markets (economic organization). It will be remembered that Park himself recognized what he described as 'a kind of cone or triangle' which has at its base the ecological organization of human beings, upon which develops economic, political and moral systems (Park, 1952: 260). The question, therefore, is not whether contemporary political and economic theories of the city are right or wrong, but whether there is another level of analysis, corresponding to Park's conception of a biotic or ecological system, which contemporary theories are neglecting altogether. If we believe there is, then the further question arises of how it can be empirically observed.

There are two possibilities we need to explore, and they correspond to the two linked aspects of Park's analysis of ecological processes. On the one hand, we find in Park's work a clear reference to the operation of 'natural' forces in human organization (this is clearest in his argument that 'biotic competition' occurs in human as well as plant and animal communities, and that it results in the formation of 'natural areas', or habitats dominated by particular 'species'). On the other hand, we also find in Park's work a strong focus on processes of evolution (the basis of his theory of urban growth and development which was further elaborated in Burgess's famous model of 'concentric rings' and in McKenzie's theory of the 'stages' of urban development). These two aspects of ecological theory are, of course, closely related, but as noted earlier, they do not necessarily entail each other. It is quite possible to identify 'natural processes' underpinning human organization without necessarily adopting an evolutionary theory, just as it is also possible to analyse human organizations as evolved systems without necessarily arguing that they are in some way rooted in biological needs. In the case of Park's human ecology, however, both elements are present, and this has not always been fully appreciated in the secondary literature.

In his review of Park's ideas, for example, Peter Dickens takes as central the concern with natural or biological processes, and he explicitly argues that Park's focus on the 'biotic' level reflected his commitment to analysing social processes in the context of human biological and psychological imperatives:

> The Chicago School of Urban Sociology . . . has much to teach us, especially as regards incorporating an understanding of instinctive behaviour into an understanding of social relations . . . Park . . . was not attempting a direct and obvious analogy between the workings of nature and those of human societies. Rather, he was arguing that there is indeed a 'biotic' level to human behaviour, one constituted by instincts of survival and competition. (1990: 29–33)

Dickens then attempts to develop the biologism in Park's theory by drawing upon more recent work in ethology, sociobiology and psychology in an attempt to identify the basic 'needs' and 'instincts' which operate through and help shape social institutions and processes, and in later work (Dickens, 1992, 1996) he has tried to outline the contours of a new approach to social science which explicitly links social action to the psychological, biological and even physical processes which come to be expressed through it.

In stark contrast with this, Savage and Warde suggest that Park's ecological theory has little to do with biologism, but is essentially an evolutionary theory:

> One should not assume that human ecology is necessarily a form of biological determinism . . . ecological ideas, rather than being biologically derived, were largely forms of evolutionary social theory . . . Rather than actually arguing that patterns of urban life were the same as those of plant life (which a moment's thought tells us is a ridiculous argument), the biological arguments are used as analogies, in order to show how comparison with biology might throw light on urban life. (1993: 15–17)

The authors are dismissive (even incredulous) of Dickens's interpretation of Park, arguing that the Chicago ethnographies explicitly rejected instinctual theories and that, when Park spoke of 'human nature', he saw it more as a social than a biological construct.

In my view, however, both biological and evolutionary interpretations of Park's ecological theory are valid. It is true, as Savage and Warde suggest, that this was an evolutionary explanation of urban change and development, but Park seems also to have been suggesting (as Dickens recognizes) that the way cities evolve somehow reflects the natural instincts and biological and psychological needs of the individuals who compete for territory within them. The evolution of urban environments cannot, for Park, be explained simply with reference to economic, political and cultural factors, for it also reflects something else, a biotic imperative beyond the level of human consciousness and reflexivity. It follows from this that, in attempting to build

upon Park's insights, we need to consider more recent developments in both biological and evolutionary theory to see whether either or both of these can help explain why conscious designs for urban living so often fail to produce the effects which their designers intended.

BIOLOGISM: NATURAL AREAS AS AN EXPRESSION OF 'NATURAL WILL'

One of the best known, yet least read, books in urban sociology is Ferdinand Toennies's 1883 study of *Gemeinschaft* and *Gesellschaft*. In this book, Toennies sought to identify 'the sentiments and motives which draw people to each other, keep them together, and induce them to joint action' (1995: 3). The strongest bonds, he argued, were those based in what he called 'natural will' which he defined (1995: 121) as the 'inborn and inherited' feelings and instincts which draw us to those with whom we share a common identity. For Toennies, the principal expressions of such natural instincts are through blood ties (the strongest example of *Gemeinschaft* is the mother–child relation), shared territory and common religious beliefs. The weakest bonds – those of *Gesellschaft* – were those based on 'rational will' involving calculations of self-interest, for these depend solely on legal contract, and relations in the *Gesellschaft* are ultimately reinforced only through the imposition of state power.

Since Toennies wrote his book, a huge amount of sociological effort has been devoted to understanding what makes a 'community' work. Toennies's central typology has perpetually been recycled in this literature, but few twentieth-century sociologists paid serious attention to his analysis of 'natural will' as the foundation of *Gemeinschaft* (this, of course, reflects sociology's rejection of instinctual and biological explanations). It is true that many community studies – such as Young and Wilmott's classic study (1957) of the working class in the East End of London in the 1950s – have pinpointed the central importance of close kin ties ('blood') and lengthy common residence (shared territory) in underpinning social cohesion, and many sociologists have also been alert to the decline of religion as a factor explaining the erosion of 'moral community' in the modern period. Few, however, have gone on to pose the obvious question of *why* social cohesion is so often grounded in common ties of blood, residence and belief.

For many nineteenth-century writers, Toennies among them, the answer was obvious. It was 'natural' for human beings to care more for those to whom they are in some way related and to seek out others similar to themselves with whom to form common areas of residence. Today, in the wake of a critical feminist onslaught against the traditional family together with 30 years of sustained argument in favour of social-constructivist explanations of social identities, such arguments seem almost heretical. Sociological orthodoxy is today wedded to the view that there was nothing 'natural' about the bonds of kinship and common territory which cemented human beings together in the past, and many sociologists seem to believe that we are free to create new forms of social organization which break radically with tradition, and to shape them according to our own designs and purposes.

It is ironic that sociology has become ever-more resistant to biological explanations of social behaviour at precisely the same time as biologists have been discovering evidence to suggest that Toennies may have been right to identify 'inborn instincts' as the source of a 'natural will'. Research in molecular biology is now beginning to map human DNA in order to pinpoint which genes appear to be responsible for which biological and behavioural characteristics. When Toennies and Park were writing, it was impossible to imagine that we could ever find empirical evidence for the existence of underlying 'instincts' and predispositions, but today this seems to be within our grasp, and this opens up the possibility of documenting some of the underlying 'natural' processes which Toennies and Park suspected were operating.

It is a key proposition of contemporary sociobiology that human beings are 'programmed' to pursue forms of behaviour which maximize the survival chances of their genes (for example, Dawkins, 1976). Such theories do not deny that cultural factors can and do modify selfish behaviour (any more than Park denied that cultural processes can and do coexist with biotic ones), for the process of socialization is precisely one in which instinctive drives are gradually harnessed and controlled by internalized norms of social behaviour. Nor do these theories deny that altruistic behaviour is common – indeed, they take as a central axiom the idea that the unit of natural selection down the ages has been the gene rather than the organism, and this means that organisms (including human beings) will be predisposed to self-sacrifice where this improves the chances of genetic survival into the next generation (the obvious example being parental altruism towards their natural offspring).

The key point of these theories for our present concerns is that, in the process of natural selection and mutation, individuals born with a genetic predisposition to favour strangers over self will have been disadvantaged in the struggle for survival relative to those born with a predisposition to favour self over others, and their genetic stock will therefore have been less likely to have survived into subsequent generations. This means that we are almost certainly programmed to favour those whom we recognize as genetically close against those whom we recognize as genetically distant. When we behave in this way, taking more care of our own children than of other people's, for example, or donating more to charities 'near to home' than to those which touch us less directly, we are not necessarily acting consciously in order to maximize the survival chances of our genes. Rather, we are responding to our 'feelings', our strongly felt sentimental predispositions. Put another way, there are good grounds for arguing for the existence of a 'natural will', inborn and inherited, which predisposes us to care more for those with whom we identify – which is precisely what Toennies argued over a hundred years ago.

What does all this have to do with human ecology and the urban environment? One illustration will have to suffice.

Regular patterns of residential segregation can readily be found in virtually all big, modern cities. Urban residents tend to cluster in certain distinctive areas on the basis of their social class, their ethnicity and their common lifestyles (shared status identities). So marked are the resulting patterns that the advertising industry in Britain can now target highly specific sub-groups of the population simply on the basis of their postcodes (the so-called ACORN system of classification of neighbourhood types).

How does such a marked 'system' of segregation come about? Social factors – politics, economics, culture – have a lot to do with it, of course. Politically, local housing authorities may allocate certain types of people to certain areas of public housing, lending institutions and real estate agencies may help steer certain kinds of buyers towards certain kinds of neighbourhoods, planning authorities may try to ensure that new developments are consistent with the 'character' of the areas in which they occur, and local pressure groups often agitate to maintain residential exclusivity against threats of 'invasion' by other social groups. Economically, the operation of supply and demand leads groups like students to gravitate towards areas of cheap rents, while developers buy up riverfront properties to convert them into luxury apartments for affluent clients, and the lower middle classes take out mortgages which buy them modest houses in the suburbs. And culturally too, people exercise choice in where to live and will often seek out areas in which they 'feel comfortable' and can best express their values and desired lifestyles – the elderly often retire to the seaside, for example, well-educated young couples tend to head for gentrifying inner city areas (Jager, 1986), and in America, a city like San Francisco has become a magnet for gays (Castells, 1983).

But is there something else going on in all this? Are these patterns of residential segregation fully explained by politics, economics and culture, or is there some fourth factor also operating, some sort of 'natural' or 'instinctive' process which leads us to seek out others like ourselves? Why, when people can exercise some degree of choice over where to live, do they tend to opt for areas where they will find others like themselves, and why, when deprived of such choice, do they often go to such extraordinary lengths to distance themselves from others who are unlike themselves?

Take, for example, the ethnic minority settlements in parts of inner city Birmingham, in England. An early study by Rex and Moore (1967) suggested that recent black immigrants were congregating in areas like Sparkbrook because they had no other choice. Local authority housing policies were effectively excluding them from predominantly white council estates in suburban areas of the city, and lending institutions were excluding them from suburban home ownership through the application of loans criteria involving security of employment and size of income which new immigrants could not hope to meet. The result was that immigrants concentrated in run-down inner city areas where large, old houses could be purchased cheaply and could then be sub-divided and let out at cheap rents. The whole study was premised on the assumption that immigrants aspired to live in predominantly white council or private housing estates but were prevented from doing so by political and economic factors.

Some years later, Rex returned to Birmingham, this time conducting a study (with Sally Tomlinson) of the predominantly black area of Handsworth. The message from this later work is very different, for they found that most residents had no desire to move out into what Rex and Moore had thought of as more 'desirable' locations. After two decades of settlement, the area had been adapted to the needs of the newcomers, and it was experienced by them as a defensible area where they could feel relatively secure and could lead a way of life which was distinctive to their needs. Translated into the terminology of the Chicago School, a newly arrived group had successfully 'invaded' inner city areas like Sparkbrook and Handsworth, had achieved 'dominance' over these areas, had 'adapted' the local environment (shops, cinemas, clubs, etc.) to their needs, and had begun to develop a distinctive 'culture' within it. Biotic competition in the early period had resulted 10 or 20 years on in the establishment of a new 'natural area' in which people felt a sense of communal belonging.

Contrast this with evidence of what happens when people are forced (normally as a result of the imposition of artificial political or administrative boundaries) to live in close proximity with others who are socially distant from them. The most dramatic examples can be found in recent events in Kosovo, in Northern Ireland, and in Burundi, but less bloody patterns of dysfunctional communities and inter-personal antagonism can be identified in other, less obvious, contexts. In his review of residential segregation in the British new towns, for example Heraud (1975) found that the original aim of achieving mixed-class neighbourhoods had to be abandoned. Those who could, voted with their feet, and segregation increased as time passed. Heraud concluded that, 'Any attempt to effect widespread social mixing on a local basis, however this is designed, will meet with little success', and he warned that 'heterogeneity at a local, or block level, seems undesirable and may promote more conflict than cohesion' (1975: 285–6). Perhaps the best example of precisely such a case where enforced mixing sparked increased conflict is that of the 'Cutteslowe Walls' in Oxford, where residents on a private housing state successfully pressed the local authority to build an imposing physical barrier between them and the tenants of a neighbouring council estate in order to enforce social segregation (see Collison, 1963; Robson, 1982).

It seems from these examples that the social organization of the city may express more than simply the outcome of economic or political processes. We can decide (as in the case of the New Towns) to encourage different people to live together, yet we often find that little or no spontaneous social organization results. By contrast, no decision was ever taken to direct ethnic minorities into inner city areas like Sparkbrook and Handsworth, yet within a relatively short space of time, these areas developed strong and distinctive cultures and residents became committed to remaining in them.

All of this suggests that there may be some connection between what Toennies saw as 'natural will' (or what sociobiologists identify as genetically programmed instincts) and the formation of what Park later identified as 'natural areas'. Natural areas of relatively high cultural homogeneity work, not only because of social factors, but because we are genetically predisposed to feel greater sympathy and concern for those we recognize as similar to ourselves. Of course, those with whom we identify in the modern urban neighbourhood may not be – and generally are not – closely related to us in genetic terms. This, however, is not the point, for as van den Berghe (1981) suggests, our disposition to cooperate with those carrying the same alleles prompts sympathy towards those categories of people whom we have learned to recognize as 'one of us'. Ethnic solidarity, for example, is a cultural phenomenon insofar as we have learned to identify certain physical signs, such as skin colour, as indicating inclusion in or exclusion from membership of a social category, but it is a biological phenomenon insofar as the sentiments of cooperation or competition which members or nonmembers feel towards each other are the products of genetic programming orienting us to favour related individuals over unrelated ones.

Residential segregation, therefore, may be understood as a natural process, culturally mediated. In addition to the economic, political and cultural forces leading us to locate in one or other area of the city, there seems to be a biotic process, grounded in genetic predispositions to seek out others like ourselves, which is also operating.

Other phenomena studied by urban sociologists may be analysed in a similar way. Elsewhere, for example, I have considered the extent to which the desire for home ownership can be explained as the culturally mediated product of an instinctual drive towards territoriality and other forms of possessiveness (Saunders, 1990), and Peter Dickens (1990) devotes much of his urban sociology text to a reanalysis of themes – urban violence, domesticity, political leadership – which he suggests can more fully be understood by bringing together sociological theories with psychological and biological ones.

There is, however, still a very strong resistance in sociology to the use of biological theories to help explain social phenomena, and critics still often assume that such theories can be discredited simply by denouncing their 'reductionism', or by asserting that they are 'ideologically motivated'. In his attack on my work on home ownership, for example, Samson treats the idea of analysing instincts as self-evidently absurd – 'Saunders seriously entertained the notion that desires for home ownership and private property may be natural instincts' – and he went on to dismiss the entire enterprise as simply 'a means of reinforcing attitudes about the nature of existing social inequality' (1994: 93–4). With a few honourable exceptions (e.g. Hirst and Woolley, 1982), sociologists remain intent on keeping their distance from the biological sciences.

Leaving such predictable criticisms on one side, it does nevertheless still have to be recognized that, like Park's original theorization of a 'biotic' level, much of the work on genetic instincts remains speculative. Despite recent staggering advances in the mapping of the human genome, we are still a long way from being able to identify the genes which are thought to be producing the sorts of behavioural tendencies considered in this chapter. It is one thing to demonstrate, as

Dawkins does, that certain kinds of genetic predispositions 'must' have been selected through processes of competition for survival, but it is another to provide the definitive empirical evidence for their existence. Unless and until such evidence is found, this kind of work is likely to remain suggestive rather than conclusive, and one of the legitimate criticisms which can be made of Dickens's recent work is that, like so many social scientific applications of realist philosophy, it ends up still begging the question of whether and how the posited 'underlying mechanisms' actually operate.

EVOLUTIONISM: A SOCIAL LOGIC OF SPATIAL FORMS

The second strand bequeathed us from Park's human ecology is the concern with evolved spatial forms. Chicago, it will be recalled, was an unplanned city which nevertheless functioned as a patterned system. We have also seen evidence that some planned communities conspicuously fail to function in this way. The question is therefore whether certain kinds of spatial arrangements have evolved in such a way as to enable social life to thrive, while more modern designed environments have generated dysfunctional outcomes because they have failed to replicate these evolved patterns.

An early attempt to raise this question was Jane Jacobs's pioneering analysis of *The Death and Life of Great American Cities*, which represented an explicit attack on the legacy of American town planning. According to Jacobs, planners have imposed grand schemes which destroy the vibrancy, spontaneity and safety of urban areas because they fail to understand the hidden dynamics which have shaped urban evolution in the past. 'It is,' she says, 'futile to plan a city's appearance, or speculate on how to endow it with a pleasing appearance of order, without knowing what sort of *innate, functioning order* it has' (1961: 24, emphasis added). Conscious design has obliterated a spontaneously-evolved 'natural' order, and the result is physical uniformity, neighbourhood demoralization and social paralysis.

Several subsequent researchers have taken up this issue and have attempted to identify the evolved patterns in human communities which best enable them to function as local systems of social order. As in the work on the instinctual basis of urban environments, however, the problem again arises of how to find evidence for an underlying process or structure which is not immediately observable, and this is reflected in a divergence in views as to the precise nature of the 'innate, functioning order' hypothesized by Jacobs.

On the one hand, studies by Oscar Newman (1972) in the United States and by Alice Coleman (1985) in Britain have suggested that evolved spatial forms succeed in maximizing security and safety for those who live in them by limiting the potential access of strangers and ensuring high levels of surveillance of surrounding space. Such arrangements encourage residents to feel and accept some degree of responsibility for monitoring behaviour in their areas. Modern, designed environments, such as high-rise and deck-access blocks of flats, fail to correspond to these elementary principles of socio-spatial order. They are generally characterized by large areas of 'confused space' for which nobody feels responsible, and by a profusion of exits and entrances which make escape by strangers relatively easy and monitoring by residents extremely difficult. Failing adequately to demarcate territorial boundaries, these designs facilitate anti-social behaviour and minimize effective surveillance of public behaviour.

Utilizing a very rudimentary methodology, Coleman shows that the frequency of occurrence of indicators of anti-social behaviour – litter, vandalism, graffiti, urine and animal faeces – is found to correlate significantly (though not always strongly) with various features of building design such as the number of dwellings per building, the height of blocks and, most crucially, the number of exits and entrances. She also demonstrates, not only that conventional housing is much less associated with anti-social behaviour then modern, purpose-built blocks, but also that housing built in Britain between the wars – bay-fronted, suburban, semi-detached homes set back from the road behind low garden walls – is much more successful than post-war housing designs involving flush facades, high screening walls and projecting garages blocking the view from the house to the street. In her view, there was, until the Second World War, an evolutionary process of 'natural selection' in vernacular building – design evolved through trial and error and builders competed for buyers by adopting proven best practice while rejecting styles which had produced problems for residents. Since the war, however, architects and builders have been responsible less to prospective buyers than to local authority planning departments which have encouraged developments that reflect current architectural fashion yet fail to embody the design principles of the earlier period. The inter-war semi was thus 'the most advanced design achieved by British mass housing before natural evolution was broken off by planning control' (1985: 103).

Coleman concluded her book with a range of design recommendations, and there is some evidence that, where these have been adopted on housing estates, crime rates and vandalism have fallen appreciably (Ballantyn, 1988). Nevertheless, her work has been criticized. Predictably, perhaps, given their lack of patience with all forms of 'reductionism' and physical determinism, sociologists have argued that the social problems in high-rise estates have more to do with the poverty of the residents than with the design of the blocks (Spicker, 1987), and architects too have argued that the problems associated with high-rise housing have been exaggerated (Glendinning and Muthesius, 1994). There are also some serious weaknesses in Coleman's data analysis which have been exposed by Hillier (1988) who goes on to suggest that her prescriptions are more likely to exacerbate than to relieve the social problems she identifies.

Hillier, it is, who represents the main alternative to Newman and Coleman. Echoing Jane Jacobs's earlier arguments, he suggests that a functioning community depends upon encouraging diverse flows of people rather than shutting strangers out, and on linking residences into the wider urban environment rather than shutting them off from it in 'defensible spaces' behind walls and fences. His arguments derive from his earlier book with Hanson, in which the authors analysed the way in which patterns in the built environment express variations in a set of simple 'syntactical rules'.

Hillier and Hanson's work (1984) is far more sophisticated and complex than anything attempted by Jacobs, Newman or Coleman. They demonstrate through a series of computer models that, by specifying some simple restrictions on a random process of aggregating cellular units, well-defined aggregate patterns emerge which resemble those found in real-life communities. For example, if we begin with cells (representing houses) of four sides, one of which is an entrance opening onto an open space, and if we specify only one rule – that cells should be added by joining the open space one to another – then we end up with what the authors call a 'beady ring' form of settlement which corresponds closely to some real-world hamlets such as those found in the Vaucluse region of France, and which can be found in more complex combinations in many larger settlements as well. Such settlements are interesting because they share a recognizably common structure (the beady ring) while varying widely in their particular forms, and this suggests that they developed not through conscious design but by some unconscious process of cumulative growth (Figures 3.1 and 3.2).

Hillier and Hanson go on to show that, by varying the initial rules in a limited number of ways, it is possible to identify a set of core principles according to which a wide variety of different building and settlement patterns are structured

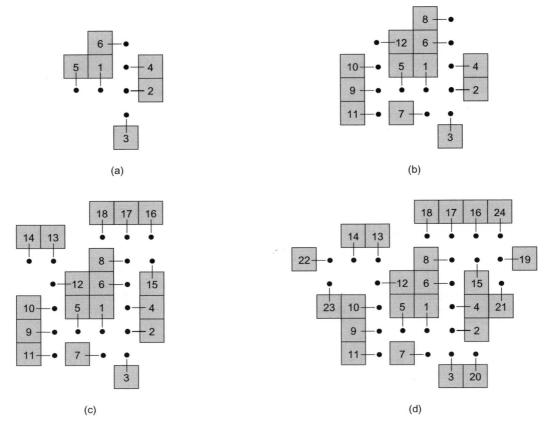

Figure 3.1 Four stages in the emergence of the 'beady ring' pattern as generated by a computer (Hillier and Hanson, 1984)

even though they develop through what are essentially random processes of aggregation. These principles can be codified into what they call a 'space syntax' – that is, spatial forms can be represented in a rule-based 'language' varying in its complexity according to the number of restrictions placed upon the process of random aggregation.

It is through the analysis of these syntactical rules that Hillier is able to assert so authoritatively that Coleman is wrong to argue for designs which limit access by strangers, for in his work with Hanson, he finds a recurring principle in evolved urban environments which combines what they call 'axial extension' of public space with intermittent 'convex' spaces along the lines (essentially the 'beady ring' pattern). Put simply, this pattern provides multiple strong routes of access for strangers while ensuring maximum connectedness of residential units to these routes, and this results in main thoroughfares being crowded with strangers while inhabitants are able to maintain surveillance. One of the crucial unin-

tended yet beneficial outcomes of the unplanned structure of such environments is, therefore, that safety is maintained by a high rate of flow of people in public spaces overseen by a high number of local residents. As Hillier and Hanson put it, 'The strangers police the space, while the inhabitants police the strangers' (1984: 18).

What is striking about all this is that these structural properties of evolved urban environments are very often missing in modern, planned housing estates, and Hillier and Hanson demonstrate that this is reflected in much lower levels of use of public spaces in even the most progressive, well-established, low-rise estates. Put crudely, the absence of spatial order in modern, designed environments is mirrored in an absence of social order, and interaction and social contact is minimized as space becomes unintelligible. This has nothing to do with population size or density, for even high-density designed environments generate and sustain much lower levels of activity than lower-density evolved ones, and the authors describe life in these new estates as 'like living in

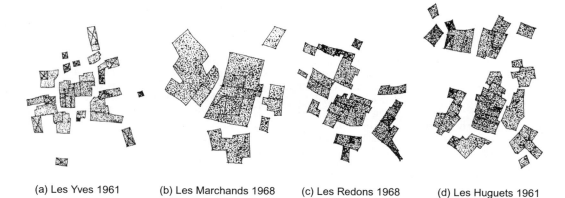

(a) Les Yves 1961   (b) Les Marchands 1968   (c) Les Redons 1968   (d) Les Huguets 1961

Figure 3.2   The 'beady ring' pattern found in four French villages (Hillier and Hanson, 1984)

perpetual night' (1984: 23). Like Newman and Coleman, therefore, Hillier and Hanson recognize that modern design has broken disastrously from the evolved wisdom of traditional spatial forms, but unlike them, they are able to identify what has gone wrong by uncovering the underlying principles of successful spatial and social order. These principles demonstrate that the problem lies, not in the lack of territorial defence and demarcation (as Coleman believes), but in the failure to integrate cores linking the interior of urban systems to the outside world.

This work on space syntax is important, not only for its substantive findings, but also for its methodology. Just as arguments about the 'natural' basis of urban social organization need to go beyond mere assertion and to demonstrate the causal properties of the underlying genetic code which is said to be producing the observed forms, so too analysis of evolved spatial forms depends upon our ability to detect the fundamental principles of urban growth and development which are generated through random processes of aggregation over time. Like analysis of the structural properties of DNA, analysis of the structural properties of space syntax holds out the promise of empirically testable insights into the unconscious and unrecognized forces shaping complex phenomenal forms.

CONCLUSION: A FATAL CONCEIT?

Through much of the Western world, the immediate post-war years were suffused with an uncritical spirit of optimism about our ability to plan a better future. In Britain, Keynesian economics offered the prospect of stable, low rates of unemployment achieved through govern-

ment intervention in the operation of the free market, the establishment of the welfare state promised a way of abolishing poverty and providing a comprehensive system of health care for all, and the Butler Education Act held out the hope of achieving effective equality of opportunity for all children. The remodelling of urban environments, presaged by the 1944 and 1947 Town and Country Planning Acts, was a further, integral element in this new spirit of optimistic grand design. Many British towns had been badly damaged by wartime bombing, and much of the remaining housing stock dated from before the First World War. For 20 years after the war, city centres were reconstructed, new towns were created and old residential neighbourhoods were obliterated as politicians, planners and architects joined forces in a concerted attempt to create a more rationally ordered urban environment better suited to the modern, post-war age of social reconstruction.

In the midst of this reformist euphoria, a few isolated voices of caution were raised warning of the dangers and difficulties inherent in attempting to impose universalistic rational blueprints on the apparent chaos of collective life. The most notable sceptics were Friedrich Hayek, whose book *The Road to Serfdom* was published in 1944, and Karl Popper, who published *The Open Society and its Enemies* just one year later. But such views were widely dismissed at the time as reactionary (Cockett, 1994), and in Britain the critics were effectively marginalized for many years by a bi-partisan commitment to expanding the scope of social and economic planning.

In retrospect, we now know that much of this concerted effort aimed at planning new forms of social organization failed to achieve the hopes and objectives of those who championed it after the war. Keynesian demand management policies eventually resulted (as many of Keynes's early critics had forecast they would) in a vicious spiral of mounting inflation and inefficient subsidy of declining industries. State welfare provision swiftly became an open-ended commitment in which ever-increasing levels of public expenditure chased ever-increasing demands for new forms of provision while 'social problems' continued to multiply. The state education system still produced poorly educated and in some cases barely literate adults and left employers complaining about the quality of the labour pool from which they were obliged to recruit. And urban planning resulted in the wholesale demolition of perfectly serviceable areas of housing redefined as 'slums', the shattering of traditional communities as people were shipped out into new estates, the blighting of large areas awaiting redevelopment, the widespread destruction of areas of architectural heritage, and the creation of new ghettos for the poor which, in the more notorious examples, have become 'no-go areas' characterized by high levels of criminality, drug abuse, unemployment, street violence, educational truancy and dysfunctional families.

There are many specific reasons why specific policies failed, but one general factor underpinning this whole period was the belief in the omnipotence and omniscience of comprehensive state planning. In the last book he published before he died in 1992, Hayek (1988) identified this belief as what he termed 'the fatal conceit'. This conceit consists in the belief that, as rational beings, we can use our intelligence and our knowledge to create institutions which will serve our purposes better than social arrangements which have evolved through time and which nobody designed, nobody controls and nobody even fully understands. It is our conceit as human beings which makes modesty in the face of evolved or natural systems so difficult to accept, for we are constantly tempted to try to improve upon them in order to turn them to our desired purposes. As Hayek suggests, intelligent individuals find 'humiliating' the very idea of being subject to forces which they do not understand and do not control, and they respond by resolving to take control of and direct these impersonal forces towards their own desired objectives.

Such a conceit is 'fatal' because we have neither the intelligence nor the knowledge required effectively to organize and manage what Hayek calls an 'extended order'. He accepts that we can learn something about the principles which underpin this order, and that we can make some limited and successful attempts to modify it, but he sees as disastrous the spirit of rationalism (epitomized in socialism) which rejects forms of organization inherited from the past if they

cannot be scientifically justified, if they cannot be fully understood, if they have no defined purpose, and if their effects cannot be fully known in advance. Irritated by the 'irrationality' of tradition, state planners have assumed that they can improve upon it through the application of intelligence, and every failure of design simply provokes a fresh attempt to do it better next time.

Hayek's argument is mainly a defence of the free market as an evolved and unconscious system of coordination of human action, but it applies equally to what this chapter has identified as the 'biotic processes' shaping human created environments. What I have tried to show is that there is an inherited 'wisdom' in evolved urban forms which, although it lacks 'purpose' and is opaque to our conscious understanding, nevertheless helps structure and order the environments in which we live. There is often a spontaneous order in the apparent chaos of the largely unplanned city just as, conversely, there is a profound absence of social organization and cohesion in many post-war planned urban environments. I have suggested that one reason for this may have something to do with the way human beings instinctively feel comfortable with and gravitate towards those whom they recognize as 'similar' (a product of the operation of the 'selfish gene'), for as the Chicago ecologists recognized, left to ourselves we tend to sift and sort ourselves into what Wirth (1938) called a 'mosaic of social worlds'. If this is correct, then attempts to re-engineer social cohesion by forcing spatial proximity upon socially distant groups have failed because they have run against the grain of our genetic inheritance. I have also suggested that unplanned urban environments often 'work' because they (unconsciously) incorporate an evolved spatial 'syntax' which is lacking in consciously designed urban blueprints. Again, Park and his colleagues seem to have been at least partially aware of this in a way that post-war planners and politicians have not, for the early work of the human ecologists points to the importance of an evolved wisdom, barely understood, which has led us down the centuries to create spatial patterns that maximize feelings of security and personal autonomy.

Park's essential insight, an insight which was once central to urban sociology yet which is today widely ignored or simply forgotten, was that urban systems are patterned partly as a result of processes which operate independently of human intention, processes which reflect both the innate instincts of human beings and the evolved logic of spatial forms. This is an insight which we continue to ignore at great cost. If many modern urban environments have today become lifeless, soulless and even dangerous places in which to live, it is in part because we have been conceited enough to believe that we could plan an environment which would work better than that bequeathed through the trial and error of the countless generations that have lived before us.

REFERENCES

Alihan, M. (1938) *Sociology Ecology: a Critical Analysis*. New York: Columbia University Press.

Ballantyn, A. (1988) 'When design is the real criminal', *Guardian*, 9 March, p. 23.

Castells, M. (1977) *The Urban Question*. London: Edward Arnold.

Castells, M. (1983) *The City and the Grassroots*. London: Edward Arnold.

Cockett, R. (1994) *Thinking the Unthinkable*. London: Harper Collins.

Coleman, A. (1985) *Utopia on Trial*. London: Hilary Shipman.

Collison, P. (1963) *The Cutteslowe Walls*. London: Faber.

Dawkins, R. (1976) *The Selfish Gene*. Oxford: Oxford University Press.

Dickens, P. (1990) *Urban Sociology*. Hemel Hempstead: Harvester Wheatsheaf.

Dickens, P. (1992) *Society and Nature*. Hemel Hempstead: Harvester Wheatsheaf.

Dickens, P. (1996) *Reconstructing Nature*. London: Routledge.

Duncan, O. (1964) 'Social organization and the ecosystem', in R. Faris (ed.), *Handbook of Modern Sociology*. Chicago: Rand McNally.

Durkheim, E. (1938) *The Rules of Sociological Method*. New York: Free Press.

Durkheim, E. (1984) *The Division of Labour in Society*. London: Macmillan.

Firey, W. (1945) 'Sentiment and symbolism as ecological variables', *American Sociological Review*, 10: 140–8.

Giddens, A. (1984) *The Constitution of Society*. Cambridge: Polity Press.

Glendinning, M. and Muthesius, S. (1994) *Tower Block*. New Haven, CT: Yale University Press.

Hannerz, U. (1980) *Exploring the City*. New York: Columbia University Press.

Harvey, D. (1982) *The Limits to Capital*. Oxford: Basil Blackwell.

Hayek, F. (1944) *The Road to Serfdom*. London: Routledge.

Hayek, F. (1988) *The Fatal Conceit*. London: Routledge.

Heraud, B. (1975) 'Social class and the new towns', in C. Lambert and D. Weir (eds), *Cities in Modern Britain*. London: Fontana.

Hillier, B. (1988) 'City of Alice's Dreams', *Architects Journal*, vol. 35, 9 July.

Hillier, B. and Hanson, J. (1984) *The Social Logic of Space*. London: Cambridge University Press.

Hirst, P. and Woolley, P. (1982) *Social Relations and Human Attributes*. London: Tavistock.

Jacobs, J. (1961) *The Death and Life of Great American Cities*. New York: Random House.

Jager, M. (1986) 'Class definition and the esthetics of gentrification', in N. Smith and P. Williams (eds), *Gentrification of the City*. Boston: Allen and Unwin.

Judge, D., Stoker, G. and Wolman, H. (eds) (1995) *Theories of Urban Politics*. London: Sage.

Marx, K. and Engels, F. (1970) *The German Ideology* (ed. C. Arthur). London: Lawrence and Wishart.

Newman, O. (1972) *Defensible Space*. New York: Macmillan.

Pahl, R. (1975) *Whose City?*, 2nd edn. Harmondsworth, Penguin.

Park, R. (1952) *Human Communities*. New York: Free Press.

Parsons, T. (1964) 'Evolutionary universals in society', *American Sociological Review*, 29: 339–57.

Popper, K. (1945) *The Open Society and its Enemies*. London: Routledge.

Reissman, L. (1964) *The Urban Process*. London: Collier Macmillan.

Rex, J. and Moore, R. (1967) *Race, Community and Conflict*. London: Oxford University Press.

Rex, J. and Tomlinson, S. (1979) *Colonial Immigrants in a British City*. London: Routledge and Kegan Paul.

Robson, B. (1982) 'The Bodley barricade: social space and social conflict', in K. Cox and R. Johnston (eds), *Conflict, Politics and the Urban Scene*. London: Longman.

Samson, C. (1994) 'The three faces of privatisation', *Sociology*, 28: 79–97.

Saunders, P. (1986) *Social Theory and the Urban Question*, (2nd edn.) London: Hutchinson.

Saunders, P. (1990) *A Nation of Home Owners*. London: Unwin Hyman.

Savage, M. and Warde, A. (1993) *Urban Sociology, Capitalism and Modernity*. London: Macmillan.

Spicker, R. (1987) 'Poverty and the depressed estates', *Housing Studies*, 2: 283–92.

Toennies, F. (1995 [1883]) *Community and Society*. New York: Harper and Row.

van den Berghe, P. (1981) *The Ethnic Phenomenon*. New York: Elsevier.

Weber, M. (1968) *Economy and Society*. New York: Bedminster Press.

Wirth, L. (1938) 'Urbanism as a way of life', *American Journal of Sociology*, 44: 1–24.

Wolfe, T. (1988) *The Bonfire of the Vanities*. London: Jonathon Cape.

Young, M. and Wilmott, P. (1957) *Family and Kinship in East London*. London, Routledge and Kegan Paul.

4

Impartial Maps: Reading and Writing Cities

HANA WIRTH-NESHER

For those who pass it without
entering, the city is one thing; it is
another for those who are trapped by
it and never leave. There is the city
where you arrive for the first time;
and there is another city which you
leave never to return. Each deserves
a different name.

Italo Calvino, *Invisible Cities*

The city is a discourse and this
discourse is truly a language: the city
speaks to its inhabitants, we speak
our city, the city where we are,
simply by living in it, by wandering
through it, by looking at it.

Roland Barthes, 'Semiology
and the Urban'

Since the turn of the century, there have been
numerous attempts to define the city's essential
characteristics, what constitutes its 'cityness'.
Beginning with Max Weber's *The City* (1958),
sociologists, anthropologists and cultural historians have offered a series of definitions of the
city as a physical and cultural phenomenon not
merely synonymous with civilization, as had been
set down by Aristotle – 'Outside the *polis* no one
is truly human.' From the German theoreticians
to the American urban sociologists who followed
them, the tendency has been to provide models
for an all-encompassing definition of urbanism in
the modern world. Constituting what has been
referred to as a German school, for example,
Weber, Simmel and Spengler claimed scientific
objectivity for their models of the metropolis. For
Weber and Spengler the contemporary city was a
degenerate form of a better model of urbanism
that preceded it, and which they described nostal-

gically. From Engels's exposé of poverty and class
relations, through Weber's and Spengler's descriptions of solitariness and fragmentation, there has
been a convention of associating the modern city
with alienation. Simmel, in contrast, called attention to the unprecedented opportunities offered
by the modern city for the development of the
intellect and for the freedom of the individual,
despite the price of blasé indifference and solipsism. Weber and Spengler were undoubtedly
made uneasy by the mass migrations and
heterogenous crowds of the modern city when
measured against the romanticized and more
homogenous city of the late Middle Ages, while
Simmel, more of an outsider as a Jew in a Christian society, recognized this unemotional ethos of
city life, but also welcomed the new freedom
made possible by impersonality.

In recent years, the coexistence of diverse
mappings of the city has become a central
concern of city planners and architects. Sceptical
about the possibility or even desirability of
universal formulations about the city, planners
have increasingly attempted to map 'affect', the
urbanite's imaginative reconstruction of his own
environment. While Kevin Lynch's pioneering
study *The Image of the City* (1971) registered
different 'readings' of the city, his aim was to
arrive at universal determinants of legibility, at a
vocabulary of cityscape elements necessary for a
city to be 'legible': path, edge, node, district and
landmark. But because Lynch reduced the urban
environment to physical features useful for
moving about the city, he ignored a symbolic level
of the image which may, in fact, determine the
recognizability of a physical feature. In other
words, recent planners have gone beyond Lynch
in their realization that 'conceptual stimuli in the
environment play a more fundamental role than

This is a revised and expanded version of the Introduction to *City Codes: Reading the Modern Urban Novel*.
(Cambridge University Press, 1996).

mere formal perception, so that physical forms are assigned certain significations which then aid in directing behavor' (Gottdiener and Lagopoulos, 1986: 6–12). Urban semioticians and planners have taken issue with Lynch's mapping of cities by claiming that his cognitive approach arrives at signification in the urban environment through the perception of its inhabitants without accounting for the way in which *symbolic worlds shape the perception of physical form* itself. The emphasis on the *reader* of the city rather than on the identification of universal features has marked recent debate about cities in a variety of fields. In their essays on urban semiotics, both Eco and Barthes have emphasized the indeterminacy of urban landmarks, pointing out the necessity for absent centres and empty signifiers, for 'meaning' which is derived from urbanites themselves (Barthes, 1971; Eco, 1973). The architect Aldo Rossi, to cite another, has insisted on understanding the individual city artefact within the construct of different collective memories (Rossi, 1989). In short, urban planners and geographers have increasingly realized the centrality of the 'reader' in the process of cognitive mapping, just as many literary critics have increasingly taken into account the phenomenology of the reading process itself (Eco, 1979; Iser, 1974, 1978; Suleiman and Crossman, 1980; Portugali, 1996, 2000).

> The obsession with the text metaphor for the city and culture is revealed in the extensive use of terms such as discourse, legibility, narrative, the vernacular, and interpretive communities, as well as in the growing interest in 'reading and writing' architecture, the city, and culture. (Ellin, 1995: 253)

Discussions of the cityscape among contemporary urban semioticians, then, finally come down to the question – whose city? The politics of difference has had a profound effect on urban research, as it has on literary and cultural studies. 'Current theorizing of the city tends to celebrate the quixotic and the flux of the urban world, and the diversity of the cityscape,' write Westwood and Williams in *Imagining Cities* (1997: 4).

When it comes to cultural models of the city based on *literary representation*, the impulse has often been to identify an essence of urbanism. Walter Benjamin, for example, has identified life in the city with the paradoxical notion of ceaseless shock, but his city is derived from his insights about Baudelaire, and is not necessarily applicable to modern urban life more generally, or to modernism as an artistic and cultural period. Franco Moretti's recent challenge to Benjamin's essays on urban experience, what he calls 'the sancta sanctorum of literary criticism', consists of his offering what amounts to the opposite claim regarding universal urban experience, 'city life

mitigates extremes and extends the range of intermediate possibilities: it arms itself against catastrophe by adopting ever more pliant and provisional attitudes' (1983: 117). But in place of Baudelaire's Paris as the ground of his observations is the world of Balzac. As propositions about the nature of urban life, there are few essays as rich as these, not because they are universally applicable as models of urban experience, but because each is grounded in an urban universe wrought by particular readers.

Yet all city dwellers do have at least one thing in common: the process of imagining the inaccessible. Insofar as cities promise plenitude, but deliver inaccessibility, urbanites are faced with a never-ending series of partial visibilities, of gaps – figures framed in the windows, partly drawn blinds, taxis transporting strangers, noises from the other side of a wall, closed doors and vigilant doormen, streets on maps or around the bend but never traversed, hidden enclaves in adjacent neighbourhoods. Because no urbanite is exempt from this partial exclusion and imaginative reconstruction, every urbanite is to some extent an *outsider*.

The *effect* of inaccessibility differs with each city dweller according to the specific experience of his or her outsiderness. 'The city dweller's life,' writes Franco Moretti, 'is dominated by a nightmare – a trifling one, to be sure, – unknown to other human beings: the terror of "missing something"' (1983: 119). But this need not be a nightmare, nor is it necessarily trifling. The city dweller learns to contend with the sensation of partial exclusion, of being an outsider, by mental *reconstruction* of areas to which he or she no longer has access, and also by *inventing* worlds to replace those which are inaccessible. The boundary between these two activities is occasionally as difficult to discern as a city's limits. And the reconstructions and inventions will depend entirely on the particular perspective of the urbanite himself, on the particular nature of his or her outsiderness. The metropolis is rendered legible, then, by multiple acts of the imagination; it is constantly invented and reinvented through narrative. The city is read by its inhabitants through the stories they remember and invent.

NARRATIVE CARTOGRAPHY: REPRESENTATIONS OF THE CITY

Sites in the 'real city' that already trigger cultural associations also function as problematic sites in fictional representations of the city. What may appear to be a 'given' geographically or what may appear to be merely a peripheral concession to fact (such as a street name or familiar landmark) can be a significant cultural locus. The narrative spaces opened in novels, for example, become arenas for cultural work (Bremer, 1992; Davis, 1987; Fisher, 1987; Gelley, 1973; Luttwack, 1984; Stallybrass and White, 1986).

Represented cities in novels perform the acts of invention and reconstruction that are endemic to metropolitan life. No matter what the effect of exclusion in the 'real' city – threatening, seductive, intoxicating, unnerving – every case will require a form of imaginative mapping. This is precisely the activity of the novelist who both reconstructs in language aspects of 'real' cities and invents cityscapes. Just as for the city dweller the city itself is a text that can never be read in its totality, the modern urban novel acts as a site for the problematic of reading cities. We have all traversed boulevards and lost our way along the streets of cities our feet have never crossed as well as those we know from physical encounters. And the spaces we have come to believe we *know*, those which we can read, are often legible through the mediation of texts about them, through the cumulative perceptions of others. Oscar Wilde remarked that London itself became foggier after the Impressionists painted their cityscapes. Narratives have had the same effect, whether historical or fictional.

When authors import aspects of 'real' cities into their fictive reconstructions, they do so by drawing on maps, street names, existing buildings and landmarks, enabling a character to turn the corner of a verifiable street on the map, to place him in a 'realistic' setting. These urban elements signify to a reader within a particular culture a whole repertoire of meanings. Dreiser assumes, for example, that the reader will recognize the significance of different locales – Fifth Avenue and the Bowery as indicators of economic status. Joyce expects his reader to be able to identify landmarks in Dublin that are signs of British imperialism. On a secondary level of signification, the novelist draws on a repertoire of urban tropes inherited from previous literature, tropes that have secured a place for themselves in the literary or artistic tradition, such as the image of the underground man, the sinister connotations of a city like Venice, or the passerby. Moreover, what I have been referring to as the 'real' city cannot be experienced without mediation as well; it is itself a text which is partly comprised of literary and artistic tropes – Hugo's sewers, Hopper's windows, Eisenstein's steps, Dickens's law courts. When Mrs Sinico, a character in Joyce's story 'A Painful Case', is run over by a train after her disappointment in love, they are Tolstoy's tracks as well as the 'real' tracks indicating an unsavoury neighbourhood as analogue for the character's emotional and social decline.

When Richard Wright's character in *The Man Who Lived Underground* (a title which identifies the work with slave narratives) escapes into a sewer, Wright situates his book and his world at a nexus between white European (Hugo, Eugene Sue, Dostoevsky) and African American literature (Frederick Douglass).

The interaction of city as text and representation of city in the literary text is most dramatic when it doubles back upon itself, when *invented* worlds themselves may be sought in the physical cityscape. Tours of London based on Sherlock Holmes and other detective or crime fiction serve as one example, as well as tourist sites originating in fictional texts: the blacksmith's house in Boston derived from Longfellow's poem or Joyce's No. 10 Eccles Street. The latter is a particularly fascinating development, for in contemporary Dublin the text of *Ulysses* has left its mark literally on the landscape: first as actual sites – Larry O'Rourke's pub or Maginni's dancing academy which have been preserved as landmarks solely because of their appearance in Joyce's novel; secondly as literal inscriptions on the city streets – bronze plaques with quotations from the text have been cemented onto the city sidewalks that mark the spots of Leopold Bloom's peregrinations. The city text is a palimpsest, therefore, of the history of its representation in art, religion, politics – in any number of cultural discourses.

I have identified four aspects of the cityscape in the representation of the city in narrative: the 'natural', the built, the human and the verbal. The *'natural' environment* refers, of course, to the inclusion or intervention of nature in the built environment, and is never nature outside the bounds of culture. The extent to which nature has been incorporated into the metropolitan imagination is evident in the *New Yorker* cartoon (Figure 4.1), in which nature is not perceived to be the surrounding ground for the man-made city, but rather as an amenity that serves the city. Parks and landscaped areas, the architectural fashioning of 'nature', fuse nature and culture in particularly interesting ways. For example, characters in Isaac Bashevis Singer's *The Family Moskat* are aware of Warsaw municipal laws that restrict caftaned Jews from entering the Saxony Gardens, a pocket of nature in the city. Even the weather can take on cultural features, such as the fog in Dickens's *Bleak House*, which is indistinguishable from the man-made gas which looms through it, or the flakes of soot which are described as snowflakes.

The *built environment* refers to city layout, architecture and other man-made objects such as trams, curtain walls and roofs. The man-made environment in an urban novel is a representation either of actual existing artefacts in 'real' cities, or of purely invented structures. The city's under-

"It's so lovely out here you wonder why they have it so far from the city."

Figure 4.1 © The New Yorker Collection 1990 Frank Modell from cartoonbank.com. All Rights Reserved

ground, for example, is an aspect of the built environment that has acquired cultural significance with the publication of novels by Eugene Sue, Victor Hugo and other writers who have represented it as the site of the collective repression of the bourgeois city dweller. Individual landmarks appear in representations of the built environment, such as Fitzgerald's Empire State Building or Dreiser's Waldorf Astoria Hotel, or precise addresses that are transformed into landmarks as a result of their embedding in a fictional text. This is the case of No. 10 Eccles Street in *Ulysses*, which I referred to earlier. Every representation that is not clearly signifying an exact location is an imaginary structure, such as the home of the child in Henry Roth's *Call It Sleep*, while architectural features of that home, like the cellar and the roof (chapter headings in the novel), refer to common features of the built environment that have acquired local significance

in the work of fiction. Aspects of the built environment do not add up to a universal lexicon of the metropolis, as they derive their significance from both the text in which they are represented and the existing repertoire of city tropes in the arts and in literature.

The *human environment* does not refer to the characters whose action or thoughts constitute the main movement of the plot, but rather to those human features which constitute setting, such as commuter crowds, street pedlars and passers-by. Although the human environment does encompass crowd scenes, it can as readily refer to types who are generic fixtures of cities in specific periods or locales: the doorman, the street musician, the beggar etc. To name a few instances from literary texts – Flaubert in *Madame Bovary* employs the organ grinder in a pivotal scene, as does Joyce in *Dubliners*, but for entirely different purposes; Dickens and Dreiser both have sharp eyes for the eccentric occupations spawned by the city, such as street car track inspectors and advertising sandwich men; Ellison supplies recurring human fixtures, such as the shoeshine boy or the street vendor, to map out the racial and social hierarchies in his representation of New York.

The *verbal environment* refers to both written and spoken language: the former includes the names of streets and places, and any other language which is visually inscribed into the cityscape – advertisements, announcements, graffiti. The giant billboard for the oculist Dr T.J. Eckleburg on the motorway to Manhattan in Fitzgerald's *The Great Gatsby*, for example, becomes an emblem of the shortsightedness of both characters and narrator in the novel. The repetition of street and building names such as Street of the Prophets or Terra Sancta in Amos Oz's *My Michael* conveys the competing claims of different peoples for the city of Jerusalem. The names of cities themselves, even if they are never represented in the novel, signify beyond their geographical referent: Moscow for Chekhov's sisters, Paris for Emma Bovary, Danzig for Oz. Moreover, the auditory rather than the written verbal environment is often an indicator of social, ethnic, or other subdivisions in the city made evident by dialect or other language usage. In Henry Roth's *Call It Sleep*, speech is an auditory landmark that identifies neighbourhood.

Each of these environments can be perceived and represented by all of the senses as the action of the novel unfolds. The reader is then put in the position of apprehending the cityscape in a visual, audial, or tactile manner, but always mediated by the written word.

Emphases shift from novel to novel and from scene to scene. *Call It Sleep* is an especially noisy book, one in which the built, human, and verbal environments are all experienced primarily by *sound*, and where fear and desire are embodied in speech. Dreiser's *Sister Carrie*, on the other hand, tends to be a visual book, one in which seeing, desire and commercialism are interlaced. In some instances, a character bypasses all of the senses in reading a city, drawing only upon mental images for which he seeks grounding in the landscape. In Henry Roth's story 'The Surveyor', a tourist lays a wreath on a spot in Seville where centuries earlier Jews were burned at the stake by the Inquisition as heretics. Nothing in the cityscape designates it as a landmark; it is identified only by historical consciousness.

Such privileging of the verbal environment may not be characteristic of other urban novelists. For Joyce in *Dubliners* and *Portrait*, inaccessibility in the urban setting results from the power of textual mediation. While there are an abundance of partly concealed windows and figures in the crowd, they tend to be analogues for the concealments wrought by language. In Bashevis Singer's novel, in contrast, strategies for the representation of Warsaw are not mediated through a literary repertoire, but rather through historical memory of the built environment; the fictional urban setting in *The Family Moskat* is a precise reconstruction of what was literally destroyed, of the Jewish community subsequently forced into a ghetto and then annihilated. Here, place can be understood only in terms of history, and is necessarily read differently by characters who cannot know the future, and readers who do. The assumption shared by these readings of settings is that the experience of exclusion and inaccessibility is shared by all city dwellers, but that the particular nature of this inaccessibility is shaped by the particular nature of the 'outsider', whether it be the result of religion, politics, class, race, gender, nationality, provincialism, or any number of other exclusionary principles.

To cite one example, Virginia Woolf's *Mrs Dalloway* overturns a whole tradition of urban literature in which the female is inscribed onto the cityscape as object of the male gaze. Woolf's protagonist reclaims her subjectivity by replacing the muse of the exotic young female in the crowd with the male passer-by and with the face of an old woman. Whereas many modern novelists summon the urban trope of the *the passer-by* who invites speculation, Woolf structures her entire book around the anticipated meeting of passers-by that never takes place, one an outsider by gender and the other by social class.

Novels reconstruct 'real' cities that contain problematic sites, and these gaps and partial views are rendered in the four aspects of the environment outlined above. The modern novel's emphasis on epistemological questions, therefore, taps into the epistemological dimension of city life. It is not surprising, then, that modernism has often been equated with urbanism. In Raymond Williams's terms, 'the key cultural factor of the modernist shift is the character of the metropolis' (1989: 45; Baumgarten, 1982; Bradbury, 1976; Howe, 1971; Pike, 1981; Sharpe, 1990; Spender, 1972). As all urban novels are not modern, what distinguishes the modern urban novel from its precursors?

THE MODERN URBAN NOVEL

Around this equation of urbanism and modernism many of the modernist platitudes revolve: loneliness, isolation, fragmentation, alienation. For the authors and cultural critics who have felt displaced and threatened by the modern city, by its influx of immigrants, its crowds, its vulgarity, T.S. Eliot's line in 'The Waste Land' is apt: 'I had not thought that death had undone so many.' I have steered away from such truisms about the city and modern culture, the Romantic insistence on the fall from rural harmony into the discord of the metropolis. I am not arguing that the city has never been an alienating milieu, but that this may not be the case for Woolf, Dreiser, or Ellison, or for any writer who may have something to gain from the modern city, or who nurtures no rural or pastoral sentiments. To reject these commonplaces about the modern city with their Spenglerian echoes, moreover, is not to deny the centrality of the city in the development of the modern novel, nor the privileged place of the city in what we loosely define as modernism. But it frees us to look at difference, to draw on alternative traditions. When a black man arrives in New York in Ellison's *Invisible Man*, he is not measuring the present city against some medieval ideal; he sees it as the promised land of freedom as represented in slave narratives. When Bashevis Singer reconstructs Warsaw in *The Family Moskat*, a novel written after the Holocaust, the pre-1939 threat of secularization and fragmentation of communal life is overshadowed by the community's subsequent annihilation and becomes material for nostalgia.

Rather than equating the city with dehumanizing features of a modernist sensibility that has increasingly been called into question as an easily recognizable period with stable features, I want to define a shift in *representation* of the city from pre-modern to modern urban novels. Novels with urban settings, after all, are as old as the novel itself. Characters in the traditional novel are on a quest for some form of self-realization that will eventually bring them to a place that provides satisfying closure. The setting in these novels serves as counterpart for character, representing the character's search for his or her 'true' identity, for an appropriate 'home'. Perhaps one of the boldest features distinguishing the modern urban novel from urban fiction in earlier periods is exactly this concept of 'home'. Unlike 'home' in the traditional English novel, which offers a refuge from the street and a resolution (even if ironic) to the plot, 'home' in the modern urban novel has been infiltrated by the 'outside' (Fanger, 1975; Welsh, 1971). The setting of the eighteenth- and nineteenth-century novel tended to be houses, with the house representing the continuity of tradition, family, social class and conventional order. These homes were not immune to violent intrusion from without, but intruders were eventually ousted or legitimized through marriage. For the servant class, of course, home could be perilous, as the employer was not perceived to be an intruder in an employee's space. While individual outsiders could occasionally make their

way into these houses through marriage or accumulation of wealth, each entry further confirmed the social order itself, on its own terms, the traditional world of houses. The novels of Austen, Fielding, the Brontes, George Eliot and Trollope are among those in the English literary tradition where house is setting, character, social order and theme. Authors like Hardy and Dickens chronicled the gradual collapse of the house, as domicile and as family. But even when the old country houses fall, as in Dickens, the ideal of the sanctity, privacy and separation of the house from the city is maintained. He affirms the necessity for clear demarcations of public and private, of the city street and the cherished private domicile.

Even in the American literary tradition, where the house as it represented an Old World social order was always suspect, the separation between the private and the public was rigorously maintained. Whether characters turned their backs on the family home, as Twain's Huck Finn or Cooper's Natty Bumpo, or whether they idealized it in *Uncle Tom's Cabin* or *Little Women*, it was sealed off from the street. Edgar Allan Poe finally demolished the ancestral home in 'The Fall of the House of Usher'; elsewhere he also gave expression to the horror of being shut out from home forever. In 'The Man of the Crowd' one city dweller is determined to trail another to his home, only to discover that neither will ever find home again, that the street is their eternal damnation.

In the pre-modern novel considered from the perspective of setting, 'home' is a private enclave, a refuge from the intensely public arena of urban life. In the modern urban novel, however, 'home' itself is problematized, no longer a haven, no longer clearly demarcated. If we borrow Bakhtin's concept of the chronotope of any genre as 'the intrinsic connectedness of temporal and spatial relationships that are artistically expressed in literature', then the home in the modern novel no longer serves as the dominant chronotope (Bakhtin, 1981: 84). For Bakhtin, the chronotope of the agora in ancient Greece represented the fully public man, the unity of a man's externalized wholeness. With the fall of public man vacating the public square, the self divides into private and public, into individual and collective identities that occupy spaces appropriate to this new subject. Each new chronotope replacing this wholly public one is intertwined with the emergence of genres – the road for the picaresque novel, the castle for the Gothic, the home for the novel. Although parlours and salons had already provided settings for novels prior to the nineteenth century, in the novels of Stendhal and Balzac, according to Bakhtin, they reached their full significance as the place where the major

spatial and temporal sequences of the novel intersect. But this increased privatization of the
modern self – traced in Bakhtin's scheme of
chronotopes, derided by Lukacs as the solipsism
of the bourgeois mind, and exposed by Jameson
as the mirage of autonomous consciousness –
gives way in the modern urban novel to the
conflation of the public and private self
(Jameson, 1981; Lukacs, 1971).

The chronotope of the modern urban novel
which I am suggesting is a space that *conflates the
public and the private* but does so in a wide variety
of ways. It is not that the street has been substituted for the parlour as the dominant setting,
although it plays a major role in these novels, but
that the opposition of parlour and street has been
eroded. Most of the action in these fictional
worlds takes place in spaces that fuse public and
private, that are uneasily indeterminate: coffee
houses, theatres, museums, pubs, restaurants,
hotels and shops. And even when the setting is an
interior 'private' home, its dwellers are exposed to
the gaze of the stranger, as Joseph K. is observed
by his neighbours in the windows opposite those
of his own room, or as Mrs Dalloway regularly
observes (and imaginatively communes) with an
old woman in the apartment facing her own. I
have chosen these two examples to emphasize that
the conflation of public and private does not
always have the same effect: in Kafka's novel it is
emblematic of K.'s paranoia, in Woolf's vision it
is emblematic of the human bonding enabled by
anonymity more readily than by social acquaintance. Woolf's *Mrs Dalloway* defies conventional
hierarchies of communication, from family, to
society, to passers-by. In other words, the urban
trope of the stranger's face in the window, of
physical proximity and mental or social distance,
can function in any number of ways, depending
upon the character of inaccessibility in that
world. The exclusion of the immigrant is not the
exclusion of the tourist, the detection of inaccessibility may differ between men and women, and
the reading of missed opportunities may differ
with age or race. What modern urban novels do
share is a predominance of these indeterminate
public and private spaces, and a construction of
self that is far more dependent on the 'street' than
it is on domestic resources. In fact, it also has the
effect of domesticating the street, of making the
city a wellspring of desire and identity.

I do not mean to suggest that 'home' has been
renounced or that 'outside' has simply been
substituted for 'inside'. It is simply that 'home' is
a shifting space, a provisional setting, an intersection of public and private that is always in
process. I would agree with Burton Pike that one
of the main shifts in representation of the city
from early to late nineteenth-century literature is
from stasis to flux, but I would argue that this

does not necessarily result in a greater emphasis
on the isolation of the individual within the city.
The effect of flux has been to undermine the existence of the private individual in the traditional
home, and to create new cultural spaces of
various mergers of the individual self and the
cityscape. The gaps in the cityscape produced by
inaccessibility and partial exclusion motivate the
city dweller to construct spaces and narratives
which constitute provisional home. For the invention of these intersections of public and private
space, the authors have drawn on the four aspects
of the 'real' city discussed above: the natural,
built, human and verbal environments. To cite
one example, the importing of landmarks into the
fictional cityscape inscribes public features into
private selves. The landmarks themselves may be
built (such as Notre Dame in the Paris of *The
Ambassadors*), human (such as the generic copper
lady in the London of *Mrs Dalloway*), or verbal
(such as the name of sites which are inaccessible
to characters, but which have cultural or historic
significance – the Hill of Evil Counsel or the
Tower of Nebi Samwil in the Jerusalem of *My
Michael*). In each case, the landmark, whether
place, person, name or all three, constitutes both
an internal and external space, an object that is
both self and other.

The problematizing of home and the indeterminacy of public and private realms affects both
the theme and the form of modern urban novels.
The private self in conflict with a public world, a
self bent on carving out a suitable private enclave,
is replaced by a subject that both constructs and is
constructed by the cityscape. At times the plot
itself unfolds as a sequence of perceptions of
place, of actual movement through the cityscape
and 'readings' of the urban environment. In the
modern urban novel, cityscape is inseparable
from self, and the specific strategies for representing the intersection of character and place are
the product of the specific form of exclusion
experienced by the character, author, or reader. In
short, setting in the modern urban novel, by its
provisional and dynamic properties, tends to
undermine the quest for a total vision, an ultimate
homecoming, lasting knowledge.

In a sense, exile in its various forms is experienced by every city dweller, and inaccessible
spaces are imagined through both memory and
desire, personal and collective. The reasons for
exile are many, and in urban novels they generate
a great variety of textual places. Isaac Bashevis
Singer can serve as one example.

Singer fled his home city of Warsaw in 1939
just before the Nazi invasion of Poland, and his
novel *The Family Moskat*, written in New York in
1947, is his reconstruction in language of the city
as yet undefiled by the ghetto and by the Holocaust. In that Warsaw from which Singer is

separated both geographically and historically, the barriers between Jew and Gentile were yet to be literalized by ghetto walls and brutally enforced by the annihilation of the Jewish population. Singer's Warsaw is a meticulous post-war reconstruction of a city by an author committed to documenting the ethnic and religious walls of a period for readers conscious of the physical walls that superseded them. His personal memory of Warsaw, combined with his historical perspective, results in a textual reconstruction that dwells on urban features of the built and human environment which retrospectively act as harbingers of death: *courtyards, janitors, and trains,* all haunted by evictions, round-ups and deportations.

The effect of such hindsight on *The Family Moskat* is twofold: movement in the narrative is toward a contraction of place, and metaphors in the fictive world are literalized by history, by a future beyond the temporal boundaries of the novel. Furthermore, certain conventional markers of the urban setting take on ominous dimensions when read retrospectively. While the city in any urban novel is experienced as a succession of partial exclusions, in this novel the partial exclusion is haunted by the subsequent *total* exclusion which the Warsaw Ghetto cruelly enforced. In effect, history subsequent to the temporal frame of the novel infiltrates both the composition and the reception of the work.

On the first few pages of the novel, the family patriarch Meshulam Moskat rides past the Saxon Gardens early in the century, while Poland was still under Russian rule.

> ... the Russian policeman standing between the two rows of car tracks, the Saxon Gardens, with densely leaved branches extending over the high rails. In the midst of the thick foliage, tiny lights flickered and died. From inside the park came a mild breeze that seemed to carry the secret whisperings of amorous couples. At the gates two gendarmes stood with swords to make sure that no long-caftaned Jews or their wives ventured into the park to breathe some fragrant air. (1950: 14)

The question of accessibility is raised immediately. Under Russian occupation, Jews with caftans and Jewish wives with wigs were barred entrance to the Saxon Gardens, under a municipal law, a legal relic from the pre-industrial city which restricted movement for visiting foreigners, certain occupational groups and Jews. This movement through the city on the opening pages of the book partakes of three time frames: the present of the fictional world which historically enforces the exclusion of certain Jews from the park; the recent past for Singer and his readers for whom this passage would be ominous; the mythic time of transhistorical archetypes. The naming of the park sets in motion historical memory: the Saxon

Gardens are a reminder of the German claim on Poland which would be repeated in 1939; the Gardens are adjacent to the Jewish ghetto, and would remain on the Other Side when the ghetto was sealed. The two gendarmes with swords, of course, are an icon of the expulsion from Eden, the exclusion that mythically paved the way for urban civilization. In Christian hermeneutics, the expulsion from Eden is a universal condition of mankind's fall from innocence, but here, in its historical context, it is limited to Jews whose apparel makes them recognizable. While the expulsion from Eden may be the first occurrence of exile in Judaeo-Christian myth, the universality of that archetype is qualified by the particular historical exile of the Jew, which in turn has acquired archetypal status in Western civilization.

Singer's personal memory of the streets invests his cityscape with a tragic dimension. His characters' routes as they walk through the city are named precisely, as the verbal environment signals restrictions in mobility. When in the space of a paragraph two characters go from Panska Street to Jerusalem Alley, and then on to Tvarda and Gzhybov streets (1950: 258), this demonstrates their mobility as they traverse major boulevards, cross intersections, move freely throughout the city. However, the streets mentioned most often, such as Twarda, Panska, Krochmalna and Leszno, are eventually divided when barbed wire barriers enforce the boundaries of the Warsaw Ghetto (see Figure 4.2). These same streets would run both inside and outside of the ghetto; they would no longer permit free movement along them for Jews. As a result, historical hindsight erects phantom barricades along the very streets that are inscribed in the book to represent freedom of movement. The literal historical ghetto walls infiltrate the routes, and hindsight lends them an ironic dimension. Jerusalem Alley, for example, is *not* in the Warsaw ghetto, while the Saxon Gardens mentioned on the opening pages of the book, with gendarmes at the gates, are exactly on the other side of the barricade.

Singer's reconstruction of Warsaw in his novel *The Family Moskat* is motivated by the desire for collective memory. The premise behind the novel is that once the map of Jewish Warsaw is inscribed in this world of fiction, it will have a chance of surviving in the collective memory of the readers. The passion for recording *that* Warsaw is motivated by its tragic loss. Despite Singer's avowal to write 'as if' the community had never ceased to exist, hindsight makes it impossible for the novel to reconstruct a world without traces of its destruction, without its historical future infiltrating its fictive present. As a result, one aspect of pre-war Warsaw, its division between Jew and Gentile, gathers into itself its

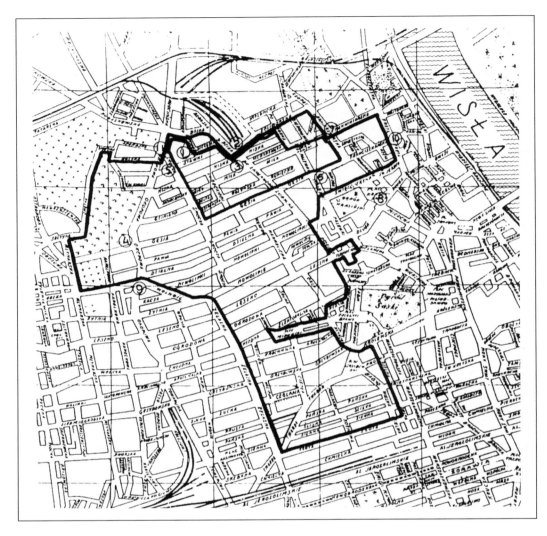

Figure 4.2   The Warsaw Ghetto (Gutman, 1982)

own horrific culmination. No matter how much the book appears to conform to the genre of the multigenerational family novel, of the young man gaining experience in the city, of the move from traditionalism to modern secular culture, the Holocaust infects the setting retrospectively, turning marginal architectural features and city descriptions into images of progressively severer forms of exclusion and doom. *The Family Moskat* re-enacts the perceptual problem and ethical dilemma of the writer committed to reconstructing a lost world as a memorial to the *living* civilization that was destroyed, but unable to shake off the knowledge of its apocalyptic end. Singer's desire to make accessible to his reader the Warsaw that was destroyed finds expression in a text that turns inaccessibility into more than a

major theme of the novel; it re-enacts an experience of inaccessibility every time it is read. The Warsaw of Jewish history overtakes the Warsaw of the survivor's painstaking reconstruction with the result that the city in the novel *is* somehow pre-Holocaust Warsaw.

While Singer was in exile from Warsaw in his personal memory, in his collective memory every Jew is in exile from Jerusalem. A novel by an Israeli novelist set in Jerusalem, therefore, will offer a variant on the exilic imagination.

To speak of Jerusalem as home seems almost paradoxical. To inhabit Jerusalem is to domesticate a vision, to literalize the most powerful urban metaphor of Western civilization. This is precisely what Zionism aimed at achieving for the Jewish people, and modern day Jerusalem is the site of history overtaking metaphor and literalizing it in a place that resists totalization. Jerusalem is the city *par excellence* in the way that it offers plenitude but is experienced as far less than that, for the very name promises transcendence, while the earthly Jerusalem is a never-ending series of exclusions.

The transformation of Jerusalem from the religious redemptive city for which an exiled nation yearned in their liturgy for over two millennia into the capital of modern day Israel is, in Zionist ideology and Jewish history, a form of reconstruction, a return to the city as a political entity. If Warsaw as depicted by Singer is 'home' in exile, 'home' through evolution and accommodation which needs to be reconstructed imaginatively lest it leave no trace in historical memory, Jerusalem is 'home' by divine decree, the city of redemption that dare not be forgotten – 'If I forget thee, O Jerusalem, let my tongue cleave to the roof of my mouth', may forgetfulness consume speech. All the more ironic, then, that the historical reconstruction of Jerusalem after the 1948 War of Independence was marred by geographical division, and that the city was divided between two warring states, between Israel and Jordan, between Arab and Jew. In the aftermath of that war, the Old City with Judaism's holy sites and its symbols of ancient nationhood were all, from the perspective of the Israeli state, on the 'Other Side', across barbed wire and stone walls. Thus, the literal reconstruction of the city continued to demand an imaginative reconstruction of that Other Side, now enemy terrain but very compellingly a part of the collective history of Israeli Jerusalemites, not only as metaphor in their civilization, but as personal memory, streets and landmarks traversed regularly before the geographical partition.

In 1968, a few months after the reunification of the city in the wake of the Six Day War, Amos Oz published a novel set in divided Jerusalem immediately after Israeli independence. In it he exposes the unease of inhabiting a metaphor, and the necessity to reconstruct in fantasy what is inaccessible in geography. The text of *My Michael* is the site of the problematics of Jerusalem as a city in the middle of the twentieth century. In contrast to *The Family Moskat* where the metaphorical always moved toward the literal, in *My Michael*, the literal cannot take root; the metaphorical haunts it.

The Jerusalem inhabited by Oz's characters in the 1950s is politically a city divided between hostile neighbours who have just fought a war. The Arab section to the east includes the ancient walled town, antiquities such as the Western Wall of Solomon's Temple and the Tower of David, holy shrines for Moslem, Christian and Jew, and historical hills such as Mount Scopus and the

Mount of Olives, famous sites in both sacred and secular scripture. The Jewish city, outside and to the west of the city's walls, is relatively new, for the most part uniform tan stone buildings covering hills of the same hue. Surrounding the city topographically are the Judean hills, vast stretches of bare, dramatic mounds that bring cool winds to the city and are covered, for the most part, with shadow and rock, not forest. During the 1950s, when this novel takes place, the city of Jerusalem occupied the tip of a narrow corridor of Israeli territory and the hills surrounding it on three sides were all Arab lands. The Israeli Jerusalemite sees those hills as not only naturally formidable, but also as politically hostile. The city itself in its totality is heterogeneous, a mixture of Jew, Moslem and Christian. Its places of worship encompass many cults and sects, and its buildings house Armenians and Poles, Germans and Greeks, Israelis and Palestinians. As in many cities, there is tolerance but little interaction among the inhabitants of different cultural or ethnic groups, but this city, from 1948 to 1967, had the additional dimension of a tangible barrier of stone and barbed wire between two enemy populations.

In Oz's Jerusalem, the names of streets and landmarks are reminders of how the literal is always poised to cross over to the metaphorical. The characters meet as students at Terra Sancta, a former convent loaned to the Hebrew University after the Independence War when the Mt Scopus campus became inaccessible due to the partition. Latin for Holy Land, the landmark establishes several motifs which will recur: (1) displacement – the university, with its aims of secular knowledge, has displaced the convent; (2) competing myths – the place of Jerusalem in Christian tradition is wholly visionary as the site of the Crucifixion and the Second Coming; it is not to be literalized in this world, as is the case for either religious or secular Zionism; (3) transience – the Christians ruled the Holy Land during the period of the Crusades, but were driven out by the Moslem Saracens, a fact which Arab historians have repeatedly used as an analogue for the fate of the modern Jewish state. During this post-war period in modern Jerusalem, Terra Sancta was merely a signifier for a literal place, the university.

The names of the streets and landmarks seem to defy all attempts to return Jerusalem to a city of mere citizens and pedestrians. In fact, the city seems to demand that the simplest actions be situated on an allegorical map. Hannah's first walk with Michael is down the Street of the Prophets. As a child when she gazed out of her window toward Arab neighbourhoods, she was literally looking at Emek Refaim, the Valley of Ghosts or Phantoms. Her first home with Michael is in

Makor Baruch, the Blessed Source. On a moonlit night she wanders down Geula Street, the Street of Redemption. When she dreams of an earthquake in Jerusalem taking place during the time of British Occupation, it occurs on Ezekiel Street. By bearing the names of prophets, poets, philosophers and scholars, the streets underscore the textuality of the city itself, its existence as language, metaphor and vision. At various times Hannah mentions streets named for Ibn Gvirol (medieval allegorist and poet of love), Maimonides (philosopher and rabbinic authority), Saadiah Gaon (medieval Jewish philosopher and translator of the Bible into Arabic). Since they all lived in the diaspora, the streets of modern Jerusalem become an ingathering of the exiles, as those for whom Jerusalem was a holy metaphor become part of the literal text of the material city, moorings for the literally lost. Emek Refaim, Valley of the Ghosts, becomes synecdoche for the city of Jerusalem.

While the street names tend to emphasize the Zionist return, the names of the hills surrounding the city have quite a different effect. During one of her early morning walks with Michael, Hannah observes: 'Shadowy hills showed in the distance at the ends of the street. "This isn't a city," I said, "it's an illusion. We're crowded in on all sides by the hills – Castel, Mount Scopus, Augusta Victoria, Nebi Samwil, Miss Carey. All of a sudden the city seems very insubstantial."' (1972: 26).

Like a palimpsest, the city names record the various claims to the city. The Castel was named by modern Israel to mark the site of the decisive battle for liberating Jerusalem in 1948, while Mount Scopus is the Greek word for 'watching', the literal translation of the Hebrew name for the hill, Har Hatsofim. An isolated Israeli enclave manned a small garrison there for 19 years during the city's partition, relieved fortnightly by a convoy under UN supervision. It too is metonymic for the entire Jewish city after the partition. Named for Kaiser Wilhelm's wife, Augusta Victoria hospital was used as a Government house by the British. The hill of Nebi Samwil is named for the prophet Samuel, but it is the Arabic name for the prophet whose mother was *Hannah*. Miss Carey is a small interfaith chapel built by an Englishwoman whose name it bears; it is at the top of what Israelis call Mount Ora. The naming of specific places in the cityscape that carry foreign and multiple names re-enacts the arbitrariness of the signs designating places in Jerusalem.

Furthermore, the non-Hebraic names for the surrounding hills are a constant reminder to modern Israelis that their city occupies a place in the ideology and teleology of other peoples. 'Villages and suburbs surround Jerusalem in a close circle, like curious bystanders surrounding a wounded woman lying in the road: Nebi Samwil, Saafat, Sheikh Jarah, Isawiyeh, Augusta Victoria, Wadi Joz, Silwan, Sur Baher, Beit Safafa. If they clenched their fists the city would be crushed' (1972: 111). In this sentence, the psychosexual, political and topographical are all conflated, with Hannah's own vulnerability and fears projected onto the cityscape.

If the stark hills are reminders of Jerusalem's Middle Eastern location, the built environment underscores this as well. Hannah Gonen lives in an urban area where dwellings are visible miles away, because they cling to bare hills, but at close range they are mysterious, because Middle Eastern architecture frequently means inner courtyards and outer walls. In fact, there are layers of walls – an outer city wall, the walls of a compound, the outer wall of a courtyard, and then the walls of a dwelling itself.

> And the walls. Every quarter, every suburb harbors a hidden kernel surrounded by high walls. Hostile strongholds barred to passers-by. Can one ever feel at home here in Jerusalem? I wonder, even if one lives here for a century? City of enclosed courtyards, her soul sealed up behind bleak walls crowned with jagged glass. There is no Jerusalem . . . There are shells within shells and the kernel is forbidden . . . . I cannot know what lies in wait for me in the monastic lairs of Ein Kerem or in the enclave of the High Commissioner's palace on the Hill of Evil Counsel. (1972: 110)

Representative of one type of Jerusalemite, the child of European immigrants, Hannah is identified by her feeling of displacement, of being exiled from the very site that marks the end of exile, of being homeless even at the centre of home. The secrecy and inaccessibility that characterize the city for its Jewish residents is conveyed in the various languages of exile: the 'monastic layers' of the presence of Christian civilization, the source of much persecution of the Jews, even here in the Jewish homeland; the closed courtyards of the Middle Eastern architecture that is alien to its predominantly European population; the jagged glass of the fear of infiltration, mainly by hostile Arab neighbours; the High Commissioner's palace of the long occupation by the British Empire. The 'shells within shells' and the forbidden 'kernel' are kabbalistic references to a spiritual Jerusalem, as it is the symbol of God's contraction and of the fragments and shards left for man whose spiritual task is the reconstitution of the divine. In general, Jerusalem is a palimpsest of historical periods and diverse longings. In this particular novel, however, it is the site of paradox, of a homecoming that feels like exile, of a Jewish European Jerusalemite uneasy in the presence of the Other at 'home'.

I have chosen these two works because they

exemplify many aspects of the urbanite as outsider and exile. Inaccessibility in these novels is historical, political, cultural and metaphysical. But examples abound of authors and characters who exemplify other types of exile. In Theodore Dreiser's *Sister Carrie* (1970) and Ralph Ellison's *Invisible Man* (1981), economic and racial divisions in New York may be transparent in the cityscape, but they are powerful forces of exclusion none the less. In Dreiser's Chicago and New York, a landscape of endless commodities offers a form of plenitude that no amount of wealth can fully embrace. In Dreiser's city, the abundance of *streets* and *windows* in the built environment challenges the separation of public and private worlds, as glimpsed interiors become nothing more than window dressing, and the street is transformed into an arena for self-display. In Ellison's New York, racial stereotyping spawns an urban doubling, a black city that parodies the white. Survival for blacks means practising strategies of invisibility, such as becoming *permanent human fixtures* of the cityscape for the whites, parading as part of the human environment as opposed to subjects who inhabit the same metropolis. How this invisibility can be achieved through mimicry and parody which safeguard the actor is one of Ellison's subjects, as the construction of inner walls mirrors outer divisions.

The tourist and traveller experience gaps in the cityscape that derive from their transience. Henry James added yet another twist to this theme by portraying what he termed 'the returning observer', the tourist who revisits a foreign city and experiences the place through both personal and collective memory. In *The Ambassadors*, his American traveller leaves New England for Paris, and in doing so retraces the complex attitudes of America toward France, both cradle of American Liberty and bloody arena of *Liberté*, as well as home of the *femme fatale, la Parisienne*.

The dissolution of the line between the public and the private gets its virtuoso expression in the writings of Franz Kafka, who, in a post-modern move, gives us the 'city' as an urban discourse that resists totalization. His characters experience perpetual discontinuity, a 'here' that can never be designated a site out 'there', a sense of urban space that is emptied of history, society, collective or even individual memory. His city is made up of signs of cityness without referents. If, as Michel de Certeau has written, 'the act of walking is to the urban system what the act of speaking, the Speech Act, is to language' (1985: 129), then in a Kafka novel, the pedestrian is walking in a foreign language, an untranslatable language. Rooms disappear from one day to the next, sites change their function without notice, institutional doors are transformed into personal paths, entrances and exits are interchangeable and inaccessible.

The consummate urban writer, he once submitted a story to a competition under the pseudonym of 'Heaven in Narrow Streets'. The sounds of city life at dawn, such as milk carts, he called 'the crickets of the metropolis'. The cityscape became for him the source of all power and creativity, as he begins a letter to Max Brod: 'Written in the street, as we shall always write to each other from now on, because the shoves you get from people passing by gives life to writing.' His diaries and letters are given over to accounts of walks, streets, buildings and scenes in public space; he is more likely to describe a passer-by or stranger than he is an acquaintance or member of his family. In his writings, the line between the public and the private has entirely disintegrated. Seated in a coffee house, he observes the interaction of husband and wife at an adjacent table, and he is embarrassed for the wife at the husband's indifference toward her, a conclusion reached from posture and facial expressions. He cannot help surmising about others, as when he is seated alone on a bench on Unter den Linden, wondering about the sad-looking ticket seller: 'How did he get into his job, how much does he make, where will he be tomorrow, what awaits him in his old age, where does he live, in what corner does he stretch out his arms before going to sleep?' Despite his tendency to conjecture about others, he is aware of the urban tendency to attribute meaning to city scenes that, upon closer examination, are devoid of meaning: 'Part of Niklasstrasse and all the bridge turns round to look sentimentally at a dog, who, loudly barking, is chasing an ambulance. Until suddenly the dog stops, turns away and proves to be an ordinary strange dog who meant nothing in particular by his pursuit of the vehicle.'

Inaccessibility becomes *the* major urban trope for Kafka. The imposing facade of apartment blocks gives rise to dreams about gaining access, walking through a long row of houses at the level of the first or second level, as if the walls dividing the houses give way to the stroller's curiosity, and all of the hidden spaces are bedrooms. 'I felt abashed to walk through people's bedrooms at a time when many of them were still lying in their beds,' he writes. In most of the entries depicting doors and stairways, there is either no access to the other side or access occurs unexpectedly, bringing with it more mystery and even greater anxiety. In one fictional sketch he opens the doors of his room at night and hears spasmodic breathing. He discerns 'yellowish glittering eyes', 'large round woman's breasts', and 'a thick yellowish tail'. The discovery that he simply entered the wrong room of his boarding house does not explain the fantastic creature that lives just on the other side of his wall. Architectural stills extend the trope of inaccessibility: 'I have enjoyed the sight from my window of the triangular piece cut out of the stone railing of the staircase that leads down on the right from the Czech bridge to the quay level. And now, over there, across the river, I see a step-ladder on the slope that leads down to the water.'

Not only are the meanings of these city encounters and scenes ever elusive, inaccessible or simply absent, but Kafka also employs non-city intertexts whose traces require re-readings of ordinary city signs, such as that of the anonymous neighbour. In one of his diary sketches a heretofore invisible neighbour bursts into the narrator's room every evening and silently wrestles with him. Even bolting his door does not prevent this singleminded neighbour from hacking it in two with an axe to perform the nightly wrestling match. A parody of Jacob's wrestling with the angel, this is no starlit Peniel, the encampment marking the place where God's face was revealed to Jacob, nor does it signal a divine renaming of Jacob into Israel as a nation's foundational myth. It is a repetitive act with an unknown neighbour who is driven to violent wrestling for no apparent reason.

The unseen neighbour constitutes a sinister yet seductive mystique summed up by this passage, in which the urban quest is a reversal of the medieval romantic quest:

> Usually the one whom you are looking for lives next door. This isn't easy to explain, you must simply accept it as a fact. It is so deeply founded that there is nothing you can do about it, even if you should make an effort to. The reason is that you know nothing of this neighbor you are looking for. That is, you know neither that you are looking for him nor that he lives next door, in which case he very certainly lives next door. (Kafka, 1965)

In the post-modernist urban novels of Paul Auster, the legacy of Kafka is felt most deeply. Auster's city is repeatedly described 'as being in a state of perpetual contingency, impermanence, ephemerality, and transience'. This is particularly true in his book *In the Country of Last Things* (1987). 'Close your eyes for a moment, turn around to look at something else, and the thing that was before you is suddenly gone, and you mustn't waste your time looking for them. Once a thing is gone, that is the end of it' (1987: 2) In Auster's city, 'Entrances do not become exits, and there is nothing to guarantee that the door you walked through a moment ago will still be there when you turn around to look for it again. That is how it works in the city. Every time you think you know the answer to a question, you discover that the question makes no sense.'

In *The New York Trilogy*, a detective named Blue is assigned the job of trailing a man named Black only to discover that Black may be trailing *him*. Auster's characters journey through space but often end up where they began. His places are devoid of collective history and even of personal history, because individual memory is of no consequence in a place where everything is perpetually shifting, where phenomenal appearances of things are not reliable. Although Auster's writings describe the cityscape in realistic detail, in some cases giving precise routes, his characters' perambulations seem as textual as they are physical. Some of his works exemplify de Certeau's concept of the pedestrian as writer, the walkers 'whose bodies follow the cursives and strokes of an urban "text" they write without reading' (1985: 124). Moreover, as the walk takes place in the mind, it is no longer dependent on the actual literal journey. 'What we are really doing when we walk through the city is thinking, and thinking in such a way that our thoughts compose a journey and this journey is no more or less than the steps we have taken' (*The Invention of Solitude*, 1982).

The walking that is not a function of body but of mind could be an apt way of describing a walk through cyberspace, as is the following observation by one of his city walkers in *In the Country of Last Things*: 'The end is only imaginary, a destination you invent to keep yourself going, but a point comes when you realize you will never get there' (1987: 183). The eternally shifting spaces, the routes that can never be repeated in exactly the same way – these are features of the cyberspace journey. In virtual space there are no boundaries, steps are never retraced in exactly the same way, and the depth of life is always on the surface, on the screen. Auster's 'real' city at times appears to be invested with the features of a virtual city. Without giving up on the streets, buildings, human density and other aspects of the conventional modern city, Auster depicts his characters' experience of that familiar landscape in terms of an electronic city as well: 'The seeming immensity of cyberspace made manifest the possibility of "always somewhere else to go", the anomic quality of cyberspace, and the absence of perception of hierarchy (for hierarchy always posits origins and ends)'.

In an essay on the poet Charles Reznikoff, Auster observes that most of his poems are rooted in the city, 'For only in the modern city can the one who sees remain unseen, take his stand in the space and yet remain transparent.' While this is true of the modern urban experience, it is equally true of the post-modern electronic city, where we are alone in front of a computer screen, hidden from sight and able to hide our identities from those with whom we communicate. Auster's characters, often portrayed in enclosed spaces, locked rooms and isolated garretts, seem to inhabit both the built and the electronic environment simultaneously, which may account for the eerie quality of his fictions.

We should be wary of turning the illusory and transient features of the post-modern city, evident in the writings of both Kafka and Auster, into definitive statements about post-modern urban experience. If we have anything to learn from these shifting landscapes which nevertheless reproduce, in great detail, the material city, it is that each urban space is a palimpsest, read in innumerable ways. 'Like words,' writes de Certeau, 'places are articulated by a thousand images.'

NOTE

This research was supported in part by a grant from the Israel Science Foundation and by the Samuel L. and Perry Haber Chair on the Study of the Jewish Experience in the United States.

REFERENCES

Auster, Paul (1982) *The Invention of Solitude*. London: Faber & Faber.

Auster, Paul (1987) *In the Country of Last Things*. London: Penguin.

Bakhtin, M.M. (1981) *The Dialogic Imagination*. Austin, TX: University of Texas Press.

Barthes, Roland (1971) 'Semiology and the Urban', first published in French in *L'Architecture d'Aujourd'hui (La Ville)* (December 1970–January 1971), 153: 11–13; reprinted in Gottdiener and Lagopoulos, 1986: 86–99.

Baumgarten, Murray (1982) *City Scriptures: Modern Jewish Writing*. Cambridge, MA: Harvard University Press.

Benjamin, Walter (1969) 'On some motifs in Baudelaire', in Hannah Arendt (ed.), *Illuminations*. New York: Schocken.

Bradbury, Malcolm (1976) 'The cities of modernism', in Malcolm Bradbury and James McFarlane (eds), *Modernism*. London: Penguin.

Bremer, Sidney H. (1992) *Urban Intersections Meetings of Life and Literature in United States Cities*. Urbana, IL: University of Illinois Press.

Davis, Leonard J. (1987) *Resisting Novels: Ideology and Fiction*. London: Methuen.

de Certeau, Michel (1985) 'Practices of space', in Marshall Blonsky (ed.), *On Signs*. Baltimore, MD: Johns Hopkins University Press.

Dreiser, Theodore (1970) *Sister Carrie*. New York: Norton.

Eco, Umberto (1973) 'Function and sign: semiotics of architecture', *VIA (Structures Implicit and Explicit)*. Philadelphia, University of Pennsylvania Press, 1: 131–53. Reprinted in Gottdiener and Lagopoulos, 1986: 55–87.

Eco, Umberto (1979) *The Role of the Reader*. Bloomington, IN: Indiana University Press.

Ellison, Ralph (1981) *Invisible Man*. New York: Vintage.

Ellin, Nan (1995) *Postmodern Urbanism*. Oxford: Blackwell.

Fanger, Donald (1975) *Dostoevsky and Romantic Realism: A Study of Dostoevsky in Relation to Balzac, Dickens, and Gogol*. Cambridge, MA: Harvard University Press.

Fisher, Philip (1987) *Hard Facts: Setting and Form in the American Novel*. New York: Oxford University Press.

Gelley, Alexander (1973) 'Setting and a sense of world in the novel', *Yale Review* 62: 186–201.

Gottdiener, M. and Lagopoulos, A.P. (1986) *The City and the Sign: An Introduction to Urban Semiotics*. New York: Columbia University Press.

Gutman, Yisrael (1982) *The Jews of Warsaw 1939–1943: Ghetto, Underground, Revolt*. Indiana University Press.

Howe, Irving (1971) 'The City in Literature', in *The Critical Point*. New York: Dell.

Iser, Wolfgang (1974) *The Implied Reader*. Baltimore, MD: Johns Hopkins University Press.

Iser, Wolfgang (1978) *The Act of Reading: A Theory of Aesthetic Response*. Baltimore, MD: Johns Hopkins University Press.

James, Henry (1986) *The Ambassadors*. London: Penguin.

Jameson, Fredric (1981) *The Political Unconscious: Narrative as a Socially Symbolic Act*. Ithaca, NY: Cornell University Press.

Kafka, Franz (1965) *Diaries 1914–1923*. New York: Schocken.

Kafka, Franz (1971) *The Complete Stories*. New York: Schocken.

Lukacs, Georg (1971) *The Theory of the Novel*. Cambridge, MA: MIT Press.

Luttwack, Leonard (1984) *The Role of Place in Literature*. Syracuse, NY: Syracuse University Press.

Lynch, Kevin (1971) *The Image of the City*. Cambridge, MA: MIT Press.

Moretti, Franco (1983) *Signs Taken for Wonders: Essays in the Sociology of Literary Forms*. London: Verso.

Oz, Amos (1972) *My Michael*. New York: Knopf.

Pike, Burton (1981) *The Image of the City in Modern Literature*. Princeton: NJ: Princeton University Press.

Portugali, Juval (ed.) (1996) *The Construction of Cognitive Mapping*. Drodrecht: Kluwer.

Portugali, Juval (ed.) (2000) *Self-Organization and the City*. Berlin: Springer.

Rossi, Aldo (1989) *The Architecture of the City*. Cambridge, MA: MIT Press.

Sharpe, William Chapman (1990) *Unreal Cities: Urban Figuration in Wordsworth, Baudelaire, Whitman, Eliot and Williams*. Baltimore, MD: Johns Hopkins University Press.

Simmel, Georg (1950) *The Sociology of Georg Simmel*. New York: Macmillan.

Singer, Isaac Bashevis (1950) *The Family Moskat*. Greenwich, CT: Fawcett.

Spender, Stephen (1972) 'Poetry and the modern city,' in Michael C. Jaye and Ann Chalmers Watts (eds), *Literature and the Urban Experience*. New Brunswick, NJ: Rutgers University Press.

Spengler, Oswald (1928) *The Decline of the West*. New York: Knopf.

Stallybrass, Peter and White, Allon (1986) *The Politics and Poetics of Transgression*. Ithaca, NY: Cornell University Press.

Suleiman, Susan and Crosman, Inge (eds) (1990) *The Reader in the Text: Essays on Audience and Interpretation*. Princeton: Princeton University Press.

Weber, Max (1958) *The City*. New York: Free Press.

Welsh, Alexander (1971) *The City of Dickens*. New York: Oxford University Press.

Westwood, Sallie and Williams, John (eds) (1997) *Imagining Cities: Scripts, Signs, Memory*. London: Routledge.

Williams, Raymond (1989) *The Politics of Modernism: Against the Conformists*. London: Verso.

Part II
THE CITY AS ENVIRONMENT

Cities form distinctive types of built environment; in turn, the growth of cities has significant effects on the physical environments in which they are located. While they are the chief loci of power and control in the world economy, as well as nationally and regionally, fundamentally cities are also environments built to meet economic as well as social and political needs, along with those of its citizens. Fulfilling its political, economic and social functions, and doing so efficiently and responsively to the plurality of demands, requires the construction and maintenance of complex infrastructures. Such infrastructures become reflected in the basic land uses into which cities are divided – residential areas, industrial and commercial zones, transportation networks – whose patterning and density maps out the distinctiveness of the urban environment.

Few attributes of the contemporary city define more clearly their difference from pre-industrial cities than does their physical form. In the wake of urban-industrialization in the core economies the form of the city was to differ markedly from its pre-industrial counterpart. In the most obvious sense the changes in urban form were the product of their size – urban-industrial cities were typically much larger than cities which had existed previously – but also in the development of mass housing for the industrial workforce, the emergent commercial areas, notably through the development of the CBD, and the introduction of novel modes of intra-urban circulation responding to the burgeoning scale of the city which resulted in a physical environment radically different from earlier cities.

While the components comprising the built form of cities may be similar, the reality is that the morphology of cities is an important measure of their uniqueness, particularly of their appearance. Similarly, while there are generic processes which are identified with the growth and change of cities, how these become reflected in morphological terms varies. In part these differences are attributable to the unique characteristics of the physical environment in which individual cities

are located. Equally important are the economic, political and cultural factors which influence the morphology of cities ranging from the impacts of different landownership systems to the cultural significance and meanings attached to home-ownership. In the advanced economies, then, the scale of urban growth is reflected in very different morphological forms, from (for example) the low density, 'spread cities' of North America and Australia to the relatively much more compact cities of Western Europe. Even within the latter there are major differences in the morphology of the city; it is not meaningful to talk of *a* West European city in morphological terms, where at the very least national differences in the physical form, and appearance, are marked.

Strangely, within the study of cities their morphological development has been less scrutinized than has their functioning. Yet, the physical form of the city reflects the interplay of a wide variety of factors – market forces, historical factors, the changing basis of state intervention including urban planning, cultural factors including lifestyle preferences for particular housing forms, as well as the role(s) attached to the delineation of public spaces, to name a few. The physical form of each city has a story to be told revealing how they have developed.

While studies of urban morphology seek to understand the physical form of the city comprehensively, the separate components comprising the urban infrastructure are worthy of individual analysis, not least because these too contribute to appreciating the differences between cities. As infrastructure, their importance is in helping to support the functioning of the city. Of the different types of infrastructure required to maintain the city, the two focused upon in this section, housing and transport, also contribute to differences in physical form. The provision of adequate housing is a basic requirement, though the manner in which housing supply is met, through the market or otherwise, the mix of tenure types and preferences, the means through which housing finance is channelled, and the modes by which particular problems, such as provision for low income as well as the homeless, are addressed defy easy generalization.

Improvements in the transportation of goods and people were critical to the development of cities, none more so than in the onset of urban-industrialization. In ways recalling historical precedent the competitiveness of cities depends on their relative accessibility and connectedness within international and national networks. The connectivity of cities within the international airline network is the contemporary equivalent of the network accessibility of cities in the developing rail nets of the nineteenth century. In spite of the development of electronic communications, cities remain the chief loci for face-to-face contact between business and government elites, so that accessibility to and within them remains a vital need to which infrastructural provision needs to respond. Yet, particularly in terms of intra-urban transportation, the success of cities as economic motors has resulted in increasing 'costs' on movement within them. In these problems the form in which cities have been built has itself become recognized as significant in exacerbating the problems of intra-urban movement. Low-density suburbia may cater for cultural and lifestyle preferences for those able to enjoy its advantages, but combined with increasing car-dependence for movement has brought into question the issue of urban sustainability.

Cities, then, are distinctive, highly varied types of built environment. But their development has in turn had deepening effects on the physical environment. While it was not until the latter part of the twentieth century that the impress of these effects had become fully recognized, awareness of them was graphically described in accounts of the emergent industrial city in the previous century. The costs of urban-industrialization, pollution for example, have probably always been disproportionately borne by the poor, in the developing cities of the early industrial nations of Europe, as much as by the citizens of Bhopal in India following the Union Carbide disaster. But the contemporary environmental problems created by continuing economic development are recognized as having global impacts. Again, how cities have developed plays a significant part here; car-borne emissions have calculable effects on pollution levels, while the ability of governments, national and local, to provide effective public transport systems not only has impacts on congestion and safety, but palpable benefits to the environment at large.

As built environments cities raise fundamental questions linked to the benefits and disadvantages of urban growth. In the core economies, in particular, cities have been the loci of, and motor for, sustained economic growth. The benefits of this growth have been accompanied by costs, externalities which are accentuated by the nature of the city and its high density. Resolution of these exposes the economic, social and political contradictions and inequities of urban growth.

5

The Physical Form of Cities:
A Historico-Geographical Approach

J.W.R. WHITEHAND

The physical forms of urban areas are one of the most obvious visual records of the societies whose environments they provide. Many of the activities of these societies evidence themselves in the urban landscape, most obviously in streets, buildings and other private and public spaces. These physical forms are much more than functional containers. Frequently surviving long after the people who shaped them have perished, they are important historical records, redolent of the cultures that have created them and occupied them in the past. Yet the urban landscape is much more than a historical artefact. It exerts an influence on subsequent generations of form creation. Indeed one of the dominant themes in urban morphology, during the period of over a century since it became a field of serious enquiry within the discipline of geography, has been the influence of the forms created in one period on those created in subsequent periods. It is not surprising therefore that it is a field in which historical approaches are deep-rooted. The most widely used of such approaches has often been referred to as morphogenetic (Whitehand, 1981a). This approach, which received great impetus during the middle years of this century from the work of the geographer M.R.G. Conzen (1958, 1960, 1962), has particularly influenced the choice of contents and perspective of this chapter.

In viewing the development of urban form historically, a number of broad topics can be recognized: first, the cyclical and secular nature of the urban growth process, and the physical forms that have resulted from it; secondly, the internal processes of adaptation and redevelopment that have accompanied the ageing of urban areas; thirdly, the characteristics and roles of the various individuals, firms and organizations that have played a part in the creation and modification of urban forms; and fourthly, the ideas concerning urban landscape management that have arisen

from the morphogenetic approach to urban growth and change.

Arguably the most important single topic to which geographical urban morphologists have devoted their attention is the process whereby urban areas have grown physically. This process, mirroring social and economic developments, has reflected both secular and cyclical change. It is the cyclical aspect of urban growth that is considered first.

CYCLES OF RESIDENTIAL GROWTH

Studies of the growth of individual cities, both within the industrial era and earlier, reveal it to be a process that has varied greatly in its speed (Adams, 1970). A broad indication of this is provided by the large fluctuations in the amounts of building activity that have occurred over time (Gottlieb, 1976). Residential building, which occupies more land than any other single land use in most cities, has been characterized by major long-term swings between boom and slump (Whitehand, 1987: 11–26). The timing of these swings in UK cities was broadly in line with that in the majority of cities in the rest of Europe, at least up to the First World War. It differed markedly, however, from that in North American cities, which tended to suffer slumps in residential building when European, especially British, cities were undergoing booms, reflecting wider economic differences between the continents of North America and Europe (Thomas, 1973).

Each boom in residential building has broadly corresponded to a zone of accretion at the edge of the city. The form of that zone has been affected by a number of factors. An important considera-tion has been the mode of transportation preva-lent at the time (Adams, 1970). Until the 1890s most movement between city centre and urban fringe was slow, the majority of it on foot or by horse-drawn vehicle. Residential accretions before that time therefore tended to be compact. However, during the residential building booms that reached their maxima at the end of the nineteenth century and early in the twentieth century (around 1890 and 1915 in most North American cities and around 1900 in most cities in the UK) the widespread extension and electrification of tramways and the development and electrification of suburban railways facilitated the development of a wider zone around the edge of the city (Figure 5.1). Ribbon development along tramway routes became accentuated and, in the case of the largest cities, separate dormitory settlements, some of which had come into existence before the electrification of suburban railway lines, now burgeoned.

The major growth in the number of motor buses and motor cars in the inter-war period reduced the constraint of railway lines and tramways on the pattern of residential development, facilitating the massive extension of cities, especially during the residential boom of the 1920s in North America and Australia and that of the 1930s in the UK. In the UK, the Tudor Walters Report of 1918 gave official blessing to low-density garden suburbs (Cherry, 1988: 83–4). But on the continent of Europe, extensions to cities were less space-consuming. It is hard to resist the inference that the difference in residential densities between almost all of the English-speaking world on the one hand and the continent of Europe on the other reflects deep-rooted differences in the predominant form of housing (Lichtenberger, 1970). During and well before the industrial era, the tenement had been the predominant residential building type in the larger cities

on the continent of Europe, although there were regional variations in the extent of this dominance. In contrast, in most of North America the single-family house was the main type of housing, although both New York and Cincinnati contained a great many tenements (Daunton, 1988: 283). In England, but not in the main cities in Scotland, the single-family house was overwhelmingly dominant. Accordingly, whereas garden suburbs of single-family houses were in the large majority in the English-speaking world, on the continent of Europe the adoption of garden suburbs generally involved higher, often substantially higher, dwelling densities (Figure 5.2). There, blocks comprised of several flats, but not always distinguishable at first sight from large detached houses, were a common form of housing, although large amounts of tenement housing, sometimes not markedly different from that built before the First World War, continued to be constructed.

Residential growth in England continued to differ in form from that on the continent of Europe after the Second World War. But there was a major change of attitude in England towards the extension of cities into the surrounding countryside. Garden suburbs were now viewed as a threat to the countryside rather than as a remedy for inner-city congestion. Green belts, within which housebuilding and most other forms of building were precluded in local authority plans, soon encircled the main cities (Elson, 1986: 24). For some two decades, during the 1950s and 1960s, multi-storey flats became a fashionable form of residential construction by local authorities. Even after this type of construction was abruptly abandoned in the early 1970s, the outward growth of cities, now constrained by

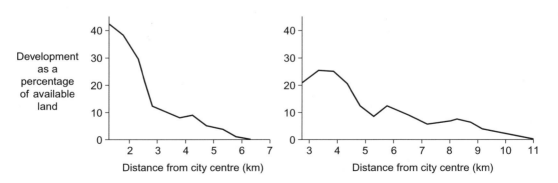

(a) 1840 - 58   **(b) 1894 - 1908**

Development as a percentage of available land

Distance from city centre (km)

Figure 5.1   Urban development around the north-west fringe of Glasgow, UK during two house-building booms: (a) 1840–58; (b) 1894–1908 (Whitehand, 1987: 49, Fig. 3.12)

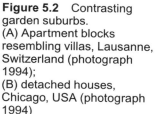

Figure 5.2 Contrasting garden suburbs.
(A) Apartment blocks resembling villas, Lausanne, Switzerland (photograph 1994);
(B) detached houses, Chicago, USA (photograph 1994)

green belts in the case of the largest cities, was much more compact than it had been in the inter-war period. Thus a major difference in the density of residential growth became evident between the United Kingdom and other predominantly English-speaking countries. In North America, cities grew outward at an unprecedented rate (Vance, 1990: 502), driven by massive growth in the number of private motor cars and unconstrained by government policies.

FRINGE BELTS

Although much of the urban morphologist's attention has understandably focused on the residential growth pulses that have had a dominant influence on the physical shape of cities, considerable attention has also been given to the intervening periods, the slumps in house building,

when other types of land use made a relatively large contribution to the emerging land-use pattern. If the conversion of rural land to urban use is recorded over time, it is apparent that many urban or quasi-urban land uses have had an incidence quite different from that of housing. For example, the number of public parks and golf courses created in England and Wales has if anything been greater in house-building slumps than in house-building booms. And conversions of rural land to many other urban uses during house-building slumps, even though they may have fallen in absolute terms, have tended to rise relative to conversions to housing (Whitehand, 1981b). This has been especially true of uses for which land has been developed at low intensity, such as for many types of institution. As a consequence, lengthy and pronounced slumps in house building have tended to be associated with the development of fringe belts comprised of a variety of often low-intensity land uses (Conzen,

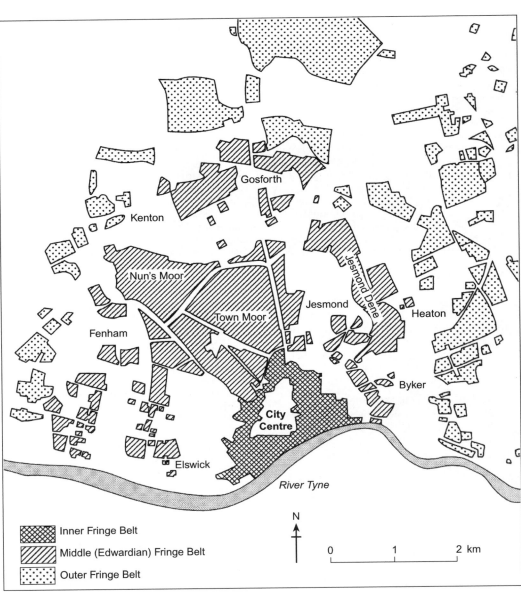

Figure 5.3 The fringe belts of Newcastle upon Tyne, UK in 1965 (Whitehand, 1967: p. 225, Fig. 1)

1960: 58–65; Whitehand, 1988). Such fringe belts tend to be most evident where a long slump in house building has occurred at the same time that some geographical limitation on the growth of the residential area has been reached. Examples of such limitations are fortification zones, which are common in the 'Old World', green belts, which are particularly significant in the case of large cities in the United Kingdom, and topographical constraints. Such fringe belts tend to

have been most distinct in conditions in which the normal expansion of the residential area has been fairly compact (Figure 5.3). The formation of fringe belts has been less evident, but by no means absent (M.P. Conzen, 1968), where outward growth has tended to be diffuse.

The tendency for urban areas to have extended outward in a series of pulsations has been associated with long-term fluctuations in the value of land (Whitehand, 1987: 39–49). Falls in land values during house-building slumps have facilitated the acquisition of urban fringe sites by land-extensive uses. Once acquired for such purposes, however, a rise in land values during a subsequent house-building boom has not generally resulted in the re-sale of these sites for house building. Thus fringe belts have not been ephemeral features but have tended to be perpetuated, embedded within the built-up area, long after the main zone of house building has moved farther out (Whitehand, 1972). The reasons for this have been numerous. Many institutions occupying fringe-belt sites have gradually developed those sites more intensively, so that a discrepancy between current use value and market value has either not arisen or has been eliminated or reduced. Sites that have remained in low-intensity use, such as public parks and certain types of sports grounds, have often been zoned in local planning documents to be retained in a land use similar to that existing. Some land users, once established on a site, have over long periods become insensitive to changes in land values, sometimes because of the lack of alternative sites that would enable them to maintain their function, perhaps a non-profit-making community function, for a specific part of the urban area.

THE CREATION OF MORPHOLOGICAL PERIODS: THE CASE OF THE ENGLISH HOUSE

Whereas the physical forms that make up fringe belts are strikingly diverse, residential accretions frequently contain numerous repetitions of similar street and plot layouts and 'standard' dwelling types (Johnston, 1969; Moudon, 1992). These layouts and dwelling types embody a succession of structural innovations and stylistic fashions that reflect the changing housing concerns of successive generations. Particular forms tend to be created continually over many years, frequently an entire house-building boom, before becoming unfashionable. Thus distinct morphological periods can be recognized on the ground (Conzen, 1960: 8). The historical geography of the English dwelling-house over the past two or three centuries illustrates how the house types and layouts that characterize these distinct periods in England have evolved.

The physical forms of residential areas in England can be seen as the products of two principal lines of influence, one from working-class housing, which had a dominant influence on urban landscapes in the nineteenth century, and the other from upper-class housing, which has been the principal influence in the twentieth century. The interaction of these two lines has

been a major process underlying the physical form of English housing.

The working-class housing line can be traced back to well before the Industrial Revolution. The basic urban property unit in medieval times in much of Europe north of the Alps was the burgage. In England it took the physical form of an elongated plot with a building at its street frontage and a garden or yard behind. In the course of time, particularly from the eighteenth century onward, many of these burgages became filled with buildings, especially working-class houses, workshops and ancillary buildings (Conzen, 1960: 66–9). Access to the burgage was generally obtained by means of an archway beneath part of the upper storey of the building that fronted the street. In some cases the entire length of a burgage became lined with buildings, apart from a narrow right-of-way along one side (Figure 5.4). The building up of the burgage was essentially a reflection of pressure on land, which became increasingly evident as the Industrial Revolution progressed. The arrangement of buildings and their physical characteristics were strongly conditioned by the original burgage shape. The dwelling-houses that were added along the length of the plot were a single room in depth and had no windows, or access, to the rear. They were thus blind-back houses or, where they shared a rear party wall with houses fronting an adjacent burgage, back-to-back houses.

Blind-back and back-to-back houses had, by the end of the eighteenth century, become widely recurrent means of increasing domestic space within existing plots. During the nineteenth century they were to become an even more prevalent form of infill (Conzen, 1962: 407; 1981: 106). However, it was in conditions unrestricted physically by the presence of burgages, on previously undeveloped sites, that the same basic structural types now became far more significant than in their original burgage environment.

Largely unconstrained by an existing urban plot pattern, back-to-back and blind-back houses were used in a variety of layouts. In Leeds, back-to-back houses were commonly set out in a rectilinear pattern of terraces in which each dwelling had a street frontage (Beresford, 1971). This meant that a very large amount of space was occupied by streets, a disadvantage that was generally avoided in Birmingham (Figure 5.5A) by arranging the dwellings along the sides of courts created within the interiors of the street blocks (Tomlinson, 1964: 51–4). In the nineteenth-century extensions to London and Newcastle upon Tyne, however, blind-back and back-to-back houses were rarely built.

Of the largest English cities to have blind-back and back-to-back houses in large numbers,

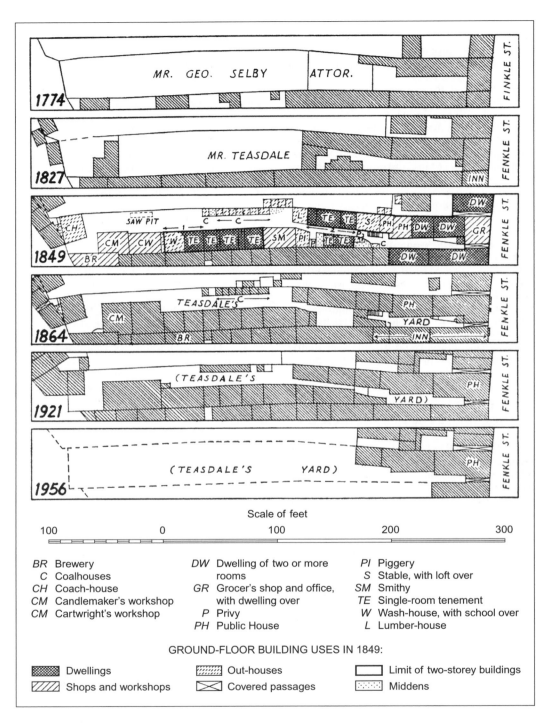

Scale of feet

| 100 | 0 | 100 | 200 | 300 |

BR	Brewery	*DW*	Dwelling of two or more	*PI*	Piggery
C	Coalhouses		rooms	*S*	Stable, with loft over
CH	Coach-house	*GR*	Grocer's shop and office,	*SM*	Smithy
CM	Candlemaker's workshop		with dwelling over	*TE*	Single-room tenement
CM	Cartwright's workshop	*P*	Privy	*W*	Wash-house, with school over
		PH	Public House	*L*	Lumber-house

GROUND-FLOOR BUILDING USES IN 1849:

Dwellings   Out-houses   Limit of two-storey buildings

Shops and workshops Covered passages   Middens

Figure 5.4   The burgage cycle: Teasdale's Yard, Alnwick, UK, 1774–1956
(Conzen, 1960: 68, Fig. 14)

Birmingham employed plans that were the most reminiscent of burgage-yard developments. There the majority of dwellings constructed during the first three-quarters of the nineteenth century were located within the interiors of street blocks. Access from the street was usually through an archway or tunnel which led into a court, usually elongated, which was flanked by blind-back and/or back-to-back houses.

In 1875, national public health legislation gave local authorities greater control over housing. The construction of back-to-back and blind-back houses soon became illegal in the majority of English cities. However, in some cities, notably Birmingham, the tunnel access from the street continued to be a common feature of new housing, but no longer as a means of access to dwellings located in the interior of a street block. It usually gave access to rear entrances, or occasionally side entrances, to terraced houses that had their own street frontages (Tomlinson, 1964: 54). Houses constructed in this way were frequently referred to as tunnel-back houses. Like the large majority of the terraced houses constructed between 1875 and the First World War, they generally had back wings (Figure 5.5B).

It was during this period that the line of influence from the housing of the upper class began to assume importance for the house types of a much larger proportion of the population. Influenced by the country houses of the landed gentry and aristocracy, the upper-middle class had begun, in the seventeenth and eighteenth centuries, to build 'villas' in private parks within a daily carriage journey of London. By the early-nineteenth century, small estates of detached and semi-detached houses for middle-class families were being developed at several points on or close to the periphery of London's built-up area (Prince, 1964: 87–104; Summerson, 1995: 17–27; Thompson, 1974: 84–5) and in the 1810s one such estate was beginning to be constructed on the south-west fringe of Birmingham (Cannadine, 1980). The single-family houses in sizeable gardens that occupied these estates were the prototypes of garden suburbs not only elsewhere in Great Britain but also on the eastern seaboard of the United States and, as the nineteenth century progressed, in the British colonies. The characteristic, compact, emphatically urban, formally arranged terraces that had characterized middle-class extensions to cities in the eighteenth century were now augmented by more open designs, in which houses set amongst trees, shrubs and lawns echoed in miniature the country parks of the English aristocracy. Eventually, in the last three decades of the nineteenth century, the detached or semi-detached house in its garden eclipsed the terraced house as the predominant form of new middle-class housing in London.

The Chicago suburb of Riverside, designed by F.L. Olmsted in 1869 (Mayer and Wade, 1969: 183–4), and the London suburb of Bedford Park, designed by Norman Shaw in 1875 (Pevsner, 1963: 402), were famous examples of what was now a much wider trend in middle-class housing towards the development of garden suburbs on the peripheries of major cities within the English-speaking world.

It was at this time that the garden suburb movement began to affect working-class housing. One of the best known examples of this is at Bournville, near Birmingham, where in 1879 the Cadburys began building houses for workers employed in their decentralized chocolate factory (Cherry, 1994: 97). Not only did the houses have front and back gardens, but semi-detached houses and short terraces of only four dwellings predominated. In the case of the short terraces, a tunnel access to the rear continued to be used. This feature was, in 1918, embodied in the recommendations of the Tudor Walters Report.

The Tudor Walters Report marked one of the great breakpoints in the evolution of the English house (Edwards, 1981: 103ff), particularly as far as the housing of the large majority if the population was concerned. It arguably contributed more than any other single factor to the contrast between the morphological periods that are often referred to as late Victorian and Edwardian on the one hand, and inter-war on the other. Semi-detached houses superseded terraced houses as the predominant type of house constructed (Figure 5.5C). Short terraces, often of only four houses, continued to be built, especially by local authorities, but they were often in a neo-Georgian 'cottage' style (Figure 5.5D) and usually part of much more open layouts than the overwhelming majority of terraces constructed before the First World War. In places a tunnel, giving access from the street to the rear of the houses located at the centre of the terrace, was a reminder of the tunnel-back terraced houses of the period 1875–1914.

The Second World War marked another significant change in the types of dwellings that were constructed, as English working-class housing was belatedly affected by the Modern Movement. In housing estates constructed by local authorities in the larger cities, terraced houses were now inter-mixed with flats. However, most recently it is clear that the break with the past has not been complete. Postmodernism has involved a return to an overwhelming predominance of single-family houses. These are often terraced but, though sometimes neo-Victorian in architectural style, their block-plans and associated street systems are quite different from those of the Victorian and Edwardian periods. There is little evidence of the back wings, uniform building

lines and rectilinear street systems that were hall-marks of those periods.

The evolution of English house types may be related broadly to fluctuations in house building. Particular types have been associated with partic-ular booms. This historical pattern has its geographical correlate (Whitehand, 1996). Diff-erent phases in the evolution of house types tended to be manifested in different geographical zones within the city, a tendency that has also been noted in the American Midwest (Adams, 1970). However, such zones are not as neat as diagrammatic representations of them tend to suggest (Whitehand, 1994: Fig. 11). Large urban areas are rarely unicentric (Whitehand, 1967: 228–9): the zones developing around one urban nucleus may merge with those growing around another. Furthermore, no sooner have physical forms been created than they become subject to a process of adaptation and renewal. In some parts of cities this process of internal change is at least as important to an understanding of the form of urban areas today as is the character of original urban development.

INTERNAL CHANGE: RESIDENTIAL AREAS

The building up of burgages, with which our outline of the development of the English house began, is a particular variant of the widespread intensification of the use of existing urban land, leading ultimately, in many cases, to redevelop-ment. Except in cases of extensive damage by fire, large-scale redevelopment, as distinct from replacement of individual buildings and modifi-cations to individual plots and buildings, was rare before the middle of the nineteenth century. From that time it became increasingly common in large cities. Within residential areas in Great Britain it was brought about in five main ways (Allan, 1965). First, there was the unfettered market mechanism in which redevelopment occurred where and when it was profitable. Secondly, there were the activities of philanthropic bodies. Thirdly, local authorities used public health acts to force private landlords to raise the standard of

their accommodation. Fourthly, local authorities purchased slum areas compulsorily, cleared them and sold them to private enterprise for redevelop-ment. Fifthly, local authorities themselves under-took redevelopment of the areas they had compulsorily purchased. In addition there was redevelopment associated with a change from residential use to some other purpose, for example by railway companies seeking to locate their termini as close as possible to city centres and by commercial firms acquiring residential land near the edge of expanding city centres. Many of these types and methods of renewal occurred simultaneously but with marked varia-tions in their relative importance, reflecting, for example, changing attitudes to the past, the adop-tion of new ideas in town planning, and swings in the popularity of different political ideologies. As well as medium- and short-term fluctuations in types and amounts of renewal, there was a long-term trend between the mid-nineteenth century and the mid-twentieth century towards greater local authority involvement, although since the 1970s this has been reversed.

Existing arrangements of streets, plots and buildings have been crucial to the process of renewal as has been shown by Mosher and Holdsworth (1992) in Pennsylvania, Coffey (1991) in Rochester (New York State) and John-ston (1968) in Melbourne, Australia. The impor-tance of the inheritance from past periods can be illustrated by the changes that have taken place to residential areas in the post-war period in Great Britain (Whitehand, 1993: 60–4). During this period some existing residential areas have come to rival 'greenfield' areas at the urban fringe in terms of the amount of change that has taken place. The type of change has been influenced heavily by prevailing fashions in building types and the density at which initial development occurred. The most important distinction is between areas initially developed as high-density working-class housing and those initially devel-oped as low-density middle-class housing. A further distinction is between areas renewed in the first two or three decades after the Second World War and those renewed since.

At the end of the Second World War the large

Figure 5.5 Nineteenth-century and inter-war house types and associated town plans in Birmingham, UK. (A) Mid-Victorian court dwellings (photograph of a court interior by permission of Jennifer Tann; plan reproduced from the Ordnance Survey 25 Inch Plan, revised 1902); (B) late-Victorian, back-wing terraced houses (photograph 1995; plan reproduced from the Ordnance Survey 25 Inch Plan, revised 1914); (C) inter-war, semi-detached houses built by private enterprise (photograph 1990; plan reproduced from the Ordnance Survey 1:2500 Plan, revised 1954; (D) inter-war, neo-Georgian, cottage-style terraced houses built by the local authority (photograph 1992; plan reproduced from the Ordnance Survey 1:2500 Plan, revised 1955) (Whitehand, 1996: 231, Fig. 17.5)

areas of housing that had accommodated the workforce of the Industrial Revolution were still largely intact. The economic return on the renewal of this huge, largely rented housing stock, whether by refurbishment or redevelopment, was generally small by comparison with alternative investments, even where units of property ownership were sufficiently extensive for large-scale redevelopment to be practicable. In the majority of cases it was in the economic interest of landlords to operate a policy of minimum maintenance. Increasing public concern over the poor housing conditions associated with this physical deterioration gave rise to a sharp increase in the acquisition of property by local authorities, often by compulsory purchase. This was followed, mainly in the 1950s and 1960s, by comprehensive redevelopment. As a result, the centres of the major industrial cities became almost surrounded by zones dominated by Modern housing, mainly in the form of flats, often of many storeys, and terraces. Brick and local stone were replaced by concrete, glass, and other materials that paid no respect to the *genius loci*. The contrast with the domestic scale of the mainly terraced dwellings that had been replaced could hardly have been sharper. It was emphasized by the multi-lane ring roads and radial roads that were simultaneously constructed. Only a small proportion of the industrial housing built before 1875 survived by the beginning of the 1970s.

Some nineteenth-century middle-class housing that had degenerated into multi-occupation suffered a similar fate to its working-class counterpart. But the majority of middle-class housing either remained relatively unchanged, save for minor adaptations, or underwent a piecemeal form of more intensive development. In the latter case, areas of large house plots in particular were gradually either redeveloped or the gardens were subdivided and additional dwellings constructed within them (Whitehand et al., 1992). Unlike the majority of redevelopments of working-class housing areas, in which local authorities played a major part, this process was largely brought about by private owner-occupiers and private developers. Each redevelopment or infilling rarely encompassed more than a few existing plots and often only a single individual garden plot was involved. Unco-ordinated, piecemeal augmentation of the street system often occurred, since it was necessary in many cases to create a new accessway, usually a cul-de-sac, to serve a particular redevelopment or infill. New building types, often flats, terraces and small detached houses, generally contrasted with existing large detached houses. Their essentially Modern, often Anglo-Scandinavian, architectural styles conflicted with the mainly traditional appearance of existing houses, in which various

historical styles, such as Tudor, were strongly represented.

Since about 1970 the pendulum of fashion has swung away from large-scale redevelopment to conservation. In the Post-Modern period local authorities are emphasizing the renovation of working-class housing. Already in the 1960s middle-class people were refurbishing working-class housing in parts of inner London. In the 1970s such renovation occurred in other major cities, greatly aided by the provision of grants from local authorities. In selected areas of Birmingham in the 1980s, the exteriors of dwellings, most of them privately owned, were entirely refurbished at the expense of the local authority. Meanwhile piecemeal infill and redevelopment continued in low-density middle-class areas. Indeed it was given a fillip by the accelerating demand for dwellings suitable for small households, both of elderly persons and young people, and the diminishing amount of building land remaining on the fringes of the larger cities, owing in particular to the strong resistance by local planning authorities to house building in the surrounding green belts. Fortunately, with the advent of Postmodernism and the return to fashion of conservation (Larkham, 1996: 31–85) has come the return to popularity of historical styles, many of them compatible with those favoured when the areas now being infilled and redeveloped were initially developed.

The existing residential areas of British cities are thus undergoing considerable change, but in working-class areas refurbishment has become much more common than redevelopment. In low-density residential areas pressure for redevelopment and infill is considerable, especially in the southern half of the country. With rising land values, developers are increasingly finding it profitable to infill even gardens of 0.15 ha or less. Rarely have areas of inter-war semi-detached housing been redeveloped. But it is not inconceivable that pressures for such redevelopment will become significant within the next decade in some parts of south-east England.

INTERNAL CHANGE: CITY CENTRES

Apart from the inner zones of the major cities, most residential areas bear a strong relation to the form they took when they were first developed, particularly in the main elements in their street plans but also, in many cases, in their buildings and plot boundaries. In contrast, the commercial cores of most cities have undergone great change. This partly reflects their far greater age and partly the fact that they have expanded into surrounding areas originally developed for non-commercial purposes.

Most present-day commercial cores in Europe were first laid out in pre-industrial times, mainly for non-commercial purposes. Even today many of their streets, and sometimes even their plots, follow medieval lines. Until the nineteenth century the large majority of central-area buildings were residential as well as commercial. Even in the late-eighteenth century most buildings in commercial cores were dwelling-houses into which shopfronts had been inserted. The shop designed exclusively for commercial purposes was, like the office and the warehouse, largely a product of the industrial era.

In the City of London, buildings designed specifically for office use were not constructed until long after specialized geographical areas, such as the financial quarter, with the Bank of England as its principal focus, had become a major feature (Thorne, 1984: 3). Edward I'Anson, architect and surveyor, who was a central figure in the rebuilding of the City of London in mid-Victorian times, recounted in 1872: 'When I first began to build on the new London Bridge approaches, previous to 1840, city offices as now constructed were not thought of; the houses were built as shops and dwellings, or as warehouses, and it was the same in Moorgate Street' (Thorne, 1984: 3). It was common even in the 1840s in the City of London for buildings to be constructed so that they could serve either as offices or as dwelling-places or as both. Although the insurance companies and banks set the pace in purpose-built offices in the late 1830s in many respects the buildings themselves involved little departure from existing residential building types (Thorne, 1984: 4). However, during the first decade or so of the Victorian period the external appearance of city centre offices underwent considerable change: the Sun Insurance Office, constructed in London in 1849, expressed a corporate solidity that set it apart from domestic buildings (Duffy, 1980: 260–2).

In the course of city growth, pressures for change have been especially evident in the city centre. The physical fabric has been adapted to demands for increased commercial space and for the incorporation of functional innovations. In the largest cities, plot boundaries have frequently been changed until the original pattern is barely recognizable. Even the street system, the most resistant to change of all the physical elements of a city, is likely to have been modified. A common feature in the nineteenth century, especially in large cities with substantial pre-industrial cores, such as London, Newcastle upon Tyne and, on the east coast of the United States, Boston, was breakthrough streets, designed as traffic improvements (Conzen, 1981: 111; M.P. Conzen, 1990: Fig. 7.3). Occasionally there were large-scale redevelopments involving the laying-out of many new streets, as in the mid-nineteenth century in Paris (Sutcliffe, 1970) and Brussels (Hall, 1986: 205–22). At this time the idea of utilizing the open space of the fringe belt beyond the city walls as a boulevard ring was adopted widely in Europe. These ring roads became lined by public buildings, promenades, parks, public utilities and transport termini (Whitehand, 1988: 48–9). In Great Britain, however, where fortifications generally fell into disuse earlier, such 'boulevards' are rare and when inner ring roads, skirting commercial cores, became fashionable in the post-war period, their construction almost invariably entailed large-scale redevelopment.

Actual buildings have often been replaced several times over in large, old-established cities. Holden and Holford (1951: 173) recorded that in the City of London, which comprised much of the commercial core of the metropolis in the nineteenth century, of all the buildings that existed in 1855 about four-fifths had already been rebuilt by 1905. Hoyt (1933: 335) suggested that there were few sites in the central business district of Chicago that had not been occupied by at least three different buildings between 1830 and 1933, and one exceptional site had undergone at least five redevelopments during that time; a massive rate of renewal, even allowing for the widespread effects of the fire of 1871. In contrast, the average building lifespan in central Glasgow, measured over the period 1840 to 1969, was about 130 years (Whitehand, 1978: 84), which is almost twice as long as the norm suggested by the tables of building depreciation of the United States Department of the Treasury (Cowan, 1963: 69–70). Van Hulten (1967: 193) reported that if the speed of rebuilding in the canal zone of the inner city of Amsterdam between 1957 and 1961 were continued a total renewal would have taken place in 50 years, but extrapolations from such short-term data are unreliable owing to pronounced short-term and long-term fluctuations in the amount of redevelopment. Such long-term information as there is, supplemented by anecdotal evidence, suggests that the lifespans of buildings in the largest American central business districts have been shorter on average than those of their counterparts in Europe. Not unrelated to this are the much greater average building heights that have become manifest in the cores of American cities in the past one hundred years. A sizeable part of lower Manhattan was transformed from buildings of mainly four to five storeys to predominantly skyscrapers in a period of a little over 20 years, 1897–1920 (Holdsworth, 1992), while European city centres retained the four to five storey building heights that they had in the nineteenth century. By 1930, when the skyscraper had spread well down the urban hierarchy in America (Bastian, 1993), it was still virtually unrepresented in Europe and it is still relatively weakly represented now.

There are also significant functional differences between major city centres on the continent of Europe and those in the English-speaking world. While the former have for the most part retained a major residential and cultural function, most of the latter have largely lost these functions. The erosion of the mixed-use character of the Western city centre has been going on since the middle of the nineteenth century and hence in many American cities virtually from their beginnings. Indeed, it could be argued that the dearth of significant pre-industrial cores in American cities, combined with the much greater diffuseness of the post-war growth of American cities, has been fundamental in the creation of city centres that lack both the social nodality and now, in certain respects, the economic nodality of those in Europe. For manufacturing and wholesaling the American city centre has almost been abandoned – more so even than its European counterpart. In most American city centres there has been considerable decline in retailing, both absolutely and relative to that in suburban centres, or indeed in so-called 'edge cities', some of which vie in size with central business districts (Knox, 1991: 190–1). Only in the case of offices has there been an absolute increase within central business districts, although this has occurred simultaneously with considerable dispersal of office activities to locations outside the central business district (Vance, 1990: 487–8).

Most of these tendencies have become particularly apparent in America since the early 1950s, more recently and less strongly in Europe. On both sides of the Atlantic they were particularly associated with the construction of large, compact office buildings, often with glass curtain walls. In Great Britain they were part of a more general reshaping of city centres on a scale unrivalled since before the First World War (Esher, 1981). A major part of this was the construction by local authorities of inner ring roads with extensive systems of underpasses and pedestrian subways. Local authorities were also frequently involved in the acquisition of large areas of land for shopping schemes, although the actual redevelopments have been largely undertaken by private enterprise. One consequence of these redevelopments was that many specialized retailers that previously occupied old premises were unable to afford the high rents in the new buildings and the tenants replacing them tended to comprise a higher proportion of shops selling high-turnover goods. Small traders were displaced and the sociocultural role of the city centre was diminished.

In contrast, in the 1980s and 1990s there have been attempts by city councils to re-establish city centres as foci of social, economic and cultural life. Beginning in America, cities in general, and city centres and their environs in particular, have been promoted as places of leisure, culture and spectacle. Waterfronts, particularly disused docklands, on which cities had frequently turned their backs in the early post-war decades, have attracted a great deal of investment, particularly in office development (Horn, 1993). Features of significance in the historical development of cities, such as industrial and transport relics have been exploited as tourist attractions, sculptures have been erected at foci of pedestrian movement and convention centres have been constructed. Decoration became a *sine qua non* for shopping centres, as did styles containing historical references in the case of office buildings. The practice of façadism, the retention of that part of the shell of an old building that fronts a street while creating a new, invariably more commodious, structure behind, was widespread (Barrett and Larkham, 1994).

AGENTS OF RESIDENTIAL DEVELOPMENT

The way in which the urban landscape has evolved needs to be seen in the light of the characteristics of the individuals, firms and organizations that played a part in its production. In the nineteenth century in Great Britain, many landowners acted as developers of their own land, laying out streets and plots, and leasing or selling plots to builders. Others sold their land outright to developers (Treen, 1982). Ninety per cent of house building in London at the end of the nineteenth century was done on speculation (Dyos, 1968: 660). Most of the building firms were small concerns, although less than 3 per cent of firms were building over 40 per cent of London's new houses. The sources of capital remained almost entirely local. Those undertaking house building in Leeds at the time still included many who were not primarily engaged in building for a livelihood (Trowell, 1983: 94–5) and the areas of operation of house builders to the immediate south-east of London appear to have been very local (Crossick, 1978: 55). Almost all of those who prepared the drawings submitted for building approval in late nineteenth-century Leeds were local (Trowell 1983: 106–8). They mostly described themselves as 'architects', but many, perhaps most, of those who prepared plans for working-class terraced houses probably lacked formal training and qualifications as architects (Kaye, 1960: 175).

There seems little doubt that the production of housing in Great Britain up to the First World War was highly fragmented. A tiny minority of sizeable building firms were emerging, but the overwhelming majority of design, building and funding was by small, local firms and individuals.

Ownership of residential property was also highly dispersed. Only a small minority of houses were owner-occupied. Most houses were owned by local people as an investment. These owners included people in a wide variety of occupations, ranging from professional men, such as solicitors, to tradesmen, such as grocers and builders, and even including people in working-class occupations (Daunton, 1977: 120–3; Dobraszczyc, 1978: 2; Holmes, 1973: 246).

After the First World War, owners of substantial estates at urban fringes were less attracted by the prospect of organizing developments themselves than their nineteenth-century predecessors had been. More commonly, estate owners sold their land to developers who divided it into building plots to sell either individually, or much more often in blocks, to builders. Alternatively, land was frequently sold direct to developer–builders, who not only prepared the land for building, by creating roads and plots, and installing basic services, but built the houses (Horsey, 1985).

The reduced role of pre-urban landowners in suburban development was accompanied by major changes in the firms responsible for residential building. A striking feature was the large number of speculative house-building firms that came into existence during the inter-war period. In the London area, they included a number that had begun as surveyors and estate agents, but had subsequently started building speculatively, usually by setting up a separate building company (Bundock, 1974: 361). Of the minority of firms that were founded before the First World War, many were new to speculative building in the inter-war period, having previously been jobbing builders or contractors. An increase in the number of firms in the building industry, especially small firms, was particularly evident during the first half of the 1930s. The continuing success of small firms reflected the fact that house building continued to be the essentially labour-intensive process it had been before 1914, the use of expensive capital equipment remaining unnecessary for most types of house building (Bowley, 1966: 388; Richardson and Aldcroft, 1968: 119). However, at the same time a small number of firms grew to a great size, one eventually employing over 20,000 men (Richardson and Aldcroft, 1968: 32). Underlying this expansion was a change in the financing of building: from a system dominated by loans from local, private individuals to one in which building societies, often with operational areas that covered virtually the entire country, became the major providers of finance (Merret and Gray, 1982: 349). A further factor was the rise to importance of local authorities as house builders. As in the nineteenth century, contrary to common belief, there is evidence that architects were much involved in suburban development at all levels in the market (Carr and Whitehand, 1996).

As had been recognized by the very largest developers before the First World War, it became evident that land ownership, building and estate agency were most effectively accomplished as facets of a single enterprise. But for the small builder, the pre-1914 system of purchasing plots, individually or in blocks, from a developer who laid out roads and installed services, continued to operate during the 1920s. However, during the 1930s small builders too were increasingly operating as developer-builders (Horsey, 1985: 150). The gradual change towards the combining of development and building in a single enterprise was also occurring at much the same time in America, although there it did not become the normal practice until after the Second World War (Vance, 1990: 459).

In many respects the types of organization responsible for the outward extension of British cities after the Second World War remained those that had been influential in the inter-war period. There were, however, a number of significant changes in the environment in which these organizations were working. First, following the Town and Country Planning Act, 1947, government assumed a major role in the development process: plans were prepared by local authorities and virtually all significant developments became subject to 'development control' in the light of these plans. Secondly, large house-building firms operating over substantial parts of the country took on much greater importance (Craven, 1969). Thirdly, demand for housing reached a new high owing in particular to the war-time hiatus in house building, the loss of a great many houses as a result of enemy action, and the rise in the number of potential households. Yet the government was simultaneously seeking to restrain development in certain areas, notably around major cities, that were 'ripe' for house building. By the later 1950s, when house building by private enterprise was regaining momentum and land values were rising in the majority of intra- and peri-urban areas, the scene was set for major conflicts among the various parties to residential development, especially landowners, developers, residents and local authority planners. These conflicts were often especially marked where attempts were made to insert additional houses on sites developed at low densities in earlier periods. Such sites often called for particularly careful blending of new with old, but unfortunately all too often the forms that developments actually took, particularly their architectural styles, were afterthoughts, when battles over density and access had been concluded (Whitehand, 1990).

AGENTS OF CITY-CENTRE CHANGE

During the nineteenth century the buildings erected in Great Britain's commercial cores were almost entirely owned and designed locally (Tate, 1983: App. I; Thompson, 1987: 89–102). For example, in central Huddersfield the work of non-local architects was confined to a few ornate, prestigious structures. The first instance of an owner-occupier from outside Huddersfield employing a non-local architect for a development in the commercial core was in 1899. In that year the Prudential Assurance Company, already a major national company, commissioned Alfred Waterhouse, the nationally renowned architect whom it had employed earlier to design its offices in central London and Glasgow, to design its local office in Huddersfield's main retailing street (Barnard, 1948; Tate, 1983: App. I; Thompson, 1987: 93). The style and materials were similar to those employed in the London building, the terracotta cladding contrasting with the predominantly local stone of existing nearby buildings.

The intrusion of outside influences in Huddersfield was much more evident in the inter-war period, with national firms such as F.W. Woolworth & Co, Marks and Spencer and Montague Burton erecting buildings in their own house styles, employing non-local architects and using materials alien to the area. Whereas before the First World War shopfronts in central Huddersfield had largely been designed by Huddersfield companies, now non-local designs predominated. Initially associated with the establishment of branches of firms based outside Huddersfield, the practice of employing non-local designers was quickly adopted by Huddersfield firms. The change from local to non-local designers occurred much more rapidly among shopfront designers than among architects. However, smaller towns than Huddersfield were less able to withstand architectural competition from nearby, big-city neighbours. Indeed, the combination of town size and distance from a major city may account for many inter-town differences in the extent to which local architects were employed in the 1930s (Whitehand, 1992b: 423). For example, Watford and Epsom, deeply within the sphere of influence of London, had low proportions of local architects employed in their commercial cores, relative to their size, and Boston in Lincolnshire, well removed from a major city, had a relatively high proportion (Figure 5.6).

A decline in the importance of local builders did not on the whole occur until the 1930s. The decline in the importance of local owners was somewhat earlier and most evident in commercial cores close to London. In all but the smallest towns, the main change in the funding of devel-

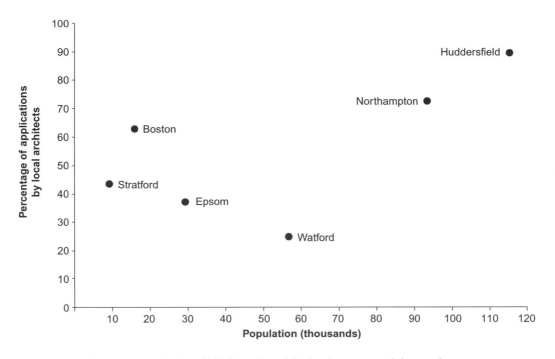

Figure 5.6 UK town populations (1931) and local design in commercial cores in the 1930s (Whitehand, 1992b: 423, Fig. 3)

opment came in the 1930s, as successful businesses, especially chain stores, were converted into
public companies (Jefferys, 1954). Property was
increasingly being viewed as an investment by the
chain stores and insurance companies. The roles
of individuals in speculative development almost
disappeared as local individuals formed development companies and London-based speculative
developers began slowly to redevelop town- and
city-centre sites (Whitehand, 1984: 25–6).

The types of funding arrangements that had
been employed in the 1930s were inadequate for
the sizes of the ventures that were envisaged after
the Second World War. This fact, combined with
a credit squeeze in 1955, just when commercial
building was reviving from the aftermath of the
war, hastened major developments in the links
between property companies and those organizations still able to provide funds on a large scale,
most notably the life assurance companies
(Whitehouse, 1964: 46–95). During the inter-war
period, insurance companies had been significant
developers of property in their own right, largely
by building premises with accommodation in
excess of their own needs and letting the surplus.
This activity was in addition to their roles as
mortgagees and shareholders. Now their property
development role was greatly expanded both as
speculative developers in their own right and
more particularly as providers of various forms of
funding for development agencies.

Speculative redevelopment in the post-war
period was heavily influenced by a handful of
companies operating on a national scale. From
the mid-1960s onward, a large proportion of
redevelopment for retailing involved the creation
of shopping precincts. Redevelopments for
owner-occupation were, like speculative redevelopments, undertaken predominantly by national
firms. Large amounts of land and property were
acquired by insurance companies and pension
funds (Barras, 1979a: 43). Individual firms tended
to spread their interests widely: for example, the
93 town- and city-centre properties owned by
Prudential Pensions (one member of the Prudential Group) at the end of 1980 were spread over
no less than 56 towns and cities (Prudential
Pensions, 1981). Local authorities also held large
amounts of land but their direct involvement in
redevelopment was comparatively small (Hillier
Parker Research, 1979: 33).

Outside London, one of the most striking
features of the post-war period has been the
shift in the locus of decision-making from
predominantly within the commercial cores
affected to predominantly outside (Whitehand
and Whitehand, 1984: 233). Inevitably this has
meant a reduction in the influence of firms with
local roots and imbued with a sense of the attributes that endow places with their particular char-

acter. At the same time there were changes in the
planning policies and activities of central government and local authorities. The precise lines of
the cause and effect relation between 'planning'
and private enterprise are difficult to disentangle.
But there is little doubt that one of the effects of
the Town and Country Planning Acts was to add
further to the scale of the already large changes
stemming from the increasing size of the organizations interested in property: the combining of
changes to the road system with redevelopment
was certainly facilitated. Nevertheless, the impression, though it is difficult to quantify, is of planners responding to the agents and processes
already mentioned more than acting as a positive,
directive force. Formal comprehensive plans for
central areas tended to be slow to emerge and in
many cases they were largely a recognition of
existing patterns and trends. There would seem to
be substance in the view of Ambrose and
Colenutt (1975: 178–9) that land-use planning
was largely treated as a matter of designating
areas in terms of land-use categories essentially to
accommodate observed trends in population and
employment. And Barras (1979b: 95) may not
have exaggerated when he suggested that the main
effect of central and local government planning
policy on office development in the City of
London was to reinforce underlying market
trends. The high proportion of new shopping
precincts located in existing commercial cores in
the 1970s would certainly seem to be attributable,
in part, to local and central government planning
policies, although as with other redevelopments
the concern of developers and chain stores to
protect their existing property interests in central
areas was almost certainly also a factor (Bennison
and Davies, 1980: 23).

URBAN LANDSCAPE MANAGEMENT

The processes at work in the development of
urban areas and the built-forms that have
emerged on the ground and seem likely to emerge
in the future must ultimately be evaluated
according to the quality of the environment, and
in particular the landscape, that they provide. The
interests that are crucial here are those of the
people using urban areas. However, despite their
fundamental importance, these interests on the
whole impinge on the processes creating urban
landscapes only indirectly: the interests of those
living and working in urban areas are filtered by
the various parties with a formal role. When these
interests are examined, however, a number of
problems come into sharper focus (Whitehand,
1992a: esp. 210–15).

Even within existing residential areas, where

individual owner-occupiers are an important force in landscape change, the character of change is not determined primarily by how it fits into the existing fabric or by the needs of the inhabitants. Moreover, the activities of non-local concerns, which tend to have less sense of place, have become increasingly important. This is especially the case in city centres. There, the incidence and type of landscape change are influenced by events far removed from what is visible to the average user of these areas, notably the investment decisions of insurance companies and, not least, the changes in the attractiveness of overseas investment to pension fund managers. Landscape problems are minor influences by comparison. This all results in, among other things, the destruction long before they are worn out physically of buildings into which enormous amounts of economic and cultural capital have been invested and, in city centres in particular, their replacement by structures out of scale with the existing landscape. These processes have led to cultural impoverishment. In city centres they have been aggravated by the enlarged scale, and impersonalization, of land and property ownership.

The main reaction to these developments has taken the form of a conservation movement. Unfortunately, the effectiveness of this movement has been hampered by the lack of a theoretical basis – a theory of urban landscape management that can give direction and coherence to the way in which conservation problems are tackled. Nevertheless, there are indications in the effects in the landscape of the processes and activities that have already been discussed that there are a number of important matters that a theory of urban landscape management needs to take into account. Most important is the fact that in addition to having considerable durability relative to the individual human lifespan, each change in the landscape embodies something of the society responsible for creating it. Thus urban landscapes embody not only the efforts and aspirations of the people occupying them now but also those of their predecessors. It is this, it has been argued, that creates the sense of place and feeling of continuity that enable individuals and groups to identify with an area. Thus fundamental to the thinking of Conzen (1966, 1975, 1988) and Caniggia (Kropf, 1993) is the view that the intelligibility of the city depends on its history. In the search for a basis for managing urban landscapes, it is a short step from this fundamental belief to regarding the city as a source of accumulated wisdom, and from there to utilizing this wisdom as the basis for prescribing change.

It is accordingly the nature and intensity of the historicity of the urban landscape that provide Conzen's main basis for devising proposals for its management. This basis is articulated in practical terms by utilizing his division of the urban landscape into three basic 'form complexes' – town plan, building forms and land use. These are regarded as to some extent a hierarchy in which building forms are contained within plots or land-use units, which are in turn set in the framework of the town plan. These three form complexes, together with the site, combine at the most local level to produce the smallest morphologically homogeneous areas, which might be termed 'urban landscape cells'. These cells are grouped into urban landscape units, which in turn combine at different levels of integration to form a hierarchy of intra-urban regions. Since the three form complexes change at different speeds, the land-use pattern being the quickest to change and the town plan the slowest, their geographical patterns within an urban area frequently differ. Particularly in old urban areas, the delimitations of cells, units and regions are complex, although the commercial core, fringe belts and certain types of residential accretion are recurrent features. The hierarchy of areal units is the geographical manifestation of the historical development of urban form and encapsulates its historicity. It is, in the Conzenian approach, the reference point for all proposals for change to the urban landscape. The approach, therefore, is essentially conservative. The accent is on the historically- and geographically-informed transformation, augmentation and conservation of what already exists.

This perspective recognizes new functional needs but requires that the ways in which they are met respect the existing landscape as a tangible record of the endeavours of past societies in a particular locale. It is founded on a long-term view of human endeavour. By emphasizing the historical and geographical context of each change in the landscape, and by laying stress on the long-term repercussions of decisions affecting the landscape, it draws attention to the responsibility that a society has to future generations. In a sense it seeks to restore that sense of geographical identity that was once taken for granted but which the Industrial Revolution and its aftermath have partially destroyed.

Whilst it is inevitable that in the foreseeable future the pace of change will largely be determined by economic conditions, in the light of the processes at work, and in particular their character and outcome in the post-war period, it would be irresponsible effectively to leave decisions about long-term social assets largely in the hands of organizations whose concern for history is slight and whose connections with the landscape effects of their actions upon those who live and work in urban areas are frequently tenuous. There is thus a large gap between the way in which the landscapes of cities have developed in recent times and Conzenian ideas concerning the principles that should govern the management of the urban landscape. Nevertheless, the mood in Western cities generally is more receptive to a concern for the visual environment than it has been for many decades. It is the responsibility of academics and practising planners to take advantage of this to harness potentially powerful, but still incipient, theory to the task of not only limiting damage to urban landscapes but, more optimistically, enhancing them.

NOTE

In a number of places this chapter draws upon the author's monograph *The Making of the Urban Landscape* (Oxford: Blackwell, 1992).

REFERENCES

Adams, J.S. (1970) 'Residential structure of Midwestern cities', *Annals of the Association of American Geographers*, 60: 37–62.

Allan, C.M. (1965) 'The genesis of British urban redevelopment with special reference to Glasgow', *Economic History Review*, 18: 598–613.

Ambrose, P. and Colenutt, B. (1975) *The Property Machine*. Harmondsworth: Penguin.

Barnard, R.W. (comp.) (1948) *A Century of Service: The Story of the Prudential 1848–1948*. London: Prudential.

Barras, R. (1979a) *The Returns from Office Development and Investment*. Centre for Environmental Studies Research Series 35.

Barras, R. (1979b) *The Development Cycle in the City of London*. Centre for Environmental Studies Research Series 36.

Barrett, H. and Larkham, P.J. (1994) *Disguising Development: Façadism in City Centres*. Research Paper No. 11, Faculty of the Built Environment, University of Central England, Birmingham.

Bastian, R.W. (1993) 'Tall office buildings in small American cities 1923–1931', *Geografiska Annaler*, 75B: 31–9.

Bennison, D.J. and Davies, R.L. (1980) 'The impact of town centre shopping schemes in Britain: their impact on traditional retail environments', *Progress in Planning*, 14: 1–104.

Beresford, M.W. (1971) 'The back-to-back house in Leeds 1787–1937', in S.D. Chapman (ed.), *The History of Working-Class Housing*. Newton Abbot: David & Charles. pp. 93–132.

Bowley, M. (1966) *The British Building Industry: Four Studies in Response and Resistance to Change*. Cambridge: Cambridge University Press.

Bundock, J.D. (1974) 'Speculative housebuilding and some aspects of the activities of the suburban housebuilder within the Greater London Outer Suburban Area 1919–39'. MA thesis, University of Kent.

Cannadine, D. (1980) *Lords and Landlords: The Aristocracy and the Towns, 1774–1967*. Leicester: Leicester University Press.

Carr, C.M.H. and Whitehand, J.W.R. (1996) 'Birmingham's inter-war suburbs: origins, development and change', in A.J. Gerrard and T.R. Slater (eds), *Managing a Conurbation: Birmingham and its Region*. Studley: Brewin.

Cherry, G.E. (1988) *Cities and Plans: The Shaping of Urban Britain in the Nineteenth and Twentieth Centuries*. London: Edward Arnold.

Cherry, G.E. (1994) *Birmingham: A Study in Geography, History and Planning*. Chichester: Wiley.

Coffey, B. (1991) 'The changing form and function of urban mansion districts: the example of Rochester, New York', *Material Culture*, 23: 15–25.

Conzen, M.P. (1968) 'Fringe location land uses: relict patterns in Madison, Wisconsin'. Unpublished paper presented to the 19th Annual Meeting of the Association of American Geographers West Lakes Division, Madison, Wisconsin.

Conzen, M.P. (1990) 'Town-plan analysis in an American setting: cadastral processes in Boston and Omaha, 1630–1930', in T.R. Slater (ed.), *The Built Form of Western Cities: Essays for M.R.G. Conzen on the Occasion of his Eightieth Birthday*. Leicester: Leicester University Press. pp. 142–70.

Conzen, M.R.G. (1958) 'The growth and character of Whitby', in G.H.J. Daysh (ed.), *A Survey of Whitby and the Surrounding Area*. Eton: Shakespeare Head Press. pp. 49–89.

Conzen, M.R.G. (1960) *Alnwick, Northumberland: A Study in Town-Plan Analysis*. Institute of British Geographers Publication 27. London: George Philip.

Conzen, M.R.G. (1962) 'The plan analysis of an English city centre', in K. Norborg (ed.), *Proceedings of the IGU Symposium in Urban Geography Lund 1960*. Lund: Gleerup. pp. 383–414.

Conzen, M.R.G. (1966) 'Historical townscapes in Britain: a problem in applied geography', in J.W. House (ed.), *Northern Geographical Essays in Honour of G.H.J. Daysh*. Newcastle upon Tyne: University of Newcastle upon Tyne. pp. 56–78.

Conzen, M.R.G. (1975) 'Geography and townscape conservation', in H. Uhlig and C. Lienau (eds), *Anglo-German Symposium in Applied Geography, Giessen-Würzburg-München*. Giessen: Lenz. pp. 95–102.

Conzen, M.R.G. (1981) 'The morphology of towns in Britain during the industrial era', in J.W.R. Whitehand (ed.), *The Urban Landscape: Historical Development and Management*. Institute of British Geographers Special Publication 13. London: Academic Press. pp. 87–126.

Conzen, M.R.G. (1988) 'Morphogenesis, morphological regions and secular human agency in the historic townscape, as exemplified by Ludlow', in D. Denecke and G. Shaw (eds), *Urban Historical Geography: Recent Progress in Britain and Germany*. Cambridge: Cambridge University Press. pp. 253–72.

Cowan, P. (1963) 'Studies in the growth, change and ageing of buildings', *Transactions of the Bartlett Society*, 1: 69–70.

Craven, E. (1969) 'Private residential expansion in Kent 1956–64: a study of pattern and process in urban growth', *Urban Studies*, 6: 1-16.

Crossick, G. (1978) *An Artisan Elite in Victorian Society: Kentish London, 1840–1880*. London: Croom Helm.

Daunton, M.J. (1977) *Coal Metropolis Cardiff*. Leicester: Leicester University Press.

Daunton, M.J. (1988) 'Cities of homes and cities of tenements: British and American comparisons, 1870–1914', *Journal of Urban History*, 14: 283–319.

Dobraszczyc, A. (1978) 'The ownership and management of working-class housing in England and Wales, 1780–1914'. Unpublished paper presented to the Annual Conference of the Urban History Group, Swansea.

Duffy, F. (1980) 'Office buildings and organizational change', in A.D. King (ed.), *Buildings and Society: Essays in the Social Development of the Built Environment*. London: Routledge and Kegan Paul. pp. 255–80.

Dyos, H.J. (1968) 'The speculative builders and developers of Victorian London', *Victorian Studies*, 11: 641–90.

Edwards, A.M. (1981) *The Design of Suburbia: A Critical Study in Environmental History*. London: Pembridge Press.

Elson, M.J. (1986) *Green Belts: Conflict Mediation in the Urban Fringe*. London: Heinemann.

Esher, L. (1981) *A Broken Wave: The Rebuilding of England 1940–1980*. London: Allen Lane.

Gottlieb, M. (1976) *Long Swings in Urban Development*. New York: National Bureau for Economic Research.

Hall, T. (1986) *Planung europäischer Hauptstädte: zur Entwicklung des Städtebaues im 19. Jahrhundert*. Stockholm: Almqvist & Wiksell.

Hillier Parker Research (1979) *British Shopping Developments*. London: Hillier Parker May and Rowden.

Holden, C.H. and Holford, W.G. (1951) *The City of London: A Record of Destruction and Survival*. London: Architectural Press.

Holdsworth, D.W. (1992) 'Morphological change in lower Manhattan, New York, 1893–1920', in J.W.R. Whitehand and P.J. Larkham (eds), *Urban Landscapes: International Perspectives*. London: Routledge. pp. 114–29.

Holmes, R. (1973) 'Ownership and migration from a study of rate books', *Area*, 5: 242–51.

Horn, J.C. (1993) 'Townscape transformations in dockland areas: case studies in the UK'. PhD thesis, University of Birmingham.

Horsey, M. (1985) 'London speculative housebuilding of the 1930s: official control and popular taste', *London Journal*, 2: 147–59.

Hoyt, H. (1933) *One Hundred Years of Land Values in Chicago*. Chicago: University of Chicago Press.

Jefferys, J.B. (1954) *Retail Trading in Britain 1850–1950*. Cambridge: Cambridge University Press.

Johnston, R.J. (1968) 'An outline of the development of Melbourne's street pattern', *Australian Geographer*, 10: 453:65.

Johnston, R.J. (1969) 'Towards an analytical study of the townscape: the residential building fabric', *Geografiska Annaler*, 51B: 20–32.

Kaye, B. (1960) *The Development of the Architectural Profession in Britain*. London: George Allen & Unwin.

Knox, P.L. (1991) 'The restless urban landscape: economic and sociocultural change and the transformation of metropolitan Washington, DC', *Annals of the Association of American Geographers*, 81: 181–209.

Kropf, K.S. (1993) 'An enquiry into the definition of built form in urban morphology'. PhD thesis, University of Birmingham.

Larkham, P.J. (1996) *Conservation and the City*. London: Routledge.

Lichtenberger, E. (1970) 'The nature of European urbanism', *Geoforum*, 4: 45:–62.

Mayer, H.M. and Wade, R.C. (1969) *Chicago: Growth of a Metropolis*. Chicago: University of Chicago Press.

Merret, S. and Gray, F. (1982) *Owner Occupation in Britain*. London: Routledge & Kegan Paul.

Mosher, A.E. and Holdsworth, D.W. (1992) 'The meaning of alley housing in industrial towns: examples from late-nineteenth and early-twentieth century Pennsylvania', *Journal of Historical Geography*. 18: 174–89.

Moudon, A.V. (1992) 'The evolution of twentieth-century residential forms: an American case study', in J.W.R. Whitehand and P.J. Larkham (eds), *Urban Landscapes: International Perspectives*. London: Routledge. pp. 170–206.

Pevsner, N. (1963) *An Outline of European Architecture*, 7th edn. Harmondsworth: Penguin Books (1st edn, 1943).

Prince, H.C. (1964) 'North-west London 1814–1863', in J.T. Coppock and H.C. Prince (eds), *Greater London*. London: Faber and Faber. pp. 80–119.

Prudential Pensions (1981) *Report to Policy Holders for the Year Ended December 1980*. London: Prudential Pensions.

Richardson, H.W. and Aldcroft, D.H. (1968) *Building in the British Economy between the Wars*. London: George Allen & Unwin.

Summerson, J. (1995) 'The beginnings of an early Victorian suburb', *London Topographical Record*, 27: 1–48.

Sutcliffe, A. (1970) *The Autumn of Central Paris: The Defeat of Town Planning 1850–1970*. London: Edward Arnold.

Tate, B. (1983) 'Alterations to the physical fabric of selected streets in central Glasgow, 1886–1905'. BA dissertation, University of Birmingham.

Thomas, B. (1973) *Migration and Economic Growth: A Study of Great Britain and the Atlantic Economy.* Cambridge: Cambridge University Press.

Thompson, F.M.L. (1974) *Hampstead: Building a Borough, 1650–1964.* London: Routledge and Kegan Paul.

Thompson, I.A. (1987) 'An investigation into the development of the building fabric of Huddersfield's CBD 1869–1939'. PhD thesis, University of Birmingham.

Thorne, R. (1984) 'Office building in the City of London 1830–1880'. Unpublished paper presented to the Urban History Group Colloquium on Urban Space and Building Form, London.

Tomlinson, M. (1964) 'The city of Birmingham: secular architecture', in W.B. Stephens (ed.), *A History of the County of Warwick*, Vol. VII. London: Oxford University Press. pp. 43–57.

Treen, C. (1982) 'The process of suburban development in north Leeds, 1870–1914', in F.M.L. Thompson (ed.), *The Rise of Suburbia.* Leicester: Leicester University Press. pp. 158–209.

Trowell, F. (1983) 'Speculative housing development in the suburb of Headingley, Leeds, 1838–1914', *Publications of the Thoresby Society*, 59: 50–118.

Vance, J.E. (1990) *The Continuing City: Urban Morphology in Western Civilization.* Baltimore: Johns Hopkins University Press.

Van Hulten, M. (1967) 'In search of the urban core of Amsterdam', in University of Amsterdam, Sociographical Department, *Urban Core and Inner City: Proceedings of the International Study Week Amsterdam 1966.* Leiden: Brill.

Whitehand, J.W.R. (1967) 'Fringe belts: a neglected aspect of urban geography', *Transactions of the Institute of British Geographers*, 41: 223–33.

Whitehand, J.W.R. (1972) 'Urban-rent theory, time series and morphogenesis: an example of eclecticism in geographical research', *Area*, 4: 215–22.

Whitehand, J.W.R. (1978) 'Long-term changes in the form of the city centre: the case of redevelopment', *Geografiska Annaler*, 60B: 79–96.

Whitehand, J.W.R. (1981a) 'Background to the urban morphogenetic tradition', in J.W.R. Whitehand (ed.), *The Urban Landscape: Historical Development and Management.* Institute of British Geographers

Special Publication 13. London: Academic Press. pp. 1–24.

Whitehand, J.W.R. (1981b) 'Fluctuations in the land-use composition of urban development during the industrial era', *Erdkunde*, 35: 129–40.

Whitehand, J.W.R. (1984) *Rebuilding Town Centres: Developers, Architects and Styles.* University of Birmingham Department of Geography Occasional Publication 19.

Whitehand, J.W.R. (1987) *The Changing Face of Cities: A Study of Development Cycles and Urban Form.* Institute of British Geographers Special Publication 21. Oxford: Blackwell.

Whitehand, J.W.R. (1988) 'Urban fringe belts: development of an idea', *Planning Perspectives*, 3: 47–58.

Whitehand, J.W.R. (1990) 'Townscape management: ideal and reality', in T.R. Slater (ed.), *The Built Form of Western Cities: Essays for M.R.G. Conzen on the Occasion of his Eightieth Birthday.* Leicester: Leicester University Press. pp. 370–93.

Whitehand, J.W.R. (1992a) *The Making of the Urban Landscape.* Institute of British Geographers Special Publication 26. Oxford Blackwell.

Whitehand, J.W.R. (1992b) 'The makers of British towns: architects, builders and property owners, c. 1850–1939', *Journal of Historical Geography*, 18: 417–38.

Whitehand, J.W.R. (1993) 'Formas de renovación urbana en Gran Bretaña: una perspectiva histórico-geográfica', *Historia Urbana*, 2: 59–70.

Whitehand, J.W.R. (1994) 'Development cycles and urban landscapes', *Geography*, 79: 3–17.

Whitehand, J.W.R. (1996) 'Making sense of Birmingham's townscapes', in A.J. Gerrard and T.R. Slater (eds), *Managing a Conurbation: Birmingham and its Region.* Studley: Brewin, pp. 226–40.

Whitehand, J.W.R., Larkham, P.J. and Jones, A.N. (1992) 'The changing suburban landscape in post-war England', in J.W.R. Whitehand and P.J. Larkham (eds), *Urban Landscapes: International Perspectives.* London: Routledge. pp. 227–65.

Whitehand, J.W.R. and Whitehand, S.M. (1984) 'The physical fabric of town centres: agents of change', *Transactions of the Institute of British Geographers*, NS9: 231–47.

Whitehouse, B.P. (1964) *Partners in Property.* London: Birn, Shaw.

6

Housing in the Twentieth Century

RAY FORREST AND PETER WILLIAMS

The production, consumption and exchange of housing is at the core of urban studies. Where we live, what we live in, with whom we live and how much it costs us to do so are among the central concerns of everyday life. With few exceptions, rental and loan payments are likely to represent a major proportion of earned incomes. Fluctuations in these costs impact in significant ways on other areas of social life and, for governments, housing costs and housing availability retained a political salience and sensitivity throughout the twentieth century. The issues may have changed and the forms of political mobilization around housing may have, in some contexts, become less overt and confrontational, but housing problems continue to have the capacity to provoke widespread social disquiet and tensions.

This chapter focuses mainly on the housing situation in the industrial or post-industrial 'North' but it is important to begin with a reminder of the global context of urbanization in order to avoid ethno- or Euro-centric views of housing problems and solutions. We cannot deal adequately in one chapter with the significant variations in social and economic conditions which shape housing problems in different parts of the world. Moreover, even within the confines of Europe, variations in culture, institutional forms, histories of urbanization and industrialization, housing stocks and policy histories have been and remain substantial and generalizations are highly problematic. With that as a cautionary note, however, it is possible to map out some of the key trends and developments which have shaped housing in the period since the Second World War.

A RETROSPECTIVE: PROGRESS IN THE TWENTIETH CENTURY?

The past hundred years has seen enormous changes in housing standards and conditions in the developed North but the reality for millions of households in developing countries remains one of absolute housing deprivation, subsistence living and marginality. At the beginning of the new millennium we are, it seems, finally entering the urban age, when the majority of people will be living in cities of varying scale. In the developing and newly industrializing countries, many of these cities will be on a mega scale, with mega problems of environmental pollution, transport congestion, poverty and shantyism. The accelerated pace of urbanization and the commercialization of land markets in the cities of the South are creating new pressures of eviction and displacement.

From a European perspective, it would appear that housing policy over much of this century has been concerned with the legacy of the rapid urban growth of the nineteenth and early twentieth century. Issues of housing conditions and their impact upon health and the economy, subsequent slum clearance and, in some countries, the associated rise of social housing, the decline of private landlordism, the growth of individual home ownership and the development of contemporary financial mechanisms and institutions are all rooted in that period. Two centuries later, the nations which were at the core of the Industrial Revolution confront very different urban conditions and housing systems. But rural–urban migration and the related housing pressures remain dominant forces in the world. As Sudjic (1996) has neatly observed,

Peter Williams has co-authored this chapter in a personal capacity. The views expressed do not necessarily reflect those of any organization with which he is associated.

Between 1810 and 1850, Manchester's population increased by 40 per cent every 10 years. It became a gigantic mechanism for the creation of wealth, and the transformation of rural migrants into city dwellers. Manchester invented the factory, the railway station, the civic university, and the back-to-back house. Asia's vast new cities are doing the equivalent for the next century. (1996: 37)

Despite all of this, it is a sobering fact that, on whatever estimates, the scale of homelessness remains significant in Europe and North America. It has been suggested, for example, that in the early 1990s there were around 18 million citizens of the European Union in some state of homelessness (FEANTSA, 1995). However, if we then include even more heroic estimates of those in living in various states of housing precariousness in other parts of the world, the scale of the global housing problem appears truly unmanageable.

Worldwide, the number of homeless people can be estimated at anywhere from 100 million to 1 billion or more, depending on how homelessness is defined. The estimate of 100 million would apply to those who have no shelter at all, including those who sleep outside (on pavements, in shop doorways, in parks or under bridges) or in public buildings (in railway, bus or metro stations) or in night shelters set up to provide homeless people with a bed. The estimate of 1 billion homeless people would also include those in accommodation that is very insecure or temporary, often of poor quality – for instance squatters who have found accommodation by illegally occupying someone else's home or land and are under constant threat of eviction, those living in refugee camps whose home has been destroyed and those living in temporary shelters (like the 250,000 pavement dwellers in Bombay). (United Nations Centre for Human Settlements, 1996: 229)

On any measure and whatever the context, we are evidently a long way from solving the housing problem. That is not to say that government interventions, rising real incomes and the development of new financial mechanisms have not produced substantially better housing conditions for the majority of households in the industrial (or post-industrial) world. If we take Britain as an example, in a relatively short period it has been transformed from a nation in which the majority of households lived as tenants of private landlords and in accommodation that lacked basic amenities such as inside toilets and fixed baths to a situation where only a minority lack such provision and home ownership is the dominant tenure form. As recently as 1950, most households were private tenants and the majority lacked exclusive access to bath, shower or inside toilet. In Greater London in 1964, more than a quarter of

households did not have exclusive access to hot water (Murie, 1974), By 1991, only 200,000 dwellings lacked basic amenities in England, representing around 1 per cent of the stock (Department of the Environment, 1993).

On such basic indicators of housing condition, similar progress can be found across Europe and North America. And even more dramatic examples of upgrading of housing standards and conditions can be found in the rapidly developing nations of South East Asia, where the pace of urban renewal and suburbanization is breathtaking by European standards. In the older capitalist core of Europe, however, as amenity provision has improved, the general physical fabric of dwellings has deteriorated and problems of disrepair and general unfitness have become more pressing policy issues. In countries such as Britain, Ireland, Belgium, France and Portugal more than a quarter of the housing stock was built before 1919. But perhaps of greater concern has been the emergence of a minority of households experiencing relatively greater housing deprivation. Robson's (1979) observations about Britain remain more generally relevant:

while the inequalities have grown less and the average provisions raised, the awareness of inequalities has grown. With a larger proportion of the population being better housed, the remaining, and still large numbers of ill-housed and homeless feel the inadequacy of their accommodation even more. (1979: 70)

While standards and expectations have risen, a minority of poor households – lone parents, poor elderly, the sick and disabled and many from ethnic minorities – find themselves in housing circumstances increasingly divergent from the majority. These minorities have, in the USA, Britain and much of Europe, become increasingly associated with low quality public housing. These poor neighbourhoods, typically in the inner city or on the urban periphery, are among the most visible signs of social polarization and social exclusion in contemporary urban society.

Housing poverty remains therefore a major issue even in the wealthiest societies. And perhaps perversely, housing-related wealth has become an increasingly important component of personal sector wealth (see Forrest and Murie, 1995). As individual ownership of dwellings has expanded in combination with periods of real house price inflation, the ownership of housing assets has come to represent a new if at times volatile division between the haves and have-nots. The home is perhaps more than ever a pivotal element in social life. The design, tenure and location of dwellings is a key dimension of social status and social stratification. And homelessness is the most overt expression of exclusion from the social and consumption norms of contemporary society. The residential dwelling is the focus of family life and the residential neighbourhood remains one of the major arenas where the mundane but important and routinized rituals of everyday life, which contribute so much to our social identity, are played out. The dwelling is the site for the storage, use and enjoyment of consumer goods. Hence in a consuming world the dwelling takes on ever-increasing significance. And the home is also, of course, the workplace for the unwaged work of women and carers, for minorities of high and low status homeworkers, where children do much of their learning and playing and where ever-increasing numbers of elderly people in increasingly ageing societies spend much of their time.

FROM RATIONAL PLANNING TO DEREGULATED MARKETS

A broad sweep of housing policy and provision in the period following the Second World War would situate changes in the context of a more general shift from the post-war period of rational planning and direct state intervention aimed at eradicating housing problems to a later phase of economic restructuring and social engineering where market processes were unambiguously reasserted. Whatever the particular phase of restructuring, and whether we are referring to shifting boundaries between the state, the market and the individual in the European capitalist core countries, (former) state socialist societies, the city states of Singapore or Hong Kong or whatever, housing has been at the forefront of broader social and economic tranformations. In the post-war period of suburbanization, mass social housing, New Towns and the development of various forms of individual home ownership have had a major impact on the everyday lives of people, on the links between housing and the wider economy, on patterns of mobility and on the relationships between home and work. With the USA as the obvious major exception, housing provision was at the heart of dominant, if contested, beliefs that states could address social needs and solve housing problems in a planned, rational manner. The British New Towns programme is perhaps one of the best expressions of state-centred, strategic rationality in addressing interrelated issues of environmental congestion, the geographic distribution of employment, higher housing standards and changing patterns of family life. Embedded in the planning of New Towns were utopian visions of healthy, cohesive communities living in land-scaped residential areas designed to maximize neighbourly interactions. And while, in the USA,

the direct provision of state housing was no more than a minor part of the policy programme, government policies were heavily geared to the promotion of home ownership as an important element of post-war social engineering and economic restructuring. Harvey (1978) linked the promotion of home ownership to a more general process of debt-financed infrastructural development and spatial reorganization in response to problems of overaccumulation and underconsumption in post-war capitalism.

> Though suburbanization had a long history, it marked post-war urbanization to an extraordinary degree. It meant the mobilization of effective demand through the total restructuring of space so as to make the consumption of the products of the auto, oil, rubber and construction industries a necessity rather than a luxury . . . It is now hard to imagine that post-war capitalism could have survived, or to imagine what it would have been like, without suburbanization and proliferating urban development. (1978: 39)

But there are other perspectives on the promotion of home ownership in post-war USA which emphasize aspects of individual behaviour, social pathology and cohesion and which resonate more widely with other national housing policies and underlying assumptions about the links between housing tenure and social action (see Winter, 1994 for an interesting exploration of these issues). For example, the influential work of Rossi (1955) on residential mobility was funded partly from a concern with the pathology of hyper-mobility – the apparent rootlessness and nomadic nature of post-war USA which was seen by some as both symptom and cause of a wider social malaise in which community ties were disintegrating and the quality of urban life was deteriorating. The mobile, affluent society was viewed as unstable. A policy solution to reduce this assumed pathological desire to move to achieve upward social mobility was the promotion of home ownership which was seen as a way of encouraging individuals and families to put down roots.

Policy and academic debates around housing have continued to reflect this cocktail of macroeconomics and social engineering. As absolute housing shortages have diminished and as fiscal pressures increased throughout the 1970s and 1980s, beliefs and ideological constructions about the supposed political and social effects of housing tenure have become intertwined with a resurgence of liberal economics and austerity policies. Whilst it would be wrong to overstate the degree of policy convergence, terms such as privatization and deregulation entered policy and popular vocabulary and progressively shaped housing policy and provision in many countries. Housing was once again in the vanguard of attempts to restructure social and economic life.

The modernist mass housing estates of the 1950s, 1960s and 1970s had become increasingly discredited and associated with concentrations of poor people in high rise flats in inner or peripheral urban locations. The solutions to the housing problems of one period had rapidly become the social and physical housing problems of the next. High-rise housing and poverty, difficult-to-let estates, social isolation, ghettoization and residential segregation, became conflated with direct state intervention in housing. Policy emphasis shifted from object (the property) to subject (the person) subsidies and increasingly towards the promotion of individual home ownership. Home ownership and social housing became progressively intertwined as privatization policies boosted the former through the disposal of the latter.

These developments took very different forms in different countries. Political cultures, institutional and subsidy structures, housing stocks and the history of housing policies vary substantially between countries. Apparently similar policies are pursued within highly varied social and economic contexts with different pressures and motivations and with very different consequences. Moreover, the pattern of policy and provision in one era conditions the possible options of the next. For the UK, the privatization of state housing represented a fortuitous coincidence of economic and social imperatives.

The creation and then sale of a substantial stock of family houses owned and managed by municipal landlords enabled a massive boost to home ownership whilst simultaneously generating a large and continuous flow of funds to government. Whilst the role of the state in housing was 'rolled back' in a number of ways, home ownership and its attendant social and political consequences were rolled forward. But governments cannot sell what they do not own. In the USA, the scope for housing privatization policies, as opposed to deregulatory measures, was limited by the already small scale and residual nature of social housing. Concentrations of poor people in heavily stigmatized housing developments offered limited scope for disposal of such properties to the tenants. And in West Germany, private landlords and other institutions received loans and subsidies to provide social housing that met specified standards and was directed at target groups but, after a stated period, the contract allowed them to withdraw this housing to use as they chose. At that point, dwellings which had served a particular social role reverted to the private market.

The 1970s and 1980s therefore saw individual households competing in an increasingly deregulated market – albeit one that in many cases was fuelled by generous tax breaks and subsidies. Governments were withdrawing from direct state provision and channelling support to individual home ownership. This trend was perhaps most evident in Thatcher's Britain but was nevertheless a perceptible policy shift in much of Western Europe. The dismantling of the Berlin Wall, the break-up of the former Soviet Union and greater marketizing pressures in China represented a new phase of social and economic transformation in which changes in housing provision, management and ownership were seen as essential elements of a new social order. The ownership of a home was emblematic of a new contract between the state and the individual, a new force for social stability and individual responsibility and, not insignificantly, the means to capture individual savings and release the state from expensive commitments for building, repair and maintenance.

At the beginning of the twenty-first century therefore, housing provision sits within a rather different ideological and economic context when compared with the decades following the end of the Second World War. The market remains ascendant and the state is to a large degree decentred. Big finance dominates and the environment of choice and deregulation presents a more complicated set of circumstances for consumers to negotiate. The polarized alternatives of the friendly bank manager or the state bureaucrat have given way to a much more fragmented system. That fragmentation of rules of access, of forms of social housing, of financial products is varied in its extent. Nevertheless, the negotiation of access to housing in a more complex and more uncertain social and economic environment is a significant element in the way in which individuals are socialized into and learn to cope with what Beck (1992) refers to as an increasingly individualized world.

POLICY PREOCCUPATIONS

The housing policy preoccupations, institutional structures and state and market responses of one period inevitably shape the problems and policy options of the next. In Europe, the Second World War was a major influence on housing priorities and policies. In the interwar years, any commitments to substantial direct state intervention in housing were generally limited and short-lived. Liberal economics dominated, housing programmes aimed at poor households were small-scale and alternative models of provision were of the self-help and cooperative variety (see Harloe, 1995 for a detailed discussion of these issues). After the Second World War, however, governments were confronted with massive housing shortages.

Almost a fifth of the housing stock in Europe had been destroyed, residential building had been halted and demobilization produced a sudden leap in demand. These chronic housing shortages required state intervention in housing provision in a variety of areas, particularly in the supply of low cost public rental housing and rent control in the private rental sector. Large-scale housing production programmes were introduced in most countries in northern Europe under a variety of subsidy systems.

> Unlike the period after the First World War, this response was not just a short term reaction to acute social unrest before a rapid return to business as usual in housing, as in the economy . . . [Moreover] now a powerful additional reason existed for state-subsidized social rental housing construction: its contribution to economic modernization and urbanization . . . Social housing investment also became one of the main tools of Keynesian-style demand management, being frequently used to help and maintain the new commitment to full employment and non-inflationary economic growth. (Harloe, 1995: 44)

In this context, Britain was rather exceptional in its almost exclusive concentration on municipal council housing as social housing. Donnison and Ungerson (1982) observed that

> in France, Austria, Germany, Scandinavia and the Netherlands, industrial, political and religious movements developed various patterns of voluntary and cooperative association which were later adopted by governments as their principal instruments for the provision of housing. (1982: 63)

Housing policy in the USA is, of course, rooted in a very different set of circumstances, which is part of the explanation for its residual approach to social housing as welfare housing and the overwhelming predominance of free market principles. Doling (1997) argues that the roots of US housing policy can be found in the profound shock of the 1930s Depression in contrast to the impact of world wars on Europe. But, he suggests that in the USA

> the concern was not primarily about the acceptance of a state responsibility for ensuring that citizens' needs were adequately met, but with finding a way out of economic difficulties through planning at the macro level. Providing an impetus to the construction industry and maintaining confidence in the mortgage market were seen as significant steps in seeking to reinvigorate a depressed economy. (1997: 18)

Attaining numerical building targets to meet substantial absolute housing shortages remained a key preoccupation of most European governments well into the 1970s. Indeed, the fiscal and material constraints of the early post-war period limited the rate of new building and most energies were directed at the repair and reconstruction of damaged stock. Continued rural–urban migration and rising birth rates meant that the housing shortage grew more acute. The pressure grew to meet rising targets by producing smaller, cheaper dwellings, and as economies recovered more reliance was placed on the private sector. Urbanization produced higher land values, further adding to problems of cost. Social housing was pushed to higher densities and often peripheral locations. New systems building techniques were adopted in the late 1950s and 1960s which lowered immediate unit costs and accelerated production but which proved in many instances to have high maintenance costs and low standards of heating and insulation.

Typically, it is these mass high-rise solutions to housing needs which are prominent among the discredited legacies of the modernist era and are also inextricably, if debatably, associated with the failures of direct state intervention. One commentator has gone as far as to link the end of modernism with the destruction of a block of social housing in the USA – the precise time and date when a version of one of Corbusier's designs was demolished because it had proved to be uninhabitable (Harvey, 1989: 40, referring to Jencks).

By the early to mid-1970s, many countries were experiencing relative economic prosperity and an apparent crude surplus of dwellings over households was emerging. Attention shifted towards more qualitative concerns and particularly towards tenure choice and tenure conversion. Major housing shortages remained in parts of southern Europe and, of course, in the rapidly developing Tiger economies of South East Asia. In general, however, political and fiscal support for mass housing provision diminished as need and provision moved closer into balance. As the urgency of the situation changed so the emphasis shifted towards more targeted assistance for localized needs, private sector solutions and questions of consumer demand. Home ownership became the dominant tenure preference, fuelled by a reorientation of subsidies. The conflation of what Kemeny (1980) has referred to as *political tenure strategies* with expressed popular desires was embodied in a variety of national and supranational policy documents. As the Economic Commission for Europe stated, 'the fact is that for most people owning their own home is a basic and natural desire' (1983: 79) and variations on that statement can be found in a variety of locations and stretching back over most of the twentieth century. Whether it is the British version of the property-owning democracy, the American or Australian Dream or Hong Kong's latest medium-term housing strategy, that familiar refrain is evident.

Fuelled by the growth and deregulation of the financial sectors in some countries, rising real incomes, changing social norms and expectations and housing policies increasingly subservient to productivist, competitive social and economic policy the momentum towards higher levels of home ownership gathered pace throughout the 1980s. At the same time, growing unemployment, poverty and increasingly concentrated spatial patterns of disadvantage emerged, often with strong ethnic, gender and class divisions. It is these concentrated spatial expressions of disadvantage which represent some of the most pressing and intractable contemporary policy preoccupations in the developed world. Moreover, mortgage-financed home ownership, now the majority tenure in many countries, sits within the more volatile and uncertain prevailing economic environment.

HOME OWNERSHIP IN ASCENDANCE?

If the industrialized world at the beginning of the twentieth century was a world in which the private landlord dominated housing provision, as a new century opens, it is individual home ownership in various shapes and forms which is ascendant. One of the more recent expressions of this pervasive ideological and policy trend came in March 1998, when China's new prime minister, Zhu Rongji, pledged to turn China into a nation of home owners by abolishing state subsidized housing. Such an announcement by the leader of a country of 1.2 billion people and where home ownership currently accommodates around 20 per cent of the urban population is further evidence of the retreat from direct housing

provision by central and local government which has been taking place throughout the world over the past 20 or so years.

International comparisons of tenure structures and rates of growth of home ownership are shown in Tables 6.1 and 6.2. These are inevitably crude comparisons and aggregate highly varied dwelling types, household circumstances and subsidy regimes under single labels. Moreover, in relation to home ownership levels, they may combine contemporary forms of loan financed urban residential property purchase with more historical rural forms and, in some cases, illegal self-build on urban peripheries. These problems are compounded when cross-national comparisons are made. Home ownership does mean different things in different countries. Home ownership might give access to land for a small holding and be the means by which a household acquires a living. Elsewhere, it is primarily a symbol of status and/or a means of accumulation. Nevertheless, such generalized aggregations serve as indications of broad trends in patterns of housing provision. With few exceptions and to varying degrees, the unambiguous trend has been towards higher levels of home ownership.

This trend has been accelerated in a number of countries in recent years by the transfer of properties from the public rental sector. The context for these transfers has varied enormously (see next section). In some cases, as with China or Russia, it was part of a fundamental reshaping of housing production, management and access; in others, it involved the addition of a new layer of owners to a highly developed private housing market. The UK's notably high growth rate through the 1980s and 1990s (from 56 per cent in 1980 to 68 per cent in 1995), for example, was fuelled to a significant degree by tenure transfers

Table 6.1   *Proportion of all households by tenure, OECD countries (%)*

Country	Home ownership	Social renting	Private renting	Others	Total	Date of source
USA	64.2	2.0	33.8	0.0	100.0	1991
UK	67.6	22.6	9.8	0.0	100.0	1994/95
Sweden	55.0	22.0	23.0	0.0	100.0	1991
Spain	78.4	1.1	13.8	6.6	99.9	1989
Netherlands	47.3	40.2	11.2	1.3	100.0	1993/94
Japan	59.8	7.0	26.4	5.1	98.3	1993
Ireland	76.8	13.9	7.7	1.6	100.0	1987
Germany	38.0	15.0	43.0	4.0	100.0	1987
France	54.4	14.5	25.1	6.0	100.0	1990
Finland	78.4	10.7	9.5	1.4	100.0	1992
Canada	62.7	5.6	31.7	0.0	100.0	1991
Australia	70.0	6.0	19.0	5.0	100.0	1994
Average	62.7	13.6	20.9	2.6	99.9	

Source: Freeman et al., 1996

Table 6.2 *Proportion of all households in owner-occupation, OECD countries, 1945/50 to 1995 (%)*

Country	1945/50	1960	1970	1980	1990	1995
USA	57	64	65	64	64	na
UK	29	42	49	56	65	68
Sweden	38	36	35	41	42	40
Spain	na	na	64	73	76	na
Netherlands	28	29	35	42	44	47
Japan	na	71	59	62	61	60
Ireland	na	na	71	76	81	na
Germany	na	na	36	40	38	na
France	na	41	45	51	54	na
Finland	56	57	59	65	67	78
Canada	66	66	66	62	63	na
Australia	53	63	67	71	70	70

Source: Freeman et al., 1996

from the public sector (and to a much lesser extent by the creation of supposedly transitional 'shared' ownership; a real indication of the drive to promote home ownership to a wide spectrum of the population). A small number of countries show modest falls in home ownership levels in recent years (e.g. Denmark, United States and Germany). Different factors have been at work, including political transformations (Germany), problems of affordability and access for some younger people and more volatile house price cycles (see below).

The growth and functioning of home ownership is strongly conditioned by the ways mortgage finance is raised and delivered in each country. Most home ownership is debt financed and there are significant differences in the types of funding instruments, the structure of interest rates, repayment mechanisms and lending practices. There are also very great differences in family structures and the capacity to save and purchase (as well as economic cycles and tax systems). Household behaviour differs greatly between countries in terms of expectations, attitudes to debt and

ownership and this impacts upon the age of entry to home ownership.

There are three main types of mortgage finance systems: the mortgage bank system, the deposit taking system and the contractual system. In the first, household funds are channelled via insurance companies and pension funds which in turn purchase mortgage bonds issued by the mortgage banks as backing to the housing loans they have granted (e.g., Sweden and Denmark). In the second, household deposits are recycled into long-term housing loans (e.g. Britain and Australia) while, in the third, potential borrowers contract to save for a specified period in order to qualify for a loan (e.g. Germany and Austria). In some countries such as Greece or Turkey there is no significant formal housing debt finance sector and informal family-based mechanisms remain important means of access to funds.

There is, in fact, little evidence that the different systems are converging. Clearly there may be significant developments in Europe following monetary union but at present systems remain doggedly national (or regional). The shape of the finance system will influence access to home ownership, its sustainability over the economic cycle and the extent to which housing market pressures impinge upon the economy as a whole. To the extent that home ownership is achievable and sustainable it can also affect public expenditure on other forms of housing provision and assistance. The structuring of mortgage finance influences who becomes owners and impacts upon the dynamics of the housing market. Higher down payments slow entry into home ownership but also allow for more rapid repayment of the mortgage debt. Fixed rate mortgages are common in some systems, allowing home buyers greater certainty regarding payments.

It is important to acknowledge the scale of state support for home ownership as a significant factor in its growth. Just as with the provision of state or public housing for rent, home ownership in many countries receives a variety of direct or

Table 6.3 *Tax and subsidy framework: owner–occupiers*

	ASL	C	FIN	F	G	IRL	J	NL	E	SW	UK	USA
Mortgage tax relief	N	N	Y	Y	N	Y	Y	Y	Y	Y	Y	Y
Imputed rental income tax	N	N	N	N	N	N	N	Y	Y	N	N	N
Capital gains tax	N	N	N	N	N	N	Y	N	Y	Y	N	Y
Subsidy for house purchasers	N	N	Y	N	N	N	N	Y	Y	N	N	Y
Subsidized interest rate on house purchase loan	N	N	Y	Y	Y	N	Y	N	Y	Y	N	Y
Eligibility for income-related housing allowance	N	N	Y	Y	N	N	N	N	N	Y	N	N
Minimum income safety net to meet housing costs	Y	N	N	N	N	Y	N	N	N	N	Y	N

ASL, Australia; C, Canada; FIN, Finland; F, France; G, Germany; IRL, Ireland; J, Japan; NL, Netherlands; E, Spain; SW, Sweden.

Source: Country researchers' questionnaires in Freeman et al., 1996

indirect subsidies (Table 6.3). Mortgage interest tax relief is the most common mechanism (ranging from 100 per cent to 10 per cent relief). The UK and Ireland are notable in that both countries have substantially eroded this very significant benefit and one which was highly regressive. In a considerable number of countries there are grants to encourage ownership or subsidized interest rates on loans for house purchase and in eight out of the twelve countries no capital gains tax is levied, giving some indication of the way home ownership has been a favoured tenure.

HOME OWNERSHIP AT RISK?

In recent years some housing markets have experienced greater price volatility and a growing number of casualties. In Britain, the rapid house price inflation of the mid- to late 1980s was followed by a severe slump and fall in nominal house prices. A variety of factors were at work in the boom, including financial deregulation which improved households' ability to borrow funds for house purchase, changing macroeconomic conditions, government policy and demographic change (Boleat, 1994; Forrest and Murie, 1994). By 1992, house-building activity in the private sector had declined to 139,000 from a peak of 199,000 in 1988 and mortgage activity was reduced from £40,111 million to £17,751 million. Home ownership became associated with mortgage arrears, repossession and negative equity.

There is, however, nothing novel about slumps in the housing market. Real house prices fell by 31 per cent between 1973 and 1977 compared to 24 per cent between 1988 to 1992 (Kennedy and Andersen, 1994). However, during the 1970s high inflation rates masked these falls and protected nominal house prices, indicating what Bootle (1996: 68) refers to as 'the power of inflation to create illusions about real values'.

This period of housing market instability was not, however, confined to Britain. The growth and support of home ownership by governments has been prevalent throughout most of the developed and developing world and this has been accompanied by financial market deregulation, the internationalization of capital markets and the integration of mortgage finance into global capital markets (Fallis, 1995). The stability of national housing finance systems evident in the 1970s has been undermined as housing finance has been exposed to global financial forces. According to Fallis, 'The deregulation and shifts in world inflation rates led to the near collapse of the housing finance system in several countries' (Fallis, 1995: 15).

Bootle (1996) documents the volatility of house prices in the world's leading industrial countries. He argues that in every one of the G7 countries 'house prices rose remorselessly during the heyday of inflation' (Bootle, 1996: 67). Between 1970 and 1992 the average annual rate of increase of house prices in Britain was 12.5 per cent, the same for Italy during 1970 to 1989. In the USA the annual rate was 7.75 per cent and in Germany 5.5 per cent. In real terms, house prices rose by an average rise of 2.5 per cent a year in Britain and Japan, 2 per cent in Canada and 1.5 per cent in Germany and the US (Bootle, 1996: 68). Kennedy and Andersen describe developments in house prices in 15 industrialized countries between 1970 to 1992, asserting that in the majority of cases volatility appears to have increased in the 1980s. In real terms, Finland, Japan and the United Kingdom show the greatest volatility both for the period as a whole and for the 1980s (1994: 13), with the Netherlands following closely behind. They also show the 10 largest house price increases, and decreases, in any single year over the period 1970–92. In the 1980s, in terms of price increases Australia is ranked third, with a peak of 38.1 per cent in 1988, Finland is second, with 3.63 per cent in the same year, and the United Kingdom is ranked seventh, with 33.0 per cent, also in the same year. The sharpest decline was experienced by Finland in 1992, when prices fell by almost 17 per cent following a decline of almost 15 per cent in the previous year. Finland is followed by the Netherlands, then Norway, Sweden, Japan and Denmark. Looking at what Kennedy and Andersen refer to as 'peak to trough movements', Finland heads the table with a 33 per cent fall in nominal house prices (40.3 per cent fall in real house prices) between 1989 to 1992. The United Kingdom is seventh, with a 10.7 per cent fall in nominal house prices (23.0 per cent fall in real house prices) between 1989 to 1992.

Inevitably, such figures mask significant local variations within countries. Fallis (1995), for example, has examined the pattern of house price increases in the United States, Britain, Australia and Canada during the latter part of the 1980s and emphasizes the regional nature of these initial increases and their subsequent volatility. He pointed particularly to London in Britain, Boston in the United States, Sydney in Australia and Toronto in Canada. These areas experienced a dramatic escalation in house prices, linked to increased demand because of employment growth, immigration and income growth within the region, and an equally dramatic decline in prices in the early 1990s.

Kennedy and Andersen (1994) also show that the ratio of mortgage debt to the value of the owner-occupied stock of dwellings was higher at the end of 1992 than in previous years and was particularly marked in Denmark, Finland, the United Kingdom, Canada and the United States. In Denmark, where 53 per cent of the housing stock was owner-occupied in 1994, estimates for 1988 indicated that 25 per cent of home owners, particularly those under 40 years old, were 'technically insolvent' with net liabilities equivalent to 100–125 per cent of property value. Kosonen (1995) and Timonen (1992) acknowledge the occurrence of negative equity in most Nordic countries but are particularly concerned about the rising incidence of housing debt problems amongst households in Finland.

Given differences in price trends and the size and characteristics of the home ownership market in different countries, the variations in the structure of mortgages and mortgage finance and the different legal arrangements, it is not surprising that the experience of mortgage arrears and possessions has also been very different. In some countries the substantial deposits put down and the relatively exclusive nature of home ownership has meant that, despite difficulties in the economy, few households have lost their homes via possession or foreclosure (in some countries, possession is not easily obtained by the lender). In contrast, in other countries, such as the UK, home ownership accommodates a wide range of households and access can be achieved with a low level of deposit. As a consequence, households are more easily exposed to a range of risks (for example, increases in interest rates, falls in income, falls in house prices).

While housing provision becomes an ever-more private responsibility, so governments are under ever-greater pressure to secure appropriate conditions in the broader environment, for example, with respect to jobs and employment, inflation and interest rates and with respect to the efficiency, effectiveness and fairness of the private housing and mortgage markets. In this regard it is no coincidence that a range of governments are now looking closer at ways of protecting home owners and home ownership from the vicissitudes of economic change.

Britain is one of the small number of countries that provides a modest state-funded safety net for home owners. As the housing market recovered in Britain in the mid-1990s the government moved to redefine the role of the state in relation to home owners in difficulty. Insurance against such risks was seen as primarily a private responsibility. Home buyers should cover their risks through private insurance schemes. Interestingly, in other countries – Germany, France and the Netherlands – the same problem has been recognized (that is, the increasing vulnerability of home owners) but the solutions sought are based on state assistance. Similarly in Australia, Canada and the United States, it is the government which

underpins the bottom end of the housing market through a system of state supported insurance-based guarantees against lender losses when making loans to low income households.

The different roles played by the state, the shape of the mortgage market and the legal structures surrounding home ownership all influence the shape and character of the sector in each country. Perhaps of even greater significance is the question of the sustainability of home ownership. Built as it is on the shifting sands of social, economic and political priorities there can be no assumption that what we have today will be a given forever. Already it is evident in some countries that private renting is undergoing a resurgence amongst younger households. This is partly a product of tighter access to home ownership, but it is also part of household risk management strategies in the context of a more flexible labour market. As yet home ownership does not have the flexibility of renting because of the substantial transaction costs and the process of property purchase and sale. Securing greater efficiencies in the home ownership market is an emerging priority for governments.

WHITHER SOCIAL HOUSING?

The construction, management, financing and role of social housing varies in important ways over space and time. Decommodification to achieve social housing goals can come in different forms. State loans and subsidies may be channelled to private agencies to serve specific social or income groups. Dwellings may be financed, owned and managed by public sector agencies and focused on the poorest sections of society on a means-tested basis or available to a wider section of the population through other means of bureaucratic allocation. Non-profit housing may be provided through a wide range of voluntary agencies and cooperatives. And the provision of housing for lower income households may be via measures to ease access into home ownership. Singapore, for example, represents a somewhat exceptional example of state built housing, being essentially built for sale into owner occupation. Typically, however, the housing is for rental (at least initially), allocated on some criteria of need and profit is not the determining factor in setting the price.

It is impossible and inappropriate within the confines of this chapter to embark on a definitional excursion (see, Doling, 1997; Oxley and Smith, 1996). The important point is to acknowledge that the degree to which the market has been the principal means of access to housing has waxed and waned within nation states and that non-market measures have varied in form and scope. Moreover, while there is an association between fiscal and social pressures and the scale of state intervention in the housing market, that relationship is contingent on a range of factors. One of Harloe's (1995) conclusions from his major study of social rented housing in Europe and the USA is that

> There is no necessary connection between crisis and restructuring on the one hand, and mass provision on the other. He points to two conditions, however, which appear to be important in determining provision, particularly mass provision. First, a situation in which the private housing market was unable, for various reasons, to provide adequate housing solutions for sections of the population. Second, when unmet housing needs among those sections of the population had a wider significance for the societies and economies in which they existed, whether in terms of heightening social tension and crisis (after the First World War and in the USA in the 1930s), or in terms of economic modernization (after 1945). (1995: 524)

Harloe also identifies two main models of social rented housing – mass and residualized. The residual form which links to periods of slum clearance and/or is targeted on the (new) urban poor is, for Harloe, the normal and most universally institutionalized form within capitalist countries. It is institutionalized to the extent that there is some acceptance by government and the wider public that the market is unable to provide affordable housing of reasonable quality for the poorest sections of society. It is residual in the sense that it is minority provision for the poor and is likely to be identifiably second-class housing and stigmatized. US social rented housing is quintessentially residual.

The mass housing forms which, in the context of capitalist Europe (as opposed to transitional economies) are most closely associated with Sweden, the Netherlands and Britain are, for Harloe, abnormal. Unlike the residual forms which have been and probably will be around for longer, these large-scale programmes of social renting, catering for a wide cross-section of the population (and to a great extent excluding the poorest) have only grown up under very specific historical conditions. In Western Europe, that period was most notably in the three decades after the Second World War. It is this form of state-provided housing which has been transformed to the greatest extent in the past two decades.

Diminishing support for substantial public housing programmes has been evident in most countries with market or mixed economies in the period since 1980. That diminished support has taken two main forms: privatization and reduced new investment. Policies of privatization have been evident in a wide range of countries –

perhaps most notably in the UK, but encompassing the USA, Ireland, Denmark, Hong Kong, Singapore, China and developing countries such as Nigeria and Egypt. State-owned dwellings have been offered for sale on the open market, on usually discounted terms, to sitting tenants. Whatever the context, there is a familiar pattern. The best dwellings are sold to the more affluent tenants. Supply to the poorest sections becomes more restricted, quantitatively and geographically. The result is a direction of change towards more residual, stigmatized sectors. It should be emphasized, however, that this occurs in its most acute form when mass privatization is combined with dramatic cuts in replacement building.

The most far reaching transformations in the state rental sectors have, however, occurred in Eastern Europe, and more recently, in China, where the transfer of ownership of properties to individuals or non-state sector organizations has been seen as pivotal to more fundamental processes of social and economic change (see Turner et al. (1992) for a detailed account of early housing reforms in Eastern Europe). One of the basic problems confronting the transitional economies of Eastern Europe in this context has been the lack of developed mechanisms (most notably loan and exchange institutions) to enable a housing *market* to take root. There has been extensive privatization but uneven and limited *commodification*.

The increased emphasis on private ownership and marketization has had very different histories in these countries. The housing system in Hungary, for example, has been shifting progressively to a more market-oriented form since the early 1980s. In Russia, however, the shift has been more recent and more dramatic. In the so-called transitional economies of Eastern Europe, various difficulties have typically complicated privatization processes: a history of low rents, poor standards of maintenance, and low incomes. These interrelated factors also have to be placed in the context of state housing stocks, which are almost invariably dominated by medium- and high-rise, system-built flats. Tenants with limited resources are therefore often being persuaded through generous discounts to buy undermaintained, obsolescent properties in a situation in which there may be no re-sale market. Governments may succeed in off-loading a major financial liability but the advantages to the new owners are at least uncertain. The carrot-and-stick approach has been further enhanced through rent increases and the introduction of various forms of rent allowances to mitigate the difficulties for those on the lowest incomes.

The inequalities, however, built into previous state-organized housing systems find new expressions in their nascent privatized forms. Most notably, the housing privileges conferred on the nomenclatura and other elite groups in the state systems which involved access to the most desirable dwellings mean that the opportunities to buy have also benefited disproportionately those same groups. For example, an assessment of housing privatization policies in Novosibirsk in Russia observed that in 1993 'the better the quality of housing the family had and the larger the floorspace of the dwelling, the more active was the family's privatization behaviour. The share of privatized dwellings in the top group of housing stratification amounted to 44 per cent, while in the bottom group it was only 2 per cent' (Bessonova et al., 1996: 125; and see Danielli and Struyk, 1996). Again this uneven pattern of take-up and of potential inequality of benefit resonates with similar policies in developed capitalist systems. Those who gained most from state rental sectors, for example the skilled labour aristocracy of post-war Britain, also gained most from a new set of subsidies to encourage private ownership. And those who follow in the aftermath often gain very little and may end up as casualties.

In the UK, where policies of mass privatization have a relatively long history, those who bought well built, state owned family houses at generous discounts in the early 1980s can be contrasted with those who purchased lower value, lower quality apartments in the late 1980s and early 1990s. The first group generally bought appreciating, saleable dwellings and experienced periods of real house price inflation. The latter bought much less desirable dwellings in a more volatile and less certain housing market. In a similar vein, Bodnar (1996) contrasts two tales of private purchase in Budapest. One owner had bought a desirable apartment, immediately rented it to a multinational corporation and had subsequently been able to buy a small villa with the rental proceeds. By way of contrast, another tenant paid roughly the same price for a system-built apartment. Maintenance costs rose, proved prohibitive and she ended up requesting a return to tenancy status.

Selling desirable, marketable apartments or family dwellings to tenants with secure and reasonable incomes in a healthy housing market is relatively unproblematic. Selling apartments to lower income tenants in a depressed or underdeveloped market may create new problems requiring later policy intervention. State disengagement may prove partial and short-lived. Again the British experience of selling high- and medium-rise state-owned flats, which gathered momentum in the late 1980s, may be salutary. Many former tenants achieved neither social nor spatial mobility and the mix of tenures which developed within single blocks created serious difficulties for housing managers (see Forrest and Murie, 1995).

The consequences of significant shifts in housing provision and policy may take considerable time to emerge – and they may emerge in very contrasting conditions compared to those in which they were introduced. Major shifts in employment levels, in the nature of employment, in demographics and in house price inflation are obvious key factors. A particularly dramatic example of changing housing pressures and social conditions came with the rapid collapse of the Soviet bloc and the reunification of Germany which produced a new wave of immigration. By the mid-1990s, some 1 million new German citizens became eligible for social housing. This coincided with a period of substantial deregulation in which a significant number of dwellings were removed from the social sector with the termination of state loans. The government was suddenly faced with major housing needs when a few years previously there was relative housing abundance (Haussermann, 1994). These events highlight a basic problem of housing. Housing stocks often have to accommodate unpredictable social changes. But the necessary adjustments in quality or quantity are inevitably slow and will lag behind, whether state-promoted or market-driven.

CONCLUDING REMARKS

There are various versions of the story of the last one hundred years of housing provision. In one version, the period of what Harloe refers to as 'abnormal' intervention, the period of mass as opposed to residual state provision, might be seen as a 'passing' moment in the history of housing provision. Market forces have been re-asserted and the opening of a new century sees a variety of forms of home ownership emerging as the type of provision best suited to contemporary industrial and post-industrial economies. The period of mass provision in the 30 or so years after the end of the Second World War, which characterized much of European housing policy, reflected merely a period of transition from private landlordism to individualized home ownership. States had to manage acute shortages and quality deficiencies alongside rising expectations. They also had a pre-eminent model of intervention. That moment has now passed, but the consequences live on.

Rising real incomes, the development of mortgage credit, greater job security and Keyensian welfarism have all contributed to what has taken place subsequently. Home ownership has achieved a dominant position, the poor are increasingly accommodated in stigmatized housing forms in enclaves of concentrated disadvantage and subsidies for home owners are progressively targeted on lower income households and, in some cases, provided a safety net for casualties of the tenure. State withdrawal proceeds not just in relation to social housing but via the privatization of risk for home owners and the reduction of general subsidies.

However, just as the economic and political context around the provision of mass social housing has been transformed so too has the context around home ownership. This is a tenure perhaps best suited to stability and predictability. The private renting which dominated at the turn of the twentieth century was embedded in a context of relative volatility in household incomes and employment. Rapid adjustments in housing costs were often necessary and possible through relatively easy residential movement. The labour market of the new twenty-first century contains some of the same features of uncertainty and unpredictability for some households as did the labour market of the early twentieth century. As one period of transition in housing begins to wane, new processes of transformation are under way which will certainly produce new and hybridized forms of home ownership and may well see a significant revival of private landlordism, albeit configured rather differently from its early twentieth-century variety.

The relationship between housing and the macro economy now appear more exposed but also more complicated. Financial deregulation, changes in the pattern of ownership of dwellings and transformations in the nature of employment and new patterns of work within households have produced a more volatile mix of circumstances. The debate is no longer so focused on how many dwellings to build, the proportion of GDP devoted to residential construction or to the employment-generating effects of building activity. Debates about what to build and where are more likely to be focused on the environmental impacts of residential development than on how close governments have come to meeting stated targets. But the growth of individual home ownership has produced a new set of associations between financial institutions, individual household budgets, employment, housing market activity and general consumer behaviour. In circumstances where the mass of households own their own homes, the vicissitudes of the economy impact more directly on individuals and their families. And the centrality of housing costs, housing debt and house values for households means that shifts in mortgage costs can impact heavily on the consumption of other goods and services and on the general health of the economy.

Many national housing policies have been constructed on assumptions of relative predictability and security in household circumstances – conditions essential to forms of housing provision requiring high levels of personal debt. But the world seems to be becoming less predictable. Jobs seem less secure, relationship breakdown more likely and the family life cycle more complicated. Climatic change too brings new threats. Thus the possibility of floods in East and Central England could threaten entire residential areas and result in rapidly rising insurance premiums. The property value rollercoaster in South East Asia has undermined household savings and entire economies. The fall-out from the collapse of the Communist bloc continues to create unpredictable housing policy problems in relation to privatization strategies, new patterns of migration or the bureaucratic complexities of property restitution.

It would, of course, be wrong to overstate the case and suggest a world in turmoil when the reality of everyday life for many is closer to unremarkable continuity rather than dramatic and unpredictable discontinuity. It would be equally wrong to exaggerate the degree of convergence of housing policy in the past few decades. At a broad level there have been parallels in housing policy discourse and direction and in patterns of social and economic transformation in a wide range of countries, as we have sought to show. However, these are always embedded in significantly different cultural contexts, specific histories of urbanization and state intervention and widely varying institutional structures. And to return to the introductory themes of this chapter, it is important to re-emphasize the inevitably rather Eurocentric view of housing change and housing problems which has been outlined. It is important to acknowledge that, if we are indeed entering the Pacific century, we are also entering the urban century. A world population that has grown enormously over the past hundred years is now predominantly an urban population. At a global level, in terms of the sheer numbers of people living in cities and the scale and severity of housing problems, it is at least arguable that *urban* housing problems are both qualitatively and quantitatively much greater than they were in 1900.

What is also very clear is that we are moving away from a concern with housing *per se*, towards a wider concept of where problems and solutions lie. In the immediate post-war period, relatively full employment was being generated as part of the reconstruction process but a decent home which met rising aspirations provided a basis for family and community building and was a trigger for, and receptacle of, domestic consumption. The challenge then was to build the appropriate residential infrastructure. To a degree this has been achieved over the subsequent 40 years, at least in terms of the majority of the population, although significant minorities remain inadequately housed in all developed countries.

The dominant housing problem has now shifted from one of numerical provision to its renewal, repair and improvement, and not just within one tenure but all tenures. This in turn is manifested spatially because areas, regardless of the distribution of tenures within them, tended to be developed at the same time. They also tend to be rooted in the same economic base and, in the 1990s, we have seen the culmination of processes that have conjoined rapid economic change, physical deterioration and social decline. The focus on area regeneration (or whatever it is called) poses a considerable policy dilemma. There is no longer a single solution; rather, it relates to the provision of better housing and infrastructure, the creation of jobs and economic prosperity and the restoration of communities.

Post-war housing solutions have become part of the problem in the sense that mass housing estates created to house wage earners and to meet rising aspirations had, in some cases, become, with the passage of years, holding camps for the unemployed and underemployed and a mechanism for destroying hope and personal esteem. Decline is, however, not simply associated with mass social housing. Residualized populations and areas cross tenures and encompass owners and tenants. Unlike previous epochs, when renewal was the order of the day and the solution was clearance and mass provision, there is no longer the political will or pressure to do this, nor is it seen as a credible way forward. In a number of developed countries, we move into the twenty-first century with an increasingly physically inadequate housing stock alongside a growing marginalized population. The complexity of the solutions is considerable and the scale of the resources required is substantial. Moreover, past models of policy intervention, whether of the *laissez-faire* or direct provision variety, now appear to lack legitimacy and have different but inherent problems. With the collapse of state socialism, the ideological baggage associated with different forms of housing provision may have diminished, but there is a danger that we end up merely wringing our hands in frustration in the face of apparently intractable housing problems. In such a context notions of empowerment and self-help as alternative models of housing provision can become little more than justifications for leaving people to fend for themselves.

For countries such as Britain, with a heritage of the Industrial Revolution and post-war recon- struction, housing is but one element of the way forward. Renewal must now encompass housing, the public infrastructure, the economic infrastructure and the communities themselves. Britain began the twentieth century with the need to bring its social infrastructure in line with its economic capacity. It begins a new century with housing as a continuing constraint, but now as part of a bigger and more complicated set of issues. But in an era of privatized individualism and home ownership, with much more fragmented forms of public provision, and with an economic base no longer driven by large enterprises offering 'permanent' mass employment, this process of renewal is more difficult. Changing economic fortunes will affect patterns of investment and disinvestment in national housing stocks. Some residential areas will be rapidly devalorized and revalorized and be unrecognizable a hundred years from now. Equally, much of the housing being built today is likely remain as a visible imprint of the past, having to accommodate new generations with very different lifestyles and expectations.

BIBLIOGRAPHY

Beck, U. (1992) *Risk Society*. London: Sage.

Bessonova, O., Kirdina, S. and O'Sullivan, R. (1996) *Market Experiment in the Housing Economy of Russia*. Moscow: Padco, Inc. for the United States Agency for International Development.

Bodnar, J. (1996) 'He that hath to him shall be given: housing privatization in Budapest after state socialism', *International Journal of Urban and Regional Research*, 20: 616–36.

Boleat, M. (1994) 'The 1985–1993 housing market in the United Kingdom', *Housing Policy Debate*, 5: 253–74.

Bootle, R. (1996) *The Death of Inflation. Surviving and Thriving in the Zero Era*. London: Nicholas Brealey.

Carter, T. (1997) 'Current practices for procuring affordable housing: the Canadian context', *Housing Policy Debate*, 8 (3): 593–631.

Danielli, J. and Struyk, R. (1994) 'Housing privatization in Moscow: who privatizes and why', *International Journal of Urban and Regional Research*, 18: 510–25.

Department of the Environment (1993) *English House Condition Survey, 1991*. London: HMSO.

Doling, J. (1997) *Comparative Housing Policy*. London: Macmillan.

Doling, J. and Ruonavaara, H. (1996) 'Home ownership undermined? An analysis of the Finnish case in the light of British experience', *Netherlands Journal of Housing and the Built Environment*, 11: 31–45.

Donnison, D. and Ungerson, C. (1982) *Housing Policy*. Harmondsworth: Penguin.

Economic Commission for Europe (1983)*Economic Bulletin for Europe*. Luxemburg: Office for Official Publications of the European Communities.

Fallis, G. (1995) *Structural Changes in Housing Markets: the American, British, Australian and Canadian Experience*. Paper presented at XXI World Congress, International Union and Housing Finance Institutions, London.

FEANTSA (1995) *Homelessness in the European Union Social and Legal Context for Housing Exclusion in the 1990s*. Brussels: European Federation of National Organisations Working with the Homeless.

Forrest, R. and Murie, A. (1994) 'Home ownership in recession', *Housing Studies*, 9: 55–74.

Forrest, R. and Murie, A. (1995) *Housing and Family Wealth: Comparative International Perspectives*. London: Routledge.

Forrest, R., Murie, A., Hawes, D., Bridge, G. and Smart, G. (1995) *Leaseholders and Service Charges in Former Local Authority Flats*. London: HMSO.

Freeman, A. (1997) 'A cross national study of tenure patterns, housing costs and taxation and subsidy patterns', *Scandinavian Housing and Planning Research*, 14: 159–74.

Freeman, A., Holmans, A. and Whitehead, C. (1996) *Is the UK Different? International Comparisons of Tenure Patterns*. London: Council of Mortgage Lenders.

Harloe, M. (1995) *The People's Home? Social Rented Housing in Europe and America*. Oxford: Blackwell.

Harvey, D. (1989) *The Condition of Postmodernity*. Oxford: Blackwell.

Harvey, D. (1978) 'The urban process under capitalism', *International Journal of Urban and Regional Research*, 2(1): 100–31.

Haussermann, H. (1994) 'Social housing in Germany', in B. Danermark, and I. Elander (eds), *Social Rented Housing in Europe: Policy, Tenure and Design*. Delft: Delft University Press.

Kemeny, J. (1980) *The Myth of Home Ownership*. London: Routledge and Kegan Paul.

Kennedy, N. and Andersen, P. (1994) *Household Saving and Real House Prices: An International Perspective*. Working Paper 21. Basle: Bank of International Settlements.

Kosonen, K. (1995) *Pohjoismaiden asuntomarkkinat vuosina 1980–1993. Vertaileva tutkimus (Nordic Housing Markets in 1980–1993. A Comparative Analysis)*. Helsinki: Palkansaasjien tutkimuslaitos.

Lunde, J. (1990) *Boligejernes formuesituation – en empirisk undersogelse*. Institute of Finance, Copenhagen Business School Working Paper 90–10.

Maclennan, D. and Williams, R. (1990) *Housing Subsidies and the Market: An International Perspective*. York: Rowntree Foundation.

Murie, A. (1974) *Housing Tenure in Britain: a Review of the Evidence*. Research Memorandum 30. University of Birmingham: CURS.

Oxley, M. and Smith, J. (1996) *Housing Policy and Rental Housing in Europe*. London: E. & FN Spon.

Robson, B.T. (1979) 'Housing, empiricism and the state', in D. Herbert and D. Smith (eds), *Social Problems and the City*. London: Oxford University Press.

Rossi, P. (1955) *Why Families Move*. Glencoe, IL: Free Press.

Sudjic, D. (1996) 'Metropolis now: Hong Kong, Shanghai, Jakarta', *City*, 1/2: 30–7.

Thomas, R. (1996) *Negative Equity: Outlook and Effects*. London: Council of Mortgage Lenders.

Timonen, P. (1992) *Asuntovelalliset – riskista kriisiin. Esitutkimus asuntovelkojen vuoksi maksuvaikeuksiin joutuneista kotitalouksista* (The housing debtors – from risk to crisis. A pilot study of households in default). Helsinki: Kuluttajatutkimuskeskus.

Turner, B., Hegedus, J. and Tosics, I. (1992) *The Reform of Housing in Eastern Europe and the Soviet Union*. London: Routledge.

United Nations Centre for Human Settlements (1996) *An Urbanizing World: Global Report on Human Settlements, 1996*. Oxford: Oxford University Press.

Van Vliet, W. (1990) *International Handbook of Housing Policies and Practices*. New York: Greenwood.

Whitehead, C. (1996) 'Trends in the provision of housing finance in sixteen countries', *European Mortgage Review*, 3: 18–22.

Winter, I. (1994) *The Radical Home Owner, Housing Tenure and Social Change*. Basel: Gordon and Breach.

7

Transport and the City

TOM HART

Relationships between transport and cities have a long and complex history, yet the greatest intensity of change has been in the period since 1870. In this period, assisted by cultural and technical change, economic growth and population growth has been concentrated on cities and their regions – with the initial focus on the 'Western' world (Western Europe, North America and Australia) spreading to Japan by 1900 and more widely since 1960. The division between an 'active' and 'responsive' role for transport in an urbanizing and commercializing world remains a controversial issue (SACTRA, 1994, 1998, 1999), yet transport has clearly interacted with cities at three levels:

1 long-distance (inter-regional/international) transport;
2 intra-regional transport (including the rise of urbanized regions);
3 transport within cities.

This chapter concentrates on transport within city regions, but no treatment of cities would be complete unless it examined the extent to which longer-distance transport and communication has been, and still is, capable of influencing their economic fortunes. After considering this topic, the chapter outlines the nature of movement within urbanized areas and relationships with urban form. It then examines the evidence relating to total urban movement and changes in modal share. The following section surveys public policy issues and places movement in the context of governance, individual preferences and cultures. Finally, there is a consideration of future prospects.

LONG-DISTANCE TRANSPORT AND THE CITY

In an influential publication, the geographer Christaller (1933) noted that significant parts of the world had already moved towards commercialized rather than subsistence lifestyles before 1800. This involved increased trade yet the constraints of land transport meant that only a few settlements – unless in exceptionally fertile areas – could exceed 10,000 inhabitants. The high cost and slow speed of transport meant that market towns were about 10 miles apart with most of their trade being with their immediate surroundings. The few higher-value items capable of being traded further afield – salt, silk, spice, precious metals – gave a boost to settlements on the relevant trade routes yet they remained of very moderate size. Using the crude definition of a city as a town of substantially larger size, the cities of 1800 were virtually confined to sites of low transport costs on navigable water or to the capitals of nation states and empires able to use state power, rather than the power of trade, to support a larger population (Chandler and Fox, 1974).

Since the seventeenth century, an enterprising Western world had been appreciating the advantages for existing, and potential, settlements of good access to navigable water and to a rapidly increasing range of roads capable of handling horse-drawn, as distinct from horseback, goods and passenger movement (Goodrich, 1961; Hart, 1983; Thirsk, 1978). River canalization and the creation of separate canals and expanding road networks were seen as a means of enhancing, and defending, the commercial status of cities – allowing them to import and export over wider areas while becoming greater centres for manufacturing and service concentration. Transportation improvements began to be used, not only to increase supplies to these cities, but to place them in a stronger position to become principal interchange hubs and specialist manufacturing centres. Inevitably, such cities could not all achieve the same level of commercial expansion. Some lagged in taking the steps necessary to place them on premier routes while others, though spending heavily on long-distance transport links, achieved lesser success than a few cities with highly favourable locations – for example New York, Chicago – or with the ability to develop manufacturing and service specialisms. Glasgow became a very successful manufacturing city *before* it had deep-water port facilities. Prior growth gave confidence to invest in improved links to deep water at Greenock and in a gradual deepening of the Clyde. The first Clyde dock in Glasgow did not open until as late as 1870, more than 120 years after a small west of Scotland town had begun to turn into a thriving commercial city.

This 'chicken and egg' relationship between cities and longer-distance transport has continued to feature in the rail, motorized road and airport eras which followed on from the early modern revival of interest in long-distance transportation as causes and facilitators of city development. Shipping changes have also been relevant as docks, steamship facilities and warehousing developed to meet trading requirements. Certain cities gained as key hubs for interchange between ocean shipping, rail and barge. Yet efforts to generate or divert demand by building expensive new facilities could have disappointing results. The Manchester economy gained from the commercially suspect Manchester Ship Canal but the net gain to the north-west of England was less (Farnie, 1980). The new 1890s canal moved activity from Liverpool to Manchester rather than boosting total activity in the region. Conversely, the much criticized Port of London remained the leading British port throughout the nineteenth century. However, sensitivity to the heavy involvement of continental cities in port improvement was a factor in the early twentieth century nationalization of the Port of London as part of a larger effort to ensure that Britain kept ahead of continental European rivals. Similar arguments on the importance of keeping ahead in airport provision are a feature of the present debate on the provision of a Fifth Terminal at London Heathrow and can also be found in the USA and Canada from the 1930s.

Such views on the importance of long-distance transport for city economies are replicated in the extensive literature concerning trunk roads and city development (Botham, 1983; Leitch, 1978) and a wider range of studies investigating cities, labour market opportunities, comparative advantage and international competitiveness (Aschauer, 1989; EU, 1993, 1997; McKinnon, 1996; SACTRA, 1998, 1999). Less prosperous cities have sometimes seen the early coming of dual carriageway or motorway links to other regions as an essential part of programmes for regeneration, yet such schemes invariably extend the 'catchment areas' and economic power of the already wealthy and well-located cities. Since the 1960s, the appearance of high-speed passenger trains able to average between 100 and 160 mph has produced parallel debates. Should such trains,

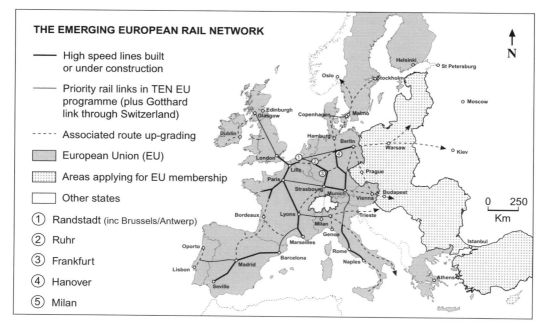

Figure 7.1 Adapted from Executive Intelligence Review (1990)

usually requiring lengthy sections of new route, be provided on well-established inter-urban corridors offering good commercial returns or should route strategy be geared to help improve access to more peripheral cities and regions? Japanese and French high-speed rail policies both reflect this dilemma. Europe offers the added complication that the continuing strength of nation states and their nationalized rail undertakings has inhibited EU efforts to open up a unified and competitive single market for rail transport.

Long-distance rail freight, to a much greater degree than road trucking, has also been inhibited by national rail traditions within Europe. It is only in the past decade that detailed attention has started to be given to developing high-quality cross-border routes on principal international corridors. These are seen as more acceptable than further expansion of trunk motorway capacity. Though some motorway, port and airport proposals remain within the Trans-European Network (TEN) Strategies developed in the 1990s, there has been increased emphasis on major rail infrastructure projects linked with other measures to raise rail efficiency and promote competition and open access on the rail network. New requirements for extensive inputs of private capital and detailed appraisal of scheme costs have led to both a slowing down and longer timescales for the most expensive rail projects. Nevertheless, of the 14 TEN Priority Projects agreed at Essen in 1995, 10 involve major improvements on key European rail corridors over the period to 2010 (see Figure 7.1). Plans now include cross-Alpine routes, expanded capacity from Germany and northern France to the Mediterranean via the Rhone Valley, new freight routes from Rotterdam and Antwerp into Germany and Eastern Europe and Scandinavian links. Work is also proceeding on the identification of further priority improvements into Eastern Europe and, though high-speed passenger route provision continues to have much greater priority than in North America, there has been recent up-rating of priorities for routes of importance for freight.

North America has seen more resilience in the expansion of high-capacity freight corridors, though rail working practices, a semi-monopoly position for many bulk traffics and the existence of strong road competition have hampered major rail inroads into the higher-value freight market. After years of planning and negotiation, it was only in 1997 that agreement was finally reached on a new freight corridor within Los Angeles allowing Pacific trade to be rail handled to and from US cities without the long delays to both road and rail traffic arising from single track access, with multiple level crossings, to the leading Pacific Coast US port.

This Los Angeles example is a particular instance of the ability of apparently congested cities to continue to grow in their world influence despite major transport problems. On balance, they have net advantages over competing and less congested cities. SACTRA has warned of the dangers and ineffectiveness of trying to use ill-considered transport mega-projects either to boost economies overall or to bring benefits to disadvantaged regions. In sharp contrast to the mega-project fascination of Aschauer (1989), the EU White Paper on Competitiveness (1993) and various national and city governments, SACTRA has raised the possibility that future economic growth and welfare may become more dependent on reducing, not increasing, the physical intensity of movement. This has important implications for future relationships between long-distance transport and competing world cities within a global strategy for sustainable development.

There is likely to be a shift of emphasis from increasing physical volumes of long-distance movement towards providing the quality of service needed to meet 'value-added' and environmental requirements – including a large increase in earnings from 'invisible' communication. Since the 1850s, longer-distance communication has included the telegraph, soon followed by the telephone and further transformed by the revolution in the quality and cost of long-distance electronic communications since the 1960s. Initially, cities were best placed to take advantage of these expanding networks but – though cities still have advantages – the widening quality networks of the 1990s have eroded some of the special communication advantages of city living and city working.

Longer-distance transport has had, and will continue to have, implications for city development yet the precise implications for particular cities depend on specific circumstances. On the one hand, improvements introduced ahead of demand can create financial/fiscal problems while damaging the economic prospects of cities with less competitive economies than others on the improved corridor *unless improvements are part of policy packages that address the issue of relative competitiveness and examine the understated total costs and quality of life perceptions of large cities.* (Hart, 1993) On the other hand, prospering cities have often been able to continue to expand with deteriorating quality in longer-distance transport and only moderate investment (relative to city wealth and population) in its improvement. Apart from special cases like the Berlin blockade and other city impacts related to political conditions, there are no examples of the costs and quality of long-distance transport becoming a severe constraint on city growth. Employment in the port areas of cities has suffered from revised labour practices, new shipping/port technology and the ability to shift considerable activity to ports on greenfield sites, but the result has been an overall improvement in distribution of benefit to city regions. Port losses have also been compensated by gains in other job sectors including airport and airport-related employment.

On a broader analysis, it is evident that 'hub and spoke' patterns have been working in favour of larger cities. Small cities cannot justify a wide range of direct links to other world cities; they must depend on interchange to international routes at recognized hubs. This applies particularly to containerized freight and to air travel, with a few international long-distance hubs becoming more prominent, for example, London, Paris, Frankfurt, Schipol (Randstad), Tokyo, Bangkok, Hong Kong, New York, Atlanta, Chicago, Los Angeles. Increasingly, airport hubs in regions of high population are being integrated with high-speed rail networks yet total growth in movement, plus regional aspirations for direct flights or status as intermediate hubs, are opening up opportunities for relative falls in the traffic share of existing major hubs (DETR Air Traffic Forecasts). City competition remains strong in seeking to gain the perceived benefits of direct services and credible hub status. One hundred years after the opening of the Ship Canal from Liverpool, Manchester – with a second runaway now open and a widening range of feeder rail services – is consolidating its position as the hub airport for northern England, leaving Liverpool Airport with the crumbs of charter flights and a handful of scheduled services. Similar jockeying for position will be a feature of the future history of ports and airports in the developing world.

THE NATURE OF MOVEMENT IN URBANIZED AREAS

The City Role in Transport Networks

While it is evident that long-distance transport had a role in city development both before and after 1870, the relationship between transport and movement within cities and within their regions began to change more rapidly from 1870. While long-distance commerce and contacts were important for the economic vitality of some cities, these activities account for a small proportion of total movement in such cities. The majority of cities were much more involved in trade with their immediate hinterlands or adjacent regions. 'Home' markets contributed far more to growth than the hype attached to international trade and, despite the emergence of a more integrated world economy since 1870, the majority of movement –

measured by tonne, passenger and vehicle miles – remains internal to city regions. This is especially true of movement by road, as shipping, air and rail is geared to longer distances.

The scale of internality is affected by geography. In urban regions adjacent to other regions and on through routes, only 60 per cent of movement may be 'internal' but this can rise to 90 per cent in urban regions at least 200 miles from other urbanized zones. Even if the definition of long-distance is relaxed to include all trips over 25 miles, the great majority of urban area movement is internal. This reflects the historical tendency, backed by the realities of travel time and cost, for most trips to be over short distances with the number of trips falling rapidly as trip length increases (and with a greater proportion of these trips being beyond urban limits).

Smaller cities and towns that lie on major corridors of regional and long-distance movement have higher volumes of through traffic with trunk traffic growth (by both road and rail) causing major problems in addition to local traffic growth. Bypasses can ease these conflicts but, the larger the city, the less likely it is that a bypass can solve problems. In hub cities, too many trips are either internal to the city or involve transit traffic which requires to interchange at airports, ports, rail terminals and other warehousing. This distinction between through traffic (using the same vehicle when moving across cities) and transit traffic (involving at least one interchange) is an important one in the nature of movement. Larger cities are characterized both as transit hubs for longer-distance goods and passenger movement and also as hubs of attraction for substantial intra-regional movement arising from patterns of economic and social activity.

Before 1870, many cities were emerging as focal points of regional commercial networks but this did not yet involve substantial commuting. In Western Europe and North America, city populations were rising rapidly but only a handful had reached the million mark. Even in these, as well as in the burgeoning industrial towns and smaller cities, walking and living close to work were the dominant characteristics.

Transport and City Form

Transport is far from being the sole determinant of city form yet human preferences and available technologies have allowed cities to be influenced by the transport options available in their principal phases of development. Early nineteenth century cities were a mixture of congestion and spaciousness with conflicting tendencies for dispersal and concentration. The lure of available greenfield or riverside sites attracted segments of industry and commerce, as well as housing, away from the congested parts of older cities yet the concentration of jobs in, or close to, the more developed parts of cities encouraged rising working-class housing densities in inner cities. These were still walking and handcart cities apart from an affluent few able to travel by other means.

This situation was changing by the 1860s. Incomes were rising, as was security of employment, while technology was allowing cities – especially in North America and Western Europe – to spread more rapidly in physical area than the rate of growth in city population and city incomes. These cities were moving from populations under 100,000 to over 250,000 and into multi-million status (Chandler and Fox, 1974; Weber, 1899) in association with incomes well above the world average. Yet this spending power did not always translate into middle-class suburbanization at low densities. Anglo-America, helped by high incomes and cultural preferences, moved in this direction but continental Europe (and Scotland) produced many examples of higher density middle-class suburbs extending out from inner zones that remained predominantly middle class and resistant for long periods to downwards filtering to lower status occupation. Working-class suburbanization also expanded with the growth of city populations, again with North America having lower densities than in most of Europe. However, these developments often included substantial local employment on new industrial sites, rather than a marked bias towards commuting to city centres.

The coming of the rail commuter, the urban tram, elevated and underground railways and the twentieth century plethora of motorized road vehicles and specialist roads saw a continuation of the tensions between city concentration and dispersal. It has been argued that the period from the 1860s to the 1920s saw the rise of the star-shaped 'mass transit' city, with the city centre and its immediate surrounding becoming the dominant centre of employment and social activity. Residences dispersed, notably in Anglo-America, but city form and patterns of movement were visibly centralizing. The centre capitalized on its high and improving level of accessibility. Building tram and trolley lines ahead of demand was speeding up the spatial expansion of some cities by the late nineteenth century, yet the form of such cities was centralizing, with much development concentrated on what became the central business district (CBD). The fact that the first New York underground (like London's Hampstead tube) had an outer terminal in a greenfield area in no way implied a decentralization of New York. Such routes, and expanding suburban railways, allowed the CBD to capitalize on a high and improving level of accessibility. Elevators and

steel frames allowed shops and offices to move upwards, increasing turnover in alliance with the speed and frequency of access permitted by suburban and regional rail and electric tram and trolley networks. Smaller cities, notably in the USA, often had sufficient existing street capacity to avoid gridlock while large cities could keep inner city congestion at bay through elevated railways, undergrounds and some selective street construction, street widening and the provision of road bridges and tunnels. Except for some sacrosanct central parks, few areas of low density survived in central areas. The norm was to intensify site development.

The weight of the evidence on spatial change and patterns of movement does support city centralization but this process was also being helped by pronounced structural shifts in urban employment away from manufacturing towards services yielding high earnings per employee and requiring effective passenger movement rather than intense flows of goods. Shops did require to receive and deliver goods in even larger volumes, but this could be accommodated within a road network having relatively minor improvements.

Yet by no means all the evidence fits a centralizing interpretation. Dispersal of jobs as well as residences was also occurring for two reasons. First, the expansion requirements of buoyant industries and buoyant city economies meant that manufacturing and distribution were looking for new sites avoiding the inefficiency of city centre and port congestion. Belt railways (built beyond the main zone of urban development) meant that existing rail routes could be linked directly and access provided to new port areas. Goods traffic expansion could be handled away from the congestion of existing facilities and could take advantage of purpose-built marshalling yards and docks employing the latest technology. Large establishments, from manufacturing plant through to hospitals and gas and electricity works, could take advantage of direct rail sidings and of the spare land available in outer areas. Even before 1900, many new private sector industrial zones were appearing in the outer areas of larger cities like Berlin, London, Glasgow, Manchester, New York and Chicago. Goods-intensive industry was already moving away from city centres (Hall, 1962) or locating from the start on the outskirts.

The second reason for dispersal was the sheer size of larger cities and the impracticability of accommodating all services in the city centre. Most basic shopping was still done within walking distance and this also applied to schools, pubs, cinemas and other leisure activities (libraries, parks, swimming pools). There were therefore growing opportunities for local service clusters (often developed from previous town centres absorbed as cities grew) as residences spread outwards. Additionally, there were desires for access not only to local employment, but to new industrial areas. This involved significant reverse-commuting (for example, from working-class housing in Glasgow to new shipyards and the massive new Singer factory seven miles down the Clyde) and growing pressure for improved non-radial links. Examination of the tram networks of larger cities by 1914 shows that they fitted web or grid patterns rather than stars. The heaviest flows were still towards city centres but other appreciable flows had also appeared. The multi-centred city was still alive, though on a scale larger than the old multi-centre cities with focal points around their city wall gateways (portals) as well as in parts of their central area.

The tension between a centralized and multi-centred city is well illustrated by pressure group politics in Chicago before the car's rise to dominance. While city centre business interests favoured upgrading of public transport with a focus on segregated rail access to the city centre, suburban interests were more concerned to improve trolley links to create a denser web (Barrett, 1983). The motor bus, the automobile and the car then proceeded to reshape these tensions in favour of the multi-centred, low density city associated with expressway/motorway grids rather than stars. Expectedly, this shift was most evident in the USA and in those US cities having their principal growth within the motor vehicle era. Nevertheless, the majority, if not vitually all cities have moved in this direction since 1950. Los Angeles, though seeing much development of the city centre between 1880 and 1920, was widely perceived as the pioneer of a new and desirable city form permitting increasingly car-based lifestyles from the 1920s. Like Chicago and other leading US cities, Los Angeles did consider taking up the option of tram subways and undergrounds to reduce congestion around the city centre, but no finance was forthcoming. The popular view, from a city already well known for low residential densities, was that undergrounds would quickly create pressures for apartment development and higher densities along their route (Fogelson, 1967). Actual decisions therefore favoured expansion of a low density multi-centred city and it is only in the 1990s that Los Angeles has come round to the view that some underground routes may be required. From around 1950, airports have emerged as important new centres for cluster development in the multi-centred city. In absolute terms, the city centres of the largest cities have continued to show considerable growth (including extensions into adjacent areas such as Canary Wharf in London, La Defense in Paris

and the area around the Yamanote loop – originally a belt railway – in Tokyo) but, relatively, they have been losing their share of employment and shopping. Smaller cities, notably in the USA, have seen absolute contraction in their city centres with dereliction and deprivation in adjacent zones even though their city regions may be at least stable, or increasing, in population and economic activity. To mix metaphors, in these doughnut cities, the spider's web has lost its centre. This brings home the intensity of a change of form away from the typical mass transit city and it has to be stressed that, even in larger cities with growing CBDs and substantial usage of mass transit, city form has also been changing away from a centralized pattern.

Definitions of Urban Areas and Intra-Urban Movement

Before 1870, the absence of commuting aided the distinction between town and country. The boundary between built-up zones and the countryside seemed self-evident even though, legally, city boundaries tended not to keep pace with physical expansion. Nevertheless, a blurring of town and country was beginning to appear with the rise of improved transport, urban sprawl and some signs of multi-centred tendencies within regions. As Weber (1899), Howard (1960) and Geddes (1915) pointed out, this blurring became a more prominent issue from 1890 as transport electrification and the subsequent expansion of the rubber-tyred petrol vehicle transformed the nature of city regions. By 1910, new patterns of conurbations and city regions accounted for some 70 per cent or more of populations in advanced economies with pronounced increases in regional mobility and wider labour market areas. The new pattern accelerated in the inter-war years and achieved even greater momentum from 1945. Conceptually, the city was replaced by metropolitan regions and travel-to-work areas covering huge physical areas compared to the compact cities of 1870. Are the cities of today to be defined on this basis or are they to be restricted to those areas having distinctly higher densities and more evident built-up features than the rest of their regions? Under the former definition, much regional movement (extending to 200 miles or more on urbanized corridors – including some cutting across national boundaries, for example, France/Germany/Switzerland; Belgium/Netherlands) becomes incorporated in urban movement (Hall, 1969); under the latter, cities would be the more compact areas within urbanized agglomerations. Even on this definition, several cities of 50 miles in width could qualify by 1990 compared to a maximum width of 10 miles in 1870.

CITY MOVEMENT AND MODAL SHARE

Despite problems relating to basic data and comparability between countries (see Appendix), there is no doubt that an explosion of urban movement has occurred since 1870. This process was well advanced by 1915 and was taken to new heights with the coming of mass motorization. Every statistic confirms that mobility was rising faster than city populations. In New York, fare-paying passengers on local transit rose from 40 million in 1860 to over 400 million in 1892 and 1,100 million by 1912. In the smaller city of Boston, the rise was from 25 million in 1869 to 320 million by 1912 (Cheape, 1980). Similar figures from Europe are recorded in McKay (1976). Buses and cars added to this pattern of growth between the wars with the recession of the 1930s slowing rather than reversing the rate of expansion. In Chicago, all trips (now mainly by car) rose by 87 per cent between 1956 and 1970 while Minneapolis saw a growth of 106.2 per cent between 1949 and 1956 (Meyer and Gomez-Ibanez, 1981). Total car miles in the USA rose by a further 60 per cent between 1970 and 1991 with Europe, from a lower base, attaining even higher growth (Pucher and Lefevre, 1996).

Tables 7.1 and 7.2 use the wide range of available data and some gap-filling assumptions to give a necessarily tentative view of the scale of the total growth in urban movement and shifts in modal share since 1870. The tables highlight differences between North America and Western Europe but no attempt was made to create tables for other world regions. Apart from Australasia and Japan, data for these areas is much less extensive, but regions such as India, China and most of Africa still had a very restricted involvement in mass urban movement by private car, even in 1990. These areas have been moving into mass transit but urban mobility is still far below the levels found in North America, Australasia, Western Europe and Japan and considerably behind levels found in Latin America, South East Asia, Eastern Europe and Russia. In preparing tabular information for 1930 and 1990, separate estimates are included for city travel (using the narrower definition of cities) and for travel within more widely defined urbanized zones. The tables reflect available knowledge of the lifestyles, cultures, real incomes, urban layouts, transport systems and travel patterns in Europe and North America in 1870, 1930 and 1990. The columns in italics estimate total movement by residents of cities and urbanized areas – an increasing proportion of which has been beyond such areas, as leisure and tourism horizons have widened. The other columns combine movement by urban residents in their urban areas with movement by visitors to these areas – also a rising proportion over the period since 1870.

These tables highlight the rise in movement in cities and urbanized regions and also the dramatic shifts in modal share. In North American cities, movement per head of population grew by 200 per cent between 1870 and 1930 and by 95 per cent between 1930 and 1990. The parallel estimates of Western European growth are 150 per cent for 1870–1930 and 80 per cent for 1930–90. In urbanized regions, movement per head doubled in North America between 1930 and 1990 while in Western Europe (from a lower base) growth averaged 140 per cent. Since the total population of urbanized regions was growing rapidly, absolute growth was even more spectacular and produced major investments in city and regional transport. The shift in modal share, boosted by the travel generation impact of the car, also brought large increases in car travel. For urbanized regions, travel per head by car rose 360 per cent in North America between 1930 and 1990 and by 1,400 per cent in Western Europe (reflecting the lower base of Western Europe).

This averaged data conceals large variations between cities. Within North America, the western and southern cities of the US were well ahead in the rise of car ownership and car-oriented regions. Several newer cities were already close to the 1990 American average by 1960. By 1990, cars accounted for some 95 per cent of all person movement in North America's urbanized regions. In 1964, Meyer, Kain and Wohl were already claiming that very few US cities had any economic need for mass transit and were adapting

Table 7.1 *Percentage modal share of person movement in cities and urbanized regions 1870–1990*

	Cities						Urbanized regions			
	North America			Western Europe			North America		Western Europe	
	1870	1930	1990	1870	1930	1990	1930	1990	1930	1990
Walk/cycle	87	18	3	91	29	10	8	1	20	4
Other private	4	27	88	4	10	71	42	95	11	81
Public transit	9	55	9	5	61	19	50	4	69	15

Table 7.2   *Average annual person miles per capita in cities and urbanized regions, North American and Western Europe, 1870–1990*

	Total movement by residents	Movement in cities			Total movement	Movement in urbanized regions		
		Residents	Others	Total		Residents	Others	Total
North America								
1870								
Walk	*640*	600	50	**650**				
Other private	*30*	24	6	**30**				
Public transit	*270*	61	9	**70**				
Totals	***940***	685	65	**750**				
1930								
Walk	*400*	360	50	**410**	*350*	320	10	**330**
Pedal cycle	*60*	55	5	**60**	*40*	35		**35**
Other private	*1200*	425	150	**575**	*2000*	1500	220	**1720**
Public transit	*2400*	1000	250	**1250**	*3000*	2000	200	**2200**
Totals	***4060***	1840	455	**2295**	*5390*	3855	430	**4285**
1990								
Walk	*140*	120	20	**140**	*120*	110	10	**120**
Pedal cycle	*30*	20	2	**22**	*20*	15		**15**
Other private	*7500*	3200	700	**3900**	*10000*	7000	900	**7900**
Public transit	*1400*	300	98	**398**	*2000*	300	70	**370**
Totals	***9070***	3640	820	**4460**	*12140*	7425	980	**8405**
Western Europe								
1870								
Walk	*640*	610	35	**645**				
Other private	*25*	20	5	**25**				
Public transit	*145*	25	5	**30**				
Totals	***810***	655	45	**700**				
1930								
Walk	*500*	450	50	**500**	*450*	400	20	**420**
Pedal cycle	*100*	90	5	**95**	*60*	55		**55**
Other private	*300*	150	25	**175**	*390*	270	30	**300**
Public transit	*2000*	900	130	**1030**	*2300*	1400	145	**1545**
Totals	***2900***	1590	210	**1800**	*3200*	2125	195	**2320**
1990								
Walk	*250*	200	50	**250**	*200*	180	5	**185**
Pedal cycle	*60*	50	5	**55**	*40*	35		**35**
Other private	*5000*	1800	400	**2200**	*6400*	4000	500	**4500**
Public transit	*1200*	500	80	**580**	*2000*	780	100	**880**
Totals	***6510***	2550	535	**3085**	*8640*	4995	605	**5600**

Notes: (1) The procedure adopted in preparing the table was to prepare estimates of total miles travelled by city residents, to adjust these downwards in respect of their travel outside home cities and to adjust upwards to take account of visitor travel within cities and urbanized regions.

(2) Reflecting higher incomes, total miles travelled per urban region residents are higher than for city residents. Due to the larger physical area of urbanized regions, total movement within these regions by residents is also considerably larger than movement within cities by city residents but with a shift in modal share away from walking, cycling and public transit. The more scattered nature of the outer parts of urbanized areas and their above average incomes gives particular impetus to travel by car, as drivers and as passengers.

(3) Private transport includes private carriages and horsedrawn cabs in 1870 widening to include non-scheduled motor coaches, private cars, motor cycles and taxicabs by 1930.

(4) Public transit includes steam railways, shipping, ferries and scheduled horse buses and trams in 1870, widening to include electric railways, trolley lines, trams and motorbuses and a small amount of air travel by 1930. Air travel (both domestic and international) was running at much higher levels by 1990, especially in North America.

to handling almost all personal movement in city regions by automobile. Commuting by car was the norm for most city dwellers, while survey after survey indicated that non-work travel had become the strongest sector in movement growth, with very high preferences for car use in urbanized regions. Longer shopping, leisure and other social trips were typical of residents moving into the phase of two and three vehicles per household. Department of Commerce data published for 1990 showed that, in most of the USA, 95–99 per cent of travel to work was by automobile. Between 1970 and 1991, the share of car travel in US personal movement (exclusive of walking and cycling) fell from 86.9 per cent to 80.7 per cent, but this is explained by the rise in domestic air travel from 10.1 per cent to 17.4 per cent. For urban travel, the automobile increased the dominance which it had already established by 1970. Lastly, within North America, Canadian cities tended to be less car-oriented than the US average, though still ahead of trends in Europe.

In Europe and Japan, the conventional view is that urban public transport remained far stronger than in the USA. The statistical evidence supports this viewpoint for 1930 and also confirms a substantial role for walking and cycling. Subsequent evidence, however, points to worldwide involvement in rising per capita car use in urban areas as soon as economic and political circumstances allowed. In Western Europe, car use accounted for almost 80 per cent of all movement in urban regions by 1990. Communism kept car ownership and use in the Soviet Union below normal economic expectations, yet political restraints were being eased from the 1960s and much more rapidly by 1990. Similarly, Communism kept car use in East Germany below the levels of West Germany, yet both areas were moving towards higher use, with East Germany no more than 15 or so years behind the West. This common direction merits more emphasis than any divergences (Hart, 1993, 1994; Pucher and Lefevre, 1996; Simpson, 1987). Despite perceptions of anti-sprawl and pro-public transit attitudes in West European town planning and transport management, income growth, personal preferences and the totality of government decisions allowed the share of car use in urban movement to rise rapidly between 1950 and 1990. Shopping by car became the norm and was associated with multi-centred development of business and retail parks (often with associated leisure and fast food facilities), replicating the American dream. Japan was somewhat more constrained by high population density and limited availability of land for urban expansion, yet here too cars made large gains in their share of movement between 1960 and 1990. The other awakening tigers of South East Asia and a reviving Latin America have

shown every inclination to follow these patterns, while India, China and parts of Africa are now experiencing high rates of growth (though still from a small base) in car ownership and usage expectations.

This process has seen many defensive reactions by city centres to protect their position through combinations of public transport and road access improvements, including relatively cheap facilities for short-term car parking. The Buchanan Report on *Traffic in Towns*, published by the British government in 1963, pointed in this direction but was soon beset by funding problems and community opposition to city motorways. Even at the time of publication, the Report was attacked for an outdated view of urban structure, seeking to perpetuate star cities when the multi-centred approach with spatial expansion and lower densities seemed preferable. Urban motorways and expressways have been built in all advanced countries, yet the costs of building in older cities and the attractions of greenfield sites ensured that most expansion of road capacity was related to the multi-centred urbanized regions.

It made little sense to build high capacity motorways into, and through, city centres when such routes would only be fully used at commuting peaks and when they could deflect lorries from preferred lines of goods movement free of city centre congestion. New alliances of business and government arose which saw greater advantages in more diffuse motorway and principal road improvements opening up greenfield and edge-of-town development prospects. This produced a new generation of beltways which intensified the change in urban structure and, in the end, brought congestion to extending suburbs. The Boston beltway (Route 28) and the M25 around London are among the best-known of these examples, but reshaped urban road strategies around the world were characterized by motorways avoiding city centres but forming part of national networks linking with other roads and with improved access to airports, key ports, greenfield manufacturing, leisure complexes and distribution parks. They became major carriers, and generators, of intra-regional movement by car and lorry.

Freight Movement in Cities and Urbanized Regions

Freight movement has followed a similar pattern to personal movement since 1870, with tonne miles per head of city region population rising and total 1990 volumes well above those of 1870 and 1930. Freight movement in urban areas has been under-researched but, as with person movement, there was some shift towards rail between

1870 and 1930. There was a considerable development of direct sidings into establishments (termed 'private' sidings since they were not available for general use by hauliers) and an expansion of public sidings – notably for coal – allowing local deliveries within urban areas. On the other hand, the inflexibility of rail and a low network density meant that the urban road network had still to bear most of the burden of collection and delivery in addition to handling the many goods trips generated entirely within urban areas. This was a factor in several road schemes, for example, the improvement of river crossings in urban areas and priorities for roads improving access to docks and major rail terminals and associated warehousing. It was significant that the inter-war Mersey Road Tunnel at Liverpool included dock branches.

By 1930, the motor truck was having the dual effect of increasing the attraction of industrial zones suited to lorry movement and of generating new opportunities to use lorries and vans for intra-city and regional movement. Rail was soon losing ground in its share of urban and regional freight movement. This became a rout from the 1960s. Local rail freight virtually disappeared, with extensive closure of both private and public sidings. More efficient and heavier lorries, assisted by improving national road networks, made deep inroads into longer-distance freight and generated their own additional flows to and from edge-of-town and out-of-town sites geared to lorry use and easy motorway access. Central congestion and the physical problems of accommodating larger and longer lorries in city lanes accentuated this process. The rundown of coal deliveries and of raw material-intensive industries (steel, heavy engineering) reduced flows in some cities, but these had been mainly by rail or water. The pressure of goods traffic on city roads therefore increased and became greater as trips lengthened, both for heavy lorries and for the lighter lorries and vans in high demand as city economies moved into higher-value added sectors. Much heavier lorries, arising from the 5–12 tonne range of 1930 to 38–50 tonnes by 1990, reduced the vehicle miles needed to carry any given volume of goods but this reduction was overwhelmed by the total increase in heavy lorry traffic, the rise of 'just-in-time' (JIT) distribution and the buoyancy of the van/estate car sector. Larger boots and the shift to estate/hatchback cars also allowed private cars to become substantial goods carriers in their own right. It became impractical to carry larger volumes of packaged and frozen shopping on foot or by public transport; cars and a new generation of 4-wheel drive private carriers began to carry shopping and items as diverse as furniture, garden rubbish, bicycles and a multiplying range of sports, hi-fi and computing equipment – not to mention babies and their attached paraphernalia! The expanding road network, always important for localized goods movement, had become integral to the vast majority of urbanized area goods movement by the 1980s.

URBAN TRANSPORT, PUBLIC POLICY AND PRIVATE MARKETS

Urban transport offers a unique interface between public policy and private markets. Interfaces exist between other aspects of public policy and markets, as in topics such as education, economic development, unemployment, housing, crime, social cohesion, health and quality of life, but transport impinges on all of these issues and has been a very visible and multi-faceted example of a subject attracting strong views from influential sections of society. The facts of the mushrooming growth of city movement and the emergence of multi-million urban agglomerations reinforced the importance of transport. In examining the resulting interface, a three-strand approach has been adopted. After an overview of shifting relationships between public policy and private markets, governmental structures for handling urban transport are considered and this is followed by categorization of current objectives, issues and responses.

The Relationship between Public Policy and Private Markets

Before 1870, there was already an extensive local government involvement in the provision and regulation of streets. In some cases, this extended to ports and contributions to rail development but it was the private sector and/or central and federal governments which had been more involved in provision of the capital for national rail and port infrastructure which was being superimposed on city transport networks. Central government interest was also extending into the sphere of railway safety and transport tariff regulation.

From 1870 onwards, however, urban governments sought a greater say as city expansion intensified transport needs and raised other issues of social concern. Just one example of the latter was the housing demolitions associated with the extension of major rail lines closer to city centres. This led to an expansion of pressures on central government, the extension of local regulation – for example, tram company franchises, bye-laws on street widths, a growing debate on housing, transit and planning – and direct involvement in tramway construction, tramway operation and improvement of the road and bridge network. Mixes of public and private finance were common but the tendency was towards stronger transport planning initiatives by local government and to a large increase in municipal ownership of local mass transit. From around 1900, the need for wider areas for transport planning became recognized and this was followed by considerable interest in regional land-use planning, further bolstered by the rise of road motorization. Prior local authority involvement in the provision or improvement of urban road arteries led to extensive and continuing involvement in road investment through to the 1980s apart from wartime gaps. Except for the USA, lower car ownership allowed extensive city public transit systems to stay viable into the 1950s and there was expansion of privately operated bus services in the outer parts of urban regions. Nevertheless, transit could not generate sufficient funds to finance investment in rail and subway systems. Operating and capital subsidies became a large element in urban transit finance from the 1950s. In the urban USA, large cuts occurred in urban services yet subsidy requirements rose. Many urbanized areas found themselves without any local transit other than school buses funded from separate sources. In continental Europe, service cuts were less extensive but support requirements rose to 70 per cent or more of total turnover. Britain occupied a middle position with some cuts, higher fares and comparatively modest levels of subsidy.

While income and expenditure accounts made increasing levels of financial support for public transit transparent, no similar accounts existed for road infrastructure. The near unanimous view was that spending on roads was beneficial and that there was no need to apply the accounting techniques used in public transit. Indeed, many local governments would have preferred to invest more in roads, being constrained only by central government rulings or insufficient sources of local income. In both North America and continental Europe, fares came to yield 40 per cent or less of the income required to maintain and improve urban public transit. For a time, there was wide political acceptance of such levels of support alongside high levels of road spending but attitudes began to change from the later 1960s. Set in the context of weakening economic growth and constraints on public spending, a more critical approach to all transport spending began to appear. Stronger proof was sought of the economic and social benefits arising from transit support while major road schemes began to be compared with alternative spending packages and/or reduced taxation. There was disenchantment with the impact of interventionist, high-spending government (both national and

local) and new interest in economic deregulation, privatization, competition and a return to mixes of public and private finance with appreciable transfers of risk to the private sector. Transport has become a leading sector in this general shift of opinion, which has also been linked with the death of blueprint planning and the rise of the partnership approach and Private Finance Initiatives (PFI). Additionally, parts of North America – notably California – became conscious of new types of auto-related air pollution while, by the 1970s, the oil price hikes and rising US imports of oil drew further attention to energy issues and transport.

Governing Structures and Urban Transport

The discussion above reveals tensions between local government and higher tiers. Two of these require elaboration. The first is the tension over finance and control of total public spending. By and large, central government aspired to a larger role in influencing the finance available to local government and actual patterns of spending. In Europe, local government became more dependent on higher proportions of revenue and capital spend coming from central government and financed from national taxation and central borrowing.

Particular urban transport projects and policies therefore became more dependent on the attitudes of national (and federal) governments. While continental European governments have been supportive of spending on all transport modes, the UK government has been less supportive and especially keen to reduce public funding for public transit and to shift major investments – as in airports, ports and the Channel Tunnel – into the private sector. There was a brief British experiment from 1909 with road taxation earmarked for road spending but the Road Fund was soon raided by the Treasury and abolished within 20 years. Following the UK, continental Europe has been moving towards tighter control over both capital and revenue subsidies for transport with a major growth of interest in opportunities for direct road pricing and other transport-related taxation (EU, 1995, 1998; *A New Deal for Transport, 1998*).

In North America, political philosophies allowed a greater role for local taxation and local referenda on borrowing and related tax increases to finance transport projects (Dunn, 1981). The US and Canadian federal governments helped some urban road schemes, treating them as part of interstate highway systems, while the US government gave limited finance towards urban public transport from the 1960s (Meyer and Gomez-Ibanez, 1981). However, most public funding for urban transport has come from state and local government gasoline and sales taxes, often with direct links to the funding of public bonds related to specific schemes.

Though several US state boundaries cut across urbanized areas, states were generally able to take wider views of transport needs than the increasingly fragmented urban government which became typical of twentieth century North America. Lesser problems of fragmentation appeared in Europe but could be eased by central government funding strategies. Even so, led by London and New York, demand increased for structures that could take an overview of transport needs in metropolitan regions. Advisory Regional Plans multiplied in the inter-war period and were sometimes associated with Executive Authorities for Public Transit. The innovative London Passenger Transport Board was created in 1933 but as an instrument of central government, not the existing London County Council. The LPTB covered a larger area than the LCC but the issues of whether such Boards should be off-shoots of central government or responsible to regionally elected councils or representative joint boards has been a recurrent theme (Smith, 1974). A less discussed, but more important, theme now appearing in urban transport debates is whether Regional Boards covering all strategic aspects of land use planning, road provision, road management and public transit are now required. Despite criticisms of planning and the dismantling of upper-tier local government in Britain, there is widespread recognition that, for transport and land use issues, a strong case exists for a strategic approach to encourage regional efficiency and confidence in partnership development (RCEP). The new consensus is that such Boards should not be directly involved as owners of operating companies but views are divided on whether they should be non-elected or imbued with democratic accountability and given direct responsibility for transport charging. Very recently, the UK has become a convert to the regional hypothecation of tax revenues and road pricing for transport purposes. In a significant policy change, Gordon Brown, UK Chancellor of the Exchequer, announced in 1999 that automatic 'escalator' increases in road fuel taxation would be replaced by increases earmarked for transport spending. Enabling powers for local road pricing have also been included in UK legislation.

Changes in structures, boundaries and financing arrangements have played a part in the competency – or lack of it – in the tactical and strategic administration of transport in urbanized regions. None the less, theoretically attractive structures cannot be seen in isolation from the vested interests of established groups; nor can they guarantee political leadership, competent staff and freedom from corruption. In any final analysis, the relationship between structures and actual performance has to be taken into account. This cannot be done without examination of policy objectives in urban transport, the effectiveness of their implementation and monitoring and their relationship with the reality of urban aspirations.

Policy Issues and Urban Transport

The policy issues affecting contemporary urban transport, and possible responses, can be summarized under the broad headings of social issues, economic issues and issues related to integration, policy-making, application and monitoring. With Europe and North America as pioneers, there has clearly been a shift from objectives primarily concerned with economic growth and equating it with increased mobility to a more penetrating assessment of the economic role of urban accessibility and its linkages with social objectives and desires for improvement in the local, regional and global environment. Within this chapter, it is only possible to give a flavour of these shifts in policy priorities and in instruments of implementation.

Economic Issues

Objective Ensure efficient movement covering costs and catering for present and future requirements.

Responses
- Leave to private competitive market.
- Adjust imperfect market by regulation (including possible support for monopoly or protected franchises/partnerships in urban transit operation), by fiscal and charging reforms and by compensating subsidy to cancel out market distortions.
- Adjust boundaries and powers of public bodies to improve quality of decision-taking on urban/regional transport and related issues, e.g. land use policies.
- Develop modelling techniques (covering all movement, not just road traffic) to test future options.
- Improve management of existing urban networks and trunk interfaces, e.g. clearways, parking regulations, priority lanes for buses/lorries/high occupancy cars, controlled entry/exit points, traffic calming, supplementary licences, congestion pricing; intensify rail and airport runway use (with extra terminal capacity where needed); improve peak-period car occupancy and encourage shifts to off-peak periods.

Objective Use transport policies to promote development or to lessen threats to established cities or CBDs.

Responses
- Research/lobbying to decide whether main economic benefit lies in using transport to give boost to region as a whole or to give priority to CBD.
- Compare transport options with other options for aiding growth or regeneration.
- Construct transport facilities and/or raise quality of service 'ahead of demand' in order to boost local/regional economy relative to competing areas.
- Examine partnerships to share costs and risk; compare benefits/costs of 'mega-projects' with larger numbers of smaller projects and promotional activities/training in sectors other than transport.

Objective Integrated transport, land use, fiscal and pricing policies to promote sustainable cities, that is, with long-term economic viability, a high quality of life and reduced emissions.

Responses
- National/international guidelines and research on patterns of urban settlement, transport pricing and taxation most likely to promote sustainable regional economies, environmental quality and increased reliance on renewable and recycled resources (contributing to targets for absolute cuts in local pollution and greenhouse gas emissions).

Social issues

SAFETY

Objective To reduce transport fatalities and injuries per head of population.

Responses
- To provide segregated routes and target blackspots.
- To reduce speeds and influence awareness.
- To improve quality of regulations and improve enforcement.
- To extend pedestrian priorities, 'safe routes to school'.

Objective To reduce transport crime, intimidation and vandalism.

Responses
- To improve lighting and encourage higher pedestrian presence.
- To eliminate blackspots and improve policing, surveillance, awareness.

- To adapt building design and street/path layouts.
- Rapid removal of evidence of vandalism and other 'zero tolerance' approaches.

LOCAL ENVIRONMENT AND HEALTH

Objective To prevent and reverse, deterioration in environmental and health standards due to increasing and more intrusive movement and reduced levels of physical exercise in normal daily routines.

Responses
- Reduce noise at source and at points of hearing, for example, altered surfaces, noise baffles, double-glazing, quieter roads and rail vehicles, quieter engines and revised take-off/landing procedures; bans on night flights, regulation of lorry routes and heavy lorry bans from designated streets; extended use of urban-area tunnels.
- Expand traffic segregation and/or reduce traffic levels and speeds.
- Reduce localized air pollution through changes in vehicle design (reducing emissions from oil-based engines and shifting to vehicles either free of emissions or free at the point of use); supplementary measures to remove gross polluters already in the fleet; effective enforcement of emission regulations; reductions in localized traffic levels (by reducing total traffic or by creation of alternative smooth-flow routes away from pollution blackspots); reduce levels of congestion.
- Reduce stress through the extension of pedestrianization, traffic-calming, park and ride.
- Avoidance of further major road construction generating extra traffic in association with creation of substantially improved conditions for the use of public transit, walking and cycling.
- Reversing health deterioration arising from the combined effects of transport-related stress/noise/air pollution and reductions in regular physical activity arising from car-based lifestyles by educational and travel awareness programmes seeking to include increased levels of walking, cycling and public transport use within daily routines for travel for work, school, shopping and leisure purposes.
- Carrot-and-stick measures to improve the relative quality and lower the perceived costs of public transit, walking and cycling.

THE GLOBAL ENVIRONMENT

Objective Increasing the contribution of urban transport and planning to reducing global damage arising from greenhouse gas emissions and other wasteful (and habitat-destroying) uses of scarce resources.

Responses

- Raise fuel efficiency in transport, expand use of renewable fuels, increase recycling and adopt policy shifts to encourage sustainable lifestyles and settlement patterns with reduced demand for intrusive movement.
- Examine trade-offs where reduced energy demand in sectors other than transport may lessen the need for reduced energy use in transport with investigation of similar trade-offs within transport (main outcome is likely to be use of absolute cuts in car and trucking use to facilitate higher energy consumption in the air sector).

EQUITY

Objective Ensure that transport policies promote equity and social inclusion.

Responses

- Adjust policies to ensure that neither low income, age, sex, location (e.g. in urban pockets difficult to serve by conventional public transit) nor physical disability is an excessive deterrent to access to facilities.
- Ensure that above groups are not disadvantaged in policies to enhance safety and environmental quality.
- Adjust vehicle, building and interchange design to ensure easy access and high standards with those having hearing and eyesight impairment.
- Adjust policies to give greater support to the provision of local facilities within easy walking distance and offering goods and services at reasonable prices.
- Adjust policies to ensure that the terms and conditions of workers in urban transport compare not unfavourably with other urban workers.
- Adjust policies to ensure that richer urban regions contribute more to programmes for national and global equity.

The listing of these policy areas and potential responses demonstrates that reactions to the problems and opportunities of increased urban movement can contain large elements of partial or 'ad hoc' responses and conflict between objectives, for example, increasing transport speeds and capacity can increase the severity of accidents while often generating so much extra use of road space that congestion reappears with traffic speeds falling at peak periods. One example from the numerous cases of the traffic generation phenomenon was the reappearance of congestion on Paris streets within 20 years of the early twentieth-century construction of the Metro system. This example was repeated on the M25 around London and the Paris Périphique in the 1990s. Other recent conflicts have included those between safety, improved air quality and energy conservation cutting CO_2 emissions from transport. In-car safety demands (and improved in-car equipment) have tended to increase car weights while catalytic convertors have reduced energy efficiency (and also have a poor performance in reducing health-threatening pollutants in those first few miles which form the majority of urban trips). Though some progress has been made in cutting units of noxious pollution per vehicle mile in urban areas, the stop–start nature of urban traffic, its continuing increase and only moderate gains in fuel efficiency have resulted in higher CO_2 emissions from urban transport. The 'simple' solution of encouraging a shift from petrol to more fuel-efficient diesel vehicles has produced its own problem of unacceptable increases from particulates damaging health.

On the whole, such conflicts were not serious public policy issues between 1870 and a varied selection of twentieth-century dates depending on the city under study. Between 1870 and the 1950s, there was both conscious and tacit consensus that the economic gains of spatially expanding urbanized areas with improving transport also brought immense social gains and desired lifestyles. Apart from safety and tendencies to congestion, city transport created few social problems. Underpinned by rising incomes, the lower residential densities and improving diet found in the expanding city regions of advanced economies greatly reduced the intensity of former health problems and offered many desirable areas of new housing. As higher-income residents moved outwards, most other residents gained from the trickle-down factor despite pockets of poor housing. In some cities, these 'pockets' were more extensive with several urban and national governments adopting more interventionist policies to provide greenfield working-class housing and accelerating the demolition and renewal of inner city slums. In certain cities, poor links between greenfield housing, ancillary facilities and city transport caused social problems but the dominant trend was to integrate transport and land use developments (Gallion and Eisner, 1963). More mobile urban residents and commercial firms released from the constraints of horse-drawn transport were the integrated expression of both economic and social gains.

With respect to safety, there was initial deterioration caused by unfamiliarity with higher speeds and the mixing of pedestrians with faster-moving vehicles which could also collide with each other. In larger cities, the ability to segregate higher speed traffic, firstly on reserved route railways and, increasingly from the 1930s, also on special purpose roads, worked to reduce accidents per mile of urban movement. But the most spectacular achievement was the marked reduction of accident rates per miles travelled on multipurpose urban roads. This reflected the adjustment of walking standards and driver behaviour to the new conditions of higher speeds plus the impact of an extended range of research and policy measures directed towards improving safety. Despite the huge increase in the volume of urban road movement since 1930, many cities have achieved – not only cuts in accident rates per miles moved – but absolute reductions in urban road deaths and injuries. The record of North American, UK and Scandinavian cities has been particularly impressive. Remaining debates affect issues such as 'could do even better' and concerns at inequity in the distribution of accident savings. Advocates of the health benefits of walking and cycling have noted that the fall in casualties for these modes has reflected a collapse in the relative attraction of walking and cycling rather than large cuts in accident rates per urban mile walked or cycled.

Safety improvement was evident by the end of the 1930s and became yet more prominent from the 1960s. Other social gains evident by 1930 included the disappearance of horse droppings from urban streets, the spread of asphalt in place of noisy and uneven stone sets and the apparently fume-free and quiet nature of motorized and rubber-tyred road traffic. Cross-subsidy also allowed regulated urban tram, rail and bus services to meet equitable objectives without need for direct subsidy.

This fortunate coincidence of economic and social objectives was to come under greater threat as car and lorry traffic intensified after the Second World War. Extensive and low-cost public transit became less feasible without direct subsidy while vocal claims were made that the apparent economic benefits of increasing mobility in urbanized regions were eroded or eradicated if proper allowance was made for the full costs of movement. The anti-car camp argued, not only for full evaluation of total costs and benefits, but also for a higher political weighting for equity and for costs and benefits not readily converted to a monetary basis. Their preference was for compact cities rather than sprawling urban regions (Sherlock, 1991). Professional support for this line of argument came in the wake of the massive oil price hike of the mid-1970s, more cautious views of the ability to apply 'technical fix' and rising national and city treasury concerns that public funding for transport was becoming unsupportable. Even where there was an apparent surplus from urban transport taxation, it did not follow that city and national economies would achieve their greatest gains from ploughback of this

funding into major roads. A green/gold alliance (Goodwin, 1992 and 1993) of environmentalists and economists meant that other policy options, including lower taxation and public borrowing, as well as redirected taxation, gained in attraction (Mulgan and Murray, 1993). Though oil prices fell back in the 1980s, the momentum for reassessment of the role of transport in urban areas was maintained by strengthened concerns about localized pollution, global warming, threats to biodiversity and the medium- to long-term economic threats to a sustainable world economy from failure to reduce population growth and raise efficiency and recycling in the use of finite physical resources (Rio Earth Summit, 1992). *Traffic degeneration* (reducing the need for intrusive movement) made its appearance as an issue (Adams, 1992) while some governments and businesses began to find political and economic benefits in being more supportive of green agendas. In 1990, an advisory report to the Dutch government (*Netherlands Travelling Clean*, 1990) urged joint improvement of the environment through absolute cuts in car use while actual changes of direction, rather than talk, began to enter transport programmes in Germany, Scandinavia, Switzerland and the UK.

The issue of whether such developments had produced radical new trends in urban transport by 1990 is much more open to debate. Driven by popular sentiment and aspirations for economic growth, developing countries seek higher mobility while North Americans have been reluctant to move from car-dependent urban lifestyles. Though some absolute revival in North American urban transit has taken place since the 1970s, the automobile continued to increase its already high share of urban movements during the 1980s. After a slight fall in the 1970s, in the USA the average annual kilometres per car rose from 14,080 in 1980 to 17,100, by 1991, while total car ownership also continued to grow. The bulk of this growth was in sprawling and decentralized urbanized regions. With increased total populations in such regions and problems in expanding expressway capacity on some corridors, US urban transport commentators became a little more optimistic about prospects for public transit growth but were highly critical of the possibility of any significant shift from cars to public transit (Meyer and Gomez-Ibanez, 1981). In particular, there was strong criticism of the ability of both revived urban rail investment and operating subsidies to have any marked influence on urban modal share. Measures to expand bus and rail usage attracted rather less criticism yet were seen as peripheral to the main urban agenda. Equity has never been a prominent consideration in US policy and the perceived central issues of tackling air pollution and congestion were seen as being made tolerable through emission legislation and pushes towards car sharing and reserved lanes for buses and high occupancy cars at peak periods. The road system continued to have ample capacity for accommodating off-peak traffic and it was such traffic which was exhibiting the highest rates of growth in car use. Proposals to hike gasoline taxes to European levels or to move towards electronic pricing on congested roads were so unpopular that they received little attention. A small revival of toll roads became apparent in the USA during the 1980s but principally in the sphere of inter-urban traffic. Substantial proportions of 'gas' taxes continued to finance road improvement and longer-term concerns over oil costs and global warming were seen as containable through big increases in automobile fuel efficiency, the development of alternative fuels and some substitution of movement by information technology.

In contrast, policy rhetoric in Europe saw greater scope for integrating all policies affecting transport in ways which could reduce the demand and need for intrusive movement while attaching greater weight to the importance of social inclusion (EU Citizens Network, 1995; UK Consultation on Integrated Transport, 1997; Transport White Paper, 1998; SACTRA, 1999; UK Consultation on Climate Change Programme, 2000; RCEP Report, 2000; Transport 2010, 2000). These views can be traced to the more interventionist traditions of Europe, higher population densities, more constrained expansion opportunities around cities and the seminal thinking of Ebenezeer Howard and other writers from the 1890s. Offended at lengthening trips to work and the social and economic inefficiency of the monster city of London, Howard argued in his *Garden Cities of Tomorrow*, that planned satellite towns for resident, work and leisure could produce a better urban region with less need for movement. In reality, his concept was hijacked and became the automobile-based vision of a multi-centred urban utopia with the relative self-sufficiency of satellite towns being replaced by extensive cross-commuting throughout urban regions. Yet the essence of his idea was to accommodate urban growth and social improvement by arranging transport and land uses to reduce the need for movement. This concept resurfaced in the 1930s when London Passenger Transport Board's plans for a general outwards expansion of commuting into London came in conflict with the notion of a Greenbelt which, as part of other objectives, would restrict the growth of trip lengths.

Post 1945, the continued desires for city expansion either involved rejection of city greenbelts or a leapfrogging of greenbelt boundaries. In both cases, the pro-mobility ethos prevailed and it was not until the 1970s oil crises that there was a revival of the view that land use and transport planning could be redirected to reduce the need

for movement (*Transport Policy*, British Transport White Paper, 1977). Nevertheless, pro-mobility views were soon in the ascendant again in Europe with acceleration of fringe-of-town and out-of-town developments and with Britain and France taking the lead. Not till the 1990s was the topic of helping the economy and society by reducing the need for urban movement revisited. Since the Rio and Kyoto Earth Summit, there has been a greater emphasis in Europe on establishing practical guidelines for transport and land use planning, reducing the need for movement and reinforced by measures to change fiscal and pricing strategies to reflect commitments to sustainable economic expansion and enhancements in urban quality of life. There has been a hardening of attitudes towards greenfield encroachment while, at the same time, attention has been given to positive measures to enhance existing city and suburban centres, bring brownfield sites back into use and encourage greater reliance on public transit, walking and cycling. The legacy of many European developments in the American direction, business attitudes supportive of space (including parking space) and many continuing incentives for individual use of cars have diluted the strength of changing policies. Nevertheless, compared to US urbanized regions with as little as 5 per cent of movement by green modes, Europe retains many cities and relatively compact regions where green modes account for 20 per cent of all person movement with higher shares in the inner areas of compact cities and higher still shares for peak travel to and from city centres. Despite considerable convergence, real differences remain in the movement patterns of North American and European cities and regions.

CONCLUSIONS

The principal conclusion from this study of urban transport since 1870 is the emergence of the feature of urban movement increasing faster than rates of growth in regional incomes in a period when cities themselves (including their surrounding zones) were forming a rising proportion of total national populations in Europe, North America and Australasia i.e. a *growth of transport intensity* within such cities and in their external links, especially in relation to personal movement. Five subsidiary conclusions can also be drawn. These are:

1 The linking of consumer preferences and technology to give an unparalleled increase in total movement, average trip lengths and in movement per head with related spatial expansion and market intensity within

integrated urbanized regions of far greater scale than in 1870.

2 The continuing importance for cities and regions of an ability to 'trade' with the wider world, including a growing emphasis on trade in tourism, other services and electronic communication.

3 The movement generation impact of the ability of several relatively backward regions of 1870 to begin to emulate, and sometimes outpace, those cities and regions already economically advanced.

4 The need to balance the evidence of city centralization (and growth of CBDs) against the substantial data, becoming more telling in the age of the auto and truck, revealing dispersal of activities within urbanized regions.

5 A shift in the balance of governmental concerns towards regional structures and away from a narrowly defined growth agenda to greater involvement in regional competitiveness and in social and environmental issues – of which many have a transport dimension.

These apparently clear conclusions conceal considerable variety within, and between, urbanized regions and considerable disputes on future trends and on the relationships between urbanization and transport of urbanization. The data on urban movement and modal share confirms substantial variation by broadly-defined world regions with peaks of car use in North America and Australasia compared with lesser levels of car intensity in Western Europe and Japan down to very low levels of car use in many African, Indian and Chinese cities. Table 7.3 indicates varied patterns in Europe and North America.

While geography and variations in real income influence Table 7.3, it also points to the importance of other factors – such as the precise nature of urban forms, culture and fiscal/pricing arrangements – in producing variations between countries. For example, Germany – with higher average incomes and higher car ownership than Britain – has appreciably lower use of cars and higher use of public transport. Higher income, but lower density Sweden also emerges with lower car use than Britain, slightly higher rail use and substantially higher bus use. Though not apparent in Table 7.3, other evidence (Pucher and Lefevre, 1996) confirms that North American cities are far from homogenous. Greater levels of public transport use, walking and cycling exist in Canadian cities and in selected US cities such as New York, Boston, Chicago, San Francisco and the Pacific North West.

Further research on these topics is needed to improve understanding of how policies might be adapted to encourage less movement and a decoupling of traffic growth from economic growth. SACTRA (1998) has described this issue as that of *reducing transport intensity* (that is, movement relative to GDP), while Adams calls it a process of *trip degeneration* (that is, a reduction in average trip lengths within, and to and from, urban areas). This thinking also has connections with reappraisals of national accounting practices to reflect resource inputs and quality of life considerations. The literature is this area has become extensive (Atkinson, 1997; Banister, 1998; Pearce and Warford, 1993; Whitelegg, 1993), with conclusions that present practices make insufficient allowance for resource depletion and quality of life evaluations. This new research agenda is receiving attention in North America as well as in other world regions. The SACTRA Final Report (1999) was more cautious about the value of the concept of 'transport intensity', but did conclude that city and city region economies (and their environmental quality) could gain from policy measures to reduce road traffic at congested times. The UK government has accepted this view (Transport 2010, 2000).

Table 7.3 *Estimated annual passenger kilometres per person, North America and selected European countries, 1992–1995*

	CANADA	USA	GB	FRANCE	GERMANY	SWEDEN	SWITZERLAND
Car	18,130	17,032	10,088	11,311	8,666	9,202	10,729
Bus	122	681	757	736	829	1,258	814
Rail	51	38	505	1,017	753	663	1,728
Population per km^2	3	28	247	106	228	20	169
Cars/1000 people	486	513	374	430	489	404	452

Note: These figures are for total movement by car, bus and train in each country but, since air has become the main means of longer-distance movement, they can be used as broad proxies for urbanized zone movement subject to some deductions to reflect rural travel.

Source: M. Bunting, 'Developments in Canadian passenger transport', *Transport Economist*, 25, Spring 1998

Concepts and Future Trends

While local factors will continue to offer variety in urban forms and urban movement, this new research is part of an evolving debate on the nature of world urbanization and the global economy. For some, there is a new logic and morality requiring a shift towards denser and less energy-intensive urban areas in the 'advanced world' bringing an equitable convergence in the transport patterns and lifestyles of sustainable world cities. Others see such a process, if taking place at all, occurring over a protracted time period with many continuing divergences between Europe, North America and the rest of the world. There is no consensus on the pace of urban transport trends within a globalization of economic, environmental and political issues.

North America

North America stands out as an area with a strong attachment to political systems, attitudes and uses of technology allowing continued high levels of urban car use. The modified American dream, to which many in Europe and the rest of the world also aspire, is of an ability of technology and open markets to respond to the challenge of sustainable development in ways allowing further spatial expansion of urbanized and commercialized zones so that the vast majority of national populations could be accommodated in decentralized, auto-based, low density urban forms far removed from the typical town and city of 1870. The 'market' would be influenced by strong doses of environmental regulation, bringing absolute reductions in emissions in America without any need to reduce movement but also including 'trade-off' permits allowing the USA to meet international targets by helping to accelerate reductions in other countries.

Despite present US (and Australian) reservations about accepting steeper targets for cuts in greenhouse gas emissions, it seems probable that, at some stage, the US will accept that its own economic advancement and world competitiveness will depend on greater advances in energy efficiency, the development of alternative fuels and resource conservation. This may involve new measures to prevent gains from fuel efficiency being lost by further growth in movement, yet such measures need not imply radical change in a US urban pattern that is already well established and can only be changed incrementally. Following the 1970s oil crises, the US did use legislation to cut oil consumption in surface transport by introducing higher standards of fuel efficiency for new cars and by enforcing 55 mph speed limits. This led to an improvement in average mpg from 13.5

in 1970 to 21.7 by 1991. Weaker policies in the 1990s and further urban sprawl eroded these gains, yet new research to expand alternative fuels and to achieve an average of 60 mph with conventional fuels has been seen as removing the need for radical action to restructure urban areas towards greener modes and less movement (Rocky Mountain Institute research quoted in OECD, 1996).

Coupled with extended car-sharing, market-based public transport improvements, IT developments reducing the need for movement, stabilized populations and some shifts to shorter work and other trips within decentralized conurbations, this scenario suggests a near stabilization of road traffic levels in most North American urbanized areas within the next 20 years but no great change in present urban forms. In any case, car ownership and use is already so high that there is no scope and no economic justification for repeating the level of North American car growth achieved between 1930 and 1990. Under the modified American dream, average modal share for cars in urbanized North American regions may fall from 95 per cent in 1990 to around 80–85 per cent by 2020, though with total urban car miles being higher than in 1990. This, however, is still a substantial change from 1930–90 trends, reinforced by a stabilizing population. Air miles, between cities and to tourism areas, would continue to rise.

Europe

In Europe, prospects for larger shifts away from urban car miles and car use seem more likely for a variety of reasons, including resurgence of a European 'urban' tradition and legacy which has remained despite the attraction of car-dominated American lifestyles. Unlike America, there is now greater interest in reducing urban vehicle miles and in reshaping land use policies to encourage shorter trips and higher shares of movement on foot, by cycle and by public transport.

While the *Netherlands Travelling Clean* target of a 75 per cent cut in car use over 25 years (1985 to 2010) looks unattainable and lacking in a dynamic view of technology, other research has confirmed the feasibility of helping city economies and raising their quality of life through 20 per cent cuts in car vehicle miles over 12 years (Sustainable Transport Study for Aberdeen, 1998). The Interim and Final Report of SACTRA (1998, 1999) to the UK government has stressed the economic and social advantages of quick action to reduce congestion by raising the marginal cost of road use. There has also been clear evidence of a new impetus within Europe to introduce higher charges for urban road use, with most of the proceeds becoming locally available for projects other than increases in road capacity. Heavy emphasis has been placed on public transport improvement, on green travel plans, brownfield development (raising urban densities) and on careful links between transport networks and greenfield development to ensure higher reliance on public transport, walking and cycling than in many developments typical of the 1960–95 period. Home delivery services and the rise of Internet shopping are also likely to affect the volume of shopping movement by car.

Lastly, the value-added rather than resource-quarrying base of the European economy and the much higher population density than in North America has created greater awareness of environmental issues and of the possibility of growth via business and government strategies stressing value-added, resource conservation (including steeper cuts in greenhouse gases) and recycling. The 'Knowledge Age', with a huge expansion of electronic communication, is eroding the pattern of ever-increasing physical movement which characterized the period from 1870 to 1990.

As in North America, urbanized regions are likely to increase their share of a stabilizing total population to around 90 per cent, but in the European case, these urban regions will account for 90 per cent or more of the land area of many countries. Yet the character of these urbanized regions seems likely to diverge, rather than converge, with North America. The regions will continue to include commuting, shopping and leisure hinterlands around the built-up areas but with a higher proportion of 'urban region' populations in relatively compact settlements and considerably greater reliance than in America on improved and expanded public transport for trips to and from the hinterlands and between overlapping urban regions. Within built-up areas, urban design and transport networks will also be much more favourable to walking, cycling and public transport use with an expected development of relatively self-contained 'city villages' (with housing, employment, shopping and leisure) within larger cities. Some expansion of 'car-free' and 'car-reduced' housing is already evident.

The extensive research conducted for the UK Climate Change Programme (2000) and the latest RCEP Report (2000) have concluded that transport is a sector of major importance in cutting CO_2 emissions. The RCEP stresses the need for cuts of around 60 per cent by 2050 but avoids specifying what this might mean for passenger miles. Further research on alternative fuels and energy efficiency will affect outcomes but indications are that it will prove easier to stabilize air travel and cut longer-distance car trips than trips within urban areas. Car mile growth in Western Europe has fallen to around one per cent a year but growth in several urbanized areas has been higher, exclusive of city centres. Even with direct road pricing applied, the car retains powerful attractions for many shorter trips with reduced car use being influenced by the quality (and perceived cost) of urban alternatives. Even so, the differing character of European and American city regions makes it more probable that the former may achieve 15 to 20 per cent cuts in car use by 2020.

While urban air quality and global commitment to greenhouse gas reductions will be significant considerations in European policies, the prime motivation in West European change is likely to be a consumer preference for a higher quality of urban life with less intrusive road traffic and an ability to deploy both private and public funding away from car (and, to a lesser extent, lorry) movement towards spending, giving greater economic and social benefits. With major new urban roads largely removed from the political agenda, measures to control and reduce existing levels of urban and inter-urban road congestion seem certain to gain more attention. Working and shopping close to home (or in the home using electronic technology) is likely to become more important as a travel substitute than in the American case while, helped by higher population densities and political attitudes, there are better prospects for attaining the quality of innovative public transport which will both generate extra usage and attract car users. Similarly, the prospects for reversing the decline in walking and cycling are better than in America.

On the other hand, Europe is already far closer to American levels of car use than most of the rest of the world. A modified car (see OECD, 1996) will retain many important advantages within towns as well as a much larger role in rural areas. Given the sustained application of policies now being applied in Western Europe, the balance of probabilities – with a favourable wind from public opinion – is that urban vehicle miles by car will stop rising within the next ten years and then begin to fall. With a shift to shorter trips and greater reliance on walking, cycling and public transport, car modal share in urban regions could fall from a West European average of 80 per cent in 1990 to around 60–65 per cent by 2020. As in North America, particular cities and their inner areas are likely to have lesser levels of car model share.

Outcomes

The complexity and variety of factors affecting urban movement make it impossible to come to precise conclusions about future trends. Differences will remain within, and between, Europe and North America while considerable increase in car use seems inevitable in the substantial parts of the urban world still having low levels of ownership. However, the great explosion of total movement, of car use and of total urbanized populations which has characterized Europe and North America since the 1870s is now over, at least in the more developed world. The future cities of the advanced world will see lesser levels of growth in urban car miles with an increasing possibility that some cities, notably in Europe, may have less car movement by 2020 than they had in 1990.

APPENDIX: DATA ON URBAN MOVEMENT AND MODAL SHARE

The amount of data relating to movement is both immense and frustrating. As with movement itself, the volume of statistical information increases rapidly from 1870, covering public transport trips and some sample surveys of road flows. Information on road vehicle ownership has also expanded since the 1920s while computerization and automatic road counts have led to a huge multiplication of data on road vehicle flows since the 1960s. However, there are major difficulties in isolating the volume and nature of movement within urban areas. These are of seven types:

1 Public transport data is often for trips and, due to the complications of flat fares and season tickets, it can be difficult to identify particular flows. Direct information on motivations behind movement is lacking or restricted by commercial confidentiality.
2 In general, data on the volume and purposes of cycling and walking has been very poor (with exceptions in countries such as the Netherlands and Denmark).
3 Road data has been concentrated on road flows (with some distinction between cars, buses and goods vehicles) with no information on load factors or journey purpose.
4 Sample surveys of origins and destinations have produced more detailed corridor results but have not been geared to provide data on movement within and beyond urban areas; samples relying on travel diaries kept by household residents also omit travel by visitors/tourists and tend to understate business travel.
5 Differences in data collection and in definitions as between countries create problems in comparisons.
6 Within countries, data do not lend themselves to adjustment to encompass total movement within different definitions of urban areas, for example, city, urban region.
7 There are considerable shortfalls in research

seeking to probe the motivations for movement and assessing the possible impact of policy changes and changing public preferences on the length of trips, trip timings and modes selected.

Since the early 1990s, the bias of research and surveys has been shifting to motivational and policy issues with a related increase in advance evaluation and subsequent monitoring of the impact of policy changes. This has led to downward revisions of overall forecasts of road traffic growth with conclusions contrasting prospects for stabilizing and reducing road traffic levels in 'traditional' towns and cities with the much greater difficulties faced in areas of recent suburbanization.

Despite these difficulties, sensitive interpretation of existing aggregate and particularized data does enable a reasonably accurate impression of changing trends in urban movement to be presented. The changing directions of research and data collection are likely to improve the quality and value of the future information base on city, regional and trunk movement.

BIBLIOGRAPHY

Adams, J.G.U. (1992) 'Towards a sustainable transport policy', in J. Roberts et al. (eds), *Travel Sickness*. London: Lawrence and Wishart.

Aschauer, D.A. (1989) 'Is public expenditure productive?', *Journal of Monetary Economics*, 23.

Atkinson, G. et al. (1997) *Measuring Sustainable Development: Macroeconomics and the Environment*. Cheltenham: Edward Elgar.

Banister, D. (ed.) (1998) *Transport Policy and the Environment*. London and New York: E & FN Spon.

Barrett, P. (1983) *The Automobile and Urban Transit: The Formation of Public Policy in Chicago, 1900–30*. Philadelphia: Temple University Press.

Botham, R. (1983) 'The road programme and regional development (in Britain)', in K.J. Button and D. Gillingwater (eds), *Transport, Location and Spatial Policy*. Aldershot: Gower.

Buchanan Report (1963) *Traffic in Towns* (official report commissioned by UK Government). London: HMSO.

Bunting, M. (1998) 'Developments in Canadian passenger transport', *Transport Economist*, 25 (Spring).

Button, K.J. (1993) *Transport, the Environment and Economic Policy*. Cheltenham: Edward Elgar.

Chandler, T. and Fox, G. (1974) *Three Thousand Years of Urban Growth*. London: Academic Press.

Cheape, C.W. (1980) *Moving the Masses: Urban Public Transit in New York, Boston and Philadelphia, 1880–1912*. Cambridge, MA: Harvard University Press.

Choay, F. *The Modern City: Planning in the Nineteenth Century*. London: Studio Vista.

Christaller, W. *Die zentralen Orte Suddeutschland*. Jena (1933). See J.H. Johnson (1967) *Urban Geography*. Oxford: Pergamon.

Climate Change Programme (2000) *Consultation Document*. DETR, London: HMSO.

DETR (Department of Environment Transport and Regions) (1997) *Revised Air Traffic Forecasts for UK*. London: DETR.

Dunn, J.A. (1981) *Miles to Go: European and American Transportation Policies*. Cambridge, MA: MIT Press.

EU (European Union formerly EC) (1990) *Green Paper on the Urban Environment*.

EU (European Union formerly EC) (1992) *The Future Development of the Common Transport Policy*. COM(92) 494 final.

EU (European Union formerly EC) (1993) *White Paper on Growth, Competitiveness and Employment*. COM(93)22.

EU (European Union formerly EC) (1995) *Citizens Network Green Paper* [dealing with urban transport revival].

EU (European Union, formerly EC) (1995) *Towards Fair and Efficient Pricing in Transport*. COM(95) 691 final.

EU (European Union, formerly EC) (1997) *The Likely Macroeconomic and Employment Impact of Investment in Trans-European Transport Networks*. Staff Working Paper.

EU (European Union, formerly EC) (1998) *Developing the Citizen's Network* [urban public transport]. COM(198) 431 final.

EU (European Union, formerly EC) (1998) *Fair Payment for Infrastructure Use: A Phased Approach to Transport Infrastructure Charging in the European Union*.

Farnie, D.A. (1980) *The Manchester Ship Canal*. Manchester: Manchester University Press.

Fogelson, R.M. (1967) *Fragmented Metropolis: Los Angeles 1850–1930*. Cambridge, MA: Harvard University Press.

Foster, M.S. (1981) *From Street car to Superhighway (USA)*. Philadelphia: Temple University Press.

Gallion, A.B. and Eisner, S. (1963) *The Urban Pattern: City Planning and Design*. London: Van Nostrand.

Geddes, P. (1915) *Cities in Evolution: An introduction to the Town Planning Movement*. Williams and Norgate: London.

Goodrich, C. (1961) *Canals and American Economic Development*. New York: Columbia.

Goodwin, P. (1992) *Transport: The New Realism*. Oxford: Transport Studies Unit.

Goodwin, P. (1993) 'Confronting traffic growth', in P. Stonham (ed.), *Local Transport Today and Tomorrow*. London: Local Transport Today.

Hall, P. (1962) *The Industries of London since 1961*. London: Hutchinson.

Hall, P. (1969) 'Transportation', *Urban Studies*, 6(3): 408–435.

Hall, P. (1988) *Cities of Tomorrow*. Oxford: Blackwell.

Hart, T. (1983) 'Transport: the historical dimension', in R.J. Button and D. Gillingwater (eds), *Transport, Location and Spatial Policy*. Aldershot: Gower.

Hart, T. (1992) 'Transport, the urban pattern and regional growth', *Urban Studies*, 29: (3/4): 483–503.

Hart, T. (1993) 'Transport investment and disadvantaged regions: UK and European Policies since the 1950s', *Urban Studies*, 30 (2): 417–435.

Hart, T. (1994) 'Transport choices and sustainability: a review of changing rends and policies', *Urban Studies*, 31.

Howard, E. (1899) *Garden Cities of Tomorrow: A Peaceful Path to Real Reform* (reprinted Faber, 1960).

Integrated Transport (1997) UK Government Consultation Paper, London: HMSO.

Knox, P.L. and Taylor, P.J. (1995) *World Cities in a World System*. Cambridge: Cambridge University Press.

Leitch Report (1978) The Report of the Advisory Committee on Trunk Road Assessment. London: HMSO.

Lloyd, P.E. and Dicken, P. (1990) *Location in Space*. London: Harper and Row.

McKay, J. (1976) *Tramways and Trolleys: The Rise of Urban Mass Transit in Europe*. Princeton, N.J: Princeton University Press.

McKinnon, A. (1996) 'Competing from the Periphery', Paper presented to Dublin Conference, Ireland.

Meyer, J. and Gomez-Ibanez, J.A. (1981) *Autos, Transit and Cities*. Cambridge, MA: Harvard University Press.

Meyer, J., Kain, J. and Wohl, M. (1964) *The Urban Transportation Problem*. Cambridge, MA: Harvard University Press.

Mogridge, M.J.H. (1990) *Travel in Towns: Jam Yesterday, Jam Today and Jam Tomorrow?* London: Macmillan.

Mulgan, G. and Murray, R. (1993) *Reconnecting Taxation*. London: Demos.

Netherlands Travelling Clean (1990) Ultrecht: Netherlands Agency for Energy and the Environment.

New Deal for Transport – Better for Everyone (1998) UK White Paper, Cm 3950. London: HMSO.

OECD (1996) Proceedings of conference 'Towards Sustainable Transportation' held in Vancouver, 24–27 March.

Pearce, D. et al. (1989) *Blueprint for a Green Economy*. London: Earthscan.

Pearce, D. et al. (1993) *Measuring Sustainable Development*. London: Earthscan.

Pearce, D. and Warford, J. (1993) *World without End: Economics, Environment and Sustainable Development*. Oxford: World Bank/Oxford University Press.

Pucher, J.R. and Lefevre, S. (1996) *The Urban Transport Crisis in Europe and North America*. London: Macmillan.

Roberts, J. et al. (eds) (1992) *Travel Sickness*. London: Lawrence and Wishart.

Royal Commission on Environment Pollution (RCEP) (1994) 18th Report: *Transport and the Environment*, Cm2674. London: HMSO.

Royal Commission on Environment Pollution (RCEP) (1997) 20th Report: *Transport and the Environment: Developments since 1994*, Cm 3752. London: HMSO.

Royal Commission on Environmental Pollution (2000) *Energy: the Changing Climate*. London: HMSO.

SACTRA (UK Standing Advisory Committee on Trunk Road Assessment) (1994) *Trunk Roads and Traffic Generation*. London: HMSO.

SACTRA (1998) *Transport Investment, Transport Intensity and Economic Growth: Interim Report*.

SACTRA (1999) *Transport and the Economy*: Final Report. HMSO for DETR: London.

Sherlock. H. (1991) *Cities are Good for Us*. London: Paladin.

Simpson, B.J. (1987) *Planning and Public Transport in Great Britain, France and West Germany*. London: Longman.

Simpson, B.J. (1994) *Urban Public Transport Today*. London: E & FN Spon.

Smith, R.G. (1974) *'Ad hoc' Governments: Special Purpose Transportation Authorities*. London: Sage.

Sudjic, D. (1993) *The 100 Mile City*. Fort Worth, TX: Harcourt Brace.

'Sustainable Development – the UK Strategy' (1994) Cm 2426. London: HMSO.

'Sustainable Transport Study for Aberdeen' (1998) Oscar Faber/ERM Report for Scottish Office Central Research Unit. London: HMSO. (see also Draft Transportation Strategy for Aberdeen, City of Aberdeen Council, 1998.)

Sutcliffe, A. (1981) *Towards the Planned City: Germany, Britain, US and France, 1780–1914*. Oxford: Basil Blackwell.

Thirsk, J. (1978) *Economic Policy and Projects*. Oxford: Clarendon Press.

Tiry, C. (1997) 'Tokyo Yamanote Line – Cityscape Mutations', *Japan Railway and Transport Review*. pp. 4–11.

Transport 2010 (2000) *The 10 Year Plan*. DETR. London: HMSO.

Transport Policy (1977) UK White Paper, Cmnd 6836. London: HMSO.

Travel Choices for Scotland (1998) Cm4010. Edinburgh: HMSO.

Weber, A.F. (1899) *The Growth of Cities in the Nineteenth Century: A Study in Statistics*. New York, Macmillan (reprinted New York: Cornell University Press, 1963).

Whitelegg, J. (1993) *Transport for a Sustainable Future: the Case for Europe*. London: John Wiley.

Yago, G. (1984) *The Decline of Transit: Urban Transportation in German and US Cities, 1900–70*. Cambridge: Cambridge University Press.

Managing Sustainable Urban Environments

ROBERTO CAMAGNI, ROBERTA CAPELLO
AND PETER NIJKAMP

8.1. THE CITY AS A SOURCE OF EXTERNALITIES

The worldwide concern about environmental quality and economic development has a history of some 25 years, starting with the UN Conference on the Human Environment (Stockholm, 1972). In the past decade, the issue has received a new focal point in the concept of sustainable (environmental) development. Since the majority (approximately 70 per cent) of the world's population is living in cities, it is clear that it is there that the consequences of a rising world population are most keenly felt. Such consequences may be positive (for example, access to education and culture, economies of scale, social contacts), but also negative (for example, congestion, concentrated pollution, criminality). If the world population rises to some 10–15 billion people in the next century, it is conceivable that all cities will be faced with major challenges and threats in the future. In various countries, cities tend to grow much faster than the average national rise in population (for example, Nairobi, Dar es Salaam and Lagos have grown sevenfold in the period 1950–1980). Thus, the general trend is one where all cities will grow and where the bigger cities may even grow faster.

Throughout the history of mankind, the social and economic agglomeration advantages of cities have stimulated urban growth. Even though suburbanization has occurred, it is clear that the city has not lost its central position as a node in a broader socio-economic network. Since awareness of the environmental aspects of urban quality of life has grown, it is increasingly being questioned whether the positive externalities outweigh the negative externalities brought about

by the city. As a consequence, much recent attention has been given to the carrying capacity of a sustainable city (see Banister, 1996a, 1996b; Breheny, 1992; Nijkamp and Perrels, 1994). The environmental carrying capacity of a city has two important aspects, namely an *intra-urban* and an *extra-urban* one. The intra-urban carrying capacity refers to the potential of the city to cope with environmental externalities within the city limits, for example, urban waste management, urban air and water pollution, traffic congestion and noise annoyance. The extra-urban carrying capacity concerns the use of land and other resources, which are necessary to ensure continuity of city life (for example, agricultural production, energy, wood, etc.). The total area needed as a life support system for a city, through the production of goods, resources or waste absorption, is often called the *ecological footprint* of a city (cf. Rees, 1992). This concept means essentially that a necessary condition for a city to survive is to import carrying capacity from the outside world.

It is interesting that the two above concerns of intra-urban and extra-urban carrying capacity are implicitly addressed in two policy documents of the European Commission. In the first, the *Green Book of the Urban Environment* (CEC, 1990a), the Commission sets out clearly that urban environmental policy should transcend a sectoral approach and focus on the social and economic choices which are the real root of the problem. In this context, a plea was made for a better coordination of urban environmental policies through more effective and integrated resource use, information provision, technological progress and use of economic stimuli.

In the second document, *Urbanization and the*

Though the chapter is the result of a common research activity by the three authors, R. Camagni has written Sections 8.3, 8.4.4 and 8.5; R. Capello Sections 8.6 and 8.7, P. Nijkamp 8.1, 8.2, 8.4.1, 8.4.2 and 8.4.3.

Functions of Cities in the European Community (CEC, 1990b), the emphasis was laid on the broader regional issues of the urbanization process (for example, spatial population distribution, networking, infrastructure and accessibility). Particular attention was given to the function of cities in regional development and vice versa, including linkages between the city and the outside world as well as the land use and environmental changes instigated by urban growth.

Urban issues appear to have come increasingly to the fore and to make up important items on policy agendas. Particular attention is often given to the problems of large cities, and these problems will be discussed in the next section.

8.2. LARGE CITIES, LARGE PROBLEMS?

Cities are a geographical concentration point of people and human activities, and are characterized by many of the problems of a modern society (see Fokkema and Nijkamp, 1996). Many cities in the developed and the developing world offer a rather depressing picture; economic growth, turbulent demographic movements, high mobility rates, poor urban housing and problematic urban public budgets put a severe stress on the urban environment and the urban habitat (see for example, Asian Development Bank, 1991; Hardoy et al., 1992; Haughton and Hunter, 1994; Pernia, 1994). Environmental degradation has apparently become a prominent feature of

Table 8.1 *Grouping of major world cities by regional/national economic performance*

High debt, high inflation, high primary export economies		Medium growth, 2–4% economies		High economic growth, 4% and over economies	
		USA	New York Los Angeles Chicago San Francisco	*Japan*	Tokyo Yokohama Osaka/Kobe
		W. Europe	London Paris Milan Rome Rhein–Ruhr Berlin Madrid		
		E. Europe	Moscow Leningrad		
Latin America		*South Asia*	Bombay	*NIEs*	Seoul
Buenos Aires	Argentina		Calcutta		Taipei
Lima	Peru		Dehli		Hong Kong
La Paz	Bolivia		Madras		Singapore
Santiago	Chile		Karachi		
Caracas	Venezuela		Dacca	*ASEAN*	Jakarta
Bogota	Colombia				Bangkok
Mexico City	Mexico	*Middle East*	Istanbul		Kuala Lumpur
Sao Paulo	Brazil		Teheran		
Rio de Janeiro	Brazil		Baghdad	*China*	Beijing
					Tianjin
Africa		*ASEAN*	Manila		Shanghai
Lagos	Nigeria				Guanzhou
Kinshasa	Zaire				
Cairo	Egypt				
Nairobi	Kenya				
Accra	Ghana				
Abidjan	Ivory Coast				
Algiers	Algeria				

Source: Lo, 1992: 198

modern city life. The continuing growth of large cities – and increasingly also of medium-sized cities – reinforces the management problems of our urbanized world (see Lo, 1992). The (socio)economic conditions in many large cities in the world are not homogeneous, but exhibit large differences (Table 8.1). Table 8.1 highlights the fact that there are clusters of cities in different regions of the world which are more uniform, but that globally many discrepancies in terms of urban economic performance can be observed. The resulting environmental problems are far reaching: a decline in air and water quality, soil pollution, waste disposal problems and fierce competition for scarce and congested urban space (cf. ESCAP, 1992; UNEP/WHO, 1992). The negative externalities of this development manifest themselves in various ways: poor health conditions, criminality, traffic insecurity, low productivity, large squatter areas, social deprivation, socio-psychological stress and ecological disturbance. A recent study of UNEP/WHO (1992) gives an illustration of the severe problems of air quality in many mega-cities. This alarming picture, however, needs some clarification. It is by no means true that big cities by necessity should have serious environmental problems. There are also several good counter-examples of cities that have developed an environmental policy which by and large serves properly the needs of the population (for example, Vancouver, Stockholm, Singapore). Large cities should not be paralysed by despair, but should try to develop effective strategies based on sound economic principles to improve the quality of the urban environment and the urban habitat.

In popular wisdom it is usual to regard cities – and especially cities in developing countries – as 'sources of evil' (in terms of environmental decay, congestion, poor health conditions etc.). Although it has to be recognized that many cities exhibit signs of decay and high social costs, it should be emphasized that the ongoing process of urbanization on a worldwide scale suggests that cities – or urban areas – all over the world exert centripetal forces which favour further city growth. Apparently, the economies of scale of a modern city far outweigh the negative externalities of urbanization. This observation sets the tone for this chapter, which takes for granted that cities are 'islands of opportunities in seas of decay'. We will argue that the city – or city size – is not necessarily the problem, but rather poor management of the city and the unprofessional organization of scarce urban space. A major reason for the low socio-economic performance of major cities in our world is institutional inertia and inefficient bureaucratic procedures which serve at best some group interests, but fail to exploit the enormous potential embodied in a

modern city. In this context, the lack of business-oriented principles for urban governance is noteworthy. Therefore, a necessary condition for urban survival – or preferably urban sustainability – is the implementation of a blend of market-based development principles and long-range public infrastructure provision which ensure urban sustainability in terms of social economic and environmental benefits for all actors in the urban space. The quality of the urban habitat will be determined decisively by its accessibility to the means of both physical and non-physical network infrastructures (see World Bank, 1994a, 1994b). This means that transportation and communication are in principle vehicles for urban sustainability, provided all social costs (and benefits) involved are charged to all users in such a way that a socially acceptable and equitable market result emerges. The recent popularity of market-based policy principles for sustainable urban development (such as, tradable area licensing schemes, or tradable car emission permits as proposed at present in, for example Mexico City) illustrates that creative policies are necessary in order to ensure that cities are – and remain – the 'home of (wo)man' (see Nijkamp and Ursem, 1998; Verhoef et al., 1996). The central role of a city in an industrialized society appears to turn into a centripetal role in a modern network society; cities are becoming 'local networks in networks of cities'. In this context, it ought to be recognized that a necessary condition for a city to survive and to ensure continuity will be a sufficient degree of accessibility via a broad spectrum of physical and non-physical networks. In this regard, communication and transportation are a *sine qua non*, as mobility (of material and non-material goods) is the necessary consequence of accessibility. However, mobility creates many negative externalities which are detrimental to urban sustainability. Rather than uncritically developing initiatives to reduce accessibility, a sound urban sustainability policy would be used on user charge principles in order to reconcile efficiency, equity and environmental quality.

Thus, urban sustainability is not a simple environmental quality objective, but is the result of a relationship between economic, social and ecological principles (see Camagni et al., 1997; Capello et al., 1999; Nijkamp, 1994; van Pelt, 1993). This means that the concept of sustainability should not be interpreted from three different angles, but is the result of a bilateral or trilateral integration between these principles (see also Figure 8.1). This distinction has implications for sustainable urban development, and for policies addressing related transportation and communication issues. Here it is advocated that the success of urban governance will depend on the professionalism of local/regional policy-making governed by sound principles drawn from business practice in corporate organizations. This seems to be a critical factor for success in also serving the needs of those living in substandard housing conditions. We will also argue that strict policies have to be developed and implemented in which the notion of urban environmental utilization space may play an important role (see Nijkamp and Opschoor, 1997). This implies that for dedicated sustainability policies threshold conditions (for example, critical loads, carrying capacity) have to be specified which have to be respected in all aspects and by all policies. We argue that the allocation of rights to use such a space may be based on fair market principles (for example, tradable permits) which should serve the needs of all citizens. The chapter concludes with some policy guidelines.

8.3. URBAN SUSTAINABILITY: THE NEW CHALLENGE

Since the publication of the Bruntland Report by the World Commission on Environment and Development (1987), the paradigm of sustainable development has gained not only widespread cultural and political recognition, but also significant theoretical acceptance among economists and environmental scientists. It is noteworthy, however, that a widespread application of the same paradigm to the particular case of cities, in spite of its acknowledged relevance for humankind in the perspective of both actual and future generations, is still largely lacking a sound theoretical foundation; consequently, policy strategies are often designed on the basis of common sense, trial and error procedures or even theologies and 'tastes'.

Two major background considerations may be mentioned here: the complexity of the theoretical enterprise, and the lack of full recognition of the nature and role of cities in the present debate. Regarding the first issue, we have to admit that the already high complexity (and consequent uncertainty) that governs the relationships between economic processes and the biosphere rises by at least an order of magnitude if the social, economic and cultural interactions that constitute the city are taken into account. But even if these latter interactions are fully accounted for in a theoretical framework, we are confronted with an intriguing paradox: we are trying to make use of theoretical tools developed from *natural* resource management in order to understand and regulate an intrinsically *non-natural* urban environment.

This is one reason for the ambiguity of much of the present literature on urban sustainability: the failure to accept that the city was born in direct opposition to the countryside, and that the city is growing as an artefact designed to attain social goals like human interaction, agglomeration economies, or effectiveness in the management of economic, cultural and knowledge processes.

The direct consequence of this is that we cannot directly transfer the theoretical tools developed in the case of natural resources to an urban environment: the city means renouncing a model of life and social organization based on human–nature integration, in favour of a model based on a human–human integration; renouncing a production function based on land and labour inputs in favour of one based upon social overhead capital, energy and information (Camagni, 1996).

In terms of the usual concepts developed in the case of natural resources and global sustainability, the city is by definition un-sustainable: it replaces non-renewable resources like fertile land by asphalt and concrete; it overcomes the carrying capacity of its territory by discharging a flow of waste water, air pollution and urban waste to the countryside; and it uses resources taken from distant territories (White and Whitney, 1992). Approaches based on 'strong' sustainability principles, allowing only a very limited substitution between natural resources and generated capital – actions that are probably the most suitable for natural resources management (Victor et al., 1994) – are almost automatically meaningless in an urban environment.

In the light of the above, it may be more meaningful to explore another pathway to the theorization of urban sustainaility. Under scrutiny should be not the city in itself – a macro-historical phenomenon that has manifested itself in all civilizations, that does not need to be justified and that only superficial romantics can reject[1] – but rather some recent trends that endanger its primordial role as the locus of social interaction, creativity and welfare. We refer here to those unlimited and chaotic growth processes that happen mainly in the phases of economic take-off and fast industrialization; in particular those recent patterns of diffused, low-density urban expansion that have been labelled as 'sprawl', 'metropolization', 'periurbanization', *'ville éclatée'*, *'ville éparpillée'*, 'megalopolis', 'edge–city development', all phenomena that blurred the conceptual distinction between city and countryside, leading to a geography of non-cities and collapsed rural environments (Boscacci and Camagni, 1994; Camagni, 1994; Camagni et al., 2000). These processes exacerbate the issue of mobility expansion and energy consumption as they lead to a car-dependent pattern of land use. But we refer also to the new processes of ghetto creation on the periphery of the big metropolises of the developed and developing world, partly linked to the recent global transformation of society and to the time lag by which government policies have come into being and tried to manage the problem. The latter issues have to be considered as part of the frame of urban sustainability, where, on the one hand, they denote imperfect or limited accessibility to the benefits of the urban environment by different groups and, on the other, they have particular impacts on the internal functioning and attractiveness of the city itself.

Further, the concept of urban sustainability should refer not to some earthly paradise in which some form of ecological equilibrium is attained, but to a multi-dimensional archetype which addresses the different major functions of the city: the functions of supplying agglomeration economies, dynamic proximity advantages, welfare, internal social interaction, proper accessibility to the external world and economies of scale in energy consumption (Camagni et al., 1997; Capello, 1997). In order to achieve the maximum welfare for the local population in the long run, the different environments which constitute the city – the economic, social, natural and built environments – have not only to interact by maximizing cross-externalities and feedbacks, but also to co-evolve in a process of virtuous dynamic adjustment (Camagni et al., 1996).

The main features of such a new conceptualization of urban sustainability may be summarized as follows (Camagni, 1996):

1   It is necessarily based on a 'weak' definition of sustainability, as far as substitution between natural and human inputs is concerned. Sustainability in an urban setting refers to the goal of meeting continuously rising (or non-declining) welfare and utility levels for the city's population (Solow, 1986), while maintaining a respect for clear environmental constraints and the long-term economic viability and attractiveness of the city for internal and external firms.

2   It is based upon a 'procedural' rationality in the sense of Simon (1972), defined as the coherence of a dynamic process of understanding and decision-making, as opposed to a 'substantive' rationality, which supposes the possibility of a never-decreasing coherence between means and goals. A procedural rationality appears as the only appropriate framework for theorization and decision-making in a condition of high complexity and widespread uncertainty regarding the fundamental relationships that shape the object of our inquiry (Faucheux and Froger, 1995; Froger, 1993; Vercelli, 1994). In fact, when complex dynamic processes are the norm, implying positive and negative feedback, synergy, network externalities and irreversibility, the possibilities of precisely anticipating the future outcomes of present conditions and policy decisions are limited, and in such circumstances a deterministic approach must be superseded by continuous monitoring, fast reaction, flexible decision-making and long-range scenario-building, these being the most suitable procedures to replace (or better accompany) static or comparative-static optimization exercises.

3   It is thus based on the principle of risk aversion and precaution (Pearce et al., 1989, 1993), implying the necessity for cautious behaviour in the presence of the possibility of coping with significant negative effects ('if pessimists were right') or of trickling irreversible trends.

4   It is necessarily based on the consideration of 'local' effects and dynamics, avoiding huge negative transborder externalities, whilst trusting that environmental virtuous behaviour will also positively affect the global equilibrium of the biosphere. Elsewhere (Camagni et al., 1996) we have argued that a 'local' approach to environmental problems presents numerous advantages with respect to a 'global' one in terms of operationality and effectiveness, due to reduced distance between polluters and victims (the 'locality theorem').

5   Once the field of inquiry is restricted to local trends and interactions, an important consequence emerges: the timespan for the full unfolding of all (negative) feedbacks and cumulative processes among the three environments that represent the city – the economic, social and physical environments – becomes much shorter than in the case of global interactions. Equally, the possibility that the present generation will suffer from present decisions becomes considerably higher.

This last element brings about two consequences. First, it means we can avoid, at least partially, the intriguing and probably (theoretically) unsolvable problem of the representation of future generations at the negotiating table of present decisions (Heister and Schneider, 1993; Pasek, 1993): in fact the concerns about urban quality of life conditions of both present and future generations are by and large the same. Secondly, from a normative point of view, it allows us to overcome the weakness of intervention processes proposed for the sake of inter-generational equity, for which the willingness to pay by the present generation is probably limited. The urban society we are going to build is one which some of us at least are bound to live in too (Camagni, 1996).

8.4. THE ECONOMY/SOCIETY/ENVIRONMENT RELATIONSHIP IN THE URBAN SETTING

8.4.1. Prologue

Modern cities are the 'home of modern (wo)man', in both the developed and the developing world. Thus, the urban habitat and its network configuration are a focal point of interest. It has been argued above that cities are not only problem areas, but also islands of new opportunities. Even though it has to be recognized that in many Third World cities a significant share of the urban population is living in substandard housing, it is also a fact that cities – through their potential economies of scale – have many more possibilities in coping with the externalities caused by their social, economic, political, technological and cultural functioning. In principle, they are in a good position to offer an urban milieu in which welfare, a good quality of life, culture and science can flourish. Both structural causes (for example, climatological or demographic conditions) and government failures caused by inertia and mismanagement are responsible for a substandard quality of urban life dominated by poverty, social stress and environmental decay (cf. Chatterji, 1984).

The ambition to reach a sustainable form of urban development means that strict measures have to be taken in order to alleviate current problem cases and to pave the way to a more acceptable urban future. Given the fact that the majority of total world population (including that in developing countries) is living in urban areas, an intensified effort has to be made to cope with the global urban challenge. As advocated in Agenda 21 of the Rio Conference, the problems of the urban habitat have to be put more at the forefront; this means that first of all sufficient information has to be collected on measurable indicators for sustainable habitats and sustainable city initiatives. By placing the problem of human settlements at the centre, other related fields (for example, technology, transportation, urban economy, social facilities) come immediately to the fore. Second, it is necessary to develop creative new types of sustainable policy, which simultaneously do justice to long-run efficiency, equity and environmental objectives. This means that inertia in urban management would have to be replaced by flexible and innovative corporate policy-making. This holds for all fields in the city, in particular transport, housing and land use. In the next subsection we will deal more specifically with urban environmental quality indicators which may impact on the city's sustainability. Then we will address urban sustainability policy,

followed by a discussion of the need for a sustainable urban transport policy. The main message is that cities are not a source of despair, but rather a window of promising development opportunity. Nevertheless, it is of primary importance to pay due attention to the negative externalities of modern cities.

It is clear that urban sustainability policy requires operational insight into environmental quality conditions, measured by means of indicators. In our discussion of urban environmental problems we will make a distinction between impacts on the natural and on the social environment of a city.

8.4.2. Environmental Problems with an Impact on the Natural Environment

Atmospheric pollution

All pollutants discharged to the atmosphere are – beyond critical concentrations – harmful to plants, animals and humans. Some are harmless in typical ambient concentrations; others have indirect effects that may be harmful. Some have effects that are local or regional, and some have global effects. In many urban areas atmospheric pollution causes severe problems. We can distinguish different emissions which pollute the urban atmosphere (cf. Nijkamp and Ursem, 1997). Examples are:

- Carbon dioxide (CO_2). Carbon dioxide emissions stem from the combustion of fossil fuels. They are seen as the main contributors to the greenhouse effect. Even relatively high amounts of carbon dioxide have no direct known detrimental effect on personal health. The problem of carbon dioxide is that it prevents heat escaping from the planet, which may generate climatic changes. Climate modelling indicates that by the year 2030 the atmospheric CO_2 concentration may result in an average temperature rise of the earth's climate of between 1.5 and 4.5 degrees centigrade. The results of global warming include a rise in the sea level, caused mainly by the thermal expansion of the oceans, with the risk of coastal area floods. When we keep in mind that a large number of big cities are located in coastal areas, the CO_2 emissions are not only a global threat but also a local threat.
- Nitrogen oxide emissions (NO_x). At transboundary levels, nitrogen oxide emissions converted to nitric acid and combined with sulphur dioxide form a significant component of acid rain, which has serious detrimental effects on many ecosystems.
- Sulphur dioxide. Sulphur dioxide can cause bronchitis and other diseases of the respiratory system. It is also the main contributor to acid

rain. The consequences of acid rain include damage to aquatic life, forests and field crops, and corrosion of structures and material. Clouds bearing acids may travel hundreds or even thousand of kilometres across several borders to precipitate acid rain.

- Carbon monoxide (CO). Carbon monoxide is especially a problem in urban areas where synergistic effects with other pollutants contribute to produce photochemical smog and surface ozone (O_3).
- Volatile organic compounds. These comprise a wide variety of hydrocarbons and other substances. They generally result from incomplete combustion of fossil fuels. When combined with nitrogen oxide emissions in sunlight, hydrocarbons and some volatile organic compounds can generate low level ozone. Ozone dims sunlight and causes watering eyes and discomfort for many people, but it normally appears not to have long-term serious health effects.
- Particulate matter. Particulate matter contributes significantly to visibility reduction and, as a carrier of toxic metals and other toxic substances, exerts pressures on human health.

Water pollution

Most people think of water pollution only in terms of water for drinking and other domestic purposes. Domestic use, however, is only a small part of the water story. The most important distinction regarding water use is between instream and withdrawal uses. The instream uses are those for which water remains in its natural channel (like commercial fishing, sports fishing, pleasure boating, swimming etc.). Withdrawal uses are those such as municipal use, industrial processing, cooling and irrigation which require water to be withdrawn from its natural channel.

The various uses of water also require different water qualities: the quality required for pleasure boating can be lower than the quality of drinking water, for example. The various uses also affect water quality differently. In using water, humans discharge an enormous variety of wastes causing water pollution. The most important ones are organic materials using dissolved oxygen in the water as they are degraded. The dissolved oxygen content influences the kind of fish and other life-supporting systems that can only survive in the water, and affects virtually every use of water.

Depletion of energy resources

Due to the high use of energy in the city by transport, houses and industry, many energy resources are overexploited. Excessive exploitation of carbon-based fuels is often seen as the major problem. Although the exploitation of the resources causes only few environmental problems in itself, the effects of overexploitation have severe negative effects on future generations. In this context, renewable energy plays a potentially important role in sustainable city initiatives.

Solid waste disposal

Solid wastes, like paper, plastics, glass and metals, which are generated in large amounts in urban areas, are still increasing annually. Roughly speaking, the weight of municipal waste in the USA generated per month is about the weight of the population that generates it (Mills and Hamilton, 1994). By far the predominant form of disposal is sanitary landfill, that is, an open space where wastes are dumped. An important problem is that cities are running out of potential places for dumping waste. One possibility in coping with this problem is to burn the waste, but the problem here is that, because of the high amount of plastics, burning may generate toxic fumes, so that a careful energy conversion system is needed. Nevertheless, this form of renewable energy is promising and deserves to be further developed.

8.4.3. Environmental Problems with an Impact on the Social Environment

Noise

The noise caused by the different economic activities of an urban area is a big problem. It has been estimated that about 110 million people in the industrialized world are exposed to noise levels in excess of 65 dB(A), a level considered as unacceptable in OECD countries. Noise has several different affects on health and well-being: it affects activities such as communication and sleep, and these effects further induce psychological and physiological disorders such as stress, tiredness and sleep disturbance.

Accident risk

In urban areas accident risk is a high social (environmental) cost. The high volumes of surface transport and the many high-risk industries (even though often located at the edge of an urban area) cause numerous accidents every year, and are therefore detrimental to urban sustainability.

Congestion

Strictly speaking, excessive traffic congestion, while an externality in an economic sense, really involves a lack of internal efficiency of transport operations rather than constituting a serious environmental problem. It is, however, closely associated and generally highly correlated with pollution and other environmental problems, which makes it a topic of concern.

Finally, it should be noted that in many cases energy indicators are appropriate tools to measure urban sustainability, as most environmental threats in the city are directly or indirectly correlated with energy use (Desai, 1990; Nijkamp and Perrels, 1994). Thus, such indicators can be used in a policy analysis to measure the impact of sustainability measures. This holds in particular for urban traffic, since the use of fossil fuels is a necessary input for mobility patterns in the city. Clearly, the economy, society and environment of a city are interrelated, calling for their holistic analysis.

8.4.4. An Integrated View on Sustainable Cities

The focus of any theoretical reflection on urban sustainability should be the relationships between the three environments or sub-systems that constitute the essence of the city: the economic, the social and the physical – natural and built – environments (Figure 8.1). In a previous paper (Camagni et al., 1997) the present authors have proposed a dual way of assessing the interaction among these environments.

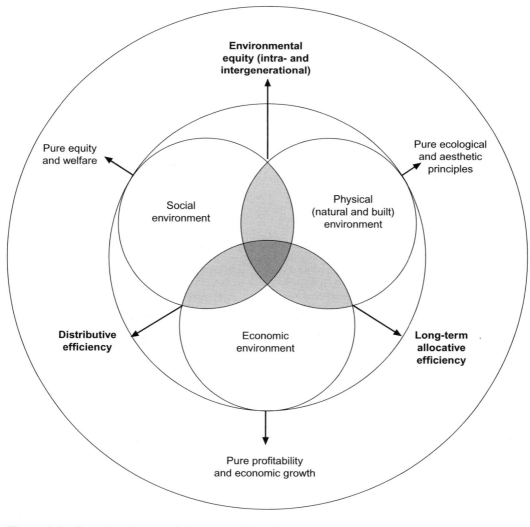

Figure 8.1 Sustainability principles and policies (Camagni, 1996; Camagni et al., 1997)

Table 8.2 *Positive and negative external effects of the interaction between the different environments in a city*

	Interaction between economic and physical environments	Interaction between economic and social environments	Interaction between social and physical environments
Positive external effects	Efficient energy use Efficient use of non-renewable natural resources Economies of scale in the use of urban environmental amenities	Accessibility to qualified housing facilities Accessibility to qualified jobs Accessibility to social amenities Accessibility to social contacts Accessibility to education facilities Accessibility to health services Diversification of options	Green areas for social amenities Residential facilities in green areas Accessibility to urban environmental amenities
Negative external effects	Depletion of natural resources Intensive energy use Water pollution Air pollution Depletion of green areas Traffic congestion Noise	Forced suburbanization due to high urban rents Social frictions on the labour market New poverties	Urban health problems Depletion of historical buildings Loss in cultural heritage

Source: Capello, 1998

1 *A static and structural approach*, focused on the study of how the external effects of the three sub-systems impinge upon each other, positively and negatively (Table 8.2). The scale and quality of the respective assets may represent:
   • positive cross-externalities for the other assets, when the presence of the former assets determines the productivity, attractiveness or marginal utility of the latter (for example, environmental assets increase the economic attractiveness of the city; economic development allows welfare policies and a wider accessibility to urban amenities, services and jobs; lower social conflict increases effectiveness of the local activities, etc.);
   • negative externalities when, due to the limited physical space in which all relationships happen, decreasing returns and bottlenecks appear both in economic terms (rising costs of factors, generating selective crowding-out effects on population and firms) and in physical terms (congestion, conflict, limited accessibility to scarce urban assets). All these effects act as positive or negative location factors.
2 *A dynamic and evolutionary approach*, focused on the assessment of the dynamic relationships among the subsystems, in the form of synergies, positive feedback effects, cumulative processes (for example, the virtuous relation-

ship between infrastructure improvement, efficiency and growth, or between rising incomes, demand for urban amenities, their supply and consequent further development), or in the form of idiosyncrasies, negative feedback effects and irreversibility (depletion or contamination of natural resources like water may irreversibly affect the economic and residential viability of the city; infrastructure improvements may generate further urbanization processes and decreasing accessibility levels within the city).

As far as the normative side of this double argument is concerned, the suggestion put forward by Camagni (1996) was to abandon the logic of pure short-term efficiency, pure equity or pure environmental principles, which was bound to lead to growing contradictions between the three subsystems, in favour of new integrated principles of:
   • Long-term allocative efficiency (taking care of the possible long-term impacts of decreasing environmental quality on the efficiency and attractiveness of the city).
   • Distributive efficiency (taking care of the long-term viability of equitable social systems).
   • Environmental equity (taking care of the negative distributional effects of environmental policies assessed in mainly economic terms).

Pure short-term, profitability principles should evolve into a *Long-term allocative efficiency*, through the internalization of negative externalities, the embedding of certain behavioural rules with respect to the environment into common business practices, and the adoption of a long-term perspective in the allocation of resources and in the definition of benefits and costs.

The resort to market principles is maintained as the most effective way of allocating resources; but this market is enriched in order to take into account – through subsidies, taxes and some regulations – the cases where a pure market fails or does not exist, or does not operate on a sufficient time horizon. The direction is towards the construction of what philosophers and theorists of justice call the '*good market*', incorporating environmental considerations in the same way as the present labour market incorporates modern working and wage conditions.

Looking at the interplay between the principles regulating the environmental and the social spheres, an *environmental equity* principle should be developed, guaranteeing both inter- and intra-generational fairness. While the former is generally underlined in many current environmental debates, opening the way to the possibility of inter-generational paternalism, the latter looks particularly crucial, in that not just the provision of environmental assets should be secured, but also that the accessibility to these assets should be fair in social terms. In the absence of this, environmental policy could become the public provision of luxury goods. Equity in terms of income distribution is quite a different matter; here we draw attention to the substantial inequalities in access to, for example, land, water, energy, environmental and sanitation facilities. In Third World cities this problem is not related only to social services, but also to the basic urban environmental services, such as clean drinking water, sanitary facilities and solid waste collection; the degree to which these services are available in all cities and all parts of the cities should be driven by environmental equity. This is especially true for the poorer segments of the population in Third World cities. Urban sustainability policies should address these differences in resource endowment by either enhancing the level of supply of public facilities (water, electricity, housing, sanitation) or by defining and (more equitably) allocating private property rights to environmental assets (Nijkamp and Opschoor, 1997).

Finally, the integration between profitability and equity principles calls for a *distributive efficiency*: this means operating through redistributive mechanisms in order to secure social stability, fair access to education and health services, wider access to options of economic upgrading and vertical societal mobility. A sustainable city is not a city of equals, but requires a wide accessibility to those basic elements that allow the continuous regeneration of its professional basis and its creative potential.

A city where distributive efficiency and environmental equity principles are established can be labelled a 'good city' in the tradition of some urban planners and urban scientists; as we mentioned above, a 'good city' is a city where the eco-dimension (both natural and built) is maintained, while progressive change is permitted (Lynch, 1981). But this is possible only when distributive efficiency as well as environmental equity principles are satisfied.

Summing up, the tentative definition of urban sustainable development which form the basis of our policy reflections is *a process of synergetic interaction and co-evolution among the basic subsystems that constitute the city – namely the economic, the social, the natural and built environment – which guarantees a non-decreasing welfare level to the local population in the long run without jeopardizing the development options of the surrounding territories, and which contributes to the reduction of the negative effects on the biosphere.*

8.5. THE ROLE OF TIME

8.5.1 Time as Irreversibility

Irreversibility is a central theme underlying sustainability. How cities develop and are planned results in outcomes which may be difficult to reverse, e.g. low density residential development. The development trajectory of urban areas is subject to very different processes and outcomes: sudden or explosive growth, sudden decline, catastrophic jumps, converging or diverging cycles, or chaotic behaviour. Most of these outcomes are characterized by strong irreversibility in the long run (see Camagni and Capello, 1996; Nijkamp and Reggiani, 1992).

This element (of irreversibility) calls for a clear distinction between a short-term and a long-term perspective, both in analysis and in policy-making. In the short term, all events happen in the neighbourhoods of the contingent historical condition (a 'local' equilibrium point), and urban sustainability policies can work by exploiting the (limited) elasticity of substitution among the inputs of the production processes (for example, stimulating energy-saving techniques) or among the transport modes in the mobility pattern of the local population (for example, stimulating the use of public transport facilities).

On the other hand, in the long run we are confronted with a radical change in the policy framework: in the production and transportation spheres technologies can change, while in terms of land-use patterns the urban form can change. But these processes of change imply in both cases huge cumulative effects via learning processes (in the case of technologies) and positive feedbacks (in the case of the transportation supply – land-use change – transport demand cycle). Once a technological or territorial trajectory gets started, usually the sunk costs encountered for changing its direction are huge and are often overlooked by the comparative static approaches based on an optimizing logic (Erdmann, 1993). If these sunk costs are high, alternative solutions or different equilibrium points may never be achieved, in spite of their possible superior efficiency, and the systems remains 'locked-in' by its historical, possibly sub-optimal, trajectory.

On the transportation technology side, the following example may clarify the message and show just how important an anticipatory and early response capability by the relevant public body can be. In Figure 8.2 the learning curves of two competing transportation technologies are drawn, an EB – environmentally benign technology – and an EA – environmentally adverse – one. In the case of an early adoption (time 0), the EB technology may need only a small amount of public subsidy in order to overcome its higher short-term cost disadvantage, but in case of a lagged adoption at time 3 the subsidy requested could easily grow bigger, as a consequence of internal learning processes on the EA technology and external investments on complementary assets (Camagni, 1996).

Irreversibility and path-dependency find another clearcut example in the territorial pattern of metropolitan expansion that has taken place in countries like the United States. As Sternlieb and Hughes (1982) rightly put it more than 15 years ago, when the issues at stake were the risk of oil shortage and the goal of energy saving: 'the U.S. has invested the bulk of its (urban) capital development since World War II in an increasingly centrifugal fashion. We cannot declare this obsolete without bankrupting the country.' This observation also highlights the importance of appropriate normative policy foundations.

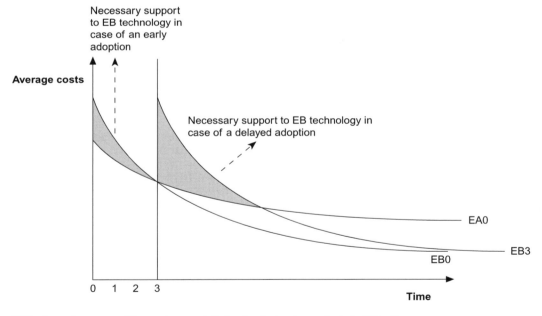

EB0 = Learning curve of the environmentally benign technology adopted at time 0

EB3 = Learning curve of the environmentally benign technology adopted at time 3

EA0 = Learning curve of the environmentally adverse technology adopted at time 0

Figure 8.2   Time trajectory of the economic advantages of alternative technologies (Camagni, 1996)

8.5.2. Time as a Positive Externality: Building Environmental Awareness

From the preceding remarks, it appears clearly that the relationship between economic development and environmental quality is much more complex, indirect and mediated than is commonly thought, especially in the urban realm where social, cultural, political and historical elements interact and co-evolve with respect to the production system and the natural environment.

The usual negative trade-off between per capita income and environmental quality is, therefore, probably also a valid relationship in a short-term *ceteris paribus* condition. In this case, it is certainly true that development, whenever it occurs, builds upon the exploitation of some natural resources: soil, energy, biomass. It impinges on the surrounding environment through the results of the manufacturing process: non-recyclable and non-degradable products, combustion gases, dirty water and waste. But 'other things' do not remain equal in the process of economic development and urbanization: infrastructure construction (and in particular sewerage and drinking water systems) and also health care, housing and social infrastructure improve at a pace that outstrips the simple effect of demographic density and agglomeration;

priorities and social values with respect to quality of life and environmental goods change, and communities are increasingly willing to allocate resources in that direction[2] (Beckerman, 1993).

In most developing countries, evidence exists to show that those elements of environmental quality that matter most, namely access to safe drinking water and sanitation, relate positively with average income levels, and show higher scores in urban than in rural areas (see Beckerman, 1993, for details). On the other hand, in the cities of developed countries, even a first glance suggests that the concentration of the more traditional forms of air pollution, sulphur dioxide and smoke, is much lower than in cities of developing countries. In Britain, for example, during the 1960s average smoke concentration in urban areas fell by 60 per cent, and concentration of sulphur dioxide fell by 30 per cent. In Greater London smoke emissions decreased by over 80 per cent in the period 1958 to 1970, in the presence of an increase of at least 30 per cent in output (Beckerman, 1993).

Our understanding of the development–environment relationship therefore is that the short-term trade-off shifts upwards as time passes, brought about by the evolution of social overhead capital of cultural and political awareness with respect to the environment, by government intervention and through economic transformation.

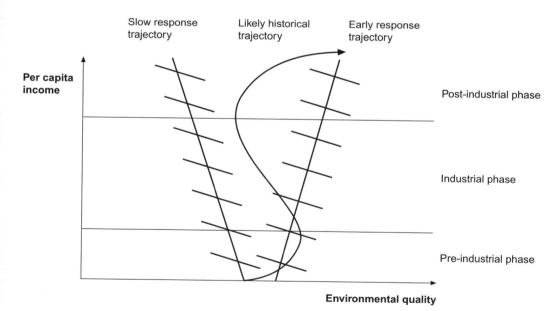

VASE: Value-driven Alternative Sustainability Evolutions

Figure 8.3   The development–environment trade-off: the VASE model (Camagni, 1996)

Through the interpolation of short-term relationships one gets the long-term trajectory of the same relationship, that may show different abstract shapes: a positive shape, in the case of an early and successful response of the local community to environmental decay, and a negative shape in the case of a slow response and low environmental awareness. These alternative outcomes are depicted in Figure 8.3, where the so-called VASE model is presented: Value-driven Alternative Sustainability Evolutions (Camagni, 1996).

While econometrics does not supply us with a definite response on the long-term shape of the development–environment relationship, mainly due to the fuzzy and subjective nature of most environmental indicators, our proposition is that the most likely relationship has an S-shape (Camagni et al., 1997), implying:

- *a positive relationship* in the early stages of development and urbanization, when social overhead capital provision show wide and increasing returns;
- *a negative relationship* in the intermediate phase of development, coinciding with rapid industrialization and metropolitan growth;
- *a positive relationship* again in the case of post-industrial societies, thanks to the emergence of new social values with respect to the environment (environmental quality is in fact a luxury good, increasingly appreciated at high income levels) and the decline of the share of polluting, manufacturing activities.

Our remarks show once again that urban evolution is not taking place in a deterministic world, in which effects follow mechanistically from causes and where trends (or 'stages') are fatally linked in time sequences defined from the beginning. Complexity of interactions and co-evolutions mean a wide spectrum of possible paths and outcomes, difficult to control and to forecast, but very much open and sensitive to discretionary practices and policy decisions, provided that they are shared by the vast majority of the local community and are implemented in a far-sighted and anticipatory way. The role of autonomous environmental values, emerging in the cultural and political spheres and embedded in grass-roots movements, research efforts, public declarations and policy engagements on both a supranational and a local scale, is clear and fundamental in this context.

8.6. DIFFERENCES AND COMMONALITIES IN DEVELOPING AND DEVELOPED COUNTRIES

Is it possible to treat sustainability problems in cities of developing and developed countries with the same logic, and with the same analytical and normative tools? As is often the case in scientific reasoning, the answer is at the same time yes and no.

As far as the goal of urban sustainability is concerned from the limited local perspective, it seems widely acceptable that this should represent a common, fundamental goal in all societies. Sustainability means the possibility of reaching and maintaining a sufficient level of well-being for the urban population in the long run, through the provision of economic advantages, social equity, and cultural and environmental assets. Provided that each community ranks differently the elements or factors of its social utility function according to its own priorities and values, and given the fact that the ways by which social utility is reached may be diverse (according to the specificities of each case), no difference exists between the urban conditions in developing and developed countries. Trade-offs on resources allocation are similar; negative feedbacks and cross-externalities among the different sub-systems are similar; and similar in character also is the role of (and the difficulties facing) the planner. Of course, local conditions in developing countries are often (but not always) much more dramatic, as the very subsistence of parts of the population is threatened.

On the other hand, a wide difference between the two cases does emerge when urban sustainability is viewed in a wider perspective, taking into consideration the role of cities in the national development context and the forces that determine their expansion. In developing countries, cities are magnets attracting human resources from outside as a consequence of their role as nodes of infrastructure and development potential. In the developing world, cities are often also the recipient of masses of desperate people, pushed out of the countryside by the crisis in the agricultural sector. In developed countries the city is a factor in the increased efficiency of the rural areas (for example, through the transfer of know-how and technology to the agricultural sector and the dispersal of industry); in developing countries it is the effect of the crisis of the non-city which stimulates much urban growth.

All this has far-reaching consequences for urban science and environmental planning. In a condition of relative demographic and economic equilibrium between city and countryside, the destiny of cities depends heavily upon the quality of their own physical lay-out, internal functioning and equilibrium between the built and the natural environment. In a condition of imbalance, on the contrary, their destiny depends upon the development process that is happening outside them; any policy intervention on urban assets is destined to amplify the perceived disparity in the development potential between the city and the countryside, fostering cumulative immigration processes that annihilate the potential effects of the initial intervention (Lo, 1992). Consequently, in developing countries, policies addressed to the sustainable development of big cities should be complemented in parallel by:

- policies focused on a balanced regional development, and in particular policies for the development of rural areas;
- policies focused on the construction of a balanced urban system, based on a creative, country-specific blending of the traditional hierarchical pattern of centres and the modern network pattern of specialized centres (Camagni, 1992; Camagni, 1994).

8.7. URBAN SUSTAINABILITY PRINCIPLES AND POLICY IMPLICATIONS

As mentioned above, the strict application of urban sustainability requires a focus on the economic, social and environmental aspects of urban life (cf. Newman and Thornley, 1994). This calls usually also for a strict urban energy conservation policy (see Banister, 1996b; Newman and Kenworthy, 1989, 1991). In many cases, targets and critical limits for various aspects (indicators) of urban sustainability have to be specified (for example, noise level, CO_2 levels, density, traffic etc.). If the actual level of negative externalities exceeds such threshold levels, a proper policy has to be designed so as to guarantee a sustainable outcome. In principle, two contrasting types of policies may be distinguished, namely standards based on regulations and market-based instruments. These measures have been discussed extensively in the literature. A more recently developed and increasingly popular concept is that of the urban environmental bubble, which defines the urban utilization space for different categories of substances which affect urban sustainability. This sustainability model presupposes two types of information: the definition of critical threshold levels for urban sustainability indicators, and indications as to the proper way of allocating the remaining constrained activity levels in terms of economy efficiency, social equity and environmental effectiveness.

In this framework, an allocation system based on market principles has in recent years gained popularity, namely the idea of tradable permits in the city, especially emission trading. Although the theory of tradable emission permits dates back to the 1960s, it has only recently become a tool in environmental policy. Emission trading is based on the objective of guaranteeing an urban environmental outcome that is in agreement with *a priori* defined critical threshold levels and which is achieved at the lowest costs possible. In a perfectly operating market these permits will be traded until the marginal abatement costs of all actors are equal to the market price of the emission permits. Various types of trading systems have been proposed, such as Ambient Permit Systems and Emission Permit Systems. Of course, there needs to be a control authority which acts as a clearing house. Given the high transaction costs of these systems, intermediate forms have emerged, notably the Pollution Offset System, which is more flexible, especially if it is accompanied by the possibility to bank credits.

Such trading systems may also serve to integrate economic and environmental aspects. Social aspects may also be covered if permits are granted to all actors in the urban space with the right to sell these rights on the market. In this respect, tradable permits are more flexible and offer more certainty than other policy systems (effluent fees for example). It should be added that there is often a strong tendency to approach urban sustainability issues from a sectoral perspective (specific industries, transport, etc.), but that an important integrating mechanism is neglected, namely land use (cf. Hayashi et al., 1992). Just like energy, land use is one of the driving forces for the city to become an 'island of sustainability'. Land use management is a *sine qua non* for proper industrial location, environmental, transport and housing policy. For example, illegal housing (such as squatters) is at odds with a policy aiming at a sustainable urban development. This is indeed reason for concern about negative urban externalities, but it ought to be recognized as well that the city also creates positive externalities. The very existence of such positive externalities warrants pro-active policy intervention based on the view that urban governance has to be driven by clear and professional management principles.

NOTES

1. Haughton and Hunter (1994: ch. 1) give an interesting list of such definitions of the city as a 'parasite on the natural and domesticated environment', a 'cancer', a 'lethal illness', 'overgrown monstrosities', 'systems of disharmony'.

2. In the USA, expenditure on PAC (pollution abatement and control) rose at an average annual rate of 3.2 per cent between 1972 and 1987, and represents a rising share in GNP; similar data are available for Germany (3.4 per cent increase during the period 1975–1985) and for Japan (with an increase of 6.1 per cent from 1975 to 1986, referring only to public expenditure); see Beckerman, 1993.

REFERENCES

Asian Development Bank (1991) *The Urban Poor.* Manila: ADB.

Banister, D. (1996a) 'Energy, quality of life and the environment: the role of transport', *Transport Reviews,* 16 (1): 23–35.

Banister, D. (1996b) 'Urban sustainability', in *The Handbook of Environmental and Resource Economics.* London: Edward Elgar.

Beckerman, W. (1993) 'The environmental limits to growth: a fresh look', in H. Giersch (ed.), *Economic Progress and Environmental Concern.* Berlin: Springer Verlag.

Boscacci, F. and Camagni, R. (eds) (1994) *Fra Città e Campagna: Periurbanizzazione e Politiche Territoriali.* Bologna: Il Mulino.

Breheny, M. (ed.) (1992) *Sustainable Development and Urban Form.* London: Pion.

Camagni, R. (1992) *Economia Urbana: Principi e Modelli Teorici.* Rome: La Nuova Italia Scientifica.

Camagni, R. (1994) 'Processi di utilizzazione e difesa dei suoli nelle fasce periurbane: dal conflitto alla cooperazione fra città e campagna', in F. Boscacci and R. Camagni (eds), *Fra Città e Campagna.* Bologna: Il Mulino.

Camagni, R. (1996) 'Lo sviluppo urbano sostenibile: le ragioni e i fondamenti di un programma di ricerca', in R. Camagni (ed.), *Economia e Pianificazione della Città Sostenibile.* Bologna: Il Mulino.

Camagni, R. and Capello, R. (1996) 'The role of indivisibilities and irreversibilities in renewable energy adoption processes: a dynamic approach', Working Paper series DEP, Politecnico di Milano.

Camagni, R., Capello, R. and Nijkamp, P. (1997) 'The co-evolutionary city', *International Journal of Urban Sciences,* 1 (1): 3246.

Camagni, R., Gibelli, M.C. and Rigamonti, P. (2000) 'Urban mobility and urban form: the social and environmental costs of different patterns of urban expansion', forthcoming in *Ecological Economics.*

Capello, R. (1998) 'Urban returns to scale and environmental resources: an estimate of environmental externalities in an urban production function', *International Journal of Environment and Pollution,* 10(1): 28–45.

Capello, R., Nijkamp, P. and Pepping, G. (1999) *Sustainable Cities and Energy Policies.* Berlin: Springer Verlag.

CEC (1990a) *Green Book on the Urban Environment,* DG XI. Brussels: European Commission.

CEC (1990b) *Urbanization and the Functions of Cities in the European Community,* DG XVI. Brussels: European Commission.

Chatterji, M. (1984) *Management and Regional Science of Economic Development.* Boston, MA: Kluwer.

Desai, A. (ed.) (1990) *Patterns of Energy Use in Developing Countries.* New Delhi: Eastern Wiley.

Erdmann, G. (1993) 'Evolutionary economics as an approach to environmental problems', in H. Giersch (ed.), *Economic Concern and Environmental Progress.* Berlin: Springer Verlag.

ESCAP (1992) *State of the Environment in Asia and the Pacific 1990.* Bangkok: ESCAP.

Faucheux, S. and Froger, G. (1995) *Decision-making under Environmental Uncertainty.* Cahiers du C3E, no. 95–1.

Fokkema, T. and Nijkamp, P. (1996) 'Large cities, large problems?', *Urban Studies,* 33 (2): 353–77.

Froger, G. (1993) *Les Modèles théoriques de développement soutenable: une synthèse des approches methodologiques.* Cahier du C3E, no. 93–19.

Hardoy, J.E., Mitlin, D. and Satterthwaite, D. (1992) *Environmental Problems in Third World Cities.* London: Earthscan.

Haughton, G. and Hunter, C. (1994) *Sustainable Cities.* London: Jessica Kingsley (Regional Studies Association).

Hayashi, Y., Tomita, Y., Doi, K. and Suparat, R. (1992) 'An international comparative study on land use – transport planning policies as control measures of the urban environment', *Land Use, Development and Globalisation* (Proceedings 6th World Conference on Transport Research, Lyon), vol. 1, pp. 255–66.

Heister, J. and Schneider, F. (1993) 'Ecological concerns in a market economy: on ethics, accounting and sustainability', in H. Giersch (ed.), *Economic Progress and Environmental Concern.* Berlin: Springer Verlag.

Lo, I.–C. (1992) 'Growth and management of the third world megalopolis', *Comprehensive Urban Studies,* 45: 189–202.

Lynch, K. (1981) *The Theory of Good City Form.* Cambridge, MA: MIT Press.

Mills, E.S. and Hamilton, B.W. (1994) *Urban Economics.* Glenview, IL: Scott, Foresman and Company.

Newman, P.W. and Kenworthy, J.R. (1989) 'Gasoline consumption and cities: a comparison of US cities with a global survey', *Journal of the American Planning Association,* 1: 24–37.

Newman, P.W. and Kenworthy, J.R. (1991) *Cities and Automobile Dependence.* Aldershot: Gower.

Newman, P.W. and Thornely, A. (1994) 'London, Paris, Berlin: economic competition and the new urban governance', *European Spatial Research and Policy,* 1 (2): 7–22.

Nijkamp, P. (1994) 'Improving urban environmental quality: socio-economic possibilities and limits', in E.M. Pernia (ed.), *Urban Poverty in Asia.* Oxford: Oxford University Press. pp. 241–92.

Nijkamp, P. and Opschoor, J.B. (1997) 'Urban environmental sustainability: critical issues and policy measures in a third-world context', in M. Chatterji and Y. Kaizhong (eds), *Regional Science in Developing Countries.* London: Macmillan. pp. 52–73.

Nijkamp, P. and Perrels, A. (1994) *Sustainable Cities in Europe.* London: Earthscan.

Nijkamp, P. and Reggiani, A. (1992) *Interaction, Evolution and Chaos in Space*. Berlin: Springer Verlag.

Nijkamp, P. and Ursem, T. (1998) 'Market solutions for sustainable cities', *International Journal of Environment and Pollution*, 10(1): 46–64.

Pasek, J. (1993) 'Philosophical aspects of intergenerational justice', in H. Giersch (ed.), *Economic Progress and Environmental Concern*. Berlin: Springer Verlag.

Pearce, D.W., Markandya, A. and Barbier, E. (1989) *Blueprint for a Green Economy*. London: Earthscan.

Pearce, D.W., Turner, R.K., O'Riordan, T., Adger, N. et al. (1993) *Blueprint 3 – Measuring Sustainable Development*. London: Earthscan.

Pernia, E.M. (ed.) (1994) *Urban Poverty in Asia*. Oxford: Oxford University Press.

Pezzey, J. (1988) 'Market mechanism of pollution control: polluter pays, economic and practical aspects', in R.K. Turner (ed.), *Sustainable Environmental Management*. London: Belhaven Press.

Rees, W. (1992) 'Ecological footprints and appropriated carrying capacity: what urban economics leaves out', *Environment and Urbanisation*, 2 (2): 121–38.

Simon, H. (1972) 'From substantive to procedural rationality', in C.B. McGuire and R. Radner (eds), *Decision and Organization*. Amsterdam: North Holland.

Solow, R. (1986) 'On the intergenerational allocation of natural resources', *Scandinavian Journal of Economics*, no. 1: 141–9.

Sternlieb, G. and Hughes, J.W. (1982) 'Energy constraints and development patterns in the 1980s', in R. Burchell and D. Listokin (eds), *Energy and Land-Use*. Brunswick, NJ: Rutgers University, Center for Urban Policy Research.

UNEP/WHO (1992) *Urban Air Pollution in Megacities of the World*. Oxford: Blackwell.

van Pelt, M.J.F. (1993) *Ecological Sustainability and Project Appraisal*. Aldershot: Avebury.

Vercelli, A. (1994) 'Sustainable growth, rationality and time', Paper presented at the International Symposium C3E-METIS on 'Modèles de développement soutenable: des approches exclusives ou compl mentaires de la soutenabilité?', Paris, 16–18 March.

Verhoef, E., Nijkamp, P. and Rietveld, P. (1996) 'Tradeable permits: their potential in the regulation of road transport externalities', Research Paper, Department of Economics, Free University, Amsterdam.

Victor, P., Hanna, E. and Kubursi, A. (1994) 'How strong is weak sustainability?', Paper presented at the International Symposium C3E-METIS on 'Modèles de développement soutenable: des approches exclusives ou complémentaires de la soutenabilité?', Paris, 16–18 March.

White, R. and Whitney, J. (1992) 'Cities and the environment, an overview', in R. Stren, R. White and J. Whitney (eds), *Sustainable Cities, Urbanization and the Environment in International Perspective*. Oxford: Westview Press.

World Bank (1994a) *World Development Report 1994*. Washington, DC: World Bank.

World Bank (1994b) *Making Development Sustainable*. Washington, DC: World Bank.

World Commission on Environment and Development (1987) *Our Common Future* (the Bruntland Report). Oxford: Oxford University Press.

Part III
THE CITY AS PEOPLE

Social difference and diversity are the very substance of cities. Their growth both reflects and indeed is often a product of social diversity. To the extent that urban growth has typically been the outcome of waves of in-migration, cities have been host to diverse ethnic groups. In the advanced economies such migration has often been international, and remains so, particularly in world cities, whose measure of globality is taken to be reflected in part by their cosmopolitan nature. Besides ethnicity, class divisions and gender constitute other bases upon which the social diversity of cities is commonly grounded, though this by no means exhausts the range of recognizable social groups within cities, which would need to include the young, the elderly, gay and lesbian groups, the disabled, the homeless and low income households including the relatively deprived among others.

Current, particularly postmodern, accounts of the city are celebratory of such difference: much of the cultural richness of cities derives from the plurality of social groups. As laudable as is such celebration, in reality the very social diversity of the city has tended to be imagined as potentially the cause of social disorganization, the 'response' to which has been the emergence of 'sorting devices', in particular different forms of segregation, which contribute to the spatial, and social, distancing of different groups within the city. Cities, then, become differentiated, usually stratified, resulting in the spatial sorting of the urban population. How these processes unfold reflects the relative power of different groups; cities have always reflected, and often magnified, the inequalities of the societies in which they are located.

Much of the sorting of the city is reflected in the way in which cities have grown. With increasing growth, following which has been the emergence of the middle classes, the development of suburbs has become a common expression of, in part, the means by which relatively advantaged groups are able to avoid the 'problems' of the central (inner) cities. While the suburb was a

Western form of urban development, it is by no means confined to the core economies. In many cities in the developing economies, affluent suburbs have emerged as a key urban form in some cases fortified, aping the development of the gated communities of North America which represent among the more extreme measures by which spatial sorting, in this case spatial enclosure, is harnessed.

While suburban development was linked to the first major outwards expansion of the city, exurban growth marked the next, while more recently counterurbanizing trends were to emerge in the advanced economies. Counterurbanization was the more remarkable precisely because it marked a reversal of the (urbanization) trends which had become taken for granted, though its significance varied between countries and is in any case contested. The onset of re-urbanization, linked in the United States in particular with the rise of New Urbanism, the rediscovery of the benefits of city living, particularly by new types of households, was to throw further doubt on the realities of counterurbanization. Each of the trends, while due to a variety of reasons, to a greater or lesser degree represented the growing importance of lifestyle preferences for residential location among those who were able to make choices.

Such trends reflect the variety of ways in which cities and metropolitan regions become segregated. Segregation, particularly reflecting class divisions, is commonplace, if not universal: even within socialism, though the evidence is not undisputed, cities were residentially segregated. Being commonplace is not the same as saying it is natural: residential segregation is indeed socially constructed, and often carefully so, so as to maintain social advantage. Its construction implies exclusion, Othering, notably of racial minorities, but also other marginalized groups. Under globalization one argument is that cities may becoming more socially polarized than they are simply socially segregated, that inequalities are becoming more pronounced, and that cities are becoming visibly more divided. If this is the case, and the debate is far from decisive on the point that cities have become more polarized over the last few decades, the celebration of difference becomes that much more ironic.

Segregation serves a multiplicity of purposes for the different groups living within the city. One, not uncontentious, function is its relation with community – that the spatial sorting of citizens into 'similar' social groups achieved by segregation becomes a (if not the) building block for community. Of course, such an argument is as value-laden as it is shot through with assumptions about the nature of communities. Perhaps a victim of its own popularity, community exaggerates consensus over conflict, romanticizing how social interaction within the city might be as opposed to what it often is. Yet, its value can be palpable, in mobilizing support and opposition to urban change, in solidifying social identity in a world undergoing considerable flux.

As much as contemporary analysis is focusing upon marginalized and excluded groups within the city, this was not always the case. In fact, with the exception of some isolated studies of such groups by human ecologists in the first half of the twentieth century, of the hobo for example, the socially marginalized were also as marginal to the study of social divisions within the city pursued by most urban academics as they were to practitioners involved with planning and restructuring the city. Some such groups remain marginalized, and sometimes even physically removed – the ongoing harassment of street traders in many cities in developing countries (in spite of the support given to the role they play in meeting consumer demand while creating employment) is matched by the admittedly less frequent cases of similar types of exclusion of activities which are viewed as at odds with the image elites wish to project of the city. Others among the previously 'invisible', women, as well as gays and lesbians, in particular, have become more visible. Even so, becoming more visible is not to be taken as implying the demise of exclusionary and discriminatory practices within the city.

Where the notion of community has (erroneously) emphasized consensus and co-operation, simply because of their pluralism, urban life is characterized more by contradiction and conflict. Among the disadvantages of the city, crime is more frequently mentioned in Quality of Life studies as a disincentive for living in cities than is any other issue. Crime, or perhaps more accurately fear of crime, permeates whole neighbourhoods, just as it does those who use it as one of the reasons explaining their move to the suburbs and beyond. Yet, urban crime is not new: the Victorian city was as much affected by crime as are contemporary cities. Attitudes to crime may too easily fall victim to moral panic. Crime is not, of course, a uniquely urban phenomenon, but combined with the imagined nature of urban social life, its alienating and anomic character, the effects of crime on its victims is easily translated into one of the, if not the, major disadvantages of living in cities.

9

Urbanization, Suburbanization, Counterurbanization and Reurbanization

TONY CHAMPION

These are confusing times for those interested in the evolution of urban systems in advanced economies and in the changing spatial structure of urban settlements, as more evidence of developments since the early 1980s becomes available. The title chosen for this chapter by the editor encapsulates the 'paradigm', or standard perspective, that by then had become the most commonly used benchmark against which the experience of individual countries and urban regions could be compared. Reflecting urban analysts' preoccupation with 'decades' of change – a very unhealthy feature of urban studies, but an understandable one bearing in mind the frequency and incidence of population censuses – a picture has emerged of 'urbanization' predominating in the 1950s and 'suburbanization' accelerating in the 1960s, with the 1970s emerging as the 'decade of counterurbanization'. The significance of the latter has remained a hotly debated topic, with past predictions for the 1980s and beyond variously suggesting a fuller development of centrifugal tendencies, a return to the 'normal' processes of metropolitan and urban concentration after the 'anomalous 1970s', and a period of 'reurbanization' associated with a natural progression through sequential stages of urban development in some form of cyclic pattern.

The aim of this chapter is to discuss the significance of the more recent developments of counterurbanization and reurbanization in the context of the ways in which urban growth in the advanced economies has traditionally taken place. The first section focuses on urbanization and, in so doing, introduces the main approaches which have been used to make sense of the urban trends of the past half-century. Urbanization has been seen variously as the increasing concentration of national populations into towns, as the increasing concentration of a country's urban population into the largest cities, and as the increasing concentration of an urban region's

population into its core at the expense of its surrounding ring. It is shown that in the 1950s all these three processes were operating side by side quite commonly across the developed world. The second section deals with the first main departure from these traditional patterns, namely the reversal of within-urban-region population shifts in favour of the suburban ring at the expense of the core, while the third section explains how counterurbanization represents the reversal of the concentration of the urban population in the largest cities and examines how far this process has proceeded. The fourth section explores the theoretical background to the 'reurbanization' hypothesis in terms of a return to traditional patterns of urban change and goes on to assess the extent to which this has occurred in recent years. It is concluded that none of these prefixed versions of 'urbanization' can adequately encapsulate the developments observed since the late 1970s, raising questions about whether it is sensible to try to impose any single model.

Throughout, the primary emphasis is on urban trends as measured by data on residential populations and on understanding these changes in terms of the direct demographic causes. Only limited mention is made of the environmental, economic, social, cultural and political factors that help to produce the observed changes in migration and population composition. These latter aspects are dealt with more fully in other chapters.

URBANIZATION AND THE STUDY OF URBAN CHANGE

Given that this chapter is entirely about patterns of urbanization, it may seem strange to begin with a separate section on urbanization. This step is, however, essential because of the variety of ways in which the term has been used in the literature. Some have conceived of urbanization in the physical sense of the increasing area of land being developed for urban use, while others view urbanization as a social process of people adopting the attitudes and behaviour traditionally associated with life in cities and towns, irrespective of where they might be living. Even when it is interpreted from the perspective of population geography so as to relate to the type of settlement in which people live, as it is for present purposes, the term has been applied in several different ways. According to which approach is taken, urbanization can be considered to be a continuing process, one that is long since over or one that is not currently very important but may undergo a resurgence, perhaps in cyclic fashion. The fact that the term has been elaborated by the attachment of various prefixes, as exemplified in the title of this chapter, suggests that our emphasis should be on 'urban change' rather than on 'urbanization' *per se*. The rest of this section, building on the discussion in Chapter 2, expands on this point and provides the context for the subsequent sections.

Traditionally, the most common way of measuring urbanization is in terms of the proportion of a national population that lives in 'urban places'. Somewhat paradoxically, it is in this sense that urbanization appears to be a steadily continuing process, yet this measure is now of virtually no value for studying urban change in advanced economies. The official figures, provided by national statistical agencies and summarized for broad regional groupings by the United Nations, indicate that around three-quarters of the population of the more developed world are considered to be urbanized, a proportion which has risen from two-thirds in 1970 and from just over half in 1950 and which is expected to increase further to five-sixths by 2030 (Table 9.1). Clearly

evident is the strong growth in 'level of urbanization' since 1950 in Eastern Europe and Southern Europe, and Japan. Even here, however, the rate of increase is now levelling off and approaching what appears to be a ceiling of 80–90 per cent – a process which was largely complete in England and Wales by 1900 and in most other parts of the more developed world by mid century. As indicated in Chapter 2, the anticipated future increases have more to do with the reclassification of existing rural settlements as a result of the outward spread of cities or their populations than with further large-scale movements of people from rural to urban areas.

Far more valuable in the study of contemporary urban change is the measure of urbanization that is based on the distribution of the population between different sizes of urban places. Its central importance was spelt out more than half a century ago in Tisdale's (1942) description of urbanization: 'Urbanization is a process of population concentration. It proceeds in two ways: the multiplication of the points of concentration and the increase in size of individual concentrations' (quoted by Berry, 1976a: 17). The most extreme examples of this latter aspect relate to those countries where one leading city has outstripped all the others, producing a primate rank-size settlement pattern, as in the case amongst the advanced economies for Vienna, Paris and London – the latter still six times the size of the UK's second city, Birmingham (Champion et al., 1987). More generally, it is seen in the high 'degree of urban concentration' found widely across the more developed world, with cities of a million inhabitants or more accounting for at least a quarter of urban dwellers in most countries and for over a third in some, these latter including Australia, Greece, the USA, Austria and Portugal (Table 9.2). Formalizing this perspective, Fielding (1982) has defined urbanization as being where there

Table 9.1   *Level of urbanization, 1950–2030*

Region	1950 (%)	1970 (%)	1990 (%)	2010 (%)	2030 (%)
More Developed World	54.9	67.6	73.7	78.7	83.7
Australia/New Zealand	74.6	84.4	85.0	85.9	88.9
Northern Europe	72.7	80.4	83.0	85.5	88.8
Western Europe	67.9	76.7	80.7	84.2	87.8
North America	63.9	73.8	75.4	79.6	84.4
Japan	50.3	71.2	77.4	80.9	85.3
Southern Europe	44.2	56.6	63.1	68.1	75.2
Eastern Europe	39.3	55.8	68.1	75.7	81.3

Figures give share of population living in urban places. Regions are ranked according to share in 1950.

Source: calculated from United Nations (1998) *World Urbanization Prospects. The 1996 Revision.* Sales No. E.98.XIII.6, Table A.2

exists a direct urban-system-wide relationship between the rate of net migratory growth of settlements and measures of their urban status (Figure 9.1a).

Table 9.2 *Degree of urban concentration, 1965–90*

Region and country	1965 (%)	1990 (%)	Change
North America			
Canada	37	39	+2
United States	49	48	−1
Eastern Europe			
Bulgaria	21	19	−2
Czechoslovakia	15	11	−4
Hungary	43	33	−10
Poland	32	28	−4
Romania	21	18	−3
Northern Europe			
Denmark	38	31	−7
Finland	27	34	+7
Sweden	17	23	+6
United Kingdom	33	26	−7
Southern Europe			
Greece	59	55	−4
Italy	42	37	−5
Portugal	44	46	+2
Western Europe			
Austria	51	47	−4
France	30	26	−4
Netherlands	18	16	−2
Australia	60	59	−1
Japan	37	36	−1

Figures refer to the proportion of the urban population that lives in cities with at least one million inhabitants in 1990.

Source: World Bank (1992) *World Development Report, 1992.* New York: Oxford University Press, Table 31

The interest in this second interpretation of urbanization arises largely because there does not seem to be the same unidirectional pattern of development as there is for the simple 'per cent urban' measure. Table 9.2 shows a widespread tendency for a reduction in the degree of urban concentration between 1965 and 1990, with 15 of the 19 advanced countries with data available for both years being estimated to have seen a fall in the proportion of their urban population living in cities of at least one million residents. Though this tendency can arise from statistical underbounding, whereby the statistical boundaries of a city fail to keep pace with its lateral expansion, the data are

suggestive of a widespread process of population redistribution down the urban hierarchy, either through the relatively faster growth of smaller urban places or through the absolute decline of the largest cities. This development has indeed been confirmed by a number of more detailed studies which have been careful to allow for underbounding; for example Fielding's (1982) analysis of France showing Paris' migratory growth rate falling behind that of the next rank of cities and eventually turning negative, the studies which have shown the USA's medium-sized and smaller metropolitan areas overtaking the growth rate of the largest ones in the 1970s (for example, Frey, 1990; Nucci and Long, 1995) and the national case studies of the present author (Champion, 1989).

This switch in the incidence of the strongest population growth away from the largest cities was termed 'polarization reversal' by Richardson (1977, 1980) and, along with observations of the revival of population growth in non-metropolitan America (Beale, 1975; Morrison and Wheeler, 1976), spawned the notion of 'counterurbanization' (Berry, 1976a). The former has been incorporated by Geyer and Kontuly (1993) into a 'theory of differential urbanization', whereby patterns of gross migration alter over time and successively favour the primate city, the intermediate-city level and ultimately the small-city level (Figure 9.2). In this approach, 'urbanization' is deemed to be taking place as long as the population is becoming increasingly concentrated in the primate city category, before the intermediate-size city category takes over as the fastest-growing of the three categories.

The seminal contribution to the rigorous analysis of this phenomenon was made by Fielding (1982). He recognized 'counterurbanization' to be taking place in a settlement system where there exists a negative relationship between migratory growth rate and urban status (Figure 9.1b). Drawing primarily on his case study of France along with statistical analyses of several other countries, Fielding visualized this new pattern as the outcome of the progressive movement of the strongest growth down the settlement hierarchy and anticipated its coming fully into being in Western Europe during the 1980s (Figure 9.1c). Thus presented as a revolution in settlement pattern trends linked to a societal shift from industrial to post-industrial eras, the nature and significance of this development has been widely disputed over the past two decades, as we will see below.

Besides these two usages of the term 'urbanization', there is a third application which overlaps the other two to a certain extent but forms a distinctive element of an alternative way of making sense of recent trends – the 'stages of

(a)

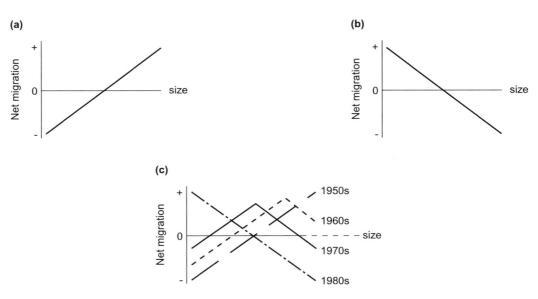

Figure 9.1 Relationship between migration and settlement size: (a) urbanization; (b) counterurbanization; (c) a possible sequence for Western Europe, 1950–80 (Fielding, 1982: 8–10) Reprinted with kind permission from Elsevier Science Ltd.

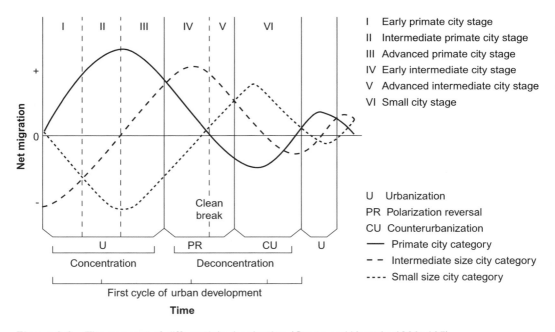

Figure 9.2 The concept of differential urbanization (Geyer and Kontuly, 1993: 165)

urban development' model. One key feature of this approach is the idea that a city has a life cycle which takes it from a 'youthful' growing phase through to an 'older' phase of stability or decline, as the benefits of the initial investment are

progressively exhausted or the original locational advantages of the site become less relevant and are changing (Birch, 1971; Rust, 1975; for a review of this literature, see Roberts, 1991). A second feature incorporated into this approach is the identification of the phasing of development through an examination of the internal patterning of growth, distinguishing between the main built-up 'core' of a city and its 'ring' or commuting hinterland. This approach was devised by Hall (1971) who suggested a four-stage model of metropolitan-area development, beginning with a period of centralization whereby people become more concentrated in the core at the expense of the ring, continuing with periods of relative and then absolute decentralization in which the core grows less rapidly than the ring and then experiences absolute loss of population to the ring, and ends up with a stage in which the metropolitan area as a whole moves into overall decline because the core's loss becomes greater than the ring's again. An additional feature added to this approach subsequently by Klaassen et al. (1981) is the idea of a recurring cycle; namely that, after the phase which Hall (1971: 118–19) refers to as 'decentralization in decline', there follows a process of reconcentration which leads on to a second cycle beginning with renewed growth overall and centralization within the core.

The main features of this cyclic model, incorporating extra elements introduced by Berg et al. (1982, 1987), are illustrated diagrammatically in Figure 9.3. It consists of four 'stages of urban development' based on whether the urban region or 'agglomeration' (core and ring together) is experiencing overall growth or decline in population numbers and on whether it is the core or the ring that is performing the more strongly. Each of

the four stages is then subdivided into two phases based on the switching of either core or ring between gain and loss (which determines whether centralization and decentralization is absolute or merely relative), giving a sequence of eight phases altogether.

In the present context, the most important point is that the term 'urbanization' is restricted to the first of the four stages, referring to the situation in which people are becoming more concentrated in a single urban region while that urban region is growing overall. Beyond this, a whole urban system can be deemed to be urbanizing if the majority of its urban regions are in this 'urbanization' stage. The next two stages, however, are clearly meant to represent movements away from what might be considered to be 'pure' urbanization, with the 'suburbanization' stage involving decentralization within the urban region and the 'disurbanization' stage signifying decentralization beyond the urban region, a process which Robert and Randolph (1983) have termed 'deconcentration' to distinguish it from within-urban-region 'decentralization' and which equates broadly with Fielding's definition of counterurbanization. Fourthly, the 'reurbanization' stage is associated with the slowing of urban-region decline, which is initiated by the core and followed by the ring and thus involves a process of renewed centralization. Finally, according to the 'cyclic' element of this model, this leads through into renewed urban-region growth and to a new period of 'urbanization'.

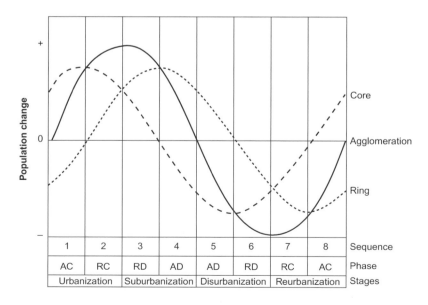

Sequence	1	2	3	4	5	6	7	8
Phase	AC	RC	RD	AD	AD	RD	RC	AC
Stages	Urbanization		Suburbanization		Disurbanization		Reurbanization	

Figure 9.3 The stages of urban development model: classification type – A, absolute; R, relative; C, centralization; D, decentralization (after Klaassen et al., 1981: 18 and Berg et al., 1982: 36)

This formulation of urban development trends, deriving from the early 1980s, is useful in summarizing what was at that time generally held to be the way in which urban population patterns had been developing over the post-war period in Western Europe and, to a considerable extent, across the more developed world as a whole. The 1950s are widely regarded as being dominated by urbanization tendencies under all three definitions outlined above, with a shift of population from rural to urban areas, with an increasing concentration of urban population in the larger cities and (more arguably) with a majority of urban regions seeing their cores growing faster than their rings. By contrast, on average across the most urbanized countries (those in Northern and Western Europe, North America and Australasia), the 1960s are seen as the heyday of 'suburbanization', associated with the rapid growth of population in the commuting hinterlands beyond the main built-up areas, or outside the 'central city' in US parlance, but this is also the period which saw the first significant weakening of Fielding's 'urbanization relationship' as the growth of some of the very large cities began to falter. The 1970s are now generally considered to be the 'decade of counterurbanization', partly because it was during this period that the larger metropolitan areas in many countries witnessed a marked reversal in net migration flows but also partly because even before the end of the decade there were signs of urban revival, notably residential gentrification, which could possibly herald the onset of 'reurbanization'.

At the same time, however, this version of events leaves a number of questions largely unresolved. In the first place, how meaningful is it to talk nowadays in terms of 'suburbanization', as if it is something separate from urbanization? Secondly, what is the real significance of 'counterurbanization' and, in particular, does it constitute a fundamental reversal of traditional urbanization processes leading to the emergence of a radically different 'post-industrial' settlement pattern or is it no more than a temporary stage in the life cycle of an urban region and its wider urban system? Thirdly and linked to this latter point, how much confidence can be placed in the idea of a 'reurbanization' stage and, even more important, in the supposition that this is not merely consolidation in decline but is the precursor of a new lease of life for the large city which introduces a new stage of 'urbanization' at the beginning of a second life cycle for that city (in the terminology of Berg et al., 1987) and leads to a new 'primate city stage' for the urban system as a whole? (in the terminology of Geyer and Kontuly, 1993). The remainder of this chapter draws on the literature that addresses these questions most directly and attempts to assess the extent to which consensus has been reached and what this means for the way in which urban areas in advanced economies are considered likely to develop in the future.

SUBURBANIZATION

Suburbanization is by no means a new phenomenon, with its origins traceable in the building of large homes by more successful entrepreneurs on the outskirts of the burgeoning centres of industry and commerce over a century and a half ago. It became significant as an urban feature during the latter half of the nineteenth century as cheap forms of mass transit loosened the ties between home and workplace for those with secure jobs and relatively 'social' hours of work. Subsequently, it accelerated as a result of further changes in transport and personal wealth, surging ahead especially strongly around mid century, to the extent that in the 'stages of urban development' model it is seen to have become the dominant element of urban change in the 1960s not only in the Anglo-American world but widely across Western Europe. Since then, however, the suburbs have, in one sense, been eclipsed by other developments, and at the same time have begun to undergo a revolution which appears to render the original terminology obsolete.

The term 'suburb' carries connotations of being something less than *urbs*, the city: 'usually residential or dormitory in character, being dependent on the city for occupational, shopping and recreational facilities' (Johnston, 1981: 331). The process has been very largely powered by the negative aspects of city cost, congestion, grime and squalor. Originally, it was dominated by the housing needs and aspirations of the family, with the emphasis very much on the male breadwinner and on healthy space in which the mother could devote her time to bringing up her children and providing for her husband. For some residents, the housing plot would be the sole concern, as everyday supplies would be delivered by firms or fetched by servants, waste would similarly be carted away and children sent off to boarding school. Even when suburban residence broadened beyond the preserve of the upper middle-class, however, the range of facilities would often be very limited: piped water (and possibly gas) supply, mains drainage in some areas (but often the cesspool), and in due course electricity, paved roads, primary school and retail outlets for everyday items (such as newsagent, grocer, post office and perhaps a garage for car maintenance), with the best services being available in strips along radial routes out of the city. The success of the suburbs in providing an escape from the problems of city living as well as allowing people to avoid the limited life chances of the deep countryside was, however, considered as definitely suboptimal by a number of commentators who decried the emerging compartmentalization of urban life, notably Ebenezer Howard, yet his 'garden city' idea – where residents would have easy access to the full range of urban facilities including jobs – had little impact until the 1950s.

Since the 1950s, suburbanization has come to be viewed more in terms of urban (or metropolitan) decentralization, reflecting the shift in patterns of development from the mainly lateral expansion of the city's built-up 'core' to much deeper penetration of house building into an extensive commuting hinterland or 'ring'. This approach to urban analysis and official statistics was adopted in the USA for the 1960 Census in the form of the Standard Metropolitan Area (SMA), which was defined on the basis of a municipality of above a certain population – called 'central city' – together with contiguous counties having metropolitan character linked to the central city by a certain level of commuting – referred to as 'suburbs' in the majority of the literature (Shryock, 1957; US Bureau of the Budget, 1964). In England and Wales this lead was followed by Hall (1971) who identified 100 Standard Metropolitan Labour Areas, each with its 'core' and 'ring' – developments of which were used for the analysis of the 1971 Census (Spence et al., 1982) and the 1981 Census (Champion et al., 1987; Coombes et al., 1982). Similar approaches have been adopted by studies of urban change in Europe, most notably the 'metropolitan areas' defined by Hall and Hay (1980) and the 'functional urban regions' used by Berg et al. (1982), while in most other advanced economies some form of metropolitan area has been identified for academic or official purposes, normally distinguishing between a central core and surrounding commuting zone. A common feature of these approaches to studying 'suburbanization' is that the 'central city' or core incorporates the

older suburbs, certainly those developed immediately adjacent to the main urban area before the First World War and often much of the motor-based extensions of the inter-war period, so what is being studied in this way is the more recent and far-flung manifestations of this centrifugal process.

Defined as such to include the physically separate settlements in the commuting hinterland that are sometimes referred to as 'exurbs', suburban areas now appear to dominate life in many countries. This would seem to be particularly true in the USA, where non-central-city parts of metropolitan areas accounted for 46 per cent of the nation's population in 1990 – up from 31 per cent in 1960, 38 per cent in 1970 and 45 per cent in 1980 (Frey, 1993a). There the 1960s do seem to have represented a landmark in this process: it was at this time that the suburban population overtook that of the central cities, its share of America's metropolitan population rising from 49 to 55 per cent on the basis of the Census definitions used in 1960 and 1970 respectively. At the same time, however, it can be seen that, while perhaps the major transformation from urban to suburban took place at this time, local metropolitan decentralization is a continuing feature of population redistribution in the USA.

Similar conclusions can be drawn from studies tracing the growth of suburban populations in other countries since mid century. For France, for instance, Boudoul and Faur (1982) have documented the particularly strong growth in 1968–75 of the parts of the *unités urbaines* lying outside the central *communes* of most of the largest cities. In Italy during the period 1971–81 the rings of the five major metropolitan areas recorded substantial population gain while their core populations declined or grew only slowly (Dematteis and Petsimeris, 1989). Data for Britain's Functional Regions (Table 9.3) reveal a strengthening and widening of decentralization from cores since the 1950s, with greatest pressures initially on the rings (the primary commuting fields) and subsequently

Table 9.3   *Great Britain: population change, 1951–91, by Functional Region zone*

Zone type	Rate for decade (%)				Deviation from GB rate			
	1951–61	1961–71	1971–81	1981–91	1951–61	1961–71	1971–81	1981–91
Great Britain	4.97	5.25	0.55	2.50				
Core	3.98	0.66	−4.20	−0.09	−0.99	−4.59	−4.75	−2.59
Ring	10.47	17.83	9.11	5.89	5.50	12.58	8.56	3.39
Outer area	1.74	11.25	10.11	8.85	−3.23	6.00	9.56	6.35
Rural area	−0.60	5.35	8.84	7.82	−5.57	0.10	8.29	5.32

Based on population present for 1951–81 and usual residents for 1981–91.

Source: Calculated from Population Censuses. Crown Copyright reserved

on the 'outer areas' (the secondary hinterlands) and the more rural labour market areas beyond.

Though comparisons between countries are bedevilled by variations in areal definitions, the differential scale and pace of decentralization tendencies in Europe can be gauged with reasonable confidence from three comparative studies. According to Hall and Hay (1980), for 15 countries taken as a whole, the proportion of population living in rings rose during the 1960s after losing ground to the cores in the previous decade and grew even more sharply in the first half of the 1970s. In more detail, however, they found a clear contrast between Great Britain, where the rings' proportion was already growing in the 1950s, and Western and Southern Europe (made up primarily of Belgium, France, Italy, the Netherlands and Spain) where the rings' proportion did not begin to increase until the early 1970s. In fact, Northern Europe (Denmark, Norway and Sweden) was the only region to see the onset of above-average ring growth in the 1960s.

Meanwhile, using the 'stages of urban development' approach, Berg et al. (1982) have demonstrated very effectively the progress of 'suburbanization' of the European urban system as a whole and of a dozen separate countries, drawn from both Western and Eastern Europe. Again, across the whole system, it is the 1960s that is definitely the decade when this process was dominant, accounting for 73 per cent of 185 Functional Urban Regions (FURs) in the study, compared with 50 per cent in the previous decade and 63 per cent for 1970–75. For individual countries, there was a profound shift in dominant stage reached in the 1950s, when only Belgium, Great Britain and Switzerland could be considered to have reached the point where the 'suburbanization' stage was dominant to that found in the early 1970s, when only Bulgaria, Hungary and Poland had not yet reached it fully.

Clearly, while originating in the nineteenth century, suburbanization emerged as the dominant process of population redistribution during the post-war period, as the exodus from the older urban cores gathered pace. There is, however, ample evidence that over time the process has grown to become very different from its original nature and has more recently been manifesting itself in novel forms. One distinction has already been drawn above, in terms of the geographical scale shifting from lateral extension of the main built-up area to decentralization over a broad commuting field. A second, equally important development is the progressive disappearance of the 'sub' element of the process, as the arrival of residential population has since the 1950s been followed by the decentralization of industrial, commercial and retail activities and lately by the growth of office and high-tech sectors, the latter

being seen as the 'third wave' of suburbanization in the USA (Cervero, 1989). Over the past ten years American commentators have been coining terms suggestive of a revolution in the form of the city, as the 'new suburbanization' (Stanback, 1991) – with its 'suburban downtowns' (Hartshorn and Muller, 1986) and 'edge cities' (Garreau, 1991) – is increasingly being seen as challenging the central cities and threatening to turn the traditional metropolitan area inside out.

In short, the past quarter of a century has seen nothing less than the urbanization of the suburbs, or – in the far-sighted words of Birch (1975: 25) – a transformation 'from suburb to urban place'. While this process appears to have progressed further in the USA than elsewhere (including Canada) and much debate is taking place even in America about the significance of these 'new suburban landscapes' (Bourne, 1989, 1994), there is no doubt that the distinction between 'urban' and 'suburban' has become increasingly blurred across the more developed world. This raises serious questions about the validity of studying 'suburbanization' as a distinctive phenomenon, whether merely in the guise of urban decentralization or as part of the 'stages of urban development' model, except insofar as the use of geographical boundaries identified in an earlier period can be considered useful for monitoring population shifts subsequently.

COUNTERURBANIZATION

It is not only at the interface between city and suburbs that blurring has occurred, but also at the fringes of the metropolitan areas and functional urban regions, as the 'suburban frontier' has moved outwards to encompass wide tracts of land and embrace formerly free-standing towns and villages. While there is a remarkable degree of correspondence between the various conceptualizations of inter-urban deconcentration, there continues to be heated discussion about whether it is really a qualitatively different phenomenon compared with suburbanization or an extension of it. Much hinges on the extent to which the growth of smaller cities and towns is the result of overspill to places that are new appendages of metropolitan territory, as opposed to arising from residents and employers seeking out these places to take advantage of their inherent characteristics, notably their smaller size and associated economic, social and environmental benefits. As we shall see, a central problem is that in more heavily populated regions it is very difficult, if not impossible, to distinguish between these two explanations purely by the examination of population redistribution patterns, but neither has more in-depth investigation of the processes as yet managed to resolve this fundamental question to general satisfaction. This apparent impasse raises the possibility that, as with urburbanization, it may be that the traditional ways of looking at these developments are now obsolete or were even misconceived in the first place.

The degree of correspondence between the three main conceptualizations of urban change introduced earlier lies in the following: whatever the actual terminology used, urban areas are defined in functional terms to include both core and ring, the process is viewed as being system-wide and, thirdly, larger places in the urban system lose out to smaller ones. As noted above in Figure 9.1, this approach is set out most explicitly in Fielding's (1982) definition of 'counterurbanization' as the situation in which the net migratory growth rate of a place is inversely related to its size and urban status, following a period of transition as the relationship switches from the positive 'urbanization' correlation. Similarly, in Geyer and Kontuly's (1993) concept of 'differential urbanization', the system goes through a transitional period of 'polarization reversal', when the intermediate-sized city category overtakes the primate city in migration growth, before reaching the 'counterurbanization' stage in which the small-city category in its turn becomes the strongest growing (Figure 9.2). The equivalent in the 'stages of urban development' model – the 'disurbanization' stage – is broadly the same, except that the stage reached is initially determined for individual urban regions and also that in this formulation not only does the urban region as a whole have to be in decline but decentralization must also be taking place from core to ring (Figure 9.3). In this approach the settlement system as a whole is reckoned to be in the 'disurbanization' stage, when the latter accounts for a majority of urban regions, including no doubt virtually all the larger cities (Berg et al., 1982).

Judging by the results of the many studies of counterurbanization, there would seem to be much evidence of the emergence of this negative relationship between growth and size of urban region, with the 1970s being a relatively common period across the more developed world. It was in the early 1970s that the 'rural population turn-around' was identified in the USA, with the non-metropolitan growth rate moving above the national figure (Beale, 1975; Berry, 1976b), while within metropolitan America the small metropolitan areas were growing faster than the medium ones and much faster than the large ones (Frey, 1989). Fielding's (1982) results on Western Europe revealed counterurbanization relationships in the 1970s for Belgium, Denmark, France, the Netherlands, Sweden, Switzerland, the UK and West Germany. Champion et al. (1987) iden-

tified a strong negative fit between 1971–81 growth rate and urban status for groupings of Britain's 280 Local Labour Market Areas. Geyer and Kontuly (1993) place France very firmly in the 'small city', i.e. counterurbanization, stage on the basis of its 1975–82 Census results.

On the other hand, other studies do not provide such a uniform picture and it would also appear that several countries were not experiencing counterurbanization at this time. Berg et al.'s work in Europe (1982) found that, while the proportion of urban regions in the 'disurbanization' stage rose steadily between the 1950s and the 1970s, it was still accounting for less than one-fifth of cases by 1970–5. Moreover, their classification of European countries by dominant stage of urban development for 1970–5 put only Belgium unequivocally in the 'disurbanization' stage, with Switzerland, the Netherlands and Great Britain rather evenly poised between 'suburbanization' and 'disurbanization'. According to Fielding (1982), Austria, Ireland, Italy, Norway, Portugal and Spain were still characterized by a positive or insignificant relationship in the 1970s. Similarly in Japan, while there was a major reduction to net migration to metropolitan areas between the 1960s and the 1970s, no clear and sustained reversal took place at this time (Tsuya and Kuroda, 1989). In related work on core/periphery differentials in migration, Cochrane and Vining (1988) identified a 'Periphery of West Europe and Japan' type, where differentials had merely narrowed, in contrast to a 'Northwest Europe' type, where periphery and core had switched over by the 1970s. A relatively cautious interpretation was also being put on events in Canada and Australia, where Bourne and Logan (1976: 136) – rather than finding clear signs of counterurbanization – concluded that 'Urbanization . . . appears to have entered a new period', with internal migration streams shifting away from the two dominant metropolitan areas in each country toward medium-sized cities and to smaller centres just outside the metropolitan regions.

This lack of uniformity in the 1970s can be explained in part by the fact that, because of differences in both history and geography, countries are clearly not all at the same stage of urban development at any particular time, nor indeed are all the regional urban systems within a single country. Mention has already been made of the uneven progress of urbanization across the more developed world and also of the progressive diffusion of the suburbanization tendency across Europe, so nothing less should be expected in relation to counterurbanization. As regards within-country variations, research on the USA traditionally draws a distinction between urban patterns in the North and those in the South and

West (for example, Frey, 1989), while in Western Europe perhaps the most marked regional contrast is the one between the heavily urbanized northwest of Italy and the much more rural Mezzogiorno (Coombes et al., 1989; Dematteis and Petsimeris, 1989). It is therefore not surprising to find some evidence from more recent studies that shows the migration reversal occurring later than the 1970s. Updating of Fielding's work on Western Europe (Champion, 1992) reveals that Austria, Ireland and Italy switched from a concentration to a deconcentration relationship between the 1970s and the early 1980s, and the counterurbanization pattern was found to intensify in Belgium, France and West Germany in the first half of the 1980s (see also Kontuly and Vogelsang, 1989).

Much more surprising for most commentators was the discovery by the mid-1980s that many of the leading counterurbanization countries were experiencing a cutback in this process or indeed some form of reversal of the original migration turnaround. The USA provided the most notable example, with not only the return of the non-metropolitan areas to below average growth in the 1980s but also the resurgence of the largest metropolitan areas to such an extent that by the latter half of the decade their growth rate exceeded the average of those with under a million people, thus reintroducing the pattern of the 1960s (Frey, 1993b). In mainland Europe some countries which had a clear counterurbanization relationship in the 1970s either switched back in the early 1980s to a positive relationship between migration growth and city size (Sweden) or saw a marked decline in counterurbanization (Denmark, the Netherlands), while others experienced a strengthening in their urbanization relationship (Norway, Portugal). Moreover, as the 1980s progressed, in common with the US experience, this slowdown in urban deconcentration and/or move towards greater concentration in large cities intensified in many cases (Finland, the Netherlands, Portugal, Sweden, Switzerland), and even Austria, Belgium and West Germany reverted to concentration in the second half of the decade (Champion, 1992). London, too, staged an impressive recovery from its high rates of population decline of the 1970s, but in terms of internal migration the dominant picture across Britain was of continued, though somewhat lower, rates of deconcentration (Champion, 1994).

This rather general swing away from urban deconcentration in the 1980s has presented a clear challenge to the understanding of recent urban-system trends, though at the same time it provides some opportunity for relating both upturn and subsequent downturn to possible causative factors. It is particularly problematic for those

many commentators who saw counterurbanization as heralding a shift to a new 'post-industrial' form of settlement based on medium-sized and smaller cities and towns and believed that they had uncovered a host of valid explanations for it (Champion, 1989: 236–7). Clearly the experience of Western Europe in the 1980s did not generally bear out Fielding's idea that this decade would see the fulfilment of the switch from an urbanization to a counterurbanization relationship between migration and settlement size, as depicted in Figure 9.1. Even more dramatic is the way in which the events in the USA appear to refute Berry's (1976b: 17) bold pronouncement that, 'A turning point has been reached in the American urban experience. Counterurbanization has replaced urbanization as the dominant force shaping the nation's settlement patterns.'

For the supporters of the counterurbanization thesis, there are at least three potential ways of making sense of the 1980s: that the process had run its course, that a chance combination of factors at that time made the decade anomalous, and that the forces of concentration and deconcentration are subject to some fairly regular cyclic rhythm.

The first interpretation does not survive the most cursory scrutiny. While it is logical that, unlike urbanization, counterurbanization is ultimately self-defeating because it erodes the differences in settlement size which power it, the scale of deconcentration experienced during the 1970s barely dented these differences. In Britain, for instance, the seemingly large-scale loss of population sustained by the 20 Dominant Functional Regions between 1971 and 1981 reduced their share of the national population by less than 3 percentage points, from 39.9 to 37.2 per cent, while in 1981 the smallest 52 of the 280 Local Labour Market Areas averaged only 54,000 people each compared to the overall LLMA average of 194,000 and together they accounted for only one-twentieth of total population but fully one-third of national territory (Champion et al., 1987). Much more fundamental change than this had been expected from the counterurbanization phenomenon.

By contrast, the second interpretation carries considerable weight. It is not difficult to identify features of any period that distinguish it from other periods. It has been suggested that the 1980s were characterized by various events that caused a temporary interruption to the fuller development of the new deconcentration pattern emerging in the 1970s. These include the strong growth of jobs in finance and business services in large cities, downturns in agricultural and energy prices, rationalization of both public sector and private consumer services in more rural areas, the switch of government policy towards inner city rejuvenation and the ageing of the baby boom cohorts into city-loving young adults (Champion, 1992). Providing empirical support for this interpretation is some evidence of a resurgence of urban deconcentration in the first half of the 1990s, notably the revival of non-metropolitan growth rates in the USA (Johnson and Beale, 1994; Nucci and Long, 1995), an acceleration of the metropolitan exodus in Britain (Champion, 1996) and the 'turnaround after the turnaround after the turnaround' in the Netherlands (Van Dam, 1996).

The third line of reasoning – that there is some cyclic rhythm in urban change superimposed on longer term trends – represents an extension of the previous one, albeit on a more formal basis. Support for a cyclic explanation emerged almost as soon as the first signs of the further reversal appeared, notably arising from Cochrane and Vining's (1988) monitoring of migration between core and peripheral regions. Mera (1988) suggested that, as a country reaches an advanced level of development, the pattern of migration becomes susceptible to a host of factors other than economic, citing political and technological factors as contributing to the return to clear urban concentration in Japan in the 1980s. Champion (1988) gave particular prominence to demographic factors, notably the generation-based cycles arising from baby boom and bust. Berry (1988) set the latest trends in the context of two centuries of change in the USA, demonstrating that the strength of urban growth follows a long-wave cycle that could include periods of metropolitan migration reversal. Building on this approach, Champion and Illeris (1990) recommended that research should recognize three separate groups of factors: those pulling towards concentration, those leading towards dispersal, and those that may have different geographical outcomes at different times. The latter include the economic forces of Frey's (1989) 'regional structuring', while the first two together are broadly equivalent to Frey's 'population deconcentration' explanation and fluctuate in their relative strengths over time and thus in their net effect on movement up or down the settlement hierarchy.

Besides these various ways of making sense of the past quarter of a century within a 'counterurbanization' perspective, there are two other main sets of reactions to the developments of the 1970s. One is to claim that there is, and has been, no such thing as counterurbanization, with Gordon (1979) being the first of many to criticize the idea of a 'clean break' from traditional urbanization and associated urban concentration. To the extent that apparent population dispersal has been taking place, this is seen either as suburbanization and local metropolitan decentralization on a wider scale outpacing the redefinition of metropolitan areas and urban regions or as the embryonic stages in the growth of major new metropolitan concentrations that will come to dominate the national settlement pattern in due course. There is plenty of evidence in support of the 'overspill' argument, most recently and comprehensively in the US context by Nucci and Long (1995) and also relating to Australia by Hugo (1994), but so far it is not clear whether the new metropolitan areas are destined to remain relatively small and independent in daily-urban-system terms or to develop into the megalopolises of the future.

The other approach is conceptually the neatest, as the general switch away from urban deconcentration in the 1980s is exactly what is predicted in the 'stages of urban development' model. Rather than forming the long-term future tendency as suggested in the 'counterurbanization' model of Figure 9.1, the equivalent 'disurbanization' phenomenon is seen as being only a temporary stage in the development of any single agglomeration and urban region, being followed by a stage of 'reurbanization' before the hypothesized onset of a second cycle (Figure 9.3). Similarly, the 'differential urbanization' approach (Figure 9.2) anticipates the fading of the net migration gains of the small city category and the re-emergence of the primate city category as the fastest growing, ushering in a new period of urbanization. The next section trawls the literature in order to discover the extent to which there has been a recovery by the larger cities and to try and establish the theoretical and longer-term significance of any such change.

REURBANIZATION

The prime issue for this section to tackle concerns the nature and significance of the recovery of larger urban areas and their inner cities since the period of widespread population exodus in the 1970s. As just mentioned, one hypothesis is that it constitutes another natural stage in the 'life' of individual urban areas and of the urban system as a whole. Alternatively, adopting the analogy of a dying star, it may merely represent a period of consolidation, as an urban area fits itself to new circumstances, or it may lead to some form of self-reinforcing 'implosion' as a 'black hole' develops, sending shock waves across the rest of the system. As we will see below, the originators of the cyclic model were extremely cautious about the prospects for large-scale reurbanization and the evidence available so far gives only rather limited support to this conceptualization. More generally, however, a number of reasons have been put forward to explain the recent recovery of larger cities, raising the question as to how long-lived this process is likely to be and whether it is significant enough to initiate a new round of urban growth and subsequent decentralization.

Compared to the great attention which has been given to it subsequently, the originators of the 'stages of urban development' model proposed the 'reurbanization' stage in a remarkably tentative manner. In the words of Klaassen and Scimeni (1981: 16), '*It is not impossible* that . . . the large town, with its economic and cultural potentials, will in the course of time win back the ground it has lost, that people once more flock towards the city, and that the town starts upon a new period of bloom, which may lead to a new cyclus' (emphasis added). They expressed reservations about the term itself, pointing out that this 'reurbanization' stage is one of continuous urban-region decline and only towards the end of it is the core meant to move into absolute population growth: 'One might wish to reserve the term "reurbanization" for the first phase of a second urban cycle' (1981: 26). They admitted that there was so far no evidence of a second cycle, that only a few cities were as yet to be found in this fourth stage of the first cycle and that these were the wrong ones from the model's point of view, all being in the small-size bracket rather than the largest cities in the most advanced stages of the cycle (1981: 26–7). Finally, while they offered the possibility that this stage might come about through market forces, their main hopes were pinned on purposeful government action – as was made even clearer in the policy agenda set out in the follow-up study which explicitly states: 'Reurbanization must be brought about fast and resolutely' (Berg et al., 1982: 44).

More recent searches for evidence of places entering the reurbanization stage have yielded what can only be described as mixed results, at best. The most systematic attempt made so far is that by Cheshire (1995), focusing on 241 FURs with at least 330,000 inhabitants in the European Union (see Figure 9.4). On the one hand, according to Cheshire (1995: 1058), among these large FURs 'the regular onward march of decentralisation appears to have faltered and, in northern Europe, it has halted, even reversed'. The proportion of urban cores gaining population was found to have recovered somewhat in Northern Europe (defined as Belgium, Denmark, the Netherlands, the UK and West Germany), reaching 47 per cent in 1981–91 after slumping to 38 and 22 per cent in 1971–5 and 1975–81 respectively, and some recovery from the later 1970s was also evident for the France/Northern Italy group. On the other hand, these changes were not enough to produce any significant increase in the number of FURs that could be classed as being in

the reurbanization stage, even in northern Europe. In fact, here the proportion of FURs in this stage fell from 15 per cent in the later 1970s to only 6 per cent in the 1980s, and instead the overall effect of the changes was to push the urban systems back into the suburbanization stage. Moreover, it was mainly the smaller FURs – particularly those with ancient cathedrals and universities – that were participating in this stage, rather than the larger and older urban regions that were meant to lead the process. The UK's experience was fairly typical, with this stage accounting for only four of its 36 large FURs: Glasgow, Oxford, Cambridge and Canterbury, with only Glasgow conforming to the model's expectations.

Nevertheless, across the advanced economies there is now much case study evidence of a marked change of fortunes for large cities from the high levels of population loss sustained in the 1970s. In the case of Glasgow, for example, Lever (1993a) has shown how the whole Strathclyde region switched from substantial loss in the 1970s to small gain in the 1980s, brought about almost entirely by the reduction in Glasgow City's annual rate of loss from 22,000 to less than 1,000. Greater London's recovery in the 1980s was even more impressive, and Inner London's particularly so (Champion and Congdon, 1988). The rate of population loss for all 280 of Britain's urban cores fell from 4.2 per cent in 1971–81 to a mere 0.1 per cent for 1981–91, while the growth rate of the rings fell back substantially, particularly in comparison with the national population trend (see Table 9.3). In Scandinavia, Copenhagen saw the stabilization of its central-city population in the 1980s (Illeris, 1994), while according to Nyström (1992) the cities of Helsinki, Oslo and Stockholm experienced actual growth, though not as strongly as in the suburbs except in the case of Stockholm and even here the pattern varied during the decade and was again favouring the outer areas by the early 1990s (Borgegård et al., 1995). The overall impression conveyed by these studies is of a reduced level of variation in population growth rates between places – a pattern which is consistent with the shrinkage in regional differentials in net migration rates across Europe since the 1970s observed by Champion et al., (1996).

The picture appears similar in the New World, though trends there are superimposed on a generally more dynamic demographic situation. In the USA, as noted above, the 1980s saw the re-emergence of the larger metropolitan areas as the fastest growing element of the urban landscape, with Frey and Speare (1992: 133) referring to 'the reurbanization of the 1980s in contrast to the 1970s pattern'. Overall, metropolitan areas with one million or more people grew by 12 per cent

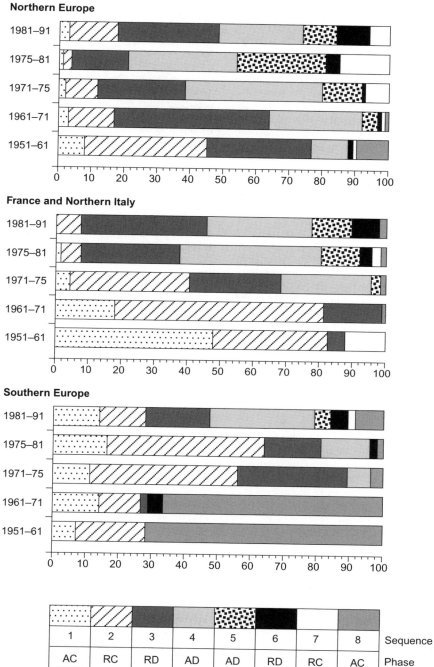

Northern Europe

France and Northern Italy

Southern Europe

								Sequence
1	2	3	4	5	6	7	8	
AC	RC	RD	AD	AD	RD	RC	AC	Phase
Urbanization		Suburbanization		Disurbanization		Reurbanization		Stages

Figure 9.4 Change in frequency distributions of Functional Urban Regions by stage of urban development, 1951–91, for three European regions; see Figure 9.3 for definitions of stage and phase (after Cheshire, 1995, Table 1: 1051)

in the 1980s compared with 8 per cent in the previous decade. Much of this growth was powered by the West and South, but even the North's large metropolitan areas switched from a decline of 0.9 per cent between 1970 and 1980 to a 2.8 per cent increase in the 1980s. New York (defined on the CMSA basis to include 18 million people in 1990) was particularly impressive, with a switch from a 3.6 per cent loss to a 3.1 per cent gain between the two decades, and there was also strong upward shifts for Philadelphia and Boston but not for Chicago and Detroit. Within metropolitan America overall, the annual average growth rate for central cities rebounded from 0.09 to 0.64 per cent between the two decades and the suburban growth rate fell back somewhat, but in the North-East and Midwest central cities still seemed to be struggling, their average annual growth rate going up from −1.09 to +0.03 per cent in the former and up from −0.85 to −0.22 per cent in the Midwest (Forstall, 1991).

Australia and Canada also provide evidence of strengthening metropolitan-area and inner-city growth in the 1980s. In the case of the former, Melbourne appears particularly dynamic in the latter half of the decade, with an overall population increase of 190,000 in 1986–91 (not far short of the 208,000 increase for the previous *ten* years), Brisbane grew by 185,000 in the five years compared to 195,000 for the previous decade, and Perth 149,000 as opposed to 218,000, while Sydney and Adelaide both continued growing at close to their earlier rate (Hugo, 1994). For Canada, the 1980s witnessed a switch in the focus of population growth back to the central provinces, notably the Toronto CMA which added almost 0.9 million people, or 30 per cent, between 1981 and 1991 – twice the 1970s growth of 15 per cent and just under 0.4 million (Bourne and Olvet, 1995). In both cases, it was the outer areas that contributed most to these gains, but inner city recovery was also common. In the case of Toronto, for instance, seven-eighths of the CMA's growth took place outside the administrative Metro area – adding over three-quarters of a million people there and almost doubling the 1981 population – and a further one-ninth was accounted for by the suburban parts of Metro Toronto, but the City of Toronto also saw a 6 per cent increase in population in the 1980s, a marked switch from its 16 per cent decline in the 1970s (Bourne and Olvet, 1995). More generally across Canada, inner cities – defined narrowly to include only the CBD and surrounding zone of mixed uses – shifted from previous decline into growth in the early 1980s (Ram et al., 1989, quoted by Bourne, 1991; Ley, 1992). A similar renaissance has been documented for Sydney, Melbourne and Adelaide (respectively, Murphy and Watson, 1990; King, 1989; Badcock, 1991).

Two somewhat contradictory sets of conclusions would seem to flow from these empirical observations. On the one hand, there are widespread signs of renewed growth or reduced decline for larger metropolitan areas and urban regions and also of a population recovery for urban cores and their inner areas. On the other, however, there appears to be no evidence of suburban ring areas losing out to core areas – not even relatively, let alone in absolute terms whereby core growth takes place alongside suburban decline, as required of the later phase of reurbanization in the 'stages of urban development' model (see Figure 9.3). The New World experience of the 1980s and early 1990s would seem to echo Cheshire's (1995) findings for Europe, namely that during this period there has been backtracking towards the pattern associated with the first phase of the suburbanization stage, involving renewed growth of both urban region and core but also with relative decentralization from core to ring.

Given the cyclic approach's apparent failure to caricature recent events, it is necessary to examine the individual lines of explanation which have been put forward to account for them. The literature falls into two rather separate camps, one dealing with the relative performance of urban regions as individual entities and the other focusing quite specifically on inner-city revitalization including gentrification. The former has already been introduced in terms of the counterurbanization debate, itself commonly being compartmentalized between the population deconcentration perspective and the effects of economic restructuring. In relation to the latter, much emphasis in the 1980s was placed on identifying 'winners' and 'losers' in the new global competition between cities. Studies of the USA (for example, Frey, 1987; Frey and Speare, 1992; Noyelle and Stanback, 1984) have shown that the greatest benefits have accrued to the larger metropolitan areas that serve as centres of finance, corporate headquarters and advanced services and to places specializing in information, high technology or knowledge-based industries, together with certain consumer-oriented areas with tourism and retirement functions. In Europe, too, various studies have pointed to the growing concentration of corporate power in particular cities, demonstrated the growing distinctiveness of cities by functional type and measured 'urban success' for different levels of the settlement hierarchy (respectively Meijer, 1993; Kunzmann and Wegener, 1991; Parkinson, 1991; see also Illeris, 1992; Lever, 1993b). Similarly, according to Bourne and Olvet, 'Canadian cities are becoming more unlike in terms of their socio-economic attributes', leading them to wonder whether there is still 'such an object as a "Canadian" city' (1995: 47).

As regards inner-city revitalization, there is now a voluminous literature on gentrification and other aspects of redevelopment (Smith and Herod, 1992). Perhaps the most fundamental point made there is that gentrification, as originally defined in terms of the social upgrading of a neighbourhood through the rehabilitation of the existing housing stock, is only one element – and by the 1980s only a minor one – of a much broader process. As set out by Bourne (1992), residential revitalization on the ground comprises four main components: intensification through the construction of infill housing or the high-density redevelopment of older neighbourhoods, implantation through housing being inserted into existing high density commercial and institutional districts, conversion of older non-residential structures such as unused warehouses, and extension through the penetration of residential uses into formerly non-residential areas such as vacant railway, port and industrial sites. This last component has been responsible for some of the most impressive changes in inner city areas, particularly where associated with waterfront redevelopment schemes such as for New York, Toronto, Vancouver and London Docklands (see, respectively, Fainstein, 1994; Bourne and Olvet, 1995; Ley, 1981; Church, 1988).

There seems to be wide agreement about the types of reasons for the residential revitalization of inner urban areas in the 1980s (see, for instance, reviews by Bourne, 1992; Badcock, 1995; Lever and Champion, 1996). As envisaged originally by Klaassen et al. (1981) and Berg et al. (1982), in most cases government action played a key role in encouraging and facilitating the process, involving the winding up of previous dispersal policies and the disbursement of often huge amounts of public money into programmes ranging from grants for improving individual homes and for local economic development incentives to massive investment in land reclamation and upgrading of infrastructure. But market forces are also seen to have played a very important role. Chief among these, as for the recovery of the whole metropolitan area, is economic restructuring, with inner-city deindustrialization running its course and with the rapid growth at this time of advanced producer services that seem less prone to decentralization than manufacturing and personal services. The greater availability and falling relative cost of space in inner areas has been cited frequently, both in terms of the 'rent-gap' thesis in explanations of the onset of gentrification and more generally in recognition of the vacant and underused land and buildings left by previous decades of population exodus and business closures and relocations, contrasting with

rising land costs and congestion in more suburban areas. Higher energy costs since the mid-1970s have also been cited as the cause of more high income people choosing to live close to CBD-based (central business district) jobs in large cities in preference to a long commute.

Demographic developments and related social changes have also been widely quoted in explanations of residential revitalization in the inner areas of large cities. Increased immigration, whether arising from growing labour market demands or as a result of the huge rise in asylum-seeking in the 1980s and 1990s, has not only been focused on the largest urban regions of destination countries but has normally impacted most strongly on their more central residential areas. Here the lower-skill migrants stand the best chances of finding older cheaper housing released by the previous exodus as well as ready access to the biggest concentrations of menial jobs created by the latest round of the economic restructuring, while high-skill international migrants – often on short-term postings for their companies – can take advantage of the cultural and entertainment attractions in the down town area. Secondly, and related to the latter, was the emergence of young, upwardly mobile professionals ('yuppies') as a significant component of society and urban geography, as these singles or partners – usually in the pre-child stage of the family life cycle and working within the financial and producer-services sectors – took over from the original 'Bohemian' gentrifiers of artists and academics as the main force in the social upgrading of inner areas in the 1980s. Thirdly, has been the much more general process of household change, involving the rapid increase in lone-parent families, in households without children and particularly in individuals living on their own as a result of population ageing, rising rates of divorce and separation and the growing tendency for young adults to leave the parental home for higher education, work, independence or escape from abuse. By and large, these groups prefer – or, because of limited financial resources, have to make do with – smaller housing units in less sought-after areas compared to the stereotypical suburban family.

What the literature is far less in agreement about, however, is the significance of the repopulation of inner city areas as a phenomenon. Part of this debate is related to whether a fundamental change is taking place in the nature of inner city populations – attitudes to which seem to vary considerably according to particular countries and cities and to the criteria adopted. In an analysis of Canadian cities, for instance, Bourne (1993a) was unable to detect any clear signs of a reversal of the traditional direct relationship between social class and distance from city centres, but Filion (1987) and Ley (1992) are much more upbeat about the achievements of residential revitalization in Canada, particularly by comparison with the changes occurring in the USA's largest cities. In the Australian context, Badcock (1993, 1995) reckons that the progress of revitalization has proceeded even more strongly than in Canada, while in continental Europe it is generally concluded that inner city areas were not as severely affected by depopulation in the 1970s as in the New World and have benefited considerably from the insertion of new residents and housing (see, for instance, Harding et al., 1994). Nevertheless, in both Old World and New, the arrival of wealthier residents has often gone hand-in-hand with the impoverishment of the remaining residents, with the emergence of the 'dual city' (Fainstein et al., 1992; Mollenkopf and Castells, 1991). Only time will tell whether the up-market housing investment will constitute more than 'islands of renewal in seas of decay' (Berry, 1985: 69).

The likely longevity of inner city revitalization, however, also appears to be a highly contentious issue (Lees and Bondi, 1995). For instance, Bourne (1993b, 1993c) argues that the conditions producing gentrification are waning and in some cases being reversed, ushering in a 'post-gentrification era'. He cites the effects of the mid-1960s 'baby bust' on the pool of potential young gentrifiers in the 1990s, allied with the shake-out in employment in financial services which began in the late 1980s and was reinforced by the early 1990s recession, and he also observes that newer immigrant groups are tending to bypass the inner city and go directly to the suburbs. By contrast, Badcock (1993, 1995) reckons that residential revitalization 'shows no signs of abating during the 1990s' (1993: 194), emphasizing the supply-side aspects relating to growing market confidence in inner city areas and the long-term nature of the trends towards advanced services employment growth, increasing household fission and greater women's labour market participation.

In conclusion, reurbanization as defined by Klaassen et al. (1981) has not yet emerged as a significant feature in the urban system of advanced economies and there is considerable disagreement about the extent to which the inner city residential revitalization which took place in the 1980s will be able to continue and lead to a fundamental change in the form of the Western city. This is a question that is taken up in more depth in other chapters, so suffice it here to say that much depends on the scale of the challenge in terms of differences between cities in the current disparities between their inner and outer areas. Also crucial, given the importance of public funds in underwriting much of the costs of inner renewal in the 1980s and 1990s, is the extent

to which housing providers and buyers can be convinced that the appropriate range of private and public services will be available in these areas in the future and that new investment will not be undermined by the further growth of 'underclass' neighbourhoods. A final, critical issue is that of densification and urban consolidation, whereby increased house building on empty or derelict land and potential green spaces may lead to increased traffic congestion and less attractive environments for residents, representing a difficult balance to be struck.

CONCLUDING COMMENTS

If there is one certainty to be drawn from this review of urban population trends, it is that nothing is the same as it was in those apparently straightforward days when 'urbanization' meant greater population concentration in any geographical framework used. First, most parts of the developed world have reached the situation where there is little scope for further increases in the proportion of people classified as living in urban places. Secondly, local decentralization, while perhaps the most enduring feature of the past few decades, is now very different from the dormitory style *sub*urbanization of the early post-war period, now involving the veritable 'urbanization of the suburbs' and some withdrawal of 'urban' facilities from the traditional cities through a form of 'deurbanization' process. Thirdly, in the light of the apparent rise and fall of counterurbanization and of the rather patchy and limited progress made so far by reurbanization, it is tempting to conclude that urban systems are currently passing through some sort of transitional period, but a transition for which no clear outcome is in sight – or, at least, none that can be presented in as simple terms as the conceptualizations discussed in this chapter.

Fundamentally, the advanced economies have moved to a situation 'beyond urbanization'. Not only is the overall concept of 'urbanization' now of very limited value – indeed one could readily argue that its continued use would be obstructive rather than merely neutral – but in retrospect it also seems not to have been very helpful to conceptualize subsequent events as variants or opposites of traditional 'urbanization' by adding prefixes to the term. If this conclusion is anywhere near correct, then maybe it is a mistake to rationalize the latest developments as 'back to the future', though this phrase does neatly encapsulate the idea of the resumption of previous trends within a new and evolving context. Perhaps it is more realistic to recognize that the present-day urban systems that form the basis for the future are very different from those which existed 50 years ago, that these are being affected by a range of individual and only partially linked processes rather than some unidirectional form of progression, and that the overall result on the ground is a mosaic of growth and decline, representing the outcome of the success and failure of separate spatial groupings of people to adjust to the opportunities and weaknesses inherent in their various circumstances. Merely to talk in terms of 'systems', however, reflects the desire to find some order out of apparent chaos and to have some certainties to build on, so the continuing talk of 'Blue Bananas', 'growth corridors', 'edge cities', 'multinodal regions', even 'return to the city', may possibly serve to influence the future course of events.

REFERENCES

Badcock, B. (1991) 'Neighbourhood change in inner Adelaide: an update', *Urban Studies*, 28 (4): 553–8.

Badcock, B. (1993) 'Notwithstanding the exaggerated claims, residential revitalisation really is changing the form of some western cities: a response to Bourne', *Urban Studies*, 30 (1): 191–5.

Badcock, B. (1995) 'Building upon the foundations of gentrification: inner-city housing development in Australia in the 1990s', *Urban Geography*, 16 (1): 70–90.

Beale, C.L. (1975) *The Revival of Population Growth in Non-metropolitan America*. Washington, DC: Economic Research Service, US Department of Agriculture.

Berg, L. van der, Burns, L.S. and Klaassen, L.H. (eds) (1987) *Spatial Cycles*. Aldershot: Gower.

Berg, L. van der, Drewett, R., Klaassen, L.H., Rossi, A. and Vijverberg, C.H.T. (1982) *Urban Europe*, vol. 1: *A Study of Growth and Decline*. Oxford: Pergamon.

Berry, B.J.L. (ed.) (1976a) *Urbanization and Counterurbanization*. Beverly Hills, CA: Sage.

Berry, B.J.L. (1976b) 'The counterurbanization process: urban America since 1970', in B.J.L. Berry (ed.), *Urbanization and Counterurbanization*, Beverly Hills, CA: Sage. pp. 17–30.

Berry, B.J.L. (1985) 'Islands of renewal in seas of decay', in P. Petersen (ed.), *The New Urban Reality*. Washington, DC: The Brookings Institution. pp. 69–96.

Berry, B.J.L. (1988) 'Migration reversals in perspective: the long-wave evidence', *International Regional Science Review*, 11 (3): 245–52.

Birch, D.L. (1971) 'Towards a stage theory of urban growth', *Journal of the American Institute of Planners*, 37 (2): 78–87.

Birch, D.L. (1975) 'From suburb to urban place', *Annals of the American Academy of Political and Social Science*, 422: 25–35.

Borgegård, L.-E. Håkansson, J. and Malmberg, G. (1995) 'Population redistribution in Sweden: long-term trends and contemporary tendencies', *Geografiska Annaler*, 77B (10: 31–45.

Boudoul, J. and Faur, J.-P. (1982) 'Renaissance des communes rurales ou nouvelle forme d'urbanisation?' *Economie et Statistiques*, 149: I–XVI.

Bourne, L.S. (1989) 'Are new urban forms emerging? Empirical tests for Canadian urban areas', *Canadian Geographer*, 33 (4): 312–28.

Bourne, L.S. (1991) 'Recycling urban systems and metropolitan areas: a geographical agenda for the 1990s', *Economic Geography*, 67 (3): 185–209.

Bourne, L.S. (1992) 'Population turnaround in the Canadian inner city: contextual factors and social consequences', *Canadian Journal of Urban Research*, 1 (1): 69–92.

Bourne, L.S. (1993a) 'Close together and worlds apart: an analysis of changes in the ecology of income in Canadian cities', *Urban Studies*, 30 (8): 1293–317.

Bourne, L.S. (1993b) 'The demise of gentrification? A commentary and prospective view', *Urban Geography*, 14 (1): 95–107.

Bourne, L.S. (1993c) 'The myth and reality of gentrification: a commentary on emerging urban forms', *Urban Studies*, 30 (1): 183–9.

Bourne, L.S. (1994) *Urban Growth and Population Distribution in North America*. Report prepared for IIED London and UN Centre for Human Settlements. Toronto: Centre for Urban and Community Studies, University of Toronto.

Bourne, L.S. and Logan, M.I. (1976) 'Changing urbanisation patterns at the margin: the examples of Australia and Canada', in B.J.L. Berry (ed.), *Urbanization and Counterurbanization*, Beverly Hills, CA: Sage. pp. 111–43.

Bourne, L.S. and Olvet, A.E. (1995) 'New urban and regional geographies in Canada: 1986–91 and beyond', *Major Report*, 33, Centre for Urban and Community Studies, University of Toronto.

Cervero, R. (1989) *America's Suburban Centers: The Land Use–Transportation Link*. Boston: Unwin Hyman.

Champion, A.G. (1988) 'The reversal of the migration turnaround: resumption of traditional trends?' *International Regional Science Review*, 11 (3): 253–60.

Champion, A.G. (ed.) (1989) *Counterurbanisation: The Changing Pace and Nature of Population Deconcentration*. London: Edward Arnold.

Champion, A.G. (1992) 'Urban and regional demographic trends in the developed world', *Urban Studies*, 29 (3/4): 461–82.

Champion, A.G. (1994) 'Population change and migration in Britain since 1981: evidence for continuing deconcentration', *Environment and Planning A*, 26 (10): 1501–20.

Champion, A.G. (1996) 'Migration to, from and within the United Kingdom', *Population Trends*, 83: 5–16.

Champion, A.G. and Congdon, P.D. (1988) 'Recent population trends for Greater London', *Population Trends*, 53: 11–17.

Champion, A.G. and Illeris, S. (1990) 'Population redistribution trends in Western Europe: a mosaic of
dynamics and crisis', in M. Hebbert and J.C. Hansen
(eds), *Unfamiliar Territory: The Reshaping of European Geography*. Aldershot: Avebury. pp. 236–53.

Champion, A.G., Mønnesland, J. and Vandermotten,
C. (1996) 'The new regional map of Europe',
Progress in Planning, 46 (1): 1–89.

Champion, A.G., Green, A.E., Owen, D.W., Ellin, D.J.
and Coombes, M.G. (1987) *Changing Places:
Britain's Demographic, Economic and Social
Complexion*. London: Edward Arnold.

Cheshire, P. (1995) 'A new phase of urban development
in Western Europe? The evidence for the 1980s',
Urban Studies, 32 (7): 1045–63.

Church, A. (1988) 'Demand-led planning, the inner-city
crisis and the labour market: London Docklands
evaluated', in B.S. Hoyle, D.A. Pinder and M.S.
Husain (eds), *Revitalising the Waterfront: International Dimensions of Dockland Redevelopment*.
London: Belhaven/Pinter. pp. 199–221.

Cochrane, S.G. and Vining, D.R. (1988) 'Recent trends in
migration between core and periphery regions in developed and advanced developing countries', *International Regional Science Review*, 11 (3): 215–44.

Coombes, M.G., Dalla Longa, R. and Raybould, S.
(1989) 'Counterurbanisation in Britain and Italy: a
comparative critique of the concept, causation and
evidence', *Progress in Planning*, 32: 1–70.

Coombes, M.G., Dixon, J.S., Godard, J.B., Openshaw,
S. and Taylor, P.J. (1982) 'Functional regions for the
population census of Great Britain', in D.T. Herbert
and R.J. Johnston (eds), *Geography and the Urban
Environment: Progress in Research and Applications*.
Chichester: Wiley. pp. 63–112.

Dematteis, G. and Petsimeris, P. (1989) 'Italy: counterurbanisation as a transitional phase in settlement
reorganisation', in A.G. Champion (ed.) *Counterurbanisation: The Changing Pace and Nature of Population Deconcentration*. London: Edward Arnold. pp.
187–206.

Fainstein, S. (1994) *The City Builders: Property, Politics
and Planning in London and New York*. Oxford:
Blackwell.

Fainstein, S., Harloe, M. and Gordon, I. (eds) (1992)
*Divided Cities: New York and London in the Contem-
porary World*. Oxford: Blackwell.

Fielding, A.J. (1982) 'Counterurbanisation in Western
Europe', *Progress in Planning*, 17 (1): 1–52.

Filion, P. (1987) 'Concepts of the inner city and recent
trends in Canada', *The Canadian Geographer*, 31 (3):
223–32.

Forstall, R.L. (1991) 'Regional and metropolitan population trends in the United States 1980–90', Paper
presented at the Association of American Geographers Annual Meeting, Miami, Florida (quoted by
Frey, 1993a).

Frey, W.H. (1987) 'Migration and depopulation of the
metropolis: regional restructuring or rural renaissance?' *American Sociological Review*, 52 (2): 240–57.

Frey, W.H. (1989) 'United States: counterurbanisation
and metropolis depopulation', in A.G. Champion
(ed.), *Counterurbanisation: The Changing Pace and
Nature of Population Deconcentration*. London:
Edward Arnold. pp. 34–61.

Frey, W.H. (1990) 'Metropolitan America: beyond the
transition', *Population Bulletin* 45 (2). Washington,
DC: Population Reference Bureau.

Frey, W.H. (1993a) 'People in places: demographic trends
in urban America', in J. Sommers and D.A. Hicks
(eds), *Rediscovering Urban America: Perspectives on the
1980s*. Washington, DC: US Department of Housing
and Urban Development, pp. 3–1–3–106.

Frey, W.H. (1993b) 'The new urban revival in the United
States', *Urban Studies*, 30 4/5: 741–74.

Frey, W.H. and Speare, A. (1992) 'The revival of metropolitan population growth in the United States: an
assessment of the findings of the 1990 Census', *Population and Development Review*, 18 (1): 129–46.

Garreau, J. (1991) *Edge City: Life on the New Frontier*.
New York: Doubleday.

Geyer, H.S. and Kontuly, T.M. (1993) 'A theoretical
foundation for the concept of differential urbanisation', *International Regional Science Review*, 15 (12):
157–77.

Gordon, P. (1979) 'Deconcentration without a "clean
break"', *Environment and Planning A*, 11: 281–9.

Hall, P. (1971) 'Spatial structure of metropolitan
England and Wales', in M. Chisholm and G.
Manners (eds), *Spatial Policy Problems of the British
Economy*. Cambridge: Cambridge University Press.
pp. 96–125.

Hall, P. and Hay, D. (1980) *Growth Centres in the European Urban System*. London: Heinemann.

Harding, A., Dawson, J., Evans, R. and Parkinson, M.
(eds) (1994) *European Cities towards 2000: Profiles,
Policies and Prospects*. Manchester: Manchester
University Press.

Hartshorn, T.A. and Muller, P.O. (1986) *Suburban Busi-
ness Centers: Employment Implications*. Washington,
DC: US Department of Commerce, Economic
Development Administration.

Hugo, G. (1994) 'The turnaround in Australia: some
first observations from the 1991 Census', *Australian
Geographer*, 25: 1–17.

Illeris, S. (1992) 'Urban and regional development in
Western Europe in the 1990s: a mosaic rather than
the triumph of the "Blue Banana"', *Scandinavian
Housing and Planning Research*, 9: 201–15.

Illeris, S. (1994) 'Why was the central city population
stabilised? The case of Copenhagen', in G.O. Braun
(ed.), *Managing and Marketing of Urban Development and Urban Life*. Berlin: Dietrich Reimer Verlag.
pp. 595–608.

Johnson, K.M. and Beale, C.L. (1994) 'The recent
revival of widespread population growth in non-
metropolitan areas of the United States', *Rural Sociology*, 59: 655–67.

Johnston, R.J. (ed) (1981) *The Dictionary of Human
Geography*. Oxford: Blackwell.

King, R.J. (1989) 'Capital switching and the role of ground rent: 2. Switching between circuits and switching between submarkets', *Environment and Planning A*, 21: 711–38.

Klaassen, L.H., Molle, W.T.M. and Paelinck, J.H.P. (eds) (1981) *Dynamics of Urban Development*. Aldershot: Gower.

Klaassen, L.H. and Scimeni, G. (1981) 'Theoretical issues in urban dynamics', in L.H. Klaassen, W.T.M. Molle and J.H.P. Paelinck (eds), *Dynamics of Urban Development*. Aldershot: Gower. pp. 8–28.

Kontuly, T.M. and Vogelsang, R. (1989) 'Federal Republic of Germany: the intensification of the migration turnaround', in A.G. Champion (ed.), *Counterurbanisation: The Changing Pace and Nature of Population Deconcentration*. London: Edward Arnold. pp. 141–61.

Kunzmann, K.R. and Wegener, M. (1991) 'The pattern of urbanisation in Western Europe 1960–1990', *Berichte aus dem Institut für Raumplanung*, 26. Dortmund: Dortmund University.

Lees, L. and Bondi, L. (1995) 'De-gentrification and economic recession: the case of New York City', *Urban Geography*, 16 (3): 234–53.

Lever, W.F. (1993a) 'Reurbanisation – the policy implications', *Urban Studies*, 30 (2): 267–84.

Lever, W. (1993b) 'Competition within the European urban system', *Urban Studies*, 30: 935–48.

Lever, W. and Champion, A. (1996) 'The urban development cycle and the economic system', in A. Bailly and W.F. Lever (eds), *The Spatial Impact of Economic Changes in Europe*. London: Avebury. pp. 204–27.

Ley, D.F. (1981) 'Inner city revitalisation in Canada: a Vancouver case study', *Canadian Geographer*, 25: 124–48.

Ley, D. (1992) 'Gentrification in recession: social change in six Canadian cities, 1981–86', *Urban Geography*, 13 (3): 230–56.

Meijer, M. (1993) 'Growth and decline of European cities: changing positions of cities in Europe', *Urban Studies*, 30 (6): 981–90.

Mera, K. (1988) 'The emergence of migration cycles?', *International Regional Science Review*, 11 (3): 253–60.

Mollenkopf, J. and Castells, M. (eds) (1991) *Dual City*. New York: Russell Sage Foundation.

Morrison, P.A. and Wheeler, J.P. (1976) 'Rural renaissance in America?' *Population Bulletin*, 31 (3): 1–27. Washington, DC: Population Reference Bureau.

Murphy, P. and Watson, S. (1990) 'Restructuring of Sydney's central industrial areas: processes and local impacts', *Australian Geographical Studies*, 28: 187:203.

Noyelle, T. and Stanback, T. (1984) *The Economic Transformation of American Cities*. Towota, NJ: Rowman & Allanheld.

Nucci, A. and Long, L. (1995) 'Spatial and demographic dynamics of metropolitan and nonmetropolitan territory in the United States', *International Journal of Population Geography*, 1 (2): 165–81.

Nyström, J. (1992) 'The cyclical urbanisation model: a critical analysis', *Geografiska Annaler*, 74B (2): 133–44.

Parkinson, M. (1991) 'European cities in the 1990s: problems and prospects', in I. Jackson (ed.), *The Future of Cities in Britain and Canada*. Aldershot: Dartmouth. pp. 67–87.

Ram, B., Norris, M. and Skof, K. (1989) *The Inner City in Transition*. Ottawa: Statistics Canada (quoted in Bourne, 1992).

Richardson, H.W. (1977) 'City size and national spatial strategies in developing countries', *Staff Working Paper*, 252. Washington, DC: World Bank.

Richardson, H.W. (1980) 'Polarisation reversal in developing countries', *Papers of the Regional Science Association*, 45: 67–85.

Robert, S. and Randolph, W.G. (1983) 'Beyond decentralisation: the evolution of population distribution in England and Wales, 1961–1981', *Geoforum*, 14: 75–192.

Roberts, S. (1991) 'A critical evaluation of the city life cycle idea', *Urban Geography*, 12 (5): 431–49.

Rust, E. (1975) *No Growth: Impacts on Metropolitan Areas*. Lexington, MA: Lexington Books.

Shryock, H. (1957) 'The natural history of Standard Metropolitan Areas', *American Journal of Sociology*, 63: 163–70.

Smith, N. and Herod, A. (1992) 'Gentrification: a comprehensive bibliography', *Discussion Paper, New Series* 1. Rutgers University, New Brunswick, NJ.

Spence, N.A. Gillespie, A., Goddard, J., Kennett, S., Pinch, S. and Williams, A.M. (1982) *British Cities: Analysis of Urban Change*. Oxford Pergamon.

Stanback, T.M. (1991) *The New Suburbanisation: Challenge to the Central City*. Boulder, CO: Westview Press.

Tisdale, H. (1942) 'The process of urbanisation', *Social Forces*, 20: 311–16.

Tsuya, N.O. and Kuroda, T. (1989) 'Japan: the slowing of urbanisation and metropolitan concentration', in A.G. Champion (ed.), *Counterurbanisation: The Changing Pace and Nature of Population Deconcentration*. London: Edward Arnold. pp. 207–29.

US Bureau of the Budget (1964) *Standard Metropolitan Statistical Areas*. Washington, DC: Government Printing Office.

Van Dam, F. (1996) 'The turnaround after the turnaround after the turnaround? The urban-rural migration wave in the Netherlands, 1973–1994'. Paper presented at the 28th International Geographical Congress, The Hague.

10

Social Segregation and Social Polarization

CHRIS HAMNETT

Any city, however small, is in fact divided into two, one the city of the poor, the other of the rich: these are at war with one another, and in either there are many smaller divisions, and you would be altogether beside the mark if you treated them all as a single State.

Plato, *Republic* IV, 422B

The existence of concentrations of rich and poor in the world's major cities is a long-standing phenomenon. Plato commented on it in his *Republic*, and the social divisions in the growing industrial cities were a perennial subject of debate in the nineteenth century from Engels's (1849) work of the condition of the English working classes onwards. Although similar concerns continued to surface in Britain during the 1930s, the baton was effectively transferred to the USA in the late nineteenth century (Ward, 1989), first to New York and Boston, and then, in the 1920s to Chicago with the work of Park and Burgess (1925). Concern about poverty, segregation and the inner city then went quiet to a large extent during the 1940s, 1950s and 1960s as sociological interest shifted to the growing suburbs. It was not until the American urban riots of the mid-1960s, and the realization of the magnitude of the racial and ethnic transformation of many American inner cities in the post-war decades, that interest resurfaced in the question of class and segregation (Beauregard, 1993a, 1993b; Fainstein, 1993). The 1960s and 1970s saw a large volume of research on ethnic segregation, particularly in the USA (Morrill, 1965; Rose, 1970), but also in Britain and elsewhere. These decades also saw a great deal of quantitative research on patterns of residential differentiation by geographers and sociologists. In the 1980s, however, two new research themes emerged, the first revolving around the existence of the 'underclass' and its structural and behavioural causes, and the second focusing on what is known as social polarization and the related issue of urban 'duality' and dual cities. These questions have been linked to issues of race, ethnicity and segregation, (Castells, 1989; Sassen, 1984, 1986) though they are by no means synonymous.

In this chapter I intend to trace the development of these concerns and issues paying particular attention to recent debates concerning polarization, duality and the underclass. The structure of the chapter is broadly historical, though the discussion of earlier work is extremely attenuated, not least because this material is already well known and well documented. My principal focus is on the work done in the past 10–15 years.

CLASS STRUCTURE AND RESIDENTIAL DIFFERENTIATION IN THE NINETEENTH CENTURY

The publication of Engels's (1849) *The Condition of the Working Classes in England, 1848* drew attention to the growth of the urban industrial working class in Britain, and to the appalling living conditions they endured. But as Glass (1968) and Steadman-Jones (1971) have shown, there was a persistent middle-class concern during the late nineteenth and early twentieth century regarding the concentration and segregation of the industrial working classes in the cities. Cooke-Taylor in 1842 commented on Manchester that:

As a stranger passes through the masses of human beings which have been accumulated around the mills and print-works in this and the neighbouring towns, he cannot contemplate those 'crowded hives' without feelings of anxiety and apprehension almost amounting to dismay. The population . . . is hourly increasing in breadth and strength. (*Notes on a Tour in the Manufacturing Districts of Lancashire*, quoted in Glass, 1968)

Sixty years later, Masterman (1904) used similar terms to describe middle-class reaction to the growth of British cities and the 'dangerous classes':

> They dread the fermenting, in the populous cities, of some new, all powerful explosive, destined one day to shatter into ruin all their desirable social order. In these massed millions of an obscure life, but dimly understood and ever increasing in magnitude, they behold a danger to security and pleasant things.

Such fears focused on the rapidly growing northern industrial towns during the first half of the nineteenth century but, as Steadman-Jones (1971) noted in *Outcast London*:

> In the period after 1850, fears about the consequences of urban existence and industrial society centred increasingly on London. For London, more than any other city, came to symbolize the problem of the 'residuum'. (1971: 12)

These fears reflected both the size of London's casual labour market (Green, 1995) and the huge size of the city itself. Victorian London was by far the largest city in the world, and during the course of the nineteenth century its population grew from just over one million to six and a half million. It also became increasingly segregated. Writing of London south of the river, Booth (1892) noted: 'the population is found to be poorer, ring by ring, as the centre is approached, while at its centre there exists an impenetrable mass of poverty.'

As Steadman-Jones (1971) clearly demonstrated, modern concerns regarding the so-called urban underclass closely parallel the concerns in the nineteenth century regarding urban 'degeneration', the 'residuum' and the 'dangerous classes'. The behaviour of the lowest classes was thought to be vice-ridden, criminal and degenerate and it was widely feared that it was pathological and self-perpetuating. Ward (1989) gives similar evidence regarding the American city and quotes a report on the conditions of the poor in Boston in 1846 which could have been written of London. It noted that there was 'a downward movement of the poorest classes . . . which if it not be checked, must sooner or later lead to a condition like that of the Old World where the separation of the rich and poor is so complete, that the former are almost afraid to visit the quarters most thickly peopled by the latter' (Ward, 1989: 15).

Concern over urban disorder and poverty somewhat faded in the first half of the twentieth century. According to Glass, suburbanization may have helped to improve working-class living conditions, to tame working-class radicalism and

soothe middle-class fears. But the point is that contemporary concerns over polarization and the rise of an urban underclass are not new. They have been around, in one form or another, for at least 150 years, rising and falling in prominence depending on changing circumstances (Beauregard, 1993a). The study of social segregation in nineteenth-century cities was an important issue in urban social geography and urban history in the 1970s and early 1980s (Dennis, 1980; Pooley, 1984), with a debate as to whether Victorian cities represented a new and distinctively modern form of social segregation, or whether they were a transitional form from pre-capitalist cities, to the modern capitalist city (Cannadine, 1977; Ward, 1975). Indeed, Ward (1980) has questioned the extent to which all Victorian cities conformed to the model of sharp class segregation suggested by Engels (see Doucet and Weaver, 1991; Harris, 1986; and Zunz, 1982 on residential class segregation in North America).

CONTEMPORARY WORK ON CLASS STRUCTURE AND RESIDENTIAL DIFFERENTIATION

Urban social geography in the 1960s and 1970s was characterized by the ever-more sophisticated quantitative analysis of urban residential structure, aided by the advent of computers and the publication of small area census data. Factorial and social area analyses of different cities proliferated in an attempt to link and test the traditional concentric (Burgess) and the sectoral (Hoyt) models of urban structure first formulated in the 1920s and 1930s. But, by the early 1970s, this type of work had reached its zenith (Johnston, 1970; Murdie, 1976). It was gradually realized that the 'game of hunt the Chicago model' (Robson, 1969) had run into an explanatory cul-de-sac, with rapidly decreasing returns to effort. More sophisticated quantitative approaches to the analysis of urban census data were producing less and less in the way of new understanding, and it was argued that a general theoretical approach to the study of urban residential structure linked to changes in class structure and other factors was needed (Harvey, 1975; Johnston, 1971; Timms, 1971).

The focus of research changed fundamentally in the mid-1970s and early 1980s with the publication of David Harvey's (1973) *Social Justice and the City*, and his work on the residential structure of Baltimore (Harvey and Chatterjee, 1974). Along with Pahl's (1970) work on urban managerialism, the nature of urban social geography and urban sociology shifted away from the study of *residential patterns per se*, and a concern with household choice and preference as the explanatory variables, towards a concern with the underlying economic and *social processes* which structured the nature of the urban housing market and, in combination with the existing class and ethnic structure, produced residential patterns. Attention shifted to study of the housing market and the *production* of urban residential space (Bassett and Short, 1980) and toward the analysis of constraints rather than choices and preferences (Short, 1978). By the mid-1980s, however, geographers had largely turned their backs on questions of segregation although, as discussed later, it is making something of a comeback in geography (Morrill, 1995; Peach, 1996) and urban sociology following Wilson's (1987) seminal work, particularly in the context of the debate over the existence of an urban 'underclass' and its links to ethnicity.

be reinforced by the sale of better council houses in better areas to skilled and higher income tenants, leaving a rump council sector dominated by high-rise flats and poor housing on marginal estates (Forrest and Murie, 1988). This reinforced some existing tendencies towards the allocation of poor council housing to ethnic minorities (Parker and Dugmore, 1977). This research has been developed in the 1990s by Peach and Byron (1993) who have shown that Afro-Caribbean tenants in Britain are overrepresented in council housing and that Afro-Caribbean women are particularly concentrated in high-rise blocks. Similar processes operate in other European cities (Hegedus and Tosics, 1994; Kovacs, 1994; Musil, 1987; O'Loughlin and Friedrichs, 1996; Pichler-Milanovich, 1994; Sykora, 1994; Van Kempen, 1994) producing a variety of forms/patterns of residential segregation of disadvantaged groups.

SOCIAL POLARIZATION AND RESIDUALIZATION IN THE HOUSING MARKET

In recent years, the focus of attention has shifted from a focus on patterns of social and spatial segregation *per se* to the broader conception of polarization and duality, particularly in global or world cities (Friedmann and Wolff, 1982; Sassen, 1991; Mollenkopf and Castells, 1991). The idea of 'polarization' in urban social structure is not new. It was current in the early 1970s in London when there was a concern that the city was becoming polarized as a result of the outmigration of skilled manual groups to the New Towns and elsewhere, and a growing concentration of the highly skilled as a result of gentrification and of the less skilled because they were caught in a housing and employment trap (Cole, 1975; Eversley, 1972; Hamnett, 1976; Harris, 1973).

In addition, in Britain the decline of private renting in the inner city and the expansion of home ownership and council renting was also believed to be leading to a growing socio-tenurial polarization and to what was termed 'residualization' of council housing. In this interpretation, the socially mixed nature of private renting was giving way to a council house tenure increasingly dominated by the less skilled, the unemployed and economically inactive, and a variety of socially marginalized groups including single parent mothers and Afro-Caribbean tenants (Hamnett, 1984, 1987; Hamnett and Randolph, 1987; Henderson and Karn, 1984; Forrest and Murie, 1988). Home ownership, on the other hand, was becoming increasingly the preserve of professionals, managers and white-collar and skilled manual workers. This process was seen to

SOCIAL POLARIZATION AND DUALITY IN WORLD CITIES

The discussion of social polarization in the early and mid-1970s was largely concerned with measurement problems and techniques concerning the use of small area census data. The theoretical debate was relatively limited (Gordon and Harloe, 1991). It was not until the early 1980s that theoretical debate on polarization in world cities took off. The idea was popularized by Friedmann and Wolff (1982), who argued that whereas major cities have long had major inequalities in income and wealth, new processes relating to the international division of labour and production are at work in contemporary world cities which were concentrating both high level business services and a professional and managerial business elite.

At the other end of the spectrum, some cities, particularly in the third world, have seen the growth of a large informal, floating or street economy as well as the service economy needed to provide for the needs of the transnational elite. They suggest that the contrast between the business elite and the third of the population who make up the permanent underclass of the world city is striking, and they pointed to the ethnic dimensions of the underclass:

> Many, though not all, of the underclass are of different ethnic origin than the ruling strata; often they have a different skin colour as well, or speak a different dialect or language. These immigrant workers give to many world cities a distinctly 'third world' aspect. There is a city that serves this underclass . . . Physically separated from and many times larger than the citadel of the ruling class, it is the ghetto of the poor. (Friedmann and Wolff, 1982: 323)

Thus, Friedmann and Wolff suggest that: 'The primary fact about world city formation is the *polarization of its social class divisions*' (1982: 322; added emphasis). But their use of the term polarization was undefined and simply implied there were sharp class divisions in world cities, which few observers would question. As I will argue, this lack of definition has characterized most of the subsequent work on polarization.

The concept of social polarization was taken up and developed by Saskia Sassen in a series of publications (1984, 1985, 1991). I have summarized Sassen's argument at some length elsewhere (Hamnett, 1994). As part of her wider thesis about the role of global cities in the world economy, she argues that the structure of economic activity in global cities, particularly the dramatic growth of financial and business services and the decline of manufacturing industry, has: 'brought about changes in the organization of work, reflected in a shift in the job supply and polarization in the income and occupational distribution of workers' (Sassen, 1991: 9). She argues that this polarization in the occupational and income structure of global cities is a result of a number of interrelated processes. First she argues that the rise of business and financial services in global cities is leading to the creation of an occupational and income structure which contrasts with the occupational structure characteristic of manufacturing industry. This new occupational structure is comprised of a mixture of highly skilled and highly paid jobs and low-skill and low-paid jobs. As she puts it:

Major growth industries show a greater incidence of jobs at the high and low paying ends of the scale than do the older industries now in decline. Almost half the jobs in the producer services are lower income jobs, and half are in the two highest earnings classes. In contrast, a large share of manufacturing workers were in the middle earning jobs during the post-war period of high growth in these industries. (1991: 9)

In addition, Sassen argues that two other processes are generating increased polarization. These are, secondly, the secondary and derivative growth in low-skilled and low-paid jobs in hotels, catering, cleaning, personal services and the like, all of which are necessary to 'service' the new global service class. Thirdly, there is the growth of what Sassen calls a 'downgraded' manufacturing sector, characterized by a high concentration of informal and sweated low-skill and low-paid work. Sassen (1991) summarizes her thesis as follows:

new conditions of growth have contributed to elements of a *new class alignment* in global cities. The occupational structure of major growth industries characterized by the locational concentration of major growth centres in global cities in combination with *the polar-ized occupational structure* of the sectors has created and contributed to a growth of a high-income stratum and a low-income stratum of workers. It has done so directly through the organization of work and occupational structure of major growth sectors. And it has done so indirectly through the jobs needed to service the new high-income workers, both at work and at home, as well as the needs of an expanded low-wage work force. (1991: 9; emphases added)

This basic thesis has been consistently reiterated and elaborated by Sassen over the past 10 years and it is clear that, although she does not define social polarization precisely, it involves absolute growth of the occupational and income distribution at *both* the *top* and the *bottom* ends combined with an absolute decline in the middle. It is therefore seen to be more than a simple increase in income inequality. It is also clear that the growth of polarization, in Sassen's view, is a product of changes in the social and spatial division of labour, which is seen to be particularly marked in global cities with their role as control and command centres and as centres of financial production. This interpretation of polarization has also received considerable support from the work of Harrison and Bluestone (1988) and Stanbach (1979). It is linked to a growing concern about the so-called 'disappearing middle' (Kuttner, 1983; Lawrence, 1984; Levy, 1987), which was influential in the USA in the 1980s and still is. In addition, Sassen has linked the expansion of low-skill and low-pay jobs to the growth of the immigrant labour force who are attracted to global cities such as New York and Los Angeles by growing job opportunities. This is a consistent element of her thesis (Sassen, 1984, 1986, 1991). But, while this may be true of New York and Los Angeles (Clarke and McNicholas, 1996), it is very questionable to what extent polarization is characteristic of all global cities as Sassen suggests. Research on the Randstad, Holland (Hamnett, 1994b), Paris (Preteceille, 1995) and London (Hamnett, 1994a), suggests that these cities have not experienced occupational polarization. On the contrary, census and other data point to a consistent picture of upwards socio-economic shift. Table 10.1 illustrates this trend for London.

Pinch (1993) has noted, however, the debate over polarization is complex, and has taken different forms in the USA and the UK with Pahl (1988) taking a very different view of polarization from Sassen and Harrison and Bluestone. Pahl's focus is on the division of work within and between households. It is not particularly urban, nor based on global cities. To this extent, the range of the thesis is quite different from that of Sassen. Pahl argues that a division is emerging between those 'work-rich' households who may have two or more

Table 10.1   *Proportionate socio-economic change in London, 1961–91, economically active males (%)*

	1961	1971	1981	1991
Managerial	11.9	13.9	16.1	20.3
Professional	4.8	6.0	6.4	8.2
Intermediate and junior non manual	23	23.5	22.0	23.7
Skilled manual	34.8	32.8	30.2	27.6
Semi-skilled manual	14.5	13.4	14.3	12.4
Unskilled	8.3	7.2	5.8	4.4
Sub-total	97.4	96.9	94.8	96.6
Armed forces and occupation inadequately described	2.6	3.1	5.2	3.4
Total	100	100	100	100

Source: Censuses of Population, 10% data

members in employment, and 'work-poor' house-holds whose members are unemployed and, because they lack the skills, contacts and income, are unable to engage in what Pahl terms 'self-provisioning'. According to Pahl, there are a range of possibilities regarding the development of social polarization, ranging from the hour-glass structure which characterizes the United States, to the onion-shaped structure which he believes may be more characteristic of the UK. What is certain, is that modern cities, like their nineteenth-century forebears, possess sharp and growing differences in wealth and poverty (Dangschat, 1994; Hamnett and Cross, 1998 a and b; Haussermann and Sackman, 1994; Kesteloot, 1994; van Kempen, 1994). The key question, however, is whether these growing inequalities are a direct and unmediated product of economic restructuring, as Sassen suggests, or whether they may arise from differences and changes in state policies or other national differences (Hamnett, 1998).

DUAL CITIES: MYTH OR METAPHOR?

The 1980s have been seen to be an era of growing inequality between rich and poor which has been manifested particularly strongly in manor cities. As a consequence, we have seen a return to Disraeli's nineteenth-century notion of the 'two nations', but this time in a specifically urban context. In their book *Dual City*, Mollenkopf and Castells (1991) comment that:

> New York remains a capital for capital, resplendent with luxury consumption and high society . . . But New York also symbolizes urban decay, the scourges

of crack, AIDS, and homelessness, and the rise of a new underclass. Wall Street may make New York one of the nerve centres of the global capitalist system, but this dominant position has a dark side in the ghettos and barrios where a growing population of poor people lives. (1991: 3)

This parallels the picture painted by Tom Wolfe in his novel *The Bonfire of the Vanities* (1988); but does this mean that New York is a 'dual city'? Mollenkopf and Castells argue that specifically in terms of poverty and income inequality the answer is 'yes' and Castells (1993) argues, more generally, that:

> the informational city is also the dual city . . . the informational economy has a structural tendency to generate a polarized occupational structure according to the informational capabilities of different social groups. Informational productivity at the top may lead to structural unemployment at the bottom or downgrading of the social conditions of manual labour. (1993: 254)

Following Sassen, Castells (1993) links also this process of dualization to immigration:

> dualization of the urban social structure is reinforced when immigrant workers take on downgraded jobs. In a parallel movement, the age differential between an increasingly older native population in European cities and a younger population of newcomers and immigrants creates two extreme segments of citizens, polarized along lines of education, ethnicity, and age simultaneously. (1993: 254–5)

Peter Marcuse (1989, 1993) has expressed serious doubts about the dual city thesis, arguing that the reality is more complex, and that the structure of the contemporary city is divided into several different groups and quarters, depending on the division of labour and on race and gender. Marcuse also argues that although the patterns have a spatial dimension, and their spatial characteristics influence their substance, they are: 'not rigid spatial patterns in the old sense in which Burgess and Park tried to describe city structure'. He also strongly disputes any idea that contemporary social divisions are new in any fundamental sense: 'The divisions of society, whether one chooses to speak of classes or socio-economic status or consumption or racial/ethnic groupings, are age-old: those derived from capitalism are hardly products of the postwar era' (Marcuse, 1993: 357). He added that:

A divided city is certainly nothing new. Never mind the slave quarters of ancient Athens and Rome, the ghettos of the middle ages, the imperial quarters of colonial cities, or the merchant sections of the medieval trading cities. At least from the outset of the industrial revolution, cities have been divided in ways that are quite familiar to us. (1993: 354)

What is new, Marcuse suggested, is the extent of homelessness, the growth of gentrification and abandonment, the role of displacement as a mechanism of expansion by the middle classes, the growth of turf allegiance and battles, the role of government in promoting gentrification and the changing form of political cleavages, most of which stem from the nature of modern capitalism.

SOCIAL POLARIZATION AND DUALITY: ALL-PURPOSE SIGNIFIERS OF INEQUALITY?

While there is little doubt that the social and spatial structure of many major cities has become more sharply divided in recent years, and that urban income inequality has grown dramatically during the 1980s, there are three main problems surrounding the uncritical use of the term. First, there is a danger that the concept of polarization is used in a variety of undefined and often contradictory ways. Pahl (1988) and Pinch (1993) have pointed to the existence of quite different conceptions of polarization, and Marcuse (1989, 1993) has queried whether the notion of duality is appropriate or valid. Thus Mollenkopf and Castells (1991) in speaking of New York state that: 'there is a process of social polarization, not just inequality: the rich are becoming richer and the poor are becoming poorer in absolute terms' (1991: 401). But why should this be termed social polarization when an increase in absolute income inequality is the issue? Likewise, in an otherwise useful study of polarization and the crisis of the welfare state in Stockholm, Borgegård and

Murdie (1994) discuss the changing position of Stockholm in the international economy, the changes in the welfare state in Sweden and impacts on income distribution, immigration and unemployment, all under the label social polarization, without anywhere defining what they mean by the term.

Indeed, the concept of social polarization is characterized more by shifting meanings than by precise definition. It has become an all-purpose general signifier of growing urban inequality and social division with the consequent disadvantages of ambiguity and lack of clarity. This issue is taken up below.

Secondly, there is the danger that the existence of social polarization is taken for granted rather than subjected to empirical analysis. Thus, polarization can easily become the received wisdom, the existence of which is simply assumed rather than problematized. While Dale and Bamford (1989) have empirically analysed aspects of Pahl's view of social polarization, and Hamnett (1994b), Hamnett and Cross (1998a, 1998b) and Buck (1994) have examined the changing income and occupational structure of London in relation to Sassen's thesis, the danger, as Fainstein et al. (1992) have perceptively pointed out, is that:

The images of a dual or polarized city are seductive, they promise to encapsulate the outcome of a wide variety of complex processes in a single, neat, and easily comprehensible phrase. Yet the hard evidence for such a sweeping and general conclusion regarding the outcome of economic restructuring and urban change is, at best, patchy and ambiguous. If the concept of the 'dual' or 'polarizing' city is of any real utility, it can serve only as a hypothesis, the prelude to empirical analysis, rather than as a conclusion which takes the existence of confirmatory evidence for granted. (1992: 13)

Finally, there is a danger that the processes thought to be generating social polarization may be inadequately or incorrectly theorized or over-generalized and empirical evidence for certain forms of social polarization may be taken as proof of the validity of these process theories. I have previously criticized aspects of Sassen's theory of social polarization and questioned whether her claim of the generality of the processes to all global cities is valid (Hamnett, 1994a, 1994b). Buck (1994) has also looked at some of the possible causes of the growing income polarization in London and New York which are not necessarily related to employment restructuring. It may be that certain forms of polarization are occurring, but not always for the reasons suggested by Sassen, Castells and others. As Levine (1992) puts it: 'To what extent is economic restructuring a "single global process"? To what extent does the social polarization of cities flow inexorably from the intrinsic nature of post-fordist production or the inherent division of the service sector into high-wage and low-wage employment?'(1992: 175).

As I have argued elsewhere (Hamnett, 1994a, 1994b), there exist major doubts as to the extent that urban social polarization is a single process with similar causes and manifestations. This is not to suggest that the concept has no value. On the contrary, my concern is that unless the term is defined and used in a reasonably precise and systematic way it will lose its descriptive/explanatory power and become no more than a catch-all term. Beauregard (1993a and 1993b), however, suggests that the ambiguity and shifting signification of certain key terms is precisely the root of their importance.

THE RHETORICAL AND REPRESENTATIONAL ROLE OF POLARIZATION

The ambiguity and shifting meaning of certain widely used concepts in social science has recently been discussed by Beauregard (1993b) in relation to representations of urban decline. The parallels with social polarization are close, and Beauregard's views are worth detailed discussion. Virtually all his references to urban decline can be replaced by polarization. He states that:

> Many of the notions that describe the social world are quite chaotic; they are concepts that bind together disparate behaviours and attributes. Commentors frequently use such terms without giving them specific meaning, knowing a focused definition would detract from their richness. They are not simple or rigid concepts and should not be treated that way. They are meant to elicit diverse impressions and resonate throughout the arguments in which they are used. Urban decline is one such concept (1993b: 188–9)

Beauregard is right in this respect and he is highly critical of approaches that seek to represent the notion of urban decline in terms of 'a range of objective indicators which produce 'a typical social science, objectivist narrative' which purports to represent the 'real' characteristics of urban decline via a series of tightly specified meanings and measurements. He argues that such narratives 'negate what is most important about urban decline; its evaluative, emotional, and symbolic content' and he states that the indeterminacy of such concepts stems from their cultural resonances rather than inadequate specification. He is strongly opposed to what he sees as the 'reductionism' implicit in trying to empir-

ically specify concepts such as urban decline which do not give greater understanding. He views the purpose of such value-laden and representational concepts as being 'mainly rhetorical' (1993b: 189)

I want to suggest that the concept of social polarization has many of the same characteristics as that of urban decline. It is inherently unstable, value-laden, symbolic and representational and its purpose is also mainly rhetorical. It is a multi-purpose signifier for urban division and growing in quality. The same is true of the dual city idea. Indeed, Mollenkopf and Castells explicity point out that: 'The dual city is a useful ideological notion because it aims to denounce inequality, exploitation, and oppression in cities, breaking with the organicist and technocratic views of cities as integrated social communities' (1991: 405). But they also point out that its 'underlying assumptions are rarely made explicit, because those who employ it tend to favour social critique over social theory. The political and emotional charge of a dualist approach and the failure to spell out its assumptions means that it cannot comprehend the complexity of urban social reality, which is certainly not reducible to a simple dichotomy' (1991: 405). They note, however, that: 'Even if the dual city metaphor can be scientifically misleading and often rhetorical or ideological, it nevertheless challenges us to explore the dimensions of growing inequality and explain the sources of the tendencies towards polarization' (1991: 11).

We therefore face something of a dilemma. While it may be politically useful to use concepts like social polarization because of their representational and rhetorical power, and implications of growing social divisions and inequality, such concepts can be empirically misleading and can divert attention away from what is happening to what we think is happening. While Beauregard raises an important issue, his anti-objectivist stance is problematic because it seems to fall victim to a rhetorical and critical trap no less powerful than the empirical one he so strongly rejects. If we do not spell out precisely what we mean and understand by social polarization we risk becoming slaves of an unexamined, imprecise or ill-defined concept. The fact that polarization is symbolic, value-laden and representational must not debar us from trying to deconstruct its uses and meanings.

Nor, having unpacked the various meanings and uses of the term, should we be debarred, as Beauregard seems to imply, from trying to assess whether the concept of polarization has any empirical validity. I disagree with his view that 'objectivist narratives' inherently negate what is most important about certain key concepts, namely their 'evaluative, emotional and symbolic content'. Whilst objectivist narratives alone are insufficient and inadequate, so are purely rhetorical narratives. Social scientists should not be restricted to the construction and deconstruction of rhetorical narratives alone. While measurement and quantitative work does not, *of itself* aid understanding, to rely on theory or conceptual deconstruction alone, without trying to assess whether postulated processes have any empirical support, is to theoretically imprison us. Theoretical development should go hand in hand with empirical analysis. The two are not mutually antagonistic. But, and here I agree with Beauregard, the primary task is to unpick the chaotic concept of polarization to try to determine what its key elements are and how they relate together (if at all). Then, and only then, can we begin to investigate the form and nature of polarization. But, if we do not undertake empirical work, we remain solely in the realm of theoretical conjecture.

UNPACKING THE CONCEPT OF SOCIAL POLARIZATION

Many commentators are convinced that we are witnessing a growth of social polarization in major capitalist cities. But the frequency with which the term is used is matched only by the general lack of a definition. Is it a synonym for growing social divisions and inequality, or does it mean something more precise and, if so, what? We must, I think, reject the notion that polarization can be simply used as a synonym for inequality, not least because inequality can take so many different forms (Gordon and Harloe, 1991: 383). Indeed, Esteban and Ray (1994), in a challenging, but extremely valuable, paper point to the major difference between inequality and polarization, showing that it is possible to have greater inequality without greater polarization and vice versa. Nor, in my view, should polarization be used to refer to increasing residential segregation by class, race, gender, etc., though this may certainly be related to growing social polarization at the city level. My reason for arguing this is that we already have a perfectly good term for this 'segregation', and we would then need to differentiate between social and spatial polarization which need not take place simultaneously.

It appears that Harloe and Gordon see polarization as a very specific form of inequality. Mollenkopf and Castells make a rather similar distinction, arguing that in New York: 'there is a process of social polarization, not just inequality: the rich are becoming richer and the poor are becoming poorer' (1991: 401). The distinction is

also made by Kloosterman (1996), who suggests that while *inequality* in the distribution of earnings (or incomes) 'refers to the extent of dispersion between given levels of earnings, polarization' refers to the phenomenon of the disappearing middle, the shrinkage of the number of middle-income jobs (or income units) and a growth (absolute or relative) at both the top and bottom ends of the income distribution. This distinction is crucial because it points to what, for me, is the key element of polarization – *a movement toward the poles of a given distribution*. Thus, the *Oxford English Dictionary* defines polarization as: 'an act of polarizing: the state of being polarized: development of poles: loosely, polarity'. This suggests that polarization can be conceived of either as a state or as a process. While it is quite possible to refer to a *state* or states of polarization, the term is commonly used to describe a *process* of polarization, where there is a movement towards the poles of the distribution. One of the best definitions is given by Marcuse (1989):

> The best image is perhaps that of the egg and the hour glass: the population of the city is normally distributed like an egg, widest in the middle and tapering off at both ends; when it becomes polarized the middle is squeezed and the ends expand till it looks like an hourglass. The middle of the egg may be defined as 'intermediate social strata' . . . Or if the polarization is between rich and poor, the middle of the egg refers to the middle-income group . . . The metaphor is not of structural dividing lines, but of a continuum along a single dimension, whose distribution is becoming increasingly bimodal. (1989: 699)

As Marcuse points out, polarization is a process whereby: 'a distribution is becoming increasingly bi-modal' irrespective of the precise dimensions along which polarization may be occurring. But polarization may be simultaneously taking place on a number of dimensions. Mollenkopf and Castells view polarization as being multi-dimensional, with distinct social and spatial dimensions:

> the tendency towards cultural, economic, and political polarization in New York takes the form of a contrast between a comparatively cohesive core of professionals in the advanced corporate services and a disorganized periphery fragmented by race, ethnicity, gender, occupational and industrial location, and the spaces they occupy. (1991: 402)

We need to specify whether we are speaking of employment, occupation or income, and whether polarization is relative or absolute. This is important because polarization may be occurring in certain respects but not in others, and the causes may be quite different.

POLARIZATION OF THE OCCUPATIONAL AND INCOME STRUCTURE

In terms of employment, polarization can be used to refer to an increase in the *number* (or proportion) of the highly skilled and the low-skilled with a decline in the number (or proportion) of the middle groups. This is the sense in which Sassen uses the term, and she links it to an increase in the *number* of highly paid and low-paid workers and to a decline in the number of middle-income workers, both of which result from the shift from manufacturing to financial and business services which is seen as particularly marked in global cities. Mollenkopf and Castells also suggest that the dual city notion

usefully emphasizes one trend – both the upper and the lower strata of a given society grow at disproportionate rates. Thus, in the perspective of the polarization thesis, the dual city becomes a simple [sic] matter of empirical testing of two basic questions:

a. Are the top and bottom of the social scale in a given city growing faster than the middle (with the key methodological issue being how to construct a scale to measure social distribution)?

b. How does such polarization, if it exists, translate into spatial distribution at the top and bottom of the local society, and how does such specific residential location affect overall socio-spatial dynamics? (1991: 407)

Mollenkopf and Castells are very clear about the outcome of the polarization process and how it should be measured, although they do not make clear whether they are speaking of absolute or relative change. But the argument about the changing *size* of the rich and poor groups is, however, very different from the argument that there is a widening *gap* between the average incomes of the rich and those of the poor: that the rich are becoming richer and the poor, poorer (either relatively or absolutely). Both theses may be correct, but the causes may be very different and it is not legitimate to use evidence for one to support claims about the existence of the other.

This may seem mere definitional nit-picking, but it has major implications for attempts to empirically determine whether polarization is (or is not) occurring in certain, specific, forms (see Buck, 1994). It is important to note here that although Sassen and Castells have argued that occupational and income polarization go hand in hand, this is not necessarily the case. I have pointed out that, in the Randstad, Holland and London (Hamnett, 1994a, 1994b) there is strong evidence of a widening income gap between rich and poor but no evidence of occupational polar-

ization. This may reflect the fact that, in the Netherlands, the official figures are based on the employed and exclude the unemployed, but Sassen's thesis relates to the changing structure of employment (see Burgers, 1996). Similarly, Preteceille (1995) found that in Paris, the occupational evidence points to professionalization rather than to polarization. As Kloosterman points out, analytically:

two concepts of polarization in urban areas can be distinguished. The first one covers the whole city population and includes, therefore, those people without work ... The second concept of urban polarization is more modest and deals only with those that have paid work. (1994: 2)

More generally, however, it appears that growing income inequality is not necessarily accompanied by a polarization of occupational structure as Sassen and Castells maintain. As Hamnett (1994a) has argued, there is evidence that, rather than polarizing, the occupational structure of Western capitalist societies appears to becoming more professionalized (Wright and Martin, 1987). This analysis is supported by Esping-Andersen (1993), who argues that occupational upgrading is inherent in the post-industrial trajectory in the United States. Indeed, Kloosterman suggests that:

During the 1970s and 80s, a decoupling seems to have taken place between the occupational level and the wage level in the United States . . . According to Esping-Andersen, a polarization of the occupational structure has been accompanied in the US by a polarization of wage structure. (1996: 468)

A similar shift appears to have occurred in Britain, in that the socio-economic structure has shifted upward whilst earnings and income inequality has risen considerably (Hamnett and Cross, 1998a, 1998b). But, this is not the result of an increase in the *number* of the less skilled and low paid as Sassen suggests, but from the impact of rising professional and managerial incomes, massive tax breaks for the rich (Hamnett, 1994a), growing unemployment and small increases in rates of government assistance for the unemployed or low paid. The key question then becomes what factors are leading to occupational depolarization and income polarization and how is the existence of the two processes to be explained and linked together? It is also necessary to examine the extent to which occupational depolarization and income polarization may, or may not, be characteristic of other Western countries, and what factors may be leading to the existence of different tendencies. Differences in welfare state regimes may be very important.

THE UNDERCLASS DEBATE

The notion of the 'residuum', a category of the poor and unemployed outside or beneath the rest of the social structure, was common in nineteenth-century London, as Steadman-Jones (1971) has shown. The existence and growth of this group was often linked to what was termed 'urban degeneration'. But, as Jones shows, the primary cause was the casualization of parts of the London labour market, and the creation of a large group of structurally unemployed or underemployed. This process of pauperization has also been discussed by Green (1995). Over the past hundred years, similar concerns have resurfaced from time to time, usually, though not always, voiced by right-wing commentators. Smith (1992) suggests that: 'the idea of an underclass is a counterpart to the idea of social classes, and acquires its meaning within that same framework'. He thus argues that the underclass, if they exist, are those who fall outside the standard class schemas in that they belong to family units who have no stable relationship with the mode of production. Indeed, Marx himself used the term 'lumpenproletariat'. Runciman argues for a similar definition of the underclass: 'those members of . . . society whose roles place them more or less permanently at the economic level where benefits are paid by the state to those unable to participate in the labour market at all' (1990: 388).

In the past 10 years the 'underclass' has become a common term in academic and media discourse. It first appeared in 1962 when Gunnar Myrdal used the term as 'a purely economic concept, to describe the chronically unemployed, underemployed, and unemployables being created by what we now call the post-industrial economy' (Gans, 1990: 271). But, as Gans (1990, 1993) and others (Robinson and Gregson, 1992; Wilson, 1987) have pointed out, the concept was subsequently hi-jacked by the radical right who employed it to encapsulate the idea of a culture of poverty, family dissolution and criminality in the inner cities, particularly in the black community.

In the 1960s, Oscar Lewis's (1968) anthropological concept of a 'culture of poverty became popular, and in the 1970s, Edward Banfield (1974) popularized the view that the problem of what he termed the 'lower class' (below the working class) was primarily the result of a pathological transmitted culture of present orientation and low expectations. He stated that:

> the lower class individual lives from moment to moment. If he [sic] has any awareness of a future, it is of something fixed, fated, beyond his control; things happen to him, he does not make them happen. Impulse governs his behaviour, either because he cannot discipline himself to sacrifice a present for a future satisfaction or because he has no sense of the future. He is therefore radically improvident: whatever he cannot use immediately he considers valueless. His bodily needs (especially for sex) and his taste for 'action' take precedence over everything else – and certainly over any work routine. (1974: 61)

> The lower-class household is usually female-based. The woman who heads it is likely to have a succession of mates who contribute intermittently to its support but take little or no part in rearing the children. In managing children, the mother is characteristically impulsive: once children have passed babyhood they are likely to be neglected or abused, and at best they never know what to expect next. A boy raised in such a household is likely at any age to join a corner gang of other such boys and to learn from the gang the 'tough' style of the lower-class man. (1974: 62)

He concluded that:

> So long as the city contains a sizeable lower class, nothing basic can be done about its most serious problems. Good jobs may be offered to all, but some will remain chronically unemployed. Slums maybe demolished, but if the housing that replaces them is occupied by the lower class it will shortly be turned into new slums. Welfare payments may be doubled or tripled and a negative income tax instituted, but some persons will continue to live in squalor and misery. New schools may be built, new curricula devised, and the teacher–pupil ratio cut in half, but if the children who attend these schools come from lower class homes, the schools will be turned into blackboard jungles, and those who graduate or drop out from them will, in most cases, be functionally illiterate. (1974: 234)

Banfield's radical cultural pessimism affords no solution but that of selective eugenics and birth control measures for the lower classes. He is of the view that welfare measures and other 'good works' are doomed to failure in that the lower classes are incapable of taking advantage of measures put forward for the improvement of their lot. A culture of low expectations and present orientation undoubtedly exists in some groups, but what Banfield critically fails to analyse are the causes and reasons why it has come into existence or grown in significance. Cultures do not simply spring into being of their own volition. As Banfield admits: 'From the beginning, the cities of the United States have had upper, middle, working and lower classes . . . [but] [T]he relative strength of the various classes have varied greatly from time to time and place to place' (1974: 63).

The question of why the relative strength of various classes and cultures has varied from time to time and place to place, and their causes, is not discussed. According to Katz (1993) the debate on the underclass in the USA accelerated in 1977 when *Time Magazine* announced the emergence of a menacing new underclass in America's inner cities. Drugs, crime, teenage pregnancy and high unemployment, not poverty, defined the 'underclass', most of whose members were young and from ethnic minority groups (see also Devine and Wright, 1993).

> Behind the [ghetto's] crumbling walls lives a large group of people who are more intractable, more socially alien and more hostile than almost anyone had imagined. They are the unreachables: the American underclass . . . Their bleak environment nurtures values that are often at odds with those of the majority – even the majority of the poor. Thus the underclass produces a disproportionate number of the nation's juvenile delinquents, school drop-outs, drug addicts and welfare mothers, and much of the adult crime, family disruption, urban decay and demand for social expenditures. (*Time Magazine*, 1977)

Katz states that with the publication of Auletta's (1982) book *The Underclass*, the term 'secured its dominance in the vocabulary of inner-city pathology'. For Auletta, the underclass was a relatively permanent minority among the poor who fell into four distinct categories, which he defined as (a) the passive poor, usually long-term welfare recipients; (b) the hostile street criminals who terrorize most cities, and who are often school dropouts and drug addicts; (c) the hustlers, who, like street criminals, may not be poor and who earn their livelihood in an underground economy; and (d) the traumatized drunks, drifters, homeless shopping-bag ladies and released mental patients who frequently roam or collapse on city streets (1982: xvi).

This right-wing portrayal of the underclass as a group of welfare-dependent, demoralized and behaviourally deviant individuals, frequently black and from single-parent families in the inner city, achieved further prominence with the publication of Charles Murray's (1984) *Losing Ground*. Murray argued that the principal cause of the growth of the underclass was the welfare programmes which had eroded the will to work and the incentives for stable family life. The right wing consistently stress culture and behaviour as the key attributes and causes of the emergence of an 'underclass', but even distinguished black analysts such as W.J. Wilson (1987) accept the term, arguing that the liberal perspective on the ghetto underclass has become less persuasive, primarily because many liberal commentators 'have been reluctant to discuss openly or . . . even to acknowledge the sharp increase in social pathologies in ghetto communities' (1987: 6). As a result, he argues that conservatives have effectively defined and dominated the debate in recent years. But, says Wilson:

> Regardless of which term is used, one cannot deny that there is a heterogeneous grouping of inner city families and individuals whose behaviour contrasts sharply with that of mainstream America. The challenge is not only to explain why this is so, but also to explain why behaviour patterns in the inner city today differ so markedly from those of only three or four decades ago (1987: 7)

> Included in this group are individuals who lack training and skills and either experience long-term unemployment or are not members of the labour force, individuals who are engaged in street-crime and other forms of aberrant behaviour, and families that experience long-term spells of poverty and/or welfare dependency. These are the populations to which I refer when I speak of the *underclass*. I use this term to depict a reality not captured in the more standard designation *lower-class*. (1987: 8)

Notwithstanding Wilson's view that it is crucial for liberals to face up to the realities of life in the black inner city ghettos (vividly depicted in films such as *Boyz 'n the Hood*), Fainstein (1993) argues (*pace* Beauregard), that the idea of the underclass has become the dominant discourse of race and class in America, and that we need to break free of the discourse if we are to make progress, both analytically and in political and policy terms. He argues that: 'Whatever their political and theoretical perspective, participants in the discourse of the underclass share a deep narrative. Like other deep narratives, that of the underclass is both explicit and implicit, saying much in its omissions' (1993: 385). Fainstein states that its logic is relatively transparent, constructed along four lines. He summarizes these as follows: underclass terminology offers a way of speaking about race in a language of class that implicity rejects the importance of race; research on the underclass tends to study the attributes or behaviours of a category of the population that is nominally separated from other groups and from processes that affect larger populations. As a result of the problematic and value-laden nature of the term, Robinson and Gregson (1992) suggest in their useful review paper that the negative connotations of the term are such that it is best not to use it, and Mingione (1996) is also sceptical although the term now seems to have acquired a life of its own which authors feel compelled to address even though they reject its validity (Lee, 1994). Wilson (1991) now substitutes the term 'ghetto poor' in the USA and makes the point that although there is a distinctive culture it is largely a response to structural changes and constraints (see Holloway, 1990). One problem of the growing use of the term in Europe is the extent to which it is valid to utilize concepts derived from one social context and apply them to another (Martiniello, 1996; Musterd, 1994; Wacquant, 1993). Murray (1990 has suggested that there is an emergent British underclass, but Lydia Morris (1993) in her work on Hartlepool concludes that whilst there are differences in kinship and friend-ship networks between the unemployed, the securely employed and the insecurely employed, they are differences of degree rather than kind, which although they add to the disadvantages of the unemployed, provide no evidence of any 'distinctive' underclass culture. She concludes that the notion of the 'underclass' is:

> an oversimplification, contaminated by its use as a tool of political rhetoric, which has been too readily applied to complex social phenomena. Its use is rarely supported by empirical research of the kind necessary to substantiate the varying claims with which it is associated. (1993: 411)

Buck (1996) argues that the difficulties with the term suggest that we should be very cautious in using it unless we can specify coherently what it means. He states that: 'The reason for using the term "underclass" is that it carries an implication that the group experiencing social and economic marginalization is . . . characterized by homo-geneity, stability and social segregation' (1996: 279). But he argues that although there has been a large increase in the number of households expe-riencing long-term inactivity, and hence poverty, in Britain, that the group displays considerable heterogeneity and there is no evidence of high levels of spatial segregation. Thus, he concludes that although it is difficult to draw firm compar-ative conclusions:

> comparison with the ideal type model of urban poverty in the USA suggests that differences in state welfare policy and in labour market regulation may lead to poverty and marginality being constituted in very different ways . . . we need to be very cautious in translating definitions of marginality and exclusion from one society to another. (1996: 297)

Unfortunately, the term has been taken over by the mass media, who employ it as a handy label without systematically examining the extent to which the term possesses analytical or empirical validity. Consequently, it is frequently reproduced willy-nilly despite the efforts of authors such as Loic Wacquant (1993, 1996a, 1996b) to analyse the term critically and the ideologies which underlie it. To this extent, the notion of an under-class as a group of people outside the economic mainstream by reason of chronic unemployment and structural change in the economy, has escaped its initial authors, aided and abetted by the new right who have a strong interest in promoting the idea of a behavioural culture of poverty which is reinforced by the evils of state aid. It seems clear that the idea of the underclass is not simply going to go away merely because the centre–left disapprove of its connotations and the

ways in which it is used to support a punitive, anti-welfare agenda. As Wilson points out, it is necessary to go on the intellectual offensive and show that the term is value-laden and empirically suspect. Alternatively, if it is thought that there are groups of people who are (semi-) permanently excluded from the labour market and in more or less permanent poverty, it is necessary to show what the consequences of this are, socially, culturally and economically, rather than uncriti-cally reproducing the idea of the underclass by conceptual repetition. The same applies both to the notion of social polarization and the 'ghetto' as Wacquant (1993, 1996a, 1997) has very power-fully argued. None the less, the term is continually reproduced via the mass media as a convenient hook. As I write, an article by John Lloyd (1996) appeared in the *New Statesman* on Labour's plans in Britain to combat social exclusion. Lloyd states that one of the New Labour thinkers, Geoff Mulgan, 'insists on the nomenclature of "social exclusion" as against the "poor" or even the newer "underclass"'. Lloyd states that this is because poverty is only one attribute of those at the bottom of the heap: they are more properly defined as excluded because they live outside the worlds of work, education and sociability. The article was termed, 'A plan to abolish the under-class'.

REFERENCES

Auletta, K. (1982) *The Underclass.* New York: Random House.

Banfield, E. (1974) *The Unheavenly City Revisited.* New York: Little Brown.

Bassett, K. and Short, J. (1980) *Housing and Residential Structure: Alternative Approaches.* London: Rout-ledge.

Beauregard, R.A. (1993a) *Voices of Decline: the Postwar Fate of US Cities.* Oxford and Cambridge, MA: Blackwell.

Beauregard, R.A. (1993b) 'Representing urban decline: postwar cities as narrative objects', *Urban Affairs Quarterly.* 19 (2): 187–202.

Booth, C. (1892) *Life and Labour of the People in London.* London: Macmillan.

Borgegård, L.E. and Murdie, R. (1994) 'Social polar-ization and the crisis of the welfare state: the case of Stockholm', *Built Environment*, 20 (3): 254–68.

Buck, N. (1994) 'Social divisions and labour market change in London: national, urban and global factors'. Paper presented for the ESRC London seminar, 28 October.

Buck, N. (1996) 'Social and economic change in contemporary Britain: the emergence of an urban underclass?' in E. Mingione (ed.), *Urban Poverty and the Underclass.* Oxford: Blackwell.

Burgers, J. (1996) 'No polarization in Dutch cities? Inequality in a corporatist country', *Urban Studies*, 33 (1): 99–105.

Cannadine, D. (1977) 'Victorian cities: how different?', *Social History*, 1 (4): 457–82.

Castells, M. (1989) *The Informational City.* Oxford: Basil Blackwell.

Castells, M. (1993) 'European cities, the informational society, and the global economy', *Tijdschrift voor Economische en Sociale Geografie*, 84: 247–57.

Clarke, W.A. and McNicholas, M. (1996) 'Re-examining economic and social polarisation in a multi-metropolitan area: the case of Los Angeles', *Area*, 28: 1.

Cole, B. (1975) 'Polarity and polarisation', in K. Dugmore (ed.), *The Migration and Distribution of Socio-Economic Groups in Greater London.* GLC Research Memorandum 443, pp. 68–80.

Dale, A. and Bamford, (1988) 'Social polarizaton in Britain, 1973–82: evidence from the General Household Survey – a comment on Pahl's hypothesis', *International Journal of Urban and Regional Research*, 13: 481–500.

Dangschat, J.S. (1994) 'Concentration of poverty in the landscapes of boom-town Hamburg: the creation of a new urban underclass', *Urban Studies*, 31 (7): 1133–48.

Dennis, R.J. (1980) 'More thoughts on Victorian cities: why study segregation?', *Area*, 12 (4): 313–7.

Devine, J.A. and Wright, J.D. (1993) *The Greatest of Evils: Urban Poverty and the American Underclass.* New York: Aldine de Gruyter.

Doucet, M. and Weaver, J. (1991) *Housing the North American City.* Montreal: McGill–Queens University Press.

Economist (1994) 'Europe and the Underclass, *The Economist*, 30 July: 19–21.

Engels, F. (1849) *The Condition of the English Working Classes in 1848.* London: Fontana.

Esping-Andersen, G. (ed.) (1993) *Changing Classes: Stratification and Mobility in Post-industrial Societies.* London: Sage.

Esteban, J.M. and Ray, D. (1994) 'On the measurement of polarization', *Econometrica*, 62 (4): 819–51.

Eversley, D. (1972) 'Rising costs and static incomes: some economic consequences of regional planning in London', *Urban Studies*, 9 (3): 347–68.

Fainstein, N. (1993) 'Race, class and segregation: discourses about African-Americans', *International Journal of Urban and Regional Research*, 17: 384–403.

Fainstein, S., Gordan, I. and Harloe, M. (eds) (1992) *Divided Cities: New York and London in the Contemporary World.* Oxford: Basil Blackwell.

Forrest, R. and Murie, A. (1988) *Selling the Welfare State: the Privatization of Public Housing.* London: Routledge.

Friedmann, J. and Wolff, G. (1982) 'World city formation: an agenda for research and action', *International Journal of Urban and Regional Research*, 6 (3): 309–43.

Gans, H. (1990) 'Deconstructing the underclass: the term's dangers as a planning concept', *American Planning Association Journal*, Summer: 271–9.

Gans, H. (1993) 'From "underclass" to "undercaste": some observations about the future of the postindustrial economy and its major victims', *International Journal of Urban and Regional Research*, 17 (3): 327–35 (reprinted in E. Mingione (ed.), *Urban Poverty and the Underclass*, Oxford: Basil Blackwell, 1996).

Glass, R. (1968) 'Urban sociology in Great Britain', in R. Pahl (ed.), *Readings in Urban Sociology.* Oxford: Pergamon Press.

Gordon, I. and Harloe, M. (1991) 'A dual to New York? London in the 1980s' in J. Mollenkopf and M. Castells (eds), *Dual City: Restructuring New York.* New York: Russell Sage Foundation.

Green, D.(1995) *From Artisans to Paupers: Economic Change and Poverty in London, 1790–1870.* Aldershot: Scolar.

Hamnett, C. (1976) 'Social change and social segregation in London', *Urban Studies*, 13: 261–71.

Hamnett, C. (1984) 'Housing the two nations: sociotenurial polarization in England and Wales, 1961–81', *Urban Studies.* 43: 389–405.

Hamnett, C. (1987) 'A tale of two cities: socio-tenurial polarization in London and the South East, 1966–81', *Environment and Planning A*, 19: 537–56.

Hamnett, C. (1989) 'Socio-tenurial polarization in London and the South-East: a reply to Berge's comments', *Environment and Planning A*, 21: 545–8.

Hamnett, C. (1994a) 'Socio-economic change in London: professionalization not polarization', *Built Environment*, 20 (3): 192–203.

Hamnett, C. (1994b) 'Social polarization, economic restructuring and welfare state regimes', *Urban Studies*, 33 (8): 1047–50.

Hamnett, C. and Cross, D. (1998a) 'Social polarization and inequality in London: the earnings evidence 1978–95, *Environment and Planning*, 16: 659–80.

Hamnett, C. and Cross, D. (1998b) 'Social change, social polarization and income inequality in London', *Geojournal*, 46(1): 39–50.

Hamnett, C. and Randolph, W. (1987) 'The residualization of council housing in Inner London', in D. Clapham (ed.), *The Future of Council Housing.* London: Croom Helm.

Harris, M. (1973) 'Some aspects of social polarisation', in D. Donnison and D. Eversley, (eds), *London: Urban Patterns, Problems and Policies.* London: Heinemann.

Harris, R. (1986) 'Home ownership and class in modern Canada', *International Journal of Urban and Regional Research.* 10 (1): 67–86.

Harrison, B. and Bluestone, B. (1988) *The Great U-Turn, Corporate Restructuring and Polarizing of America.* New York: Basic Books.

Harvey, D. (1973) *Social Justice and the City.* London: Edward Arnold.

Harvey, D. (1975) 'Class structure in a capitalist society and the theory of residential differentiation', in R. Peel, et al. (eds), *Processes in Physical and Human Geography.* Bristol: Colston Papers.

Harvey, D. and Chatterjee, L. (1974) 'Absolute rent and the structuring of space by financial and governmental institutions', *Antipode*. 6 (1): 22–36.

Hausermann, H. and Sackman, R. (1994) 'Changes in Berlin: the emergence of an underclass?', *Built Environment*, 20 (3): 231–41.

Hedegus, J. and Tosics, I. (1994) 'The poor, the rich and the transformation of urban space', *Urban Studies*, 31 (7): 989–94.

Henderson, J. and Karn, V. (1984) 'Race, class and the allocation of state housing in Britain', *Urban Studies*, 21: 115–28.

Holloway, S.R. (1990) 'Urban economic structure and the urban underclass: an examination of two problematic phenomena', *Urban Geography*, 11 (4): 319–46.

Johnston, R.J. (1970) 'On spatial patterns in the residential structure of cities', *Canadian Geographer*, XIV (4) 361–77.

Johnston, R.J. (1971) 'Towards a general model of intra-urban residential patterns: some cross-cultural observations', *Progress in Geography*, 2.

Katz, M.B. (ed.) (1993) *The 'Underclass' Debate: Views from History*, Princeton, NJ: Princeton University Press.

Kesteloot, C. (1994) 'Three levels of socio-spatial polarization in Brussels', *Built Environment*, 20 (3): 204–17.

Kloosterman, R. (1994) 'Three worlds of welfare capitalism? The welfare state and the postindustrial trajectory in the Netherlands after 1980', *West European Politics*, 17 (4): 166–89.

Kloosterman, R. (1996) 'Double Dutch: polarization trends in Amsterdam and Rotterdam after 1980', *Regional Studies*, 30 (5): 467–76.

Kovacs, Z. (1994) 'A city at the crossroads: social and economic transformation in Budapest', *Urban Studies*, 31 (7): 1081–96.

Kuttner, B. (1983) 'The declining middle', *Atlantic Monthly*, July: 60–72.

Lawrence, R.Z. (1984) 'Sectoral shifts and the size of the middle class', *The Brooking Review*, Fall: 3–11.

Lee, P. (1994) 'Housing and spatial deprivation: relocating the underclass and the new urban poor', *Urban Studies*, 31 (7): 1191–210.

Levine, M. (1992) 'The changing face of urban capitalism', *Urban Affairs Quarterly*, 28 (1): 171–80.

Levy, P. (1987) 'The middle class: is it really vanishing?' *The Brookings Review*, Summer: 77–122.

Lewis, O. (1968) 'The culture of poverty', in D. Moynihan (ed.), *Understanding Poverty: Perspectives from the Social Sciences*. New York.

Lloyd, J. (1996) 'A plan to abolish the underclass', *New Statesman*, 28 August: 14–16.

McNichol, J. (1987) 'In pursuit of the underclass, *Journal of Social Policy*, 16 (3): 293–318.

Marcuse, P. (1989) '"Dual City": a muddy metaphor for a quartered city?' *International Journal of Urban and Regional Research*, 13 (4): 697–708.

Marcuse, P. (1993) 'What's so new about divided cities?', *International Journal of Urban and Regional Research*, 17 (3): 355–65.

Martiniello, M. (1996) 'The existence of an urban underclass in Belgium', *New Community*, 22 (4): 655–70.

Masterman, C.F. (1904) *The English City*. Publisher unknown.

Mingione, E. (1996) 'Urban poverty in the advanced industrial world: concepts, analysis and debates', in E. Mingione, (ed.), *Urban Poverty and the Underclass*. Oxford: Basil Blackwell.

Mollenkopf, J. and Castells, M. (eds) (1991) *Dual City: Restructuring New York*. New York: Russell Sage Foundation.

Morrill, R.L. (1965) 'The negro ghetto: problems and alternatives', *The Geographical Review*, 55 (3): 339–61.

Morrill, R.L. (1995) 'Racial segregation and class in a liberal metropolis', *Geographical Analysis*, 27: 22–41.

Morris, L. (1993) 'Is there a British underclass?', *International Journal of Urban and Regional Research*, 17 (3): 404–12.

Murdie, R.A. (1976) 'Spatial form in the residential mosaic', in D.T. Herbert and R.J. Johnson (eds), *Spatial Processes and Form*. Chichester: J. Wiley.

Murray, C. (1984) *Losing Ground: American Social Policy, 1950–1980*. New York: Basic Books.

Murray, C. (1990) *The Emerging Underclass*. London Institute of Economic Affairs.

Musil, J. (1987) 'Housing policy and the socio-spatial structure of cities in a socialist country: the example of Prague', *International Journal of Urban and Regional Research*. 11 (1) 27–36.

Musterd, S. (1994) 'Arising European underclass', *Built Environment*, 20 (3): 185–90.

O'Loughlin, J. and Friedrichs, J. (eds) (1996) *Social Polarization in Post-industrial Metropolises*. Berlin: Walter de Gruyter.

Pahl, R. (1970) *Whose City?* Harlow: Longman.

Pahl, R.E. (1988) 'Some remarks on informal work, social polarization and social structure', *International Journal of Urban and Regional Research*, 12: 247–67.

Park, R.E. and Burgess, E.W. (1925) *The City*. Chicago: University of Chicago Press.

Parker, J. and Dugmore, K. (1977) 'Race and allocation of council housing: a GLC Survey', *New Community*, 6: 27–41.

Peach, C. (1996) 'Does Britain have ghettos?', *Transactions of the Institute of British Geographers*, 21 (1): 216–35.

Peach, C. and Byron, M. (1993) 'Carribbean tenants in council housing: "race", class and gender', *New Community*, 19 (3): 407–23.

Pichler-Milanovich, N. (1994) 'The role of housing policy in the transformation process of Central–East European cities', *Urban Studies*, 31 (7): 1097–115.

Pinch, S. (1993) 'Social polarization: a comparison of evidence from Britain and the United States', *Environment and Planning A*, 25: 779–95.

Pooley, C. (1984) 'Residential differentiation in Victorian cities: a reassessment', *Transactions of the Institute of British Geographers*, NS 9: 131–44.

Preteceille, E. (1995) 'Division sociale de l'espace et globalisation: le cas de la metropole parisienne', *Sociétés Contemporaines*. 22/23: 33–67.

Robinson, F. and Gregson, N. (1992) 'The underclass: a class apart?', *Critical Social Policy*, 34: 38–51.

Robson, B. (1969) *Urban Analysis: A Study of City Structure*. Cambridge: Cambridge University Press.

Rose, H.M. (1970) 'The development of an urban subsystem: the case of the Negro ghetto', *Annals of the Association of American Geographers*, 60 (1): 1–17.

Runciman, W.G. (1990) 'How many classes are there in contemporary British society?', *Sociology*, 24 (3): 377–96.

Sassen, S. (1984) 'The new labour demand in global cities', in M.P. Smith (ed.), *Cities in Transformation*, vol. 26, *Urban Affairs Annual*, Beverly Hills, CA: Sage.

Sassen, S. (1985) 'Capital mobility and labour migration: their expression in core cities', in M. Timberlake (ed.), *Urbanization in the World Economy*. New York: Academic Press.

Sassen, S. (1986) 'New York City: economic restructuring and immigration', *Development and Change*, 17: 85–119.

Sassen, S. (1991) *Global City: New York, London and Tokyo*. Princeton, NJ: Princeton University Press.

Short, J. (1978) 'Residential mobility in the private housing market of Bristol', *Transactions of the Institute of British Geographers*, 3 (4): 311–47.

Smith, D.J. (ed.) (1992) *Understanding the Underclass*. London: Policy Studies Institute.

Stanbach, T.M. (1979) *Understanding the Service Economy*. Baltimore, MD: Johns Hopkins University Press.

Steadman-Jones, G. (1971) *Outcast London*. Harmondsworth: Penguin.

Sykora, L. (1994) 'Local urban restructuring as a mirror of globalisation processes: Prague in the 1990's', *Urban Studies*, 31 (7): 1149–66.

Time Magazine (1977) 'The American Underclass', *Time*, 29 August: 14–15.

Timms, D. (1971) *The Urban Mosaic: Towards a Theory of Residential Differentiation*. Cambridge: Cambridge University Press.

van Kempen, E. (1994) 'The dual city and the poor: social polarization, social segregation and life chances', *Urban Studies*, 31 (7) 995–1015.

Wacquant, L. (1993) 'Urban outcasts: stigma and division in the black American ghetto and the French urban periphery', *International Journal of Urban and Regional Research*, 17 (3): 366–84.

Wacquant, L. (1996a) 'L'underclass urbaine dans l'imaginaire social et scientifique Americain,' in S. Paugam (ed.), *L'exclusion: l'etat des savoirs*, Paris.

Wacquant, L. (1996b) 'Three pernicious premises in the study of the American ghetto', *International Journal of Urban and Regional Studies*, 21 (2): 341–54.

Wacquant, L. (1997) 'Negative social capital: state breakdown and social destitution in America's urban core', *Netherlands Journal of the Built Environment*.

Ward, D. (1975) 'Victorian cities: how modern?' *Journal of Historical Geography*, 1 (2): 135–51.

Ward, D. (1980) 'Environs and neighbours in the "two nations": residential differentiation in mid-nineteenth century Leeds', *Journal of Historical Geography*, 6 (2): 133–62.

Ward, D. (1989) *Poverty, Ethnicity, and the American City, 1840–1925: Changing Conceptions of the Slum and Ghetto*. Cambridge: Cambridge University Press.

Wilson, W.J. (1987) *The Truly Disadvantaged: the Inner City, the Underclass and Public Policy*. Chicago: Chicago University Press.

Wilson, W.J. (1991) 'Studying inner-city dislocations', *American Sociological Review*, 56: 1–14.

Wolfe, T. (1988) *The Bonfire of the Vanities*. London: Jonathan Cape.

Wright, E. and Martin, B. (1987) 'The transformation of the American class structure, 1960–1980', *American Journal of Sociology*, 93: 1–29.

Zunz, O. (1982) *Poverty, Ethnicity and the American City, 1840–1925*.

11

Race Relations in the City

JOE T. DARDEN

CONCEPTUAL FRAMEWORK

In multi-racial societies, racial conflict often occurs in cities. However, it is not the fact that multi-racial groups are living in cities that gives rise to conflict. It is conceivable that multi-racial groups could live in harmony. Thus, to understand why racial conflict occurs, one must probe deep into the *ideology* and behaviour of the dominant racial group.

Smith (1989: 4) argues that ideology is a fundamentally political, prescriptive medium through which the popular legitimacy of iniquitous social and economic arrangements are secured. Ideology involves collective decision-making and serves as a way to gain popular support for strategies adopted by the dominant group, particularly in democratic societies (Smith, 1989: 4).

Ideology provides a means by which one group of people carve out a certain vision of how things are and ought to be (Reeves, 1983: 3a). Giddens (1981) believes that ideology refers to the modes in which exploitative domination is legitimized. It is the dominant racial group that makes most decisions and establishes the racial climate, policies, practices and structure of the society. Cities are merely organizational constructs designed to accommodate the ideology of the dominant group. The role of the dominant group is examined in this chapter in two countries – the USA and Britain. Although spatial separation has resulted in some minor societal differences, the dominant group in the two countries share a common belief when the issue of race is involved. Whether in the USA or Britain, there is a widespread belief among the dominant group in the ideology of white supremacy.

In order to understand race relations in cities in Britain and the USA, it is necessary to understand this ideology which is held by most people of white European descent – that is, the dominant group in the two multi-racial industrial countries.

The ideology of white supremacy holds that in any *relations* involving people of colour, the white race must have the superior position (Rose and Associates, 1969: 68).

Thus, in economic relations, political relations and social relations, most whites believe that they must have the advantage. Any other arrangement, such as racial equality or control by people of colour, would threaten the 'white comfort zone' and therefore be unacceptable to most whites. The intensity to which most whites in Britain and the USA hold these views varies from very intense to passive (Figure 11.1). Indeed, there is a segment of the white population in each country that does not hold these views at all, but rather detests them and is actively working in the area of race relations to bring about racial equality. It should be clear, however, that such views of white supremacy are not merely a fringe element of white society in Britain and the USA; the belief in white superiority is held by a large segment of white society.

Depending on events in cities and the society involving people of colour, the degree of white support for racial equality or white supremacy may shift from a greater percentage of supporters for racial equality to a greater percentage of supporters for maintaining white supremacy (Figure 11.1). Events that may influence such shifts may involve black immigration, black movement into predominantly white areas, changes in the structure of the economy resulting in white insecurity, and black competition for white-held jobs. Such changes threaten the nature of race relations. The core of race relations in cities of Britain and the USA has been and continues to be that of inequality between whites and people of colour. Thus, changes which may result in reducing racial inequality or threatening the extent of white control or demographic dominance often lead to a shift in the percentage of supporters for maintaining white supremacy.

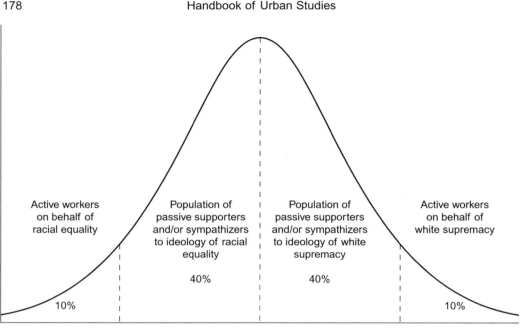

Figure 11.1   Model of race relations in a multi-racial society: conflict between two opposing ideologies

Among the changes in cities of Britain and the USA which have shaped race relations are international immigration and internal migration of people of colour. To maintain white supremacy, such movement of people of colour has been controlled by immigration policies, racial segregation and discrimination in the housing and labour markets in each country.

It is the belief in white supremacy and the extent to which it is acted upon in policy and/or practice that has shaped the nature of racial conflict in the cities examined in this chapter. When the belief is strong and widespread, it results in policies and/or practices that deny equal access to jobs and housing and other amenities to people of colour. Such denial of equal access based on race results in observable racial inequality. Through policies and practices, whites are advantaged and people of colour are disadvantaged. Such inequalities along racial lines are observable in cities through examination of the quality of life of racial groups. Thus, underlying most racial conflicts in cities examined in this chapter is the factor of racial inequality, which is perpetuated by the dominant group. This chapter will demonstrate how such behaviour impacts on racial minority groups and their responses to such behaviour. Although racial minorities are impacted by the behaviour of the dominant group in all spheres of life, this chapter will focus primarily on the housing and labour markets.

It is important to note that although racial inequality provides the potential for racial conflict, such conflict does not occur until there is a degree of racial consciousness on the part of the minority group which leads to a struggle for racial equality. Such efforts are often manifest through a protest of racial injustices, a statement of grievances, or competition for better housing or jobs which are disproportionately held by the dominant group. It is the reaction of a dominant group towards minority demands for equality that racial conflict in cities of the USA and Britain must be understood. Given that the dominant group in each country shares a common racial ideology, their reaction to demands for change by people of colour is often more similar than different.

Although forms of blatant discrimination have been outlawed in each country, a second more subtle form of discrimination is increasingly apparent. This type of discrimination, which is practised in Britain and the USA, can be referred to as *institutional racism*. It has its roots in the ideology of white supremacy and can be defined as a process created by the dominant white population to keep persons in a position or status of inferiority by means of laws, policies, or actions or institutional structures which do not use colour itself as the subordinating mechanism, but instead use other mechanisms indirectly related to colour to get the desired results (Bullock and Rodgers, 1976; Dawns, 1970: 79).

Unlike overt racism, which involves individual

whites acting against individual blacks or other people of colour, which may result in death, injury, or destruction of property and often involves a fringe element of white society, institutional racism originates and operates by and with the approval of the established and respected forces in society. It thus receives far less opposition and condemnation (Carmichael and Hamilton, 1967: 4). Yet, institutional racism is the primary mechanism by which racial inequality is sustained and white supremacy is maintained. With these concepts clearly understood, I turn now to a discussion of the many ways whites and people of colour came in closer proximity and the subsequent change in the nature of race relations.

RACE RELATIONS IN BRITISH CITIES

Although the concept of 'people of colour' consists of those population groups that are not white, this chapter will focus on those people of colour of African descent. Relations between people of European and African descent have been historically the most challenging. To be sure, such difficult race relations remain down to the present.

Black Immigration to Britain

Although blacks had lived in Britain for centuries (Freyer, 1984), due to post-war labour shortages the British recruited more blacks from the West Indies in the late 1940s. Prior to 1948, few white Britons came in contact with blacks. Although Britain has never been a 'promised land' for blacks, relations between the races appeared on the surface to be relatively peaceful and hospitable (see Holmes, 1982). There was no apparent threat to white dominance and sense of white control. After 1948, more white Britons came in contact with black people than ever before, especially in the cities of London, Nottingham, Birmingham and Manchester (Pilkington, 1988).

From the very beginning, black immigrants to Britain constituted a relatively marginal complement of labour located at the bottom of the occupational ladder of the highly industrialized, predominantly white society (Freeman, 1987). The West Indian immigrants found low-income jobs in the struggling textile and manufacturing industries. They were used as a replacement workforce, filling jobs vacated by upwardly mobile white workers and they were largely denied positions in the skilled and managerial workforce (Peach, 1968). The relatively low income of the West Indian immigrants (which reflected their

lack of employment opportunities, not their skills and experiences) obviously limited their housing options. However, these options were further constrained by discrimination in access to housing in both the private and public sectors (Smith, 1989: Chap. 4).

Coming in contact with white Britons, they were viewed as ex-colonial dependants, black immigrant workers of subordinate status. Although few in number at first, the arrival of these immigrants of colour was perceived as a threat to white society. Thus, a debate ensued about the status and future of black people in Britain. Some white Britons wondered whether their country would develop a 'colour problem' similar to the American South, where blacks and whites lived separately and unequally.

Anti-black racism in Britain had emerged by 1951 and was clearly evident within the political arena. The Labour Party undertook policies and propounded ideas which significantly influenced the creation of a racist Britain within the present context (see Joshi and Carter, 1985: 69–70). The intrusion of anti-black racism into domestic politics prior to 1951 was necessarily limited by the small size of the settled black population. With the onset of relatively large scale black immigration during the post-war period, racism flowered everywhere, poisoning the entire body politic (Jacobs, 1985–7). Black immigrants to Britain clearly entered an intensely hostile white world where racism was recent only in its domestic application (Segal, 1967: 300). The absence of overt racism in social legislation has been misinterpreted as evidence of British tolerance, when in reality, black people encountered discrimination on a massive scale (Jacobs, 1985: 11). From the British perspective, it was black labour that was wanted, not black presence (Sivanandan, 1976–352).

Racism, even when acknowledged, was almost exclusively explained as individual acts of prejudice by landlords, state agents, vendors and others. The powerful institutions, which control the property market, were rarely criticized even though building societies, for example, invented 'blue zones' to specifically exclude areas of black settlement from the mortgage loan market (Segal, 1967). In other words, white society chose to ignore the stark realities of racism.

The vast majority of the black immigrants from the Caribbean are today permanent settlers in Britain. They constitute about 500,000 people and represent 56 per cent of the total black population. Almost half of the African-Caribbean population was born in England (Owen, 1994: 1). However, when the figures are adjusted for census under-enumeration, the black-Caribbean population increases to 517,000 (see Owen, 1995: 46). African-Caribbean blacks represent only 1.7 per cent of the total British population.

Segregation and Discrimination in Employment

Although many whites came in contact with blacks for the first time in the 1950s, they had already held preconceived negative views about blacks. The stereotype view of blacks was that they were unskilled, poorly educated, lazy and stupid, but had extraordinary sexual powers (Pilkington, 1988). These prejudices affected the way white people *related* to black newcomers. Employers refused to hire black workers because they prejudged them to be lazy and unreliable. Yet, these ideas were not supported by the facts. The characteristics of the black immigrants to Britain reflected wide and diverse strata. Indeed, 46 per cent were skilled workers, 24 per cent had professional or managerial experience, 5 per cent were semi-skilled, and contrary to the perception of most whites, only 13 per cent of the West Indian blacks were unskilled, manual workers (Pilkington, 1988: 23).

Although unskilled West Indians had no difficulty finding manual jobs in manufacturing and service industries, skilled black workers who expected to find work equivalent to the abilities based on their merit were excluded by the British, qualifications not withstanding. For most blacks from the West Indies this was the first 'reality check', namely, that racial discrimination was a fact of British life. Such knowledge of the 'real Britain' began to spread as black teachers, carpenters, mechanics, tailors, doctors and journalists were all rejected for employment in Britain.

Such rejection was particularly shocking to black ex-servicemen, who had been accustomed to operating according to the principle of merit. However, upon entering Britain as civilians, they too learned that colour, not merit was the basis for employment. Not only did the private sector discriminate, but discrimination also occurred in the British government (Wrench, 1995).

The British government did not assess candidates for the Civil Service based on their competence alone. Race mattered as a critical factor in employment selection (Harris, 1991: 9). Whites were clearly given preference over people of colour.

A subtle means of discrimination was to mark the black candidate down so that he or she would have difficulty qualifying. Hard evidence that governmental recruitment officers unjustifiably marked qualified Afro-Caribbean applicants down can be found in memos of governmental officials cited by Harris (1991: 11). Persons were excluded, therefore, not based on merit, but based on colour. Discrimination was targeted primarily at black men due to the white stereotypes of black male sexuality. Blacks were particularly excluded from industries which brought them in contact with the public or placed them in positions of authority over white people (Harris, 1991).

Lindsey (1993) reveals how West Indians became marginalized in Britain's second largest city between 1948 and 1962. Specifically, racial discrimination was a major factor responsible for the split labour nature of the British job market. Such structure ensured that white British workers would maintain the higher level jobs and the West Indian immigrants, who were considered inferior, would occupy lower level ones. Such racial discrimination has continued in Britain. Blacks during the 1980s were more likely than whites to be unemployed whether in prosperous regions or depressed regions (Smith, 1989: 39).

Also, similar to the USA, is the degree of racial disparity in unemployment rates. Newnham (1986) found that unemployment rates among black workers are consistently at least twice those experienced by whites, irrespective of their qualifications and language skills. These discrepancies are even worse for young blacks, where it is not uncommon to have a third of those aged 16–24 years unemployed compared with only 17 per cent of young whites.

Similar to the occupational structure in the USA, black workers in Britain typically occupy lower status jobs than their white counterparts and are paid lower average wages (Cross and Johnson, 1989). Since such lower occupational status of blacks has not been attributed to their length of time in Britain, researchers have contributed it to persistent racial discrimination (Smith, 1989).

The post-war phase of non-interventionism on the part of policy-makers was a critical factor in the persistent inequality faced by blacks in Britain. In the absence of strong government directives, firms have instituted racial quotas and limited highly qualified blacks' access to skilled and supervisory jobs.

Such segregation of black immigrants in a limited number of jobs as a result of discrimination based on race has denied blacks economic mobility. Moreover, it has perpetuated racial inequality. Such actions by white British private and public sectors have planted the seeds for black grievances, which has inevitably led to racial conflict as blacks strive for racial and social justice and the whites refuse to yield to black demands.

Racial Segregation and Discrimination in Housing

The residential concentration of blacks is another dimension of race relations leading to racial conflict. In general, people of Afro-Caribbean origin have been heavily concentrated in the South-East and West Midlands regions of Britain. Within these areas, 58 per cent of the black population reside in Greater London and 16 per cent live in metropolitan areas of the West Midlands. They live in the largest urban areas, and within such large areas, they are much more concentrated in inner and south London and Birmingham (Owen, 1995). Smaller concentrations of Afro-Caribbeans are located in Yorkshire and Humberside (4 per cent) and Greater Manchester (3 per cent).

Clearly segregated within metropolitan areas, this pattern of population concentration, which started during the period of black immigration to Britain, had been essentially set by 1961 (Smith, 1989: 27). However, Peach (1996: 227) found that Afro-Caribbean segregation in Greater London declined between 1961 and 1991. He also found that average levels of Afro-Caribbean segregation in British cities was much lower in 1991 than those for Bangladeshi, Pakistanis and Indians.

Even though the overall numbers of blacks in Britain are small, because of colour and the fact that they are heavily concentrated in certain cities, their numbers appear larger than they actually are. Eighty-one per cent of the Afro-Caribbean population lived in metropolitan counties in 1989 (Office of Population Censuses and Surveys, 1991: 25). Moreover, 43 per cent of Afro-Caribbeans lived in inner city zones of London, Birmingham and Manchester, while only 6 per cent of the white British population lived there (Small, 1994: 64; Smith, 1989: 26). Wherever blacks have lived, most have been relegated to the most deprived urban areas (Brown, 1984).

In the beginning, most black immigrants were relegated to housing in the private market. Thus, most lived in inner city slums, condemned to the worst housing in the least desirable areas where high rent and inflated prices were charged for property that was on average inferior to that occupied by whites (Jacobs, 1985: 14). Moreover, during the early post-war period, the relatively few blacks who managed to acquire council housing were invariably housed in the poorest properties also (Jacobs, 1985: 19).

Rapid building in the 1960s, often below standard and based on new techniques, had led to the development of 'new slums'. This polarization in the quality of housing was to allow the 'public sector urban gatekeeper' to allocate housing on the basis of 'deserving' and 'non-deserving' categories. The non-deserving were often allocated poor quality housing – these included the homeless, lone parents, large families, the unemployed and so-called 'problem families'. In most cases, blacks were overrepresented in the 'non-deserving' category (Luthera, 1988: 23). Race was a key factor in the overrepresentation of blacks in this category. Such a pattern of unjust treatment based on race, which emerged during black immigration in Britain, has been sustained, resulting in a continued potential for racial conflict. While black residential segregation in Britain is not at the same level as in cities in the USA, residential location, nevertheless, has been a critical factor in determining the subordinate status of blacks (see McKay, 1977).

Such segregation seems to be related to discrimination. For example, in 1967, when it was not illegal to refuse people accommodation on racial grounds, Daniel (1968) revealed that Afro-Caribbeans were often told that furnished accommodations had been rented already in 63 cases out of 120 properties in which 'phone enquiries were made. In five cases, the Afro-Caribbean person was quoted a higher rent than the white British person. When enquiring in person, the Afro-Caribbean person was told the properties had been rented already on 40 occasions out of the 60 properties in which enquiries were made. On six occasions, the Afro-Caribbean person was quoted a higher rent. Unfurnished accommodation was hardly ever made available to blacks (Daniel, 1968: 155–9). A repeat of the test six years later showed that discrimination had been reduced (Smith, 1977: 287), but had not disappeared.

Follow-up interviews with the discriminating landlords and real estate agents showed that two stereotypes existed. The first was that blacks were, amongst other things, 'unclean and noisy' and would 'upset' the existing white tenants, and secondly, blacks would be bad payers, break the conditions of their tenancy or contract of repayment and reduce the area to a slum (with a consequent fall in property/rental values) (Luthera, 1988: 110).

In fact, Patterson (1963) found just the opposite in Brixton (a heavily black section in London). There, not only did managing estate agents report black settlers to be prompt payers of rent, but they also paid at a level above the average rent paid by all dwellers. In the early years, this was regarded as a 'colour tax' paid by blacks to induce landlords to rent to them. It was not as if blacks were receiving any better facilities in return. Instead, blacks, like other immigrants, enjoyed the least in terms of access to facilities, such as hot water (Davison, 1964: 107).

Many of the early surveys reported white home owners who refused to sell to blacks and estate agents who either refused to show them any properties or explicitly limited them to the least saleable ones on their books (PEP, 1967: 170–6; PEP, 1977). Some estate agents in Brixton reported sales to West Indians at inflated prices (Patterson, 1969: 194). During the 1970s, private sector racial discrimination in housing took a more sophisticated form with estate agents giving limited information to blacks and *steering* them into existing areas of concentration (PEP, 1976).

Thus, due to white British fear of black competition and the widespread belief by most whites that blacks have no rights in the labour and housing market that whites are bound to respect, white discrimination against blacks has occurred in both employment and housing. Such patterns of discrimination, which have led to continued racial inequality, were the seeds that would inevitably lead to racial conflict in British cities. Discrimination also leads to poverty, which is in turn related to crime.

Race, Poverty and Crime

Poverty and crime are highly correlated. When the poor perceive their condition as related to racial injustice and discrimination, the potential for violent rebellion heightens. Although crimes have been disproportionately higher among blacks, those who often are arrested for crimes are also disproportionately poor, unemployed and disadvantaged. It is this condition that has led to what I call 'political crimes'. Such crimes can be characterized as a form of rebellion against an oppressive system. Such rebellions occurred in cities in the USA during the 1960s. Rebellion against racial oppression occurred in British cities during the 1980s. During 1980 and 1981, urban unrest was observed in the St Paul's area of Bristol, the Brixton area of London, parts of Liverpool, Manchester and the Handsworth area of Birmingham. There were also disturbances in Leeds, Bradford and other cities.

Researchers who investigated the underlying reasons for the disorders related them to the reaction of Afro-Caribbeans and Asians to persistent racial discrimination, high unemployment, deprivation, political exclusion and powerlessness. Moreover, there was hostility and mistrust towards the white police officers who harassed, abused and assaulted them (Benyon, 1987; Scarman, 1981). Whether in cities in Britain or the USA, the behaviour of white police officers towards black male residents has been a key factor that has triggered urban rebellions.

Behaviour of the Police as a Factor in Racial Conflict

In the 1960s in Britain, written evidence began to emerge related to the role of the police in racial conflict. In 1965, the West Indian Standing Conference published a study of police–black relations. The Committee concluded that racial prejudice was interwoven into the fabric of police and black immigrant relationships. Moreover, the police were considered malicious and hostile (Hunte, 1965).

Black males, particularly young black males, have a higher chance of being apprehended by the police than do young white males (Willis, 1983). Moreover, black males have been subjected to general harassment from the police, which has resulted in black mistrust of the law enforcement system in general. Similar to the behaviour of the police towards blacks in the USA, the behaviour of the British police has been the immediate precipitant that has triggered rebellions by blacks in British cities.

Police behaviour is usually influenced by political power. If most whites prefer that police stop blacks unjustly, or engage in differential behaviour based on race, such police behaviour is likely to continue. Therefore, change in race relations in Britain will be slow since blacks have little influence on the political process.

Prospects for Black Political Power

Unlike the USA, Britain's small black population has little political influence. Blacks in Britain remain grossly underrepresented as MPs and at local levels (Commission for Racial Equality, 1999). Even where there are substantial, homogeneous black communities, political leadership often remains under white control (Norris and Lovenduski, 1993). Therefore, policy-makers largely ignore the needs of poor urban black residents. It is not surprising that the social and economic disadvantage experienced by urban blacks has not changed (Thomas and Krishnarayan, 1994).

Policy Implications and Implementation

In the years immediately after the end of the Second World War, the British provided Afro-Caribbeans freer access to Britain in order to fill a labour shortage. Fearing increased immigration by non-whites, policy-makers in 1962 reversed their policy and established a quota system, which strictly controlled and reduced the number of Afro-Caribbeans and Asians who could settle in the country.

At least since the 1960s, strong opposition to black immigration has been on the rise in Britain. However, policy-makers set out early to formulate immigration policies that would reduce black immigration, but not appear racially discriminatory. One such invention was the concept of partiality. This allowed entry into the country if one could demonstrate a historical family link. Therefore, a white-skinned immigrant from

Africa or India could obtain entry, while black-skinned immigrants could not. The British introduced such policies, while at the same time stating that such policies were not racist since race was not specifically mentioned.

Yet, as is typical of those who hold the ideology of white supremacy, black people in and of themselves were being defined by the British as 'a problem'. In other words, the problem was not immigration but 'black immigration'. Since Britain's immigration control policies have been inseparable from race, such policies have been a factor in establishing a climate for racial conflict in British cities. Most British policy-makers have had a history of side stepping the issue of race in Britain. Such a history is reflected in their approach to anti-discrimination policies.

British housing and labour market anti-discrimination policies have evolved from a mistaken belief that discrimination did not exist in Britain and that Britain was an example for the world of 'good race relations'. Insofar as racial segregation was a problem, in Britain it was seen as only temporary, resulting from the period of immigration (Smith, 1989: 113). Most Britons adopted the myth that segregation was merely a cultural adaptation experienced by people who possessed the same rights and opportunities as white Britons.

Given such beliefs that racial discrimination did not exist in Britain, it was not surprising that many policy-makers felt that no anti-discrimination policies were necessary. As has been documented, however, racial discrimination became widespread in both the housing and labour markets (Harris, 1991; Pilkington, 1988).

Until inhibited by the 1968 Race Relations Act, local authorities persistently evaded their responsibilities to house black applicants and made no bones of their intentions to give priority to the white population (Jacobs, 1985: 20). While the house allocation system was not specifically designed to discriminate against blacks, in practice it clearly did so at every turn (Jacobs, 1985: 23). All involved in the allocation process, from housing managers and senior councillors to clerks and housing vectors, would have a common set of values about black inferiority (Jacobs, 1985: 24).

British policy-makers have taken a non-interventionist approach to the discriminatory treatment of blacks in Britain. Policies have been weak. As a result, some employers have continued to institute race-based quotas that limited black access to high-level jobs regardless of their qualifications. This denial of opportunity in the labour market impacts on their status in the housing market. Discrimination has also occurred by which blacks in Britain received virtually none of the benefits associated with

governmental subsidized council housing following the Second World War (Smith, 1989: 115). This exclusion, based on race, forced blacks to seek housing in the decaying private housing sector in the inner cities.

It was not until 1965 that the first major piece of legislation to address racial discrimination was passed in the form of a Race Relations Act (1965). (The Act was updated in 1968 and 1976.) The Race Relations Act of 1965 was passed in response to racial violence against West Indians and a mounting tide of racist speeches made at well-publicized meetings in the early 1960s (Lester, 1987: 22).

The first act of 1965 was extremely narrow in scope. It failed to touch the real problems of discrimination in housing. Three years later, it was replaced by the Race Relations Act of 1968. This act was much more ambitious in its scope. It made it unlawful to discriminate on racial grounds in employment, housing, education and the provision of goods, facilities and services to the public. The act, however, continued to be weak in the area of enforcement. The emphasis was upon conciliation and friendly settlement rather than legal rights and remedies (Lester, 1987: 23). Only the Race Relations Board could bring legal proceedings and only after elaborate conciliation procedures had been exhausted.

Another weakness of the Act was that it was confined only to deliberate or intentional discrimination. Traditional practices that were not intentional (but had discriminatory effects) were not included (Lester, 1987: 23). Furthermore, the definition of discrimination was not flexible enough to permit 'positive action' to be taken to overcome the effects of past discrimination or disadvantage. By action is meant special training programmes to equip people of colour for areas in which they have been underrepresented.

Finally, the 1965 and 1968 Acts dealt only with individual acts of discrimination and with direct discrimination. According to Wilson, the Acts:

> assumed that the main problem was that of active discrimination against individuals (generated by prejudice), that this could and would be curbed by a widespread positive reaction to the law by those in a position to discriminate and that any residual discrimination would be brought to light and dealt with as a result of complaints by people concerned to assert their rights. (Wilson, 1973: 4)

The results showed that the conciliation procedure did not remedy the wrongs, and individual complaints as a procedure were ineffective in eliminating pervasive and entrenched discriminatory practices (Lester, 1987: 23).

More importantly, according to Smith (1989), the Act came too late to guard against institutionalized racism in housing allocations. Other researchers have pointed out the lack of enforce-ment and lack of coordination of anti-discrimination measures (Cross, 1982). Similar problems were found with the Equal Employment Opportunity Commission in the USA, which will be discussed later in this chapter.

However, one should not overlook a major obstacle to eliminating discrimination which is the persistent and strong opposition by white males to equal opportunity and racial equality. Jenkins and Solomon (1987: 215) have pointed out how the lack of commitment and hostility towards equal opportunity occurs among management, trade union representatives and among the average white British workers. Most see genuine racial equality for black workers as reducing their own mobility. Such views are very common among those who fear competition with people of colour and is related to their belief in the ideology of white supremacy.

RACE RELATIONS IN CITIES IN THE USA

Interregional Migration – South to North

Unlike Britain, the forces underlying racial divi-sions and racial conflict in the USA must first be understood through an examination of black migration from the rural South to the urban North. Until recently, immigration to the USA of people of colour has played less of a role in racial conflict.

Black migration to the North had two waves. The first wave followed the First World War and the second wave came after the Second World War. Together, these mass movements of blacks changed race relations in the North. When blacks were few in number and relatively 'invis-ible' to whites, race relations appeared calm. As black migration increased, Northern whites viewed black migration with alarm and fear. Such fear of black competition for jobs was espe-cially prevalent among working-class whites, many of whom were first or second generation immigrants (Massey and Denton, 1993: 29). Although many were as unskilled as blacks, they found satisfaction in their 'whiteness' in a society where white colour gave immediate advantages in the area of employment and housing over people of colour.

From 1930 to 1940, about 400,000 blacks moved from the rural South to the urban North (Farley and Allen, 1987: 113). When they arrived, they faced not the 'promised land', but a severe housing shortage combined with housing discrimination. Blacks found that Northern whites, like whites in the South, did not want racial residential integration. One mechanism for

avoiding racial integration was the construction of black ghettos, that is, areas where the popula-tion was majority black and separate from white areas. The black ghetto became an institutional tool to subordinate and exploit blacks in cities. Such ghettos were usually not only racially sepa-rate, but also unequal. In the South, laws passed by state governments kept blacks separate and unequal, whereas in most Northern cities the primary mechanism for achieving the same results was the construction and maintenance of black ghettos. According to a report of a National Advisory Commission appointed by former Pres-ident Lyndon Johnson, 'what white Americans have never fully understood, but what the Negro can never forget, is that white society is deeply implicated in the ghetto. White institutions created it, white institutions maintain it, and white society condones it' (National Advisory Commission on Civil Disorders, 1968: 2).

Over the years, black migration to cities has occurred at an accelerating rate. As a result, the black population has become more urbanized, and more metropolitan than the white popula-tion. Almost all black population growth is occurring within metropolitan areas, primarily within central cities.

On the other hand, the vast majority of white population growth is occurring in suburban portions of metropolitan areas. As a result, central cities are becoming more heavily black, while the suburban areas surrounding them are remaining heavily white. Moreover, the black population reside disproportionately in the largest metropolitan areas; those with popula-tions of one million or more.

It is important to note that due primarily to discrimination based on race, the black pattern of settlement and spatial expansion in metropolitan areas diverge sharply from that typical of the white immigrant. As white immigrants arrived in cities, most were integrated into the larger society obtaining better jobs and housing, that is, social and spatial mobility in the suburbs. On the contrary, black migrants were generally not inte-grated into the larger predominantly white society. Instead most blacks remained segregated in central cities.

Such divergent patterns led the National Advisory Commission on Civil Disorders to conclude that 'our nation is moving toward two societies, one black, one white – separate and unequal'. Such a statement incorrectly assumed that there had once been unity and equality between the races, when in fact there has always been a situation of 'separate and unequal' between blacks and whites in cities in the USA. We will examine later whether these patterns of separation and inequality are increasing or decreasing.

Recent Immigration and Changing Racial Diversity in the USA

The USA is now experiencing the largest wave of immigration in the country's history. Whereas early immigration consisted of people predominantly from Europe, immigration since the 1970s has consisted of people predominantly from Asia and Latin America. Mexico, which borders the USA, supplies the largest share of both legal and undocumented immigrants. Within the Caribbean and Central America, the Dominican Republic and Jamaica supply a high proportion of the legal immigrants. The impact of recent immigration by people of colour can have a profound effect on race relations in metropolitan areas.

Debate is already intensifying over the issue of immigration. Some Americans are arguing that the 'new immigrants of colour' will not integrate as easily or move up the economic ladder as quickly as earlier white immigrants from Europe and will therefore place strain on US public resources. Others have argued that immigrants take jobs away from American citizens and lower wages overall (Beck, 1996: A11). In spite of the evidence that immigrants have little or no impact on wages or jobs, efforts to control immigration (legal and illegal) continued during 1995. These efforts are discussed later in the section on immigration controls.

One reason for the aggressive increased concerns over immigration is the perceived *threat* of an increasing population that is different in characteristics from the majority white population. An important variable in understanding race relations is the 'dissimilarity factor'. In general, the greater the dissimilarity of the minority group from the majority group, the greater the discrimination, segregation and denial of equal access to jobs and housing imposed on the minority group.

Segregation and Discrimination in Housing

Racial residential segregation is a tool in which the dominant white population excludes people of colour in cities from the benefits of society. The ultimate outcome of racial residential segregation is racial inequality. The National Advisory Commission on Civil Disorders (1968: 10) recognized that fact when it concluded that 'pervasive discrimination and segregation in employment and housing have resulted in the continuing exclusion of a great number of blacks from the benefits of economic progress'.

Blacks are also excluded more often than other groups, making them the most residentially segregated racial minority group in the country. Support for this statement is provided through calculations of indexes of dissimilarity between the dominant white population and three racial minority groups in the largest metropolitan areas with a population over one million (Table 11.1). The three racial groups are blacks, Asian/Pacific Islanders and Hispanics. The results show that blacks, the most racially visible group, are the most segregated from the white population, with an average level of segregation of 66 per cent compared with 42 per cent for Hispanics and only 38 per cent for Asian/Pacific Islanders (Li, et al., 1995).

Thus, residential segregation in US metropolitan areas occurs along a colour continuum. Whites exclude some racial minority groups more often than others. Darden's (1989) study of multiple minority groups revealed that Asians experience the smallest amount of residential segregation from whites, have the highest level of suburbanization, and have the greatest opportunity for social, economic and spatial mobility. Blacks, on the other hand, experience the greatest amount of residential segregation, the lowest level of suburbanization and have the least opportunity for social, economic and spatial mobility of all racial/ethnic minority groups.

Race, more than ethnicity or class, drives residential segregation in American housing markets. Due to persistent discrimination in housing based on colour, blacks are residentially segregated and largely excluded from the suburbs, regardless of their level of education, income and occupation. Thus, future improvements in the social and economic status of blacks will not necessarily lead to residential integration and greater suburbanization. Such a pattern differs from that of any other minority group and reinforces the significance of colour – that is, black colour – in explaining the unequal status of blacks compared with members of the other minority groups. This conclusion is also supported by others.

After examining an overwhelming amount of evidence, Massey and Denton (1993) came to three conclusions. First, black residential segregation continues unabated in the nation's largest metropolitan black communities, and this spatial isolation cannot be attributed to class. Secondly, although whites now accept open housing in principle, they have not come to terms with its implications in practice. Whites in practice are unwilling to tolerate more than a small percentage of blacks in their neighbourhoods. Thirdly, discrimination against blacks is widespread and continues at very high levels in urban housing markets (Massey and Denton, 1993: 109).

In general terms, racial discrimination in housing exists whenever individuals, in this case blacks, are prevented from obtaining the housing

Table 11.1   *Racial residential segregation indices in the largest metropolitan areas, 1990*

MSA/PMSA	Black	% in pop.	Asian	% in pop.	Hispanic	% in pop.
Los Angeles–Long Beach, CA PMSA	72.97	10.5	46.19	10.2	61.07	37.8
New York, NY PMSA	82.18	23.2	48.11	6.2	65.78	22.1
Chicago, IL PMSA	85.78	21.7	43.36	3.6	63.19	12.1
Philadelphia, PA–NJ PMSA	77.14	18.8	43.07	2.1	62.57	3.6
Detroit, MI PMSA	87.63	21.4	43.45	1.3	39.71	1.9
Washington, DC–MD–VA MSA	66.07	26.2	32.33	5.0	40.91	5.7
Houston, TX PMSA	66.80	18.1	46.14	3.7	49.30	21.4
Boston, MA PMSA	78.32	6.0	36.81	3.0	36.37	5.0
Atlanta, GA MSA	67.83	25.8	40.26	1.8	34.48	2.0
Riverside–San Bernardino, CA PMSA	43.75	6.5	33.30	3.6	35.84	26.5
Dallas, TX PMSA	63.10	15.8	40.82	2.5	49.54	14.4
San Diego, CA MSA	58.12	6.0	47.91	7.4	45.20	20.4
Minneapolis–St Paul, MN–WI MSA	61.95	3.6	41.24	2.6	35.42	1.5
St Louis, MO–IL MSA	76.99	17.2	38.48	0.9	22.88	1.1
Anaheim–Santa Ana, CA PMSA	37.49	1.6	33.16	10.0	49.92	23.4
Baltimore, MD MSA	71.36	25.7	38.31	1.8	30.16	1.3
Phoenix, AZ MSA	49.89	3.3	27.44	1.6	48.09	16.3
Oakland, CA PMSA	67.75	14.2	39.23	12.4	38.78	13.1
Tampa–St Petersburg–Clearwater, FL MSA	71.24	9.1	35.10	1.1	46.80	7.0
Pittsburgh, PA PMSA	71.03	8.1	50.58	0.8	29.77	0.6
Seattle, WA PMSA	56.23	4.0	36.56	6.7	20.37	2.8
Miami–Hialeah, FL PMSA	71.76	19.1	26.88	1.2	50.32	49.2
Cleveland, OH PMSA	85.11	19.3	36.76	1.1	55.31	1.9
Newark, NJ PMSA	82.50	22.4	29.48	2.8	66.74	10.3
Denver, CO PMSA	64.47	5.7	29.02	2.2	46.47	13.0
San Francisco, CA PMSA	63.93	7.4	50.22	20.0	49.81	14.5
Kansas, MO–KS MSA	72.55	12.7	33.13	1.1	39.35	2.9
San Jose, CA PMSA	42.79	3.5	38.39	16.8	47.80	21.0
Sacramento, CA MSA	55.69	6.7	48.11	7.4	36.98	11.6
Cincinnati, OH–KY–IN PMSA	75.75	13.1	40.25	0.8	25.39	0.5
Milwaukee, WI PMSA	82.76	13.6	42.23	1.3	56.35	3.6
Norfolk–Virginia Beach–Newport, VA MSA	49.44	28.2	34.27	2.4	29.57	2.3
Columbus, OH MSA	67.33	11.9	44.20	1.5	27.55	0.8
Fort Worth–Arlington, TX PMSA	61.90	10.6	40.30	2.2	45.06	11.3
San Antonio, TX MSA	54.22	6.5	30.20	1.1	53.75	47.6
Bergen–Passaic, NJ PMSA	76.77	7.5	34.48	5.1	58.80	11.6
Fort Lauderdale–Hollywood– Pompano Beach, FL	72.22	15.7	27.18	1.4	30.73	9.1
Indianapolis, IN MSA	74.28	13.7	37.25	0.8	25.73	0.9
Portland, OR PMSA	66.43	3.1	29.99	3.7	26.87	3.6
New Orleans, LA MSA	68.80	34.4	49.72	1.6	31.00	4.3
Charlotte–Gastonia–Rock Hill, NC–SC MSA	53.44	19.9	42.99	0.9	32.09	0.9
Orlando, FL MSA	60.45	12.1	25.32	1.8	26.46	9.0
Salt Lake City–Ogden, UT MSA	47.12	17.2	31.05	2.3	32.23	5.8
Middlesex–Sommerset–Hunterdon, NJ PMSA	53.96	6.5	36.36	5.4	49.88	7.0
Rochester, NY MSA	67.36	9.1	40.70	1.4	55.39	3.1
Means	66.33	13.5	38.23	3.9	42.35	10.8

Source: Computed by the author from data obtained from the US Department of Commerce, *1990 Census of Population and Housing. Summary Tape File 1A*. Washington, DC: Data User Services Division, 1991

they want in the location they prefer for reasons of race. Such discrimination by race hampers the possibility of integrated neighbourhoods, despite improved white attitudes towards blacks and increases in the socioeconomic status of blacks (Foley, 1973).

Prior to the 1968 Fair Housing Act, white real estate brokers simply refused to show or sell blacks homes in predominantly white areas. The most effective, subtle and widespread discriminatory technique still used today by white real estate brokers is 'racial steering'. This is a practice by which a real estate broker directs buyers toward or away from particular houses or neighbourhoods according to the buyer's race (Aleinikoff, 1976: 809; Openshaw, 1973; Saltman, 1975: 43–5; US Commission on Civil Rights, 1971: 60–1). Black homeseekers are steered away from white areas, while whites are directed to them. Conversely, white homeseekers are steered away from black areas while blacks are directed to them.

Widespread discrimination in housing has continued to affect various racial and ethnic groups. However, the most serious and persistent discrimination has been against blacks. It is no accident that blacks are the most residentially segregated and the least suburbanized minority group.

Discrimination typically is uncovered through the use of an investigative procedure called paired testing. Two individuals, one white and one from a minority racial group, are trained to pose as homeseekers and are given fictional identities which render them virtually identical in terms of income, family size, preferences etc. Both team-mates separately contact a real estate agent or landlord and attempt to acquire housing. The testers independently complete reports following their contacts, and analysis of dozens of such testing studies conducted during the 1980s has revealed the alarming frequency with which housing discrimination occurs in the sale and rental of housing (Galster, 1990a, 1990b, 1990c). More recently, the results of the Housing Discrimination Study (HDS) have been released by the Department of Housing and Urban Development (Turner et al., 1991). Thirty-eight hundred paired tests of housing discrimination against blacks and Hispanics were conducted in 25 metropolitan areas in 1989. On average, both black and Hispanic renters and would-be owners were likely to be discriminated against 50 per cent of the time when they responded to advertisements in major metropolitan newspapers.

Such restrictions on black spatial mobility due to discrimination have important social and economic consequences for the black population. Blacks, more than any other group, are disproportionately concentrated in central cities. Moreover, within central cities, blacks are disproportionately concentrated in poor areas away from economic opportunity.

Race, Lack of Economic Opportunity and the Emerging Underclass

In the period since the end of the Second World War, many blue-collar jobs that once constituted the economic backbones of cities and provided employment opportunities for their poorly educated residents, have either vanished or been moved to the suburbs (Kasarda, 1989: 4). Thus, newer and better job opportunities are locating further away from the places of black residence, forcing black families to pend more time and money commuting to work or looking for work (Darden, 1986: 112).

Given that blacks have more restricted residential location choices than members of other racial minority groups, the cost associated with distance reduces access to some jobs. In addition, there is a tendency for employers to hire workers who reflect the racial character of the area in which they are located – resulting in an indirect effect of housing segregation on employment opportunities (Kain, 1968; McDonald, 1981: 28).

The decline of jobs in central cities has been most pronounced in certain cities of the North Central Region, which includes such metropolitan areas as Detroit and Chicago (Darden et al., 1987: 2–3).

Discrimination in the Labour Market

In addition to a lack of economic opportunity due to housing segregation and discrimination, blacks also continue to face discrimination in the labour market regardless of residential location resulting in higher rates of unemployment and lower wages.

With rare exception, black men and women experience higher unemployment than whites at all levels of schooling. Thus black men who have graduated from college, are more than twice as likely to be unemployed as white college graduates. Moreover, blacks experience higher unemployment regardless of their metropolitan residence, although inequalities tend to be greatest in central cities (Darden et al., 1992).

In addition to higher unemployment and underemployment, blacks are more likely than whites to be in jobs offering fewer opportunities for career growth. Corcoran and Duncan (1979) found that blacks receive less on-the-job training than whites. Controlling for characteristics such as education and marital status, Boston (1990) concluded that the probability of black men and women moving from secondary jobs (jobs characterized by low levels of training) to primary sector jobs offering more training is about one-half the corresponding probability for whites.

Beyond the statistical record, controlled experiments using paired testers have investigated hiring discrimination. In these experiments, minority job applicants are paired with non-Hispanic white applicants. The applicants are given similar backgrounds and are chosen and trained to be as similar as possible in job-related characteristics such as appearance, articulateness, and apparent energy level. How the minority applicants are treated in job applications can then be observed and compared with the treatment received by their non-Hispanic white 'twins'.

A study targeted at entry-level jobs concluded that in one out of five paired tests the white applicant was able to advance further through the hiring process than his or her equally qualified black counterpart (Turner et al., 1991: 31). Such experiments provide irrefutable evidence of pervasive hiring discrimination in US labour markets.

Racial discrimination in employment contributes to blacks' inferior rates of employment and wage levels. This occurs both directly and through discriminations' effect upon occupational segregation. Segregation of blacks into occupations associated with low pay, minimum on-the-job training or chances for advancement, and cyclica instability intensifies interracial economic disparities and enhances the probability of racial conflict.

Thus, by limiting physical access to work, the means by which people became productive workers, and the occupations through which economic success is promoted, discrimination and segregation in housing and employment systematically create and perpetuate a host of socioeconomic inequalities between blacks and whites. These disparities manifest themselves in many ways, but perhaps most dramatically in higher black rates of poverty, unemployment, school and labour force dropouts, welfare dependency, crime and substance abuse (Darden et al., 1992).

These economic and social inequalities reinforce the stereotypes held by some whites and serves as a rationale for continuing white supremacy. It legitimizes white prejudices about the 'undesirable characteristics' supposedly possessed by all members of the black population.

Such reinforced prejudices make it more likely that white households will want to perpetuate all dimensions of segregation. Some may be unwilling to remain in their neighbourhoods once blacks begin to move in or attend local schools. When seeking different accommodation, they may be reluctant to search in areas where blacks live or go to school. These actions directly reinforce housing segregation. Finally, white employees' hostility toward potential black workers might encourage employers to discriminate for the sake of maintaining work place tranquillity, or they may discriminate due to their own ideology. Black households bear crushing costs from the perpetuation of this vicious circle. Elimination of employment discrimination should close nearly half the gap in incomes between white and black males (Price and Mills, 1985), and reduce the high proportion of blacks in poverty and living in high poverty areas.

The Spatial Concentration of Poverty: New or a Continuing Trend?

Researchers are continuing to debate whether the spatial concentration of black poverty is a new or a continuing trend. Wilson (1987) argues that the concentration of black poverty is not only a post-Second World War phenomenon, but it occurred in the 1980s as civil rights laws provided new opportunities for middle- and working-class blacks to move out of the ghetto in large numbers leaving behind an isolated and truly disadvantaged black poor population lacking the institutions and human resources necessary for success. Massey and Denton (1993), on the other hand, have taken the opposite view. They argue that the geographic concentration of black poverty is not new – poverty was just as concentrated in the ghetto of the 1930s as in the black underclass communities of the 1970s. They also do not attribute the causes of black poverty concentration to civil rights laws. Instead, they argue that black poverty is attributed to a high degree of segregation from the rest of society and a great deal of hardship stemming from larger economic changes (Massey and Denton, 1993: 117–18).

Whatever the cause, the recent studies suggest a growing trend in the number of poverty areas. Kasarda (1993) analysed census tracts in the 100 largest central cities and found that the percentage of all central city poor persons living in extreme poverty areas increased from 16.5 per cent to 22.5 per cent in 1980 to 28.2 per cent by 1990. Moreover, the spatial concentration of blacks was the most severe compared to other groups.

Using 1990 census data, Jargowsky (1994) analysed changes in ghetto poverty among blacks from 1980 to 1990. Defining 'very poor' to mean census tracts with poverty rates in excess of 40 per cent, Jargowsky concluded that ghetto poverty among blacks increased both in terms of the number of blacks living in ghettos and as a percentage of the black population.

Whether the concentration of ghetto poverty is new or a continuing fact of black life in America's cities, it is clear that the National Advisory Commission on Civil Disorders (1968) recognized its existence in the late 1960s and warned white Americans to act to avoid a permanent separation along the lines of race and class. Since most white Americans ignored the warnings, by the end of the 1970s many large American cities were populated by an increasingly black, unemployed, poorly educated population with single-family households forced to live in endless poverty with no hope of social mobility. Observing this phenomenon, researchers began to coin the phrase 'the urban underclass' (Katz, 1989). Although researchers disagree over what constitutes an 'urban underclass', and whether it is growing, the concept in its most general form refers to a population that is highly segregated and poor with few tools or skills to change their condition. Thus, a feeling of hopelessness is a common characteristic (Jencks and Peterson, 1991; Ricketts and Sawhill, 1988; Ruggles and Marton, 1986). Several authors have attributed the black underclass condition directly to white racism, residential segregation and discrimination, and the systematic failure of white American institutions to address the needs of poor blacks in America's cities (Glasgow, 1980; Massey and Denton, 1993).

The continuous existence of an underclass has serious and costly consequences for America's metropolitan areas. Among such consequences is the potential for rising crime and racial conflict.

Race, Poverty and Crime

The National Advisory Commission on Civil Disorders (1968) noted more than 30 years ago that concentrating the poor in certain sections of cities may lead to rising crime. In America's largest cities, the most concentrated poor population is disproportionately black. They reside in areas where crime rates have been historically higher than anywhere else.

The National Advisory Commission mentioned two facts that are still true today. First, most crimes in these poverty-ridden, inner city areas are committed by a small number of the residents, and secondly, the principal victims are the residents themselves. In terms of race relations, most everyday crimes committed in America's large cities are not crimes committed against a member of another race. Throughout America's metropolitan areas, most crimes committed by blacks involve other blacks as victims, just as most crimes committed by whites involve other whites as victims.

Such crimes, however, do have implications for race relations. Although most crimes in the black ghetto are committed by a small percentage of young black males (aged 14–24), they have a tendency to create fear in the minds of many whites, i.e., fear of blacks in general and black males in particular. Such fear reduces interracial interaction and the opportunity for productive race relations. Moreover, high black crime rates provide a rationale for those whites who wish to rationalize segregation by any means necessary. Instead of focusing on the poverty which is related to the crimes, they focus instead on race, which leads to such behaviour as 'keeping the blacks in their place'. Many blacks, however, view the relationship between blacks and crime with a much broader and different meaning, which inevitably speaks to the racially unequal, unjust and oppressive American metropolitan system.

The concept that these concentrated areas of poverty are analogous to white controlled internal colonies emerged following the civil disorders of the 1960s. Staples (1975), for example, viewed black crime as a function of a group's power to define what behaviour can be classified as criminal or legitimate.

In a society such as the USA, blacks share many of the characteristics of natives in colonial countries, such as political and economic oppression, cultural subjugation, and control of their community by an alien group. Race is the major factor why nearly one in four young black men in the USA were in prison or jail or on probation in 1990 (Mauer, 1990). Five years later, the number of black men in their twenties, who were imprisoned or on probation or parole, had risen to one in three. Mauer, the report's author, warned that if one in three white men were under criminal justice supervision, the nation would declare a national emergency (Butterfield, 1995).

Miller (1992), who has studied the damaging impact of the US criminal justice system on blacks, considers the pervasive criminalization of black youth today as analogous to the white slave owners' practice of crippling young black male slaves to prevent them from escaping to freedom. Moreover, white collar crimes, which are mostly committed by the white population, are rarely or lightly punished. The real relationship between crime and race is the systematic exploitation of blacks, reducing them to levels of poverty below the standard of living of most whites, a situation which produces a disproportionate number of economic crimes among the black population. Some blacks in prison are defined as political prisoners because their status is determined by the economic conditions imposed upon them by the political state, not their criminal activity.

As long as blacks remain segregated and denied equal opportunity to leave poor areas of cities, the potential for violent racial conflict is always present. Although there are periods of remission, such areas remain potential powder kegs ready to explode. The lesson for such prediction can be derived from past civil disorders in America's cities. The National Advisory Commission on Civil Disorders (1968: 10) concluded that the primary factors responsible for the 'mood of violence' among many blacks in urban America was the attitude and behaviour of white Americans towards blacks.

The Behaviour of the Police in Racial Conflict

Most of the black violent protest in America's cities has been set off by the behaviour of white police officers in particular who are viewed by many poor urban blacks as the oppressor. They are sent into black neighbourhoods to *control* blacks by any means necessary. Thus, many blacks firmly believe that police brutality and harassment occur disproportionately in black neighbourhoods. This belief, which is frequently supported by documented incidents, is one of the major reasons for intense mistrust and resentment against the police (Hawkins and Thomas, 1991).

The role of the police, as representatives of the dominant group, is critical. Depending upon the definition of their role as 'keepers of the peace' or as enforcers of 'law and order', they may inhibit a propensity toward violence or, in an atmosphere of suspicion and mistrust, they become the catalyst, their actions serving as a principal precipitating event. When a reputation for brutality or the frequent harassment of racial minorities occurs, even routine investigations and arrests may become occasions for onlookers to react violently. The same may be said of court hearings and sentencing decisions if these are perceived as racially unjust. Such was the reaction following the not guilty decision in the Rodney King beating. To many blacks, the decision represented the reality of those infamous words spoken by Supreme Court Justice Roger Taney in the Dred Scott Decision, 'Blacks have no rights that whites are bound to respect'.

Changing Black Political Power and Consequent White Reaction

Debate continues over whether there has been much improvement in the condition of life in the black ghetto since the disorders of the 1960s. Although one may question progress in the social and economic arenas, one area where there has been clear and measurable progress has been in the political arena. When the civil disorders occurred in the late 1960s, only two major cities had black mayors – Cleveland, Ohio, and Gary, Indiana. Since then, blacks have served as mayor of several major American cities including New York, Chicago, Los Angeles, Philadelphia, Detroit, and San Francisco.

Moreover, the number of black elected officials has steadily increased. From 1990 to 1993, the total number of black elected officials increased from 7,370 to 8,015 (Joint Center for Political and Economic Studies, 1994). Although blacks still represent only 1.6 per cent of all elected officials, their presence in positions of decision-making, particularly in America's cities, may be a key factor that explains why some cities have remained calm even though the social and economic conditions of some urban blacks may have not improved. Blacks in decision-making roles seem to have a positive and moderating effect on racial relations in cities. Moreover, city governments with black elected officials have been found to be more responsive to the needs of urban blacks. This has been especially true in the promotion of black business development (Bates and Williams, 1993) and employment of blacks in city government.

Policy Implementation

As the USA continues to undergo a profound racial demographic shift, it is moving to tighten its policy on immigration and loosen its enforcement of anti-discrimination legislation. Important to race relations were changes in US immigration laws enacted in 1965. Those changes made it easier for immigration to occur resulting in increased numbers and changes in the racial composition of the American population. However, larger numbers of immigrants from non-European countries coupled with a large illegal flow from Mexico and other developing countries, has prompted policy-makers to re-examine immigration policy.

Because of disproportionate immigration of people of colour, non-Hispanic whites constituted 73.6 per cent of the population in 1995. The remaining percentages were: blacks 12 per cent, Hispanics 10.2 per cent, and Asians 3.3 per cent. If the present trends continue, the white non-Hispanic population will constitute 61 per cent of the population by 2030; Hispanics will be 19 per cent, blacks 13 per cent and Asians 6.6 per cent (US Bureau of the Census, 1996). However, these projections are fuelling the racial flames of fear, which are leading to changes in immigration legislation.

Republican members of Congress are sponsoring bills to change immigration laws. These bills are aimed at reducing immigration and stopping the increased racial diversity. As America moves to implement new policies to restrict immigration, it is becoming more loose concerning enforcement of anti-discrimination laws in the housing and labour markets.

Moreover, similar to the efforts in Britain, past federal efforts to eradicate housing market discrimination have been hampered by weak legislation and lack of enforcement since the Fair Housing Act of 1968. The present system is fundamentally flawed because it continues to rely on individual victims detecting discrimination against them and then filing suit. As a result, current anti-discrimination efforts provide little deterrence for prospective discriminators because they rightly believe that their actions are unlikely to be challenged. Victims typically cannot detect such illegal acts and, even when they do, their prospective benefits from seeking redress often fall short of their prospective costs of doing so (Darden et al., 1992). Moreover, relatively few victims of housing discrimination realize that they have been victimized because discriminatory techniques have become more subtle (Turner et al., 1991).

Current civil rights enforcement policy in the labour market has its origins in the 1964 Civil Rights Act (Withers and Winston, 1989). Title VII of the Civil Rights Act prohibits discrimination in all aspects of employment and compensation on the basis of race, colour, religion, gender or national origin. Title VII also established the Equal Employment Opportunity Commission (EEOC) to investigate complaints of such discrimination. The EEOC receives and investigates discrimination charges, resolving them through conciliation and, if necessary, court action. As such, current enforcement policy generally relies upon individual reported discrimination. In its authority to file 'pattern and practice', or systematic suits against private employers, the EEOC also may initiate investigations without a specific charge being filed.

Unlike the more limited measures in Britain, affirmative action programmes (established to implement Executive Order 11246 in 1965) are another measure to aid American minorities in their employment. As implemented by the Office of Federal Contract Compliance Programs (OFCCP) of the Department of Labor, this policy requires companies with $50,000 or more in federal contracts and 50 or more employees to take 'affirmative action'. Due to a white male 'backlash', however, 'affirmative action' is presently under attack.

In sum, the USA, like Britain, is becoming increasingly racially tense. Evidence suggests that racial incidents involving attacks against people of colour are on the rise. Yet, policies to address the issue are increasingly retrogressive or not enforced. It appears that the forces of white supremacy are gaining strength which further weakens the prospects of progress towards racial equality in cities in the USA.

CONCLUSIONS

The major thesis of this chapter has been that race relations, including racial conflict, have been formed and motivated by a common ideology, that is, white supremacy. Such an ideology has been deeply ingrained in the white British and white American cultures. It is this ideology that has shaped immigration policy related to people of colour. It is also this ideology that has caused similar patterns of racial residential segregation in cities and occupational segregation in the workplace.

The ideology of white supremacy, which holds that in any relations involving whites and people of colour the white population must have the advantage, has led to racially discriminatory immigration policies in both countries. It has also led to racial discrimination in housing and employment. Such discrimination occurs due to white fear of losing control over the better jobs, housing and neighbourhoods to an increasing population of blacks.

The ultimate results of the implementation of this ideology have been the perpetuation of racial inequality. African-Americans in the USA and Afro-Caribbeans in Britain have experienced a common fate of exploitation, segregation and deprivation. In both Britain and the USA it has been the denial of equal treatment of blacks by whites in housing, employment and in the criminal justice system that has been most responsible for racial conflict in cities.

REFERENCES

Aleinikoff, Alexander (1976) 'Racial steering: the real estate broker and title viii', *Yale Law Journal*, 6: 808–25.

Bates, Timothy and Williams, Darrell, L. (1993) 'Racial politics: does it pay?', *Social Science Quarterly*, 74 (3): 507.

Beck, Roy (1996) 'The pro-immigration lobby', *New York Times*, 30 April, A11.

Benyon, J. (1987) 'Interpretations of civil disorder', in J. Benyon and J. Solomons (eds), *The Roots of Urban Unrest*. Oxford: Pergamon.

Boston, T. (1990) 'Segmented labor markets: new evidence from a study of four race-gender groups', *Industrial and Labor Relations Review*, 44: 99–105.

Brown, C. (1984) *Black and White Britain*. London: Heinemann.

Bullock, Charles and Rodgers, Harrell (1976) 'Institutional racism: prefrusites, freezing, and mapping', *Phylon*, 37: 212–23.

Butterfield, Fox (1995) 'More blacks in their 20s have trouble with the law', *New York Times*, 5 October.

Carmichael, Stokely and Hamilton, Charles V. (1967) *Black Power*. New York: Vintage Books.

Commission for Racial Equality (1999) 'Ethnic minorities in Britain', *CRE Fact Sheets*, revised. London.

Corcoran, M. and Duncan, G. (1979) 'Work history, labour force attachment, and earnings differences between the races and sexes', *Journal of Human Resources*, 14: 497–520.

Cross, M. (1982) 'Racial equality and social policy: omission or commission?', in C. Jones and J. Stevenson (eds), *The Year Book of Social Policy in Britain, 1980–81*. London: Routledge & Kegan Paul. pp. 77–88.

Cross, M. and Johnson, M. (1989) *Race and the Urban System*. Cambridge: Cambridge University Press.

Daniel, W. (1968) *Racial Discrimination in England*. Harmondsworth: Penguin.

Darden, Joe T. (1986) 'Accessibility to housing: differential residential segregation for blacks, Hispanics, American Indians, and Asians', in Jamshid Momeni (ed.), *Race, Ethnicity and Minority Housing in the United States*. Westport, CT: Greenwood.

Darden, Joe T. (1989) 'Blacks and other racial minorities: the significance of colour in inequality', *Urban Geography*, 10: 562–77.

Darden, Joe T., Duleep, Harriet O. and Galster, George (1992) 'Civil rights in metropolitan America', *Journal of Urban Affairs*, 14 (3&4): 469–96.

Darden, Joe T., Hill, R.C., Thomas, J. and Thomas, R. (1987) *Detroit: Race and Uneven Development*. Philadelphia: Temple University Press.

Davison, R.B. (1964) *Commonwealth Immigrants*. IRR, Oxford: Oxford University Press.

Downs, Anthony (1970) *Urban Problems and Prospects*. Chicago: Markham Publishing Co.

Farley, Reynolds and Allen, Walter (1987) *The Color Line and the Quality of Life in America*. New York: Russell Sage.

Foley, Donald L. (1973) 'Institutional and contextual factors affecting the housing choices of minority residents', in Amos H. Hawley and Vincent R. Rock (eds), *Segregation in Residential Areas*. Washington, DC: National Academy of Sciences. pp. 85–147.

Freeman, Gary P. (1987) 'Caribbean migration to Britain and France: from assimilation to selection', in Barry B. Levine (ed.), *The Caribbean Erodus*. New York: Praeger. pp. 185–203.

Freyer, P. (1984) *Staying Power: The History of Black People in Britain*. London: Pluto Press.

Galster, G. (1990a) 'Racial steering by real estate agents: a review of the audit evidence', *Review of Black Political Economy*. 18: 105–29.

Galster, G. (1990b) 'Racial discrimination in housing markets in the 1980s: a review of the audit evidence', *Journal of Planning Education and Research*, 9: 165–75.

Galster, G. (1990c) 'Racial steering by real estate agents: mechanisms and motivations', *Review of Black Political Economy*, 19: 39–63.

Giddens, A. (1981) *A Contemporary Critique of Historical Materialism*, vol. 1. London: Macmillan.

Glasgow, Douglas. (1980) *The Black Underclass: Poverty, Unemployment and the Entrapment of Ghetto Youth*. New York: Vintage.

Harris, Clive (1991) 'Configurations of racism: the civil service, 1945–60', *Race and Class*, 33 (1):1–29.

Hawkins, Homer and Thomas, Richard (1991) 'White policing of black populations: a history of race and social control in America', in Ellis Cashmore and Eugene McLaughlin (eds), *Out of Order*. London: Routledge.

Holmes, C. (1982) 'The promised land? Immigration into Britain, 1870–1980', in D.A. Coleman (ed.), *Demography of Immigrant and Minority Groups in the United Kingdom*. London: Academic Press.

Hunte, J. (1965) *Nigger Hunting in England*. London: West Indian Standing Committee.

Jacobs, Sidney (1985) 'Race, empire and the welfare state: council housing and racism', *Critical Social Policy*, 13: 6–28.

Jargowsky, P.A. (1994) 'Ghetto poverty among blacks in the 1980s', *Journal of Policy Analysis and Management*, 13: 288–310.

Jencks, Christopher and Peterson, Paul (eds) (1991) *The Urban Underclass*. Washington, DC: Brookings Institution.

Jenkins, Richard and Solomon, John (1987) *Racism and Equal Opportunity Policies in the 1980s*. Cambridge: Cambridge University Press.

Joint Center for Political and Economic Studies (1994) 'Political Trendletter', *Focus* 21: 14.

Joshi, S. and Carter, B. (1985) 'The role of labour in the creation of a racist Britain', *Race and Class*, 25 (3): 53–70.

Kain, J. (1968) 'Housing segregation, negro employment, and metropolitan decentralization', *Quarterly Journal of Economics*, 88: 513–19.

Kasarda, J.D. (1989) 'Urban industrial transition and the underclass', in W.J. Wilson (ed.), *The Ghetto Underclass*. (Annals of the Association of Political and Social Science.) Beverly Hills, CA: Sage.

Kasarda, J.D. (1993) 'Inner-city concentrated poverty and neighbourhood distress: 1970 to 1990', *Housing Policy Debate*, 4: 253–302.

Katz, Michael (1989) *The Undeserving Poor: From the War on Poverty to the War on Welfare*. New York: Pantheon.

Lester, Anthony (1987) 'Anti-discrimination legislation in Great Britain', *New Community*, 14 (1, 2): 21–31.

Li, Chun-Hao, Bagaka's, J. and Darden, J. (1995) 'A comparison of the US Census Summary Tape Files 1A and 3A in measuring residential segregation', *Journal of Economic and Social Measurement*, 21: 145–55.

Lindsey, Lydia (1993) 'The split-labour phenomenon: its impact on West Indian workers as a marginal working class in Birmingham, England, 1948–1962', *Journal of Negro History*, 78 (2): 83–110.

Luthera, M.S. (1988) 'Race, community, housing and the state – a historical overview', in A. Bhat et al. (eds), *Britain's Black Population: A New Perspective*. Aldershot: Gower. pp. 103–46.

Massey, D.S. and Denton, N.A. (1993) *American Apartheid: Segregation and the Making of the Underclass*. Cambridge, MA: Harvard University Press.

Massey, D. and Mullan, B.P. (1985) 'Commentary and debate: reply to 'Goldstein and White', *American Journal of Sociology*, 91: 396–9.

Mauer, Marc (1990) 'Young black men and the criminal justice system: a growing problem', *The Sentencing Project*. San Francisco: Center on Juvenile and Criminal Justice.

McDonald, J.F. (1981) 'The direct and indirect effects of housing segregation on employment opportunities for blacks', *Annals of Regional Science*, 15: 27–38.

McKay, D.H. (1977) *Housing and Race in Industrial Society: Civil Rights and Urban Policy in Britain and the United States*. London: Croom Helm.

Miller, Jerome (1992) *Hobbling a Generation: Young African American Males in the Criminal Justice*

System of America's Cities. Baltimore: National Center on Institutions and Alternatives.

National Advisory Commission on Civil Disorders (1968) *Report of the National Advisory Commission on Civil Disorders*. New York: Dutton.

Newnham, A. (1986) *Employment, Unemployment and Black People*. London: Runnymede Trust.

Norris, P. and Lovenduski, J. (1993) 'If only candidates came forward: supply-side explanations of candidate selection in Britain', *British Journal of Political Science*, 23 ((3): 373–408.

Office of Population Censuses and Surveys (1991) *Labour Force Survey, 1988–1989*. London: HMSO.

Openshaw, H. (1973) *Race and Residence: An Analysis of Property Values in Transitional Areas, Atlanta, Georgia, 1960–1071*. Atlanta, GA: Georgia State University School of Business Administration. Monograph No. 53.

Owen, David (1994) *Black People in Great Britain: Social and Economic Circumstances*. 1991 Census Statistical Paper No. 6. Coventry: Centre for Research in Ethnic Relations.

Owen, David (1995) 'Growth, development, size and structure of ethnic minority populations, origins and geographical location', in David Coleman and John Salt (eds), *General Demographic Characteristics of Ethnic Minority Populations*. London: OPCS.

Patterson, S. (1963) *Dark Strangers*. London: Tavistock.

Patterson, S. (1969) *Immigration and Race Relations in Britain, 1960–1967*. London: Oxford University Press.

Peach, Ceri (1968) *West Indian Migration to Britain: A Social Geography*. London: Oxford University Press for the Institute of Race Relations.

Peach, Ceri (1996) 'Does Britain have ghettos?', *Transactions of the Institute of British Geographers*, 21(1): 216–35.

PEP, 1967 (1968) *Racial Disadvantage in England*. Harmondsworth: Penguin.

PEP, 1976 (1976) *Extent of Discrimination*. Harmondsworth: Penguin.

PEP, 1977 (1977) *The Facts of Racial Disadvantage*. Harmondsworth: Penguin.

Pilkington, Edward (1988) *Beyond the Mother Country: West Indians and the Nottinghill White Riots*. London: Tauris.

Price, R. and Mills, E. (1985) 'Race and residence in earnings determination', *Journal of Urban Economics*, 17: 1–18.

Race Relations Act (1965) *Public General Acts*. Elizabeth II, Chapter 73. London: HMSO.

Race Relations Act (1968) *Public Acts*. Elizabeth II, Chapter 71. London: HMSO.

Race Relations Act (1976) *Public General Acts*. Elizabeth II, Chapter 74. London: HMSO.

Reeves, F. (1983) *British Racial Discourse*. Cambridge: Cambridge University Press.

Rex, John (1988) *The Ghetto and the Underclass: Essays on Race and Social Policy*. Aldershot: Gower.

Ricketts, Errol and Sawhill, Isabel (1988) 'Defining and measuring the underclass', *Journal of Policy Analysis and Management*, 7: 316–25.

Rose, E.J.B. and Associates (1969) *Colour and Citizenship*. London: Oxford University Press.

Ruggles, Patricia and Marton, William P. (1986) *Measuring the Size and Characteristics of the Underclass: How Much Do We Know?* Washington, DC: Urban Institute.

Saltman, J. (1975) 'Implementing open housing laws through social action', *Journal of Applied Behavioural Science*, 11: 39–61.

Scarman, Lord (1981) *The Brixton Disorders* (The Scarman Report). London: HMSO.

Segal, R. (1967) *The Race War*. Harmondsworth: Penguin.

Sivanandan, A. (1976) 'Race, class and the state: the black experience', *Race and Class*, 17 (4): 347–68.

Small, Stephen (1994) *Racialized Barriers: The Black Experience in the United States and England in the 1980s*. London: Routledge.

Smith, D. (1977) *The Facts of Racial Disadvantage*. Harmondsworth: Penguin.

Smith, Susan (1987) 'Residential segregation: a geography of English racism', in Peter Jackson (ed.), *Race and Racism*. London: Allen & Unwin. pp. 25–49.

Smith, Susan (1989) *The Politics of Race and Residence*. Cambridge: Polity Press.

Staples, R. (1975) 'White racism, black crime, and American justice: an application of the colonial model to explain crime and race', *Phylon*, 36: 14–22.

Thomas, Huw and Krishnarayan, Vijay (1994) *Race and Equality and Planning*. Aldershot: Avebury.

Turner, M., Struyk, R. and Yinger, J. (1991) *Housing Discrimination Study: Synthesis Report*. Washington, DC: Urban Institute.

US Bureau of the Census (1996) 'Population projections of the United States by age, sex, race and Hispanic origin, 1995 to 2050', *Current Population Reports*. Washington, DC: US Government Printing Office.

US Commission on Civil Rights (1971) *The Federal Civil Rights Enforcement Effort: One Year Later*. A report of the Commission. Washington, DC: US Government Printing Office.

Willis, C. (1983) *The Use, Effectiveness and Impact of Police Stop and Search Powers*. London: Home Office Research Unit.

Wilson, W.J. (1973) *Power, Racism and Privilege: Race Relations in Theoretical and Sociohistorical Perspectives*. New York: Macmillan.

Wilson, W.J. (1987) *The Truly Disadvantaged: The Inner City, the Underclass, and Public Policy*. Chicago: University of Chicago Press.

Withers, C. and Winston, J. (1989) 'Employment', in R. Gavan and W. Taylor (eds), *One Nation Indivisible: The Civil Rights Challenge of the 1990s*. Washington, DC: Citizens Commission on Civil Rights.

Wrench, John (1995) 'Racism and the labour market in post-war Britain: the second generation and the continuance of discrimination', in Marcel Van Der Linden (ed.), *Racism and the Labour Market: Historical Studies*. Berne: Peter Lang.

12

Communities in the City

RONAN PADDISON

In the analysis of urban social life the notion of the community has had a long, if somewhat chequered, history. Intuitively, its appeal is not difficult to identify: as a means of counteracting the size and potential alienation of the city, community has typically been interpreted as the means by which the individual is able to develop a sense of belonging and identity with at least part of it. Further, as researchers from the Chicago School sought to trace out so carefully, social interaction within the city frequently takes on a spatial expression; it was not just that the city could be divided into areas which were more or less distinguished by the patterning of residential segregation, but that this could be linked often to patterns of social interaction and bonding (Park, 1929/1952). Community, then, became intricately involved with how social behaviour within the city was organized, and indeed could be understood.

The intuitive understanding that urban communities exist belies the epistemological problems that have arisen both in their definition, and empirically in their identification (Bennett, 1989; Burns et al., 1994); Johnston, et al., 1994). Definitionally, community is among a number of 'slippery' social science terms, alongside others such as integration and development, which have proved sufficiently difficult to define to the point that the validity of the term has been questioned by some social scientists (Stacey, 1969). Yet, if as part of its chequered history, the study of communities has been in the past all but abandoned by urban sociologists and others, more recently since the 1980s it has enjoyed something of a renaissance, not only because of its seeming importance as a local counter to the socially erosive effects of globalization, and its ideological handmaiden, neo-liberalism, but also to its common appeal to politicians and policy-makers. This renaissance does not mean that the epistemological problems of definition and identification have been overcome; indeed, not only do they

remain, but they have been added to by new concerns with the ways in which community can be used as a totalizing construct, obscuring the differences which characterize individuals and groups within the city, its very diversity (Young, 1990).

It is the very ubiquity with which the term has been and is used, stemming no doubt from its intuitive appeal, that is at least in some measure the source of such problems (Hoggett, 1997). It is not only academics who have used the term so commonly; from public policy-makers, and particularly by politicians, its usage has gained widespread currency. In the contemporary British city community policing along with community health care, community development and community schools are among the numerous public policies which have invoked the term, paralleling similar trends in the United States and elsewhere. It is not hard to see why the term community is being used adjectivally to qualify public policy: not only does it convey connotations of empathy and localness, themselves implied values of community, but it also harnesses new relationships between the state and civil society, notably of partnership and shared responsibility, itself one of the increasingly distinctive trends of the shift towards governance in post-Fordism. By the 1990s, then, notions of community empowerment had become central to the methods of regenerating the city in the advanced economies, as they had in the goal of fostering the upgrading of squatter settlements in cities in less developed countries. But multiple usage conceals the multiple meanings which can be attached to the term, not to mention the rhetorical uses (as in its linking to empowerment and partnership) to which community can be directed, particularly by politicians and government agencies.

Part of the reason why the notion of community has become such common currency is that it has served two ends – as a means of describing social life and behaviour (within the city) and also a normative tool by which to legitimize relationships and processes designed to meet political ends. Any critical understanding of the place of the community needs to draw out an appreciation of both. Nor are these two aspects unrelated. Indeed, where the 'benefits' of community have become harnessed by political debate, through in particular the recent resurgence in interest in communitarianism (Etzioni, 1995), its advocacy may be seen in part a reaction to the multiple social dislocations within the city resulting from global and local restructuring. Its proclamation, too, serves as a rebuttal to the claim that in late capitalism urban communities no longer exist.

Such a loss has become expressed in the doubts held that in the present-day city the notion of community in the traditional sense of attachment to neighbourhood, to place, has any meaning, or at least as much meaning as it had in the past. Urban social life, it is argued, has become increasingly atomistic, in which, spurred on by neo-liberalist ideas, individualism has gained currency over collectivist solutions (Walzer, 1995). Neither argument is accurate in any universal sense. Spatially defined communities, characterized by neighbouring and supportive social networks, persist in the city. Yet, few would argue that they are universal, that the city is comprehensively divided into a network of territorial communities in which place, social interaction and identity are strongly fused to give common social purpose. But such communities do persist, perhaps most visibly in the class and/or ethnically based enclaves which form distinctive zones in virtually all cities.

Much as territory and place continue to be important bases for community in the contemporary city, to limit the meaning of community to its territorially-based form is to deny the diversity of forms it can assume. Interest-based groups reflecting the social diversity of the city's population, the elderly, disabled, gay and lesbian groups, the homeless, represent a different order of urban communities functioning often at a city-wide level, as well as, in many cases, identified with particular parts of it. Further, as Webber (1963) identified more than a generation ago, the increased mobility of the more affluent resulted in much social interaction taking place extra-locally within the 'non space urban realm'.

These introductory remarks imply that community is problematic as a concept, and that its mapping out needs to take account of the different forms it can assume, and the different purposes it can serve as giving meaning to urban social life as well as to its employment as a mobilizing agent for social change. This chapter looks at the different meanings of community, but before doing so examines how it is located within the wider study of urban social analysis, as it has been developed by urban sociologists, in particular. Here a paradox is readily identifiable, between the existence of communities and the alienating effects of city living. Communities, it is frequently argued, are important precisely because they act as a cushion to the otherwise alienating nature of city life. The argument presumes community to be 'local', either encompassing a small territory of the city or a well-defined subset of its population. Whatever its precise nature, it leaves open the question of whether the city is (or can be) an imagined community.

COMMUNITY STUDIES

One of the paradoxes of urban sociological analysis is that while empirical studies have claimed to identify the existence of communities within the city, orthodox social theories of the city have emphasized the negative attributes of social life in the city (Flanagan, 1995). In the classic work of Tönnies (1883/1995), Simmel (1905/1950) and others the over-riding image of the city was that it was alienating, corrosive of pre-urban forms of community, centring on interaction where, if there was greater interdependence between individuals, contact was instrumental and, as for Simmel, based on calculative reasoning. This tradition of thought, so much a product of nineteenth century urban-industrialization, was to continue to have a major influence on attitudes towards the city, and the interpretation of urban social life and behaviour, in the following century. As much as it was a product of the development of urban-industrialism in the nineteenth century, the speed of social change helps explain why theorists grappled to come to terms with its meaning and were drawn to make comparisons with pre-urban social conditions.

This is nowhere clearer than in Tönnies's still frequently quoted distinction between *Gemeinschaft* and *Gesellschaft*, mapped out in a work written in the 1880s. If the distinction was not explicitly devised to counterpose rural social life with that of the urban, typically this is how the argument has become interpreted. In the pre-urban/pre-industrial world *Gemeinschaft* mapped out a social order in which interaction was intense, personal and based on primary social relations. It was marked by strong social solidarity – effectively the individual was part of an integrated social collectivity. By contrast in the city *Gesellschaft* conditions predominated, characterized by transitory, fleeting social relationships. Individuals had become more interdependent on one another

(because of the division of labour), but (inter)
action was founded on its calculated exchange
value (because of the deepening effects of capitalism). Neither word, *Gemeinschaft* or
Gesellschaft, is directly translatable into English,
though usually they are equated with 'community'
and 'society'.

Tönnies's analysis represents an important
tradition within urban social theory, that the twin
processes of urbanization and industrialization
were in one way or another dystopian, and that a
sense of community was lost. Within this depiction community was linked to a world which had
been and as belonging to an idealized past. Even
if it is possible to interpret Tönnies's dualistic
analysis too literally – actual societies would be
located along a spectrum whose end points were
represented by the ideal images of *Gemein-
schaften* or *Gesellschaften* – the location of these
within the continuum was largely fixed by the
nature of place, rural or urban.

One of the more obvious ways in which urban
society was to differ from pre-urban societies
arose from the sheer numbers of people that
constituted the city. Inevitably, this would deper-
sonalize social relations. For Simmel, interested in
explaining the experience of urban life, the scale
of the city and the intensity of the city, meant
inevitably that the individual would need to
become insensitive to the myriad changes and
people which are part of the daily encounter.
Alienation too was fostered by the money
economy which provided the framework for so
much of the interpersonal relations which took
place, and which by definition meant that relationships were both based on calculation and
potentially conflictual.

Some of these negative qualities of urban life
were to be expanded upon by Wirth (1938), to
whom the distinctiveness of social life in the city
was the product of three main factors, size,
density and heterogeneity. As for Tönnies, Wirth's
methodological approach was to posit the ways in
which the urban differed from the non (or pre-)
urban, though in Wirth's formulation the latter
was left implicit. Size was critical, and was linked
to the rapid growth of the city through immigrant
groups from different cultural backgrounds. For
Wirth, as for Tönnies, Simmel and Weber, the
sheer massing of people within the city meant
that social relations were likely to be superficial,
competitive rather than solidaristic and instrumental. These relationships were buttressed by
the effects of density and heterogeneity, in which
case Wirth emphasized how diverse was the
network of interaction between individuals but
how superficial they were. In short, living in the
city was disorganizing, depersonalizing, alienating, and competitive.

Set against these visions of city life empirical
studies in both the United States and Britain,
conducted mainly in the 1950s and 1960s, were to
show how the city comprised a mosaic of social
worlds characterized more by social cohesion and
organization than by alienation and disorganization. Indeed, many of these studies were not only
to refute the classical theorists but were to celebrate the persistence of communities, some of
which were to lend support to the idea that
Gemeinschaft-type relations were persistent
within the city. A territorial definition of what
constituted the community, together with a
shared sense of belonging were important
markers, within which, as empirical analysis was
able to demonstrate, social cohesion would be
reflected in the sense of neighbourliness and in
the patterning of friendship and kinship
networks. A precursor of the trend was Whyte's
(1943) study, *Street Corner Society*, following
which there was a plethora of analyses in US
cities demonstrating the viability of community
as the setting of convivial relations, and, variously, as the territorial base on which the support
and defence of (local) collective values could be
founded.

Whyte's study became a classic in showing how
social life, far from being disorganized, was
ordered in ways that gave meaning to its residents.
The argument was all the more striking because
of the type of neighbourhood in which he was
working, a poor working class area within
Boston, one which to the outsider would be more
likely to be 'read' as socially dysfunctional. To the
outsider, then, the prevalence of street gangs and
racketeers, characteristic of the neighbourhood,
was likely to be interpreted in terms of its moral
disorder. But in a tightly-knit community which
had its own codes of behaviour, what appeared
dysfunctional could actually be supportive, as
where, given the difficulties of entry into the
formal labour market, informal (albeit often
illegal), activities provided the means of access to
income generation.

It was within poor neighbourhoods that the
research indicated there was a strong sense of
community. Suttles's (1968) work in Chicago, in
the 'Addams Area', a multi-ethnic area in which
growing numbers of blacks, Mexicans and Puerto
Ricans were displacing an older Italian community, argued for the importance of territorial identification. Conflict was restricted to involving
groups considered to be outsiders, such was the
strength of identification with the Addams area
that it transcended ethnic differences. In fact,
such an identification was necessary, as being the
only effective means by which to defend the area
from external threat. Community became a
coping strategy.

In Britain, too, empirical studies in the 1950s
and 1960s were to emphasize the importance of

community life within cities and towns. Detailed case studies, drawing methodologically from anthropology, were conducted in a variety of different environments, in Featherstone, a mining town in Yorkshire, Swansea, Gosforth, and perhaps most famously in the East End of London in Young and Wilmott's classic work, *Family and Kinship in East London*. To a generation of (British) urbanists, Young and Wilmott's work was to emphasize how *Gemeinschaft* conditions remained in the inner city, in which the supportive roles of kinship and extended family were to underwrite a strong sense of community, although they were to prove vulnerable when confronted by urban renewal.

Though the geographical locations analysed in these studies varied, in fact there were strong similarities between them. Attention was focused on the working-class community, sometimes in areas threatened with industrial decline or the onset of urban renewal. In other words they were areas which were characterized by social change, sometimes rapid, which threatened the traditional communities which had been built with the spread of urban-industrialism. Such communities were portrayed as relatively stable, in spite of the conflictual environment against which they were set; in South Yorkshire Featherstone was depicted as a community in which the traditions of social life – the strong sense of solidarity, the marked sexual division of labour, and the penetrating influence of trade unionism in daily life and in basic values – were of long establishment (Dennis et al., 1956).

Yet the tightly knit world of the working-class community of the inner city would not be able to resist the progress of urban renewal, as Young and Wilmott were to demonstrate graphically in London. In the East End (of London) redevelopment was to sever the ties of community through the breakup of kinship and extended family networks. Once displaced to the newly-constructed peripheral suburbs, supportive social networks frequently became severed, an observation which was to be made elsewhere in British cities as relocation to peripheral estates accompanied the process of inner city renewal. What was lost was the sense of community which had pervaded the old working-class areas, in spite of the conscious attempt to rebuild communities in the mass housing estates and, particularly in the neighbourhoods of the New Towns.

It has become commonplace to criticize these studies – that they were overly-empirical, that they tended to emphasize the consensual nature of community over the fact that communities are often conflictual, and that they romanticized the community (Hoggett, 1997). Many of the so-called communities which were to fall victim to the bulldozer of post-war urban renewal were

characterized by poor housing conditions and poverty, far from the romanticized image subsequent analysis may have given them. Perhaps most critical was their not infrequent failure to define adequately the meaning of community. As Stacey (1969) was to argue, the 'community studies' literature had been predicated on identification with a geographical area and social coherence, yet on neither count did they achieve sufficient rigour. The territorial extent of the community was under-specified as was the construct of social coherence, who precisely was 'inside' the group, and what constituted a 'sense of community'. Paradoxically, a generation later, by which time the term community was being rehabilitated, concepts such as social coherence and a sense of community were to become even more commonplace, yet still contested over as to their definition and empirical interpretation.

If anything such continuing difficulties serve to emphasize the slippery nature of community, rather than denying that, however they are defined, empirical analysis would not be able to identify individuals living within cities claiming a sense of belonging to a community. Nor is this to deny the different forms community can assume, where for some citizens this is deeply rooted in territory and a sense of place, while for others it can be expressed more in terms of interests and lifestyles emphasizing social interaction which extends beyond their immediate locality.

In differing ways, then, a sense of community helps give social meaning to urban life in spite of those urban social theorists who have emphasized the alienating nature of city life. What needs to be emphasized here is that most of these urban social theorists are concerned with accounting for the nature of social interaction and behaviour within the city as a whole. Yet, typically, for the individual, everyday experience does not involve negotiating the city as a whole, but rather occurs within particular, often relatively small parts of it. Indeed, such a recognition was to be given a direct expression in the rapidly developing urban-industrial cities of the nineteenth century through the development of the suburb. Above all the suburb provided the means of withdrawing into the more private spheres of home and the local community as coping strategies to the alienation of the city.

Such an argument is not gender-neutral (Wilson, 1991). In the distinction between space and place, the one transactional and instrumental, the other associated with home and its immediate environs, the emotive, more personal, needs of the individual and households, how men and women relate to the city varies. The stereotype of women's attachment to place is married to the suburb, and to home, while men are the more likely to have to negotiate space, both predicated fundamentally on the sexual division of labour.

Of course, such a stereotype bears only partial testimony to reality, not least in the more developed economies because of the feminization of the labour force, and of the gains made by feminist movements in challenging patriarchical practices. Yet, the negotiation of space by women has (re)affirmed the dangers of urban life, of the unexpected, not least through apprehension over personal safety. At the same time, for both women and men coping with urban space is addressed by their interaction with only a limited part of it, with large parts of the city being typically 'unexplored territory'.

Outside the geography of mundane routines, for the citizen much of the rest of the city, then, is imagined space, certainly infrequently encountered, perhaps, even for long-term residents of the city, there are areas which have never been visited. But, some parts of the city which are habitually negotiated, which are not part of the local area, such as the city centre, are encountered in everyday experience. Even where the citizens feel affinities towards such spaces, the city centre is above all where the stranger is encountered, where the massing of people in the city has its most obvious effects on the alienating of the individual. Indeed, where in modernity the nature of urban social life has tended to steadily corrode the meaning of public space, more recent interpretations of it have emphasized the threats it poses for the individual. Benefits, it is suggested, follow from the privatization and increased surveillance of spaces where by necessity citizens, strangers to one another, are brought into close proximity. Apart, then, from the commercial reasons which help explain the development of shopping malls and out-of- or edge-of- town centres, the sanitized spaces which they offer the individual are a counter to the potentially more dangerous environments of the city centre, an argument which again has gender implications.

The distinction between the imagined city and the everyday experience of the individual focusing on only a small part of it is suggestive of the scalar qualities of urban life, and of how they give meaning to urban living. This is particularly the case for communities defined by a strong sense of place in which the local area is both familiar and supportive. Little wonder, then, that such communities will be defended should they be threatened with redevelopment or some other change which is perceived as resulting in the erosion, or worse, loss of community. The paradox of communities within the city which is otherwise alienating and dysfunctional in social terms reflects the different scales at which we live in cities.

Yet, if in this sense community becomes a means by which to overcome the scale of the city, as Fischer (1975) has suggested it is the very size of the city which renders the potential for subcultural communities, not necessarily tied to the local neighbourhood, to emerge. Cities offer critical mass within which individuals are able to express their interests collectively, in which the heterogeneity of the city creates opportunities for subcultures to develop. The argument stands Wirth's conclusion on the heterogeneity of the city on its head: that whereas to Wirth it tended to weaken social life, to Fischer increasing differentiation was enriching, fostering contact and allowing for groups to establish cultural identities.

COMMUNITY FORMS AND PROCESSES

The recognition that communities do exist, and are empirically identifiable, in cities needs to be accompanied by an acceptance that the form they assume, and the processes which underpin them, will vary. It would be naive to assume that cities are subdivided neatly into discrete communities. Cities may comprise a mosaic of social worlds, but these do not conform to a template in which (for example) residents of different parts of the city necessarily identify with the local area in which they live.

Even if, as it was argued in the earlier discussion, the defining of community has been something of a semantic minefield, there is a broad measure of agreement as to its basic attributes. Fundamentally, communities 'are not merely territorial units but consist in the links that exist between people sharing common interests in a network of social relationships' (Hill, 1994: 34; Karp et al., 1991). Effectively the community occupies the intermediate ground between the institutions of wider society, the city and its external environment, and the family and other immediate institutions with which the individual has daily social interaction.

In a study examining the social life within American neighbourhoods the Warrens (1977) have identified three factors which help distinguish between different types of community:

1 Identity – assessing the sense of place residents have, including the extent to which they consider they have shared values and interests with neighbours;
2 Interaction – what is the pattern of neighbouring within the area, the strength of the ties linking residents;
3 Linkages – the degree of closure of the local area from other parts of the city, adjacent or otherwise, and the purposes served by these linkages, such as to establish political leverage within external institutions operating at a city-wide level so as to gain some advantage for the local area.

Gottdiener (1994) has adapted the Warrens' framework to define six types of community, ethnic villages, interactive middle-class neighbourhoods, diffuse, anomic and transitory communities and, finally, the defended neighbourhood. While giving a brief outline of these, the discussion here will focus on the last of them, the defended community.

At opposite ends of the 'spectrum' are the interactive middle-class neighbourhood and the anomic community. In the interactive middle-class community performance on each of the three criteria mentioned earlier is high. These are highly integrated communities in which there is a strong sense of belonging, extensive social interaction and which are able to use the professional skills of the residents to build external political linkages so as to buttress community advantage. In the anomic community scores on each of the three factors is low; typically these are low income neighbourhoods characterized by social breakdown and high crime rates. Compounding their social disorganization, local institutional capacity to counteract on some collective basis social problems is low; resident support of both the formal political process, in voting, and of the informal sector, through voluntary organizations, is low. Diffuse communities are closer to the interactive middle-class neighbourhood in the sense that while there is a low degree of social interaction within the former, nevertheless it functions as an effective community. Typically the residents of such communities will show strong linkages with areas outside the immediate locality. Like the interactive middle-class neighbourhood diffuse communities are able to use external linkages to defend the area's status; conversely it is in anomic communities where the need for such linkages is the greatest, and where community capacity building is a priority to local regeneration.

Ethnic urban villages are identifiable as communities with a strong sense of identity, including that of place in that, typically, they occupy distinct territories, and in which there is a high degree of social interaction. They are also identified by a high degree of closure from other areas of the city, while at the same time some of their external linkages may be strong; where they comprise immigrant populations connections with their country of origin may be maintained, and indeed important, while in other cases as immigrant populations political power bases within City Hall may have been forged so as to ensure status identity. Where the collective identity is so critical to group integrity, restricting the social interaction with those from other parts of the city, particularly from neighbouring areas, is itself an important mechanism ensuring boundary maintenance; the ethnic urban village

provides the most obvious illustration of Park's description of the city as comprising 'little worlds that touch but do not interpenetrate' (Park, 1925/1952: 40).

In his study *Urban Enclaves*, of which the ethnic urban village is one example, Abrahamson (1996) demonstrates the supportive role such communities give to its ethnic members. In San Francisco's Chinatown residents' answers to the question as to why they like living there

> focus on factors that are the advantages of *any* enclave. They like being close to the Chinese restaurants and shops where they can obtain goods and services available nowhere else. They also like the proximity to their relatives and friends and are comfortable living in a relatively homogenous community. (Abrahamson, 1996: 69)

These advantages differed between members of the group – for the more elderly the benefits were measured in terms of the proximity of the supportive network, for the young in terms of the fact that Chinatown could function as a place of work as well as of residence. This supportive capacity reflects the self-reliance of such communities; if, because of the hostility of the host population, self-reliance is borne out of necessity, it is also the means of safeguarding the integrity of the group.

Where one of the characteristics of the urban ethnic village is its 'relative closure' from much of the rest of the city in which it is located, particularly as a cultural expression, the paradox is that such communities are identified often in terms of a strong global sense of place. Immigrant ethnic villages such as San Francisco's Chinatown, the Cuban community of Miami's Little Havana or the ethnoburbs of Los Angeles (Li, 1998) reflect the globalization of the city, in which the development of local communities reflects the development of global diasporas which maintain strong cultural links with the homeland in the new environment. As globalization deepens, the arrival of new immigrant communities is leading to the territorial repartitioning of the city. Nor has their location been confined to the inner city as was so often the case in immigrant-fed urban growth in the past, where in cosmopolitan world cities such as Los Angeles, ethnoburbs such as 'Little Taipei' have been established in less than two decades in a process of invasion and succession not unlike that described by the Chicago ecologists.

Defended communities function as a special category where the threat to the local neighbourhood is sufficient to foster collective local action. The sources of threat are manifold; from capital with the pressure to redevelop a local neighbourhood involving its densification and/or a change in the type of land use; from the state, central or local, intent on implementing some infrastructural improvement, a new road or a new landfill site for waste disposal for example, which has benefits for the city (or fractions of it), but which is a clear threat to the local area; from residential populations in adjacent areas 'spilling over' into the neighbourhood. Precisely because of its high density the city comprises a complex weave of spatially-defined externalities generated by individual as well as collective action, many of which pose threats to the *in situ* residential population. While Gottdiener (1994) identifies the defended community as a separate type the ubiquity of conflict within the city means that defending the community often cross-cuts the other categories described, particularly as the threats to them ebb and flow.

Defending the community is not new; the creation of suburbs and their incorporation as separate legal entities in the nineteenth century, the forerunner of the jurisdictional jigsaw characterizing the US city (particularly in northern US states, for example) and institutional practices of zoning-in and -out are well-established methods by which the community has sought to safeguard its (territorial) base. Yet, for several reasons the defence of community has become more salient in the city in late capitalism. Neoliberal globalization has exacerbated the competitiveness of urban life; conflict, ever a defining feature of urban life, has become more salient as restructuring has (re)affirmed disparities of power and wealth. Threats to the local neighbourhood, as we have seen, are commonplace; the processes of restructuring, economic and physical, characterizing contemporary cities have increased the likelihood of threat. Simultaneously, and linked to the overarching processes of globalization in complex ways, inherited, relatively stable social formations linked to the 'centred subject' have been steadily eroded. Where self-identity is partly a reflection of the wider relationships of which the individual is part, the erosion of stable collective identifies – working-class communities centring on some long-standing manufacturing base now redundant through global economic restructuring, for example – destabilizes the sense of self. Globalization, then, has contributed in part to the 'decentred subject' through the de- and re-territorialization of social life. Old, previously relatively stable communities have been undermined through devalorization. New identities have formed around race, ethnicity, gender, sexual preference, ablement and other interest-based groups as urban societies have become more fragmented and 'fluid'. Further, through formal or informal political means the demands of different groups have been voiced, and even those who were relatively 'invisible' within urban society, the disabled for instance,

have joined the vocal interest-based communities in demanding a fairer city.

These trends have been associated with a rapid rise in defended communities in the city. Often their development has sought to establish or reaffirm the territorial basis of the community; clearly this is so in the case of neighbourhoods seeking to defend their community in the face of some threat. The rapid spread of gated communities, initially in the United States, but which are now found in cities in both the advanced and less developed economies, are among the more visible reminder of how the defence of personal privilege has become a feature of cities which are both more unequal and divided. Interest-based groups, too, may seek to establish a territorial claim, as in the case of gay communities which have developed in a number of cities in the advanced economies. The paradox would seem to be that while in the advanced economies lifetime mobility for the individual has increased, attenuating the geographical scale in which personal interaction takes place, there is if anything an increasing need to reaffirm the existence of the bounded community.

How such communities defend themselves, what resources they are able to muster varies; in the interactive middle-class neighbourhood, mentioned previously, local institutional capacity is high, whereas in the anomic community such capacity needs building, often through pro-active measures aimed at empowerment. In the middle-class community the professional skills and resources which can be mustered to confront threat can also be used to broker political linkages.

The development of the so-called Green Bans in Sydney in the 1970s demonstrate the ability of interactive middle-class neighbourhoods to successfully resist threat (Roddewig, 1978). What was distinctive about the Green Bans was the ability of neighbourhood groups to broker powerful linkages with other groups, notably with local trade union movements, to halt developmental changes affecting the neighbourhood. One of the early Green Bans, in the affluent, middle-class Sydney suburb of Hunter's Hill, sought to prevent the development of native bushland in an area which was only a few miles from the city centre. The community sought the support of the major union involved in the development, the Builder's Labor Federation, whose leader was persuaded to join forces with the local activists. Green Bans were a relatively short-lived form of resistance, and were the product of particular circumstances. Nevertheless they were successful in demonstrating the power of neighbourhood defence once mobilized. They exemplify a powerful source of local action; even if they are episodic, and are often stimulated by NIMBYism to maintain local privilege, they are commonplace in the city as significant reminders of the power of place-based attachments. The Green Bans were important too in reflecting one of Castells's (1983) criteria for urban social movement, the ability of local (community) activism to transform power relations, although these effects were not to have a more general or durable impact.

The development of gay communities in San Francisco provided Castells (1983) with a grassroots movement which was more successful in establishing and defending a territorial community protecting the rights and interests of an 'outsider' group. It was one of the earlier of such communities now found in cities as different as London, Paris, Madrid, Manchester and Sydney. While gay communities are defended in the sense that their establishment needed to overcome hostility, their development and maintenance has much to do with the increasing role played by consumption in self-identities in postmodern societies, and particularly in the city. Mort (1998) has traced the development of the gay community in London, specifically within Soho, an area which has been linked historically with unconventional, sometimes outsider, groups. The transformation of Soho in the 1980s exemplified a coming together of a number of changes centring on the interplay between consumption and identity, its commercialism through advertising which sought to promote consumerism alongside the image of the city (London) as a major site of consumption, processes which could be targeted at specific markets, such as young men. As in other gay communities territoriality provides a source of power, enabling members to adopt social practices which would be the less acceptable in the city beyond.

COMMUNITY: INSIDE/OUTSIDE

Earlier it was suggested that community is a problematic term because of the difficulties arising from how it should be defined. But there are other reasons why community is problematic arising from the assumption implied by the term that it is inclusionary. How the boundaries of the community are defined determines those who are included, and those who are excluded. It is precisely because of their inclusionary nature that we tend to think of communities as being somehow a 'good thing', of having, like the term decentralization, apple-pie qualities which defy denial as a value in its own right. Yet, the rhetoric of inclusion acts as a mask to the divisions within communities, and their ability to act as a (comm)unity. Of course, within the city as a whole, communities frequently find themselves in conflict, over the allocation of resources, zero-sum competitions to attract or avoid a particular land use and the like, reflecting the scale and diversity which characterize the city. But at the more local level, within the individual (territorial) community functioning as a unit is for the most part something of an ideal, concealing differences and divisions within it.

In short, then, community is a totalizing construct which assumes a degree of internal coherence which is rarely the case, and is in all likelihood unattainable. Even taking one of the most basic indicators upon which the territorially-defined community is predicated, the existence of a distinct local place, it is clear that how this is perceived will differ between community members. They may not even be able to agree on what are the geographical limits of the place, its boundaries. Where there have been attempts to define the territorial boundaries of the community frequently there is at best only limited agreement as to the limits (Paddison, 1981). But also, by its nature, place is not some neutral container in which the social interaction of everyday life occurs; individuals and groups imbue spaces with different meanings. As Staeheli and Thompson (1997) show in their work in Boulder (Colorado) different groups can prioritize the needs of the same area in radically different ways. Attachments to the community too will vary between members of it. Adapting Relph's (1976) insider/outsider terminology residents will differ in their 'insideness' with some members of the community feeling a much greater degree of attachment with local place than others. As Relph expressed it, 'to be inside a place is to belong to it and to identify with it, and the more profoundly inside you are the stronger is the identity with the place' (1976: 49).

There are other ways than the different attachment to place in which community glosses over internal difference. To assume that the territorial community could itself be a common bond presumes that the other identities which residents have – of ethnicity, gender, for example – can somehow be either subsumed under the umbrella of community, or that the latter itself will be able to supercede such difference. Clearly many communities are defined in terms of these characteristics. But the act of drawing a boundary around the community implies a degree of internal homogeneity which may be more myth than reality.

Even so, the drawing of boundaries inevitably becomes entwined with community. By its nature boundary demarcation tends to be brutal, notably in defining who is inside and who is outside. That is, community is about similarity, but it is also about difference, about the exclusion of 'others'. Physically, such communities become expressed in the search for the homogenous community. In his book *The Uses of Disorder: Personal Identity and City Life*, Sennett (1970) argues that the quest for the 'purified community', represented by the homogenous neighbourhood, is a desire for collective defence against the hostile and the unpredictable. In fact, such a community does not necessarily function in a highly interactive way; the residents of them do not need to engage with one another. The purified community is an imagined community; 'the image of the community is purified of all that might convey a feeling of difference, let alone a conflict, in who "we" are'. Sennett's argument of purified community had wider application to cities than to residential areas: in the sense that it became translated into the need to establish a stable social order, its place was apparent in the wider history of city planning. But it is in suburbia, and particularly in the more affluent suburbs, that the twin objectives of asserting control and of avoiding disorder and difference along with the unexpected can be achieved.

Postmodern accounts tend to be especially critical of the myth of community for failing to recognize its exclusionary nature. As Sandercock (1998) argued within the territorially-based community 'there may well be a dominant group with a distinct set of values and lifestyle, but . . . there are always those who are regarded as not belonging, as inferior, deviant, or threatening – whether it's the poor who live "on the other side of the tracks", the Aborigines who dwell on the fringes . . . or the gays who live in the night, in the shadows' (1998: 191). Community becomes a dangerous construct, precisely because it hides the process of making individuals and groups behind the facade of its inclusionary rhetoric.

THE APPEAL TO COMMUNITY

While the idea(l) of the community has come under sustained attack, paradoxically its appeal as a means by which to achieve social change has, if anything, increased. How cities are restructured, and particularly how the more deprived neighbourhoods of the city can be regenerated, makes recourse to the need for community participation. More ambitiously the recent spread of communitarian ideas has sought to re-invent a political philosophy based on the multiplicity of communities which exist to foster communal life, and fundamentally to bring about the re-moralization of society. Such ideas have become influential in moulding how local urban regeneration will be more likely to succeed where it can build up(on) social capital.

Bringing the community into the practice of urban regeneration is a demonstrably more fraught exercise than governments have assumed it to be. In the more top-down forms of participation employed to discover opinions on an urban planning proposal (for example), involving the community might be little more than an exercise in tokenism (Arnstein, 1969). But even where participation was limited to a public meeting in which local views were being sought it was often clear that 'the community' did not hold a single view. Eliciting a single community opinion was implausible, unless the multiplicity of different views was to be ignored in favour of the dominant voice. Such an outcome is hardly surprising: communities, as we have argued, are invariably divided.

The incorporation of community participation in urban regeneration in the United States and Britain (for example) has been on a much more ambitious scale than is represented by the single public meeting designed to gain local reaction to a land-use change affecting the neighbourhood. In Britain the introduction of the City Challenge and Single Regeneration Budget programmes have been built upon the notion of partnership between local government, the private sector, voluntary bodies and local communities. In some cases other government agencies and, quangos, might also be involved in the local regeneration project. Within such programmes, the funding for which is competitive, the role of the local community is problematic, often marginalized to the more significant objectives of the programme ensuring the leverage of private sector investment or to the need of meeting tight deadlines for the submission of bids.

This is not to deny that community involvement in the task of urban regeneration is argued as being of mutual benefit both to the local community and to the development agency charged with the task of regenerating the local area. The experience of Tyne and Wear Development Corporation (England) is a case in point. As an Urban Development Corporation set up in 1987, its chief task was 'to bring land and buildings (in a riverside area) back into effective use . . . and to encourage the development of existing and new industry and commerce' (Russell, 1998: vii). It was a major challenge. Covering an area of some 6,000 acres, the area had experienced pronounced economic decline, and with it social dislocation. Selective outmigration in the 1980s had left a residual population, largely unskilled, and communities which had become fractured as a result of the economic and social changes. The Development Corporation set in tow a community development strategy aimed at increasing the ability of the population to compete in the labour market, fostering community business, widening social housing opportunities and increasing tenant management, encouraging participation and building up the social and cultural facilities of the area. Though not atypical as a set of objectives aimed at regenerating inner city/old industrial areas, what was different was that the Corporation built community participation into the process of agenda-setting, i.e. at an early stage, and sought to use innovative techniques (such as monitoring panels) to ensure that implementation was meeting the intended targets.

In an assessment of community involvement Russell (1998) argued that it proved to be of mutual benefit to both the governmental agency and to the local population. For the Development Corporation, it was more than a device for legitimating the agency's programme, though where the UDCs, unlike local government, lacked a democratic base, gaining legitimacy in the eyes of the local population was important. As bonuses, community participation 'gave a better product by the application of on-the-ground knowledge and pressure for what was appropriate and workable' (Russell, 1998: 45). Equally, by extending ownership, vandalism and other problems were reduced. Though, as Russell demonstrated, there were shortcomings to the ways in which community participation was both structured and used, its employment was largely beneficial to the overarching aim of establishing sustainable local regeneration.

Compared to experience elsewhere – in, for example, the earlier years of the flagship UDC established by the Thatcher Government in London's Docklands – community involvement in Tyne and Wear was to be more carefully crafted into the mechanics of urban regeneration. As an instrumental device, it is a moot point whether in the north-east England case, participation was able to involve the multiple publics inherent in urban communities. The practice of community participation underlines how complex and time-absorbing it is, how it needs to demonstrate awareness of the different groups and interests represented in any community, and how it needs to be built into the different stages of the process of regeneration (Atkinson and Cope, 1997).

CONCLUSIONS

Community remains a useful concept by which to understand how urban social life has, and is given, meaning, but only insofar as its limitations and complexities are acknowledged. Critically, communities are complex: rather than unitary, they are multi-dimensional, harbouring within them a diversity of different interests and voices. Rather than searching for the unity of community, as Abu-Lughod (1994) said working in the

East Village of Manhattan, the point should be to focus on controversies within the community as 'there is no single "authorial" [authoritative] image of the neighbourhood' (1994: 195). Recognizing the community as multi-dimensional points the way forward to recognizing a sense of diversity within a shared sense of belonging.

Communities, then, are political, not in the sense of partisan politics, but in the sense of containing within them competing demands and the potential for conflict as well as harmony. The harnessing of community participation within urban governance, to the extent that it has sought to give real expression to local demands has a tendency to gloss over the political nature of the community. Participation is too easily co-opted for specific governmental ends based on the belief that communities are more unified than is actually the case.

Communities, too, are divisive. They are about the constructing of boundaries between who is defined as inside and who outside. Being inside gives both position and status to the individual, cocooning them from the social diversity of the city, providing a means of support and a (territorial) base upon which to maintain individual and group identity. But boundaries are exclusionary, a means of othering those to be excluded, and though in a liberal democratic society people have a right to form exclusivist types of association, their almost natural tendency is to promote self-interest towards outsiders. Bell (1998) questions, then, the development of the 'Residential Community Association', embracing the lives of some 32 million Americans, as being more 'privatistic enclaves' than communities, aimed at maintaining the privileges of the middle and upper income groups.

The exclusivity of community may of course be beneficial for individuals and groups which are otherwise more vulnerable, or even the subject of discrimination, within society at large. For immigrants to the city, and particularly for the ethnically distinctive immigrant, group association brings a suite of benefits which can help adjustment to the new (urban) environment. Community can in this sense be defensive, and protective, as well as sometimes helping to provide a means of gaining access to labour and other urban markets. Thus, ethnic economies characterize cities in the advanced economies (Barrett et al., 1996) as they do cities in less developed economies (Speece, 1990).

In a globalizing world economy, and particularly in its world cities, exclusivist ethnic communities are reaffirming themselves in cities such as London, New York and Los Angeles. The nature of such communities is changing: globalizing processes are altering the spatiality of such communities. Community (in the global era) is in the process of being disembedded from its purely local basis, the local intersecting with the global; thus 'second-generation Bangladeshis in the East End of London . . . engage in lively, diverse commentaries of belonging which range across numerous boundaries of space and time' (Eade, 1997: 24).

Rather, then, than abandon the notion of community, as earlier urban analysts have suggested, providing there is adequate recognition given to its complex and contested nature, it has continued (analytic) value in understanding social life in late modern cities. Clearly though, as the lessons of the Residential Community Associations, gated communities and other exclusivist associations demonstrate, the danger is that community can be used to protect the interests of the more privileged at a cost to the city as community.

REFERENCES

Abrahamson, M. (1996) *Urban Enclaves: Identity and Place in America*. New York: St Martin's Press.

Abu-Lughod, J.L. (1994) *From Urban Village to East Village: The Battle for New York's Lower East Side*. Oxford: Blackwell.

Albrow M., Eade J., Dürrschmidt, J. and Washbourne, N. (1997) 'The impact of globalization on sociological concepts', in J. Eade (ed.), *Living the Global City*. London: Routledge pp. 20–36.

Atkinson, R. and Cope, S. (1998) 'Community participation and urban regeneration in Britain', in P. Hoggett (ed.), *Contested Communities*. Bristol: The Policy Press. pp. 201–21.

Arnstein, S. (1969) 'A ladder of citizen participation', *Journal of the American Institute of Planners*, 35: 216–24.

Barrett, G.A., Jones, T.P. and McEvoy, D. (1996) 'Ethnic minority business: theoretical discourse in Britain and North America', *Urban Studies*, 33 (4/5): 783–809.

Bell, D.A. (1998) 'Residential community associations: community or disunity', in A. Etzioni (ed.), *The Essential Communitarian Reader*. Lanham: Rowman and Littlefield Publishers Inc. pp. 167–76.

Bennett, R. (ed.) (1989) *Territory and Administration in Europe*. London: Pinter.

Brent, J. (1998) 'Community without unity', in P. Hoggett (ed.), *Contested Communities*. Bristol: The Policy Press. pp. 68–83.

Burns, D., Hambleton, R. and Hoggett, P. (1994) *The Politics of Decentralisation: Revitalising Local Democracy*. Basingstoke: Macmillan.

Castells, M. (1983) *The City and the Grassroots: A Cross-Cultural Theory of Urban Social Movements*. London: Edward Arnold.

Dennis, N., Henriques, F. and Slaughter, C. (1956) *Coal is Our Life*. London: Tavistock.

Eade, J. (ed.) (1997) *Living the Global City: Globalization as a Local Process*. London: Routledge.

Etzioni, A. (1995) *The Spirit of Community: Rights, Responsibilities and the Communitarian Agenda*. London: Fontana.

Fischer, C.S. (1975) 'Towards a subcultural theory of urbanism', *American Journal of Sociology*, 80: 1319–41.

Flanagan, W. (1995) *Urban Sociology: Images and Structure*. Boston: Allyn & Bacon.

Gottdiener, M. (1994) *The New Urban Sociology*. New York: McGraw-Hill.

Hill, D.M. (ed.) (1994) *Citizens and Cities: Urban Policy in the 1990s*. New York: Harvester Wheatsheaf.

Hoggett, P. (1997) 'Contested communities', in P. Hoggett (ed.), *Contested Communities*. Bristol: The Policy Press. pp. 3–16.

Johnston, R.J., Gregory, D. and Smith, D.M. (eds) (1994) *The Dictionary of Human Geography*. 3rd edn. Oxford: Blackwell.

Karp, P.A., Stone, G.P. and Yoels, W.C. (1991) *Being Urban: a Sociology of City Life*. 2nd edn. New York: Praeger.

Klug, G.E. (1999) *City Making: Building Communities without Building Walls*. Princeton, NJ: Princeton University Press.

Li, W. (1998) 'Anatomy of a new ethnic settlement: the Chinese *ethnoburb* in Los Angeles', *Urban Studies*, 35 (3): 479–501.

Mort, F. (1998) 'Cityscapes: comsumption, masculinities and the mapping of London since 1950', *Urban Studies*, 35 (5–6): 889–907.

Paddison, R. (1981) 'Identifying the local political community: a case study of Glasgow', in A.D. Burnett and P.J. Taylor (eds) *Political Studies from Spatial Perspectives*. Chichester: John Wiley. pp. 341–56.

Park, R.E. (1925/1952) 'The city: suggestions for investigations of human behaviour in an urban environment', in R.E. Park, *Human Communities: The City and Human Ecology*. Glencoe, IL: The Free Press. pp. 13–51.

Park, R.E. (ed.) (1929/1952) *Human Communities: The City and Human Ecology*. Glencoe, IL: The Free Press.

Relph, E. (1976) *Places and Placeness*. London: Pion Ltd.

Roddewig, R.J. (1978) *Green Bans: The Birth of Australian Environmental Politics*. Sydney: Hole & Iremonger.

Russell, H. (1998) *A Place for the Community? Tyne and Wear Development Corporation's Approach to Regeneration*. Bristol: The Policy Press.

Sandercock, L.A. (1998) *Towards Cosmoplis*. Chichester: John Wiley & Sons.

Sennett, R. (1970) *The Uses of Disorder: Personal Identity and City Life*. New York: Knopf.

Simmel. G. (1905/1950) 'The metropolis and mental life', in K.H. Wolff (ed.), *The Sociology of George Simmel*. New York: The Free Press. pp. 409–24.

Speece, M. (1990) 'Ethnodomination in marketing channels revisited', in A.M. Findlay, R. Paddison and J.A. Dawson (eds), *Retailing Environments in Developing Countries*. London: Routledge. pp. 138–55.

Stacey, M. (1969) 'The myth of community studies', *British Journal of Sociology*, 20(2): 134–47.

Staeheli, L.A. and Thompson, A. (1997) 'Citizenship, community and struggles for public space', *The Professional Geographer*, 49: 28–38.

Suttles, G. (1968) *The Social Order of the Slum*. Chicago: University of Chicago Press.

Tönnies, F. (1883/1995) *Community and Society*. New York: Harper & Row.

Waltzer, M. (1995) 'The communitarian critique of liberalism', in A. Etzioni (ed). *The New Communitarian Thinking: Persons, Virtues, Institutions and Communities*. Virginia: University of Virginia Press.

Warrens, R. and D.I. (1977) *The Neighbourhood Organizer's Handbook*. Notre Dame, IN: Notre Dame Press.

Webber, M.M. (1963) 'Order in diversity: community without propinquity', in L. Wingo (ed.) *Cities and Space*. Baltimore, MD: Johns Hopkins University Press. pp 23–56.

Whyte, W.F. (1943) *Street Corner Society*. Chicago: University of Chicago Press.

Wilson, E. 91991) *The Sphinx in the City: Urban Life, the Control of Disorder, and Women*. London: Virago.

Wirth, L. (1938) 'Urbanism as a way of life', *American Journal of Sociology*, 40: 1–24.

Young, I.M. (1990) *Justice and the Politics of Difference*. Princeton, NJ: Princeton University Press.

Young, M. and Wilmott, P. (1957) *Family and Kinship in East London*. London: Routledge and Kegan Paul.

13

Women, Men, Cities

LINDA M. McDOWELL

For our house is the corner of the world. As has often been said, it is our first universe, a real cosmos in every sense of the world.

Gaston Bachelard, *The Poetics of Space*, 1969, p.2

I have a terrible fear of the suburbs: I cannot bear the provinces, and especially the edges of conurbations. Just like home. The safe familiarity of the bay windows, the neat gardens, safety like a trap, ready to ensnare in its enfolding arms.

Walkerdine, 1985: 66

The city was the source of enlightenment and the fountain of culture for my parents and their friends who were city born and bred. No shtetls for them. They believed in the city; 'peasant' was a word which indicated ignorance, dirt and prejudice. They looked to the city above all for its liberating intelligence, for education, for ideas, as a cultural and political centre.

Amirah Inglis, *A Tale of Three Cities*, 1989, p.79

INTRODUCTION: STOCKPORT, LONDON, CAMBRIDGE

I was . . . one of the children of the post war boom, who would leave the safe innocence of the suburbs for the stripped pine promises of the new middle class, for the glamour of the metropolis and the desperate lure of the academy. (Walkerdine, 1985: 63–4)

These are the words of Valerie Walkerdine, born two years before me, in Derby rather than in Stockport where I was born, but her trajectory was mine too. A child of the post-war settlement, a first generation university student, I – like so many of the feminist academics currently teaching in British universities – was a grammar school girl in the 1960s, escaping marriage and motherhood, at least for a while or as a sole pursuit, through a scholarship and later an academic position in an urban university. And now so many of us are urban dwellers, some of those women to whom writers about the city as a liberating arena for women (Wilson, 1991a), or about women's growing significance as inner area gentrifiers (Bondi, 1992; Butler and Hamnett, 1994; Lyons, 1996; Rose, 1989; Warde, 1991) must have been referring. But these women, like me, are now middle class and have relatively little in common with other women who live in the same inner areas of cities. These women are more likely to be the tenants of private landlords or else to live in increasingly poorly maintained council property, rather than in the gentrified terraces that are our homes. And, in the uncertain urban labour markets of the 1990s and the new century, they are more likely to be unemployed or in poorly paid casual and temporary work than to have permanent well-paid employment as academics (or other professional occupations into which educated women have made successful, if too small, inroads) or to be able to afford the commodified domestic services that keep our lives running smoothly. Indeed, it may be that we meet these women only in our homes as they supplement their incomes by cleaning or ironing for us. Although we are all women, as Virginia Woolf so perceptively argued three-quarters of a century ago, our needs and demands may be very different. Thus at a meeting of the Women's Cooperative Guild in 1915 Woolf noted that 'If every reform they demand was granted this very instant it would not touch one hair of my comfortable capitalistic head' (Woolf, 1929 in Davies, 1977: xxxi). Class divides women as much as gender unites them.

So what exactly, if anything, do women *per se* and women in cities have in common at the end of the twentieth century rather than at the beginning? Is gender a fundamental axis of urban analysis? Do women as a group have different perceptions of, and experiences in the city from men? Or are the differences among women, divided as they/we are not only by class, but also by race and ethnicity, life-cycle stage, age and fitness, as marked as the similarities between them/us? Certainly having money and health makes a huge difference to the 'urban experience'. Does this mean that the project that feminists interested in urban divisions have been engaged in since 'round about 1973' (Harvey, 1989) – that significant date in the transition to postmodernity – has been at best too single-minded, at worst misguided? These are difficult questions to raise as they seem to deny the political significance of feminism, the policy implications of gender needs, and the common cause between women but they have been asked in an increasingly insistent form, both in the political arena and in the academy for the past decade or so. One resolution has been to hold to a form of strategic essentialism (all women are united by their gender) for political purposes (and here I mean far more than party politics for one of the key successes of feminist politics has been to extend the definition of what is political into the everyday, to 'ordinary' issues, many of them urban) while at the same time unpicking the differences and diversity that distinguish women from each other.

In this chapter I want to address the significance of the recognition of difference and its implications for urban analyses and policies, while also giving a flavour of the excitement that has stimulated a vast array of studies about gender divisions and the city – as a built form, as a spatial arena, as an artefact and an image – in the twenty and more years since the geographer Pat Burnett published what was probably the first paper in an English language journal about gender and the city, a paper published in *Antipode* in 1973. In the intervening years, not only has feminist scholarship burgeoned and shifted in its emphasis from a primary focus on material social relations to the inclusion of symbolic meaning and systems of representations (McDowell, 1993a, 1993b, 1999; Rose, 1993; Segal, 1999) but urban planners and the policy-makers no longer ignore what are too often referred to as 'women's issues'. It is important, however, to establish from the outset that the intersection of gender and urban policy is not solely about women's issues – men are gendered

too and, like women, their experiences in and of the city are affected by their age, ethnicity, class, sexuality and other social characteristics. Indeed the changing position of men in urban areas and the significance of the visible effects of what has been termed a crisis of masculinity is one of the most noticeable shifts in research that focuses on the connections between gender and urban issues (Campbell, 1993). This parallels a wider focus on masculinity in recent scholarship about gender relations (see, for example, Connell, 1995).

While I hope to give some, inevitably brief, history of these debates that have altered – indeed transformed may not be too strong a word – analyses of urban issues and problems in the past two decades, I also want to suggest from the start that it is impossible to discuss gender and urban policy in the abstract, without an idea of the type of society or city that we might be aiming towards. At the most general level, therefore, I want to situate my chapter in the context of a search for urban policies that will result in or at least move us towards a more socially just society. In the particular context of this chapter, of course, the reduction of inequalities based on gender is the principal aim. And here we have to answer some difficult questions about what is meant when or whether we talk about women's needs, the empowerment of women, gender issues or equal opportunities policies. As I have already intimated, it has become increasingly clear that it is difficult to separate injustices based on gender from those of class, race or ethnicity, age, sexual preference and so on. The overall aim of an urban policy whose goal is greater social justice must be to close the increasing gulf between the powerful and the powerless, the haves and the have-nots in contemporary cities, whether the have-nots are defined by gender, race, age or class or, as is usually the case, positioned by a complex cross-cutting of all these social divisions, and others.

These are large issues: the philosophical literature about social justice, for example, is enormous, and while I can only touch on it in this chapter, I want to end with a consideration of the alternative definitions of justice that inform equal opportunities policies in the urban context – indeed it would be impossible to consider gender relations without such a discussion. Urban policies must be based on some normative notion of what equality between women and men might look like, of how this affects urban form and access to and the distribution of urban goods and resources.

Before I either review the 'urban studies and gender' literature or assess alternative conceptions of social justice, however, I want to raise some questions about the press coverage of what might be defined as urban policy and gender in action. It is an example of community action (or is it?) which *Observer* columnist Melanie Phillips first raised in the summer of 1994 and then again, almost two years later, in the winter of 1996. I found that these were articles and an issue that made me think about the significance and appropriateness of the topic 'women and cities' that the editor of this book had asked me to write about.[1]

BALSALL HEATH, BIRMINGHAM: COMMUNITY ACTION OR VIGILANTISM?

The inner city suburb of Balsall Heath in Birmingham, in common with many similar areas in British cities, has a history of housing decline, class change and in-migration from different areas of Britain, and, since the Second World War but particularly from the 1970s, from other Commonwealth or former Commonwealth parts of the empire. It is areas like Balsall Heath that Rex and Moore (1967), in their classic study of urban social structure based on research in Birmingham, dubbed 'twilight areas' or, after Burgess's 1920s work on inner Chicago 'the zone in transition'. But as critics of these classic sociological analyses of urban structure have argued, these areas are also home for many people who have a strong attachment to and pride in their area, who have invested in owner occupation and in a range of improvements and who resent not only the negative labelling of their community but also, more problematically, the shifting population of less fortunate and less-rooted people who also find some sort of a home here. Such areas are thus cut through the social divisions – those of ethnicity, of class, of income, of different types of lifestyle – and with potential conflicts of interest between local residents. In Balsall Heath another form of conflict became an increasing source of tension for the local residents. In the past few years the area had become a place where sex was for sale: for sale in some of the small Victorian houses where women sat under a light in the window, often in a state of semi-undress and, more obviously, for sale on the streets. Groups of women prostitutes, and some young men, would gather each evening, and during the day, on street corners in order to attract the attention of punters who cruised the area in their cars. It was not only moral outrage, but also physical dangers to their children – from the passing cars, from the used condoms and syringes junked in the small front gardens – that stirred the local community into action. Headed by a group of mainly Asian men, a rota was worked out of men who would sit in full visibility on the street corners and patrol the local pavements each evening through into the night in an attempt to shame the punters and destroy the trade for the 'working girls'.

Now what is the relevance of this? It is, you might think, a fine example of community action, a gesture of gender solidarity between 'decent' men and women. Indeed the social commentator Melanie Philips, who seems in her column in the *Observer* to have set herself up as the voice of middle (class) England, saw it in exactly those terms, declaring that the action was giving people a 'sense of communal self-respect'. Although she mentioned that insoluble problem of local community action – the displacement of whatever use or activity is disliked from one into another area – and worried about vigilantism in a passing reference, her overall conclusion was that the actions by these Muslim men had 'pointed social values in a more healthy and realistic direction' (*Observer*, 17 July 1994: 12)

I want to suggest that there is a more complicated story to tell here that reveals some of the tensions and divisions between the residents of this area: differences that are based on class and gender lines. Indeed, I think that this story is an exemplary illustration of the way in which feminist scholars and policy-makers have come to recognize over the past decade or so that it is problematic to talk about gender or gender interests *per se*. We should no longer assume a simple dichotomy or division of interests along gender lines, with men on one side and women on the other, nor that there is a commonality of interests that unites either all women or all residents of a small area. Returning to Balsall Heath illustrates this point.

In an article in *The Guardian* (23 July 1994: 25) published a few days after Melanie Phillips's piece, Maggie O'Kane revealed the ways in which 'the local community' was divided along ethnic and gender lines. She interviewed several women who lived in the area who told her that they felt intimidated by the gangs of men patrolling the streets. These women included an older white woman who had been verbally abused on returning from a party, and a woman in her twenties who lived opposite a house used for prostitution who explained the effect of the patrols on her attitudes and behaviour: 'I find it intimidating walking past the vigilantes. They don't have a perspective on who might or who might not be a prostitute. I used to talk to the girls but now I am even wary of doing that.' The prostitutes themselves, although more used to verbal abuse, also suffered. One 'girl' (actually a woman of 44) queuing in a local shop reported: 'I was just standing there when a gang of them surrounded me and started shouting at me "Clear out, you slut".' The attitudes of Asian women in the area, married or single, young or elderly, were not considered by either of the two reporters. However, as the work of an Asian women's community action group in London, the Southall Black Sisters, has shown there are often tensions in Muslim families, not only between the generations but also between young men and women who have different attitudes about women's position in contemporary Britain.

Phillips clearly was not convinced by O'Kane's journalism nor by the academic feminist literature that increasingly has emphasized the diversity of women's interests. In February 1996, she wrote a follow-up article in which she revealed the success of the community picket in cleaning up Balsall Heath with the exception of a single housing estate, where, she reported, a new alliance of the housing department, the police and the Muslim-led street watch group was beginning to have success in evicting not only prostitutes, but also drug dealers, pimps and gangsters. Phillips quotes only a single voice against what she terms 'the people power that cleaned the streets' – a minister of the United Reform Church, David Clark, who is quoted as suggesting: 'It's almost fascist; it's the sort of thing that was done to the Jews and the Gypsies, when the police and local groups get together like this. It's treating people like objects, like the ethnic cleansing in Bosnia.' His unease clearly did not worry Phillips who wrote, 'But these "victims" are drug dealers and pimps.'

In the conclusion to her article Phillips raises an interesting issue. She writes: 'At the root of the controversy is a fundamental collision of values, between those who believe there are absolutes of right and wrong and those who believe that all values are relative and no-one should judge anyone else' (*Observer*, 4 February 1996: 7).

I think Phillips is wrong here. While it is perfectly appropriate to make judgements – urban planners and policy-makers do so every day, as do scholars who comment on city questions – we still need to ask ourselves who it is who is making these judgements and on what basis. How are the different interests of women who live in Balsall Health, in Birmingham, or indeed in any other urban area, to be judged against the women who work there, albeit in an occupation that many feel is morally repugnant? How are we positioned as academics or policy-makers? How does our own race, gender or religion influence our arguments? And are the absolutes that Phillips feels are so clear unchangeable or are the views we express about urban problems actually dependent on our own positions or the particular circumstances of a particular city at that time? These are some of the issues that are at the centre of contemporary debates about urban policy and social justice and I shall return to them at the end of the chapter. This recognition of the fragmentation or diversity of interests among women and the acceptance that decisions and even knowledge often is context-dependent or 'situated' has, however, been the main change of emphasis in

theoretical work on gender issues undertaken over the past two decades. It also raises difficult issues for urban policy and politics that I shall also want to return to.

First, however, and perhaps paradoxically given this claim, I want to establish the extent to which women in Britain today do have certain common interests and face common issues in cities.

FROM THE 1970s TO THE 1990s: KEY CHANGES IN THE POSITION OF WOMEN IN BRITAIN

These two decades in Britain were ones, of course, dominated by Conservative governments; a period in which, since 1979 and to some extent before, in the words of Brian Robson, 'urban policy in Britain has been faced with political ambivalence and declining expenditure' (Robson, 1994: 131).[2] There have been cuts not only in earmarked urban budgets, but also in local and central government spending that is most crucial in cities. Direct investment in housing is the most obvious example where expenditure has declined but also in a whole raft of welfare services, both in direct provision such as the health service where there has been a shift of resources away from the cities to the shires, and in financial support for the vulnerable. The abolition of income support for 16-18-year-olds, for example, is visible in the shaming increase of young homeless on city streets. Moreover, the declining real value of unemployment benefit and state pensions, over two decades in which unemployment has affected over three million at any one time and the numbers of the elderly have increased, have condemned many to lives of private desperation, not only in the inner areas of the largest cities which we traditionally regard as areas of deprivation, but also on suburban council estates in smaller towns such as Oxford. The unrest that broke out there on the Blackbird Leys estate in 1991 was a challenge to the too-easy associations between poverty, unrest and the inner areas of large conurbations. Further, as recent research published from the ESRC panel survey and by the Joseph Rowntree Foundation (1995) shows, geographic patterns of poverty are changing. Smaller towns and cities in the southeast, Chelmsford as well as Oxford for example, have experienced growing numbers of households in poverty over the 1980s (Green, 1994). While these are not necessarily gender-specific issues, they do have a differential effect on men's and women's lives. The elderly poor, for example, are disproportionately female, and single parent mothers are disproportionately represented among the poor (Glendinning and Millar, 1992).

The overall effect of changes over the past two decades has been a marked increase in the extent of inequality between the richest and the poorest members of society. While trends in urban policy have exacerbated, or rather perhaps not alleviated injustices, wider changes, at a national and indeed international level are also part of the explanation of greater inequality. I do not want to labour these points but rather to draw out the implications for women's positions in society and for changing gender relations, and to examine some of the consequences for urban policy.

Waged Work

Perhaps the most significant change has been in the structure of the economy. The shift to the service sector, common in advanced economies in the past two decades, has been accompanied by a rise in women's participation in waged labour to such an extent that some commentators argue that we are witnessing a feminization of the economy (Wilkinson, 1994). Whether or not this is so (and see Hakim, 1993, 1996, for a dissenting view), it is clear that in Britain there has been the withdrawal of some two million men from the labour force since 1971 and a concomitant increase of a similar number of women workers – although, as is well known, the figures for women's labour-market participation are notoriously inaccurate. Working-class men have been affected by the decline in skilled manual employment and, more recently middle-class men by so-called downsizing and by the shakeout of middle management. There are several thousand middle-aged men in their forties and fifties who are 'discouraged' workers, often ineligible for unemployment benefit by virtue of golden handshakes and redundancy payments and also failing to register for work as little that is appropriate to their age and skills is available. The impact of unemployment on men took a spatially uneven form. In the heavy decline in the male manufacturing industries in the 1970s and 1980s, men in the industrial urban areas of the north of England, Scotland and Wales were most severely affected, whereas in the recession of the early 1990s in which service sector employees also lost their jobs, the urban impacts were more evenly spread both across the regions and between the larger urban conurbations and the smaller cities and towns (Lawless et al., 1996; Martin, 1995; Philo, 1995). The rise of an 'underclass' of alienated young men, especially in Britain's northern cities, who, excluded from the labour market, enact rites of masculine identity through car theft and joy-riding, is a new aspect of gender problems in cities (Campbell, 1993). In long summers (1991, 1994) sporadic unrest may turn into 'riots',

in the most recent of which Asian youth, regarded as less prone to riot than either white or African-Caribbean men, have also been involved in towns such as Bradford and Luton. This urban crisis of masculinity (McDowell, 2000) has parallels in a wider concern about the educational achievements of young men and the gap that is appearing between the results of girls and boys at the statutory school-leaving age. In Britain it seems that girls are now more successful in examinations at 16 than boys. The gap is particularly wide for African-Caribbean young people.

The more than two million women who have entered the labour market are in very different occupations and jobs from the men who have left or who never join. Predominantly in the service sector and in part-time employment, they are low paid and have poor terms and conditions of employment. The net effect has been an overall reduction of the amount and value of paid work being done in the economy, in terms of the number of hours and the total remuneration with the rapid growth of low-paid workers. This casualization or peripheralization of the labour market has been further exacerbated by the privatization of public sector employment, especially under the compulsory competitive tendering programme of the successive conservative governments to 1997. Although the vast majority of women currently in the labour market are both in the service sector and in part-time employment, there are also growing numbers of women who have successfully entered professional and white collar employment, often on equal terms with men, at least in terms of salary levels at particular positions in the hierarchy. (There is still significant occupational segregation, both horizontal and vertical, in the British labour market; Scott, 1994.) Women's growing possession of educational and professional credentials, however, has opened up class divisions between women that are reflected in a growing income divide between women (McDowell, 1991). White middle-class women in professional occupations, especially those without children (and an increasing number of women are remaining childless) have more in common in terms of their social characteristics, lifestyles and income with men in similar positions in the labour market than with their working-class sisters.

There are also important divisions between women from different ethnic backgrounds in terms of their pattern of labour market participation and their position in the economy. Women from African-Caribbean backgrounds, for example, are both more likely still to work in manufacturing or in the public sector of services and full time than white women. And whereas women from the Asian subcontinent who are Muslim are least likely to be in employment, other women from India and Bangladesh have high participation rates. These differences affect women's position in the family, their independence and their purchasing power and so place them differentially when urban policies that are 'gender-sensitive' are being considered. Despite these class and ethnic differences, however, feminist economist Anne Phillips (1989) has argued that the lives of most women now are amazingly homogeneous in the sense that they are for a considerable period of their life both mothers and waged workers and do their own domestic labour. So in two senses women are the proletariat, the new working class in the labour market and, in the main, domestic labourers in their homes.

While some of the women who have entered the labour market over the past two decades have become the sole income earners in their household, over this period there has been an increase both of dual earning households and of households without a wage earner. Thus there has been a growing gap created between (Nelson and Smith 1999; Pahl, 1984) work-rich and work-poor households. In households headed by a married couple where the man has lost his job, the social security regulations act as a disincentive for women who are in low paid employment and so paradoxically at first sight women may choose to leave the labour market if their husband loses his job. The work-rich households are predominantly middle-class households whom, as Gregson and Lowe (1994) have shown, are increasingly employing what they have termed a new service class of domestic workers – cleaners, housekeepers, nannies and child minders to service their domestic needs – a class composed primarily, of course, of other women. This commodification and privatization of domestic work may be a possible area for future intervention. For the majority of less affluent dual income households, women's entry into the labour market has entailed a speed-up or intensification of their overall workload, as two people do at least three jobs – two in the labour market and the domestic labour. This speed-up is exacerbated by the declining standards or withdrawal of a range of state-provided services – school meals for children for example, meals on wheels and home helps for elderly dependants, poor and expensive public transport, higher costs and increased waiting times for dental treatment and so on. As research has demonstrated (Sassoon, 1987), it is predominantly women who do not only the domestic labour and care of dependants but who undertake all the organization and transport required to patch together the range of services used by their families but widely distributed across the urban area. For these women, as feminist scholars and activists have long argued, a city designed on

the basis of full-time work for men, the separation of so-called non-conforming uses, the home from work, retail from industrial uses, a transport system geared around peak time flows and periphery–centre links ignores the realities of an increasing number of women's lives. There are clear policy implications here that hardly need spelling out. However, as the implications are, in fact, no less than a redesign of the principles of urban planning as we know them – Hayden published an article questioning whether a nonsexist city is possible almost two decades ago (1984) – the type of planning interventions that come under the heading of gender planning are a lot smaller-scale and somewhat more pragmatic. I shall outline some of them below. We might do well, however, to remind ourselves of research that shows that women who are full-time housewives express greater dissatisfaction with suburban living than employed men who only return to these areas in the evening for R&R (Van Vliet, 1988).

Household Structure and Single Mothers

As well as entry into waged work, perhaps the most significant, or at least most discussed, change of the past two decades had been the rise in the number of single women living without a man, especially unsupported mothers. Indeed in the early 1990s the degree of political hysteria around this was such as to justify the label a 'moral panic'. A wide range of social ills have been attributed to single mothers, especially but not only by commentators from the radical right of the political spectrum. Few have reached the heights of absurdity of US commentator Charles Murray (1994), who with particular relevance to this chapter seemed to suggest that inner city decline in general, and urban crime in particular, was the sole responsibility of single mothers, if not individually then because of their moral fecklessness in inadequately socializing the type of wild young men that Campbell (1993) has studied. We might variously add to the list teenage violence, drug-taking and family breakdown, all associated by urban commentators on the right of the political spectrum. While the debate about the causes or consequences of the growing number of unsupported mothers is beyond the scope of this chapter, it is interesting to compare their labour market participation rates with comparable rates for married mothers of the same age. Participation rates are significantly lower for single mothers, many of whose earning capacity does not allow them to purchase the necessary child care, although survey research has found that many single mothers would prefer to be employed than to rely on state benefits

(Glendinning and Millar, 1992). Here a number of possible interventions from income support to more specifically urban policies relating training schemes and new job opportunities to child care provision and affordable housing in inner areas and outer urban estates are important.[3] In her innovative work in Los Angeles the architect and academic Dolores Hayden designed such a scheme for the inner city. In Britain, the Women's Design Service in London is involved in the design of a number of projects for women including an Asian women's centre and housing for single women but these examples are few and far between.

Urban Violence

While I have emphasized both homogeneity and heterogeneity in women's position in the contemporary city, it is perhaps urban violence, and the threat of it, that most obviously unites women, whatever their circumstances. Interesting work by, among others, Gill Valentine (1989, 1993), has shown how the fear of attack restricts women's access to certain areas in the city at particular times of the day or night. A smaller proportion of women than men are able to drive, and not all of those with a licence have unrestricted access to a car, although as the 1991 census figures on journey to work reveal, the gender differentials here are now beginning to close. Although more women than men still go to work by bus (15 per cent compared to 6 per cent in 1991 but down from 24 per cent and 11 per cent respectively in 1981), 52 per cent of women compared with 67 per cent of men now go to work by car. (In 1981 the figures were 37 per cent and 58 per cent respectively.) The combination of fear of violence and the problems of a public transport system that is either unsafe (because of, for example, cuts in platform staff on the underground in Greater London, and the increase in single operator buses in many cities) or not available is one reason for the reduction of the gender gap in travel patterns. It is also a reason, in my view, to look sceptically at green urban policies to reduce car usage in city centres. However, this conclusion is directly opposed to the need to reduce pollution levels and to make city streets safer for children. A comparison of the figures for children's journies to school over the past two decades reveals a staggering reduction in the percentage of children who walk to school – a decline from 91 per cent in 1971 to only 9 per cent in 1991. In the long term, of course, advocating increased car usage by women is merely adding to the spiral of declining public transport and streets empty of pedestrians.

Fear and Delight?

While the city streets are too often places of insecurity and fear for women, and for others too whose appearance might distinguish them from the norm – gay men, people of colour and homeless people are frequent victims of street violence – the urban crowd is also a place of escape and anonymity where conventional boundaries might be transgressed and the familial lifestyle of the suburbs denied (Mort, 1995; Sennett, 1977, 1993; Walkowitz, 1992). A growing literature, primarily by and about gay men but latterly including lesbians too, has documented the significance of urban bars and clubs and areas of gay residential gentrification in the establishment of an alternative sexual identity (Adler and Brenner, 1992; Bell and Valentine, 1995; Knopp, 1992; Lauria and Knopp, 1985). While the earliest urban research that was from an explicitly feminist focus tended to emphasize the negative implications of urban living, and especially the disadvantages of current urban land use planning, there is also a strand of feminist research on the city that celebrates the city's significance for large numbers of women, from the early pioneers of women's education and employment, for the women involved in the suffrage movement, for lesbians and for women who just want to be independent.

Elizabeth Wilson's writing on the city – books with such evocative titles as *Hallucinations: Life in the Post-modern City* (1988) and *The Sphinx and the City* (1991b) – are good examples of this genre that draws on, and feminizes, the celebration of the city in the work of that great modernist poet and writer, Charles Baudelaire. Writing about mid-nineteenth century Paris, Baudelaire identified as the key urban figure the *flâneur*, an urban voyeur who, in Baudelaire's immortal phrase, 'botanized on the asphalt', dawdling, gazing and observing the fantastic urban spectacle laid out for his delectation. Although Baudelaire took for granted that the *flâneur* was a masculine figure, in an interesting exchange with Janet Wolff (1985), Wilson (1991a) has asserted not only the existence of the *flâneuse* (although George Sand had to adopt male dress to wander Parisian streets at night in the 1870s) but also impugned the very masculinity of the *flâneur* himself: after all, these men not only did not work – and 'breadwinning' has been the key defining characteristic of modern masculinity – but they also enjoyed window shopping!

Wilson convincingly suggests that the anonymity and excitement of the late nineteenth- and twentieth-century city was crucial for the development of feminist politics, as urban living often brought freedom from patriarchal control, in both the literal sense of an escape from the father and in the more general sense. For the 'New Woman' of the twentieth century, challenging the Victorian sanctity of hearth and home, the city brought a welcome anonymity. Wilson is on less certain ground, however, in my view, in her arguments that the origins of modern town planning lie in men's fears of the female urban crowd and so a desire to restrict and restrain women's activities and keep them in the private sphere. This desire was surely restricted to 'virtuous' women, the domestic angel of the hearth of Victorian ideology, whereas the eroto-feminization of the streets, the city as an arena of female sexuality ('loose' women of the street; after all street walkers is a common euphemism for female prostitutes), there for the male gaze and for taking, is an all-too-familiar masculine representation of cities (Walkowitz, 1992). The binary construction of the madonna and the whore, the good woman and the bad woman, is a classic division between women. It also seems to me that the fear of the 'otherness' of the urban crowd (the mob?) is beyond gender in the sense that it is a long-understood reaction of authority in whatever guise – the police or the town planners – to fears of the lack of control of crowds of all sorts: the working class, poll tax protesters or football fans. It is the city of course, especially in the late nineteenth and early twentieth century, that facilitated conglomeration in numbers previously unknown, leading to productive as well as challenging chaos. As Marx recognized, the city was a crucial arena for the forging of class consciousness and class politics.

THE IMAGINARY CITY: MEAN STREETS

The association of cities with outsiders, and with public forms of association, has been a dominant theme of representations of urban life in literature and in films. Indeed, the social critic Raymond Williams (1989) has argued that the city itself was the crucible of modernity. In the vast migrations, especially from Europe to North America, that fed the late nineteenth and twentieth century cities of the 'advanced' world, writers, painters, musicians were thrown together in a form of creative chaos. Among the products of the modern movement were the great social realist novels that dealt with urban poverty and inequalities, with the consequences of migration and ambition. Dickens in Great Britain, Dreiser and Sinclair in the USA, produced sweeping panoramic visions of the effects of urban life on the men and women of the time. Like the urban sociologists of the time – Simmel, Spengler, later Wirth, Park and Burgess – and of course their predecessor, Baudelaire, these novelists were

fascinated by the arcane and the unusual, the outsider and the outlaw. Their novels featured the criminal and the cheat, the drug-takers and law-breakers, risking their reputations in the anonymity of the city. The urban sociologists tended to ignore women, seduced by the glamour of the male outsider, although Baudelaire recognised the fallen woman as a crucial urban outsider and novelists from Dreiser's *Sister Carrie* onwards have seen the city as an arena for women's seduction and downfall. Donald (1992) has argued that this emphasis on the outsider heralded the classic protagonist of urban fiction – the detective who, with faith in rational science and equally rational but bleak view of the venality of his fellow men and women, was able to pierce the gloom of the urban underworld and restore relative order to urban chaos (see also Willett, 1996). It is significant that this figure, at least until the rise of a feminist genre in the 1980s, is male. Here, in detective fiction, we see a clear distinction between the masculine, public world of the streets and the feminine private world of the home, that feminist scholars have long recognized. Indeed, not only is the detective a man in a male urban world but he is specifically portrayed as unencumbered by the ties of domesticity. He is a loner, single or divorced, forever caught in the bleak internal neon light of the open fridge door, whose empty interior parallels his own interior life.

The fictional female private eye had to wait for the feminist deconstruction and blurring of the binary associations of man/public/city/streets and woman/private/rural/home that ordered not only Western fiction but also the whole of Western social theory from the Enlightenment (see Pateman and Gross, 1987). Similarly, the gaze of modernity – whether in literature, film or painting – was that of the male; women were the object of the gaze rather than the 'eye' doing the gazing (Pollock, 1988; Rose, 1993). The female private eye seemed to have to wait too for the development of significant numbers of real 'material girls' – single, independent and employed. The feminine and feminist heroine of the new genre of hard-boiled fiction is, like her male predecessor, single and unencumbered; she changes her man as often as her gun, has a liking for spirits and is as at home on the mean streets as any men. Paretsky's V.I. Warshawski and Grafton's Kinsey Millhone epitomize these new urban heroines. I have searched in vain for a female detective with children in the literally dozens of female fictional detectives that now exist but I was pleased to welcome Spring's Laura Principal, with a fictional base in Cambridge, to the list. It is a nice feeling when a favourite female protagonist actually walks down your own mean street! The 1980s also saw the development of a whole range of films starring tough and independent urban women, that the critic Judith Williamson dubbed the SWW genre (single working women). *Fatal Attraction* (1987) was perhaps the apotheosis of this genre with its central fight for control over a man between the wicked SWW and the good suburban-based wife. Its disturbed central character (played by Glenn Close) lived in inner New York in an area that seemed often to be lit by flames like a charnel house. Her final action that was meant to signify her insane challenge to the domestic virtues of suburbia was to boil the family bunny in a cooking pot!

While female dicks of detective fiction may be the ideal, as well as an idealized, urban character, counterposed to the mad SWW, feminists interested in the urban and in cities have also analysed the significance of the city to women in literature as well as the portrayal of different aspects of urban life in novels and films. Heron's *Streets of Desire* (1993) is an excellent example of the first set of analyses. Here she illustrates the central theme of rootlessness and displacement that lies at the heart of city novels by twentieth century women writers, drawing examples from each of the decades of this century. For all women, city fictions are narratives of self-discovery, and as Heron argues, have a special significance for Black women for whom the city is a historical stage where migrants converge and 'women come into their own voice: the voice that articulates historical experience as urban' (Heron, 1993: 3). While Heron's collection is dominated by the great fictional cities of Western literature – Paris, London and New York – a collection by Rieder (1991), *Cosmopolis*, includes urban stories by women based in cities as diverse as Bombay, Beirut, Beijing, Soweto and Sao Paulo. In many of these stories, women appear as outsiders, city heroines created by women writers who themselves are strangers to the city they are writing about, whose fictional lives reflect alienation and loneliness, as well as opportunities for greater freedom. In *Inner Cities*, Modjeska (1989) straddles the boundaries of 'fact' and fiction in a series of essays by Australian women, the majority of whom are migrants both to the country itself and to Australia's cities. On another continent, and moving even further towards 'reality' from these tales of imaginary and remembered cities, Hayden (1995) has reclaimed the hidden history of migrant women (and in this case men), Black and Asian, in the early urban development of Los Angeles, providing a stimulating example of how to uncover the *Power of Place* – the title of her book – revealing the significance of women pioneers in urban iconography, street layouts and monuments.

POLICY RECOMMENDATIONS AND RESEARCH AGENDAS

Turning now specifically to urban policy, and back to Britain, what might be included on a shopping list of urban policies addressed to the changing circumstances of women in contemporary British cities?

A glance at any recent book, article or policy document that takes 'gender sensitivity' (the current policy euphemism for inequality or discrimination against women) seriously, reveals a consensus around a number of key issues. The reviews by Little (1993) and Greed (1994) are not atypical of the sort of debates and recommendations that have been common in feminist circles, both among academics and practitioners, for at least a decade and probably longer. For example, Little includes a range of recommendations from women's representation in a whole number of planning fora from the neighbourhood to the national level to specific policies for transport (increased running times, shelters, ramps, guards, women only parking spaces, arrangements between women students and women taxi drivers in university cities and so on), for shopping (extended opening times, rearrangement of stores), housing (low income provision, mixed development, sheltered accommodation) and so forth. Greed (1994) has identified a similar range of policies, as well as developing work on a more specific issue – namely public conveniences – that bane of life for all women in the city, for those who habitually use urban leisure facilities or shops and, worst of all, for women who are accompanied by a toddler or two on any of their trips. A commendation should be reserved for those few urban planners who have included nappy changing facilities in *men's* public lavatories. Despite the so-called crisis of single parenthood, well over 80 per cent of children in this country have two parents and not insignificant numbers of men occasionally take their children shopping. Garber and Turner (1994), in a review of urban research (rather than policy) focused on gender issues also included a similar range of issues. Thus they had chapters on rape programmes, child care provision, refuges for battered women, affordable and public housing, central city redevelopment programmes, as well as chapters on women's involvement in collective local action and in formal local politics. Here, as with most of the policy literature, gender is used synonymously with 'women's issues'.

While I have no argument with the sorts of shopping lists identified in these and similar texts on urban policy and research and can only agree that they would improve the quality of life for many women, I want to conclude by turning to the philosophical foundations of equal opportunities policies, or gender-sensitive practices. When we, as academics, planners, policy-makers or politicians, argue for gender-sensitive planning, or for policies to meet women's needs – and these are now common terms not only in the documents of British urban policy but also in Third World planning – what exactly do we mean? As an aside before pursuing this theme in the context of contemporary Western cities, in the women in development and gender and development (WID/GAD) policy literatures (see Pearson, 1992 for a review) there are two proposals that urban planners might consider: namely that all policy documents should include a gender impact statement and that the evaluation of policies should include a gender audit. It is also interesting to compare the rhetoric of development policy, where women are seen as a resource and serious considerations of measures to increase women's empowerment are now conventional in national and international policy documents, with at best, the silence about so-called women's issues, or at worst the construction of women as a problem, that is usual in British urban policy documents.

To return to the more general issue: what vision of a good city lies behind the calls for gender equality; on what conception of social justice, of greater equality between men and women, might it be based? Further, should politicians and practitioners be planning at all to meet policy-defined needs (which so often are based on stereotypical views of women's primary role as helpmeets, domestic provisioners or temporary workers), or rather should there be an attempt to create a wider range of choices for the diverse groups of people who live in cities? As Robson saw need to remind us, 'cities are made up of people' (1994: 132).

In a series of talks on BBC Radio 4 in 1993, David Harvey, then Professor of Geography in Oxford, now at Johns Hopkins in Baltimore, made a passionate plea for the reinstatement of a utopian vision of a good city. It is a plea with which I have a great deal of sympathy, as it seems important to hold on to a progressive notion of social change. But whose vision of a good city are we to plan for? The rise of a postmodernist discourse has challenged the rational modernist assumptions on which town planning practice and urban policies are based (Dear, 1986; Watson and Gibson, 1995). It is no longer clear that there is a singular view of 'the good city' that Donnison and Soto (1980) were concerned to identify many years ago. Increasingly, urban planning is, or should be, concerned to establish the diversity of interests in contemporary cities.

DIVERSITY AND SOCIAL JUSTICE

As I argued from the example of Balsall Health at the start of this chapter, it is difficult to conceive of a set of interests that might unite the women who live and work in this small area of a single city, let alone to plan for all women with their diverse interests in the range of cities and towns that make up urban Britain. The recognition of diversity among women is paralleled by contemporary critiques of normative versions of social justice, and indeed by the postmodern critique of what have rather dismissively been lumped together as meta-narratives – whether they are the liberal notion of the individual that lies behind the ideal of equal citizens and conventional equal opportunities policies or the class-based prescriptions of collective interests in Marxist theory. Instead a number of philosophers have argued for a group-based notion of rights, in which the desires and needs of different interests are not compared to a singular norm, whether it be an idealized worker, parent or way to live. As feminists, and more latterly postmodern critics, have pointed out what is set up as a universal norm or an ideal to aim for, whether in legislation or in urban policy, too often embodies masculinist or class-specific assumptions about the way to be or live (England, 1994; Laws, 1994; McDowell, 1994; Pateman, 1988; Phillips, 1987; Young, 1990).

Looking at the specific dilemma that has long faced feminist politics and equal opportunities policies clarifies the general argument. When claims are made about the necessity for women's equality with men, what exactly is it that is being suggested? As many feminist philosophers have pointed out, feminists are faced with what has become known as the equality/difference dilemma (Phillips, 1987). In the workplace, for example, equal opportunity policies have tended to focus on a liberal notion of equality and to be based on the presumption that strategies should be aimed at women's equality with men within the existing structures of workplace practices. But this presumption denies the differences between women and men. As Rhode has argued, 'Women ought not to have to seem just like men to gain equal respect, recognition and economic security' (Rhode, 1990: 7).

The alternative strategy is to argue for policies based on an acceptance of women's difference from men – and it is this pole of the dichotomy that has tended to inform urban policy. Women have particular or special needs – for safety, for example, or in relation to their particular responsibilities for dependants. However, the danger of emphasizing, even celebrating as some radical feminists do, women's differences, resulting as many of them do from the oppressive structures of a patriarchal society, are well known. If women are weaker, or predominantly carers, then it is too easily argued, as many judges do in rape cases for example (Smith, 1989), that they should be restricted to the home, not out on the city streets celebrating their independence. As Jaggar has pointed out, 'measures apparently designed for women's special protection may end up by protecting them primarily from the benefits that men enjoy' (Jagger, 1990: 244). But I have shown above that the city streets, albeit often dangerous, have also been a significant crucible of women's fight against patriarchal and familial oppression, a place of possibilities to transgress conventional norms of femininity and acceptable female behaviour.

If both equality with or difference from masculinist norms – and it might be added in the context of urban policy particular (white middle class?) ways of living in the city – are unacceptable as the normative basis for urban planning, what then should inform policy? Here I want to return to my initial point about the diversity between women and suggest that the debate about gender must be recast in different terms, emphasizing multiple differences among women (and indeed men) rather than a binary distinction between men and women. Thus there should be an effort to move away from policies based on either women's similarity to or difference from men, although the problems this raises for a movement based on demands for justice for women should not be minimized.

In attempting to think about the implications for urban policy of multiple differences, about people's membership of a whole range of diverse and cross-cutting interest groups, some of them spatially or locally based, and others not, the work of Iris Marion Young is extremely provocative. In her 1990 book *Justice and the Politics of Difference*, she argues for an urban politics based on what she has termed the celebration rather than the hierarchization of differences. 'Minorities' of whatever sort should be seen merely as different from each other and not inferior to a (white, male) norm. Young suggests that different interest groups should come together in democratically elected locally based groups to formulate and negotiate their demands which would then be transmitted to a strategic level planning authority that is city-wide. Her ideal is the construction of a locally based community that does not exclude those with whom the majority does not identify, but is open to what she terms 'unassimilated otherness', that 'takes account of and provides voice for the different groups that dwell together in the city' (Young, 1990: 227) without forming a community in the conventionally understood sense of a unitary interest. Instead of notions of justice being based on a community of rights and enti-

tlements as in liberal theory, what moral philosophers have termed the 'generalized other' (Benhabib, 1986), or even 'the view from nowhere' (Haraway, 1991), claims for justice must be based on the point of view of the 'concrete other', and be secured through *complementary reciprocity*, that is each person 'is entitled to expect and to assume from the other forms of behaviour through which the other feels recognized and confirmed as a concrete, individual being, with specific needs, talents and capacities' (Benhabib, 1986: 341, quoted by Young, 1990: 231).

Thus, in the Balsall Heath case, the community interests would include, rather than exclude, the prostitutes who worked in the area as well as local residents, whose interests anyway, as we have seen, diverge. Lest the question arises of the participation of groups whose membership is based on views that are, for example, based on racist or sexist grounds, Young suggests that who should participate on what grounds and which issues are allowable should be limited by three principles. Individuals and group actions and their consequences must be judged, first, not to do harm to others, secondly, not to inhibit the ability of individuals to develop and exercise their capacities within the limits of mutual respect and cooperation, and finally not to determine the conditions under which other agents are compelled to act (Young, 1990: 251).

Young attempts to distinguish her proposals from the usual interpretation of decentralized decision-making based on local autonomy, which she believes reproduces the problems of exclusion that the ideal of community poses. Instead she insists that local negotiation of needs must be in the context of what she terms 'democratic empowerment in large-scale regional government' (1990: 227). It is crucial that a forum exists in which all participating groups acknowledge and are open to listening to others. Thus Young's view of this wider forum is emphatically not one in which 'the public' is conceived as a unity transcending group differences.

Young recognizes that her proposals are at present utopian. Indeed she herself dubs them as 'laughably utopian', reminding us that it is in the US cities about which she was writing that social injustice is most visible at the present time: 'homeless people lying in doorways, rape in parks, and cold-blooded racist murder are the realities of city life' (1990: 241). But she also reaffirms the alternative view of cities as exciting, as erotic, arenas where strangers meet, where new freedoms are possible. Clearly her vision of the good city has parallels with the work of urban commentators such as Jane Jacobs and Richard Sennett who respectively celebrate or mourn lost urban diversity and the chaos in the streets that has been identified as the key theme of urban literature. The potential for conflict between these diverse urban dwellers when competing to use a single space – Harvey's (1992) analysis of Tompkins Park in New York is a thought-provoking counter-example – is ignored. Zukin (1995) has shown how middle-class views of diverse urban culture exclude other inner-city users and her analysis of small-scale urban entrepreneurial policies to 'reclaim' New York's parks, for example, reveals the chilling underside of cleaning up that city. The parallels with Balsall Heath are clear even though Zukin's New York example of 'community' action involves businesses and the state authorities rather than local residents as in Birmingham's inner city.

Rather than pursue these specific cases or Young's romantic utopian view of cities, I want to conclude this chapter by suggesting that Young's ideas about local democracy and the structures needed to extend it are worth consideration. Young develops a case for the extension, not only of strategic planning authorities for large cities, but also the introduction of regional governments with mechanisms for representing immediate neighbourhoods and towns, interest groups from a variety of spatial scales. The new Scottish and Welsh parliaments are a beginning. The Labour government, elected in 1997, has also reintroduced a strategic authority for London, headed by Ken Livingstone. Critical to the success of this authority will be its inclusion of community participation.

While this chapter might be seen to have strayed from the brief to look at women and cities, and contemporary urban policies, behind the discussion about social justice lies the belief that gender issues cannot and should not be considered in isolation but are a fundamental aspect of all urban policies. The emphasis on diversity is also a plea to develop more varied and sensitive gender-related policies, not to cast gender off the agenda, as some scholars anxious about the dangers of taking diversity into account seem to fear. Unlike Melanie Phillips, I do not believe that the renouncement of absolute values leads to the slippery path of relativism (McDowell, 1995), but rather that only careful attention to difference will produce urban policies to match the diversity of the late twentieth century urban condition. At the end of the millenium, for the first time in history, more than half the world's population lived in a city: the destination for migrants and exiles, a place of dreams and desires, of hopes and fears, and, in fiction at least, if not in policy, ultimately ungraspable and unknowable in its diversity.

NOTES

1. I have also used this example as a teaching aid in a chapter in an Open University course text (Allen, J., Massey, D. and Pryke, M. (eds) (1999) *Unsettling Cities*. London: Routledge).

2. At the election in May 1997 'new' Labour forced a majority government but as it continued the previous government's policy of restraining public spending, this comment remains valid. In July 2000, a rise in public spending was announced. The urban impacts remain to be seen.

3. The New Deal for Single Parents introduced in 1999 includes a range of employment-related policies, but these are, as yet, separate from specifically urban initiatives.

REFERENCES

Adler, S. and Brenner, J. (1992) 'Gender and space: lesbians and gay men in the city', *International Journal of Urban and Regional Research*, 16: 24–34.

Bell, D. and Valentine, G. (eds) (1995) *Mapping Desire*. London: Routledge.

Benhabib, S. (1986) *Critique, Norm and Utopia*. New York: Columbia University Press.

Bondi, L. (1992) 'Gender divisions and gentrification', *Transactions, Institute of British Geographers*, 16: 190–8.

Burnett, P. (1973) 'Social change, the status of women and models of city form and development', *Antipode* 5: 57–61.

Butler, T. and Hamnett, C. (1994) 'Gentrification, class and gender: some comments on Warde's "Gentrification as consumption"', *Environment and Planning D: Society and Space*, 12: 477–93.

Campbell, B. (1993) *Goliath: Britain's Dangerous Places*. London: Virago.

Connell, R. (1995) *Masculinities*. Cambridge: Polity.

Davies, M. (1977) *Life as We Have Known It*. London: Virago (first published 1915).

Dear, M. (1986) 'Postmodernism and planning', *Environment and Planning D: Society and Space*, 4: 367–84.

Donald, J. (1992) 'Metropolis: the city as text', in R. Bocock and K. Thompson (eds), *Social and Cultural Forms of Modernity*. Cambridge: Polity.

Donnison, D. and Soto, P. (1980) *The Good City: A Study of Urban Development and Policy in Britain*. London: Heinemann.

England, K. (1994) 'From "Social justice and the City" to women-friendly cities? Feminist theory and politics', *Urban Geography*, 15: 628–43.

Garber, J. and Turner, R. (eds) (1994) *Gender in Urban Research*. Urban Affairs Annual Review 42. London: Sage.

Glendinning, V. and Millar, J. (ed.) (1992) *Women and Poverty in Britain: The 1990s*. London: Harvester Wheatsheaf.

Greed, C. (1994) *Women and Planning*. London: Routledge.

Green, A.E. (1994) *The Geography of Poverty and Wealth*. Coventry: Institute of Employment Research, University of Warwick.

Gregson, N. and Lowe, M. (1994) *Servicing the Middle Classes: Class, Gender and Waged Domestic Labour in Contemporary Britain*. London: Routledge.

Hakim, C. (1993) 'The myth of rising female employment', *Work, Employment and Society*, 7: 97:120.

Hakim, C. (1996) *Female Heterogeneity and the Polarisation of Women's Employment*. London: Athlone.

Haraway, D. (1991) *Simians, Cyborgs and Women: The Reinvention of Nature*. London: Free Association Books.

Harvey, D. (1989) *The Condition of Postmodernity*. Oxford: Blackwell.

Harvey, D. (1992) 'Social justice, postmodernism and the city', *International Journal of Urban and Regional Research*, 16: 588–601.

Hayden, D. (1984) 'What would a non sexist city be like?', in C. Stimpson, E. Dixler, M. Nelson and K. Yatrakis (eds), *Women and the North American City*. Chicago: University of Chicago Press. pp. 167–84.

Hayden, D. (1995) *The Power of Place*. Cambridge, MA: MIT University Press.

Heron, L. (ed.) (1993) *Streets of Desire: Women's Fictions of the Twentieth Century City*. London: Virago.

Jaggar, A.M. (1990) 'Sexual difference and sexual inequality', in D. Rhode (ed.), *Theoretical Perspectives on Sexual Difference*. London: Yale University Press. pp. 239–54.

Joseph Rowntree Foundation (1995) *Income and Wealth*, vols I and II. York: Joseph Rowntree Foundation.

Knopp, L. (1992) 'Sexuality and the spatial dynamics of capitalism', *Environment and Planning D: Society and Space*, 10: 651–70.

Lauria, M. and Knopp, L. (1985) 'Towards an analysis of the role of gay communities in the urban renaissance', *Urban Geography*, 5: 152–69.

Lawless, P., Martin, M. and Hardy, S. (1996) *Unemployment and Social Exclusion: Landscapes of Labour Inequality*. London: Jessica Kingsley.

Laws, G. (1994) 'Social justice and urban politics: an introduction', *Urban Geography*, 15: 603–11.

Little, J. (1993) *Gender Planning and the Policy Process*. London: Pergamon.

Lyons, M. (1996) 'Employment, feminisation and gentrification in London, 1981–93', *Environment and Planning A*, 28: 341–56.

McDowell, L. (1991) 'Life without father and Ford: the new gender order of postfordism', *Transactions of the Institute of British Geographers*, 16: 400–20.

McDowell, L. (1993a) 'Space, place and gender relations: part 1 – feminist empiricism and the geography of social relations', *Progress in Human Geography*, 17: 157–79.

McDowell, L. (1993b) 'Space, place and gender relations: part 2 – identity, difference, feminist geometries and geographies', *Progress in Human Geography*, 17: 305–18.

McDowell, L. (1994) 'Social justice, organisational culture and workplace democracy: cultural imperialism in the City of London', *Urban Geography*, 15: 661–80.

McDowell, L. (1995) 'Understanding diversity: the problem of/for "theory"', in R. Johnston, P. Taylor and M. Watts (eds), *Geographies of Global Change*, Oxford: Blackwell. pp. 280–94.

McDowell, L. (1999) *Gender, Identity and Place: Understanding Feminist Geographies*. Cambridge: Polity.

McDowell, L. (2000) 'The trouble with men?', *International Journal of Urban and Regional Research*, 24: 201–9.

Martin, R. (1995) 'Income and poverty inequalities across regional Britain: the North–South divide lingers on', in C. Philo (ed.), *Off the Map: The Social Geography of Poverty in the UK*. London: Child Poverty Action Group. pp. 23–44.

Modjeska, D. (1989) *Inner Cities: Australian Women's Memory of Place*. Victoria, Australia and Harmondsworth: Penguin.

Mort, F. (1995) 'Archaeologies of city life: commercial culture, masculinity, and spatial relations in 1980s London', *Environment and Planning D: Society and Space*, 13: 573–90.

Murray, C. (1994) *Losing Ground: American Social Policy 1950–80* (2nd edn). New York: Basic Books.

Nelson, M. and Smith, J. (1999) *Working Hard and Making Do*. Berkeley: University of California Press.

O'Kane, M. (1994) 'Cruising, abusing or on the game?', *The Guardian* 23 July, p. 25.

Pahl, R. (1984) *The Division of Labour*. Oxford: Blackwell.

Pateman, C. (1988) *The Sexual Contract*. Cambridge: Polity.

Pateman, C. and Gross, E. (1987) *Feminist Challenges: Social and Political Theory*. Boston: Northeastern University Press.

Pearson, R. (1992) 'Gender matters in development', in T. Allen and A. Thomas (eds), *Poverty and Development in the 1990s*. Oxford: Oxford University Press. pp. 291–312.

Phillips, A. (1987) *Feminism and Equality*. Oxford: Blackwell.

Phillips, A. (1989) *Divided Loyalties: Dilemmas of Sex and Class*. London: Virago.

Phillips, M. (1995) 'When a community stands up for itself', *Observer*, 17 July: 25.

Phillips, M. (1996) 'Bye hooker, bye crook', *Observer*, 4 February: 7.

Philo, C. (ed.) (1995) *Off the Map: The Social Geography of Poverty in the UK*. London: Child Poverty Action Group.

Pollock, G. (1988) *Vision and Difference: Femininity, Feminism and Histories of Art*. London: Routledge.

Rex, J. and Moore, R. (1967) *Race, Community and Conflict*. London: IRR Press.

Rhode, D. (ed.) (1990) *Theoretical Perspectives on Sexual Difference*. London: Yale University Press.

Rieder, I. (1991) *Cosmopolis: Urban Stories by Women*. Dublin: Attic Press.

Robson, B. (1994) 'No city, no civilization', *Transactions of the Institute of British Geographers*, 19: 131–41.

Rose, D. (1989) 'A feminist perspective on employment restructuring and gentrification: the case of Montreal', in J. Wolch and M. Dear (eds), *The Power of Geography*. London: Unwin Hyman. pp. 118–38.

Rose, G. (1993) *Feminism and Geography*. Cambridge: Polity.

Sassoon, A.S. (ed.) (1987) *Women and the State*. London: Hutchinson.

Scott, A.M. (1994) *Gender Segregation and Social Change*. Oxford: Oxford University Press.

Segal, L. (1999) *Why Feminism?* Cambridge: Polity.

Sennett, R. (1977) *The Fall of Public Man*. New York: Knopf (reissued 1993, London: Faber).

Sennett, R. (1993) *The Conscience of the Eye: The Design and Social Life of Cities*. London: Faber.

Smith, J. (1989) *Misogynies*. London: Faber.

Valentine, G. (1989) 'The geography of women's fear', *Area*, 21: 385–90.

Valentine, G. (1993) 'Negotiating and managing multiple sexual identities', *Transactions of the Institute of British Geographers*, 18: 237–48.

Van Vliet, w. (ed.) (1988) *Women, Housing and Community*. Aldershot: Avebury.

Walkerdine, V. (1985) 'Dreams from an ordinary childhood', in L. Heron (ed.), *Truth, Dare or Promise: Girls Growing up in the Fifties*. London: Virago. pp. 63–78.

Walkowitz, J. (1992) *City of Dreadful Delight*. London: Virago.

Warde, A. (1991) 'Gentrification as consumption: issues of class and gender', *Environment and Planning D: Society and Space*, 9: 223–32.

Watson, S. and Gibson, K. (1995) *Postmodern Cities and Spaces*. Oxford: Blackwell.

Wilkinson, H. (1994) *No Turning Back: Generations and the Genderquake*. London: Demos.

Willett, R. (1996) *The Naked City: Urban Crime Fiction in the USA*. Manchester: Manchester University Press.

Williams, R. (1989) 'Metropolitan perceptions and the emergence of modernism', in *The Politics of Modernism*. London: Verso. pp. 37–48.

Wilson, E. (1988) *Hallucinations: Life in the Postmodern City*. London: Radius.

Wilson, E. (1991a) 'The invisible flâneur', *New Left Review*, 191: 90–110.

Wilson, E. (1991b) *The Sphinx and the City*. London: Virago.

Wolff, J. (1985) 'The invisible *flâneuse*: women and the literature of modernity', *Theory, Culture and Society*, 2(3): 37–46.

Woolf, V. (1929) *A Room of One's Own*. London: Hogarth.

Young, I.M. (1990) *Justice and the Politics of Difference*. Princeton, NJ: Princeton University Press.

Zukin, S. (1995) *The Cultures of Cities*. Oxford: Blackwell.

14

Urban Crime in the USA and Western Europe: A Comparison

PAULA D. McCLAIN

New York, London, Rome, Paris – all cities that bring to mind art, culture, sophistication and international finance. Cities are important, yet many have not fared well over the past several decades. Since the 1960s, the fortunes of US cities, both big and small, have waxed and waned with changes in the domestic political climate, domestic and international economic downturns and upswings, and increasing national and local debt burdens. Many forces affecting the well-being of urban centres have been beyond the control of local elected officials. Yet many problems have been the result of neglect or outright malfeasance of local officials. Many of these problems have led to the use of the word 'crisis' to preface discussions of urban problems.

The 1960s saw many US cities in flames and communities disrupted by social unrest among racial minorities excluded from the political process, oppressed and abused by local police forces, and abandoned by the political system. The 1960s was a time of 'crisis of civil disorders'. The 1970s brought 'fiscal crisis' to many major cities. New York City and Cleveland, among others, went bankrupt, drastically reduced city services, and were saved by federal government-backed loans and guarantees. The 1980s brought massive urban development, with a booming economy, only to see real estate prices plummet with the massive collapses of savings and loan institutions. The urban 'crisis' of the 1990s was that of urban crime and violence, particularly homicide (McClain, 1992, 1993; Rose and McClain, 1990). Despite the national trend downward in homicide and urban crime in the USA, the extent and increasing frequency of the violence overwhelmed many cities.

While acute in some American cities, violent crime has also ebbed and flowed in many European cities. Many, while not experiencing the decline of the urban core nor the levels of violent crime that are characteristic of many American cities, have nevertheless experienced increases in urban violence. Sometimes many of the same factors that have influenced increases in American cities, for example, drugs, firearms and gangs, have played a role in rising crime levels in European cities also.

This chapter focuses on issues of urban crime in both the United States and Western Europe, with special emphasis on the similarities and/or differences in the types of crimes committed and the levels of violence. The chapter is divided into six sections. The first analyses crime and crime rates in the USA and fear of crime, while the second examines crime and crime rates in selected Western European countries overall, and fear of crime in Europe, but in less depth than the discussion of the subject in the USA. The following two sections discuss the concept of dangerous places and crime (crime and crime rates in selected US and European cities are compared) and the concept and use of the term 'underclass'. The fifth section examines the two paradigms through which crime and crime control policies have been shaped, and while emphasis is on the debate in the USA, examples from the European context are included. Finally the chapter concludes by looking at policies and programmes designed to control and reduce violent crime in Europe and the USA.

CRIME AND CRIME RATES IN THE USA

How much violent crime is there in the USA? The Federal Bureau of Investigation (FBI) defines violent crime as consisting of four offence categories – murder and non-negligent manslaughter, forcible rape, robbery and aggravated assault. All violent crimes involve force or threat of force (Federal Bureau of Investigation, 1997: 10). In 1992, based on crime victimization surveys, there were 6.6 million violent victimizations, including 23,760 homicides (55 per cent killed by handguns), 141,000 rapes, 1.2 million robberies and 5.3 million assaults.[1] Five per cent of all US households had a member victimized by violence and Americans had a greater chance of being a victim of a violent crime than of being injured in a car accident (Bureau of Justice Statistics, 1994). While overall aggregate crime rates were decreasing, for some segments of the population, crime was not decreasing. In 1992, the violent crime rate for blacks was the highest it has ever been, 50 per 1,000 persons, compared with a decreased 29.9 per 1,000 persons for whites. Juvenile victimization rates are increasing. The overall aggregate victimization rate for those aged 12–15 reached its highest level in 1992, as did the victimization rate for those aged 16–19. Teenage black males have the highest victimization rates for violent crime (113 per 100,000), while elderly white females had the lowest rate (3 per 100,000). Teenagers overall had very high rates, 90 per 1,000 for teenage white males, 55 per 1,000 for teenage white females, and 94 per 1,000 for teenage black females (Bureau of Justice Statistics, 1994).

In 1991, homicide was the tenth leading cause of death for all Americans, but for black males and females aged 15–24, homicide was the leading cause of death. For whites aged 15–24, homicide was the third leading cause of death, exceeded only by accidents and suicide (Bureau of Justice Statistics, 1994). Black males had the highest homicide rate (72 per 100,000 population), followed by black females (14 per 100,000), white males (9 per 100,000), and white females (3 per 100,000). For all age groups, however, black males aged 15–24 had the highest homicide rate (159 per 100,000).

By 1994, the total number of homicides nationwide had dropped slightly to 23,305, a 5 per cent decrease from 1993. (By 1995, the number of homicides had dropped even more to 21,597, which was down 7 per cent from 1994 and 13 per cent from 1991). Nevertheless, the number of violent victimizations increased to 10.9 million – 430,000 rapes/sexual assaults, 1.3 million robberies, more than 2.5 million aggravated assaults and 6.6 simple assaults (Bureau of Justice Statistics, 1997.) However, aggregate figures are deceptive. Fox argues that there are two crime trends in the USA – one for the young, one for the mature – which are moving in opposite directions (Fox, 1996). As with earlier years, violent deaths and violent crime were increasing among certain segments of the US population, as were offending rates. Black males aged 15–24 still had the highest homicide rate and were most often homicide offenders in these incidents. Yet, rates of homicide offending and victimization continued to increase substantially between young white and black males.

The rates of violent victimization for white and black teenage males aged 12–15 had become more similar, a rate of 135.6 victimizations per 1,000 persons for white teens and 141.6 victimizations for black teens in this age group. White male teens aged 16–19 actually had a higher victimization rate, 146 per 1,000 persons, than did black male teens in the same age category, 124.8 per 1,000 persons (Bureau of Justice Statistics, 1997: 15). The rate of killing by white male teenagers has doubled since 1985 and the rate by black male teenagers has tripled (Fox, 1996). Overall, the rate of homicide committed by teenagers aged 14–17 has increased 172 per cent from 7.0 per 100,000 in 1985 to 19.1 per 100,000 in 1994 (Fox, 1996). Juvenile violence is on the upswing and, if not checked, according to Fox (1996), the number of 14–17-year-olds who will commit murder could increase from 4,000 per year in 1994 to nearly 5,000 or more annually by 2005 because of changing demographics. The arrest in 1997 of six white teenagers in Mississippi for planning and committing the murder of the mother of one of the youths and the shooting up of their school which left two of their classmates dead is but one illustration of the daily stream of youth violence committed in the USA.

Crime in the USA is primarily intra-racial; one is more likely to be victimized by someone of the same race than someone of another race. In 1996, 93 per cent of black murder victims were slain by black offenders, and 85 per cent of white murder victims were slain by white offenders (Federal Bureau of Investigation, 1997). The reality of crime in the USA is that the risk of criminal victimization is race-specific, with the risk to blacks, who principally reside in urban centres, being substantially higher than the risk to whites. Table 14.1 shows the black homicide risk in 14 major cities in 1980 and 1990, compared with the risk for the total population. In all 14 cities, the black homicide risk is higher than the risk to the total population in both years. In several cities, for example, Washington, DC, Baltimore, Dallas, Houston, and Atlanta, black homicide risk increased substantially over the period.

Blacks are six times more likely than whites to be victims of homicide (Rose and McClain, 1990) and two-and-a-half times more likely than whites to be victims of rape. For robbery, the black victimization rate is three times that for whites, and the black rate for aggravated assault is one-and-a-half times that for whites (Rose, 1981; Silberman, 1978). In an article published in 1989 the Federal Bureau of Investigation said, based on 1987 data, that a non-white male born in the USA in these times had a one in 38 chance of being murdered, while the risk to the aggregate population was one in 177. For men already 20 years of age, another report stated that

Table 14.1 *Total risk and black homicide risk in major US cities, 1980 and 1990*
(per 100,000 persons in the population)

City	Total risk			Black risk		
	1980	1990	% diff.	1980	1990	% diff.
Atlanta	38.6	60.7	+57.3	46.3	78.0	+68.5
Baltimore	28.6	42.7	+49.3	40.6	65.2	+60.6
Chicago	28.6	NR	–	45.1	NR	–
Cleveland	45.7	31.6	–14.1	76.8	53.1	–30.9
Dallas	33.6	42.5	+26.5	55.3	91.6	+65.6
Detroit	45.7	62.3	+36.3	59.3	70.2	+18.4
Houston	37.1	37.7	+1.6	45.9	71.2	+55.1
Los Angeles	33.8	30.2	–10.7	87.3	54.7	–37.3
Memphis	24.2	33.4	+38.0	37.7	55.9	+48.3
New Orleans	37.5	49.9	+33.1	56.8	73.4	+29.2
New York	23.9	26.6	+4.2	46.8	45.8	–2.1
Philadelphia	26.4	32.9	+24.6	51.8	60.8	+17.4
St Louis	46.8	48.8	+4.7	91.6	89.7	–2.0
Washington	36.1	79.3	+119.7	39.9	113.9	+185.5

diff., difference; NR, not reported.

Sources: 1980: Harold M. Rose (1986) 'Can we substantially lower homicide risk in the nation's larger black communities?', *Secretary's Task Force on Black and Minority Mental Health* (Washington, DC: US Department of Health and Human Services). 1990: calculated by Steven Tauber, research assistant, from Supplemental Homicide Reports, 1990.

non-whites had a one in 41 chance of being murdered, while whites had a one in 224 chance (*Arizona Republic*, 1989). In 1996, there were an estimated 19,645 murders, with contributing agencies providing data on 18,108 of the victims. Of the latter figure, 49 per cent of the victims were black (Federal Bureau of Investigation, 1997: 14). On the other side, in 1996 blacks represented slightly more than one-half, 52 per cent, of those arrested for murder and non-negligent manslaughter (Federal Bureau of Investigation, 1997: 14); yet blacks comprise only 12 per cent of the population.

What accounts for the disproportionate representation of blacks as both victims and offenders of violent crime? Moreover, what accounts for the increase in black and white juveniles engaged in violent crime? Two divergent explanations have emerged. On the one hand, some conservative scholars suggest these increases in black violence are the result of the liberal policies of the Great Society era (Murray, 1984), and individual personality attributes that are more important contributors to escalating crime rates than are external factors (Wilson and Herrnstein, 1985). On the other hand, others argue that the problem stems from external forces that affect black urban residents, such as centuries of racism, persistent poverty and the structure of the economy of urban environments in which most blacks reside (Brenner and Swank, 1986; Rose and McClain, 1990; Turner et al., 1981).

The reasons most frequently offered as explanations for increasing youth vulnerability are:

1 an increasing rate of gang involvement;
2 the growth of the crack cocaine epidemic;
3 the easy availability of guns.

Others attribute these behaviours, that make violent victimization more likely, to the rise of a youth street culture and/or an oppositional culture that rejects selected mainstream values (see Anderson, 1994; Fordham, 1996). Fox (1996) attributes the increase in black and white juvenile crime to drugs, greater access to firearms, a more casual attitude about violence, and the negative socializing forces of gangs and the media. Coupled with these negative influences is the breakdown of positive socializing forces, such as family, school, religion and neighbourhood. Fox also recognizes that part of the breakdown in the positive socializing forces are attributable to funding cuts in programmes that support these forces.

Fear of Crime

Fear of crime among US residents has decreased slightly over the years as violent crime has declined. A 1997 Gallup poll identified that 38 per cent of Americans were afraid to walk alone at night, which represents a decrease from 45 per cent in 1983 and 44 per cent in 1992 (*Sourcebook of Criminal Justice Statistics Online*, 1995: Table 2.32). Yet, despite this overall decline, fear of crime, like victimization rates, has racial dimensions. Blacks and Latinos are more worried about crime than are whites. Approximately two-fifths (40.9 per cent) of black Americans and more than one-third of Latino Americans very frequently, or somewhat frequently, worried about being murdered. In contrast, only one-fifth of white Americans very frequently, or somewhat frequently, worried about being murdered. About the same proportion of black Americans (42.8 per cent) and close to half of Latinos (47.4 per cent) very frequently, or somewhat frequently worried about being knifed or shot. Less than one-third (29.1 per cent) of white Americans worried about being knifed or shot. Clearly, black and Latino Americans worry about different levels of violent victimization more than do white Americans.

Racial differences, although not as acute, also exist regarding fear of home burglary. Two-thirds (66.6 per cent) of Latinos and more than half (53.6 per cent) of blacks very frequently or somewhat frequently worry about their homes being burgled. While still substantial, less than half of whites (46.5 per cent) frequently, or somewhat frequently, worry about home burglary. Racial differences abate, however, on the question of sexual assault and home burglary. Whites, blacks, and Latino Americans are worried in similar proportions about someone in their family being sexually assaulted. Among whites, 45.3 per cent are worried, as are 47.3 per cent of Latinos and slightly more than half, 51.8 per cent, of blacks (*Sourcebook of Criminal Justice Statistics Online*, 1995: Table 2.0026).

Fear of crime is present within the US population, but in the aggregate fear of crime has been decreasing. Nevertheless, those at greatest risk, blacks and Latinos, are more fearful than are white Americans. One could argue that the level of fear is an accurate reflection of the race-specific victimization rates, and we should expect to see these differences.

CRIME AND CRIME RATES IN EUROPEAN COUNTRIES

While most Europeans may think of violent crime as a US phenomenon, violent crime is also problematic for several European countries. Most European nations have experienced a sharp increase in registered crime over the past three decades (van Dijk, 1991). According to van Dijk, between 1985 and 1986 registered crime in France, England and Wales, the Federal Republic

of Germany (FRG) and the Netherlands reached a rate of 7,000 registered crimes per 100,000 inhabitants. (Van Dijk estimates that the actual increase is probably higher because many crimes may not have been reported to the police.) Property crimes were particularly high in Sweden and the Netherlands during this period, and recorded crimes of violence were higher in Sweden, the FRG and England and Wales. European countries with victimization rates of about 25 per cent were the Netherlands, Spain and the FRG. Countries with victimization rates of about 20 per cent were Scotland, England and Wales, France and Belgium. Those with rates around 15 per cent were Northern Ireland, Switzerland, Norway and Finland (van Dijk, 1991). Countries with the lowest rates were characterized by lower levels of urbanization, meaning that many residents live in villages and small towns and few in cities of over 100,000 persons (van Dijk, 1991).

In addition to the countries already mentioned, France also experienced increases in crime. Between 1972 and 1984, recorded crimes in France increased by 119 per cent. Crime rates grew steadily, with cities experiencing a doubling in burglaries and a quadrupling of robberies involving female victims (de Liège, 1991). In addressing increases in European crime rates overall, van Dijk concludes;

> Crime has gone up from 1955 onwards across western Europe. In Sweden the upward trend started immediately after the war. At the end of the 1980s crimes like theft, burglary, simple assault and indecent assault place a heavy burden on the inhabitants of the larger European cities in particular . . . (1991: 31)

Table 14.2, using data from Interpol, displays a comparison of violent crime rates for the USA and selected European countries for 1992.[2] While the United States homicide rate is among the highest (9.3 per 100,000), others exceed it by some way. Northern Ireland, the Netherlands and Scotland have higher homicide rates, 30.64, 24.88 and 11.41 per 100,000 persons respectively. Sweden, with 8.39 homicides per 100,000 persons, is close to the USA and ranks fifth overall.

Within the European context, Northern Ireland, the Netherlands, Scotland and Sweden have the highest homicide rates. Sweden has the highest rape rate (19.42 per 100,000), followed by Denmark (10.77 per 100,000), the Netherlands (10.82 per 100,000), and Scotland (10.19 per 100,000). Spain has the highest robbery rate (163.91 per 100,000), followed by the Netherlands (124.74 per 100,000), France (121.79 per 100,000) and Scotland (121.77 per 100,000). England leads in serious assaults (362.10 per 100,000), with the Netherlands (191.75 per 100,000). Denmark (169.23 per 100,000), Luxembourg (131.05 per 100,000) and Scotland (123.41 per

Table 14.2   *Crime rates per 100,000 persons for violent crime offence categories for the USA and selected European countries, 1992*

Country	Homicide	Rape	Robbery	Aggravated/ serious assault[a]
USA	9.30	42.80	263.60	441.80
Austria	2.61	7.09	55.08	2.52
Belgium	2.70	7.81	98.71	117.91
Denmark	4.59	10.77	87.91	169.23
England/Wales[b]	2.52	7.98	89.36	362.10
Finland	0.57	7.30	46.29	38.81
France	4.70	9.31	121.79	96.67
Germany	4.12	7.82	70.40	104.77
Greece	2.54	2.69	14.80	66.42
Ireland	0.71	3.61	72.70	18.04
Italy	5.77	NR	55.29	36.20
Luxembourg	1.32	5.26	71.58	131.05
Northern Ireland[b]	30.64	9.87	117.71	30.19
Norway	2.06	8.89	23.24	44.95
Netherlands	24.88	10.82	124.74	191.75
Portugal	4.30	1.30	4.10	7.04
Scotland[b]	11.41	10.19	121.77	123.41
Spain	2.29	4.01	163.91	23.76
Sweden	8.39	19.42	71.55	42.51
Switzerland	2.64	4.58	35.66	53.59

[a]Aggravated assault is the term used in US reporting, serious assault is the term used in European reporting. Definitions are similar.
[b]Figures are for 1991.

Sources; Federal Bureau of Investigation, *Crime in the United States, 1996* (Washington, DC: US Government Printing Office, 1997); International Criminal Police Organization, *International Crime Statistics 1991–1992*. (Lyons, France: International Criminal Police Organization, Interpol General Secretariat, 1992)

100,000). What is clear from these data is that Scotland, while not the highest on every violent crime dimension, is consistently high in each category. Moreover, although several European countries have homicide rates higher than the USA, the latter has higher victimization rates for rape, robbery and aggravated assault.

What accounts for the high homicide rates in Northern Ireland, the Netherlands and Scotland, the high rape rates in Sweden, Denmark, the Netherlands and Scotland, the high robbery rate in the Netherlands, and the high serious assault rate in England? Clearly, countries differ in many respects, consequently the reasons may be country-specific. Moreover, the causes of violent crime are complex and simple answers are not sufficient. However, the available literature may provide some insight and tentative answers.

Norway and Denmark have experienced an increase in motorcycle gang violence associated with control of territory and the distribution of drugs. The violence has begun to spill over into Sweden and Finland as well (Zuzevich, 1997). Several members of rival gangs have been murdered, and each murder often generates a retaliation murder. Organized crime is growing at a fast pace throughout the European Union. The collapse of the Berlin Wall, the increases in migration from Eastern European countries to Western Europe, and the loosening of borders within the EU have created an environment in which organized crime syndicates flourish (Solomon, 1995). The result has been increases in drug trafficking, prostitution, importation of illegal workers and other associated crimes (Bovenkerk, 1993).

Research on crime among ethnic minorities and other immigrants from other European countries in the Netherlands and Sweden has come to similar conclusions. While crime and victimization rates are higher among ethnic minority and other immigrant populations, these groupings are also the poorest, have a greater proportion of youth in their populations, have disproportionately higher unemployment rates, have lower education levels and are the objects of societal and police discrimination. Moreover, most of the crime is intra-ethnic or intra-immigrant. Ethnic minorities and other immigrants are more likely than others to be the victims of crime and the perpetrators are more likely to be members of their own group (Junger-Tas, 1997; Martens, 1997).

Organized crime, sometimes associated with terrorist activities, has also been prominent in Northern Ireland (Maguire, 1993). Northern Ireland also has higher levels of unemployment, a weaker private sector, and higher levels of low-paid jobs than other areas in the UK (Maguire, 1993), and many children grow up in poverty (Kilpatrick et al., 1995). In 1993, more than one-quarter (27 per cent) of Northern Ireland's population was under the age of 18, and many of these children were born outside marriage and lived in single-parent homes. Moreover, most economically deprived and socially disadvantaged areas were those that experienced high and increasing levels of political conflict and violence (Kilpatrick et al., 1995). Research suggests a close relationship between growing up in areas with high levels of violent political conflict and criminal behaviour and juvenile offending (Farrington, 1995; Kilpatrick et al., 1995).

Some countries have also experienced an increase in juvenile crime, particularly among second-generation immigrant populations. In Germany, the Netherlands and Switzerland, the number of young people under the age of 15 increased from 217,000 in 1960 to 1,459,000 in 1981 (Sun and Reed, 1995). The problems of youthful delinquency, disadvantageous living conditions, prejudice and discrimination, stigmatization and marginalization in urban slums have been identified as contributing to increasing crime rates (Junger and Polder, 1992; Sun and Reed, 1995). Garland (1996) estimates that over the past 30 years, high crime rates have become a part of everyday life in Britain. While experiencing a downturn in juvenile crime, England has seen an increase in offending among those aged 17–20 years of age (Matthews, 1995). This category of youth is increasingly involved in opportunistic armed robberies and organized drug dealing. Anti-immigrant violence and prejudice coupled with the rise of right-wing extremism have also contributed to increases in violent crimes against immigrants in several countries, particularly Germany, France and Britain (Bovenkerk, 1993; Roberts, 1994).

Fear of Crime

The literature on the fear of crime in European countries is more diffuse than in the United States, as is the survey data on the question. Nevertheless, the literature suggests that fear of crime among residents has been on the increase in several European countries. The 1993 Scottish Crime Survey found that the percentage of people who said they felt 'a bit' or 'very' unsafe walking alone after dark (41 per cent) had risen significantly from the 1988 survey in which 30 per cent felt 'a bit' or 'very'

unsafe (Monaghan, 1997). A 1994 MORI poll in England showed that 77 per cent of the respondents were afraid of having their home or possessions burgled and 56 per cent feared having their home or possessions vandalized (Loveday, 1994). The 1992 British Crime Survey identified that the highest victimization rates occurred in the most deprived areas, usually large urban public housing estates, and that the fear of crime among its residents may not exaggerate the likelihood of victimization (Loveday, 1994). Other literature suggests that the fear of crime is also on the increase in France (de Liège, 1991).

Despite the focus on national trends and rates in both the USA and Western Europe in this section, violent crime occurs at the local level, most often in urban environments. This is the case for both the USA and Europe. The next section examines the relationship between urban environments and crime, and examines crime and crime rates in selected US and European cities.

URBAN ENVIRONMENT AND DANGEROUS PLACES

While not exclusively an urban phenomenon, most crime and violence occur in cities and urbanized areas. Considerable scholarly research has been undertaken in the field of urbanism and urbanization and their impacts on individual level behaviour. These studies have developed a variety of findings and conclusions that are instructive to the understanding of patterns of urban violence. This section will review the literature on the effects of urbanism on behaviour, particularly as it relates to criminal behaviour.

Urban areas are characterized as more prone to violence and less safe than either suburban or rural areas. The question of what makes some neighbourhoods more dangerous places to live in is difficult to answer, but several attempts have been made. The classic works of Wirth (1938), Shaw and McKay (1942, 1969) and Simmel (1951) suggest that certain aspects of the urban environment promote social disorganization and individual alienation. These two conditions may lead to deviant social behaviour that could result in urban areas being more dangerous places to live than less urban environments. Included among the elements that contribute to deviant behaviour are size, density, heterogeneity, low economic status, residential mobility and family disruptions.

Sampson and Groves (1989) tested Shaw and McKay's hypothesis on data from England and Wales and found that the social-disorganization theory provided a good explanation of teenage crime rates in urban areas in those areas of Britain. The fact that the theory was a good explanation of juvenile crime in a country other than the USA suggests that the effects of some aspects of urban environments on individual level criminal behaviour may not be country-specific. Others have also tested Shaw and McKay's theory of a relationship between social disorganization and crime (Bottoms and Wiles, 1986; Bursik and Grasmick, 1993; Ingram, 1993; Reiss, 1986; Taylor and Covington, 1993).

While taking issue with the contention that urbanism results in isolation and anonymity, Fischer (1976a, 1976b) argues that the concentration of large numbers of people who share the same attitudes, including the acceptance of illegal activity, contributes to the development of subcultures which sanction many types of deviant behaviour. Fischer argues that the 'deviance and disorganization' found in urban areas are the result of the development of distinctive subcultures that encourage or tolerate behaviours that the broader society considers to be deviant, rather than the results of individual alienation. However, Fischer does emphasize that ecological variables – population density, heterogeneity and size – have great influence on social life.

A body of literature has examined the consequences of urbanism on crime and criminal victimization. Cohen and Felson (1979) view an individual's susceptibility to criminal victimization as stemming both from the victim's characteristics and behaviours, and the context in which they are found. It has also been suggested that neighbourhoods (communities) are not random aggregates of individuals, but are 'collective representations' of people and environment (Hunter, 1974: 178; Rees, 1970). In support of this position, Kornhauser (1978) specifies that the reasons why some areas are more violence-prone than others may depend on the sociological characteristics of communities, or on the characteristics of individuals selectively aggregated into communities; in effect, it may be contextual. In the same vein, Roncek (1981) concluded that environmental factors or residential areas are important to explaining where crimes occur, suggesting that changes in the environments of residential areas can affect crime. Overall, these studies suggest that there are elements of urban environments that may promote criminal behaviour among individual residents.

US AND EUROPEAN CITIES: A COMPARISON

In 1991, four of the nation's largest cities – San Diego, Dallas, Phoenix and Los Angeles – had record high numbers of homicides. New York City, Chicago and San Antonio all recorded their second-highest ever homicide totals, and while not its highest, Detroit experienced a 5 per cent increase over its 1990 homicide level (*New York Times*, 1992). Among smaller cities, Washington, DC experienced one of its most violent years in history in 1991, with recorded homicides of 490, up from 483 in 1990. Washington, DC., until recently, had the dubious distinction of being the nation's 'homicide capital'. In the USA, homicide is now the leading cause of death among black males ages 15–24, and black females 25–44.

Over the past 25 years, the large urban centres of the USA have experienced an increase in the incidence of violent crime, particularly homicide. Explanations for the increase in the latter are varied. Some attribute the rise to the widespread availability of handguns (Zimring, 1979), weakening attachment to traditional values (Waldron and Eyer, 1975), a demographic structure that tends to support violent behaviour (Wolfgang, 1978), cyclical behaviour (Archer and Gartner, 1976), increasing social isolation in urban neighbourhoods (Wilson, 1987), and urban ecological factors that contribute to violent behaviour (Rose and McClain, 1990). There is a consensus, however, that blacks tend to be disproportionately represented as both victims and offenders of criminal homicide (Rose and McClain, 1990). Typically, blacks reside in the city centre areas in which homicides often occur (Rose and Deskins, 1980). Studies also show that the homicide rate tends to increase with the size of place and density of population (Mayhew and Levinger, 1976). Homicide is also more frequent in areas with large concentrations of low socio-economic status persons, poor housing stock and high unemployment (Shin et al., 1977).

The reasons for this tremendous increase in urban violence in the 1990s were the introduction of crack cocaine in US cities beginning in 1985 and the development of entrepreneurial gangs associated with the distribution of the drug (Baumer 1994; Rose and McClain, 1990, 1998). Along with this drug enterprise came the introduction of massive fire power in automatic assault weapons. While much of the violence was aimed at rival drug gangs, many innocent bystanders were also killed. Drive-by shootings killed many neighbourhood residents who happened to be in the proverbial wrong place at the wrong time. In 1992, overall violent crime rates began to inch downward and have continued to decline. Yet, despite the decrease in violence overall, in some neighbourhoods in American cities violence levels have intensified.

Table 14.3 shows the number of violent crimes committed in 1989 and 1996 in 15 large urban centres and the changes in the number of violent crimes committed. Also, detailed in the notes to

Table 14.3 *Number of violent crimes in 15 large urban centres, 1989 and 1996*

1989

City	Population	Murder	Rape	Robbery	Assault
Atlanta	426,482	246	691	6,796	9,119
Baltimore	763,138	262	541	7,966	6,849
Chicago	2,988,260	742	NR	31,588	37,615
Cleveland	523,906	144	837	4,045	2,939
Dallas	996,320	351	1,185	9,442	10,250
Detroit	1,039,599	624	1,424	11,902	11,006
Houston	1,713,499	459	1,152	9,820	8,097
Los Angeles	3,441,449	877	1,996	31,063	43,361
Memphis	651,081	141	781	3,781	3,327
New Orleans	528,589	251	388	5,449	4,115
New York	7,369,454	1,905	3,254	93,377	70,951
Newark	313,839	107	376	5,310	4,547
Philadelphia	1,652,188	475	784	10,233	6,562
St Louis	405,066	158	330	4,220	7,936
Washington, DC	604,000	434	186	6,541	5,775

1996

City	Population	Murder[a]	Rape	Robbery	Assault[b]
Atlanta	413,123	196	392	4,805	8,306
Baltimore[c]	716,446	328	641	10,393	8,145
Chicago	2,754,118	789	NR	26,860	37,097
Cleveland	496,049	103	643	4,062	2,823
Dallas	1,060,585	217	740	6,122	9,201
Detroit	1,002,299	428	1,119	9,504	12,188
Houston	1,772,143	261	1,002	8,276	12,917[d]
Los Angeles	3,498,139	711	1,463	25,189	35,477
Memphis[c]	631,626	161	789	5,970	5,615
New Orleans[c]	488,300	351	390	5,700	4,580
New York	7,339,594	983	2,332	49,670	45,674
Newark	261,909	92	179	4,219	4,271
Philadelphia	1,528,403	414	704	15,485[d]	6,764[d]
St Louis	374,061	166	269	4,062	5,682
Washington DC	543,000	397	260[d]	6,444	6,310

NR, not reported.

[a]Rates calculated per 100,000 for four cities – New York (13.19); Los Angeles (20.32); Dallas (20.46); and Philadelphia (27.08).

[b]Rates calculated per 100,000 for this column – New York (622.29); Chicago (1,346); Detroit (1,216); Philadelphia (442.55); Los Angeles (1,014.16); Washington, DC (1,162.06); Houston (728.89); Baltimore (1,136.86); New Orleans (937.94); Memphis (888.97); Atlanta (2,010.53); Dallas (867.54); Cleveland (569.09); St Louis (1,519); and Newark (1,630.71).

[c]Lost population since 1989, but increased incidents in all violent crime categories.

[d]Increased since 1989

Source: Federal Bureau of Investigation, *Crime in the United States – 1989* (Washington, DC: US Government Printing Office, 1990a), pp. 69–117. Federal Bureau of Investigation, *Crime in the United States – 1996* (Washington, DC: US Government Printing Office, 1997), pp. 84–156

the table are the calculated assault rates in 1996 for all cities, and homicide rates in 1996 for selected cities. While crime in many cities has declined, several cities, for example, Baltimore, New Orleans and Memphis, despite population loss, have experienced an increase in all categories of violent crime since 1989. Also, Philadelphia has experienced significant increases in robberies and assaults, and Houston has experienced significant increases in aggravated assaults.

While not completely comparable because of possible reporting differences, Table 14.4 shows the number of violent crimes, as reported to the United Nations in its crime survey, in 11

Table 14.4   *Number of violent crimes in selected European cities, 1990–1994*

City	1990	1991	1992	1993	1994	1994 rate per 100,000
Vienna						
City population	1,487,577	1,508,120	1,536,442	1,561,333	1,589,052	
Homicides	82	68	76	75	84	5.28
Major assaults	32	57	50	81	77	4.84
Total assaults	8,053	8,785	9,537	9,081	9,950	626.15
Rapes	198	189	202	208	180	11.32
Robberies	1,558	1,579	1,847	1,659	1,696	106.73
Copenhagen						
City population	470,000	470,000	470,000	470,000	470,000	
Homicides	46	41	42	62	41	8.72
Major assaults	202	235	200	193	237	50.42
Total assaults	1,184	1,257	1,437	1,320	1,350	287.23
Rapes	51	77	76	70	52	11.06
Robberies	NR	719	671	716	610	129.78
Helsinki						
City population	490,629	492,400	497,542	501,514	508,588	
Homicides	75	61	90	67	74	14.55
Major assaults	425	368	276	206	239	46.99
Total assaults	4,898	4,708	4,070	3,674	3,879	762.69
Rapes	78	84	64	65	79	15.53
Robberies	1,019	916	696	570	691	135.86
Paris						
City population	2,152,423	2,152,185	2,158,330	2,155,593	2,157,459	
Homicides	NR	NR	NR	NR	NR	—
Major assaults	4,436	4,466	4,779	4,715	5,491	254.51
Total assaults	NR	NR	NR	NR	NR	—
Rapes	402	375	405	389	398	18.44
Robberies	1,425	1,348	1,480	1,751	1,460	67.67
Rome						
City population	2,791,354	2,773,889	2,723,327	2,687,881	2,667,052	
Homicides	158	115	87	101	109	4.08
Major assaults	NR	NR	NR	NR	NR	—
Total assaults	163	126	146	352	448	16.79
Rapes	35	23	24	27	31	1.16
Robberies	2,979	2,362	2,356	2,428	2,141	80.27
Madrid						
City population	3,120,732	3,120,732	3,084,673	3,084,673	3,084,673	
Homicides	NR	NR	NR	NR	NR	—
Major assaults	4,319	3,544	3,117	2,908	2,471	80.10
Total assaults	[Included in above figures]					
Rapes	195	275	248	273	242	7.84
Robberies	15,851	16,197	17,127	17,986	16,135	523.07
Stockholm						
City population	673,320	676,909	681,970	688,766	NR	[1993 rate]
Homicides	119	114	136	144	NR	20.90
Major assaults	236	238	300	267	NR	38.76
Total assaults	6,723	6,736	7,553	8,196	NR	1,189.95
Rapes	207	207	250	367	NR	53.28
Robberies	2,219	2,270	2,054	1,857	NR	269.61

Table 14.4   *cont.*

City	1990	1991	1992	1993	1994	1994 rate per 100,000
Zurich						
City population	341,300	343,300	345,200	343,000	342,900	
Homicides	35	46	30	34	39	11.37
Major assaults	15	20	34	46	49	14.28
Total assaults	543	548	682	641	551	160.68
Rapes	58	63	46	67	43	12.54
Robberies	563	796	925	906	702	204.72
London						
City population	7,227,344	7,264,052	7,373,100	7,391,139	7,423,777	
Homicides	184	184	175	160	169	2.27
Major assaults	959	640	855	712	657	8.84
Total assaults	33,720	36,025	35,797	37,475	43,782	589.75
Rapes	982	1,158	1,193	1,316	1,406	18.93
Robberies	18,156	21,927	23,617	24,537	25,547	344.12
Glasgow						
City population	689,210	688,600	684,260	681,470	680,000	
Homicides	169	183	235	195	203	29.85
Major assaults	NR	NR	NR	NR	NR	—
Total assaults	1,481	1,803	1,935	1,649	1,729	254.26
Rapes	83	102	146	112	122	17.94
Robberies	2,413	3,167	3,469	2,562	2,342	344.41

NR, Not reported.

Sources: The Fifth United Nations Survey of Crime Trends and Operations of Criminal Justice Systems, data set. Rates calculated by the author

European cities. Also shown is the calculated rate per 100,000 for each offence category for 1994. (The rates are likely to be affected by how tightly the boundary of the city is defined, e.g. whether or not it includes middle-class residential suburbs.) (Several cities that should be represented, such as Berlin, did not report their statistics.) Glasgow, Scotland has the highest homicide rate (29 per 100,000 in 1994) of the 10 European cities listed in Table 14.4. In fact, its homicide rate has been consistently higher than the other cities for the 1990–1994 period. As a comparison, London, England, with a population of more than seven million, had a 1994 homicide rate of 2.27 per 100,000. Putting Glasgow's homicide rate within the US context, its 1994 homicide rate was higher than that of New York City (13.19 per 100,000)[3], Los Angeles (20.32 per 100,000), Dallas (20.46 per 100,000) and Philadelphia (27.08 per 100,000), all cities associated in the public mind with high rates of violence.[4]

In 1984, Glasgow's homicide rate was 4.4 per 100,000 (McClintock and Wikström, 1990); thus in the decade that followed, the homicide rate in Glasgow increased more than seven times. This seven-fold increase occurred as Glasgow's population dropped from 850,000 in 1984 to 680,000 in 1994. A study of violent crime in Edinburgh, Scot-

land found that offending rates and victimizations were related to the proportion of the population that was foreign-born, number of vacant residences, single parents and unemployed residents (McClintock and Wikström, 1992). Although cities differ in size, regional location, manufacturing base, population distribution and other dimensions, it is possible that similar factors influence offending and victimization rates in Glasgow also. A study of crime in Scotland overall found a strong relationship between unemployment and crime rates (Reilly and Witt, 1992).

While 1994 data were not reported, Stockholm, Sweden has the second highest rate of violent crime (20.9 per 100,000), based on 1993 data. An analysis of crime trends in Scotland and Sweden for 1950 to 1984 identified that violent crime was on the rise in both countries, but the rates were increasing more sharply in Scotland (McClintock and Wikström, 1990). Helsinki, Finland ranked third in 1994, with a homicide rate of 14.55 per 100,000. Of surprise is Zurich's homicide rate. While Switzerland's popular image is of a relatively crime-free country, Zurich's 1994 homicide rate (11.4 per 100,000) placed it fourth among the ten European cities, with a rate close to that of New York City in 1996.

Glasgow's and London's 1994 robbery rates

were virtually identical, 344.41 and 344.12 per
100,000 respectively, which were the second and
third highest among the ten cities. Madrid, Spain
had the highest robbery rate at 523.07 robberies
per 100,000. Stockholm leads in the assault
victimization rate, an incredible 1,189.95 per
100,000. Not only is this the highest among the
European cities (London is a distant second at
589.75 per 100,000), but it is higher than most of
the US cities listed in Table 14.3. For example,
Stockholm's rate is higher than rates of New York
City (622.29 per 100,000), Chicago (1,345 per
100,000), Detroit (1,216 per 100,000), Los
Angeles (1,162 per 100,000), St Louis (1,519 per
100,000) and Newark (1,630 per 100,000). The
only city with a higher assault rate than Stockholm is Atlanta (2,010.53 per 100,000).

Within the US context, the question of who
commits most of these crimes is raised often in
the literature and the media. The answer for many
is a group that some have labelled the urban
'underclass'. The use of this term in the urban
violence literature is prominent, and many have
accepted its usage uncritically. An assumption
also exists that its definition and meaning have
universal agreement. Therefore, a discussion of
the term, its origins and its usage is important in
any discussion of urban crime.

THE CONCEPT OF THE UNDERCLASS

One of the current debates raging within American academic circles (and also discussed by
contributors to this present volume) concerns the
definition and use of the term 'underclass'.
Unfortunately, while the debate rages, the concept
has been readily adapted as part of the American
lexicon, and is used as if it had a distinct and
discrete meaning. Conservatives, eager to
discredit the social programmes of the Great
Society era and the expansion of the welfare state,
argued that welfare and other 'liberal' policies
had 'created a new caste of American – perhaps
as much as one-tenth of this nation – a caste of
people free from basic wants but almost totally
dependent upon the state, with little hope or
prospects of breaking free' (Anderson, 1979: 56;
Gilder, 1980; Murray, 1984).

While admitting that there is little agreement
on a definition of the term, William Julius Wilson
has advocated the use of the concept of the
underclass (Wilson, 1976, 1987) – although more
recently, Wilson (1990) acknowledged the confusion and debate within academic circles over the
use and misuse of the term. In *The Truly Disadvantaged* (1987), Wilson suggests that present-day
ghetto neighbourhoods consist almost exclusively of the most disadvantaged segments of the

black urban community who are outside the
mainstream of the American occupational
system. He defined the underclass as consisting of
the following elements:

> ... individuals who lack training and skills and either
> experience long-term unemployment or are not
> members of the labour force, individuals who are
> engaged in street-crime and other forms of aberrant
> behaviour, and families that experience long-term
> spells of poverty and/or welfare dependency.
> (1987: 8)

Expanding further on his definition, Wilson
suggests that these current occupants possess
behaviour patterns that differ markedly from
ghetto residents of three or four decades ago.
These behaviour patterns coupled with the flight
of middle- and working-class blacks from the
urban core have contributed to what Wilson calls
the social isolation of ghetto residents. This isolation has made it impossible for present day ghetto
residents to continue to sustain the traditional
black community institutions, such as churches
and businesses, that earlier ghetto communities
maintained. For these reasons, and others, Wilson
argues strenuously for the use of the underclass
label because it is meant to depict a reality not
captured in the standard designation of 'lower
class' (Wilson, 1985: 546). Moreover, he argues, to
employ a more neutral descriptor represents a
failure and/or refusal to come to grips with the
severity of the plight of the inner-city poor (Rose,
1988).

Wilson's definition of the underclass has been
the subject of controversy. Lowi (1988: 853)
states that, although Wilson attempts to extricate
himself from the conservative label applied after
the publication of *The Declining Significance of
Race* (1976), he failed to do so when he adopted
the rhetoric of President Reagan about an undefined group he called the 'truly disadvantaged'.
Lowi contends that the term 'is a throwback to
the 19th century concept of deserving poor, as
distinct from the "undeserving poor"'. Lowi
further questions the need for concepts such as 'a
"culture of poverty" and "underclass" for the
ghettos when occupied by blacks when we did
not need such concepts when the ghettos were
occupied by Jews or Irish or Italians or non-
Jewish Poles' (1988: 855). The major defining
trait of the underclass is persistent poverty
(Kelly, 1985), which has led Rose (1988) to argue
that this lack of precision in the definition of the
underclass is a major weakness in the concept.
Gans, arguing for the discontinuation of the use
of the term, states 'that the term underclass has
taken on so many connotations of undeserving-
ness and blame-worthiness that it has become
hopelessly polluted in meaning' (Gans, 1990:
272).

Underclass Defined

In a less scholarly form, Auletta (1982) defined the underclass as consisting of four distinct categories of people (a) the passive poor, usually long-term welfare recipients; (b) hostile street criminals; (c) hustlers who earn a living in the irregular economy but rarely commit violent crimes; and (d) traumatized drunks, drifters, homeless shopping-bag ladies and released mental patients (Auletta, 1982: xvi). Auletta's view of the underclass, as well as those of Lemann (1986) and previous underclass proponents, identified personal inadequacy and behavioural defects as the cause of this apparently new phenomenon of intergenerational welfare dependency. Their position at the bottom of the socio-economic ladder was of their own doing, an approach that is reminiscent of William Ryan's 'blaming the victim' thesis (Ryan, 1971). Moreover, Auletta, like Wilson, lumps together a variety of human conditions into one category – a welfare mother and a criminal supposedly exhibit similar behaviour patterns. Although his evidence is anecdotal and impressionistic, Auletta estimates that there are nine million people who are unable to assimilate into the American mainstream (1982: xvi).

Macnicol (1987: 314) argues that proponents of the underclass concept are unaware of or ignore its many conceptual flaws and are 'completely ignorant of its long and undistinguished pedigree'. The origin of the term evolved from eugenic studies during the 1920s in Britain aimed at defining the residuum or social problem group – those with mental defects. Macnicol states that in order to do justice to the definitional debate of the underclass, the concept must be related to wider definitions of social class and theories of stigma, labelling and deviance. He argues:

> At an obvious level the concept has been sustained by a large measure of simple class prejudice (involving the fetishization of middle class social mores) legitimized, in the inter-war years, by a 'biologization' of class through theories of heredity and, thereafter, by psychological models of personal inadequacy. Yet many of the concept's proponents have seen the underclass as distinct from the working class – in effect, a rootless mass divorced from the means of production – definable only in terms of social inefficiency, and hence not strictly a class in a neo-Marxist sense. (1987: 299)

Apparently during this earlier period, as often happens today, there were definitional problems with the concept (Macnicol, 1987). First, populist versions of the concept had been espoused and internalized by ordinary working-class people as the behavioural obverse of respectability – analogous to Lowi's distinction between the 'deserving' and 'undeserving' poor in the nineteenth century. The literature on social stratification in the USA, as exemplified in the work of W. Lloyd Warner, suggests a similar perception of social class. The element that distinguishes the 'levels above the common man' (upper-upper, lower-upper and upper-middle classes) and the 'level of the common man' (lower-middle and upper-lower classes) from the bottom of society, 'the level below the common man' (lower-lower class), is the perceived notion of respectability. As Kornhauser, in a synthesis of Warner's research, explained:

> The upper-lower class are the 'honest workmen' and the clean poor . . . Though they live in the less desirable sections of town and have lower incomes than the classes above them, they are still 'respectable'. Respectable is what the lower-lower class is not. This class forms the 'level below the common man'. It is composed of semi-skilled and unskilled workers . . . Lower-lower class people live in the worst sections of town and are thought by the other classes to be immoral. (1953: 231)

Warner suggests that other classes perceive the lower-lower to lack the '. . . cardinal virtues in which Americans pride themselves'. Moreover, their sexual behaviour differs from that of the other class. However, Warner suggests that 'their reputation for immorality often is no more than the projected fantasy of those above them' (quoted in Kornhauser, 1953: 231).

Second, there was difficulty in separating the specific underclass concept from wider assumptions about the inheritance of intelligence, ability and positive social qualities (Macnicol, 1987). The concept of the underclass, then and now, is linked with the separate question of intergenerational transmission. If this connection was not present, the underclass would essentially be those at the 'bottom' at any point in time. However, the emphasis is on the transmission of social inefficiency rather than structural inequality (Macnicol, 1987: 315).

Finally, although the concept has been popularized by conservatives, the general notion of an underclass has also been employed by those on the left to describe the casualties of capitalism (Macnicol, 1987). Marx and Engels used the phrase 'lumpenproletariat' to describe what they called the '"dangerous class", the social scum, that passively rotting mass thrown off by the lowest layers of old society . . .' (Marx and Engels, 1948: 20). Included in this group were casual labourers and the unemployed. Macnicol (1987: 300) lamented that the term has been used so loosely that 'Ralf Dahrendorf has even applied it to the problem of soccer hooliganism'.

Although the concept suffered from definitional problems, Macnicol says that the concept had enormous symbolic importance as part of a broader reformist strategy within conservative social thought. A 1929 Report of the Mental Deficiency Committee, as reported by Macnicol (1987: 302), stated that if the underclass could be identified it would consist of '. . . insane persons, epileptics, paupers, criminals (especially recidivists), unemployables, habitual slum dwellers, prostitutes, inebriates and other social inefficients than would a group of families not containing mental defectives'. Additionally, the lowest 10 per cent of the social scale make up the underclass. The solution to the 'problem' of the underclass was, of course, sterilization.

Some would argue that the current American conservative use of the term is not far removed from the early twentieth-century British usage. The responsibility for the position of the urban underclass is placed squarely on their own shoulders, with little recognition of the contribution of events that occur in the larger environment to the development of the group (Gilder, 1980; Murray, 1984; Wilson, 1976). Additionally, the 10 per cent figure has also been adopted as the estimated size of the underclass (Anderson, 1979:56). Moreover, some ascribe the intractability of poverty to the attitudes, values and behaviour of the city poor (Reed, 1998).

The principal question also seems to be – What should be done with these people? In spite of these similarities, however, there is one major and significant difference. In the most recent US variant of the term, the underclass consists mainly of minorities, is concentrated in the older industrial cities of the Northeast, and is socially, culturally and geographically isolated from mainstream society (Ricketts and Sawhill, 1988).

Considering the negative origins of the concept and its conceptual problems, we should be suspect of its usage, and its supposed meaning. Ricketts and Sawhill (1988: 316) raise three questions concerning the problems in defining the underclass namely:

- Is the underclass a new label for the poor or does it represent a distinct group?
- Assuming the two groups are different, is the underclass a subset of the poor, are the poor a subset of the underclass, or do the two groups overlap?
- What characteristic most distinguishes the underclass from the poor: (a) length of time they remain poor, (b) their geographic concentration and/or isolation, (c) their attitudes, or (d) their behaviour?

The term as used in the USA has become a surrogate for race – poor, black, urban residents. Conservatives, and liberals, combine a variety of disparate individuals into this group simply on the basis that they share the same geographic space. The assumption is made that all these individuals share the same values and outlooks simply because they live in the same neighbourhoods. The concept of the underclass is evidently wrought with problems, and should be used cautiously and only with an explanation of how it is being used.

How do American and European societies get a handle on urban violence? What are the answers and where are they to be found? Part of the answer is in how the issue is conceptualized and defined for policy-making purposes. The following section examines the two dominant paradigms through which urban violence is viewed and the directions each paradigm suggests for policy responses.

TWO CONCEPTUAL PARADIGMS OF VIOLENT CRIME

Much debate is taking place in both the United States and many European countries, but particularly in the United States, over the most appropriate approach(es) to reducing levels of violent crime. Within the US context, two paradigms are debated, each taking a different view of crime and victimization. The two are, broadly defined, a criminal justice and a public health paradigm. Each begins with a different definition of crime and presents a different set of solutions. One paradigm, criminal justice, is in vogue, the other, public health, is fighting for recognition. The literature suggests that a similar paradigm conflict is present in several European countries. For example, in France the debate is over an emphasis on welfare policies, more specifically policies that target and provide activities for youth, or 'repression' policies by criminal justice agencies. In France, repression means strict enforcement of the criminal code by police and the judiciary and the widespread use of imprisonment as punishment for crimes (de Liège, 1991).

The prevailing approach in the USA to homicide and violent crime is a criminal justice conceptualization, which uses the legal definition as the basic conceptualization of the problem for policy formulation purposes. Homicide is defined as a criminal act and therefore is deemed a criminal justice problem. From this perspective, the emphasis is placed on the perpetrator and the institutions involved in solving the problem are the police, prosecutors, prisons and parole system. Thus the policy recommendations call for the death sentence, more police, longer jail terms, mandatory sentencing for firearms crimes and stricter furlough and parole standards. The emphasis is on punishment, and the assumption is made that if the individuals who perpetrate these crimes are locked up, the crimes will not occur. Unfortunately, given the structure of the problem, there are others in the queue ready to replace those who are arrested and incarcerated. Moreover, the policies only take effect *after* someone (or a number of someones) has been murdered.

The other less popular paradigm says that in order to develop policies that will address urban violence the prevailing idea of urban violence must be challenged and American society persuaded to perceive and approach the problem differently. The only way to begin to ameliorate the situation, as a society, is to reconceptualize the problem so that the policies developed address the risk of victimization that confronts urban residents. The criminal justice conceptualization places the emphasis on the offender *after* the homicide has occurred. This orientation to the problem makes it virtually impossible to discuss or develop homicide-prevention strategies. If homicide *prevention* and the *preservation* of the lives of urban citizens are reasonable policy goals, then an alternative conceptualization is in order.

The public health conceptualization views homicide not as a criminal act but as a cause of death (O'Carroll and Mercy, 1986). The term 'homicide' in this context refers to victimization, rather than perpetration. Additionally, this reconceptualization places homicide within the framework of a public health response, rather than a criminal justice response. The emphasis, therefore, is on prevention and control rather than on punishment. The time frame is before the event, rather than after.

Mercy and O'Carroll (1988) argue that violence in general, and homicide in particular, have enormous public health implications. Several factors imply that homicide should be viewed from a public health perspective. First, as other causes of death decline in importance because of innovations in medical treatment, homicide is becoming more prominent as a leading cause of death. In fact, in 1985 the Centers for Disease Control listed homicide as the number one cause of death of black males of 15–34 years of age (O'Carroll and Mercy, 1986). Second, there is increasing recognition within the public health field of the importance of behavioural factors in the etiology and prevention of disease. For example, prevention of the three leading causes of death in the United States – heart disease, cancer and stroke – have been attributed to behavioural modifications such as exercise, changes in diet and non-smoking (Mercy and O'Carroll, 1988). Public health perspectives and practices hold promise for beginning to address the seriousness of the problem of urban violence.

The primary objectives of public health are to preserve, promote and improve health (Last, 1980: 3). The philosophy of public health has several dimensions directed toward attaining these goals (Mercy and O'Carroll, 1988). First, public health emphasizes prevention of disease or injury from occurring or recurring. This parallels nicely the differences between homicide defined as a cause of death (preventable) and homicide defined as a criminal act (after the fact). Second, public health interventions are concentrated on those at greatest risk of disease or injury – those in greatest need of intervention (Mercy and O'Carroll, 1988). This implies that homicide policies would focus on those at greatest risk of victimization – urban black residents – rather than exclusively on those who are the victimizers (as in the criminal justice approach).

The public health model views violence as a learned response to environmental sources of stress (Spivak et al., 1989). Public health literature argues that national problems require multiple solutions. Clearly, this is the case with urban violence. Multi-tiered strategies address different segments of the population. Within the public health framework, the types of interventions are known as primary, secondary and tertiary prevention strategies, each addressing specific groups of individuals. Primary prevention strategies are designed to reduce health problems in the general population. Prevention strategies as the primary level take the form of educational and public information campaigns aimed at teaching the masses of the public about risk factors. For example, primary prevention strategies to combat heart disease include programmes that raise the consciousness of the general public to the dangers of eating fatty foods, or smoking, or having a high cholesterol count. Secondary prevention strategies are interventions aimed at people who are at risk. For heart disease, secondary prevention strategies include efforts targeted at those who are at risk for developing heart disease because they smoke, have high blood pressure or high cholesterol or have a family history of the disease. Tertiary prevention strategies encompass all the strategies designed to prevent those who are already ill with the disease, say heart disease, from becoming more sick (Prothrow-Stith and Weissman, 1991: 130–43).

Reducing the homicide risk among urban Americans will require a set of long-term strategies. The US governmental system and its citizens are not enthusiastic about long-term strategies; the system is only geared to short-term solutions. The polity wants to see results from its policies and programmes, and if things do not change quickly, the commitment begins to wane. Any set of strategies devised will have marginal utility in the short run, but over time may do some good. The desire for short-term results promotes the punishment (criminal justice) response to the problem. It is only in public health problems (for example, cancer and AIDS) that the political system is inclined to make long-term policy commitments, realizing that results will occur for, and benefits will be reaped by, future generations.

If homicide is viewed within the context of a public health model, policy options for attacking, and hopefully controlling the problem are substantially broadened. The current stream of criminal justice approaches is nicely subsumed under the tertiary prevention strategies of the public health approach as part of an overarching solution to urban violence, but it is not seen as the only approach. Moreover, in the public health framework the criminal justice machinery can become an active participant in the development of prevention strategies, because the broadened definition of the problem allows these agencies to move beyond their arrest and punishment responsibilities, which a few agencies are beginning to see as essential to reducing levels of urban violence.

In the public health paradigm, we begin to focus on preventing the development of behaviour patterns and world-views in at-risk children that put them in danger in later years of becoming victims of lethal violence. As Allen (1981) has argued, potential victims can be made aware of their own dangerous behaviour patterns that make them susceptible to being murdered. The concern, therefore, is with helping people learn and practise alternative behaviour patterns (Spivak et al., 1989). Furthermore, it means that we begin to recognize that other social problems (for example, teenage pregnancy) are connected to the issue of lethal violence. Homicide is a very complex phenomenon that will require a complex set of solutions. The public health model offers us a vastly superior policy framework for tackling the issue than does the narrowly focused criminal justice model.

While the political system and politicians in the USA have emphasized the criminal justice approach, one federal agency, the Centers for Disease Control and Prevention (CDC), and many local and community-based initiatives have adopted the public health approach to violent crime. On the international front, on 13 May 1997 more than 1,200 delegates from 191 member states of the World Health Organization (WHO) unanimously endorsed an international plan of action to deal with violence as a public health problem (WHO, 1997). More recent survey data also suggest that the American public may be shifting its emphasis from criminal justice to an alternative approach. In 1995, only 30 per cent of the American public said that the USA should

attack the problem of crime by spending more money on police, prisons and judges. Some two-thirds (63 per cent) of Americans felt that crime should be attacked by spending more money on social and economic problems (*Sourcebook on Criminal Justice Statistics Online*, 1995).

A similar pattern has been observed in British public opinion. Despite the government's emphasis on criminal justice responses to crime, a 1994 MORI poll found that most Britons blamed drugs and unemployment as the key causes of crime, with unemployment the most frequently cited major cause of crime. Parental discipline, lenient sentencing and alcohol as explanations of crime were rarely identified, and only 16 per cent believed that absent fathers or single motherhood were causes of crime. More significantly, in view of the various law and order measures implemented, especially the Criminal Justice and Public Order Act of 1994, only 14 per cent felt that jailing more offenders would cut crime (Loveday, 1994). The concluding section looks briefly at crime prevention strategies, many of which use the public health paradigm, in the USA and Europe.

CONCLUSION: CRIME PREVENTION AND REDUCTION STRATEGIES

Crime is a concern for all countries and their citizens. As this chapter has shown, violent crime is not just a problem in the United States and its cities. Many European cities are grappling with the problem. Despite decreases in aggregate levels of violence, in both US and European cities the age of victimization and offending is decreasing and offending and victimization rates among juveniles are on the rise. Finally, the question is what to do to control and reduce the levels of violence. Those who study violent crime recognize that it is a complex phenomenon. There are multiple reasons and causes for violence. As such, no one policy or one programme will provide the complete solution. It is a problem that requires multiple approaches at all levels of society. While space does not allow an extensive listing of crime prevention and reduction strategies in the USA and European countries, several examples from different countries may be illustrative of the approaches being taken to address the problem. (Table 14.5 lists a small selection of crime prevention strategies, most of which are not discussed in this section.)

Many governments, particularly those of the USA and Britain, have emphasized a criminal justice approach to reducing crime that is manifest in the Omnibus Crime Control Act of 1994 in the USA and the Criminal Justice and Public

Table 14.5 *Selected crime prevention strategies in the USA and Western Europe*

1 **Neighbourhood Watch**: This is a programme for involving citizens in the monitoring of crime in their neighbourhoods. Neighbourhood residents watch homes when residents are away, report suspicious individuals in the neighbourhood, and other related activities. Many programmes distribute signs and window stickers to alert people that a Neighbourhood Watch programme is in effect in an area. This programme is used in both the USA and some Western European countries

2 **Drug Abuse Resistance Education (DARE)**: Most police departments in the USA participate in DARE. This programme sends officers into schools to talk with students about the hazards of drug use and the importance of resisting peer pressure to use drugs. The assumption is that crime reduction will follow if young people are dissuaded from using drugs

3 **Take Back the Night**: These programmes are usually sponsored by women's organizations and domestic violence groups in the USA to highlight domestic violence and rape. Many groups sponsor the marches held at night and make the statement that women will now allow crime to keep them from walking in their neighbourhoods

4 **Students Against Violence Everywhere (SAVE)**: This is an organization that promotes non-violence in schools and communities: SAVE provides education about the effects and consequences of violence and provides safe extracurricular activities for students. SAVE chapters are established in schools in the USA.

5 **Save Our Sons and Daughters (SOSAD)**: An organization founded by a mother who lost her 16-year-old son to violence in Detroit, and which has expanded nationally. It provides grief counselling and violence prevention activities, and its goal is to reduce violence in the USA

6 **Community policing**: This approach to policing has been adopted by many US and European police forces. The aim is to put police in neighbourhoods and to interact with residents in their daily lives. This may take the form of a substation located within a housing project and/or with the department and its officers attending community meetings, overseeing sports activities, and walking patrols through neighbourhoods. Neighbourhood residents are more likely to cooperate with officers who are a part of and concerned about the community

7 **Safe Cities and Crime Concern**: Primarily a programme in the UK, its aim is to link crime prevention with economic development. Urban regeneration in connection with business partnerships would provide a mechanism for reducing crime levels and the fear of crime among urban residents. In many cities this is associated with the growth of surveillance techniques, e.g. CCTV

8 **Youth Violence Prevention Programmes**: This is a broad umbrella term for many programmes in the USA and several Western European countries aimed at reducing current levels of youth violence and to combat and prevent the development of behaviour among urban youth that leads to violent behaviour. Many of these programmes are based on the public health approach to urban violence. For example, France has Mission Locales, which are youth training centres that provide a place for young people to meet, discuss and resolve problems associated with employment, training, finance and counselling (Graham and Bennett, 1995). Norway has introduced a national campaign to reduce bullying in schools, as has France through the School Safety Project (Graham and Bennett, 1995)

Order Act of 1994 in Britain. These acts call for more police, stiffer sentences, more jail time, the abolition of parole and other criminal justice related penalties. Some scholars attribute the reduction in aggregate rates to these tough criminal justice measures (Dilulio, 1995; Wilson, 1995). Yet, others feel that the approach is limited in its ability to reduce crime (McClain, 1992, 1993; Scheingold, 1995).

Many police departments and local governments and groups have taken alternative approaches, many, but not all, within the context of the public health approach. Some jurisdictions, such as Glasgow, Scotland, have experimented with improvements in street lighting to prevent crime and reduce the fear of crime among neighbourhood residents. Some early evaluations found a reduction in fear of crime, especially among the elderly and women, and a reduction of

criminal victimization in some areas. Other areas, on the other hand, experienced an increase in criminal victimization. Nevertheless, later evaluations of the approach found that for many residents the increased lighting had the opposite effect – they were more fearful when the lights came on then they were before the lights were installed (Nair et al., 1993). Many US cities also experimented with street lighting as an approach to crime reduction, also with the same mixed results (Painter, 1996).

Another approach to crime control taken by police departments is the concept of community policing. This approach has taken hold in the USA and several European countries, for example, Great Britain (Monaghan, 1997). Community policing puts the police and police substations into the communities they serve. The departments and officers participate in social activities, community meetings and youth sports leagues. The purpose is to make the police less an occupation force and more a participant in the community. The view is that citizens are more likely to cooperate with people they know are concerned about the community. Moreover, a constant police presence in the community is a deterrent to crime. Officers also advise citizens on crime prevention activities, such as Neighbourhood Watch schemes.

The failure of many crime control policies in Britain led the Thatcher government reluctantly to focus on crime prevention strategies. The government premised its change of direction on the theme that crime could not be controlled by the institutions of the criminal justice system alone; every citizen had a responsibility to fight crime and prevent criminal victimization (Monaghan, 1997). According to Monaghan (1997), the central government no longer claimed, as it had in the election campaign of 1979, that its policies would reduce the crime rate. The Safe Cities and Crime Concern programmes were initiated in 1988 in England, Scotland and Wales to get local business communities and community residents involved in crime prevention. The common aims of the Safe Cities programmes were to reduce crime, to reduce the fear of crime and to create safer cities in which businesses and community life could develop. Crime prevention was set within the context of urban regeneration (Monaghan, 1997). However, in 1994 when the Home Office began to end funding for local Safe Cities programmes, few local business communities were prepared to assume responsibility for the continuation of the programmes (Loveday, 1994). Some suggest that the Safe Cities and Crime Concern programmes have failed due to a lack of cooperation among the various participants.

In the USA, most states and many cities have established a network of violence prevention programmes, based on the public health model, to combat and reduce violent behaviour. Many are based on the successful Violence Prevention Project of the Health Promotion Program for Urban Youth run by the Boston Department of Health and Hospitals. The programme is focused on reducing levels of violent behaviour and associated social and medical hazards for adolescents. It is a community-based outreach and educational primary prevention programme that focuses on changing individual behaviour and community attitudes around violence (Spivak et al., 1989). The primary prevention strategies are supported by a network of secondary therapeutic clinical treatment services.

The primary prevention programme consists of a violence prevention curriculum in the schools, but that is also taken outside the traditional classroom to other settings such as alternative schools, recreational programmes, public housing developments, Sunday schools, boys' and girls' clubs, neighbourhood health centres, and other local organizations. Ministers, police officers and other community leaders and activists are recruited to spread the violence reduction message, to, in effect, 'saturate' the community with the message (Spivak et al., 1989). Connected with this primary prevention programme is a mass media campaign of public service announcements, posters, billboards and T-shirts addressing the problems of violent behaviour. Secondary level therapeutic treatment services are available for those youths who are already exhibiting violent behaviour patterns.

Presently violence prevention programmes of this type are being run in major cities and in many states. The Centers for Disease Control is involved in piloting several youth violence prevention programmes around the country (Rosenberg, 1995). While evaluations are ongoing on the success of these efforts, what is clear is that the multiple strategy approach to the prevention of violence is the only approach that will allow countries to get a handle on increasing levels of violence and to control and reduce these levels. The criminal justice paradigm as the sole approach is short-sighted; the public health model subsumes the criminal justice approach while at the same time providing mechanisms for the development of preventive strategies.

NOTES

I would like to thank Paul C. Jacobson for his insightful and helpful comments on earlier drafts of this chapter. I am also indebted to Stacy Nykios, my research assistant, for finding much of the literature and crime statistics cited in this chapter, and Steve C. Tauber, my former research assistant, for calculating some of the crime rates.

1. Considerable differences exist between official crime statistics reported by the Federal Bureau of Investigation and the crime victimization surveys conducted by the Bureau of Justice Statistics. The survey data suggest higher levels of victimization in the USA than do the official statistics. Certainly, official statistics need interpreting with caution.

2. I must caution that crime data are subject to measurement error because many rely on victim reporting to the police, and police department reporting to another agency. Data source differences are problematic also. These rates are based on data collected by the International Criminal Police Organization and published by Interpol General Secretariat. Literature examining data for the same year in England and Wales has a rate per 100,000 of 1.1, which is lower than the Interpol rate (Golding, 1995). Golding uses public health data, but

definitional and reporting differences between the two sources may be similar to the differences in homicide statistics maintained by local health departments and those reported to the Federal Bureau of Investigation in the USA. Local health departments maintain statistics on resident homicides, while the Federal Bureau of Investigation counts all homicides.

3. As Table 14.3 shows, New York's homicide rate has dropped dramatically from 1989 to 1996.

4. I must express a note of caution about the data year differences. It is possible that 1996 data for Glasgow are below the figures for the US cities in 1996 which I have used for comparison. It is equally possible that Glasgow's rate could still be above that for the four US cities. While either of these may be the case, the point is to provide a comparative view of homicide in US and European cities.

BIBLIOGRAPHY

Allen, Nancy H. (1981) 'Homicide prevention and intervention', *Suicide and Life Threatening Behavior*, 11: 167–79.

Anderson, Elijah (1994) 'The code of the streets', *Atlantic Monthly*, May: 81–94.

Anderson, Martin (1979) *Welfare*. Palo Alto, CA: Hoover Institute.

Archer, D. and Gartner, R. (1976) 'Violent acts and violent times: a comparative approach to postwar homicide rates', *American Sociological Review*, 41: 937–63.

Archer, D., Gartner, R., Akert, R. and Lockwood, T. (1978) 'Cities and homicide: a new look at an old paradox', *Comparative Studies in Sociology*, 1: 73–95.

Arizona Republic (1989) 'Non-white men face greatest murder risk', 13 January: A–10.

Auletta, Ken (1982) *The Underclass*. New York: Random House.

Baumer, Eric (1994) 'Poverty, crack and crime: a cross-city analysis', *Journal of Research in Crime and Delinquency*, 31 (August): 311–27.

Bottoms, Anthony, E. and Wiles, Paul (1989) 'Housing tenure and residential crime careers in Britain', in Albert J. Reiss, Jr and Michael Tonry (eds), *Communities and Crime*. Chicago: University of Chicago Press.

Bovenkerk, Frank (1993) 'Crime and the multi-ethnic society: a view from Europe', *Crime, Law and Social Change*, 19 (April): 271–80.

Brenner, M. Harvey and Swank, Robert T. (1986) 'Homicide and economic change: recent analysis of the joint economic report of 1984', *Journal of Quantitative Criminology*, 3 (March): 69–80.

Bureau of Justice Statistics (1994) *Violent Crime*. Washington, DC: US Department of Justice. April, NCJ–147486.

Bureau of Justice Statistics (1997) *Criminal Victimization in the United States, 1994*. Washington, DC: US Department of Justice. May, NCJ–162126.

Bursik, Robert J. Jr and Grasmick, Harold G. (1993) 'Economic deprivation and neighbourhood crime rates, 1960–1980', *Law and Society Review*, 27: 263–83.

Cohen, L.E. and Felson, M. (1979) 'Social change and crime rate trends: a routine activity approach', *American Sociological Review*, 44: 588–608.

Dilulio, John J. Jr (1995) 'Arresting ideas: tougher law enforcement is driving down urban crime', *Policy Review* (Fall): 12–16.

Farrington, D.P. (1995) 'The development of offending and antisocial behavior from childhood: key findings from the Cambridge Study in Delinquent Development', *Journal of Child Psychology and Psychiatry*, 360: 929–64.

Federal Bureau of Investigation (1987) *Uniform Crime Reports*. Washington, DC: US Government Printing Office.

Federal Bureau of Investigation (1990a) *Crime in the United States – 1989*. Washington, DC: US Government Printing Office.

Federal Bureau of Investigation (1990b) *Supplemental Homicide Reports*. Washington, DC: US Government Printing Office.

Federal Bureau of Investigation (1997) *Crime in the United States – 1996*. Washington, DC: US Government Printing Office.

Fischer, Claude S. (1976a) 'Toward a subcultural theory of urbanism', *American Journal of Sociology*, 80: 1319–40.

Fischer, Claude S. (1976b) *The Urban Experience*. New York: Harcourt Brace Jovanovich.

Fordham, Signithia (1996) *Blacked Out, Dilemmas of Race, Identity and Success at Capital High*. Chicago: University of Chicago Press.

Fox, James Alan (1996) 'Trends in juvenile violence: a report to the United States Attorney General on Current and Future Rates of Juvenile Offending', Washington, DC: United States Department of Justice.

Gans, Herbert J. (1990) 'Deconstructing the underclass: the term's danger as a planning concept', *Journal of the American Planning Association*, 56(3): 271–8.

Garland, David (1996) 'The limits of the sovereign state', *British Journal of Criminology*, 36 (Autumn): 445–71.

Gilder, George (1980) *Wealth and Poverty*. New York: Basic Books.

Glasgow, Douglas G. (1980) *The Black Underclass: Poverty, Unemployment, and Entrapment of Ghetto Youth*. San Francisco, CA: Jossey-Bass.

Golding, A.M.B. (1995) 'Understanding and preventing violence: a review', *Public Health*, 109: 91–7.

Graham, John and Bennett, Trevor (1995) *Crime Prevention Strategies in Europe and North America*. Helsinki: European Institute for Crime Prevention and Control.

Hunter, Albert (1974) *Symbolic Communities*. Chicago: University of Chicago Press.

Ingram, A. Leigh (1993) 'Type of place, urbanism, and delinquency: further testing the determinist theory', *Journal of Research in Crime and Delinquency*, 30 (May): 192–212.

International Criminal Police Organization (1991–92) *International Crime Statistics 1991–1992.* Lyons, France: International Criminal Police Organization, Interpol General Secretariat.

Junger, Marianne and Polder, Wim (1992) 'Some explanations of crime among four ethnic groups in the Netherlands', *Journal of Quantitative Criminology*, 8: 51–78.

Junger-Tas, Josine (1997) 'Ethnic minorities and criminal justice in the Netherlands', in Michael Tonry (ed.), *Ethnicity, Crime, and Immigration: Comparative and Cross-National Perspectives.* Chicago: University of Chicago Press.

Kelly, Robert F. (1985) 'The family and the urban underclass: an integrative framework', *Journal of Family Issues*, 6 (June): 159–84.

Kilpatrick, Rosemary, Trew, Karen and Young, Valerie (1995) 'Juvenile crime in Northern Ireland: psychological research and social policy', *Irish Journal of Psychology*, 16: 356–65.

Kornhauser, Ruth R. (1953) 'The Warner approach to social stratification', in Reinhard Bendix and Seymour Martin Lipset (eds), *Class, Status and Power.* Glencoe, IL: Free Press.

Kornhauser, Ruth R. (1978) *Social Sciences of Delinquency.* Chicago: University of Chicago Press.

Last, John M. (1980) 'Scope and methods of prevention', in John M. Last (ed.) *Public Health and Prevention Medicine.* 11th edition. New York: Appleton-Century Crofts.

Lemann, Nicholas (1986) 'The origins of the underclass', *The Atlantic Monthly* (June): 31–55.

de Liège, Marie-Pierre (1991) 'Social development and the prevention of crime in France: a challenge for local parties and central government', in Frances Heidensohn and Martin Farrell (eds), *Crime in Europe.* London: Routledge. pp. 121–32.

Los Angeles Times (1989) 'L.A. outrage makes little impact on gang epidemic', 30 January: 1.

Loveday, Barry (1994) 'Government strategies for community crime prevention programmes in England and Wales: a study in failure?', *International Journal of Sociology of Law*, 22: 181–202.

Lowi, Theodore J. (1971) *The Politics of Disorder.* New York: W.W. Norton.

Lowi, Theodore J. (1988) 'The theory of the underclass: a review of Wilson's *The Truly Disadvantage*', *Policy Studies Review*, 7 (Summer): 852–8.

Macnicol, John (1987) 'In pursuit of the underclass', *Journal of Social Policy* 16 (3): 293–318.

Maguire, Keith (1993) 'Fraud, extortion and racketeering: the black economy in Northern Ireland', *Crime, Law and Social Change*, 20 (November): 273–92.

Martens, Peter L. (1997) 'Immigrants, crime, and criminal justice in Sweden', in Michael Tonry (ed.), *Ethnicity, Crime, and Immigration: Comparative and Cross-National Perspectives.* Chicago: University of Chicago Press.

Matthews, Roger (1995) 'Crime and its consequences in England and Wales', *Annals of the American Academy of Political and Social Services*, 539 (May): 169–82.

Marx, Karl and Engels, Frederich (1948) *The Communist Manifesto.* New York: International Publishers.

Mayhew, B. and Levinger, R. (1976) 'Size and the density of interaction in human aggregates', *American Journal of Sociology*, 82: 86–110.

McClain, Paula D. (1992) 'Reconceptualizing urban violence: a policy analytic approach', *National Political Science Review* 3: 9–24.

McClain, Paula D. (1993) 'Urban violence: agendas, politics, and problem redefinition', in Paula D. McClain (ed.), *Minority Group Influence: Agenda Setting, Formulation, and Public Policy.* Westport, CT: Greenwood Press.

McClintock, F.H. and Wikström, Per-Olaf H. (1990) 'Violent crime in Scotland and Sweden', *British Journal of Criminology*, 30 (Spring): 207–28.

McClintock, F.H. and Wikström, Per-Olaf H. (1992) 'The comparative study of urban violence: criminal violence in Edinburgh and Stockholm', *British Journal of Criminology*, 32 (Autumn): 505–20.

Mercy, James A. and O'Carroll, Patrick W. (1988) 'New directions in violence prediction: the public health arena', *Violence and Victims*, 3: 285–301.

Monaghan, Bernadette (1997) 'Crime prevention in Scotland', *International Journal of Sociology of Law*, 25: 21–44.

Morison, John and Geary, Ray (1993) 'Studying crime and conflict: the Northern Ireland example', *Northern Ireland Legal Quarterly*, 44 (Spring): 65–70.

Murray, Charles (1984) *Losing Ground.* New York: Basic Books.

Nair, Gwyneth, Ditton, Jason and Phillips, Samuel (1993) 'Environmental improvements and the fear of crime: the sad case of the "Pond" area in Glasgow', *British Journal of Criminology*, 33 (Autumn): 555–61.

New York Times (1992) 'Preliminary 1991 Figures Show Drop in Homicides', Daniels, Lee A., January 3, Section B, Page 3, Column 5, Metropolitan Desk.

O'Carroll, Patrick W. and Mercy, James A. (1986) 'Homicide trends in the United States', in Darnell F. Hawkins (ed) *Homicide Among Black Americans.* Lanham, MD: University Press of America.

Painter, Kate (1996) 'The influence of street lighting improvements on crime, fear and pedestrian street use, after dark', *Landscape and Urban Planning*, 35 (1996): 193–201.

Prothrow-Stith, Deborah and Weissman, Michael (1991) *Deadly Consequences: How Violence is Destroying Our Teenage Population and a Plan to Begin Solving the Problem.* New York: Harper Collins.

Reed, Adolph Jr (1988) 'The liberal technocrat', *The Nation*, 6 February: 167–70.

Rees, P.H. (1970) 'Concepts of social space', in B.J.L. Berry and F.C. Horton (eds), *Geographic Perspectives on Urban Systems.* Englewood Cliffs, NJ: Prentice-Hall.

Reilly, Barry and Robert Witt (1992) 'Crime and unemployment in Scotland: an exconometric analysis using regional data', *Scottish Journal of Political Economy*, 39 (May): 213–28.

Reiss, Albert, J. Jr (1986) 'Why are communities important in understanding crime?', in Albert J. Reiss Jr and Michael Tonry (eds), *Communities and Crime*. Chicago: University of Chicago Press.

Ricketts, Erol R. and Sawhill, Isabel V. (1988) 'Defining and measuring the underclass', *Journal of Policy Analysis and Management*, 7: 316–25.

Roberts, Geoffrey K. (1994) 'Extremism in Germany: sparrows or avalanche?', *European Journal of Political Research*, 25: 461–82.

Roncek, Dennis W. (1981) 'Dangerous places: crime and residential environment', *Social Forces* 60: 74–98.

Rose, Harold M. (1971) *The Black Ghetto: A Spatial Behavioral Perspective*. New York: McGraw-Hill.

Rose, Harold M. (1981) 'The changing spatial dimension of black homicide in selected American cities', *Journal of Environmental Systems*, 11: 57–80.

Rose, Harold M. (1986) 'Can we substantially lower homicide risk in the nation's larger black communities?', *Secretary's Task Force on Black and Minority Mental Health*. Washington, DC: US Department of Health and Human Services.

Rose, Harold M. (1988) 'Are non-race specific politics the key to resolving the plight of the inner-city poor?', *Policy Studies Review*, 7 (Summer): 859–64.

Rose, Harold M. and Deskins, D.R. Jr (1980) 'Felony murder: the case of Detroit', *Urban Geography* 1: 1–22.

Rose, Harold M. and McClain, Paula D. (1981) *Black Homicide and the Urban Environment*. Washington, DC: National Institutes of Mental Health.

Rose, Harold M. and McClain, Paula D. (1990) *Race, Place, and Risk: Black Homicide in Urban America*. Albany, NY: State of University of New York Press.

Rose, Harold M. and McClain, Paula D. (1998) '*Race, Place and Risk* revisited: a perspective on the emergence of a new structural paradigm', *Homicide Studies: An Interdisciplinary and International Journal*, 2 (May): 101–29.

Rosenberg, Mark. L. (1995) 'Violence in America: an integrated approach to understanding and prevention', *Journal of Health Care for the Poor and Underserved*, 6: 102–10.

Ryan, William (1971) *Blaming the Victim*. New York: Pantheon.

Sampson, Robert J. and Groves, Byron W. (1989) 'Community structures and crime: testing social-disorganization theory', *American Journal of Sociology*, 94 (January): 774–802.

Scheingold, Stuart A. (1995) 'Politics, public policy, and street crime', *Annals of the American Academy of Political and Social Sciences*, 539 (May): 155–68.

Shaw, C.R. and McKay, H.D. (1942) *Juvenile Delinquency and Urban Areas*. Chicago: University of Chicago Press.

Shaw, C.R. and McKay, H.D. (1969) *Juvenile Delinquency and Urban Areas* (2nd edn). Chicago: University of Chicago Press.

Shin, Y., Jekicka, D. and Lee, E.S. (1977) 'Homicide among blacks', *Phylon*, 39: 398–407.

Silberman, Charles (1978) *Criminal Violence, Criminal Justice*. New York: Random House.

Simmel, George (1951) 'Metropolis and mental life', in Kurt H. Wolff (ed.), *The Society of George Simmel*. New York: Free Press.

Solomon, Joel S. (1995) 'Forming a more secure union: the growing problem of organized crime in Europe as a challenge to national sovereignty', *Dickinson Journal of International Law*, 13 (Spring): 623–48.

Sourcebook of Criminal Justice Statistics Online (1995). Washington, DC: Bureau of Justice Statistics.

Spivak, Howard, Hausman, Alice J. and Prothrow-Stith, Deborah (1989) '"Practitioners" forum: public health and the primary prevention of adolescent violence – "The Violence Prevention Project"', *Violence and Victims*, 4: 203–12.

Sun, Hung-En and Reed, Jack (1995) 'Migration and crime in Europe', *Social Pathology* 1 (Fall): 228–52.

Taylor, Ralph B. and Covington, Jeanette (1993) 'Community structural change and fear of crime', *Social Problems*, 40 (August): 374–95.

Turner, Charles W., Fenn, Michael R. and Cole, Allen M. (1981) 'A social psychological analysis of violent behavior', in Richard B. Stuart (ed.), *Violent Behavior*. New York: Brunner/Mazel. pp. 31–67.

United Nations (1997) *The Fifth United Nations Survey of Crime Trends and Operations of Criminal Justice Systems*. Data set.

US Public Health Department (1979) *Vital Statistics*. Washington, DC: US Public Health Department.

van Dijk, Jan (1991) 'More than a matter of security: trends in crime prevention in Europe', in Frances Heidensohn and Martin Farrell (eds), *Crime in Europe*. London: Routledge. pp. 27–42.

Waldron, I. and Eyer, J. (1975) 'Socioeconomic causes of the recent rise in death rates for 15–24 year olds', *Social Science and Medicine*, 9: 383–96.

Washington Post (1990a) 'Homicides by teenagers boosted 1989 death toll', 1 January: D–1.

Washington Post (1990b) 'Big cities breaking homicide records', 9 December: A–20.

Willie, Charles V. (1988) 'Rebuttal to a conservative strategy for reducing poverty: a book review of William Julius Wilson, *The Truly Disadvantaged*', *Policy Studies Review*, (Summer): 865–75.

Wilson, James Q. (1983) *Thinking about Crime* (rev. edn). New York: Basic Books.

Wilson, James Q. (1995) 'What to do about crime: blaming crime on root causes', *Vital Speeches*, 61 (1 April): 373:7.

Wilson, James Q. and Herrnstein, Richard (1985) *Crime and Human Nature*. New York: Simon & Schuster.

Wilson, William Julius (1976) *The Declining Significance of Race*. Chicago: University of Chicago Press.

Wilson, William Julius (1985) 'Cycles of deprivation and the underclass debate', *Social Service Review* (December): 541–59.

Wilson, William Julius (1987) *The Truly Disadvantaged: The Inner-City, the Underclass, and Public Policy*. Chicago: University of Chicago Press.

Wilson, William Julius (1990) 'Social theory and public agenda research: the challenge of studying inner-city social dislocations', Presidential address, annual meeting of the American Sociological Association, 12 August.

Wirth, Louis (1938) 'Urbanism as a way of life', *American Journal of Sociology*, 64: 1–24.

Wolfgang, Marvin E. (1958) *Patterns in Criminal Homicide*. Philadelphia: University of Pennsylvania.

Wolfgang, Marvin E. (1978) 'Perceived and real changes in crime and punishment', *Daedalus*, 86 (Winter): 143–58.

Wolfgang, Marvin E. and Ferracuti, Franco (1967) *The Subculture of Violence*. London: Tavistock – Social Science Paperbacks.

World Health Organization (WHO) (1997) 'World Health Assembly endorses plan of action to deal with violence as a public health issue', Press Release (13 May).

Zimring, F.E. (1979) 'Determinants of the death rate for robbery: a Detroit time study', in H.M. Rose (ed.), *Lethal Aspects of Urban Violence*. Lexington, MA: Lexington Books.

Zuzevich, Debra (1997) 'Nordic biker wars spread to continent', *Crime and Justice International*, 13 (5). http://www.acsp.uic.edu/oicj/pubs/cjintl/1305/130511.shtml

Part IV

THE CITY AS ECONOMY

It is only in the modern era that economic change has become so significant in explaining urban development, the growth of cities, and, following shifts in the spatial patterning of production, the relative decline of others. While production and trade characterized the growth of cities in the pre-modern (pre-industrial) period, urban centres developed as often for political, administrative or symbolic (including ecclesiastical) reasons as they did for primarily economic ones. While these functions remain important to cities, the development of industrial capitalism shifted the balance between rural and urban, fostering the industrialization–urbanization nexus. Where previously urban centres functioned within societies which remained largely rural, increasingly the balance tilted towards urbanization. Both the scale of production and international trading were unprecedented, as was the scale of urban growth in the core nations of the world economy.

The current phase of economic and urban restructuring linked to the shift towards the emergence of post-Fordist, post-industrial economies has brought into question those very aspects of cities which had become perhaps taken-for-granted, but which nonetheless had been vital to their success as the main loci of accumulation. The success of the city lay in its ability to support industrial capitalism through the economies it was able to harness. Cities became uniquely efficient motors of capital accumulation because of the economies – agglomeration and external – and the innovatory and entrepreneurial environments they created. Yet the benefits of the city were not inevitable; the increasing scale of the city could lead to agglomeration diseconomies, while shifts in the spatial division of labour, brought about through technological advance or through the search by capitalism for cheaper loci for production, could undermine their economic raison d'être. In the advanced economies, and particularly in those industrial metropolitan areas which have become casualties within the ongoing round of economic restructuring, the 'search' is on for

the regeneration of those attributes which had fostered their initial growth. Not surprisingly, this involves the revisiting of old concepts, Marshall's ideas on the creation of 'industrial atmosphere' for one, in an attempt to create the conditions which had led to their early growth.

Contemporary restructuring is questioning the economic role of cities in other ways. Theories explaining urban growth which enjoyed considerable support, particularly those which prioritized manufacturing, have an increasingly anachronistic feel, particularly in the post-industrial cities whose economies are largely, if not wholly, based upon services. According to the basic: non-basic model, cities grew by virtue of their ability to export manufactured goods, an ability which depended upon the development of a wide range of supportive services, producer and consumer. In the post-industrial economy information processing services become the motor of urban economic growth rather than merely supportive of it.

Changes in the international spatial division of labour and the deepening process of globalization have become centrally entwined with urban change. Among the urban shifts linked to the new international division of labour, the emergence of the Asian tiger (and 'cub') economies, with their primate mega-cities are among the more obvious changes. In other new industrial economies, including Mexico and Brazil, industrialization has spawned rapid urban development as it had done in the first industrial nations.

Whereas the new international division of labour has resulted in the shift of urban growth towards particular economies in the former Third World, globalization has tended to emphasize the dominance of cities in the core economies, and in particular the dominance of the world cities. Among contemporary ideas of urban growth the notion of the world city has few equals. In the now classic accounts the hypothesis has focused on the upper rungs of the world city hierarchy, and indeed on the uppermost, comprising New York, London and Tokyo. In fact, to varying degrees virtually all cities have become enmeshed within the world economy, though most function at a national and/or regional rather than a global scale. Emphases on the world city, or indeed on those cities which aspire to such a role, has had the effect of crowding out analyses of how cities of different sizes, function and location are incorporated within the world economy – in what sense, if any they are 'world cities'. Nor, of course, is economic growth simply a product of globalization or global position. The national framework, operating through macro-economic policies, and factors more particular to cities individually, their legacy, and local social and political structures, for example are important too in influencing city economic growth.

Indisputably, globalization, with its emphasis of 'one-ness' in a world undergoing rapid economic and technological change, has (re)kindled a sense of competition between cities for economic growth. How such competition unfolds for individual cities depends on their position within global and regional networks, their assets in being able to attract new economic activities, their quality of life amongst a range of different factors. Globalization has reaffirmed the uneven development between cities characterizing earlier phases in the evolution of the world economy, but the patterning of such unevenness has altered.

Further, where competition and the competitive ability of cities have become virtual canons in the understanding of how the development of new urban economies is unfolding, it is accompanied by the emergence of cross-national networking collaborative networks between cities aimed too at bolstering the urban economy.

Accompanying the processes of urban economic change, in both the advanced and the developing economies, has been the growth of the informal economy. The mechanisms underpinning the development of informalization differ between the core and developing economies, as does their significance; in many African cities the informal economy dominates over the formal, providing not only a vital means of income generation for the urban poor, but also in ensuring cities are adequately provisioned. In the advanced economies restructuring has fostered the duality of urban labour markets, the spread of informal forms of employment being an essential support to those otherwise marginalized by restructuring. Whether in the developing or advanced economies, a broad consensus has been reached that the urban informal economy (except where it is associated with criminal activity) should be viewed positively by policy and is to be fostered.

15

Urban Scale Economies

J. VERNON HENDERSON

Scale economies are the basis of urban agglomeration – the reason we have cities. The magnitude and nature of scale economies are critical in determining individual city sizes, what cities do, the size distribution of cities, the possibility of multiple equilibria in location patterns, the growth process of cities, inequality among residents of different cities, and the efficiency of a variety of public policies trying to influence the geographical organization of economic resources.

Based on a selective review of the literature, this chapter investigates three sets of issues concerning urban scale economies. To try to help us understand their source and basis, the micro-foundation models of urban scale economies are first presented. The next section examines the nature of these scale economies – what measures of local scale matter – focusing on empirical evidence. The third section then turns to recent key questions. To what extent are urban scale externalities dynamic rather than static and how does the notion of dynamic externalities contribute to our understanding of urban growth, equilibrium and location processes?

MICRO-FOUNDATIONS

The original models of urban agglomeration either treated scale economies as being at the level of the factory (e.g., Mills, 1967) or as black-box external economies of scale to the firm (e.g., Henderson, 1974). Given the desire to move beyond consideration of factory towns to modern diverse cities, the external economy of scale approach dominated work until the early 1980s (e.g., Hochman, 1977; Kanemoto, 1980).

In the external economy of scale approach firm output x is written as

$$x = G(N)x(k) \qquad (1)$$

where $x(\cdot)$ represents the firm's own constant returns to scale technology, with a vector of private inputs, k. $G(N)$ is a Hicks's neutral shift factor external to the firm; its arguments are local scale measures such as total city workforce. With positive externalities $G' > 0$, indicating that as city scale rises, firm output rises, holding its own inputs fixed. That is the basis for urban agglomeration for firms and hence workers. Hicks's neutrality is not critical but empirical evidence suggests it holds (see below).

With Hicks's neutrality, the technology in equation (1) can be rewritten in terms of unit cost functions and profit functions as

$$c = G(N)^{-1}c(p) \qquad (2)$$

$$\pi = G(N)\pi(p) \qquad (3)$$

where c is the firm's unit cost of production, p is a vector of input and output prices, and $c(p)$ represents the firm's constant returns to scale technology. In equation (3), π is the firm's profits; but now, while $\pi(p)$ represents the firm's technology, such technology must exhibit decreasing returns to scale (to, say, fixed entrepreneurial inputs) in order to be well behaved. Equation (3) can be used also to determine firm size, unlike (2), through the partial derivative with respect to output price. Use of (2) and (3) in empirical work is common, since all right-hand side arguments are exogenous variables to the competitive firms (for example, in particular, prices); while in (1) the vector k is endogenous, raising problems of simultaneity bias in estimation.

There is strong empirical evidence on the existence, nature and magnitude of agglomeration economies, reviewed later. However, the issue emerging in the 1980s concerned what really are the sources of these black-box external economies of scale – what are the micro-foundations? In thinking about that issue, early urban economists drew from Alfred Marshall, who wrote in 1890 about the benefits of agglomeration. Marshall discussed:

1   Information spillovers: 'the mysteries of the trade become no mystery; but are, as it were, in the air . . . inventions, and improvements in machinery, in processes, and in the general organization of business have their merits promptly discussed; if one man starts an idea it is taken up by others and combined with suggestions of their own; and thus becomes the source of yet more new ideas.'

2   Labour market externalities: 'Employers are apt to resort to any place where they are likely to find a good choice of workers with the special skill which they require; while men seeking employment naturally go to places where there are many employers.'

3   Urban diversity and Adam Smith specialization by plants: 'subsidiary trades grow up in the neighbourhood, supplying it with materials, organizing its traffic, and in many ways conducing to the economy of its material . . . for subsidiary industries devoting themselves each to one small branch of the process of production, and working it for a great many of their neighbours, are able to keep in constant use machinery of the most highly specialized character' (1890: 332).

With this cue from Marshall, urban theorists in the last 15 years have worked on developing models of these and related micro-foundations of scale economies. I turn to these now.

Information Spillovers

Equation (1) can be treated as directly modelling information spillovers, where N is a measure of the number of people in the locality. In this formulation we measure the volume of local spillover communications by the number of communicators. Communications concern information about available technology, qualities and prices of suppliers and buyers, market conditions in general and about product specifications, and interaction with regulators and government officials. Moving beyond equation (1), since communication involves space, the first attempt at more specificity in micro-foundations involves introducing urban space into the formulation. We follow Fujita and Ogawa's (1982) path-breaking work.

Consider a linear city, as pictured in Figure 15.1. The (endogenous) business district lies, in the monocentric case, between b_1 and b_2 with the centre at $(b_1 + b_2)/2$. Residents commute to the business district from the 'surrounding' residential area. Firms operate with the same fixed labour/land ratio normalized to 1. Then output of a firm at location y is

$$x(y) = G(\int_{b_1}^{b_2} g(|y - s|)ds) \cdot 1 \tag{4}$$

$|y - s|$ is the distance between plants at y and s. $g' < 0$ so the value of messages, or the interaction between y and s, decays with distance. That is the spatial component to spillovers. $\int_{b_1}^{b_2} g(|y - s|)ds$ is the total undecayed message volume for producer y, which gives a shift factor $G(\cdot)$ on the firm's technology (1), assuming $G' > 0$.

The common way to proceed is to assume G is linear or log linear and to give $g(\cdot)$ one of two forms. With expotential decay, $g = e^{-\alpha|y - s|}$, firms always remain in communication. With linear decay, where

$$g = \max[0, a - \alpha|y - s|],$$

once $|y - s|$ exceeds a/α, firms cease to communicate. Once we consider multi-centred cities with developers who operate imperfectly competitive centres, the linear form presents direct strategic options for different clusters of firms to move strategically out of communication, while with the expotential form communications diminish but never cease (Henderson and Mitra, 1996). For purposes here we utilize the expotential following Fujita and Ogawa. Total employment in the monocentric city is given by the measure $N = b_2 - b_1$. Integrating over s in (1) we get firm output equal to

$$x(y) = \int_{b_1}^{b_2} e^{-\alpha|y - s|}ds$$
$$= \frac{1}{2}[2 - e^{-\alpha(y - b_1)} - e^{-\alpha(b_2 - y)}]. \tag{5}$$

Then integrating over $x(y)$ we get city output, X, where

$$X = \int_{b_1}^{b_2} x(y)dy - 2\alpha^{-1}[N - \alpha^{-1}[1 - e^{-\alpha N}]] \tag{6}$$

for $N \equiv b_2 - b_1$. In terms of urban agglomeration, the critical feature is the second derivative of (6),

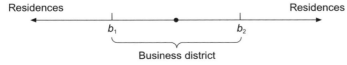

Residences   Residences

b_1 b_2

Business district

Figure 15.1   Information flows in a linear city

where given $dX/dN = 2\alpha^{-1}[1 - e^{\alpha N}] > 0$ for $N > 0$, then

$$\frac{d^2X}{dN^2} = 2e^{-\alpha N} > 0. \tag{7}$$

Output increases at an increasing rate with increasing urban scale, or average output per worker rises with N. Note, despite the introduction of space, the reduced form result in (5) and (6) is a special case of (1). There is 'structure' under the black box.

A dissatisfaction with (4) is that communications are exogenous. People do not choose to communicate, with each having an endogenous message level. Kim (1988) shows how to endogenize the relationship. Suppose net output, or profits for a firm at y in Figure 15.1 are given by

$$\pi(y) = \int f(V(s,y))\, ds - \alpha \int |s - y| \cdot V(s,y) ds. \tag{8}$$

In (8) $V(s,y)$ is the endogenous number of messages firm y chooses to receive from s, where there is a cost per message from s of $\alpha|s - y|$, which is increasing in distance. A firm has to work harder to hear what is being said at s, as $|y - s|$ increases. Alternatively, say, messages are received by face-to-face interaction and each message received by y involves a commuting cost of $\alpha|s - y|$. Now suppose $f(V(\cdot))$ takes the functional form

$$f(V) = V - V\log V, 0 < V < 1$$

where $f' = -\log V$. Then maximizing (8) with respect to the V for each s, y pair, the first-order condition is

$$-\log V = \alpha|s - y| \quad \forall s \tag{9}$$

noting $V < 1$. Thus

$$V = e^{-\alpha|s - y|}$$

and $\pi(y)$ in (8) reduces to

$$\pi(y) = \int_{b_1}^{b_2} e^{-\alpha|s - y|} ds$$

which gives us the same form as (5). Thus underlying the black box in (1) can be a simple reduced form as in (6) with endogenously chosen communications, where city output per worker rises with city size.

Two types of extensions come to mind. First, Fujita and Ogawa (1982) explore the message relationship in multicentred linear cities with two or three business districts. The result with the expotential form to $g(\cdot)$ is ever-increasing scale benefits to urban size since firms in very distant centres still receive messages from others and always benefit from more firms and centres. With a linear form to $g(\cdot)$, at some distance, centres cease to communicate. In Fujita and Ogawa (1982) centres form through 'self-organization' of firms (Krugman, 1993), and there are typically multiple-equilibria urban configurations (any configuration that is locally 'stable' to local perturbations of business district and residential borders). In Henderson and Slade (1993) and Henderson and Mitra (1996), developers form centres choosing business district locations strategically to enhance or diminish communications.

The second extension is to consider the nature of face-to-face communication versus telephone or electronic communication. Gasper and Glaeser (1996) ask whether reduction in the effective cost of electronic/telephone communications reduces the demand for the face-to-face communications upon which urban information spillovers are based (the conversations among workers and managers of different local firms in the same industry and with buyers, suppliers and regulators). In this model, endogenous initial contacts are made by telephone, and if the relationship seems very productive, more contacts are made face-to-face (as opposed to contact ceasing or to more contact by phone for moderately productive matches). If the cost of telephone calls drops, more initial contacts are undertaken leading ambiguously to more face-to-face contacts (as opposed to increased usage of telephone for further contact). In general, telephone and face-to-face contacts may be complements, so modern telecommunications can lead to enhanced productivity of face-to-face meetings and urban agglomeration.

Search and Matching Externalities

Workers are heterogenous in skills and abilities and firms collectively in an industry need to choose production techniques to match the skills of their workers. (Alternatively, heterogenous firms need to find workers with specific skills.) But there is always uncertainty in any market about the availability of different skill workers. In a search-matching model, the larger the market the more likely a firm is to find matches. Helsley and Strange (1990) formalize these notions as follows.

Each city gets a drawing of N heterogenous workers, from a universe of workers distributed uniformly by skill over the unit circle (no ranking *per se* of skills here – just pure heterogeneity). Each worker in the city has then an ex post address y on the unit circle, but those addresses are unknown to firms at the time they make decisions.

The m firms in a city must choose their own address s and implied labour market segments they hire in, so as to maximize expected profits. For a worker with address y and a firm with address s the value of a match, or the output from the match is

$$\max[0, \alpha - B|s - y|].$$

The value of matches declines with distance between the worker and firm, reflecting decreasing quality of the match. With m identical firms on the unit circle, only knowing that city workers are drawn from a uniform distribution, a Nash equilibrium in choice of s's will have the firms evenly spread over the unit circle, so each labour market area is $(s - (2m)^{-1}, s + (2m)^{-1})$, as depicted in Figure 15.2.

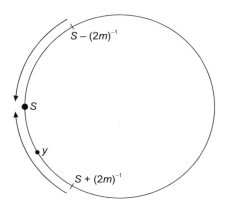

Figure 15.2 Matching firms and workers

For any firm, the expected distance to any worker in the firm's market area is

$$\int_{s-(2m)^{-1}}^{s} |s - y|(2m)dy = (4m)^{-1} \qquad (10)$$

where we are looking at one side (or the other) of the market (symmetrical about s) and $1/(2m)^{-1} = 2m$ is the density function for possible locations of workers on that side of the firm's market. Given expected distance $(4m)^{-1}$, expected output for any worker is

$$\alpha - B(4m)^{-1}.$$

The firm's fixed cost of production is C (apart from labour costs) and its total expected employment is N/m. Thus expected net output per firm $x(s)$ is

$$x(s) = \frac{N}{m}(\alpha - B(4m)^{-1}) - C. \qquad (11)$$

Total city output, X, is $m*x$, or

$$X = N(\alpha - B(4m)^{-1}) - Cm. \qquad (12)$$

Solving for the equilibrium number of firms in a city (given N workers with unknown addresses) is a little complicated (see Helsley and Strange, 1990 for a solution), so here I just solve for the

optimum. Maximizing X, with respect to m, we see there is a trade-off between increasing expected match productivity by increasing m and incurring more fixed cost increments of C. Increasing m increases plant productivity by lowering market segments a firm hires in and reducing chances of getting poorly skilled workers for that firm. That is, each firm 'searches in' or actually hires in a more homogeneous sub-market. Maximizing (12) with respect to m gives

$$m = B^{1/2}(2C^{1/2})^{-1}N^{1/2}.$$

Substituting for m back into (12) gives us expected city output

$$X = \alpha N - B^{1/2}C^{1/2}N^{1/2}. \qquad (13)$$

In (13), as in (6), $dX/dN = \alpha - \frac{1}{2}B^{1/2}C^{1/2}N^{-1/2} > 0$ for α or N large enough and thus

$$d^2X/dN^2 = \frac{1}{4}B^{1/2}C^{1/2}N^{-3/2} > 0. \qquad (14)$$

Thus expected city output rises at an increasing rate with city size, N. Or, average output per worker rises with city size. Again the black box in (1) corresponds to the reduced form of a micro-foundations model.

In a related paper, Helsley and Strange (1991) model another matching problem, beyond the scope of Marshall. I am going to modestly respecify their model, without changing the essence. In period 1 entrepreneurs enter a city and undertake local projects involving committed investment. For any entrepreneur, the quality of the match of the project with his or her specific skills is unknown at the time they commit to the project. If the quality of the match is good, the entrepreneur continues with the project in the next period. If the quality is bad, he or she wants to sell the local project (of now known character-istics to just local residents) to another local resi-dent looking for a project. For example, in the second period, all those with bad matches would want to trade. The expected value of match improvement and the opportunities to trade up are both improved with increases in the size of the local market, or city size.

The basic principle is related to concepts from the insurance literature. On projects whose risk (of a bad match) cannot be insured against, oper-ating in a larger market provides better insurance (through improved opportunities for resale) against poor matches. More generally, in words from Hoover (1948: 288), where the idea seems to originate:

> Diversification affords some insurance against total collapse of the economic life of an area . . . The greater the variety of industries in the area the more remote is the possibility that all or a major part of the area's industries will hit the skids at the same time.

Adam Smith Specialization and Local Diversity

Efficiency in production of, say, a city's export good is improved by diversity in the local (non-traded) intermediate goods sector. Such inputs include business services, special-order materials, parts and machinery, repair services, and the like. Increased 'specialization' in this sector, or increased diversity of the array of products to use as intermediate inputs, increases efficiency of export producers, since they can find the specific part, service or material they want (implicitly, they get better matches on precise inputs they want, versus what is offered).

There are two ways to model this. Abdel-Rahman and Fujita (1990) use the mechanical formulation in Dixit and Stiglitz (1977) as adapted to production to model diversity *per se*. Becker and Henderson (1996) adapt the Becker and Murphy (1992) model of Adam Smith specialization among firms, which is more directly connected to Marshall's notion. However, as I will point out, the two approaches are intimately linked, with the latter being a particular, more complex version of Dixit and Stiglitz specifications of diversity.

In Abdel-Rahman and Fujita (1990), final city output Z is produced by competitive firms using inputs of labour n_z and diverse intermediate inputs x_i, where x_i is the ith variety of x which is available, for $i = 1, \ldots m$, with m endogenous. For a given m, competitive firms in the Z sector have constant returns to scale, so we may aggregate and work with the industry production function

$$Z = N_Z^\alpha (\sum_{i=1}^{m} X_i^\rho)^{(\frac{1-\alpha}{\rho})}, 0 < \rho < 1. \quad (15)$$

In this Dixit and Stiglitz formulation ρ is a measure of substitutability among varieties. As $\rho \to 1$, inputs become more perfectly substitutable (are almost additive in the production function), so diversity of inputs is unimportant to Z efficiency. Diversity is measured by the magnitude of m, indicating the number of varieties available, where in this simple formulation all varieties enter (15) symmetrically. If $\rho \to 1$, having more X through greater m (more X_i varieties) is almost the same as having more of any one X_i. If $\rho \to 0$, inputs become more complementary. For small ρ, greater m is very different from, and more valuable than increases in any X_i.[1] What limits the number of varieties lies on the cost side, where the amount of labour N_i needed to produce any amount of X_i is

$$N_i = f + cX_i. \quad (16)$$

In (16), each new variety involves a fixed cost (in labour units), f, as well as constant marginal costs.

Total employment in the city $\bar{N}$ is

$$\bar{N} = N_z + \sum_{i=1}^{m} N_i. \quad (17)$$

In a Dixit and Stiglitz model, given X's enter (15) symmetrically, producers are perfectly symmetrical in choices, and the equilibrium *and* optimal employment and output for any X_i are respectively

$$N_i = \frac{f}{1-\rho} \quad \text{(a)}$$

$$X_i = \frac{\rho f}{(1-\rho)c} \quad \text{(b)} \quad (18)$$

Substituting (18a) into $\bar{N} = N_z + mN$ for N (where $N = N_i = N_j \forall i, j$) and then substituting that for N_z and (18b) for X_i into (15) and optimizing with respect to m gives

$$m = f^{-1} \bar{N} \frac{(1-\rho)(1-\alpha)}{1-\alpha+\alpha\rho} \quad (19)$$

Note the entire optimal solution in (18) and (19) can be directly solved by substituting (17) for N_z in (15), substituting (16) for X_i in (15) and maximizing the resulting expression with respect to m and X_i. The resulting two first-order conditions with rearrangement give (18a) and (19) directly.

In contrast to the optimum, the equilibrium m takes several pages to solve, because of the complexity of solving for demand relationships under monopolistic competition. While N and X are given in (18), m is smaller in the equilibrium (too few varieties). Abdel-Rahman and Fujita (1990) discuss urban subsidization policies (undertaken by local governments) to achieve optimal varieties, given the externality involved (entrants to the X industry do not recognize the benefit to Z producers of the increased local diversity).

If we substitute for X_i from (18b), N_z from (17) given N_i in (18a), and then for m into (15), the result is

$$Z = A\bar{N}^{(\frac{1-\alpha+\alpha\rho}{\rho})} \quad (20)$$

where A is a parameter collection. In (20) given

$$\frac{dZ}{d\bar{N}} = A\left(\frac{1-\alpha+\alpha\rho}{\rho}\right)\bar{N}^{(\frac{1-\alpha-\rho+\alpha\rho}{\rho})},$$

$$\frac{d^2Z}{d\bar{N}^2} = A\frac{(1-\alpha+\alpha\rho)}{\rho}\frac{(1-\alpha-\rho+\alpha\rho)}{\rho}$$

$$\bar{N}^{(\frac{1-\alpha-\rho+\alpha\rho}{\rho})-1} > 0,$$

given $1-\alpha-\rho+\alpha\rho > 0$ (noting $(1-\alpha) > \rho(1-\alpha)$, for $\rho < 1$). As in other cases, the black-box technology in equation (1) has micro-foundations yielding a reduced form, where average output per worker rises with city size.

Abdel-Rahman and Fujita (1990) is a diversity

version of agglomeration economies. Becker and Henderson (1996) present an adaptation of Becker–Murphy specification of Adam Smith specialization. Let us look at that, and then see why it really reduces to a Dixit and Stiglitz type formulation, albeit perhaps richer than that in equation (15).

There is a continuum of tasks all on the unit circle which firms in the Y industry must complete in order for final city output to result. But firm specialization in a limited segment of tasks will improve efficiency, through limiting the diversity of tasks any entrepreneur faces and allowing the entrepreneur to devote more time to mastering and administering any one task. Specifically, if tasks on the unit circle are indexed by s, for the entrepreneur undertaking s, potential output from s is

$$Z(s) = ET(s)^{\gamma}n(s)^{\delta}$$
$$0 < \gamma, \delta < 1; \gamma + \delta > 1. \qquad (21)$$

$T(s)$ is the time the entrepreneur devotes to learning about task s and supervising labour at task s, where $n(s)$ is the quantity of labour employed at s. Without loss of generality, we assume each entrepreneur spans a contiguous set of tasks of width w on the unit circle. Since it is efficient for the entrepreneur to split his or her time (and hire labour) equally across tasks (equation (21) is the same at all tasks), we drop the s index so $T = T(s)$. Each entrepreneur has one unit of time, so each is constrained by

$$Tw = 1. \qquad (22)$$

Now for the tricky and uncomfortable part of Becker–Murphy. Given all tasks on the unit circle 'must be performed', firm output is

$$z = w \cdot \min_{0 \le s \le 1}\{z^{(s)}\}. \qquad (23)$$

In this Leontief-type world, the minimum $y(s)$ of any firm on the circle gives us all other $y(s)$'s; and firm input is its span of tasks w, multiplied by this minimum $y(s)$. Imposing a best Nash equilibrium (see below) where all firms are symmetrical, for m firms in the city and industry

$$wm = 1. \qquad (24)$$

From (21) and (23) firm output is $z = wET^{\gamma}n^{\delta}$, but from (22) and (24) $T = w^{-1}$ and $w = m^{-1}$ so

$$z = Em^{\gamma-1}n^{\delta}.$$

City output in mz or

$$Z = Em^{\gamma}\bar{N}^{\delta}, \gamma + \delta > 1. \qquad (25)$$

Note total city employment $\bar{N}$ is $n(s)$ times the circumference of the unit circle which is 1. Now, in (25) there are two factors of production, with joint agglomeration economies given $(\gamma + \delta) > 1$. If we fix m/N at the national endowment ratio, we can rewrite (25) as

$$Z = EA^{\gamma}\bar{N}^{\gamma+\delta}, \qquad (26)$$

in which $\partial Z/\partial \bar{N} > 0$ and $\partial^2 Z/\partial N^2 > 0$, our agglomeration economy requirement.

The problem with this Becker–Murphy formulation lies in the Leontief specification, requiring all tasks on the unit circle to be completed. That means there will be a coordination failure problem. For example, if firms divide the unit circle non-symmetrically, that is a Nash equilibrium since if any firm deviates and unilaterally changes its span of output it loses. Either firm and city output goes to zero (if it reduces its span) or firm and city output drops (if it increases its span, duplicating some of another firm's tasks and reducing $y(s)$ at existing tasks for the same firm labour force, given the entrepreneur's time is further splintered). The problem is we want the city's span of tasks to be endogenous, not just the firm's. To endogenize the city's span of tasks takes us back to Dixit and Stiglitz.

In particular, assume city output (again as in Abdel-Rahman and Fujita) is produced by competitive firms aggregated up so city output is simply

$$Z = (\int z(s)^{\rho}ds)^{1/\rho}.$$

Each firm produces a span of products, as indexed by s, of width w, so $Z = (\int_0^{w_1} y(s)^{\rho}ds + \ldots + \int_{w_{m-1}}^{w_m} y(s)^{\rho}ds)^{1/\rho}$ for m firms. $z(s)$ has the same form as in (21) so $z(s) = ET(s)^{\gamma}n(s)^{\delta}$ subject to $\int T(s)ds = 1$ for each entrepreneur. To limit each firm's w, there are coordination costs to the firm, of Bw^{β}, increasing in w. Firms have to choose w as well as n and T. Under symmetry total city output is $Z = (mw^{1/\rho}z = m^{(1-\rho\delta)/\rho}w^{(1-\rho(\gamma+\delta))/\rho}\bar{N}^{\delta}$ (noting $nwm = \bar{N}$ with symmetry so $n(s) = n$ and $T = w^{-1}$), before coordination costs of $mw^{\beta}B$. There is no simple reduced form to this problem, which is why the Becker–Murphy technology is used. Note that, for a given $m/\bar{N}$ ratio of A, controlling for w, Z is increasing at an increasing rate in $\bar{N}$ (given $Z = A^{(1-\rho\delta)/\rho}w^{(1-\rho(\gamma+\delta))/\rho}\bar{N}^{1/\rho}$).

THE NATURE OF AGGLOMERATION ECONOMIES FOR SPECIFIC INDUSTRIES

A big issue that emerged in the 1960s (e.g., Chinitz, 1961), following on Hoover's (1948) early work, concerned the nature of scale economies for any specific industry. So far, we have really been emersed in a one final good world. Turning to the world of a myriad of final products, what is the scale argument in equations (1), (6), (13), (20) and (25) above for any specific industry? For any industry, in particular, is it own industry local employment (N), is it total local employment in all industries ($\bar{N}$), or is it some measure of the diversity of the total local environment? And how local is local – is it neighbourhood, county/district, basic metropolitan area, urban conglomeration or mega-city region? In part these are empirical issues, but one conceptual issue stands out.

Urban economists debate whether external scale effects for a plant are ones of *localization*, deriving from the scale of just the *own* industry locally, or ones of *urbanization*, deriving from the overall scale/diversity of total local employment. For the former, the notion is that a plant is most concerned with learning from others doing similar things. So, for example, a plant producing a standardized product is most concerned with the details of precise specifications of machines to buy, exactly who to buy inputs from, how to access unfamiliar markets locally or further away, how to deal with local implementers of regulations and the like. This information comes from others in your own industry who have the same concerns. On the one hand, Jacobs (1969), argues that, for the latter, plants need new ideas which can only come from being in diverse (and hence large, see later) environments. So, for example, a plant engaged in R&D learns from a diversity of inventive activity and needs a diverse labour market. A high fashion apparel producer needs to be at the centre of the action in a big city of sophisticated consumers, publishers and high profile purchasers.

The distinction as to the nature of scale economics has enormous implications for the spatial organization of production into cities. As Henderson (1974) theorized, if economies are ones of localization, cities will tend to specialize in production of just one traded good (or one set of interrelated goods). Then there will be different types of cities in the economy, each type specialized in a different good, where different types of cities tend to have different typical sizes. Since to some large extent this is what we see in an economy (see Alexandersson, 1959; Bergsman et al., 1972; Henderson, 1988), it is important to understand why.

Despite urban agglomeration economies, all production in most economies does not occur in just one mega-city. City sizes are limited because of other diseconomies to sizes of cities. The typical modelling of diseconomies in the new urban economics is based on escalating commuting, congestion and land rent costs (Henderson, 1974; Mills, 1967). The older location literature (Beckmann, 1968, as well as also in Mills, 1967) and the new economic geography literature (Fujita, et al., 1995), stress the limitations (due to shipping costs) to the extent of the agricultural and regional market of any one city. The commuting congestion cost argument is based on the notion (in Figure 15.1) that firms must cluster together in a city to enhance information spillovers (and, in the old days, ship from the same central transport node). Workers then commute from the surrounding residential area. As the city size grows, commuting distances and rent costs per worker rise. Breaking a city into multi-centres as it grows is a strategy to limit this problem (by again shortening commuting distances), but it is a strategy which also limits agglomeration benefits since information spillovers are diminished. Regardless, overall at some size in either a single-centred (Henderson, 1974; Mills 1967) or multi-centres (Henderson and Slade, 1993) city, the marginal scale benefits to increasing city size are met by and then exceeded by the marginal commuting–rent–environmental costs.

Presuming these limitations to city sizes, if economies for an industry are ones of localization, cities will most certainly specialize. Why? There are no scale benefits to be realized from putting two unrelated industries together, and only scale diseconomies in the commuting–residential sector. If two industries are together, splitting them into different cities would lower commuting–rent–environmental costs, with no loss of productivity.

Economies being ones of urbanization does not eliminate specialization *per se*. If the degree of urbanization economies differs across industries in a simple function like equation (1), then different industries will be best off in cities of different sizes. However, respecifying the form of scale in (1) to involve diversity, not just pure size, means efficiency for industries experiencing urbanization economies would demand diverse environments *per se*. Abdel-Rahman and Fujita (1990) model mixed systems of cities, where there are both specialized and non-specialized cities.

The empirical evidence tends to support all this reasoning and we turn to basic results now. There is a substantial literature from the late 1970s and mid-1980s estimating the magnitude of scale economies and their nature, most notably Sveikauskas (1975, 1978), Nakamura (1985) and Henderson (1986). The focus concerns the elasticity, ε, of firm output with respect to the local scale of the own industry. In equation (1)

$$\varepsilon \equiv \frac{N}{X}\frac{\partial X}{\partial N} = \frac{dG(\cdot)}{dN} \cdot \frac{N}{G(\cdot)} \tag{27}$$

which measures the percentage change in firm output with respect to a change in own industry local scale, with no change in firm behaviour (inputs). To estimate the elasticity in equation (27) one can work with a trans-log specification of the production functions in (1), instrumenting to try to control for endogeneity of firm inputs (Nakamura, 1985). Alternatively, we can apply

Shepard's Lemma to the cost function in (3) where $\partial c/\partial w = n/x$, for w the wage rate and n/x the ratio of firm labour to output. Then, the estimating equation in logs is

$$log\,(x/n) = \log g(N) + \log [(\partial c(\cdot)/\partial w)^{-1}].\quad (28)$$

The estimation procedure is to approximate the second term on the right hand by a polynomial in prices (local wages and gross capital rental rates, including industrial property tax rates), so output per worker is a function of this polynomial and scale measures in $G(\cdot)$. Two common functional specifications for $G(N)$ are

$$G(N) = N^\varepsilon \qquad\qquad (a)$$

$$G(N) = e^{-\phi/N}, \text{so } \varepsilon \equiv \phi/N. \quad (b) \qquad (29)$$

The first formulation is a constant elasticity and the second a declining one. If estimated at the industry level for the city, then certainly endogeneity and instrumenting issues arise in the relationship between N and the error term also affecting X/N.

Sveikauskas, Nakamura and Henderson all conclude that for most standardized manufacturing products the appropriate measure of scale in (27)–(29) is local own industry employment. Table 15.1 reports summary results for the USA, Brazil and Japan based on 1970s data, on the

Table 15.1 *Degree of economies of scale: estimates of elasticity of firm output with respect to total local own industry employment*

	USA[a]	Brazil[b]	Japan[c]
Primary metals	0.12	0.10	0.06
Non-electrical machinery	0.07	0.04	0.08
Apparel	0.07	0.02	0.06
Textiles	n.a.	0.06	0.04
Pulp and paper	0.05	0.07	0.01
Food processing	0.10	0.02	0.03
Electrical machinery	0.11	n.a.	0.08
Transport equipment	n.a.	0.07	0.04

(a) For the USA, the underlying formulation implies declining elasticities. The elasticity quoted is evaluated at 1,000 employees. From Henderson, 1988.

(b) For Brazil, the formulation implies declining elasticities. The elasticity quoted is evaluated at 1,000 employees. From Henderson, 1988.

(c) The estimates for Japan are for a constant elasticity formulation. They are from Nakamura, 1985.

The USA and Brazil results are based on OLS regressions, while for Japan 3SLS was used. For the USA, based on Hausman tests, endogeneity is an issue for textiles, apparel and food, but I report OLS coefficients. 2SLS coefficients were double in magnitude. For textiles the results appear unstable.

magnitude of these localization economies. In Table 15.1, a typical elasticity is, say, 0.07, which says a 10 per cent increase in own local industry employment increases firm output by 0.7 per cent which is a good return to the firm for doing nothing. Henderson (1988) demonstrates for these products that there is a high degree of urban specialization so that there are steel, textile, pulp and paper, food processing, auto, diverse manufacturing etc., type cities and that average sizes of these cities in the USA and Brazil are related to both degrees of localization economies and capital and skill intensity in these industries.

For industries producing non-standardized products, such as portions of the apparel industry, of furniture, and of publishing, evidence is stronger for urbanization economies. This is also true of high-tech industries (Henderson et al., 1995) and we suspect for R&D activities (Fujita and Ishii, 1994) and portions of the financial and business service sector. Certainly these are the industries more concentrated in large diverse metro areas.

More recent work tends to support these earlier results but focuses on some other related issues. Ciccone and Hall (1996) worry that the appropriate measure of scale ought to be employment density (as in Figure 15.1), or how closely packed together own industry workers are. They find evidence supporting this notion. Another related issue concerns the question of what is 'local'? Jaffe et al. (1993), in exploring patterns of patent situations, find evidence suggesting that information spillovers and hence localization economies diminish sharply over space. Nakamura (1985) experimented with sorting out scale effects from localized employment in the political city versus the bigger prefecture geographic unit, and finds in favour of the former.

Another issue concerns the measure of urbanization scale. While earlier authors simply use a measure such as total local populations or total local employment, Henderson and Kuncoro (1996) experiment with direct diversity measures such as a Hirschman–Hefindahl index of sum of squared employment shares of different local industries. The Hirschman–Hefindahl index for city j would be

$$HHI_j = \sum_{i=1}^{l}(S_{ij})^2, S_{ij} = \frac{N_{ij}}{N_j} \qquad (30)$$

where N_{ij} is employment in industry i in city j, Nj is total local employment in city j and l is the number of industries. With perfect diversity so all S_{ij} are equal, $HHI_j = l^{-1}$. With total lack of diversity $HHI_j = 1$, so all local employment is in one industry. An increase in HHI lowers diversity and should hurt productivity. Ellison and Glaeser (1997) suggest refining (30) to normalize for national employment shares of an industry $N \bar{S}_i$ ($= \bar{N}_i/\bar{N}$ for $\bar{N}_i$ national employment in industry i and $\bar{N}$ total national employment). This is especially important in comparisons over time where $\bar{S}_i$ may change, driving (as a trend) some of the changes in S_{ij}. A refined diversity index could be

$$g_j = \sum_{i=1}^{I} (S_{ij} - \bar{S}_i)^2.$$

Henderson and Kuncoro (1996), using (30), in a cross-sectional context for base period existing firms (with appropriate instrumenting), find strong evidence of diversity effects for new manufacturing plants in Indonesia, along with localization economies. Their finding of fairly universal diversity effects, in contrast to results in earlier work, could be because of a refined urbanization measure (although diversity and urban size are highly correlated). However, it could be the context: new industrial plants in a developing country (in the non-corporate sector, moreover).

The final topic of recent empirical focus concerns another issue about scale economies – are they static or dynamic? This question derives from theoretical work of the past decade or so, and is so important that I devote the third section of this chapter to it.

DYNAMIC EXTERNALITIES

The externalities in (1) are specified to be static ones, in the sense that only N today in $G(N)$ affects firm output today. So only immediate external scale matters. But there are reasons to believe that history, or the patterns of prior scale, matters for some dimensions of Marshall's externality notions. In considering information spillovers, there are two conceptual considerations.

First, there may be a transmission and testing mechanism for new ideas. One firm learns of a new technique or new supplier and that information only gets imparted to others through casual contacts among workers, information supplied by observers such as suppliers and purchasers who visit that plant, or through regulators. That transmission takes time. Moreover, there may be a testing period where other firms watch to see if the idea works out. So it is not today's scale of information flows, but last year's, the year before's, and so on that matters, as information goes through a transmission and filtering process. Also relevant is the notion that all this communication requires an established functioning social communications network, something which takes time to develop in newly occupied locations.

The second conceptional consideration involves notions of local stocks of knowledge. Producers in a location individually learn and information spills over among them, creating a stock of local knowledge about how to produce and market a product. These are Romer (1986) knowledge accumulation notions, applied to local areas. While in Romer knowledge accumulations are universal or transmission of information across locations is instantaneous, here, given evidence in Jaffe et al. (1993), some accumulation is entirely local. Localities accumulate a stock of local trade secrets, that a firm can only learn in a reasonable period of time by joining the location, and which may or may not be of use if a firm exits and takes that knowledge elsewhere. So there is a general versus location-specific knowledge distinction here, quite apart from how quickly information spreads across locations. Some local trade secrets about who to buy inputs from locally, how to deal with local regulators, perhaps who to bribe at what price, how to negotiate tax and regulatory breaks, what consultants to hire, etc., are really location-specific knowledge, that gets built up over time.

Quite apart from intellectual curiosity about the nature of scale effects, dynamic externalities matter critically for several reasons. First, if lagged rather than just current scale effects matter, the potential for multiple equilibria based on different histories is greatly enhanced. Secondly, because history matters, it becomes more difficult for locations to change production patterns. In terms of policy, attempts successively to decentralize production or help lagging regions in a country may be doomed because locations with little history of producing a product are so information-deficient. Finally, in urban growth models, local knowledge accumulations may enhance, or fuel city growth and may also lead to dynamic inequality across cities over time.

In this section, we examine the theory of dynamic externalities and then turn to empirical evidence. Since these concepts are so new, there is little published literature on the subject. Some of what I discuss are ideas still in the formation stages. But it is clear for agglomeration economies, dynamic considerations will dominate analyses for the next several years.

Theory

I review two different theoretical approaches here. The first is from Rauch (1993), with significant refinements by Mitra (1995). Then I turn to notions from the endogenous growth literature. Rauch[2] specifies a simple model where equation (1) could be rewritten as

$$x(t) = G(N(t-1)) \cdot x(k(t)) \tag{31}$$

or, scale effects lagged one period $(t - 1)$ affect output in time t. Equation (31) can be generalized to

$$x(t) = G(N(t-2), N(t-2)...N(t-n)) \cdot x(k(t)).$$

Rauch then considers the following experiment. Suppose we have two regions, A and B, where A is inherently more productive than B. For example, in (31) we could have $G(N_A(t-1), F_A)$ and $G(N_B(t-1), F_B)$ where $\partial G/\partial F > 0$ and $F_A > F_B$, for F being fixed locational amenities. Suppose as a historical accident all production starts in B. The questions are, under what conditions will all production shift to A, how will the shift occur, and what will the transition look like.

The problem in shifting production from B to A is that, in the first period in which producers are in A, they will experience no scale benefits (i.e., $N_A(t-1)$ will be zero), so production (within limits on the shape of G and $F_A - F_B$ magnitude) will be initially very inefficient in A compared to B with its historical advantage. That has two implications. First, it may not be efficient to switch production from B to A. The discounted gain in productivity over time from an improvement in F may not be worth the transition costs, or productivity loss from operating initially with poor information. Note this would not be the case with static externalities: if $F_A > F_B$ the efficient solution is to shift all producers at once, immediately getting full externalities in A. Secondly, even if it is efficient, production may not shift. In particular, no firm in a simple Nash game would want to deviate from the current locational choice of B to move to A. Doing so would involve a big initial production loss given $N_A(t-1) = 0$. So there is a first-mover problem . . . 'who will bell the cat?'. Both problems indicate that multiple equilibria based on history can exist. If all production starts in B, it will stay there. If it starts in A, it will stay there also.

In considering transition possibilities, Rauch (1993) and the more formal and complete treatment in Mitra (1995) consider several issues. The first concerns the efficient transition process. In general, in moving production from B to A one does not want to do the move all at once – the loss for everyone from A's initial deficient history is too great. Instead, the idea is to stage the transition, first building N_A up slowly and then to move all producers remaining in B to A at some point. For the latter, for example, if total scale is $\bar{N}$, as soon as $G(N_A(t-1), F_A) \rightarrow G(\bar{N} - N_A(t-1), F_B)$, you would want to move all remaining producers.

The second issue concerns under what conditions the transition from B to A will occur, if it is efficient. If there are 'large agents' such as either local government or land developers who own/control activity in A, generally by subsidizing initial movers and later collecting high compensating rent differentials (reflecting the benefits of being in A once history is established there, over returning to B, or going elsewhere), large agents can (1) ensure transition will occur if it is efficient and (2) orchestrate an efficient transition process. If there are no large agents but only atomistic agents, where any moves depend on Nash-consistent behaviour of (small) firms, we have the classic first-mover problem. Mitra (1995) shows that not all is lost in this case. In more sophisticated games, first movers from B to A recognize that, if they move, that may trigger successive moves by potential stayers (since the initial move will improve $N_A(t-1)$ for later movers). Given that, it may pay any firm to move first because the later-triggered moves will in the long term increase that firm's discounted profits. Then some initial number of firms may deviate and move from B to A creating a transition process.

The second type of dynamic externality formulation is based on human capital formation in the endogenous growth literature (Romer, 1986). Here human capital formation is quite general – on-the-job experience in local production, as well as formal education. In this case, equation (1) gets rewritten as

$$x_{ij}(t) = G(h_i(t), \bar{h}_j(t), N_j(t))x(k(t)). \tag{32}$$

$x_{ij}(t)$ is output of firm i in city j in time t. $h_i(t)$ is the human capital of the entrepreneur in firm i in time t, reflecting an accumulation of private investment in h_i over time. $\bar{h}_j(t)$ is the average level of human capital of entrepreneurs in city j, reflecting knowledge spillovers. A firm's productivity is improved if all firms around it are more experienced. Finally, $N_j(t)$ reflects either or both traditional static agglomeration economies and the fact that total human capital stocks, as well as averages in cities may improve any firm's productivity.

Henderson and Becker (1996) show that, in a simple endogenous growth model in an urban system, if h_i and $\bar{h}_j$ (which will be equal in a representative (identical) entrepreneur model) grow endogenously over time, individual city sizes will also increase over time. Growth in $\bar{h}_j$ accentuates urban scale economies – for example, the information that spills over (proportional to $N_j(t)$) is richer in quality (proportional to $\bar{h}_j$). Expanding scale benefits relative to urban diseconomies in commuting leads to increasing efficient and equilibrium city sizes. And, of course, technological improvements will lower commuting diseconomies, spurring city growth. Consider the impact of intra-city rail transit, the car and intra-city highways on urban form and size in the past hundred years.

Equation (32) does not specify whether knowledge is transportable across cities and/or industries. Based on an important paper by Benabou (1993), we know this can be a critical issue. Adapting Benabou, consider an inter-generational model in which all people are born equal but must choose one of two occupations. One occupation requires considerable human capital investment and the other little. However, the productivity of human capital investment is improved by neighbourhood peer group effects in an inter-generational model, in which a child's human capital formation process is enhanced by parents' skill level and by the skill composition of adults in the child's neighbourhood – children of highly skilled workers will be more efficient in human capital formation, and that will be enhanced by living in high-skill neighbourhoods. Under general conditions this model can lead to neighbourhood stratification between high- and low-skill people, and evolving inter-generational inequality.

In equation (32), if there are neighbourhood or city-specific effects, entrepreneurs become committed to a city by accumulating a stock of location-specific human capital, where desired accumulation may vary across types of cities (given different human capital spillover and private benefits across industries). That will lead to different human capital accumulations across cities; and, in an inter-generational context (given differential spillovers like neighbourhood peer group effects), can lead also to evolving inequality.

Evidence

Completed works by Glaeser et al. (1992), Miracky (1992), Henderson et al. (1995) and Henderson and Kuncoro (1996) try to establish the existence of dynamic externalities and to some extent their magnitude. In pioneering work Glaeser and co-authors examine whether 1987 employment in cities is affected by 1956 industrial environment conditions (localization, urbanization measures, and the like). They draw a sample of the six largest industries in each city in 1956 and regress 1987–56 employment growth for those industries collectively on base 1956 conditions. While they find a positive answer for some 1956 conditions, their work is flawed by the sample drawn, which is mostly wholesale, retail and construction activities, and, even within manufacturing, mostly in declining industries once strong in the USA and now of more minor importance. As stated earlier, we might expect certain externalities to be most important for firms in rapidly developing and evolving industries, as opposed to firms looking for a graceful exit.

Miracky (1992) and Henderson et al. (1995) focus on the sample for any industry of all cities and separate out industries. Henderson et al. in particular focus on key traditional manufacturing industries and on key new high-tech industries, looking at employment growth and location choice for the period 1970–87. They find strong evidence of only 'dynamic' localization externalities (MAR: Marshall–Allen–Romer externalities) from traditional manufacturers, but evidence of both dynamic localization and dynamic urbanization (Jacobs externalities) for new high-tech industries. In a related paper on Indonesia, Henderson and Kuncoro (1996) find evidence of both base period localization and urbanization measures affecting location decisions and inferred profitability for new industrial plants in the (smaller plant size) unincorporated sector. They also find that maturity of the local environment (average age of existing other manufacturing plants) positively and strongly affects profitability, perhaps a better indicator of dynamic influence.

All these studies suffer from several problems. The first concerns whether employment today is affected by industrial environment conditions in a base period *per se*; or whether other unobserved time invariant locational attributes are driving both, producing the observed correlation. To try to deal with this classic simultaneity problem, Henderson et al. (1995) and Henderson and Kuncoro (1996) experimented with instrumenting, and also with trying to control for all locational attributes (natural resource deposits, weather, right-to-work laws, etc.). Doing so still leaves externalities intact. But the instrumenting problem is severe – trying to find a long enough list of truly exogenous (for example, geographic/climate variables and potential conditions in neighbouring areas) variables which are still sufficiently correlated with right-hand side variables to have efficiency in estimation.

A second problem is that these estimations based on cross-sectional growth between two distant time periods do not allow us to infer the lag structure in equation (29b) and answer the question of how long history matters. Third, the regressions look just at employment, rather than the productivity growth desired from (29b). While the two are related, employment growth involves its own adjustment mechanisms, with employees reacting to better production conditions with lagged adjustment, including considerations of capital irreversibility, renegotiated union contracts and the like. To sort all this out, what is needed is a panel data set on plant productivity as it relates to plant inputs and a long evolving history of the local industrial environment.

CONCLUSIONS

This chapter has reviewed key concepts and issues concerning urban scale economies. I started with models of micro-foundations for scale economies, which motivated the traditional 'black-box' approach and which also give shape to the notions expressed in older writings by Marshall and Hoover and, later, Chinitz, Jacobs and Mills. We examined single good models of information spillovers with both exogenous spillovers and endogenous acquisition, of search and matching externalities, and of intra-industry specialization and local diversity. All solved for reduced forms in which output per worker rises with urban scale; and each had its own source of market failure and dilemmas in modelling.

In the second section I turned to a discussion of the nature of externalities in a multi-good world, distinguishing localization (MAR: Marshall–Allen–Romer) and urbanization (Jacobs) economies. The nature of scale economies as weighed against urban dis-economies and dis-amenities has strong implications for urban form, implying patterns of urban specialization among smaller cities and diversity among larger metro areas. This part of the chapter reviewed international empirical evidence from the USA, Brazil, Japan and Indonesia, suggesting that most industries benefited from localization economies, but some high-tech and service industries may benefit from urbanization economies.

In the third section I turned to the latest wave of work on urban scale economies, which attempts conceptually and empirically to distinguish static from dynamic externalities. Dynamic externalities suggest that past conditions through slow transmission of information or contribution to a stock of local trade secrets affect productivity today. In contrast to static externalities, even in the presence of 'large agents' such as land developers and city governments who can internalize urban scale economies, with dynamic externalities there is strong persistence in historical location patterns, slowing industrial relocation in the presence of altered comparative advantage. Dynamic externalities also contribute to local urban growth. The empirical evidence on dynamic externalities is at an early stage, but, so far, supports the notion of their existence and strength.

NOTES

1. It is also common to write (15) as

$$Z = N_Z^\alpha (\int_1^m X(i)^\rho di)^{(1-\alpha)/\rho}.$$

Note that

$$\partial Z/\partial m = \frac{(1-\alpha)}{\rho} \ X(i)^\rho. \ (Z(\int X_i^\rho di)^{-1} \to \infty \text{ as } \rho \to 0.$$

2. Rauch, in fact, makes the argument in G the highest N ever recorded up to and including time $t-1$.

REFERENCES

Abdel-Rahman, H. and Fujita, M. (1990) 'Product varieties, Marshallian externalities and city sizes', *Journal of Regional Science*, 2: 165–83.

Alexandersson, G. (1959) *The Industrial Structure of American Cities*. Lincoln: University of Nebraska Press.

Becker, G. and Murphy, K. (1992) 'The division of labor, co-ordination costs and knowledge', *Quarterly Journal of Economics*, 107: 1137–60.

Becker, R. and Henderson, J.V. (1996) 'Notes on city formation', Brown University, mimeo.

Beckmann, M.J. (1968) *Location Theory*. New York: Random House.

Benabou, R. (1993) 'Workings of a city: location, education, and production', *Quarterly Journal of Economics*, 108: 619–52.

Bergsman, Greenston, P. and Healy, R. (1972) 'The agglomeration process in urban growth', *Urban Studies*, October.

Chinitz, B. (1961) 'Contrasts in agglomeration: New York and Pittsburgh', *American Economic Review*, 51: 279–89, 299–302.

Ciccone, A. and Hall, R.E. (1996) 'Productivity and the density of economic activity', *American Economic Review*, 86: 54–70.

Dixit, A.K. and Stiglitz, J.E. (1997) 'Monopolistic competition and optimum product diversity', *American Economic Review*, 67: 297–308.

Ellison, G. and Glaeser, E. (1997) 'Geographic concentration in US manufacturing: a dartboard approach', *Journal of Political Economy*, 105: 889–927.

Fujita, M. and Ishii, T. (1994) 'Global location behavior and organizational dynamics of Japanese electronic firms and their impact on regional economies'. Paper prepared for Prince Bertil Symposium on the Dynamic Firm, Stockholm.

Fujita, M. and Ogawa, M. (1982) 'Multiple equilibria and structural transition of non-monocentric urban configurations', *Regional Science and Urban Economics*, 12: 161–96.

Fujita, M., Krugman, P. and Mori, T. (1995) 'On the evolution of hierarchical urban systems', University of Pennsylvania, mimeo.

Gasper, J. and Glaeser, E. (1996) 'Information technology and the future of cities'. Stanford University, mimeo.

Glaeser, E., Kallal, H., Scheinkman, J. and Shleifer, A. (1992) 'Growth in cities', *Journal of Political Economy*, 100: 1126–52.

Helsley, R.W. and Strange, W.C. (1990) 'Matching and agglomeration economies in a system of cities', *Regional Science and Urban Economics*, 20: 189–212.

Helsley, R. W. and Strange, W.C. (1991) 'Agglomeration economies and urban capital markets', *Journal of Urban Economics*, 29: 96–112.

Henderson, J.V. (1974) 'The sizes and types of cities', *American Economic Review*, 643: 640–56.

Henderson, J.V. (1986) 'Efficiency of resource usage and city size', *Journal of Urban Economics*, 19: 47–70.

Henderson, J.V. (1988) *Urban Development: Theory, Fact and Illusion*. New York: Oxford University Press.

Henderson, J.V. and Becker, R. (1996) 'City formation'. Brown University, mimeo.

Henderson, J.V. and Kuncoro, A. (1996) 'Industrial decentralization in Indonesia', *World Bank Economic Review*, 10: 513–50.

Henderson, V. and Mitra, A. (1996) 'The new urban landscape: developers and edge cities', *Regional Science and Urban Economics,* 26: 613–43.

Henderson, V. and Slade, E. (1993) 'Development games in non-monocentric cities', *Journal of Urban Economics*, 34: 207–29.

Henderson, J.V., Kuncoro, A. and Turner, M. (1995) 'Industrial development in cities', *Journal of Political Economy*, 103: 167–90.

Hochman, O. (1977) 'A two-factor three-sector model of an economy with cities: a contribution of urban economics and international trade theories'. Ben Grurion University, mimeo.

Hoover, E.M. (1948) *The Location of Economic Activity*. New York: McGraw-Hill.

Jacobs, J. (1969) *The Economy of Cities*. New York: Vintage Press.

Jaffe, P., Henderson, R. and Tratjenberg, M. (1993) 'Geographic localization of knowledge spillovers as evidenced by patent citations', *Quarterly Journal of Economics*, 108: 577–98.

Kanemoto, Y. (1980) *Theories of Urban Externalities*. Amsterdam: North-Holland Press.

Kim, H.S. (1988) 'Optimal and equilibrium land use pattern in a city: a non-monocentric approach'. PhD Dissertation, Brown University.

Krugman, P. (1993) 'First nature, second nature and metropolitan location', *Journal of Regional Science*, 34: 129–44.

Marshall, A. (1890) *Principles of Economics*. London: Macmillan.

Mills, E.S. (1967) 'An aggregation model of resource allocation in a metropolitan area', *American Economic Review*, 57: 197–210.

Miracky, W.F. (1992) 'Technological spillovers, the product cycles and regional growth'. MIT, memo.

Mitra, A. (1995) 'Dynamic externalities and industrial location'. PhD thesis, Brown University.

Nakamura, R. (1985) 'Agglomeration economies in urban manufacturing industries: a case of Japanese cities', *Journal of Urban Economics*, 17: 108–24.

Rauch, J. (1993) 'Does history matter when it only matters a little: the case of city-industry location', *Quarterly Journal of Economics*, 108: 843–68.

Romer, P. (1986) 'Increasing returns and long-run growth', *Journal of Political Economy*, 94: 1002–37.

Sveikauskas, L. (1975) 'The productivity of cities', *Quarterly Journal of Economics*, 89: 393–413.

Sveikauskas, L. (1978) 'The productivity of cities', US Bureau of Labor Statistics, mimeo., J-94.

16

Cities in the Global Economy

SASKIA SASSEN

The dispersal capacities emerging with globalization and telematics led many observers to assert that cities would become obsolete. Indeed, many of the once-great industrial centres in the highly developed countries did suffer severe decline. The off-shoring of factories, the expansion of global networks of affiliates and subsidiaries, the move of back offices to suburbs and out of central cities also left their mark on the more service-centred cities. But against all predictions, a significant number of major cities also saw their concentration of economic power rise. While the decline of industrial centres as a consequence of the internationalization of production beginning in the 1960s has been thoroughly documented and explained, until recently the same could not be said about the rise of major service cities in the 1980s. Today we have a rich new scholarship, replete with debates and disagreements, on cities in a global economy.

What explains the new or sharply expanded role of a particular kind of city in the world economy since the early 1980s? It basically results from the intersection of two major processes. One is the sharp growth of the globalization of economic activity, particularly some very specific features of its implementation which have received far less attention than the dispersal aspects. It is a fact that globalization has raised the scale and the complexity of economic transactions, thereby feeding the growth of top-level multinational headquarter functions and the growth of services for firms, particularly advanced corporate services. The second is the growing service intensity in the organization of the economy, a process evident in firms in all industrial sectors, from mining to finance.

The key process from the perspective of the urban economy is the growing demand for services by firms in all industries and the fact that cities are preferred production sites for such services, whether at the global, national or regional level. The growing service intensity in economic organization generally and the specific conditions of production for advanced corporate services, including the conditions under which information technologies are available, combine to make many cities once again key 'production' sites, roles they had lost when mass manufacturing became the dominant economic sector. That is to say, the strategic space for mass manufacturing is the large, vertically integrated factory; the city is, at most, the space for administrative activities. If there is another strategic space for mass manufacturing it lies in the realm of government – the site for the execution and legitimation of many of the features of the social contract that is a part of the regime characterized by the dominance of mass manufacturing, especially, but not only, in the highly developed countries.

The growing service intensity in the organization of all industries brings with it a newly strategic role for cities as production sites of these necessary service inputs. This holds for cities at many levels of the urban hierarchy, including cities that service sub-national regional economies. In the case of some cities, these servicing functions have global reach; in these cities also, the dominant economic engine tends to be highly specialized service industries. These are the world cities or global cities that are the focus of a new research literature. How many there are, what is their shifting hierarchy, how novel a development do they represent – all are subjects for debate. But there is growing agreement about the fact of a network of major cities cutting across the North/South divide, cities that function as centres for the coordination, control and servicing of global capital.

The first section of this chapter examines the key components in the new narrative that has emerged from the research literature on world or global cities. The second reviews the evolution of this literature. In this review I confine myself to a fairly narrowly defined field of scholarship grounded in the notion that the contemporary forms assumed by globalization over the past two decades have specific organizational requirements and political possibilities and that the new technologies produce specific opportunities and capacities. This does not preclude the existence of enormous continuities with past periods – a subject of considerable debate also in this literature – but it does posit the specificity of the current era and hence of the role of cities. The third section examines some of the themes that are emerging as an agenda for research and theory.

It is impossible in such a short piece to do full justice to the many scholars who have and are contributing to this new literature. Because of the diversity of variables that can be incorporated – from finance to immigration – the subject of cities in the global economy has contributed not only a research literature in the social sciences but also in cultural studies and in anthropology, and most recently in certain aspects of political science, notably questions of citizenship, governance and the participation of sub-national units in international relations. It is also a subject that has been given modernist and postmodernist treatments, highly theorized and highly empirical treatments. Finally, it is a subject with an older literature on capitals of empires and world cities – from Braudel (1984) to Peter Hall 1996) – and on urban hierarchies in world systems analysis (Chase Dunn, 1984).

TOWARD A NEW NARRATIVE: RECOVERING PLACE IN THE GLOBAL ECONOMY

The massive trends towards the spatial dispersal of economic activities at the metropolitan, national and global level which we associate with globalization have contributed to a demand for new forms of territorial centralization of top-level management and control operations *because* this dispersal is happening under conditions of concentration in control, ownership and profit appropriation (Sassen, 2000a). National and global markets as well as globally integrated organizations require central places where the work of globalization gets done (Friedman and Wolff, 1982).[1] Further, information industries require a vast physical infrastructure containing strategic nodes with hyperconcentrations of facilities (Castells, 1989; Graham and Marvin, 1996); there is a distinction to be made between the capacity for global transmission/communication and the material conditions that make this possible. Finally, even the most advanced information industries have a work process that is at least partly place-bound because of the combination of resources it requires, even when the outputs are hypermobile.

This type of emphasis allows us to see cities as production sites for the leading information industries of our time and it allows us to recover the infrastructure of activities, firms and jobs, that is necessary to run the advanced corporate economy.[2] Advanced information industries are typically conceptualized in terms of the hypermobility of their outputs and the high levels of expertise of their professionals rather than in terms of the work process involved and the requisite infrastructure of facilities and non-expert jobs that are also part of these industries.

A central proposition here is that we cannot take the existence of a global economic system as a given, but rather need to examine the particular ways in which the conditions for economic globalization are produced. This requires examining not only communication capacities and the power of multinationals, but also the infrastructure of facilities and work processes necessary for the implementation of global economic systems, including the production of those inputs that constitute the capability for global control and the infrastructure of jobs involved in this production. The emphasis shifts to the *practice* of global control – the work of producing and reproducing the organization and management of a global production system and a global market place for finance, both under conditions of economic concentration. The recovery of place and production also implies that global processes can be studied in great empirical detail.

The specific forms assumed by globalization over the past decade have created particular organizational requirements. The emergence of global markets for finance and specialized services, the growth of investment as a major type of international transaction, all have contributed to the expansion in command functions and in the demand for specialized services for firms. Much new global economic activity is not encompassed by the organizational form of the transnational corporation or bank. Nor is much of this activity encompassed by the power of such firms, a power often invoked to explain the fact of economic globalization. The spatial and organizational forms assumed by globalization and the actual work of running transnational operations have made cities one type of strategic place and producer services a strategic input.

In brief, the combination of geographic dispersal of economic activities and system integration which lies at the heart of the current economic era has contributed to new or expanded central functions and the complexity of transactions has raised the demand by firms for highly specialized services. Rather than becoming obsolete due to the dispersal made possible by information technologies, cities (a) concentrate command functions, (b) are post-industrial production sites for the leading industries of this period – finance and specialized services, and (c) are transnational market places where firms and governments from anywhere in the world can buy financial instruments and specialized services.

Such a focus allows us to conceive of globalization as constituted through a global grid of strategic sites which emerges as a new cross-border geography of centrality.

The New Urban Economy

This is not to say that everything in the economy of these cities has changed. On the contrary, there is much continuity and much similarity with cities that are not global nodes. It is rather that the implantation of global processes and markets has meant that the internationalized sector of the economy has expanded sharply and has imposed a new valorization dynamic, often with devastating effects on large sectors of the urban economy. High prices and profit levels in the internationalized sector and its ancillary activities, for example, restaurants and hotels, made it increasingly difficult in the 1980s for other sectors to compete for space and investment. Many of the latter have experienced considerable downgrading and/or displacement; or have lost economic vigour to the point of not being able to re-take their economic space when the recession weakened the dominant sectors. Illustrations are neighbourhood shops catering to local needs replaced by up-scale boutiques and restaurants catering to new high-income urban elites. The sharpness of the rise in profit levels in the international finance and service sector also contributed to the sharpness of the ensuing crisis. These trends are evident in many cities of the highly developed world, though rarely as sharply as in major US cities (see , for example *Le Debat*, 1994 and Veltz, 1996 for Paris; Todd, 1995 for Toronto; Levine, 1995 for Montreal; Brake, 1991 and Noller et al. 1994 for Frankfurt; and generally, *Wissenschafts Forum* 1995; Peraldi and Perrin, 1996; Short and Kim, 1999; Allen et al., 1999).

Though at a different order of magnitude, these trends also became evident towards the late 1980s in a number of major cities in the developing world that have become integrated into various world markets: São Paulo, Buenos Aires, Bangkok, Taipei, Mexico City are but some examples (Cicolella and Mignaqui, 2000; Schiffer, 2000; Kowarick et al., 1991; Santos, 1993; see various chapters in Cohen et al., 1996; see various chapters in Gilbert, 1996 and in Rakodi, 1997; Hardoy and Satterthwaite, 1989;

see various chapters in Henderson and Castells, 1987; Landell-Mills et al. (1989). Central to the development of this new core in these cities were the deregulation of financial markets, ascendance of finance and specialized services, and integration into the world markets, real estate speculation and high-income commercial and residential gentrification. The opening of stock markets to foreign investors and the privatization of what were once public sector firms have been crucial institutional arenas for this articulation. Given the vast size of some of these cities, the impact of this new economic complex is not always as evident as in central London or Frankfurt, but the transformation has occurred.

Accompanying these sharp growth rates in producer services was an increase in the level of employment specialization in business and financial services in major cities beginning in the late 1980s. There is today a general trend towards high concentration of finance and certain producer services in the downtowns of major international financial centres around the world: from Toronto and Sydney to Frankfurt and Zurich we are seeing growing specialization in finance and related services in the downtown areas (Brake, 1991; Gad, 1991; Hitz et al., 1995; Todd, 1995). These trends are also becoming evident in major Southern cities, notably Mexico (Parnreiter, 2000), São Paulo (Schiffer, 2000) and Buenos Aires (Ciccolella and Mignaqui, 2000).

These cities have emerged as important producers of services for export, with a tendency towards specialization.[3] New York and London are not only leading producers but also exporters of financial services, accounting, advertising, management consulting, international legal services and other business services. For instance, out of a total private sector employment of 2.8 million jobs in New York City in 1995, almost 1.3 million were export-oriented (both national and international) (see generally Crahan and Vourvoulais-Bush, 1997). Cities such as New York are among the most important international markets for these services, with New York and London the world's largest sources of service exports.

There are also tendencies towards specialization among different cities within a country. In the USA, New York leads in banking, securities, manufacturing administration, accounting and advertising. Washington leads in legal services, computing and data processing, management and public relations, research and development, and membership organization (Markusen and Gwiasda, 1994). New York is more narrowly specialized as a financial and business centre and cultural centre. Some of the legal activity concentrated in Washington is actually serving New York City businesses which have to go through legal and regulatory procedures, lobbying, etc. These are bound to be in the national capital.[4]

It is important to recognize that manufacturing remains a crucial economic sector in all of these economies, even when it may have ceased to be so in some of these cities. I return to this in a later section.

A NEW THEORETICAL FRAMEWORK EMERGES

We now know from unpublished papers and a variety of publications that by the early 1980s a number of scholars had begun to study cities in the context of globalization.[5] But it is one article in particular, 'World city formation' by Friedmann and Wolff (1982), that marked a new phase. This article took a variety of elements that were emerging in the research literature on cities, on the global economy, on immigration, and a number of other subjects, and sought to formalize these into several propositions about the meaning of the global economy for cities. The key elements in this framework were the emergence of several cities as basing points for global capital, a hierarchy (albeit a shifting one) of such cities, and the social and political consequences for these cities of being such basing points.

By the mid- and late 1980s we see the beginnings of a research literature (concerned with one or another of these propositions, and often proceeding quite autonomously from the World City hypothesis notion). This literature makes a number of specific contributions to comparative analyses of cities, to international trade in services and its impact on cities, the impact of economic globalization on the social and spatial structure of major cities, etc.[6] At this time we also see more explicit research on the subject of cities in the global economy, including an elaboration of the framework presented in the World City hypothesis.[7] But we also see work quite independent from that hypothesis, notably the work by Thrift (1987) on the formation of an intermediate economy, and the research on producer services (e.g. Noyelle and Stanback, 1984) and the office economy (Daniels, 1991). All of these make extremely important contributions to questions about the organization of the economy and its spatial implications; they lay a groundwork for much of the subsequent literature.

With the books by Castells (1989), King (1990a) and Sassen (2000d), what had been a hypothesis in the early 1980s became a full-fledged theorization and empirical specification. These three books add important propositions to the general framework: Castells's proposition that globalization as constituted today has engendered a space of flows that reconfigures economic and political power; Sassen's proposition that it is not simply a matter of global coordination but one of the *production* of global control capacities and that an examination along these lines allows us to understand the role of global cities; King's emphasis on the historical transformation of the link between cities and internationalization.

It is important to distinguish what is different about this literature from a broader, earlier literature on world cities prominently represented by the work of Peter Hall already in the 1960s, and a new literature on mega-cities especially focused on Latin America and Asia (e.g. Dogan and Kasarda, 1988; Gilbert, 1996; Rakodi, 1997; Fuchs et al., 1995; Lo and Yeung, 1996). These literatures do not have the fact of globalization and the centrality of crossborder networks connecting cities as crucial variables. The earlier literature on world cities is closer to the notion of capitals of empires: one city at the top of the power hierarchy. In the current literature on global cities the determining factor is a crossborder, global network of cities that function as strategic sites for global economic operations. There is no such entity as a single global city – as there is with the capital of an empire; by definition, the global city is part of a network of cities (Sassen, 2000d). Similarly, an older literature focused on past world cities, as in the work of Braudel, and earlier studies of major centres of world commerce and banking, as well as more recent work on urban hierarchies in the world system (Chase Dunn, 1984), are to be differentiated from the current literature if we historicize the world economy and specify what is distinct today. Finally, we need to distinguish between a narrowly specified literature on global and world cities today and various literatures that directly or indirectly contribute to our understanding of these cities, notably the research on producer services.

By the mid-1990s the subject was clearly a large field for research, with scholars in many different disciplines and countries working on particular aspects of global cities. This was reflected in the variety of authors and themes in several state of the art collections, notably by Fainstein et al. (1993), Knox and Taylor (1995), Brotchie et al. (1995), Noller et al. (1994), and several others that elaborate, critique, expand the empirical base and generally advance this theoretical and methodological project.[8] We can also see it in several new important books that set the stage for highly focused research on particular variables, notably Fainstein (1993), Keil (1993), Hitz et al. (1995), Pozos (1996), among others. During this period several book series were created by various publishers in different countries: the series on World Cities edited by Knox for Belhaven Press, the series edited by Milton Santos and his colleagues in São Paulo for Hucitec, the series edited by Martin Wentz for Campus Verlag (Frankfurt), the series edited by Ciccolella for EUDEBA (Buenos Aires), and the series organized by Fu-chen Lo for the UN University Press (Tokyo) are just some.

In the late 1990s two types of scholarly work added, directly and indirectly, a set of new dimensions to global city research. One of these is the renewed interest in the nature of localized growth in the spatial economy in a variety of disciplines. The observed coincidence between spatial industrial clustering and regional specialization is central to the principle of increasing returns to scale in urban and regional economic analysis. This was stimulated by the development of formal research techniques and the specific changes in the organization of economies over the last two decades, notably the fact that external scale economies assume a new importance over internal scale economies. Under these conditions, localization economies become important. While economic geographers have long focused on these issues, we are now also seeing this within neo-classical economics where the interest in locational questions and spatial clustering is related to applications of modern trade theory to problems of spatial allocation. Much of the research on spatial clustering has focused on the manufacturing sector; but this is, nonetheless, a useful set of contributions to the understanding of the specific types of service industries that concentrate in global cities (see Sassen 2000a).

A second important development in the late 1990s was the work by Scott et al. (2000) on global-city regions. This relocates some of the dynamics and issues about cities and globalization on the territorial scale of the region and is far more likely to include a cross-section of a country's economic activities than the scale of the city. It is likely, for instance, to include as key variables manufacturing and basic infrastructure. This, in turn, brings with it a more benign focus on globalization. The concept of the global city introduces a far stronger emphasis on strategic components of the global economy, and hence on questions of power than that of the region. Secondly, the concept of the global city will tend to have a stronger emphasis on the networked economy because of the nature of the industries that tend to be located there: finance and specialized services. And, thirdly, the concept of the global city will tend to have more of an emphasis on economic and spatial polarization because of the disproportionate concentration of very high and very low income jobs in the city compared with what would be the case for the region. In contrast, the concept of the global-city region is

more attuned to questions about the nature and specifics of broad urbanization patterns, a more encompassing economic base, more middle sectors of both households and firms, and hence to the possibility of having a more even distribution of economic benefits under globalization. There is a strong emphasis on competition and competitiveness in the global-city region construct. In contrast, the nature itself of the leading industries in global cities strengthens the importance of cross-border networks and specialized division of functions among cities in different countries and/or regions, beyond inter-national competition *per se*.

It is not only the growth of the research literature but also the growth of a body of critical responses and analyses that signals the strength and vigour of this field of enquiry. There is only space here for the briefest mention, a sort of guide to criticisms: Logan and Swanstrom's (1990) critique of the excessive weight given to global structural processes in comparing internal versus external factors that shape a city's economic development; Waldinger's (1988) and Hamnett's (1994, 1996) critiques of Sassen's proposition that globalization has contributed to social and economic polarization in global cities; Markusen and Gwiasda's (1994)) critique of the notion that New York is at the top of the US urban hierarchy and how a comparison with Washington shows that the latter has a higher level of specialization than New York in many advanced specialized services, notably in legal services; Michael Peter Smith's critique of the literature for its neglect of grassroots transnationalism and the new kinds of politics and identity formation that this entails; Beauregard's (1991) critique of the explanatory variables for changes in the built environment and the real estate industry; Simon's (1995) critique of the neglect of the periphery, notably Africa; the debate in *Urban Affairs* (March 1998); the special issue on 'Ségrégations Urbaines' in *Sociétiés Contemporaines* (1995); the multi-year project on New York, London, Tokyo and Paris sponsored by the SSRC (Social Science Research Council) where many of the authors criticize what they consider an excessive emphasis on the impacts of globalization (Mollenkopf, forthcoming); and many others.

There are two types of scholarly literature that intersect with this body of research on cities and the global economy, and indeed often invoke or use it to develop their arguments. They are on the one hand a literature in anthropological and cultural studies on transnationality, globalization and identity formation.[9] The other is the scholarship on the global economy by regional economic geographers, who have also focused on cities (e.g. Moulaert and Scott, 1997; Shachar, 1990; and Veltz, 1996).

In terms of method, a number of strategies have been developed. Perhaps the most ambitious quantitative method is the effort by Smith and Timberlake (2000), who conceptualize urban areas as central nodes in multiplex networks of economic, social, demographic and information flows. They use the methodological logic of sociological network analysis, using particularly two measures. One of these is structural or relational equivalence between actors (i.e. cities) in a network; the second measure is centrality. Both of these measures relate to a number of propositions developed in the literature on cities in the global economy. The necessary data on inter-city flows may exist but it will take a lot of work to constitute the requisite data sets (see also Sassen, 2000c). Hill, drawing on the work of Wallerstein, McKenzie and Hymer (1972), calls for a focus on 'global production systems' linking places across the globe in an increasingly integrated vertical division of labour. David Meyer (1991; 2000) has developed a quantitative analysis of the distribution of international bank branches linking cities in various regions of the world to the world's financial centres and to each other.

Several very recent efforts will make a strong contribution to the possibility of empirical specification of key hypotheses describing the global city. One of the most important developments for research on global cities is the work by Taylor et al. (2000), GAWC (2000) and Beaverstock et al. (2000) who have constructed a new and pioneering data set that makes it possible to map the global networks of affiliates of the leading firms worldwide in accounting, law, advertising and finance. These networks of affiliates can be used to classify cities in terms of their participation in cross-border networks. At this time, the data set shows the networks of affiliates of 46 major advanced producer services firms in 55 cities involved in cross-city networks. The data can be analysed using a variety of hypotheses and statistical techniques. A second important contribution is by Drennan et al. (1996) who have developed a measure that allows us to establish the impact of high levels of international trade in producer services characteristic of global cities on median income families compared with that of families in cities with low levels, if any, of international trade in producer services. A third contribution is by Elliott (1999) who ranked metropolitan areas along variables that allows for a specification of global city status and developed a measure of the impact of global city status on earnings inequality. The work by Hamnett and Cross (1998) and by Klosterman (1996) also makes methodological contributions in this same domain.

Castells (1989) and Sassen (1991) developed several techniques of analysis which range from methods to understand the place of cities in global markets to expanding the representation of the global. In *The Informational City* and *The Global City* each of these authors sought to establish rather broadly what is the array of data sets that can be brought into analyses of this subject – from international flows of capital and information to very localized social effects. This was an effort to resist the simplification in mainstream accounts that emphasize the global dispersal of activities and telecommunications and exclude most social issues. For instance, Sassen's formulation of urban 'spatial circuits of economic operations' seeks to capture the diversity of firms, workers and work cultures that constitute the leading information industries in that portion of a sector that is urban. Urban space can be mapped in terms of these circuits for a whole range of global market-oriented industries. Techniques for data analysis traditionally used by economic geographers can also be helpful: for instance, Wheeler's (1986) examination of the dispersion of higher-order financial services throughout the US urban hierarchy – which he found had proceeded at a much slower rate than the headquarters of other large corporations. This allows him to posit that corporations tend to proceed up the urban hierarchy for their advanced service and banking needs.

THE RESEARCH AND THEORY AGENDA

There are a number of emerging issues for research and theorization. Some of these are narrowly technical while others are broad and need to be specified more rigorously, both theoretically and empirically. I will discuss a few at some length and just name a few others.

Spatial Agglomeration in a Global Economy

Among the more empirical questions are several concerning locational issues in a context of globalization and telematics. A central one concerns the locational needs and options of different kinds of headquarters and producer services, and the extent to which their mutual dependence has a spatial dimension. It is common in the general literature and in some more scholarly accounts to use headquarters concentration as an indication of whether a city is an international business centre. The loss of headquarters is then interpreted as a decline in a city's status. The use of headquarters concentration as an index is actually a problematic measure given the way in which corporations are classified.

Which headquarters concentrate in major international financial and business centres depends on a number of variables. First, how we measure or simply count headquarters makes a difference. Frequently, the key measure is size of firm in terms of employment and overall revenue.[10] In this case, some of the largest firms in the world are still manufacturing firms and many of these have their main headquarters in proximity to their major factory complex, which is unlikely to be in a large city due to space constraints. Such firms are likely, however, to have secondary headquarters for highly specialized functions in major cities. Further, many manufacturing firms are oriented to the national market and do not need to be located in a global or any city. If we change the measure, the results can change dramatically. For instance, in the case of New York City, 40 per cent of US firms with half their revenue from international sales have their headquarters in the New York City region.[11] Secondly, the nature of the urban system in a country is a factor. Pronounced urban primacy will tend to entail a disproportionate concentration of headquarters no matter what measure one uses. Thirdly, different economic histories and business traditions may combine to produce different results. Further, headquarters concentration may be linked to a specific economic phase. We need more research and more complex measures to understand the impact of globalization and telematics on locational patterns of firms and their implications for the future of global cities.

Another locational issue with urban implications concerns the conditions for agglomeration, long a central feature in urban economies. For instance, the production process in advanced corporate services benefits from proximity to other specialized services, especially in the leading and most innovative sectors of these industries. Complexity and innovation often require multiple highly specialized inputs from several industries. The production of a financial instrument, for example, requires inputs from accounting, advertising, legal expertise, economic consulting, public relations, designers, and printers. The particular characteristics of production of these services, especially those involved in complex and innovative operations, explain their pronounced concentration in major cities. Routinization has long been and is also here a force reducing the importance of agglomeration. Yet, the acceleration of economic transactions and the premium put on time, have created new forces for agglomeration. Routine operations can easily be dispersed; but where time is of the essence, as it is today in many of the leading sectors of these industries, the benefits of agglomeration are still extremely high.

Can we detect significant changes in the importance of innovation in these sectors and hence in the advantages of agglomeration? There is a strong suggestion in current locational and organizational patterns that the agglomeration of the leading sectors of producer services in major cities actually constitutes a production complex. This producer services complex is intimately connected to the world of corporate headquarters; they are often thought of as forming a joint headquarters–corporate services complex. But in my reading we need to distinguish the two. Although it is true that headquarters still tend to be disproportionately concentrated in cities, many have moved out over the past two decades. Headquarters can indeed locate outside cities, but they need a producer services *complex* – a single firm in each speciality won't do for the more complex operations – in order to buy or contract for the needed specialized services and financing. Further, headquarters of firms with very high overseas activity or in highly innovative and complex lines of business tend to locate in major cities. In brief, firms in more routinized lines of activity, with predominantly regional or national markets, appear to be increasingly free to move or install their headquarters outside cities. Firms in highly competitive and innovative lines of activity and/or with a strong world market orientation appear to benefit from being located at the centre of major international business centres, no matter how high the costs.

Both types of firms, however, need access to a corporate services complex. Where this complex is located is probably increasingly unimportant from the perspective of many, though not all, headquarters. Such a specialized complex is most likely to be in a city rather than, for example, a suburban office park. The latter will be the site for producer services firms but not for a services complex. And only such a complex is capable of handling the most advanced and complicated corporate demands. There is a large literature in the US on the spatial distribution of top level corporate functions and corporate services across the urban system; though there are theoretical and empirical disagreements, most studies show considerable growth of these activities beginning in the 1980s and continuing in the 1990s at various levels of the urban system (Daniels and Lever, 1996; Noyelle and Stanback, 1984; Wheeler, 1986). In the case of cities that are major international business centres, the scale, power and profit levels of this new core of economic activities are vast. In this context, globalization becomes a question of scale and added complexity in a process that is also taking place at lower levels of the urban hierarchy, where a national or regional, rather than global, orientation may prevail.

Space and Power: the New Centrality

A more theorized version of these locational questions can be posited in terms of centrality. Cities have historically provided economic actors with something we can think of as centrality – agglomeration economies, massive concentrations of information on the latest developments, a market place. How do economic globalization and the new technologies alter the role of centrality and hence the key comparative advantage of cities as economic entities?

Telematics and globalization have emerged as fundamental forces in the reorganization of economic space. This reorganization ranges from the spatial virtualization of a growing number of economic activities to the reconfiguration of the geography of the built environment for economic activity. Whether in electronic space or in the geography of the built environment, this reorganization involves institutional and structural changes. One outcome of these transformations has been captured in images of geographic dispersal at the global scale and the neutralization of place and distance in a growing number of economic activities.

But is a space economy that lacks points of physical concentration possible in an economic system characterized by significant concentration in ownership, control and the appropriation of profits? Another way of formulating this, one that captures both the physical, organizational and power dimension, is in terms of centrality: can such an economic system operate without centres? And, further, how far can forms of centrality constituted in electronic space go in replacing some of the functions commonly associated with geographic/organizational forms of centrality?

The Topoi of e-space:
Global Cities and Global Value Chains

The vast new economic topography that is being implemented through electronic space (e-space) is one moment, one fragment, of an even vaster economic chain that is in good part embedded in non-electronic spaces. There is no fully virtualized firm and no fully digitialized industry. Even the most advanced information industries, such as finance, are installed only partly in e-space. And so are industries that produce digital products, such as software designers. The growing digitalization of economic activities has not eliminated the need for major international business and financial centres and all the material resources they concentrate, from state of the art telematics infrastructure to brain talent.

None the less, telematics and globalization have emerged as fundamental forces reshaping the organization of economic space (see, for example, Special Issue of *Journal of Urban Technology*, Spring 1997; Castells, 1996; Graham and Marvin, 1996; Rotzer, 1995; Salomon, 1996). This reshaping ranges from the spatial virtualization of a growing number of economic activities to the reconfiguration of the geography of the built environment *for* economic activity. Whether in e-space or in the geography of the built environment, this reshaping involves organizational and structural changes. Telematics maximizes the potential for geographic dispersal and globalization entails an economic logic that maximizes the attractions/profitability of such dispersal.

The transformation in the spatial correlates of centrality through new technologies and globalization engenders a whole new problematic around the definition of what constitutes centrality today in an economic system where (1) a share of transactions occurs through technologies that neutralize distance and place, and do so on a global scale, and (2) centrality has historically been embodied in certain types of built environment and urban form, that is, the central business district. Further, the fact of a new geography of centrality, even if transnational, contains possibilities for regulatory enforcement that are absent in an economic geography lacking strategic points of agglomeration.

There are at least two sets of issues on which we need more research. First, leading economic sectors that are highly digitalized require strategic sites with vast concentrations of infrastructure, the requisite labour resources, talent and buildings. This holds for finance but also for the multimedia industries that use digital production processes and produce digitalized products. What is the range of articulations and their spatial expression between the virtual and the actual organizational component? What are the implications for urban space?

Secondly, the sharpening inequalities in the distribution of the infrastructure for e-space, whether private computer networks or the Net, in the conditions for access to e-space and, within e-space, in the conditions for access to high-powered segments and features, are all contributing to new geographies of centrality both on the ground and in e-space. What does this mean for cities?

The Place of Manufacturing in the New Urban Service Economy

Another subject for research and debate is the relation between manufacturing and producer services in the advanced urban economy. The new service economy benefits from manufacturing because the latter feeds the growth of the producer services sector, but it does so whether located in the particular area, in another region, or overseas. While manufacturing, and mining and agriculture for that matter, feed the growth in the demand for producer services, their actual location is of secondary importance in the case of global level service firms: thus whether a manufacturing corporation has its plants off-shore or inside a country may be quite irrelevant as long as it buys its services from those top-level firms.

Secondly, the territorial dispersal of plants, especially if international, actually raises the demand for producer services because of the increased complexity of transactions. This is yet another meaning of globalization: the growth of producer service firms headquartered in New York or London or Paris can be fed by manufacturing located anywhere in the world as long as it is part of a multinational corporate network. It is worth remembering here that as GM was off-shoring production jobs and devastating Detroit's employment base, its financial and public relations headquarters office in New York City was as dynamic as ever, indeed busier than ever.

Thirdly, a good part of the producer services sector is fed by financial and business transactions that either have nothing to do with manufacturing, as is the case in many of the global financial markets, or for which manufacturing is incidental, as in much of the merger and acquisition activity which was really centred on buying and selling rather than the buying of manufacturing firms. We need much more research on many particular aspects in this relation between manufacturing and producer services, especially in the context of spatial dispersal and cross-border organization of manufacturing.

Not unrelated to the question of manufacturing is the importance of conventional infrastructure in the operation of economic sectors that are heavy users of telematics. This is a subject that has received little attention. The dominant notion seems to be that telematics obliterates the need for conventional infrastructure. But it is precisely the nature of the production process in advanced industries, whether they operate globally or nationally, which contributes to explain the immense rise in business travel we have seen in all advanced economies over the past decade. The virtual office is a far more limited option than a purely technological analysis would suggest. Certain types of economic activities can be run from a virtual office located anywhere. But for work processes requiring multiple specialized inputs, considerable innovation and risk-taking, the need for direct interaction with other firms and specialists remains a key locational factor. Hence the metropolitanization and regionalization of an economic sector has bound-aries that are set by the time it takes for a reasonable commute to the major city or cities in the region. The irony of today's electronic era is that the notion of the region and conventional forms of infrastructure re-emerge as critical for key economic sectors. This type of region in many ways diverges from older, geographic notions of region. It corresponds rather to a type of centrality – a metropolitan grid of nodes connected via telematics. But for this digital grid to work, conventional infrastructure – ideally of the most advanced kind – is also a necessity.

New Forms of Marginality and Polarization

The new growth sectors, the new organizational capacities of firms, and the new technologies – all three interrelated – are contributing to produce not only a new geography of centrality but also a new geography of marginality. The evidence for the USA, Western Europe and Japan suggests that it will take government policy and action to reduce the new forms of spatial and social inequality.

There are misunderstandings that seem to prevail in much general commentary about what matters in an advanced economic system, the information economy and economic globalization. Many types of firms, workers and places, such as industrial services, which look as if they do not belong in an advanced, information-based, globally oriented economic system are actually integral parts of such a system. They need policy recognition and support: they can't compete in the new environments where leading sectors have bid up prices and standards, even though their products and labour are in demand. For instance, the financial industry in Manhattan, one of the most sophisticated and complex industries, needs truckers to deliver not only software, but also tables and light bulbs; and it needs blue-collar maintenance workers and cleaners. These activities and workers need to be able to make a decent income if they are to stay in the region.

Yet another dimension not sufficiently recognized is the fact of a new valuation dynamic: the combination of globalization and the new technologies has altered the criteria and mechanisms through which factors, inputs, goods, services are valued/priced. This has had devastating effects on some localities, industries, firms and workers.

The Global City and the National State

Globalization has transformed the meaning of and the sites for the governance of economies (see, for example, Jessop, 1999; Sassen, 1998:

Chap 10). One of the key properties of the current phase in the long history of the world economy is the ascendance of information technologies, the associated increase in the mobility and liquidity of capital, and the resulting decline in the regulatory capacities of national states over key sectors of their economies.

One of the characteristics of the current phase in the world economy is a reassertion of the importance of sub-national units: whether global cities or strategic regions such as Silicon Valley in California. This signals the possibility that the impact of globalization cannot simply be reduced to the notion of the declining significance of the national state, as is so often asserted, but rather a triangulation: national state, global economy and strategic localities – typically major international financial and business centres. The strategic relationship is no longer the diad national state–global economy. Taylor's (1995) work on the changing nature of territoriality in the modern world-system, and his observations on national states and cities (2000) set an agenda for research (see also Jessop, 1999; Sassen 1996: Chap. 1).

The excessive emphasis on the hypermobility and liquidity of capital is a partial account which tends to obscure the relation between foreign policy, local policy and the global economy. It is an account that excludes, for instance, the possibility of a *de facto* participation of global cities in international economic policy and practice and hence in foreign policy insofar as economic policy has become a growing concern in foreign policy. It also has the effect of excluding a variety of global processes that are really about the re-territorializing of people, economic practices, cultures. The immigrant communities and the neighbourhood sub-economies they often form are an instantiation of this (Basch et al., 1994; Smith, 1997; see also Holston, 1996). The formation of transnational bonds and communities through immigration raises a whole series of additional issues that have the effect of displacing certain political functions away from the international relations among national states and onto privatized spheres of individuals, households and communities.

We are also seeing the formation of cross-border regions that are increasingly acting as units, not because they are cohesive but because of a shared spatial and organizational terrain (see, for example, Chen, 1995b; Sum, 1999 on East Asia; Herzog, 1993 and Alegria, 1992 on the Tijuana–San Diego region). The formation of a new 'transnational class' of managers and professionals represents yet another dimension of triangulation. More broadly defined, several scholars are exploring transnationality from below, a process particularly evident in global

cities (Portes et al., 1999; Smith, 1995). Finally, there is an emergent debate about the return of the city-state given the conditions characterizing global cities – strong articulation with global markets, multiple forms of transnationality and weakened articulation with the national economy and national state.

These transformations in key aspects of the modern state and the modern inter-state system signal a conceptual and a practical opening for the inclusion of cities that are strategic in the global economy and thereby contribute to triangulate what was once a partnership of two.

CONCLUSION

An examination of globalization through cities, more specifically global cities, introduces a strong emphasis on strategic components of the global economy rather than the broader and more diffuse homogenizing dynamics we associate with the globalization of consumer markets. As a consequence, this also brings an emphasis on questions of power and inequality. And it brings an emphasis on the actual work of managing, servicing and financing the global economy. Second, a focus on the city in studying globalization will tend to bring to the fore the growing inequalities between highly provisioned and profoundly disadvantaged sectors and spaces of the city, and hence such a focus introduces yet another formulation of question of power and inequality.

Third, a focus on cities and globalization brings a strong emphasis on the networked economy because of the nature of the industries that tend to be located there: finance and specialized services, the new multimedia sectors, and telecommunications services. These industries are characterized by cross-border networks and specialized divisions of functions among cities rather than inter-national competition *per se*. In the case of global finance and the leading specialized services catering to global firms and markets – law, accounting, credit rating, telecommunications – it is clear that we are dealing with a cross-border system, one that is embedded in a series of cities, each possibly part of a different country. It is a *de facto* global system.

Finally, a focus on networked cross-border dynamics among global cities also allows us to capture more readily the growing intensity of such transactions in other domains – political, cultural, social, criminal.

Global cities around the world are the terrain where a multiplicity of globalization processes assume concrete, localized forms. These localized forms are, in good part, what globalization is about. Recovering place means recovering the multiplicity of presences in this landscape. The large city of today has emerged as a strategic site for a wide range of new types of operations – political, economic, 'cultural', subjective. It is one of the nexi where the formation of new claims, by both the powerful and the disadvantaged, materializes and assumes concrete forms.

Economic globalization and telecommunications have contributed to produce a spatiality for the urban which pivots on cross-border networks and territorial locations with massive concentrations of resources. This is not a completely new feature. Over the centuries cities have been at the cross-roads of major, often worldwide, processes. What is different today is the intensity, complexity and global span of these networks, the extent to which significant portions of economies are now dematerialized and digitalized and hence the extent to which they can travel at great speeds through some of these networks, and, thirdly, the numbers of cities that are part of cross-border networks operating at vast geographic scales.

The new urban spatiality thus produced is partial in a double sense: it accounts for only part of what happens in cities and what cities are about, and it inhabits only part of what we might think of as the space of the city, whether this be understood in terms as diverse as those of a city's administrative boundaries or in the sense of a city's public imagery Yet, while partial, it is strategic.

NOTES

1. The producer services, and most especially finance and advance corporate services, can be seen as industries producing the organizational commodities necessary for the implementation and management of global economic systems. Producer services are intermediate outputs, that is, services bought by firms. They cover financial, legal and general management matters, innovation, development, design, administration, personnel, production technology, maintenance, transport, communications, wholesale distribution, advertising, cleaning services for firms, security and storage. Central components of the producer services categories are a range of industries with mixed business and consumer markets; they are insurance, banking, financial services, real estate, legal services, accounting and professional associations. (For more detailed discussions see, e.g., Daniels, 1991; Noyelle and Dutka, 1988.)

2. Methodologically speaking, this is one way of addressing the question of the unit of analysis in studies of contemporary economic processes. 'National economy' is a problematic category when there are high levels of internationalization. And 'world economy' is a problematic category because of the impossibility of engaging in detailed empirical study at that scale. Highly internationalized cities such as New York or London offer the possibility of examining globalization processes in great detail, within a bounded setting, and with all their multiple, often contradictory aspects.

3. All the major economies in the developed world display a similar pattern towards sharp concentration of financial activity in one centre: Paris in France, Milan in Italy, Zurich in Switzerland, Frankfurt in Germany, Toronto in Canada, Tokyo in Japan, Amsterdam in the Netherlands, and Sydney in Australia. The evidence also shows that the concentration of financial activity in such leading centres has actually increased over the past decade. Thus in Switzerland, Basle used to be a very important financial centre which has now been completely overshadowed by Zurich. And Montreal was certainly the other major centre in Canada two decades ago, and has now been overtaken by Toronto. Similarly in Japan, Osaka was once a far more powerful competitor with Tokyo in the financial markets in Japan than it had become by the late 1980s.

4. The data on producer services is creating a certain amount of confusion in the US. For instance, the fact of faster growth at the national level and in medium sized cities is often interpreted as indicating a loss of share and declining position of leading centres such as New York or Chicago. Thus one way of reading these data is as decentralization of producer services: NY and Chicago losing share of all producer services in the US. Another way is to read it as growth everywhere, rather than a zero sum situation where growth in a new location ipso facto is construed as a loss in an older location. In my reading these patterns point to the growing service intensity in the organization of the economy nationwide.

5. Browning and Roberts, 1980; Cohen, 1981; Glickman, 1984; Hill and Feagin, 1984; Portes and Walton, 1981; Rodriguez and Feagin, 1986; Ross and Trachte, 1983; Sassen, 1982, 1984; Soja et al., 1983; Thrift, 1987.

6. Beauregard, 1991; Brake, 1991; Kunzmann and Wegener, 1991; Mollenkopf and Castells, 1991; Noller et al., 1994; von Petz et al., 1992; Sassen, 1984, 1988; Savitch, 1988.

7. Alegria, 1992; Cybriwsky, 1991; Douglass, 1993; Drennan, 1992; European Institute of Urban Affairs, 1992; Friedmann, 1986; Fujita, 1991; Fujita and Hill, 1993; Knight and Gappert, 1989; Kowarick and Campanario, 1986; Logan and Swanstrom, 1990; Machimura, 1992; Meyer, 1991; Rodriguez and Feagin, 1986; Sachar, 1990; Smith and Feagin, 1987; Stren and White, 1989; Teresaka et al., 1988; Timberlake, 1985.

8. Notable are the edited collections by Fujita and Hill, 1993; King, 1995; Wentz, 1991; *Wissenschafts Forum*, 1995; Yeung, 1996. It should also be noted that other impressive collections, while not as focused on global cities, have multiple chapters dedicated to the subject of cities and the global economy and its literature: see e.g. Cohen et al., 1996; LeGates and Stout, 1996.

9. See for instance several chapters in Basch et al., 1994; Eades, 1996; Holston, 1996; McDowell, 1997; Palumbo-Liu, 1999; Yaeger, 1996.

10. For instance, using information on the corporate headquarter location of the largest 250 US corporations for 1974, 1982 and 1989, Lyons and Salmon (1995) tested several of the central hypotheses/propositions about the nature of global cities. They found a changing concentration of corporate headquarters, with a group of smaller diversified regional cities gaining the losses of cities at the top, notably New York City. Among the gainers were Atlanta, Dallas/Fort Worth, Philadelphia and St Louis. The highest concentration of such headquarters continues to be in New York, Chicago, Los Angeles and San Francisco, which together account for almost half of the top 250 headquarters (see also Sassen, 2000b: ch. 4).

11. One of the patterns that has become clear is that global integration of markets and deregulation have not necessarily had the effect of dispersing firms and holdings to the point of eliminating their concentration in particular localities. For instance, while there was a lot of dispersal, it was not enough to eliminate a disproportionate concentration of certain kinds of firms in the growing network of global cities. For evidence see Sassen (2000b), Short and Kim (1999), GAWC (1999) and Taylor (2000).

BIBLIOGRAPHY

Abbott, Carl (1996) 'The internationalization of Washington, D.C.', *Urban Affairs Review*, 31(5): 571–94.

Abu-Lughod, Janet L. (1999) *New York, Los Angeles, Chicago: America's Global Cities*. Minnesota: University of Minnesota Press.

Alegria, Tito Olazabal (1992) *Desarrollo urbano en la frontera Mexico-Estados Unidos*. Mexico: Consejo Nacional para la Cultura y las Artes.

Allen, John, Massey, Doreen and Pryke, Michael (eds) (1999) *Unsettling Cities*. London: Routledge.

Amin, A. and Thrift, N. (1992) 'Neo-Marshallian nodes in the global networks', *International Journal of Urban and Regional Research*, 16(4): 571–87.

Amin, A. and Thrift, N. (1995) *Globalization, Institutions, and Regional Development*. Oxford: Oxford University Press.

Asher, François (1995) *Metapolis ou l'avenir des villes*. Paris: Editions Odile Jacob.

Basch, Linda, Schiller, Nina Glick and Szanton-Blanc, Cristina (1994) *Nations Unbound: Transnationalized Projects and the Deterritorialized Nation-State*. New York: Gordon and Breach.

Baum, Scott (1997) 'Sydney, Australia: A global city? Testing the social polarisation thesis', *Urban Studies*, 34(11): 1881–901.

Beauregard, Robert (1991) 'Capital restructuring and the new built environment of global cities: New York and Los Angeles', *International Journal of Urban and Regional Research*, 15(1): 90–105.

Beaverstock, Jonathan V. and Smith, Joanne (1996) 'Lending jobs to global cities: skilled international labour migration, investment banking and the City of London', *Urban Studies*, 33(8): 1377–94.

Beaverstock, J.V., Smith, R.G., Taylor, P.J., Walker, D.R.F. and Lorimer, H. (2000) 'Globalization and world cities: some measurement methodologies', *Applied Geography*, 20: 43–63.

Berner, Erhard and Korff, Rudiger (1995) 'Globalization and local resistance: the creation of localities in Manila and Bangkok', *International Journal of Urban and Regional Research*, 19(2): 208–22.

Body-Gendrot, Sophie, Mung, Emmanual Ma and Hodier, Catherine (eds) (1992) 'Entrepreneurs entre deux mondes: les creations d'enterprises par les etrangers: France, Europe, Amerique du Nord', Special Issue, *Revue Europeenne des Migrations Internationales*, 8(12): 5–8.

Bonamy, Joel and May, Nicole (ed.) (1994) *Services et mutations urbaines*. Paris: Anthropos.

Bonilla, Frank, Melendez, Edwin, Morales, Rebecca and Torres, Maria de los Angeles (eds) (1998) *Borderless Borders*. Philadelphia: Temple University Press.

Brake, Klaus (1991) *Dienstleistungen und Raumliche Entwicklung Frankfurt*. Oldenburg: Universität Oldenburg, Stadt- und Regionalplanung.

Braudel, Fernand (1984) *The Perspective of The World*, Vol. III. London: Collins.

Brotchie, J, Barry, M., Blakely, E., Hall, P. and Newton, P. (eds) (1995) *Cities in Competition: Productive and Sustainable Cities for the 21st Century*. Melbourne: Longman Australia.

Browning, Harley I. and Roberts, Bryan (1980) 'Urbanisation, sectoral transformation and the utilisation of labour in Latin America', *Comparative Urban Research*, 8(1): 68–104.

Brunn, Stanley, D. and Leinbach, Thomas R. (eds) (1991) *Collapsing Space and Time: Geographic Aspects of Communications and Information*. London: HarperCollins Academic.

Budd, Leslie (1995) 'Globalization, territory and strategic alliances in different financial centres', *Urban Studies*, 32(2): 345–60.

Budd, Leslie and Whimster, Sam (eds) (1992) *Global Finance and Urban Living: A Study of Metropolitan Change*. London: Routledge.

Burgel, Galia and Burgel, Guy (1996) 'Global trends and city politics: friends or foes of urban development?', in Cohen et al. (eds), *Preparing for the Urban Future. Global Pressures and Local Forces*. Washington, DC: Woodrow Wilson Center Press. (Distributed by The Johns Hopkins University Press.)

Carrez, Jean-François (1991) *Le developpement des fonctions tertiaires internationales á Paris et dans les metropoles regionales*. Rapport au Premier Ministre. Paris: La Documentation Francaise.

Castells, M. (1972) *La Question Urbaine*. Paris: Maspero.

Castells, M. (1989)*The Informational City*. London: Blackwell.

Castells, M. (1996) *The Networked Society*. Oxford: Blackwell.

Castells, M. and Aoyama, Yuko (1994) 'Paths towards the informational society: employment structure in G-7 countries, 1920–90', *International Labor Review*, 133(1): 5–33.

Castells, Manuel and Hall, Peter (1994) *Technopoles of the World: The Making of Twenty-First Century Industrial Complexes*. London: Routledge.

Chase-Dunn, C. (1984) 'Urbanization in the world system: new directions for research', in M.P. Smith (ed.), *Cities in Transformation*. Beverly Hills, CA: Sage.

Chen, Xiangming (1995a) 'Chicago as a global city', *Chicago Office*, 5: 15–20.

Chen, Xiangming (1995b) 'The evolution of free economic zones and the recent development of cross-national growth zones', *International Journal of Urban and Regional Research*, 19(4): 593–621.

Ciccolella, Pablo and Mignaqui, Iliana (2000) 'Buenos Aires', in S. Sassen (ed.) *Cities and their Cross-Border Networks*. Tokyo: UNU Press.

Clark, Terry Nicholas and Hoffman-Martinot, Vincent (eds) (1998) *The New Public Culture*. Oxford: Westview Press.

Cochrane, Allan, Peck, Jamie and Tickell, Adam (1996) 'Manchester plays games: exploring the local politics of globalization', *Urban Studies*, 33(8): 1319–36.

Coffey, William (1996) 'Forward and backward linkages of producer-services establishments: evidence from the Montreal metropolitan area', *Urban Geography*, 17(7): 604–32.

Cohen, Michael, A., Ruble, Blair A., Tulchin, Joseph S. and Garland, Allison M. (eds) (1996) *Preparing for the Urban Future. Global Pressures and Local Forces*. Washington, DC: Woodrow Wilson Center Press. (Distributed by The Johns Hopkins University Press.)

Cohen, R.B. (1981) 'The new international division of labor, multinational corporations and urban hierarchy', in Michael A. Dear and A.J. Scott (eds), *Urbanization and Urban Planning in Capitalist Society*. New York: Methuen.

Copjec, Joan and Sorkin, Michael (eds) (1999) *Giving Ground*. London: Verso.

Corbridge, Stuart, Martin, Ron and Thrift, Nigel (eds) (1994) *Money, Power and Space*. Oxford: Blackwell.

Cox, Kevin R. (1993) 'The local and the global in the new urban politics: a critical view', *Environment and Planning D*, 11: 433–48.

Cybriwsky, R. (1991) *Tokyo. The Changing Profile of an Urban Giant*. (World Cities Series, eds. R.J. Johnson and P.L. Knox.) London: Belhaven Press.

Daniels, Peter W. (1985) *Service Industries: A Geographical Appraisal*. London and New York: Methuen.

Daniels, Peter W. (1991) 'Producer services and the development of the space economy', in Peter W. Daniels and Frank Moulaert (eds), *The Changing Geography of Advanced Producer Services*. London and New York: Belhaven Press.

Daniels, Peter W. and Lever, W.F. (1996) *The Global Economy in Transition*. Essex: Longman.

Le Debat: Le Nouveau Paris. Special Issue, Summer 1994. (Paris: Gallimard.)

Diken, Bulent (1998) *Strangers, Ambivalence and Social Theory*. Aldershot: Ashgate.

Dogan, M. and Kasarda, J.D. (eds) (1988) *A World of Giant Cities*. Newbury Park, CA: Sage.

Douglass, M. (1993) 'The "new" Tokyo story. Restructuring space and the struggle for place in a world city', in K. Fujita and R.C. Hill (eds), *Japanese Cities in the World Economy*. Philadelphia: Temple University Press. pp. 83–119.

Drennan, Mathew P. (1992) 'Gateway cities: the metropolitan sources of US producer services exports', *Urban Studies*, 29(2): 217–35.

Drennan, Matthew P., Tobier, Emanuel, and Lewis, Jonathan (1996) 'The interruption of income convergence and income growth in large cities in the 1980s', *Urban Studies*, 33(1): 63–82.

Dunn, Seamus (ed.) (1994) *Managing Divided Cities*. Stafford, UK: Keele University Press.

Eade, John (ed.) (1996) *Living the Global City: Globalization as Local process*. London: Routledge.

Edel, Matthew (1986) 'Capitalism, accumulation and the explanation of urban phenomena', in Michael Dear and Allen Scott (eds), *Urbanization and Urban Planning in Capitalist Society*. New York: Methuen.

Elliott, James R. (1999) 'Putting "global cities" in their place: urban hierarchy and low-income employment during the post-war era', *Urban Geography*, 20(2): 95–115.

Eurocities (1989) *Documents and Subjects of Eurocities Conference*. Barcelona: 21–22 April 1989.

European Institute of Urban Affairs (1992) *Urbanisation and the Functions of Cities in the European Community: A Report to Commission of the European Communities Directorate General for Regional Policy (XVI)*. Liverpool: Liverpool John Moores University.

Fainstein, Susan (2001) *The City Builders*. Lawrence: University of Kansas Press. Originally published in 1993 by Oxford: Blackwell.

Fainstein, S., Gordon, I. and Harloe, M. (1993) *Divided City: Economic Restructuring and Social Change in London and New York*. New York: Blackwell.

Featherstone, Mike (1993) 'Global and local cultures', in J. Bird, B. Curtis, T. Putnam, G. Robertson and L. Tickner (eds), *Mapping the Futures: Local Cultures, Global Change*. London: Routledge, pp. 169–87.

Fincher, Ruth and Jacobs, Jane M. (eds) (1998) *Cities of Difference*. New York: Guilford Press.

Findlay, A.M., Li, F.L.N., Jowett, A.J. and Skeldon, R. (1996) 'Skilled international migration and the global city: A study of expatriates in Hong Kong', *Transactions of the Institute of British Geographers*, 21(1): 49–61.

Friedmann, John (1986) 'The world city hypothesis', *Development and Change*, 17: 69–84.

Friedmann, John (1995) 'Where we stand: A decade of world city research', in P.L. Knox, and P.J. Taylor, (eds), *World Cities in a World-System*. Cambridge: Cambridge University Press. pp. 21–47.

Friedmann, John and Wolff, G. (1982) 'World city formation: an agenda for research and action', *International Journal of Urban and Regional Research*, 6: 309–44.

Frost, Martin and Spence, Nigel, (1992) 'Global city characteristics and Central London's employment', *Urban Studies*, 30(3): 547–58.

Fuchs, Gotthard, Moltmann, Bernhard and Prigge, Walter (eds) (1995) *Mythos Metropole*. Frankfurt: Suhrkamp.

Fujita, Kuniko (1991) 'A world city and flexible specialization: restructuring of the Tokyo metropolis', *International Journal of Urban and Regional Research*, 15(1): 269–84.

Fujita, Kuniko and Hill, R.C. (eds) (1993) *Japanese Cities in the World Economy*. Philadelphia: Temple University Press.

Futur Anterior (1995) Special issue: *La Ville-monde aujourd'hui: entre virtualité et ancrage.* (eds Thierry Pillon and Anne Querrien), Vols 30–32. Paris: L'Harmattan.

Gad, Gunter, (1991) 'Toronto's financial district', *Canadian Urban Landscapes 1*. 203–7.

Garcia, Soledad (1996) 'Cities and citizenship', *International Journal of Urban and Regional Research*, 20(1): 7–21.

Gilbert, A. (1996) *The Mega-City in Latin America*. Tokyo and New York: United Nations University Press.

Glickman, Norman J. (1984) 'Cities and the international division of labor'. Presented at the Second World Congress of Regional Science Association, Rotterdam, Netherlands, 10 June 1984.

Graham, Stephen and Marvin, Simon (1996) *Telecommunications and the City: Electronic Spaces, Urban Places*. London: Routledge.

Gravesteijn, S.G.E., van Griensven, S. and de Smidt, M.C. (eds) (1998) *Timing Global Cities, Nederlandse Geografische Studies*, 241. Utrecht.

Hall, Peter (1996) *The World Cities*. New York: McGraw Hill.

Hamnett, Chris (1994) 'Social polarisation in global cities: theory and evidence', *Urban Studies*, 31(3): 401–24.

Hamnett, Chris (1996) 'Why Sassen is wrong: a response to Burgers', *Urban Studies*, 33(1): 107–10.

Harris, Nigel and Fabricius, I. (eds) (1996) *Cities and Stuctural Adjustment*. London: University College London.

Harvey, David (1985) *The Urbanization of Capital*. Oxford: Blackwell.

Harvey, David (1989) *The Condition of Postmodernity*. Oxford: Blackwell.

Henderson, Jeff and Castells, Manuel (eds) (1987) *Global Restructuring and Territorial Development*. London: Sage.

Hepworth, Mark (1991) *Geography of the Information Economy*. London: Belhaven.

Hill, Richard Child and Feagin, Joe R. (1984) 'Detroit and Houston: two cities in global perspective'. Presented at 79th Meeting of the ASA, San Antonio, Texas, 29 August 1984.

Hitz, Keil, Lehrer, Ronneberger, Schmid, Wolff (eds) (1995) *Capitales Fatales*. Zurich: Rotpunkt Verlag.

Holston, James (ed.) (1996) 'Cities and citizenship', *Public Culture*, Special Issue, 8(2). (Re-issued 1999, Baltimore, MD: Johns Hopkins University Press.)

Hymer, Stephen and Rowthorn, Robert (1970) 'Multinational corporations and international oligopoly', in Charles P. Kindleberger (ed.), *The International Corporation*. Cambridge, MA: MIT Press.

Jessop, Robert (1999) 'Reflections on globalization and its illogics', in Kris Olds et al. (eds) (1999) *Globalization and the Asian Pacific: Contested Territories*. London: Routledge. pp. 19–38.

Journal of Urban Technology (1995) 'Information technologies and inner-city communities', Special Issue, 3(1).

Katznelson, Ira (1992) *Marxism and the City*. Oxford: Clarendon Press.

Keil, Roger (1993) *Weltstadt- Stadt der Welt: Internationalisierung und lokale Politik in Los Angeles*. Munster: Westfaelisches Dampfboot.

King, A.D. (1990a) *Global Cities: Post-Imperialism and the Internationalization of London*. London: Routledge.

King, A.D. (1990b) *Urbanism, Colonialism, and the World Economy: Culture and Spatial Foundations of the World Urban System*. The International Library of Sociology. London and New York: Routledge.

King, A.D. (ed.) (1995) *Representing the City: Ethnicity, Capital and Culture in the 21st Century*. London: Macmillan.

Klopp, Brett (1998) 'Integration and political representation in a multicultural city: the case of Frankfurt am Main', *German Politics and Society*, 49, no. 4. pp. 42–68.

Klosterman, Robert C. (1996) 'Double Dutch: polarization trends in Amsterdam and Rotterdam after 1980', *Regional Studies*, 30(5): 467–76.

Knight, Richard V. and Gappert, Gary (eds) (1989) *Cities in a Global Society*. Newbury Park: Sage.

Knox, Paul L. and Taylor, Peter J. (eds) (1995) *World Cities in a World-System*. Cambridge: Cambridge University Press.

Kowarick, L. and Campanario, M. (1986) 'São Paulo: the price of world city status', *Development and Change*, 17(1): 159–74.

Kowarick, L., Campos, A.M. and de Mello, M.C. (1991) 'Os Percursos de Desigualdade', in R. Rolnik, L. Kowarick and N. Somekh (eds), *São Paulo, Crise e Mudança*. São Paulo: Brasiliense.

Kunzmann, K.R. and Wegener, M. (1991) *The Pattern of Urbanisation in Western Europe 1960–1990*. Report for Directorate-General XVI of the Commission of the European Communities as part of the study 'Urbanisation and the Function of Cities in the European Community'. Dortmund, Germany: Institut fur Raumplanung, 15 March 1991.

Landrieu, Josee, May, Nicole, Spector, Therese and Veltz, Pierre (eds) (1998) *La Ville Eclatee*. La Tour d'Aigues: Editiones de l'Aube.

Lazzarato, M., Moulier-Boutang, Y, Negri, A. and Santilli, G. (1993) *Des entreprises pas comme les autres: Benetton en Italie, Le Sentier à Paris*. Paris: Publisud.

LeGates, Richard T. and Stout, Frederic (eds) (1996) *The City Reader*. London and New York: Routeledge.

Lo, Fu-chen and Yeung, Yue-man (eds) (1996) *Emerging World Cities in Pacific Asia*. Tokyo: United Nations University.

Logan, John R. and Swanstrom, Todd (eds) (1990) *Beyond the City Limits: Urban Policy and Economic Restructuring in Comparative Perspective*. Philadelphia: Temple University Press.

Low, Setha M. (1999) 'Theorizing the city', in S.M. Low (ed.), *Theorizing the City*. New Brunswick, NJ: Rutgers University Press. pp. 1–33.

McDowell, Linda (1997) *Capital Culture*. Oxford: Blackwell Publishers.

Machimura, Takashi (1992) 'The urban restructuring process in the 1980s: transforming Tokyo into a world city', *International Journal of Urban and Regional Research*, 16(1): 114–28.

Machimura, Takashi (1998) 'Symbolic use of globalization in urban politics in Tokyo', *International Journal of Urban and Regional Research*, 22(2): 183–94.

Marcuse, Peter and van Kempen, Ronald (2000) *Globalizing Cities: A New Spatial Order*. Oxford: Blackwell.

Markusen, A. and Gwiasda, V. (1994) 'Multipolarity and the layering of functions in the world cities: New York City's struggle to stay on top', *International Journal of Urban and Regional Research*, 18: 167–93.

Meyer, David R. (1991) 'Change in the world system of metropolises: the role of business intermediaries', *Urban Geography*, 12(5): 393–416.

Meyer, David (2000) 'Hong Kong', in S. Sassen (ed.), *Cities and their Cross-Border Networks*. Tokyo: UNU Press.

Mitchell, Mathew and Sassen, Saskia (1996) 'Can cities like New York bet on manufacturing?', in *Manufacturing Cities: Competitive Advantage and the Urban Industrial Community*. Conference sponsored by the Harvard University Graduate School of Design and The Loeb Fellowship. Cambridge, MA, 8 March.

Mitchelson, Ronald L. and Wheeler, James O. (1994) 'The flow of information in a global economy: the role of the American urban system in 1990', *Annals of the Association of American Geographers*. 84(1): 87–107.

Mollenkopf, John and Castells, Manuel (eds) (1991) *Dual City: Restructuring New York*. NY: Russell Sage Foundation.

Moss, Mitchell (1991) 'New fibres of urban economic development', *Portfolio: A Quarterly Review of Trade and Transportation*. 4(1): 11–18.

Moulaert, F. and Scott, A.J. (1997) *Cities, Enterprises and Society on the Eve of the 21st Century*. London and New York: Pinter.

Mozere, Liane, Peraldi, Michel and Rey, Henri (eds) (1999) *Intelligence Des Banlieues*. La Tour d'Aigues: Editiones de l'Aube.

Murie, Alan and Musterd, Sako (1996) 'Social segregation, housing tenure and social change in Dutch cities in the late 1980s', *Urban Studies*, 33(3): 495–516.

Nijman, Jan (1996) 'Breaking the rules: Miami in the urban hierarchy', *Urban Geography*, 17(1): 5–22.

Noller, Peter, Prigge, Walter and Ronneberger, Klaus (eds) (1994) *Stadt-Welt*. Frankfurt: Campus Verlag.

Noyelle, T. and Dutka, A.B. (1988) *International Trade in Business Services: Accounting, Advertising, Law and Management Consulting*. Cambridge, MA: Ballinger Publishing.

Noyelle, T. and Stanback, Thomas Jr (1984) *The Economic Transformation of American Cities*. Totowa, NJ: Rowman and Allanheld.

Olds, Kris, Dicken, Peter, Kelly, Philip F., Kong, Lilly and Yeung, Henry Wai-Chung (eds) (1999) *Globalization and the Asian Pacific: Contested Territories*. London: Routledge.

Palumbo-Liu, David (1999) *Asian/American*. Stanford: Stanford University Press.

Peraldi, Michel and Perrin, Evelyne (eds) (1996) *Reseaux Productifs et Territoires Urbains*. Toulouse: Presses Universitaires du Mirail.

Persky, Joseph and Wievel, Wim (1994) 'The growing localness of the global city', *Economic Geography*, 70(2): 129–43.

von Petz, Ursula and Schmals, Klaus M. (eds) (1992) *Metropole, Weltstadt, Global City: Neue Formen der Urbanisierung*. Dortmund: Dortmunder Beitrage zur Raumplanung, Vol. 60. University of Dortmund.

Pirez, Pedro (1994) *Buenos Aires Metropolitana*. Buenos Aires: Centro.

Portes, A. and Lungo, M. (eds) (1992a) *Urbanización en Centroamerica*. San José, Costa Rica: Facultad Latinoamericana de Ciencias Sociales – Flacso.

Portes, A. and Lungo, M. (eds) (1992b) *Urbanización en el Caribe*. San José, Costa Rica: Facultad Latinoamericana de Ciencias Sociales – Flacso.

Portes, A. and Walton, J. (1981) *Labor, Class and the International System*. New York: Academic Press.

Portes, A., Guarnizo, Luis and Landolt, Patricia (1999) 'The study of transnationalism: pitfalls and promise of an emergent research field', *Ethnic and Racial Studies*. (March).

Pratt, A.C. (1997) 'The cultural industries production system: a case study of employment change in Britain, 1984–1991', *Environment and Planning A*, 29 (11): 1953–74.

Pryke, M. (1991) 'An international city going global: spatial change in the City of London', *Environment and Planning D: Society and Space*, 9: 197–222.

Rakodi, C. (ed.) (1997) *The Urban Challenge in Africa: Growth and Management of its Large Cities*. Tokyo and New York: United Nations University Press.

Razin, Eran and Light, Ivan, (1998) 'Ethnic entrepreneurs in America's largest metropolitan areas', *Urban Affairs Review*, 33(3): 332–60.

Reboud, Louis (ed.) (1997) *La relation de service au coeur de l'analyse economique*. Paris: L'Harmattan.

Robertson, Roland (1997) 'Social theory, cultural relativity and the problem of globality', in A.D. King (ed.), *Culture, Globalization and the World-System: Contemporary Conditions for the Representation of Identity*. Minnesota: University of Minnesota Press. pp. 69–90.

Rodriguez, Nestor P. and Feagin, J.R. (1986) 'Urban specialization in the world system', *Urban Affairs Quarterly*, 22(2): 187–220.

Ronneberger, Klaus and Noller, Peter (1994) 'Globalisierte Oekonomie und regionale Identitaet: Neue Dienstleister', in Martin Wentz (ed.), *Die Zukunft des Staedtischen: Region*. Frankfurt: Campus Verlag. pp. 26–33.

Rosenblat, Celine and Pumain, Denise (1993) 'The location of multinational firms in the European urban system', *Urban Studies*, 30(10): 1691–709.

Ross, Robert and Trachte, Kent, (1983) 'Global cities and global classes: the peripheralization of labor in New York City', *Review*, 6(3): 393–431.

Rotzer, Florian (1995) *Die Telepolis: Urbanität im digitalen Zeitalter*. Mannheim: Bollmann.

Roulleau-Berger, Laurence (1999) *Le travail en friche*. La Tour d'Aigues: Editiones de l'Aube.

Sachar, A. (1990) 'The global economy and world cities', in A. Sachar and S. Oberg (eds), *The World Economy and the Spatial Organization of Power*. Aldershot: Avebury. pp. 149–60.

Salomon, Ilan (1996) 'Telecommunications, cities and technological opportunism', *The Annals of Regional Science*, 30: 75–90.

Sanchez, Roberto and Alegria Tito (1992) 'Las cuidades de la frontera norte', Departemento de Estudios Urbanos y Medio Ambiente, El Colegio de la Frontera Norte, Tijuana, Mexico.

Santos, Milton, De Souze, Maria Adelia A. and Silveira, Maria Laura (eds) (1994) *Territorio Globalizacao e Fragmentacao*. São Paulo: Editorial Hucitec.

Sassen, Saskia (1982) 'Recomposition and Peripheralization at the Core', *Contemporary Marxism*, 5: 88–100. (Reissued in S. Jonas and M. Dixon (eds), *The New Nomads: Immigration and Changes in the New International Division of Labor*. San Francisco: Synthesis Publications.)

Sassen, Saskia (1984) 'The new labor demand in global cities', in M.P. Smith (ed.), *Cities in Transformation*. Beverly Hills, CA: Sage.

Sassen, Saskia (1994) 'The informal economy: between new developments and old regulations', *The Yale Law Journal*, 103(8): 2289–304.

Sassen, Saskia (2000a) *Cities and their Cross-Border Networks*. Tokyo: United Nations University Press. London: Blackwell.

Sassen, Saskia (2000b) *Cities in a World Economy*. Thousand Oaks, CA: Pine Forge/Sage Press. (New updated edition; originally published in 1994.)

Sassen, Saskia (2000c) *Cities and their Cross-Border Networks*. United Nations University Press (Tokyo).

Sassen, S. (2000d) *The Global City: New York, London, Tokyo*. Princeton: Princeton University Press. Updated edition.

Savitch, H, (1988) *Post-Industrial Cities*. Princeton, NJ: Princeton University Press.

Savitch, H.V. (1996) 'Cities in a global era: a new paradigm for the next millennium', in M. Cohen et al. (eds), *Preparing for the Urban Future*. Washington, DC: Woodrow Wilson Center Press. pp. 39–65.

Sayer, Andrew and Walker, Richard (1992) *The New Social Economy: Reworking the Division of Labor*. Cambridge, MA and Oxford, UK: Blackwell.

Schiffer, Sueli Ramos (2000) 'Sao Paulo', in S. Sassen (ed.), *Cities and their Cross-Border Networks*. Tokyo: UNU Press.

Scott, A.J., Storpoer, M., Soja, E. and Agnew, J. (eds) (2000) *Global-City Regions*. Oxford: Oxford University Press.

Shank, G. (ed.) (1994) *Japan Enters the 21st Century. A Special Issue of Social Justice*, 21(2).

Short, John R. and Kim, Y. (1999) *Globalization and the City*. Essex: Longman.

Simon, David (1995) 'The world city hypothesis: reflections from the periphery', in P.L. Knox and P.J. Taylor (eds), *World Cities in a World-System*. Cambridge: Cambridge University Press. pp. 132–55.

Singelman, J. and Browning, H.L. (1980) 'Industrial transformation and occupational change in the US, 1960–70', *Social Forces*, 59: 246–64. Cambridge: Cambridge University Press. pp. 132–55.

Skeldon, R . (1997) 'Hong Kong : colonial city to global city to provincial city?', *Cities*, 14(5): 265–71.

Sklair, L. (1991) *Sociology of the Global System: Social Changes in Global Perspective*. Baltimore, MD: Johns Hopkins University Press.

Smith, M.P. and Feagin, J.R. (1987) *The Capitalist City: Global Restructuring and Territorial Development*. London: Sage.

Smith, D. and Timberlake, M. (2000) 'Cities in global matrices', in S. Sassen (ed.), *Cities and their Cross-Border Networks*. Tokyo: UNU Press.

Social Justice (1993) 'Global Crisis, Local Struggles', Special Issue, 20(3–4).

Sociétés Contemporaines (1995) 'Segregations Urbaines', Special Issue, 22/23 (June–September). (Paris: L'Harmattan.)

Soja, Edward, Morales, Rebecca and Wolff, Goetz (1983) 'Urban restructuring: an analysis of social and spatial change in Los Angeles', *Economic Geography*, 59(2): 195–230.

Stanback, T.M. and Noyelle, T.J. (1982) *Cities in Transition: Changing Job Structures in Atlanta, Denver, Buffalo, Phoenix, Columbus (Ohio), Nashville, Charlotte*. Montclair, NJ: Allenheld, Osmun.

Stren, Richard (1996) 'The studies of cities: popular perceptions, academic disciplines, and emerging agendas', in M. Cohen et al. (eds), *Preparing for the Urban Future*. Washington, DC: Woodrow Wilson Center Press, pp. 392–420.

Stren, R.E. and White, R.R. (1989) *African Cities in Crisis: Managing Rapid Urban Growth*. Boulder, CO: Westview Press.

Suro, Roberto (1998) *Strangers Among Us: How Latino Immigration is Transforming America*. NY: Alfred A. Knopf.

Tabak, Faruk and Chrichlow, Michaeline A. (eds) (2000) *Informalization: Process and Structure*. Baltimore, MD: Johns Hopkins University Press.

Tardanico, Richard and Lungo, Mario (1995) 'Local dimensions of global restructuring in urban Costa Rica', *International Journal of Urban and Regional Research*, 19(2): 223–49.

Taylor, Peter J. (1995) 'World cities and territorial states: the rise and fall of their mutuality', in P.L. Knox and P.J. Taylor (eds), *World Cities in a World-System*. Cambridge: Cambridge University Press. pp. 48–62.

Taylor, Peter J. (2000) 'World cities and territorial states under conditions of contemporary globalization', *Political Geography*, 19(5): 5–32.

Teresaka, Akinobu et al. (1988) 'The transformation of regional systems in an information-oriented society', *Geographical Review of Japan*, 61(1): 159–73.

Thrift, N. (1987) 'The fixers: the urban geography of international commercial capital', in J. Henderson and M. Castells (eds), *Global Restructuring and Territorial Development*. London: Sage.

Thrift, Nigel and Leyshon, Andrew (1994) 'A phantom state? The de-traditionalization of money, the international financial system and international financial centres', *Political Geography*, 13(4): 299–327.

Timberlake, M. (ed.) (1985) *Urbanization in the World Economy*. Orlando, FL: Academic.

Todd, Graham (1995) '"Going Global" in the semi-periphery: world cities as political projects. The case of Toronto', in P.L. Knox and P.J. Taylor (eds), *World Cities in a World-System*. Cambridge: Cambridge University Press. pp. 192–214.

Torres, Rodolfo D., Inda, Jonathan Xavier and Miron Louis F. (1999) *Race, Identity, and Citizenship*. Oxford: Blackwell.

Veltz, Pierre (1996) *Mondialisation, villes et territoires: l'economie d'archipel*. Paris: Presses Universitaires de France.

Vogel, David (1993) 'New York City as a national and global finance center', in Martin Shefter (ed.), *Capital of the American Century: The National and International Influence of New York City*. New York: Russell Sage Foundation. pp. 49–70.

Walker, Richard (1996) 'Another round of globalization in San Francisco', *Urban Geography*, 17(1): 60–94.

Warf, Barney and Erickson, Rodney (1996) 'Introduction: globalization and the U.S. city system', *Urban Geography*, 17(1): 1–4.

Wentz, Martin (ed.) (1991) *Stadtplanung in Frankfurt: Wohnen, Arbeiten, Verkehr*. Frankfurt: Campus Verlag.

Wheeler, James O. (1986) 'Corporate spatial links with financial institutions: the role of the metropolitan hierarchy', *Annals of the Association of American Geographers*, 76(2): 262–74.

Wissenschaft Forum (1995) 'Global City: Zitadellen der Internationalisierung', Special Issue 12(2).

Yeung, Yue-man (1996) 'An Asian perspective on the global city', *International Social Science Journal*, 147: 25–32.

Yeung, Yue-man (2000) *Globalization and Networked Societies*. University of Hawaii Press.

17

The Post-fordist City

W.F. LEVER

The rationale for the rapid growth of industrial towns and cities in Europe and North America was the comparative advantage of scale economies. The widespread application of steam-power, rather than the earlier waterpower, facilitated the development of large-scale industrial plants such as steelworks and shipyards. These large plants required the assembly of large labour forces, much of them drawn from rural areas, or from immigration in the case of North America, and the rates of population growth were rapid, often 10 per cent per year at peak growth rates. The implementation of Fordist regimes, based on Taylorist ideas of labour specialization, scientific management and the optimal use of time, extended the process of factory growth into assembly-line production systems for the manufacture of consumer goods such as automobiles. Although more capital-intense, these production systems still required large quantities of labour. Fordism, however, was substantially more than a system of production, in that it came to describe a whole economic and social system. In Lipietz's (1989) formulation the Fordist regime of accumulation was based in the final analysis on the idea of mass production, mass consumption and a Keynesian system of state regulation (Friedmann, 1995).

The key economic features of Fordism include production systems based on monopolistic or oligopolistic capitalism increasing concentration of capital, growth in output and worker productivity, especially in consumer products, and an expansion of demand for, and supply of private and public services. It assumes almost full employment, although there is a transition from employment in manufacturing to employment in services from the mid-1960s; more of the employed workforce is female, and there is an increasing skill division of labour. The replacement of human capital by fixed capital in manufacturing increasingly separates the workforce into a primary and secondary labour forces. Mass consumption becomes increasingly predominant, especially in standardized household durables such as electrical goods and automobiles. Production enjoys economies of scale in the form of mass production, which is functionally decentralized and often multinationally organized and controlled (Pacione, 1997; Wallace, 1990).

This Fordist economic system is paralleled by a socio-institutional structure with a collectivistic character (Mandel, 1980). This social structure is organized mainly by occupation, but with a tendency towards homogenization and experiencing income conversion. Within the labour market, the increased use of trade unions and collective bargaining brings increasing income conversion. Politics are closely aligned with occupation and organized labour, and the regional and class dimensions are important. The degree of state intervention is Keynesian–Liberal collectivist in character, with markets regulated against problems such as monopolistic price fixing, levels of demand maintained by public expenditure and the expansion of the welfare state. In terms of the space economy, pronounced regional specializations of early industrialization become overlaid by new spatial divisions of labour based on functional decentralization and specialization: regional unemployment disparities remain relatively stable, although industrial and economic structures may converge (Gordon, 1980; Martin, 1988).

In urban terms, Fordism could be equated with the success of large cities and large city systems. The predominant modes of production required locations in large cities, not just as the homes of large industrial workforces but as the providers of the most advantageous sets of externalities. Large cities meant large local markets and an extensive array of advanced producer services, including data processing, financial and legal services, education, personal and ancillary services, access to political decision-makers. As the world economy globalized, the large cities remained the key locations in corporate structures and on informational networks (Clark, 1996). The success of Fordist production systems was equated with the success of large cities as economies, and debates on 'the urban problem' revolved around the most effective ways of slowing their growth, which was causing congestion, high space costs, high labour costs, system capacity problems in transport and services and environmental–ecological problems. The policy response in Europe and to a lesser extent in North America, was to filter out some of the growth to smaller urban places and lagging regions through land use planning, licensing of economic development, space controls and population movement (Hall, 1988).

POST-FORDISM

This paradigm of continuous growth and development within a global economy in ways favourable to the expansion of the largest cities, however, was not sustainable. Fordism came to be replaced by post-fordism, with some scholars arguing that the transition was automatic, inevitable, or path-dependent (Bonefield and Holloway, 1991; Clarke, 1990; Rustin, 1989) while others have argued that there has been no sharp distinctive break between the two stages (Lovering, 1991; Sayer, 1989). Nevertheless, there has been a transition to post-fordism. Sternberg (1993) lists some eight characteristics of postfordist, or postmodern, economies and societies. First, there is a high value placed on knowledge or information within the process of wealth creation. Secondly, the postmodernist trend will extend consumerism into all areas of private and social life, including aesthetics, art, leisure and pleasure. Thirdly, it is characterized by global interdependence on production, finance, distribution, migration and trade. Fourthly, Sternberg identifies a new mercantilism in which national coalitions between industry, government and labour seek to develop strategic comparative advantage as a basis for national prosperity. Fifthly the growth of multinational enterprises and financial institutions run by a new class of global executives and professionals will shape consumption and production patterns. Sixthly, flexible specialization, characterized by new principles of production, specialist units of production, decentralized management and versatile technologies and workforces, will become the new system of production. Seventhly, new social movements will come into being, humanizing capital with greater concerns for ethnic groups, for women and for the environment. Lastly, there is increasing rejection of the technocracy and consumerism which so characterized Fordism, and the growth of communitarian, social and religious values and traditions by way of replacement.

Whilst most observers recognize these elements of the post-fordist world, there are three theoretical positions at the core of the post-fordist debate. These are the regulation approach, the neo-Schumpeterian approach and the flexible specialization approach. Each has developed a separate explanation of the process by which the era of Fordist mass production is being replaced by new systems (Amin, 1994). The regulation approach was pioneered in France in the 1970s and refined in the 1980s by political economists trying to explain the dynamics of long-term, 50-year, cycles of economic stability and change (Aglietta, 1979; Lipietz, 1985). The objective was to develop a theoretical framework that could explain the apparent inconsistency between capitalism's inherent tendency towards instability, crisis and change and its ability to stabilize around a set of institutions, rules and norms (regulations) to secure a relatively long period of economic stability. There was concern that the recession of the 1970s was not just another lull within a recurrent cycle but a more generalized crisis of the institutional structures which had organized the post-war global economy. It was important, the regulationists argued, to recognize the historical processes that underpinned economic change. Two key concepts at the heart of regulation approach are the 'regime of accumulation' and the 'mode of regulation'. The regime of accumulation refers to a set of regularities at the level of the macroeconomy which facilitates a coherent process of capital accumulation, including the organization of production and work, relationships and forms of exchange between sectors, common rules of economic management and norms of income-sharing (wages, profits) and norms of consumption in the market place. The mode of regulation refers to the institutional framework (laws, contracts, etc.) and a nexus of cultural habits which secure capitalist reproduction, and comprise laws, state policy, political practices, industrial codes, governance and social expectations (Nielson, 1991). Fordism is therefore a distinctive type of labour process involving large plants, mass production, scale economies, rising incomes linked to productivity and mass consumption. The slow-down and successive recession of the 1980s and 1990s are seen as the 'crisis of Fordism'. Critics of the regulationist arguments for Fordism and post-fordism have argued that the dominant Fordist mode of production is not ubiquitous, that there are several national variants of the paradigm and that the regulation approach tends to generate a systemic functionalist and logical coherence to history which it in fact does not possess (Clarke, 1988; Hirst and Zeithin, 1991).

The neo-Schumpeterian view of Fordism and post-fordism stems from Kondratiev's work on long-wave (50-year) cycles of growth, of an alternation between boom and recession, in capitalist economies. Schumpeter extended this work by identifying the key role of path-breaking entrepreneurs whose innovations led to new technological paradigms. For Freeman and Perez (1988), the successful transition from one long wave to another is dependent upon 'quantum leaps' in industrial productivity achieved by the diffusion of major innovations throughout the macroeconomy. These technical innovations, however, need to be matched by socio-institutional innovations. In neo-Schumpeterian analysis, the passing of the age of mass production, termed the fourth Kondratieff or long wave, is claimed to have been underpinned by electro-mechanical technologies, the products of mass consumption industries and oil and petrochemicals as the basic sources of cheap energy. In common with the regulationist school, it identifies standardization, massification, scale economies, oligopolistic competition and mass consumption of cheap goods as the distinctive features of the fourth Kondratiev arranged around vertically integrated, hierarchical corporations. In terms of the socio-institutional context, the neo-Schumpeterian view focuses on state policy – for education, housing, welfare and Keynesian interventions into the macroeconomy. Post-fordism begins to come about with the competitive failure of oligopolistic structures in maturing technologies. Productivity gains are reduced as wages and prices rise. The major problem or crisis occurs as new techno-economic paradigms come into increasing conflict with the enduring socio-institutional structures of the fourth Kondratieff. The benefits of innovation are slow to diffuse as inertia due to the reluctance of management and labour, political inertia and legislative drag impede the spread of new ideas (Nielson, 1991).

The flexible specialization approach draws a sharp distinction between mass production involving the use of special purpose, product-specific machines operated by semi-skilled workers to produce standardized goods and flexible specialization, or craft production, in which skilled workers produce a variety of customized goods (Piore and Sabel, 1984). The two systems are thought to co-exist, but at different points in history one system comes to predominate over the other. With the crisis of recession from the 1970s, the problem of declining profits, growing uncertainty emanating from the breakdown of post-war economic systems such as the Bretton Woods

agreements, and the threat to mass production represented by rejection of standardized products, the flexible specialization model offers a better understanding of post-fordism at the present time. The rise of non-specialist and highly flexible manufacturing and design and flexible work practices characterize the new system, which favours small-batch production without losing large-scale economies. Thus, the Fordist emphasis on large plants and large corporations gives way to the small and medium-sized enterprise which in its turn may not require location in large urban centres accommodating large workforces (Amin and Robins, 1990; Harrison, 1994; Leborgue and Lipietz, 1992). Criticism of the flexible specialization approach has focused on its duality, which it is argued caricaturized the two extremes of mass production and flexible specialization, failing to recognize the heterogeneous nature of both systems. Piore and Sabel (1984) have also been criticized for their naivety in assuming that much of industrial product can be generated in a craft paradigm, when elements of Fordism are too deeply embedded to be replaced to such an extent.

The key constituents of the post-fordist city therefore stem from economic changes which have seen the reduction in importance of scale economies and hence the need for large plants, in large cities. This has been accompanied by the growth of the small enterprise sector, requiring less labour employed more flexibly, and the transition from employment in manufacturing to employment in services. Higher levels of information, managerial changes such as just-in-time systems, and disintegration of vertical production 'filieres' or chains of production in single or multiple establishments will impact on the urban hierarchy in different ways. The Fordist town was characterized by strongly agglomerative processes, the standardization and industrialization of construction, the nuclearization of the family and far-reaching processes of social disintegration. Supported by the large-scale use of the car, extreme spatial-functional differentiations developed, characterized by suburbanism, the formation of satellite towns, the depopulation of the inner cities, the loss of smaller industrial and service enterprises, and the growth of hypermarkets and trading estates. Life in the nuclear family, standardized labour, television and cars became the basis of a new model of life and consumption and structured urban space. The 'incongeniality' of the standardized towns, whose spaces were differentiated according to function, became a central issue for critical urban sociology. State and local government supported this process through traffic development, housing policies and subsidies. Serious social conflicts were caused by the process by which residential areas near the city were turned into slums, often by planning blight or uncertainty, as a preliminary step towards commercial use for predominantly tertiary functions, by the loss of infrastructure, the downgrading of local public services and the expulsion of the population from deep-rooted residential areas and the drastic reduction of the quality of life (Esser and Hirsch, 1994). The crisis of Fordism rapidly or simultaneously became the crisis of the Fordist city.

POST-FORDIST ECONOMIES AND THE URBAN HIERARCHY

Fordist globalization based on multinational enterprises, which often had more power than the national economies which hosted them, together with the international agencies such as the International Monetary Fund, the World Bank, the Single European Market and the North American Free Trade Association, are often argued to have reduced the power of national governments. However, the crisis of Fordism has enabled regions and urban regions to take greater control over their economies. Some, such as Sabel (1989), have argued that this represents a return to the nineteenth-century pattern of flexible specialization in specific sectors, which predominated in local urban economies: Lyons for silk, Sheffield for cutlery and St Etienne for cast iron hardware and arms, for example. New urban and regional economies have emerged in which small and medium-sized enterprises are networked by flows of goods and services in a model based on collaboration rather than competition. The network is sufficiently flexible to achieve scale economies without the diseconomies of large plant inflexibilities. Locationally such an industrial network may be positioned on a network of small and medium-sized towns. Again, the intention is to achieve agglomerative or locationalization economies without the agglomerative diseconomies of congestion, high rents, high efficiency wages and urban problems such as crime (Scott, 1988). Earlier models might have taken the form of the Black Country in the English Midlands, where a complex of metal producing and metal products formed over about a dozen urban centres, and the Ruhrgebiet in Germany where a similar coal–steel–metalworking complex developed over about 20 towns.

In the post-fordist era, the first and best-known case was that of the Third Italy, identified by Bagnasco (1977) and so called because it contrasted with the poor, agricultural South and the old heavy-industrial North focused on Genoa, Turin and Milan. It is a string of industrial districts extending from the Venetian provinces in the north through Bologna and Florence to Ancona in the south, producing a wide range of goods, many of which have a high fashion or design content. These include knitwear (Carpi), special machinery (Parma, Bologna), ceramics (Sassuolo), textiles (Como, Prato), mechanical pumps and shoes (both Modena) and musical implements (Ancona). This networked development, however, is now spreading into southern and northern Italy, with new industrial districts in Turin (robotics) and the Canavese region of Piedmont (computers and software), around Milan in Lombardy (furniture and machine tools) and around Bari (textiles) (Leone, 1994).

Other examples of new industrial districts with poly-nuclear urban structures are to be found elsewhere in Europe. Jutland in Denmark has recently developed a network of textile, clothing, furniture, machine tool and even shipbuilding firms as a countermagnet to the overconcentration of economic activity in Copenhagen. In Germany, the *Land* of Baden-Wurttemberg has a number of textile, clothing, textile machinery, machine tool, electrical machinery and automobile component districts. In Scotland, the emergence of 'Silicon Glen' may be seen as the development of a new industrial space in which the smaller urban centres such as the New Towns (East Kilbride, Glenrothes and Livingston) and older towns such as Dunfermline have become the major foci, whereas the major urban centres such as Glasgow have not succeeded in attracting the industry. This, in a sense, typifies issues of flexible production systems in which locations that have no long history of Fordist or pre-fordist industry with its adversarial labour relations are preferred to the older industrial cities and towns (Pike and Sengenberger, 1992).

In the United States the two best-known high-technology industrial districts are the centre of semi-conductor production in Silicon Valley, south-east of San Francisco, and the concentration of mini-computer producers along Route 128 circling Boston (Saxenian, 1985). A third area of industrial networking appears to be developing in Orange County, south of Los Angeles, based on the film, television, video and music recording industries, but also including aerospace and automobiles. An interesting feature of this industrial district is the fact that it has spread over the border into Mexico where production occurs in 'maquilladores' – US-owned plants in low labour-cost Mexico (Storper and Christopherson, 1986).

If post-fordism equates with small units of production located in smaller urban centres, then what of the large units of production in the 1990s? As early as the 1970s evidence was emerging that the location of large Fordist plants in large urban centres, or at their peripheries, was no longer the most profitable locus of production.

The advantage of access to large workforces, large local markets and the range of positive externalities was being offset by the growth of diseconomies such as high wages, traffic congestion, negative externalities and high rents. Researchers such as Tyler, Moore and Rhodes (1984) by the 1980s were able to demonstrate that profit rates in manufacturing were inversely correlated with urban size, and in the case of London were positively correlated with linear distance from the centre of London measured along the motorways that radiated from the city. Researchers such as Massey and Meegan (1979, 1982) were able to show that as the 'crisis of Fordism' hit manufacturing firms in Britain and Europe, they were most likely to remove their excess capacity in the largest cities – often where the capital vintage was oldest – and refocus production, usually with substantial labour reductions, in smaller urban places.

The consequence of these processes in terms of the urban hierarchy has been described as the urban–rural shift (Keeble et al., 1983). Thus, within Europe in the 1970s the highly urbanized regions' share of manufacturing output fell by 1.7 per cent, that of the urbanized regions fell by 0.3 per cent, that of the less urbanized regions grew by 0.9 per cent and that of the rural regions grew by 1.1 per cent. In terms of gross value-added the relative share shifts were -2.9 per cent, +0.2 per cent and +1.1 per cent and +1.6 per cent. The same relationship holds at the nationally disaggregated level, as well as the European-wide level. Thus in Germany the figures were -1.8 per cent, +0.3 per cent and +0.6 per cent and +0.9 per cent, and in France they were -1.6 per cent, -0.6 percent, +1.1 per cent and + 1.1 per cent. In locational terms this means that urban regions such as the Paris Basin and the Ruhr performed poorly and rural/small urban centre regions such as Brittany, the Auvergne and Thuringia performed well. A comparison between employment change and output change shows how productivity improved much more markedly in the rural and small urban centres than it did in the large cities.

Although much of the analysis indicating the movement of output and employment to smaller urban centres has focused on the manufacturing sectors, there is evidence to believe that the same processes hold good in services, although analyses of output necessarily relate to tradable services (Daniels, 1993). Whilst sectors such as retailing may have concentrated upon large units, in the form of hypermarkets on out-of-town sites, other sectors such as insurance, and leisure and recreation, have seen a dispersal to small urban centres in small units linked in some cases by advanced telecommunications. In the 1990s, however, post-fordist consumerism has become refocused on inner city and city centre developments which offer wide choice and greater specialization in retailing, entertainment and culture. The balance of these two trends, however, mirrors the urban–rural shift of manufacturing. In the United Kingdom, for example, employees in retailing between 1981 and 1991 fell by 6.6 per cent in London and by 5.4 per cent in the other principal cities, whereas in other cities it rose by 5.5 per cent, in older towns it rose by 16.4 percent and in smaller towns and rural areas it rose by 25.5 per cent (Townsend et al., 1996).

Whilst industry and services have relocated themselves across the urban hierarchy, and indeed within cities between the core and the periphery, in post-fordist systems, the demand for labour has been similarly adjusted. Flexible systems of production in both manufacturing and services have had several impacts on the construction of labour demand. The most obvious single effect has been growing polarization within the urban labour force. This duality is most clearly expressed as a widening gulf between a stable core of high-waged workers (typically white/male/educated) and an unstable periphery of low-waged workers (typically female/black/poorly qualified) (Piore, 1980). The core–periphery divide within the labour market reflects the division within the post-fordist production system between core producers and their periphery of subcontractors, franchisees, suppliers and surplus/periodic/seasonal capacity. Workers in the latter sector are likely to find employment uncertain, periodic and subject to marked swings between labour surplus and overly tight labour demand. The uncertainties experienced by workers in the secondary labour market are not only reflected in their economic circumstances but have implications for their position in the housing market as they are unlikely to be mortgagable, fixing them within the private or social rental sector. There are also correlates between occupying a position in the secondary labour market and poorer health, poorer access to services such as education and leisure, and less political voice. Such groupings are now often termed the underclass in the US and the socially excluded in Europe.

Whilst much of the theory pertaining to the separation of primary and secondary labour markets has been developed in the context of the manufacturing sector, similar distractions may be found in the service sector. Here, the primary labour force comprises those workers engaged in high-level information transfers such as corporate managers, educationalists, administrators, financial and legal service workers and strategic information processors. The secondary labour market in services comprises routinized information processors such as keyers and call centre staff and personal service sector workers. In terms of location within the post-fordist city many of the latter still occupy Fordist locations at the city edge or on out-of-town sites such as retail parks, whereas many of the former remain, despite changes in information technology, located close to the city centre (Budd and Whimster, 1992). The exception would be the growth of inner city employment associated with the development of urban economies based on the leisure, consumption and culture industries, either as a long-term process or in the form of hallmark, one-off, events.

Theorists have rejected the view that Fordism and post-fordism are separated by a sharp break (Elam, 1990) and now tend to regard them as a continuous process. In terms of the city, however, the break is more sharp. Fordism, certainly in its latter stages, has been equated with suburbanization and decentralization, whereas post-fordism is often invoked as a cause of recentralization and reurbanization.

POST-FORDIST CITY POLITICS

Under Fordist systems of production, larger and larger units of production made it increasingly difficult for the local state, and even national government, to control economic development. At the national level, governments found their attempts to manage the macroeconomy through interest rates, the exchange values of currencies and public sector borrowing could be frustrated by the large-scale hypermobility of capital (Leyshon, 1992). The more recent trend, over the past two decades, however, has seen greater opportunity for local government and other locally based institutions to play a role in economic development. Whilst under Fordism local modes of regulation played a minor and subordinate role in assuring the coherence of the overall regime, when the central state and other large-scale modes of regulation played the crucial roles, efforts to respond to the crises of Fordism have involved a shift in the division of labour. The specific local conditions of production and reproduction required by globally mobile capital cannot be orchestrated by the central state. Hence local political organizations, their skills in negotiating with supraregional and multinational capital, and the effectiveness with which they tailor the particular set of local conditions of production have become decisive factors in shaping a city's economic profile as well as its place in the international urban hierarchy.

The greater autonomy of the local state has been termed a 'perforated sovereignty' (Mayer, 1994), by which nation states become more exposed to trans-sovereign contracts by local agencies, not always democratically elected and accountable. Regional and local forces become more active in advancing their locational policy strategies orientated directly to the world market and many observers see this as contributing to the greater salience of the local state and to the 'hollowing-out' of the nation state. Ironically, as Page (1993) points out, this greater local salience has at times occurred when there has been a clear intention on the part of the national government to reduce the power of local government, as in the case of Britain in the 1980s and early 1990s. In other cases, however, such as France and Spain, the enhanced powers enjoyed at the local regional and urban scale have been willingly conferred by the central state as part of a deliberate policy of decentralization (Jessop, 1994).

This phase of the post-fordist local politics dates, according to some, from the economic crises of the late 1970s, when the macroeconomic shocks caused by sharp increases in the price of crude oil forced central governments to reduce their levels of grant finance to local government, who in turn were left with diminished resources with which to tackle increasing unemployment and deprivation. Local authorities were thereafter more likely to engage in anti-unemployment policies and local labour market interventions such as planning agreements, the use of local authority reserves such as pension funds to support local industry, and procurement policies aimed at the local economy. Paralleling these direct interventions were less direct approaches, such as programmes of urban marketing designed to enhance a city's image in the hope of attracting inward-investment, the development of tourism (both business and leisure) and cities competing for hallmark events to provide a one-off financial boost or to upgrade infrastructure and to enhance the quality of life. In European cities, these less direct approaches are typified by the boost given to Barcelona's economy by its hosting of the 1992 Olympic Games, the improvement of Glasgow's image achieved by its designation of European City of Culture in 1990, and the growth associated with the EuroLille project in the Nord-Pas de Calais (Logan and Swanstrom 1990: Stöhr, 1990). Some local authorities and city councils seem to be aware of the increasingly polarized occupational and class structure of their local economies and attempt to counteract the attendant social disintegration with consciously chosen strategies to stimulate growth and to target job creation on particular sectors of the labour market such as the long-term unemployed of former manual workers. From case studies, it is possible to demonstrate that some urban leaders who were engaged in local economic development activities were often far from certain as to how exactly an improvement in the course of urban development might be brought about, except in agreeing that 'industry and employment matters should be important'

(Cochrane, 1992: 122). Gradually these activities have consolidated into a more systematic economic development policy strategy oriented explicitly to fostering growth, which in its turn is supposed to create employment, although policies to foster efficiency and competitiveness may not always have this desired effect (Lever, 1997).

This increased level of local economic intervention is expressed not merely in the quantitative growth of local government spending for economic development, but, and often more importantly, in qualitatively different approaches to economic intervention which seek to make use of indigenous skills and entrepreneurship, which emphasize innovation and new technologies, and which involve partners from outside the public sector in the organization of conditions for local economic development. Traditionally, local authority actions have been aimed at attracting mobile capital through incentives such as financial and tax breaks, infrastructural provision or assistance with site selection. A shift in the approach to local economic development has now emerged. Subsidies are now targeted towards industries promising innovation and growth; more public resources are focused on stimulating research and development, consulting and the transfer of technology, as well as on constructing alliances with universities, polytechnics, chambers of trade and commerce, and labour unions. The planning and preparation of land has now become of great strategic significance, especially in the larger cities, which may be administratively underbounded and heavily reliant on the redevelopment of 'brownfield' land (Bennett et al., 1990; Cooke and Imrie, 1989).

The focus now has been switched to creating 'milieux innovatrices' (Aydalot, 1994), which encourage new firm formation. Such milieux require to be information-rich, with a benign fiscal regime and good external and agglomerative economies. In addition to publicizing the virtues of the local business climate, local authorities are likely to stress the quality of life, the availability of good services (especially education) and good image. Finally, new development strategies frequently include employment strategies which involve the third or alternative 'not-for-profit' sector.

These attempts to engender local economic development have had the effect of gradually breaking down the traditional sharp distinction between different policy areas. This is especially true in the case of labour market and social policy areas but also in educational, cultural and environmental policies. These have all become more integrated with local economic development measures. In addition, the new efforts have led to institutional changes: new departments and inter-agency networks have been created within the administration and new institutions that contribute in significant ways to shaping local politics have been established outside the local authority (for example, urban development corporations, training and enterprise councils, technology centres, growth alliances, local business fora and roundtables). What has been described as institutional thickness has emerged as a measure of this overlay of agencies, some public, some private (McLeod, 1997). 'Thicker rafts' of institutions are felt to be more flexible and more responsive to the needs of inward investors of existing businesses, and to contribute to the emergence of the 'entrepreneurial city'.

This greater economic autonomy of post-fordist cities has been described as 'delinking' from the national economy and the state. Over time, perhaps as a consequence of post-fordism, the rate of economic change of city regions has increasingly diverged from the national rate of growth (or decline) (Lever, 1997) of their host state. This delinking is generally expected to be positive; in other words, entrepreneurial and dynamic cities will outperform their relevant national economies, and cities such as Barcelona, Rotterdam, Munich and Turin in Europe are cited as examples. The reality is, however, that the pursuit of distinctive local economic policies may cause an urban economy to lag significantly behind the national performance: examples from Europe in 1986–96 include Marseilles, Hamburg and Amsterdam.

A second aspect of the politics of post-fordist cities which is distinctive is the restructuring and subordination of social consumption or welfare. The pressures exerted by economic restructuring and mass unemployment and by shrinking subsidies from central government, and the prioritization of the economic development function have forced local authorities to reduce their commitment to one of the formerly central functions of local state politics, namely the provision of social consumption goods and welfare services. Not only has local government spending for social consumption declined as a proportion of overall expenditure but a qualitative restructuring has taken place involving an increase in the importance of non-stable (private and not-for-profit voluntary sector) organizations or quangos. In several policy fields where the state used to be the sole provider of a service, non-governmental agencies have been created or private markets have emerged. Sometimes these now involve more cooperation with neighbourhood-level initiatives or other social movements (Evers, 1991). On the positive side, these movements are seen as having the virtues of greater efficiency and of greater community responsibility and involvement. Negatively, however, as in the case of Great Britain, they can be seen as an attempt by central government to disempower local authorities, often controlled by a political party different from that of central government under the guise of making local government more accountable to the local electorate. In some cases the set of urban welfare policies have been collapsed into a single programme (for example, the Single Regeneration Budget), grants from which are awarded on a competitive basis, rather than one of objective need. The criterion for success in such competition may be the ability of such schemes to attract private capital, and they thus are linked to the economic success of a local area.

This means that social welfare measures and assistance which need to be relatively universal and guaranteed by the national welfare state (but delivered by the local state) are now an arena of conflict and are implemented in a fragmented fashion. This movement away from service provision by elected authorities to an 'enabling' role by which services are offered by voluntary sector agencies, the private sector and others has generated 'growth coalitions'. The range of services now offered outwith local government but formerly in its control is large and extends from school catering and policing to the provision of subsidized housing and refuse disposal. These new public–private–voluntary forms of cooperation in the area of social consumption are part of the structural changes in the repertoire of municipal action in the post-fordist city. Whether the local struggles and bargaining processes result in more egalitarian and accountable models responsible to broad local needs, nor in division models enforcing processes of polarization, marginalization and social exclusion, one of the certain new characteristics of the emerging local 'welfare state' that distinguishes post-fordist cities from the past is its role in enabling negotiation with outside actors.

POST-FORDIST URBAN CULTURE

The mass production systems of Fordism can be equated with mass consumption. Fordist techniques made available goods and services in large quantities at prices which fell in real terms whilst offering higher wages to households whose consumption patterns since the 1940s were increasingly shaped by the desire to own mass housing, electrical goods, cars, furniture and sports equipment. Mass produced food, often pre-prepared, sold through decreasing numbers of large retailers, grew in importance as larger numbers of females entered, or re-entered employment. Even services were subject to the same pattern, in the form of mass package holidays and branded eating experiences such as McDonald's.

Socioculturally, post-fordism can be argued to represent a rejection of mass consumption. Improved education and more sophisticated advertising, some of it relating to the informationalizing of retailing, has led to more selective and discerning patterns of consumption. Mass production for many in the post-industrial city was rejected in favour of patterns of consumption that involved goods and services which were distinctive, bespoke and, in the case of goods, had high levels of design and craft embedded in them. The emphasis swung to clothing, footwear, furniture and jewellery, which were produced in small batches, or individually. Mass services were supplanted by much more customized offerings in the form of esoteric restaurants, often with obscure ethnic origins, customized or small group holidays, and more individualistic leisure activities. Even goods such as cars, the stereotypical products of Fordist systems, used new techniques such as Computer Aided Manufacturing (CAM) to break up long production runs by producing 'limited editions' marginally different one from another.

At the urban scale, while the mass production of suburban housing for owner occupance continues, and would appear to be the majority choice for most nuclear families in the child-rearing stage of the life cycle (Lever and Champion, 1998) new types of households and the new economic order are generating a greater variety of residential communities and a new spatial structure. Authors such as Harvey (1982, 1989) and Christopherson (1994) have drawn attention to the relationship between the new international division of labour and the spatial restructuring of the city. Business and financial service activities have become concentrated in a small number of countries and urban centres whilst manufacturing has moved to low wage-cost countries, removing the demand for blue collar workers in the cities of the developed world. Sassen (1994) has described the 'hollowing out' of city economies with growing social polarization which, in its turn, exacerbates polarities in the housing market and in residential districts. Most studies have stressed the extent to which high income households have been able to displace inner city working-class communities through gentrification and the residential conversion of formerly non-residential buildings such as dockside warehouses to obtain housing close to CBD jobs. To protect themselves from the more serious social pathologies of crime, assault and invasion high-income communities are now becoming gated communities, with surveillance and security services, a reflection of the growing polarization (Christopherson, 1994; Judd, 1994; Marcuse, 1995). A further manifestation of the desire for protection from a polarizing society is the (largely North American) creation of residential communities of specific populations, particularly the elderly, where the encroachment of non-conforming populations would immediately be obvious.

It is not only the international business class who are creating special inner city communities using their high incomes to protect themselves. An interesting study of international administrators within the European Community (Papadopoulos, 1997) describes how Eurocrats in Brussels have been able to acquire particular residential neighbourhoods for their use by driving out poorer households.

Much of the work on polarization in post-fordist cities has examined processes within the residential housing market, but there are parallel studies of the privatization of formerly public spaces by high income households to exclude 'undesirable populations' – the poor, the homeless, young children and teenagers. Perhaps the best example of this type of development is the enclosed shopping mall in North America and increasingly in Europe. Loukaitou-Sideris (1993), in a study of privately controlled public space in Los Angeles, describes how such spaces are inwardly orientated, with high enclosing walls, blank facades, distanced from the public street and with obscured street level access. These public open spaces are effectively discontinued from the surrounding city. Activities therein are severely restricted, lacking children's playspace for example! All the spaces are aimed at a specific clientele, the workers in the vicinity and shoppers, thus allowing the owners of adjacent commercial space to capture a particular segment of the market and to orient other marketing strategies to that segment.

The exclusion of populations from areas formerly open to the public is usually justified on commercial grounds. Excluding 'undesirable' individuals and activities is felt to guarantee higher levels of investment in job creation by removing the risk of problems such as theft, arson, damage and assault, which would in turn alienate customers, clients and visitors. This reworking of public and private spaces in North American, and increasingly also European, cities is a matter of stimulating consumption. Business improvement districts, such as midtown Manhattan, reflect the intervention of private companies in what they regard as a context where a purely public service, such as cleansing, police, urban design or transportation, has failed.

CONCLUSION

The post-fordist city is less easy to define or describe than the Fordist city. Fordist cities from the late nineteenth century and the first half of the twentieth century were quite a homogeneous group, with scale economies predominant, mass production and consumption, and a clear socio-economic hierarchy. Post-fordist cities, on the other hand, are a more heterogeneous set. They are, however, characterized by smaller scales, by greater internal heterogeneity and polarization. To an extent, individualism has replaced communitarianizm and state socialism. It suited the political moves of the right-wing national governments of the 1980s to emphasize entrepreneurialism and self-reliance, and to erode the supportive mechanisms of city governments. Ironically, one of the consequences, at the urban and at the enterprise level, of this increased emphasis on self-reliance has been the emergence of partnership and collaboration, rather than enhanced competition, as the paradigm within the relationships are conducted.

REFERENCES

Aglietta, M. (1979) *A Theory of Capitalist Regulation*. London: New Left Books.

Amin, A. (1994) 'Post-fordism: models, fantasies and phantoms of transition', in A. Amin (ed.), *Post-Fordism*. Oxford: Blackwell.

Amin, A. and Robins, K. (1990) 'The re-emergence of regional economies? The mythical geography of flexible accumulation', *Environment and Planning D: Society and Space*, 8(1): 7–34.

Aydalot, P. (1994) *Milieux innovateurs en Europe*. Paris: GREMI.

Bagnasco, A. (1977) *Tre Italie: la problematica territoriale dello sviluppo italiano*. Bologna: IR Mulino.

Bennett, R., Krebs, G. and Zimmerman, H. (eds) (1990) *Local Economic Development in Britain and Germany*. London: Anglo-German Foundation.

Bonefield, W. and Holloway, J. (1991) *Post-Fordism and Social Form*. London: Macmillan.

Budd, L. and Whimster, S. (1992) *Global Finance and Urban Living*. London: Routledge.

Christopherson, S. (1994) 'The fortress city: privatized spaces, consumer citizenship', in A. Amin (ed.), *Post-Fordism*. Oxford: Blackwell. pp. 407–27.

Clark, D. (1996) *Urban World, Global City*. London: Routledge.

Clarke, S. (1990) 'New utopias for old: Fordist dreams and post-fordist fantasies', *Capital and Class*, 42: 131–55.

Clarke, S. (1988) 'Overaccumulation, class struggle and the regulation approach', *Capital and Class*, 36: 59–92.

Cochrane, A. (1992) 'Das veränderte Gesicht der städtischer Politik in Sheffield: vom "municipal labourism" zu "public-private partnership"', in H. Heineld and M. Mayer (eds), *Politik in europaischen Städten*. Basel: Birkhäuser. pp. 119–36.

Cooke, P. and Imrie, R. (1989) 'Little victories: local economic development in European regions', *Entrepreneurship and Regional Development*, 1: 313–27.

Daniels, P. (1993) *Services in the World Economy*. Oxford: Blackwell.

Elam, M. (1990) 'Puzzling out the post-fordist debate: technology, markets and institutions', *Economic and Industrial Democracy*, 11: 9–37.

Esser, J. and Hirsch, J. (1994) 'The crisis of fordism and the dimension of a post-fordist regional and urban structure', in A. Amin (ed.), *Post-Fordism*. Oxford: Blackwell. pp. 71–97.

Evers, A. (1991) 'Pluralismus, Fragmentierung und Vermittlungsfährgkeit', in H. Heinheld and M. Mayer (eds), *Brennpunkt Stadt*. Basel: Birkhäuser. pp. 299–311.

Freeman, C. and Perez, C. (1988) 'Structural crisis of adjustment, business cycles and investment behaviour', in G. Dosi, C. Freeman, R. Nelson, G. Silverberg and L. Soete (eds), *Technical Change and Economic Theory*. London: Frances Pinter.

Friedmann, J. (1995) 'Where do we stand: a decade of world city research', in P.L. Knox and P.J. Taylor (eds), *World Cities in a World-System*. Cambridge: Cambridge University Press. pp. 21–47.

Gordon, D. (1980) 'Stages of communication and long economic cycles', in T. Hopkins and I. Wallerstein (eds), *Processes of the World System*. London: Sage. pp. 127–159.

Hall, P. (1988) *Cities of Tomorrow*. Oxford: Blackwell.

Harrison, B. (1994) *Lean and Mean: the Changing Landscape of Corporate Power in the Age of Flexibility*. New York: Basic Books.

Harvey, D. (1982) *The Limits to Capital*. Oxford: Blackwell.

Harvey, D. (1989) *The Condition of Postmodernity*. Oxford: Blackwell.

Hirst, P. and Zeithin, J. (1991) 'Flexible specialisation and the competitive failure of UK manufacturing', *Political Quarterly*, 60(3): 164–78.

Jessop, B. (1994) *From Keynesian Welfare to the Schumpeterian Workforce State*. Lancaster Regionalism Working Group. Working Paper 45.

Judd, D. (1994) 'The rise of the new walled city', in H. Liggett and D. Perry (eds), *Representing the City*. Newbury Park, CA: Sage. pp. 127–51.

Keeble, D., Owens, P.L. and Thompson, C. (1983) 'The urban–rural manufacturing shift in the European Community', *Urban Studies*, 20: 405–18.

Leborgue, D. and Lipietz, A. (1992) 'Conceptual fallacies and open questions on post-fordism', in M. Storper and A.J. Scott (eds), *Pathways to Industrialisation and Regional Development*. London: Routledge.

Leone, V. (1994) *La Rivalorizzazione Territoriale in Italia*. Milan: Franco Agneli.

Lever, W.F. (1997) 'Policies to improve the efficiency of urban areas', in C. Jensen-Butler, A. Shachar and J. van Weesep (eds), *European Cities in Competition*. Aldershot: Avebury. pp. 357–84.

Lever, W.F. and Champion, A.J. (1998) 'The urban development cycle and the economic system' in W. Lever and A. Bailey (eds), *The Spatial Impact of Economic Changes in Europe*. Aldershot: Avebury. pp. 204–27.

Leyshon, A. (1992) 'The transformation of regulatory order: regulating the global economy and environment', *Geoforum*, 23: 249–68.

Lipietz, A. (1985) *The Enchanted World: Inflation, Credit and the World Crisis*. London: Verso.

Lipietz, A. (1989) *Mirages and Miracles: The Crisis of Global Fordism*. London: Verso.

Logan, J. and Swanstrom, T. (eds) (1990) *Beyond the City Limits*. Philadelphia: Temple University Press.

Loukaitou-Sideris, A. (1993) 'Privatization of public open space', *Town Planning Review*, 64: 139–67.

Lovering, J. (1991) 'Theorising post-fordism: why contingency matters', *International Journal of Urban and Regional Research*, 15: 298–301.

McLeod, G. (1997) '"Institutional thickness" and industrial governance in lowland Scotland', *Area*, 29: 299–311.

Mandel, E. (1980) *Long Waves of Capitalist Development: the Marxist Interpretation*. Cambridge: Cambridge University Press.

Marcuse, P. (1995) 'Not chaos but walls; postmodernism and the partitioned city', in S. Watson and K. Gibson (eds), *Postmodern Cities and Spaces*. Oxford: Blackwell. pp. 243–53.

Martin, R. (1988) 'Industrial capitalism in transition: the contemporary reorganisation of the British space-economy' in D. Massey and J. Allen (eds), *Uneven Redevelopment*. London: Hodder and Stoughton. pp. 202–31.

Massey, D. and Meegan, R. (1979) 'The geography of industrial reorganisation', *Progress in Planning*, 10: 155–237.

Massey, D. and Meegan, R. (1982) *The Anatomy of Job Loss*. London: Methuen.

Mayer, M. (1994) 'Post-fordist city politics', in A. Amin (ed.), *Post-Fordism*. Oxford: Blackwell. pp. 316–38.

Nielson, K. (1991) 'Towards a flexible future – theories and politics', in B. Jessop, H. Kastendiek, K. Nielsen and O. Pedersen (eds), The *Politics of Flexibility*. Aldershot: Edward Elgar. pp. 21–34.

Pacione, M. (1997) *Britain's Cities: Geographies of Division in Urban Britain*. London: Routledge.

Page, E.C. (1993) 'The future of local government in Britain', in V. Bullman (ed.) *Die Politik der dritten Elsen Regionen im EG-Integrationsprozess*. Baden-Baden: Nomos. pp. 214–37.

Papadopoulos, A.G. (1997) *Urban Regimes and Strategies*. Chicago: University of Chicago Press.

Pike, F. and Sengenberger, W. (1992) *Industrial Districts and Local Economic Regeneration*. Geneva: International Institute for Labour Studies.

Piore, M. (1980) 'The technological foundations of dualism and discontinuity', in S. Berger and M. Piore (eds), *Dualism and Discontinuity in Industrial Societies*. Cambridge: Cambridge University Press. pp. 21–35.

Piore, M. and Sabel, C. (1984) *The Second Industrial Divide: Possibilities for Prosperity*. New York: Basic Books.

Rustin, M. (1989) 'The politics of post-fordism, or the trouble with "New Times"', *New Left Review*, 175: 54–78.

Sabel, C.F. (1989) 'Flexible specialisation and the reemergence of regional economies', in P. Hirst and J. Zeitlin (eds), *Reversing Industrial Decline? Industrial Structure and Policy in Britain and Her Competitors*. Oxford: Berg. pp. 74–101.

Sassen, S. (1994) *Cities in a World Economy*. Thousand Oaks, CA: Pine Forge.

Saxenian, A. (1985) 'Silicon Valley and Route 128: regional prototypes or historical exceptions', in M. Castells (ed.), *High Technology, Space and Society*. Beverly Hills, CA: Sage. pp. 81–105.

Sayer, A. (1989) 'Post-fordism in question', *International Journal of Urban and Regional Research*, 13: 666–95.

Scott, A.J. (1988) *New Industrial Spaces*. London: Pion.

Sternberg, E. (1993) *Transformations: the Eight New Ages of Capitalism*. Buffalo, NY: Department of Planning and Design.

Stohr, W.B. (ed.) (1990) *Global Challenge and Local Response: Initiatives for Economic Regulation in Contemporary Europe*. London: Mansell.

Storper, M. and Christopherson, S. (1986) 'Flexible specialization and regional agglomeration: the case of the US motion picture industry', *Annals of the Association of American Geographers*, 76: 104–17.

Townsend, A., Sadler, D. and Hudson, R. (1996) 'Geographical dimensions of the UK retailing employment change', in N. Wrigley and M. Lowe (eds), *Retailing, Consumption and Capital*. Harlow: Longman. pp. 208–19.

Tyler, P., Moore, B. and Rhodes, J. (1984) *Geographical Variations in Industrial Costs*, Discussion Paper 12. Cambridge: University of Cambridge, Department of Land Economy.

Wallace, I. (1990) *The Global Economic System*. London: Unwin Hyman.

18

The Post-Industrial City

DOUGLAS V. SHAW

The 'post-industrial' city is an emerging set of urban forms and functions that appears to be sufficiently different from the industrial city of the past two centuries to warrant a separate definition. At the same time, it is not yet sufficiently articulated to justify a name and nomenclature of its own that defines it in a manner other than as a reaction to the cities of the recent past. The use of 'post-industrial', then, represents both a sense on the part of observers that we have crossed a significant development boundary, and that the precise nature of the terrain on the other side is still largely unknown.

INDUSTRIAL ORIGINS, POST-INDUSTRIAL OUTCOMES

As the post-industrial city is in many ways a continuation of the industrial city, it is appropriate to begin with a brief review of those aspects of prior urban development that shed light on the transitions now in progress. Much of the context of the post-industrial city may be discovered in its predecessors, in what Harvey S. Perloff has aptly called 'the inheritance component of the future' (Perloff, 1980: 279).

Cities are usually classified by their major purpose, and throughout the past 6,000 years of human settlement commerce and exchange have formed the dominant functions of most cities, most of the time. Cities located at strategic points gathered the goods of their hinterlands and organized them for transfer to other places, usually other cities, which functioned as distribution points for the goods of distant places. Cities were the focal points in trade systems that varied in their extent from regional to trans-continental. In the ancient world, cities such as Tyre, Rhodes, and Byzantium served as crucial nexus points for the expanding water-borne trade of the Mediterranean basin.

Later, medieval cities like Venice, Genoa, Barcelona and Constantinople carved out similar roles, while Damascus became a staging area for the great caravans that crossed the mountains and deserts of central Asia, tenuously linking Europe and the Middle East to China and India (Hohenberg and Lees, 1985: 59–73, 106–36; Mumford, 1961: esp 321–8, 410–20).

During the later Middle Ages and Early Modern era production methods slowly improved in many areas of manufacture, with consequent improvements in both the quantity and quality of goods available. Much basic production, fine craft work excepted, took place outside of cities, with cities serving their traditional commercial roles, often enhanced by greater production and a general quickening of economic activity. In the late eighteenth century the rate of change in production methods increased markedly, and with the application of power sources such as coal and steam, the nature of manufacturing labour changed as well. Requiring relatively large and coordinated labour forces, the new factories became the engines of unprecedented urban growth and of changes in the basic functions of both existing and newly formed cities. The most successful of these 'post-commercial' cities grew with unprecedented speed, providing the means for the creation of wealth through control of industrial development at a pace that astonished contemporary observers, while at the same time, to the dismay of many of those same observers, adding new accretions of social and economic ills (Chudacoff and Smith, 1994: 78–107; Cowen, 1998: 3–31; Hohenberg and Lees, 1985: 179–214).

As social historian E.P. Thompson has observed, 'There is no such thing as economic growth which is not, at the same time, growth or change of a culture' (Thompson, 1967: 97). The dynamic of the Industrial Revolution challenged the moral, cultural, organizational and hierarchical structures of every society it touched. The ethics of capitalism, the time-discipline of the factory and the unprecedented economic and political power accruing to those who controlled productive resources all worked to undermine long-standing assumptions, habits and relationships. The intrusion of industrialism altered the structure and sources of economic rewards, and hence created new sets of socio-economic beneficiaries and victims (Gutman, 1976: 19–76).

Large-scale manufacturing depended on the distribution of production through cheap and efficient transport. Perfected in the second quarter of the nineteenth century, the steam railroad met that need, and the revolution in production advanced in tandem with an equally profound revolution in distribution. Railroads also worked as centralizing forces, gathering activity in dense settlements around their terminals and way stations. As rail-based transport lessened the historical cost advantage of water-borne transport, industrial cities became almost by definition railroad cities as well. In the second half of the nineteenth century improvements in internal movement inside cities, both horizontal (street railways) and vertical (elevators), allowed a level of concentration of people and activity unknown in earlier times. Innovation after innovation worked to create or accommodate concentration and density (Chudacoff and Smith, 1994: 78–89; Taylor, 1951: 74–103).

The thrust of innovation in systems affecting cities in the twentieth century, however, worked primarily in the opposite direction, as the telephone, internal combustion engine and the use of electricity as a factory motive force allowed some activities to move outside of high-density settings while retaining commercial competitiveness. The scale of activity made possible by 'modern' industrial methods led to management and commercial innovations and to large increases in non-production personnel. In twentieth century industrial economies, in contrast to those of the previous century, managerial and clerical workforces grew at much faster rates than factory labour (Chandler, 1990: 31–46; Hohenberg and Lees, 1985: 248–74).

The scale of urban population growth during the industrial era was astonishing. Manchester, England, the shock city of the English Industrial Revolution, doubled in population between 1801 and 1831, from slightly above 70,000 to 142,000 (Briggs, 1965: 88–9). In the rapidly growing United States, Chicago exploded from 30,000 in 1850 to 1.1 million in 1890. Detroit, the citadel of 'Fordist' production, grew from 80,000 in 1870 to 465,000 forty years later, and then between 1910 and 1920, the first decade of mass automobile production, more than doubled again, to 994,000

(Philpott, 1991: 6; Sugrue, 1996: 23). Cities tended to grow faster physically than institutionally, frequently overwhelming their lagging infrastructures and policy-making processes. Much of the built environment reflected the need to house and serve thousands upon thousands of hopeful migrants quickly rather than well, and the resulting construction deficiencies formed much of the physical landscape in these cities until well beyond the industrial era. Conversely, cities outside industrial mainstreams or lacking good transport connections languished. Industrialization remade urban hierarchies, thrusting industrial cities toward the forefront (Chudacoff and Smith, 1994: 111–46, 177–201; Hohenberg and Lees, 1985: 248–330).

From their earliest incarnations, cities functioned as cultural repositories of learning and civilization, as places where accumulated knowledge was nurtured, preserved and passed from generation to generation through means both formal and informal. Ancient temples, medieval monasteries and early modern universities served similar purposes in their respective times. Preservation and transmission, particularly of religious lore, took precedence over seeking new truth: Galileo's forced recantation in the 1630s exemplified the alarm aroused by non-traditional views. In the twentieth century the conscious search for new secular knowledge through funded research and development accelerated the pace of innovation and enhanced the ability to solve specific problems through focused empirical research. The Rockefeller Institute for Medical Research in New York City, founded in 1901, provided an early model of a 'pure research' institution, an organizational form subsequently extended by others into virtually every discipline (Chernow, 1998: 467–79). Indeed, the institutionalized production of knowledge and invention as commodities are two hallmark innovations of the twentieth century, and are among the more important bridges linking the industrial to the post-industrial city.

In the final third of the twentieth century the engine of economic growth – heavy industry built around the twin symbols of industrial might, iron and steel – began to decline in the economies of developed countries in both absolute and relative importance. Various types of services replaced heavy industry as growth centres, from highly specialized producer services such as corporate law, banking and accounting, to low-paying consumer services. The proto-typical post-industrial city, then, is a city in which traditional industry maintains a significant but decreasing share of economic activity, replaced as an engine of economic growth by the production of various types of services, from producer services, to medical, educational and governmental services, to consumer services. None of these service areas was new, but as primary beneficiaries of the twentieth-century knowledge revolution, they all expanded and became more specialized. When traditional industry began to decline, however, growth in these areas not only continued but accelerated, and they rapidly increased both their relative and absolute shares in evolving urban economies (Bluestone and Harrison, 1982; Drennan, 1991; Hall, 1996: 402–22).

THE TERM 'POST-INDUSTRIAL'

First used in 1919 by Indian cultural reformer A.K. Coomaraswamy (Gappert, 1979: 31), the label 'post-industrial' did not become a term of everyday use until popularized by Harvard sociologist Daniel Bell in his influential *The Coming of Post-Industrial Society: A Venture in Social Forecasting* (1973). According to Bell,

> The concept of the post-industrial society is a large generalization. Its meaning can be more easily understood if one specifies five dimensions, or components, of the term:
>
> 1 Economic sector: the change from a goods-producing to a service economy.
> 2 Occupational distribution: the pre-eminence of the professional and technical class.
> 3 Axial principle: the centrality of theoretical knowledge as the source of innovation and of policy formulation for the society.
> 4 Future orientation: the control of technology and technological assessment.
> 5 Decision making: the creation of a new 'intellectual technology'. (Bell, 1973: 14)

For Bell, the United States's labour force reached a critical juncture in 1956 when, 'for the first time in the history of industrial civilization', the number of white-collar workers exceeded the number of blue-collar workers. He also noted that in the 1950s and 1960s the greatest growth in white-collar workers was in the professional and technical categories, areas demanding at least some post-secondary education, and that the fastest growing sub-groups were those of scientists and engineers. The result, he posited, would be a future society structured around the kind of knowledge created by these technocrats. To emphasize his point, Bell drew a sharp distinction between industrial and post-industrial societies:

> Industrial society is the coordination of machines and men for the production of goods. Post-industrial society is organized around knowledge, for the purpose of social control and the directing of innovation and change; and this in turn gives rise to new social relationships and new structures, which have to be managed politically. (Bell, 1973: 20)

Although Bell's conceptualization has hardly been accepted uncritically (see Gappert, 1979: 31–7), the entry of 'post-industrial' into the common language of social and economic discourse, and its continued use over three decades basically as Bell defined it, seems to confirm his 1973 forecast of emerging fundamental structural changes in the nature of economic organizations and social relations.

GLOBAL CITIES

At the same time that most older industrial societies began to experience the impact of industrial levelling off or decline and the growth of post-industrial service sectors, a series of changes in the world economy began to reshape urban networks and to hasten the development of a small number of world cities that possessed the resources to exploit and benefit from a new, more internationalized economic order. Just as the Industrial Revolution strengthened the prospects of some cities at the expense of others, the dynamic changes that shaped the global, post-industrial economic order also led to differential outcomes for different places (Kantor, 1987: 506–7).

The key components of late-twentieth-century economic globalism included the continued elaboration and growth of multinational corporations with worldwide interests; the weakening of national restraints on the free flow of capital between countries and regions, and the development of new types of investment instruments to accommodate those flows; and the movement of manufacturing away from the older core of industrial nations to Third World countries in the early stages of industrial development. An important consequence was a reduction in the effectiveness of strictly national barriers to investment and trade, and an expansion of markets for the transfer of capital and goods. More than ever, markets seemed to transcend the borders and interests of nation states, and the ability of individual countries to direct their internal economies and shape the manner in which they interacted with external structures and forces declined accordingly.

These changes have affected the dynamics of urban networks worldwide. While cities tied firmly to smokestacks and factories frequently suffered social and economic distress after 1970, those cities positioned to provide the services required by the new global order grew in affluence, commercial importance and economic power. Decisions made in these centres of post-industrial growth and change, frequently termed global cities, disproportionately affected the course of economic and technological developments in distant parts of the world. This emerging trans-national economy, described by Saskia Sassen as 'spatially dispersed, yet globally integrated', (1991:3) required a number of highly specialized producer services in such areas as law, banking, finance and accounting. These services, while hardly new in kind, were in form more advanced, specialized and innovative than their more traditional antecedents. Sassen posited four enhanced roles for cities at the centre of post-industrial international trade and finance. Global cities would function, she wrote:

> first, as highly concentrated command points in the organization of the world economy; second, as key locations for finance and for specialized service firms, which have replaced manufacturing as the leading economic sectors; third, as sites of production, including the production of innovations, in these leading industries; and fourth, as markets for the products and innovations produced. (Sassen, 1991: 3–4)

Those able to provide these service products required a highly specialized international clientele, and that clientele had to be available in sufficient numbers to provide a flow of work capable of supporting a critical mass of highly trained, highly paid, providers. Thus service producers and service consumers found each other in abundance only in a relatively small number of places. As the global economy expanded, and the demand for these services increased, such post-industrial 'command points' as New York City, London, Paris, Tokyo and Hong Kong, and to a greater or lesser extent their surrounding metropolitan areas, experienced expansion both in corporate headquarters employment and in the services multinational corporations required as adjuncts to their global business development (Sassen, 1991: 19–34, 90–125).

In a major study of New York City, London and Tokyo, Sassen theorized that global cities constituted a new type of urban development, and that cities at the centre of post-industrial change would develop in similar, parallel ways as they responded to the demands of a new age; and that the interests of these cities and their business elites had the potential to run in directions counter to the internal and external interests of the nations in which they were situated. The financial industry in particular, a key to multinational corporate growth, had become increasingly concentrated in these three cities. Sassen contended that as the global economy became ever more integrated, the position and economic power of these three financial centres would correspondingly be enhanced as demand for their specialized producer services grew. A spatially dispersed global economy, driven by large corporations exercising control over worldwide production and distribution systems by means of modern telecommunications, would at the same time experience additional concentration in the cities best equipped to provide the specialized services required by those corporations (Sassen, 1991: 3–167).

The decline of manufacturing in the industrial West and the growth in producer services has tended to rearrange the distribution of opportunity and income in cities affected by these changes, regardless of their degree of participation in the global economy. Factory jobs that paid middle-class wages have been a primary casualty of the transition to a post-industrial economy, with industrial workers frequently pushed into lower paying service jobs or out of the labour force entirely. Meanwhile, managers and the new professionals on whom they depend receive, year by year, increasingly lucrative salaries, bonuses and benefits. Thus the broad middle of urban societies may be shrinking, while the income gap between the well paid and the working poor widens. The post-industrial city, then, is also likely to be a dual city, in which rich and poor draw further away from each other spatially within the urban region, as well as in terms of differential access to economic resources. While the extent of this dichotomy and its meaning is a subject of continuing scholarly debate, few observers deny that troubling social consequences are to a greater or lesser extent a constituent accompaniment of the post-industrial urban order (Davis, 1992b; Marcuse, 1997; White, 1998).

The extent to which this perceived dualism will prove to be a temporary or permanent condition is as yet unknown. White (1998) notes that the decline of manufacturing in the West is directly responsible for economic growth and personal opportunities in those parts of the world where industry has relocated, and in some places economic change has also increased distributive equity. It is possible that the dualism currently experienced by older industrial societies is transitional, with the dislocations created by industrial decline likely to lessen as a replacement order takes form. It was Benjamin Disraeli, after all, who portrayed the social and economic upheavals of mid-nineteenth-century industrializing England in equally stark dualistic terms in his classic novel of social criticism, *Sybil, or The Two Nations* (Disraeli, [1845] 1995).

Certainly the relative stability and predictability that characterized the industrial decades seem to have come to a close. Peck and Tickell (1994) pick up on the theme of transition in a provocative essay that argues that the present era represents not a new era of stability, but a time of instability between an ordered, regulated industrial system and a social and economic order of a new type, its full shape and direction as yet unknown. The present global economy, they hold, constitutes a disordered interregnum that possesses neither adequate regulation nor the basis for a new coherence. Renewed stability can only come from 'a new institutional fix . . . at the global scale'. This global fix, however, will likely rely on the enduring assertive might of the traditional nation state. 'The global financial institutions in particular must be harnessed and reformed,' they write, 'a process which will doubtless require concerted action through nation states. It is consequently important that the nation state is not written off as a key site in this regulatory struggle, for this is likely to remain the principal scale at which democratic control and political power can be (re)coupled' (Peck and Tickell, 1994: 311). Nation states, then, might seek to recreate stability, perhaps collectively, by attempting to impose a new trans-national order on the global services economy. National assertiveness in these arenas may have the potential to harness and restrain trans-national economic activity, and by extension, to reduce the independent influence of the global cities that facilitate its development.

POST-INDUSTRIAL URBAN NETWORKS

It is an over simplification to think of cities as either 'global' or 'non-global'. The degree to which cities have become providers of the specialized services that have become the moving force in post-industrial urban economic development moves along a continuum. Some cities, such as New York, London and Paris, have retained their positions as global financial centres, adapting their institutional structures to changing markets, methods and opportunities. Other cities, most notably Tokyo, have emerged as leading financial centres during the transition to a post-industrial urban order.

Tokyo, as a financial and global services centre, has grown in importance as Japan has advanced to the status of an industrial power and technological leader in the decades following the Second World War. Tokyo has become the most important centre in a trading region often referred to as the Pacific Rim, which includes the Asian and North American nations and city-states of the Pacific basin. After Japan developed its modern industrial capacity at home, its corporations began to invest in production and distribution facilities in other countries. Japanese banks and corporations became important links in global networks (Tabb, 1995: 86–111).

As a result, Tokyo has increased its hinterland to reach far out into the Pacific basin. The city of Los Angeles in the United States, for example, has become in recent years a major centre in a Pacific-based urban network, and an important port of entry for Asian-based commerce and migration into the United States. Japanese investment in the California economy has proceeded rapidly, and on a sufficient scale to lead one critic to refer to 'the contemporary "Nipponization" of the Southern California economy', and to claim that so much real estate in downtown Los Angeles has passed to Japanese owners as to constitute 'a *tsunami* of Japanese capital' (Davis, 1992: 135). Los Angeles, then, a large industrial centre that experienced a restructuring of its factory economy in the 1980s and 1990s away from heavy industry and toward apparel and electronics, has maintained its prosperity at least in part through increased participation in a Pacific-rim post-industrial urban order (Ong and Bloomenberg, 1996: 311–14; Scott and Soja, 1996: 19–20; Tabb, 1995: 255–84).

Urban networks are forever in a state of flux. Over time, cities have risen and fallen in importance in response to changes in trade patterns, to innovations in communication and transport technology, and to the efforts of urban elites consciously working to develop perceived locational advantages that will 'boost' their places over others. New York City in the 1820s supported a canal connecting its port with the inland Great Lakes, and consequently emerged decisively as the principal entrepôt in North America, effectively eliminating several strong rivals (Taylor, 1951: 194–8). That tradition of support for innovation has continued. After 1970, local officials in New York City supported improvements in infrastructure, particularly in telecommunications, and changes in land uses that allowed the construction of additional Manhattan high-rise office towers. These policy responses have aided and encouraged post-industrial development, and that in turn has enhanced New York City's status as the leading centre for financial and producer services in the Western Hemisphere. Municipal and related institutional structures in such cities as London, Paris and Tokyo have responded in similar fashion, with comparable results (Savitch, 1988: 284–307; Tabb, 1995: 169–97). Savitch (1998) emphasizes the role of urban governments in facilitating developmental growth and change.

Virtually all cities in almost every nation have been reshaped to a greater or lesser degree by post-industrial processes. As the 'industrial' world lost much of its factory base, non-industrial countries became locations of industrial growth. Industrial corporations frequently moved factory production to new, lower cost locations, most frequently in Asia and Latin America. In addition, local entrepreneurs in these regions began to enter markets and to export goods to the more prosperous countries of North America and Europe, sometimes to the detriment of industry in those countries. This movement of industry to the world's economic periphery has been a boon to such nations as Mexico, Argentina, Brazil, Korea, Taiwan, Singapore and Hong Kong. It has meant new opportunities, rising standards of living for some if not all, and greater integration into the world economy (Dicken, 1998: esp. 115–43; Haggard, 1990: 223–53; 1995: 46–99). The port of Los Angeles provides a dramatic illustration of these changes. Its most important export by volume in the late 1980s consisted of cargo containers that had carried Asian manufactured goods to the United States and, for want of American exports to Asia, returned home empty (Davis, 1992a: 135).

Industrial outposts in less developed nations are frequently tied to and dependent upon the capital and institutional structures emanating from global centres. They are part of a worldwide restructuring of the location of work and the organization of economic activity. Peter Hall writes of 'a new division of labour' taking place globally,

> a division based not on product (Lancashire cotton, Sheffield steel) but on process (London and New York global finance, Berkshire and Westchester back offices, Leeds and Omaha direct telephone sales). In so far as an activity can be decentralized to a lower-cost location, it will be; and while manufacturing moves out from the advanced economies to Thailand and China, so services now move out to suburban or provincial locations, limited so far by linguistic and cultural barriers, but doubtless soon overleaping those too. (Hall, 1998: 404)

These changes, then, work to reshape not only urban networks, but also the nature of work and the organization of space within individual cities and urbanized areas as well. Industry moves beyond borders, clerical and data processing tasks move to more remote locations, and command and control functions migrate to global centres. All are held together by air travel and telecommunications. Traditional locational advantages lose their historic value, and the local begins to fade imperceptibly into the global (Amin and Thrift, 1994: 5–16; Kantor, 1987: 506–7, 514–16).

One consequence is that governments at every level have found the new post-industrial order difficult to grasp and hold. Local governments in particular have found the global mobility of capital and jobs disruptive of their traditional relationships with private business interests. The relative 'placelessness' of many business and industrial functions has diminished the ability of cities and regions, which are by definition place-specific, to

control their economic destinies. According to Savitch,

> The immensely decentralized composition and flexibility of global forces stand in marked contrast to hierarchical and fixed behavior of government institutions. Particularly acute are the contrasting modes of operation between markets and governmental institutions. Markets are flexible, prolific, and immensely responsive to mass demands, whereas government institutions are steeped in formalism and routine. (Savitch, 1998: 257)

Savitch, like Amin and Thrift, advocates 'thickening' institutional structures, both public and private not-for-profit, by promoting institutional 'diversity, pluralism, and autonomy', particularly at the local level, in order to increase institutional capacities to deal with the consequences of rapid, frequently unpredicted, social and economic changes (Amin and Thrift, 1994; Savitch, 1998: 260–5; see also Swanstrom, 1996).

The role and status of localities in general, and cities in particular, has become more tenuous. Kantor holds that cities in the post-industrial era are increasingly dependent on corporate decisions that generally are not subject to public accountability. 'In effect,' he writes, 'cities have become captives of a highly competitive urban economic environment in which traditional factors – such as geography, physical infrastructure, and transportation – that once tied business to a specific community matter less than ever.' Thus for executives considering where to locate new corporate functions, or where to relocate existing ones, 'cities are almost interchangeable', to be played off one against another, forced to compete from positions of comparative weakness for the capital investment and accompanying job creation that corporations dispense (Kantor, 1987: 506–7; Warren, 1990: 541–7; Yago, 1983: 124–7).

The transition to robust post-industrial economies for cities lacking strong command and control functions has, therefore, frequently proved difficult. Philadelphia, a former centre of manufacturing in the north-eastern United States, has successfully restructured its economy in ways that allow participation in the post-industrial order, but without directly challenging or competing with primary post-industrial global cities such as New York, London and Tokyo. During the 1980s Philadelphia lost about 20 per cent of its manufacturing jobs, but gained 26 per cent in non-manufacturing employment; producers' services grew by about 50 per cent. Philadelphia's services, however, were primarily produced for, and consumed in, local and regional markets. The impact and reach of these services did not extend, as in the case of those produced in global cities, to the world at large, and the services available were less specialized.

Those seeking the more esoteric specialized services of the modern age had to seek them elsewhere in the global urban network (Stull and Madden, 1990: 45–66, 126–32).

To encourage service-based economic growth, successive Philadelphia administrations after 1950 supported large-scale urban renewal that demolished numerous businesses on the margins of downtown and several neighbourhoods of mostly poor residents. In their place rose the glimmering office towers of the post-industrial service economy, along with upscale retail establishments and luxury housing targeted for the affluent professionals who would inhabit those steel-and-glass skyscrapers by day. As so often in the course of urban renewal in the United States, the poor and the marginal, who dared to live in the path of someone else's definition of progress, were summarily bulldozed out of the way (Kantor, 1988: 256–8; see also Beauregard, 1989; Guinther, 1996: 204–35).

DIFFICULT TRANSITIONS: DETROIT, USA, BIRMINGHAM, UK

Some cities have failed to find a place in the post-industrial order commensurate with their previous positions as industrial centres. These cities have been passed over by the innovations and changes of the last quarter century and have as a consequence frequently experienced considerable economic contraction and social distress.

Detroit, Michigan, in the United States, represents a city that has not made a successful transition from an industrial to a post-industrial economic base. Growing spectacularly during the early decades of the twentieth century when the city was synonymous with the technical innovations that created the automobile boom, Detroit tripled its population from 465,000 to 1.5 million between 1910 and 1930. Population peaked in 1950 at 1.8 million and then began to decline, to 1.2 million in 1980 and to just over a million in 1990. Recent estimates project continued decline into the next century.

Detroit's robust pre-1960 economy drew first European immigrants and then African Americans from the American south. Like Europeans before them, blacks found physically demanding but comparatively remunerative work on the assembly lines of the automobile factories. Racial geography quickly became a pressing issue. As the African American population expanded into neighbourhoods occupied by whites, residents first resisted black settlement, and when resistance failed, fled to the suburbs. As a result, Detroit from the 1920s onward regularly experienced great racial tensions which occasionally rose to the level of collective violence, while at the same time becoming one of the nation's most segregated industrial cities. From 1920 until the present, Detroit has had one of the highest segregation indexes in the United States. Racial clashes, most frequently involving conflicts over residential space, exploded in several major riots, culminating in a devastating conflagration in July 1967 in which 43 people died and more than 7,000 were arrested (Sugrue, 1996: 231–68, 273; Thomas, 1997: 127–31).

A recent study contends that Detroit's industrial economy had already been slowly crumbling for at least two decades. Thomas Sugrue, in *The Origins of the Urban Crisis: Race and Inequality in Postwar Detroit* (1996) argues that post-war automation not only reduced employment in existing automobile factories, but also eliminated independent smaller producers, such as Packard and Hudson, that were unable to raise the vast capital sums automation demanded. Thus the 'Big Three' American producers – General Motors, Ford and Chrysler – not only increased their domestic production and sales dominance, but did so in ways that shifted production to capital intensive 'labour saving' automated processes that required ever fewer workers. Additionally, Detroit automobile makers chose to reduce their dependence on outside suppliers and began producing components previously produced under contract by local independent vendors. These new functions were often assigned to newly built suburban-type automated plants scattered throughout the Midwest. Thus the post-war restructuring and decentralizing of production further reduced the level of employment in Detroit. As a result of all of these factors, Detroit's manufacturing employment base fell by 38 per cent between 1947 and 1967 (Sugrue, 1996: 125–44).

The resulting unemployment rippled through the community, affecting neighbourhood retail and service sectors and creating a class of permanently unemployed or underemployed older males. African Americans suffered more severely than whites, with longer bouts of unemployment after layoffs and with sharply lower wages than whites when commencing alternative employment. Sugrue argues that the development of the city's African American 'underclass', with increasing numbers of adult males outside the workforce, began in the 1950s. By 1980 an astonishing 56 per cent of African American males over the age of 16 were without jobs (Sugrue, 1996: 143–52).

The large-scale African American riot in 1967 intensified the movement of people and capital investment from city to suburbs. Unlike more strategically situated cities, Detroit lost existing functions in addition to production, such as downtown retail and consumer service activities, without attracting those new functions and investments associated with successful post-industrial urban economies. Further, the national, and then global, decentralization of the automobile industry, as well as the increasing success of non-American name plates in the American market, continued to reduce local employment possibilities in the city's traditional economic mainstay (Sugrue, 1996: 259–71).

The result was a city in chronic fiscal distress, with a large and growing population of minority poor, and a depressed economic base that had lost much of its historic underpinning without developing the means by which to replace those lost functions with a sufficient quantity of activities linked to the innovations of the post-industrial era. As elsewhere, members of minority groups, and African Americans in particular, experienced high levels of distress (Massey and Eggers, 1993). Strategic economic planning efforts involving city officials and members of the business elite produced few results, beyond two new stadiums for the city's professional sports teams (DiGaetano and Lawless, 1999: 559–63). A 1998 study of urban policy in the United States by urban analyst Robert Waste labelled Detroit a 'point of no return city', a place with a degree of economic and social distress of such immense proportions as to preclude revival within existing policy paradigms (Waste, 1998: 12–16).

Detroit is a conspicuous example of a city that gave up considerable ground in the transition from industrial to post-industrial, losing its former leadership position as a place of technological innovation, and declining in importance in both national and international urban networks. With presumably intended hyperbole, Peter Hall described Detroit's decline as 'an astonishing case of industrial dereliction; perhaps, before long, the first major industrial city in history to revert to farmland' (Hall, 1998: 499).

Like Detroit, Birmingham, UK, was an important centre of manufacturing and technological innovations, with strong ties in the twentieth century to the automobile industry. Differing somewhat from Detroit, its economy never became quite as thoroughly identified with and dependent upon a single industry. Birmingham developed as a centre of iron and steel production, metal work, engineering and specialized manufacturing, with a prosperous, sophisticated, machine-driven economy built around a strong manufacturing core. One of several cities in the English north and midlands that rose as symbols of economic strength and technological prowess during the nineteenth century, Birmingham played an integral role in the British Industrial Revolution. Its perpetual prosperity seemed sufficiently assured in the mid-twentieth century that following the passage of the Town and Country Planning Act of 1947, planning policy for the city and region acted to discourage additional industrial expansion in order to steer economic development toward areas of the country where prosperity lagged (Briggs, 1965: 184–240; Carter, 1977: 128; Hohenberg and Lees, 1985: 287).

After 1970 Birmingham began to undergo an economic restructuring similar to those of other industrial cities. Its factories began to close, and its industrial jobs began to disappear. The pace of deterioration quickened in the 1980s. Birmingham lost 70,000 manufacturing jobs between 1981 and 1991, and during those years the manufacturing proportion of the labour force dropped from 39 per cent to 27.5 per cent. Unemployment peaked at an astonishing 21 per cent in 1986, then declined gradually to 11 per cent in 1991, still almost twice the national average (DiGaetano and Lawless, 1999: 551–3; DiGaetano and Klemanski, 1993: 68–9).

As in Detroit, unemployment led to social dislocation and distress. Birmingham's growing non-white population suffered higher levels of unemployment, and, consequently, more extreme distress. 'British blacks,' Hall writes, 'like American blacks, have remained heavily concentrated in the inner and middle rings of the big cities. Relatively few have entered the ranks of the middle class.' During the 1980s collective violence occasionally erupted in the ghetto-like neighbourhoods of several British cities. Birmingham experienced a riot in 1985, when local unemployment approached its peak. The root causes were similar to those of the Detroit eruption of 1967. Comparing these experiences, Hall concludes that British urban violence in the 1980s involved 'the same pre-history of mounting, barely controllable tension among the young ghetto blacks as they spar with police; the same small triggering incident of an arrest, followed by the spreading of rumour like wildfire; the same almost immediate conflagration.' According to an official inquiry into a similar riot in Brixton (London), the cause was not race directly; instead, 'it was a clash of cultures, exacerbated by the fact that the black subculture was built on deprivation and disadvantage' (Hall, 1996: 397–8). A report prepared by the Church of England tied collective violence to job loss: '*It is the national decline in the number of manual jobs, and the concentration of manual workers in the UPAs* [Urban Poverty Areas] *that lies at the heart of the problem*' (quoted in Hall, 1996: 399; emphasis in original). Few British cities were as hard hit by the loss of unskilled and semi-skilled jobs as was Birmingham.

In response to decline, Birmingham's political and business leadership came together in the early 1980s to promote diversified non-industrial economic development, and to protect, if possible, the remnants of its former industrial base. Through a series of innovative public–private partnerships, the city and several private developers joined forces to build office space and industrial parks, one of which was devoted to high-technology research. The city added a large convention centre, office space and an upscale retail mall to its downtown (DiGaetano and Klemanski, 1993: 68–73). While Birmingham has hardly recovered its former stature as an engine of economic growth, it seems to have confronted the challenges of the post-industrial age somewhat more successfully than had Detroit.

These two case studies represent something of an extreme. While virtually all industrial cities suffered serious employment loss as they attempted to adapt to post-industrial circumstances, few suffered as long or as hard as Birmingham and Detroit, and the scars left on the landscape and labourers alike will take decades to heal. Other cities, like Philadelphia and Los Angeles, seem to have retooled more effectively and moved on; and a few, such as New York City, London, Paris, Tokyo and Singapore, have successfully ridden the crest of the post-industrial wave.

CITIES AS INFORMATION AND KNOWLEDGE CENTRES

Modern global networks depend on rapid, low-cost communication. The ability of large organizations to function effectively across several continents with no physical connection between command and control centres, back office operations, manufacturing, distribution and sales is wholly a product of modern telecommunications. Thus, technological advance has led to rapid organizational evolution, allowing far-flung activities to be closely coordinated, and lessening the importance of location for many types of activity (Fathy, 1991: 1–7, 99–103; Malone and Rockart, 1993: 37–9). In addition, modern air travel has made face-to-face contact – when that is required – a matter of journeys that are likely to be short in time regardless of the distance to be travelled.

From earliest times, communication between two human beings took the form of either unamplified speech or the transport of documents by animal-drawn vehicles or wind-driven ships. Cities served as communication and market centres, providing places in which people could come together and speak to one another, trading such items as information, raw materials and finished products (Mumford, 1961: 70–3). Not until the second quarter of the nineteenth century did this begin to change. The communications revolution began with the application of steam power to land and sea transport in the form of railroads and steamships. These were soon joined by the telegraph, which allowed the transmission of coded information in the form of dits and dots over extensive distance by wire. The opening of the transatlantic cable in 1866 closed the yawning communications gap between North America and Europe, linking local markets into the first semblance of a global network (Taylor, 1951: 73–113, 152).

Advances since have largely elaborated on the telegraph and the cable. The telephone and the radio added transmission of human voices, and the gradual perfection of long-distance telephony increased the convenience and lowered the cost of this form of communication. The recent introduction of personal computers, facsimile transfers, electronic mail networks, high-speed data transmission lines, fibre-optic cable, and cellular telephones has further lowered costs and increased both the range and the speed of communication (Hall, 1998: 460–78). As costs fell and linkages improved, the advantages of gathering and transmitting data over long distances increased, and the communication expenses and hurdles incurred by locating production functions away from command and control dropped sharply (Casson, 1997: 274–97). Thus cities with corporate headquarters have increasingly become centres of information flows (Drennan, 1991: 33–8; Hall, 1996: 402–4).

The use of computers and computer networks to collect and analyse data is at the heart of the information revolution, affecting processes great and small. A study of the apparel industry, for example, found that information gained by retailers in tracking sales and inventory not only increased their responsiveness to consumer-driven market trends, but also gave them new-found leverage in their negotiations with manufacturers, altering long-standing relations within the industry (Hammond, 1993: 185–8). Control of critical information, then, may lead to systems changes beyond those initially envisioned: here, information gained by service sector retailers strengthened their position with industrial sector apparel makers, providing a small yet pertinent example of post-industrial shifts.

From their earliest pre-historic beginnings, cities served as centres of learning and accumulated lore; what Mumford refers to as the city as 'a storehouse, a conservator and accumulator'. From this knowledge base, cities developed as centres of creativity and innovation, 'and by the command of these functions . . . the city served its ultimate function, that of transformer', which Mumford believed approached a zenith in Golden Age Athens (Mumford, 1961: 97, 119–82).

The factors that came together to produce a golden age of intellectual ferment and progress in that particular place and time have defied precise definition. Times of rapid innovation, intellectual, artistic and technological, have occurred periodically in more recent human civilization, and each one has redirected our ways of thinking or doing. Hall devotes a section of *Cities in Civilization*, titled 'The Innovative Milieu', to the nature of technological innovation across time and space. He looks at textiles in Manchester, ship building in Glasgow, electrical engineering in Berlin, automobiles in Detroit, computers in San Jose, and electronic innovation in Tokyo–Kanagawa (Hall, 1998: 291–499).

He finds a number of commonalities. The first is 'the overwhelming strength of the heroic tradition. Most of the innovations were created by individual entrepreneurs in the garret, later garage, tradition' (Hall, 1998: 493). If individuals lit the spark, however, it was the culture in which they lived that provided the dry tinder. 'None stumbled on a discovery by pure accident or serendipity,' Hall writes. Instead, each was part of a local network of 'competitor–cooperators' working toward similar goals. Participants steadily expanded their technical knowledge and skills, which then became a kind of 'shared intellectual property' on which all could build (Hall, 1998: 493–4). The larger cultures in which they worked tended to be, for their times, relatively egalitarian, non-hierarchical, open, free of strong barriers to innovation, ready to reward talent regardless of origin, and receptive to commerce and money-making (Hall, 1998: 493–7).

In looking at cities that are no longer places of innovation, Hall finds that each past era of ferment led to one of consolidation, followed eventually by ossification and decline (Hall, 1998: 499–500). Kindleberger (1996) posits a similar life cycle for nations that have achieved economic primacy; they grew rapidly through a series of innovations, then successfully consolidated those methods into institutional structures that in time became resistant to further change, only to be overtaken in turn by a nation more open and receptive to further innovation.

How to nurture a culture of continuous innovation is a long-standing goal of economic planners and urban policy-makers. As past innovations seem to have emerged from networks of creative individuals, Hall suggests that metropolitan areas are the places most likely to produce future innovations, and that rare phenomenon, an innovative milieu. Certainly, the conscious development of cities as knowledge centres, as Knight and others have suggested, is a form of economic development that has the potential to renew many obsolescent industrial cities. With the advent of rapid, low-cost, worldwide communication,

global economic networks have become the new driving force of economic growth, and in this context, information and knowledge have become critical to the creation of new wealth (Knight, 1987: 196–208; Sassen, 1991).

Cities remain, as they have always been, places where people come together to live in dense settlements. In looking at the technologically creative environments of San Jose, California, and Tokyo–Kanagawa, Japan, Hall writes, 'They resemble nothing so much as huge and complex ecosystems, which must be constantly nourished if they are not to wither and die; and that is what they are, human ecosystems which contain a disproportionate number of the world's most creative individuals' (Hall, 1998: 500). This could also serve as a fairly reliable depiction of a city. It is in cities, where knowledge is most often created, and information traditionally stored, at least in its industrial era paper format, and where networks of inspired individuals have the opportunity to come together to achieve creative synergies, that the majority of future post-industrial innovations, and innovative milieus, are likely to occur (see Fathy, 1991: 81–91).

CONCLUSION

The ancient world lasted for 3,000 years, the medieval age for less than 1,000 years, and the industrial era for about 100 years. Our post-industrial revolution has occurred in just 25 years, and its pace is quickening. In one great blow, the new revolution has remade the industrial fabric of society, radically altered the behavior of capital, broken down national boundaries and is remodeling government. (Savitch, 1998: 249)

In three well-written sentences, Savitch captures both the scope and speed of the process by which the industrial has given way to the post-industrial. Few systems in our society are untouched by the changes unleashed. It is the pace of change, more than the nature of the changes themselves, that sets the transition to the post-industrial apart from earlier transitions. Further, the present changes intrude into almost every corner of the planet inhabited by human beings, leaving few individuals untouched.

Like most transitions, there are differential costs and benefits. Industrial workers in developed countries, who had spent decades building a work culture that gave them entrance to the middle class, witnessed the rapid destruction of their jobs, their cultures and their expectations for safe and secure economic futures. At the same time, the corporate executives who ordered the plants shut, along with the high-level providers of producers' services to those corporations, managed to extract from the misery they brought to others considerable wealth and comfort for themselves. Whole cities fell from prosperity and industrial leadership to advanced states of decay in less than a generation. Detroit, USA and Birmingham, UK illustrate the drastic impacts of these post-industrial changes on physical environments and human lives alike.

In those parts of the world to which industry has moved, the deindustrialization of the West has provided economic development, employment and entry into the global economy. The prosperity of some newly industrial states has contributed to world migration flows. Some industrializing countries require more workers than their population produces; the availability of work in those places attracts temporary migrants or permanent immigrants from poorer countries. The role and status of migrants in the societies to which they move is a continuing source of cultural and political tension (Massey et al., 1998: 4–7, 275–94).

The problems of migrants are closely related to the problems of distressed urban populations in developed countries, which are frequently disproportionately made up of racial or ethnic minorities. Nightengale raises the possibility of a 'global inner city' that has similar characteristics regardless of location. The related problems of racial discrimination and poverty, so apparent in Birmingham and Detroit, are, he suggests, compounded by the nature of the global economy and by '[g]lobally linked racial logics' (Nightengale, 1998: 230–5).

Numerous dislocations and social problems have accompanied the transition to a post-industrial order. In a time of transition, however, it is virtually impossible to know which conditions are temporary social adjustments and which will congeal into a new set of long-lasting inequalities, firmly embedded in a new stability. How the new wealth created by the engines of post-industrial economic processes is ultimately distributed, both within and among societies, is a political and policy issue of great importance. The nature of that distribution will, ultimately, help to shape the health and vitality of a more globalized world and of the post-industrial cities that have become its engines of economic growth and social change.

REFERENCES

Amin, Ash and Thrift, Nigel (1994) 'Living in the global', in Ash Amin and Nigel Thrift (eds), *Globalization, Institutions, and Regional Development in Europe*. Oxford: Oxford University Press.

Bell, Daniel (1973) *The Coming of Post-Industrial Society: A Venture in Social Forecasting*. New York: Basic Books.

Beauregard, Robert (1989) 'The spatial transformation of postwar Philadelphia', in Robert Beauregard (ed.), *Atop the Urban Hierarchy*. Totowa, NJ: Rowman and Littlefield.

Bluestone, Barry and Harrison, Bennett (1982) *the Deindustrialization of America: Plant Closings, Community Abandonment, and the Dismantling of Basic Industry*. New York: Basic Books.

Briggs, Asa (1965) *Victorian Cities*. New York: Harper Colophon Books.

Carter, Christopher (1977) 'The changing pattern of industrial land use in the West Midlands conurbation', in James Joyce (ed.), *Metropolitan Development and Change, The West Midlands: A Policy Review*. Birmingham: The University of Aston.

Casson, Mark (1997) *Information and Organization: A New Perspective on the Theory of the Firm*. Oxford: Clarendon Press.

Chandler, Alfred (1990) *Scale and Scope: The Dynamics of Industrial Capitalism*. Cambridge, MA: Harvard University Press.

Chernow, Ron (1998) *Titan: The Life of John D. Rockefeller*. New York: Random House.

Chudacoff, Howard and Smith, Judith (1994) *The Evolution of American Urban Society*, 4th edn. Englewood Cliffs, NJ: Prentice Hall.

Cowen, Alexander (1998) *Urban Europe, 1500–1700*. London: Arnold.

Davis, Mike (1992a) *City of Quartz: Excavating the Future in Los Angeles*. New York: Vintage Press.

Davis, Mike (1992b) 'Fortress Los Angeles: the militarization of urban space', in Michael Sorkin (ed.), *Variations on a Theme Park: The New American City and the End of Public Space*. New York: Hill and Wang.

Dicken, Peter (1998) *Global Shift: Transforming the World Economy*. New York: Guilford Press.

DiGaetano, Alan and Klemanski, John S. (1993) 'Urban regimes in comparative perspective: the politics of urban development in Britain', *Urban Affairs Quarterly*, 29: 54–83.

DiGaetano, Alan and Lawless, Paul (1999) 'Urban governance and industrial decline: governing structures and policy agendas in Birmingham and Sheffield, England, and Detroit, Michigan 1980–1997', *Urban Affairs Review*, 34: 546–77.

Disraeli, Benjamin (1995) *Sybil, or the Tale of Two Nations*. Ware: Wordsworth. (Originally published 1845.)

Drennan, Matthew (1991) 'The decline and rise of the New York economy', in John Mollenkopf and Manuel Castells (eds), *Dual City: Restructuring New York*. New York: Russell Sage Foundation.

Fathy, Tarik (1991) *Telecity: Information Technology and its Impact on City Form*. New York: Praeger Publishers.

Gappert, Gary (1979) *Post-Affluent America: The Social Economy of the Future*. New York: Franklin Watts.

Goldsmith, William and Blakely, Edward (1992) *Separate Societies: Poverty and Inequality in U.S. Cities*. Philadelphia: Temple University Press.

Guinther, John (1996) *Direction of Cities*. New York: Penguin Books.

Gutman, Herbert (1976) *Work, Culture and Society in Industrializing America: Essays in American Working Class and Social History.* New York: Alfred A. Knopf.

Haggard, Stephan (1990) *Pathways from the Periphery: The Politics of Growth in the Newly Industrializing Countries.* Ithaca, NY: Cornell University Press.

Haggard, Stephan (1995) *Developing Nations and the Politics of Global Integration.* Washington, DC: The Brookings Institution.

Hall, Peter (1996) *Cities of Tomorrow.* Oxford: Basil Blackwell.

Hall, Peter (1998) *Cities in Civilization.* New York: Pantheon Books.

Hammond, Janice (1993) 'Quick response in retail/manufacturing channels', in Stephen Bradley et al. (eds) *Globalization, Technology, and Competition: The Fusion of Computers and Telecommunications in the 1990s.* Boston, MA: Harvard Business School Press.

Hohenberg, Paul and Lees, Lynn (1985) *The Making of Urban Europe, 1000–1950.* Cambridge, MA: Harvard University Press.

Kantor, Paul (1987) 'The dependent city: the changing political economy of urban economic development in the United States', *Urban Affairs Quarterly,* 22: 493–520.

Kantor, Paul (1988) *The Dependent City: The Changing Political Economy of Urban America.* Boston, MA: Scott, Foresman and Company.

Kindleberger, Charles (1996) *World Economic Primacy, 1500–1990.* New York: Oxford University Press.

Knight, Richard (1987) 'Knowledge and the advanced industrial metropolis', in Gary Gappert (ed.), *The Future of Winter Cities.* Newbury Park, CA: Sage.

Malone, Thomas and Rockart, John (1993) 'How will information technology reshape organizations? Computers as coordination technology', in Stephen Bradley, et al . (eds), *Globalization, Technology, and Competition: The Fusion of Computers and Telecommunications in the 1990s.* Boston, MA: Harvard Business School Press.

Marcuse, Peter (1997) 'The Enclave, the Citadel and the Ghetto: What has changed in the Post-Fordist city?', *Urban Affairs Review,* 33: 228–64.

Massey, Douglas and Eggers, Mitchell (1993) 'The spatial concentration of affluence and poverty during the 1970s', *Urban Affairs Quarterly,* 29: 299–315.

Massey, D., Taylor, J.E., Arango, J., Kouaouci, A., Pellegrino, A. and Hugo, G. (1998) *Worlds in Motion: Understanding International Migration at the End of the Millennium.* Oxford: Clarendon Press.

Mumford, Lewis (1961) *The City in History: Its Origins, Its Transformations, and Its Prospects.* New York: Harcourt, Brace and World.

Nightengale, Carl (1998) 'The global inner city: toward a historical analysis', in Michael Katz and Thomas Sugrue (eds), *W.E.B. Dubois, Race, and the City: The Philadelphia Negro and its Legacy.* Philadelphia: University of Pennsylvania Press.

Ong, Paul and Bloomenberg, Evelyn (1996) 'Income and racial inequality in Los Angeles', in Alan Scott and Edward Soja (eds), *The City: Los Angeles and Urban Theory at the End of the Twentieth Century.* Berkeley, CA: University of California Press.

Peck, Jamie and Tickell, Adam (1994) 'Searching for a new institutional fix: the after-fordist crisis and the global–local disorder', in Ash Amin (ed.), *Post-Fordism: A Reader.* Oxford: Basil Blackwell.

Perloff, Harvey S. (1980) *Planning the Post-Industrial City.* Chicago: Planners Press.

Philpott, Thomas (1991) *The Slum and the Ghetto: Immigrants, Blacks and Reformers in Chicago, 1990–1930.* Belmont, CA: Wadsworth Publishing Company.

Sassen, Saskia (1991) *The Global City: New York, London, Tokyo.* Princeton, NJ: Princeton University Press.

Savitch, H.V. (1988) *Post-Industrial Cities: Politics and Planning in New York, Paris and London.* Princeton, NJ: Princeton University Press.

Savitch, H.V. (1998) 'Global challenge and institutional capacity: or how we can refit local administration for the next century', *Administration and Society,* 30: 248–73.

Scott, Alan and Soja, Edward (1996) 'Introduction to Los Angeles: City and Region' in Alan Scott and Edward Soja (eds), *The City: Los Angeles and Urban Theory at the End of the Twentieth Century.* Berkeley, CA: University of California Press.

Stull, William and Madden, Janice (1990) *Post Industrial Philadelphia: Structural Changes in the Metropolitan Economy.* Philadelphia: University of Pennsylvania Press.

Sugrue, Thomas (1996) *The Origins of the Racial Crisis: Race and Inequality in Post-War Detroit.* Princeton, NJ: Princeton University Press.

Swanstrom, Todd (1996) 'Ideas matter: Reflections on the new regionalism', *Cityscape: A Journal of Policy Development and Research,* 2: 5–21.

Tabb, William (1995) *The Postwar Japanese System: Cultural Economy and Economic Transformation.* New York: Oxford University Press.

Taylor, George Rogers (1951) *The Transportation Revolution, 1815–1860.* New York: Harper and Row.

Thomas, June (1997) *Redevelopment and Race: Planning a Finer City in Postwar Detroit.* Baltimore, MD: Johns Hopkins University Press.

Thompson, E. (1967) 'Time, work-discipline, and industrial capitalism', *Past and Present,* 38: 56–97.

Warren, Robert (1990) 'National urban policy and the local state: paradoxes of meaning, action, and consequences', *Urban Affairs Quarterly,* 25: 541–61.

Waste, Robert (1998) *Independent Cities: Rethinking US Urban Policy.* New York: Oxford University Press.

White, James W. (1998) 'Old wine, cracked bottle? Tokyo, Paris, and the global city hypothesis', *Urban Affairs Review,* 33: 451–77.

Yago, Glenn (1983) 'Urban policy and national political economy', *Urban Affairs Quarterly,* 19: 113–32.

19

The New Urban Economies

DONALD MCNEILL AND AIDAN WHILE

Silicon Valley, Docklands, Barcelona . . . all essential namedrops in recent urban research and all examples of responses to changing patterns of economic activity in recent years. As recession and unemployment have bitten deep into old industrial areas, more and more local authorities and city governments are pursuing a policy agenda that draws on a set of exemplars and paradigms which – taken together – add up to the emergence of 'new urban economies'. However, these new directions are far from being value-free. Our argument here is that they are politically loaded, and that we need to understand the political tendencies which accompany these new economies. The growing interest in issues of 'power, discourse, culture and institutions' in political economy (Barnes, 1995) suggests that, far from being inevitable responses to processes of economic change, these new models are adopted, accepted and legitimized by a range of political actors, be they politicians, journalists, academics, or civil servants and officers. There is nothing inevitable about these policy agendas, for as Beauregard (1993: 22) has pointed out:

> An ideology that celebrates 'newness' and 'growth', and that portrays investors as risk-takers bringing prosperity to all and strength to the nation, legitimizes uneven spatial development. By hailing restlessness [of capital] as a positive virtue, and by focusing attention on the creative rather than the destructive aspects of urban development . . . complex new patterns of investment and disinvestment, bearing significant costs for many groups, are justified.

Our intention, then, is to critically review the *political* economy of recent urban change. We begin by briefly outlining the main manifestations of these 'new urban economies', arguing that these have been associated very closely with a 'global shift' (Dicken, 1992) in economic activity. Secondly, we review the substantial body of work which suggests the emergence of a 'new urban politics' (NUP) which has developed as political actors attempt to comprehend and adapt to these changes.[1] Thirdly, we identify several discourses by which the embrace of such 'newness' is justified and discussed, and which often contain highly contestable politico-economic ideas.

NEW ECONOMIC SPACES

The emergence of the new economies has sparked a spirited debate within economic geography as to their salience and, crucially, their universality. In general, they are linked to grander theorizations of changes in the global economy and the way production is organized and regulated, and are usually discussed in relation to regulation theory, post-fordism, or related theories of economic restructuring. These are grounded in a profound transformation of state–society–economy relations which has occurred since the mid-1970s (see Amin, 1994 for a summary). As these are theories of change, some commentators have expressed concern over the strength of their descriptions of transition. Thrift (1989), pointing to the 'perils of transition models', has expressed dismay at the lack of historical context and sociological rigour in the construction and use of some of these theoretical structures. Rustin (1989) has suggested that the examples used may not be generalizable, and may reflect only the most visible 'leading edge' of economic development. Without wishing to jump into our critique too early, it is worth noting that the novelty of some of these 'new' economies can be overplayed, as can their universal relevance. None the less, and without wishing to be overly rigid in our categorization, we provide a fourfold typology of themes which have been exciting academics in recent years.

1   Agglomeration economies: the growing interest in locally based agglomerations of production can be seen as a response to Fordist crisis. In light industrial sectors such as electronics and clothes production, certain areas have emerged which have been resistant to economic crisis (Sabel, 1994: 106–16). The success of these 'industrial districts' clustered together in regional economies such as the 'Third Italy', Baden-Wüttemburg and the Los Angeles area, has been due, it is suggested, to the notion of flexible production, where factors of trust, knowledge, actual physical proximity and strong buyer–supplier relationships lead to a synergy, whereby more wealth is produced by the interactions between firms than would have been the case had each ploughed their own furrow. The importance and density of transactions between firms, in many cases combined with partnership between public and private sector organizations and a strong emphasis on innovation, has been the defining characteristic of these new economies. Morgan (1992) argues that the emergence of these areas shows the importance of networking, of the horizontal relationships between different actors that occur between large manufacturers, smaller suppliers, training agencies, banks and local state policy-makers.

Industrial districts owe their success to other factors than firm organization alone, however. Scott (1988: 58) suggests that the Third Italy owes its success to a combination of the new and the old, high technology co-existing with home-working, labour-intensive firms communicating through telematics, and traditional skills being used to produce goods which are competitive in global 'niche' markets.

2   The growing importance of information and knowledge-rich economies has encouraged the emergence of *informational cities*. As manufacturing has declined, local strategists have sought to attract financial services, the headquarters of transnational corporations or government agencies. Debate on this area tends to speak in terms of a hierarchy of informational cities. At the top, there are the 'global cities' such as London, New York, and Tokyo, centres of financial trading with associated financial services such as media, legal services and accountancy, and housing the headquarters of most large corporations (Sassen, 1991). Further down, there are national capitals, with a cluster of government offices, national or regional headquarters or major corporations, and associated businesses linked to the national economy. This is followed by regional financial and industrial centres, and finally by cities which are marginal to global information flows.

Fainstein (1994) suggests three reasons for the emergence of such informational cities: first, the growth of international trade and the deregulation of financial markets have altered world capital flows. Firms have sought space to respond to the new demands, and have largely found them in the global cities where the agenda and discourse of international finance is constructed (Thrift, 1994). Secondly, the decentralization of production has brought an 'increased need for command and control posts' (Fainstein, 1994: 27), which need not be within the global city, thus opening up possibilities for smaller cities to attract office investment. Thirdly, the increasing sophistication of computer and telecommunications technology has had a tendency to split company headquarters from routine information processing, or 'back office' functions. This has led to major cities reinforcing their strengths as command centres, while regional or non-core cities may take on more routine tasks.

The future of the informational city is not easy to predict, and has long been a source of controversy among urbanists:

> On the one hand, utopianists and futurologists herald telecommunications as the quick-fix solution to the social, environmental or political ills of the industrial city and industrial society more widely. On the other, dystopians or anti-utopians paint portraits of an increasingly polarized and depressing urban era dominated by global corporations who shape telematics and the new urban forces in their own image. (Graham and Marvin, 1996: 5)

However, telematics – 'services and infrastructures which link computer and digital media equipment over telecommunication links' (Graham and Marvin, 1996: 2–3) – offer a substantial range of policy options to local authorities which allows them to attract innovative companies to their areas. Through facilities such as teleports, 'satellite links associated with property developments and links to local telecommunications networks', local authorities can specialize in attracting particular sectors to their areas, such as media in Cologne, financial services in New York, or maritime industries to Le Havre and Bremen (Graham and Marvin, 1996: 350).

3   The *technopole* shares many of the characteristics of the industrial district, and is also based upon economies of agglomeration. Technopoles are centres for the promotion of high-technology industry, frequently private sector initiated but often with a considerable degree of cooperation with the public sector. They include industrial complexes linking manufacturing with research and development, as in Silicon Valley or Route 128 in Boston; they can be strictly research-based, with no link to manufacturing; or they can be used to attract high-technology manufacturing firms as a means of stimulating economic growth

in a particular local space (Castells and Hall, 1994: 10–11). They tend to be semi-urban, located on the fringes of major urban areas but – crucially – easily accessible from road networks.

4   The urban leisure economy: the deindustrialized cities of North America and Europe have usually found it easier in recent years to build a new opera house or art gallery than attract major manufacturing companies. While the new industries of flexspec and high technology have largely taken up residence in green-field or semi-rural sites, local governments have been left with the problem of what to do with the areas of dereliction in their cities. Possible solutions have not been slow in coming. Urban regeneration has become a growth industry in itself as a variety of options have opened for local urban leaders seeking to rebuild their cities. Most are now totally committed to developing their economies along the lines of arts, culture and entertainment, both as a means of attracting tourists and as recognition of the importance of the arts as an employer. Furthermore, attempts to persuade major corporations or government agencies to locate in their cities can be aided if that city contains good arts and leisure facilities. Thus cities such as Barcelona and Baltimore have become models of urban regeneration, targeting investment in business and leisure tourist infrastructure, improving the urban landscape, and subsidising the arts and culture. These leisure economies are usually property-led, reflecting both the increased profitability of property holdings for major investors such as pension funds (Fainstein, 1994: 30), and the limited scope for local government to pursue alternative development options.

Governments pursue such development with at least one of the following aims in mind.

1   To attract high-spending tourists into the city through a programme of hotel and transport infrastructure building. Business tourism is seen as a particularly lucrative activity, and most cities now have at least one purpose-built convention centre to hold trade fairs or conferences. The projected influx is also felt to have beneficial effects on spending in local shops, restaurants and services.

2   To create a cultural infrastructure of theatres, museums, galleries and performance spaces, what Zukin (1995) calls the 'symbolic economy'. This is not only linked to attracting tourists, as it can also act as a powerful magnet for corporations seeking to provide good 'lifestyle' opportunities for its employees, or can encourage potentially footloose residents to stay in the city. The presence of such institutions also enhances place image, explaining why the Basque region mortgaged

itself to attract a European branch of the Guggenheim art gallery to Bilbao, housing it in a sparkling showpiece building designed by an internationally renowned 'trophy' architect (see McNeill, 2000).

3   Such developments provide jobs which – in old industrial cities – may help to tackle long-term structural unemployment. Policymakers are now more aware of the employment potential of the arts, and the increasing importance of the newer media industries in urban economies (Bianchini and Parkinson, 1993).

4   It allows governments to embark upon redesigning urban spaces, particularly those devalorized by capital flight and manufacturing decline. Derelict industrial sites have been turned into heritage parks, old canals or waterfronts have become housing or restaurant areas, and warehouse conversions have helped build urban living into something chic. As well as improving the quality of the urban environment, the use of big-name architects or innovative urban designers also allows for the redefinition of city images, again important for place-marketing strategies.

The new leisure economies are now widespread (Sudjic, 1992 provides a readable introduction), and are perhaps the most visible manifestations of changing economic activity in cities.

These, then, are the main manifestations of economic novelty in urban areas. Many of these developments have become established features of the urban (and ex-urban) landscape. Yet these spatial developments conceal very important power relations within the city: they may involve the destruction of old ways of life, and endow particular social groups over others with the chance of sharing in wealth creation. They are, then, rooted not only in pure economic change, but are determined very strongly by local political actors.

THE NEW URBAN POLITICS, AND THE ECONOMICS OF THE 'NEW' CITY

So, it is clear that there are a range of activities which can be grouped together as being somehow 'new', 'leading edge', 'dynamic', 'creative', or 'entrepreneurial'. This ties with Barnes's (1996) review of '"post"-prefixed' economic geographies, academic accounts that provide explanations for such change based on ideas of progress, which in turn relies on analysis elaborated from traditions of historical materialism (Leyshon, 1997). However, this is to imply the teleological inevitability of such processes, discarding the fact that they are legitimated politically. Leyshon (1997) thus argues for more attention to discourse, to the frameworks of understanding used by 'geo-economic intellectuals' such as Robert Reich and Will Hutton, which in turn have had a significant impact on how the national and international/global economy is interpreted. Similarly, Beauregard's (1993) *Voices of Decline* traces through the shifting place of the city in the public discourse of politicians, journalists and policy-makers, justifying his approach as follows:

> To the degree that commentators aspire to provide a factual basis for their interpretations and evaluations, representations of decline are grounded in an empirical reality. Yet the relation between discourse and material conditions is increasingly problematic, and that indeterminacy is unavoidable. The 'real world' provides material for our representations but these representations then become subject to social construction through language and discourse. The reality of urban decline is always mediated by our representations of it. (Beauregard, 1993: 36)

And so, discourse is now being taken more seriously by economic and political geographers, usually at a basic level that examines the stories and scripts which predominate in public debate on urban economic change.

Many of those involved have been quick to identify globalization as a crucial structuring force in these futures, and it is here that the focus on discourse has importance. Lovering (1995) has articulated a forceful criticism of the role of academics in constructing a one-sided 'new localist' agenda which ties in with the city government as entrepreneur, freed from the shackles of centralized government, surfing the globalization wave in a competitive race for prosperity or survival. The question, after Beauregard, is this: whose reality is being articulated, and which version of the truth should we listen to? To answer this we first have to account for *why* we should be suspicious of representation and discourse.

The impact and implications of processes of 'globalization' (Swyngeduow, 1992), where globalization is entwined with an apparent growth in the power of the local, sub-national or regional level, are captured in Cox's notion of the 'new urban politics' (Cox, 1993). The NUP is based on evidence that the restructuring of global and national space economies has transformed the options open to urban leaders, characterized by an increasing involvement in local economic strategies. For Cox and others (see Cox and Mair, 1988; Harvey, 1989; Hay, 1994) the emergence of the NUP reflects growing competition between cities in the search to tap into growth opportunities at varying spatial scales, linked with the rise in more entrepreneurial forms of city leadership (Hall and Hubbard, 1996; Harvey, 1989; Mollenkopf, 1983).

This marks a shift from traditional 'managerial' approaches based on the provision of collective services, to a more growth-oriented polity, in which the public sector assumes 'characteristics once distinct to the private sector – risk-taking, inventiveness, promotion and profit motivation' (Hall and Hubbard, 1996: 153).

One of the defining characteristics of the new modes of local governance is a more visible role for urban coalitions or public–private partnerships in driving strategies for urban development and regeneration (Bassett, 1996; Elkin, 1987; Harding, 1994; Logan and Molotch, 1987; Peck and Tickell, 1994a; Ward, 1996). The political rationality of these developments has been explored in detail by USA-inspired urban political theory, notably urban regime theory and growth coalition approaches. While many commentators have suggested that US local government is peculiarly dependent on business and development in order to secure its tax base (Harding, 1994; Ward, 1996), regime theory has been proposed as a useful tool in illuminating various aspects of urban governance at a more general level (Stoker and Mossberger, 1994).

Regime theory (Elkin, 1987; Stone, 1989) points to the limited power of public officials to bridge the gap between state and market in capitalistic societies. Lacking direct economic control, the local state essentially relies on cooperation with private actors if it is to induce market activity. Regime theorists are primarily interested in the process of coalition or regime building between state and non-state interests. By their nature urban regimes rely on voluntary participation and operate through informal negotiation and bargaining, as each side tries to move towards a common position. What is interesting here is that although coalitions can, and sometimes do, pursue *progressive* agendas, a regime polity inevitably privileges business interest, which tend to have a regressive effect on urban policies. As Keating (1993: 391) argues, the ability of city governments 'to integrate social and economic concerns is weakened everywhere by the alliance of political and business interests in development coalitions'.

The implications by this are clear: at any given level, regimes or growth coalitions will have to present essentially partial strategies as being in the general interest. They have to pursue a variety of mechanisms to aid in perpetuating this hegemony, which will include the solidification of the regime itself. Stoker and Mossberger's (1994) review of regime theory identifies various areas where territorially based power elites seek to enhance, or extend or deepen a 'capacity to act' (1994: 197) through the building of regimes based upon identification of goals (for example, image-building, project realization), motivation of participants (which could be business, community organizations, technocrats or voters), the development of 'a common sense of purpose' (p. 204) through use of symbols or invocation of tradition, and a relationship with the 'wider political environment' (p. 207). Discourse is of crucial significance in the establishment of 'scripts' for understanding social change. Whether or not one agrees with regime theory's somewhat schematic view of urban politics, we are moving beyond an interest in material 'reality', and are beginning to engage with how such materialist analyses are also discursive. It is time to identify what some of these discourses may be.

THE DISCOURSES OF THE NEW ECONOMIES

To recap, a focus on meaning, on discourse, is important when we consider the stake local 'mediators' may have in the urban economy. Politicians depend on votes, local newspapers frequently have an interest in urban growth (Thomas, 1994), and academics will benefit from research grants (Logan and Molotch, 1987). In this section, we discuss five discourses that have emerged from academics, politicians and policymakers who have sought to justify the pursuit of new economic sectors. They are as follows: cities as agents, or the 'new localism'; the competitive city; the informational city; the just city; and urban renaissance. We explore each of these in turn.

The 'New Localism'

It is now commonplace for accounts of urban restructuring to be couched in terms of a global–local interplay (Dunford and Kafkalas, 1992). Such an interplay, it is argued, originates in the shift in global economic activity and breakdown in the international economic order of the period after the Second World War (Swyngedouw, 1992). However, it has taken on a particular form in individual nation states as they respond to neo-liberal hegemony at the level of international economic institutions, such as the EU. As a consequence, national governments have sought to regulate the local, albeit in highly specific ways. In the UK, successive Conservative governments were able to redefine the powers of elected local governments to influence economic development through the creation of new institutions and the empowerment of business elites (Peck, 1995). In Spain, Germany, France and Italy, by contrast, the rise of powerful regional movements and prominent urban leaders has

suggested an alternative architecture of European integration, giving us a vision of a 'Europe of the Regions' which sits within a federal structure.

This has placed a lot of emphasis on the emergence of the territorially clustered agglomeration economies discussed above, with urban and regional governments promoting policies to reposition their economies in a post-industrial world. However, it has been pointed out that the striking thing about these 'local' economies is just how 'unlocal' such strategies are, being part of the 'prevailing orthodoxy of neo-liberal economic policy' (Peck and Tickell, 1994b: 281). Overconcentration on the local, they argue, ignores the growing globalization of economic life, manifested in supra-national trading blocs such as the European Union and the North American Free Trade Agreement, which prevents national governments from pursuing their desired economic destinies. Within this discourse, localities are only in control of their destinies to the extent that superior levels of the state allow.

Nevertheless, the experiences of Birmingham (Martin and Pearce, 1992) and Barcelona (Marshall, 1996), combined with the greater autonomy of North American cities (Stone, 1989) has suggested that local leaders are taking seriously an entrepreneurial script that leaves little room for dissent. Lovering (1995) suggests that British urban policy has been dominated by a 'new localism', whereby academics actively construct an agenda that focuses on competitiveness, disregarding the highly unequal effects such policies will have on local labour markets. Similarly, the participation of business interests in economic development policy-making has a highly politicized character, with the creation of discursive and institutional 'platforms' (such as urban development corporations) to empower the private sector (Peck, 1995). Opposition can be difficult to articulate in such situations: Sadler (1993: 183), describes actual cases of 'attacks upon the motives or personal integrity' that alternative commentaries on local development have attracted from politicians or local agencies. The enthusiasm for localism, then, has to be seen as a construct, a means of creating a favourable climate for potentially controversial development strategies.

Competitive cities

The emphasis placed on local proactivity tends to be manifested in elite-led civic boosterist regimes, as is usually the case in North American cities, with a varying balance of power between city government and other institutions. In these cases, business groups, local government, trade unions, newspapers, sports teams, universities and property developers are likely to be the main players in the locality and are thus likely to share spin-offs from prosperity. Consequently, these growth coalitions will try to foster 'value-free development' (Logan and Malotch, 1987) in their attempts to persuade local voters that success for the locality will be shared generally. Here, 'success' is generally equated with 'competitiveness', which in turn is often meshed with discourses based upon local pride, identity, or traditions.

Localities are – to state the obvious – in markedly different structural positions within a national economy, and are in turn visited by restructuring in different ways (Cooke, 1989; Massey, 1995). Central government has considerable powers at its disposal to influence the spatial distribution of growth, thus putting some regions in a favourable position in national accumulation strategies. The governments of Reagan and Thatcher – despite their rhetoric about the sovereignty of the market – favoured particular regions with a *de facto* spatial policy. Tickell and Peck's (1995) analysis of Thatcherism in the South East of England suggests that local efforts were considerably aided by a very deliberate funding regime based around London's importance as a global financial node. This clearly implies that other UK regions suffered in relative terms. In the absence of a post-fordist regional policy, it could be argued that the promotion of certain fractions of capital within a national economic space – as happened with finance capital under Thatcherism – will negatively affect the competitiveness of localities within less favoured regions. Lovering and Boddy (1988) illustrate the relationship between defence spending and formal regional aid in the UK, noting the contribution made to employment by government military contracts. Similarly, the hype surrounding technopoles and science parks has been challenged by Massey et al. (1992), who argue that they are active in perpetuating uneven development, though sucking in government investment to already advantaged regions.

Furthermore, measures to stimulate competitiveness – such as localized bargaining, curbs on trade union activity, and technological innovation – are presented as being self-evidently beneficial for the local economy. This may be the case for newly created jobs in the service sector, but leads to the creation of clusters of unemployment in areas dependent on a single industry. Appeal is often made to other euphemisms – 'rationalization' or 'modernization' – to conceal the complex effects that restructuring has for local economies (see Berman, 1983 on how modernization has been exploited by power elites throughout the history of modernity).

Accepting that competition is not an inevitable response to the logic of capital, but rather an agenda constructed by local and national political actors, we can begin to examine how particular strategies are selected and developed. This poses various questions for urban analysts. Who constructs the economic development agenda? How is it represented? Why does it take one form and not another? Harvey (1989) describes various strategies open to governments wishing to reposition their economies, to compete for investment or resources: first, there is competition within the spatial division of labour, which can be achieved by improving technology, infrastructure and cutting labour conditions; secondly, they can compete within the spatial division of consumption through attracting and retaining tourists and high income residents; thirdly, they can seek to become financial, governmental, or information centres with all the spin-offs that this can entail; fourthly, they can compete for government redistribution, be this through formal urban policy or state investment in military technology. Harvey suggests that this signifies a shift to entrepreneuralism in urban governance, with the public sector taking on the role of risk-taker to smooth the way for private investment. It should be clear, then, that the creation of competitive strategies may well involve the destruction of local traditions of radicalism, the over-riding of community groups in the pursuit of development opportunities, or the co-optation of local developers and business elites into decision-making processes. This latter process is not known for strengthening municipal financial probity.

Along with the stridency of appeals to the 'need to compete' comes a corresponding vagueness as to what amounts to 'success'. Leaving aside the distributional consequences of these policies, there seems to be a chronic inability to determine what actually constitutes continuing economic growth. The proliferation of league tables and quality-of-life indices give policymakers ammunition to defend (or question) their strategies, but beyond that their use is limited. This leaves local regimes to follow what Harvey (1989) calls the 'serial reproduction' of convention centres, sports facilities and waterfront developments. Taylor et al. (1996) note the destructiveness of Manchester and Sheffield's competing attempts to become 'sport cities' through stadia construction and hosting major international events such as the Olympics and the World Student Games (Loftman and Nevin, 1996). Jauhiainen (1995) shows how the Baltimore model of redeveloping derelict waterfronts has been copied by European urban planners in Barcelona, Cardiff and Genoa, as the symbolic and economic possibilities from the redevelopments became apparent. Such projects usually involve a combination of similar elements: an aquarium, waterside promenades, restored ships, and so on. The appearance of such developments in virtually every city with developable waterfront space puts into doubt the competitive edge that can be gained from such strategies.

Occasionally, competition between places is more tangible. Cases such as the contest over the siting of the German parliament between Bonn and Berlin (Haussermann and Strom, 1994), or for the National Gallery of Scottish Art between Edinburgh and Glasgow (Buxton, 1994) bring other discourses into play. Here, the notion of civic identity is often aligned with the 'need to compete' to attract events and institutions which may ultimately have highly stratified effects on local social groups. This is encapsulated by the four-yearly scramble for the Olympic nomination, or the staging of a World Fair or Expo. Such major events can have a logic of their own, which can help local elites establish hegemonic control over the terms of debate (see Ley and Olds, 1988 on the local state and Vancouver's 1986 Expo). Manchester's urban leaders have stated that even a failed Olympic nomination bid can be worthwhile in terms of the publicity and central government funding generated. Civic boosterism is the urban variant of nationalism: less likely to end in war and bloodshed, perhaps, but with the same logic of territorial competition, and the same appeals to turf and vertical loyalty between rulers and ruled. Competition, success, modernization and civic pride (and rivalry): all can be mixed into the boosterist message as elites seek to restructure not only the local economy, but the balance of social forces which exist in a particular place.

The Informational City

As we noted above, there now exists a substantial body of work detailing the importance of informational industries in the urban economy. Cities now seek to attract high technology industry, to try to drop anchor in a fast-flowing global system of invisible financial trading, corporate strategy and knowledge exchange. There is little room for the faint-hearted here, and it has been suggested by Borja and Castells (1997) that cities are better adapted to global change than nation states:

> For one thing, they enjoy greater representativeness and legitimacy with regard to those they represent: they are institutional agents for social and cultural integration in territorial communities. For another, they have much more flexibility, adaptability and room for manoeuvre in a world of cross-linked flows, changing demand and supply and decentralized, interactive technological systems. (Borja and Castells, 1997: 6)

Cities are thus duty-bound to develop mechanisms by which they can try to manage such information flows. Borja and Castells argue that cities must develop strategic plans, public–private partnerships, concrete infrastructure targets, to be enhanced by devolving greater powers from central government, and developing powerful municipal leadership. The essence of their argument is that only with strong civic societies can a more equitable informational future be achieved. In addition, the informational city is seen as entailing a qualitative progression for localities in terms of how they compete spatially. Jessop (1998: 79) argues that the whole idea of competition should be thought of in terms of 'strong' and 'weak' forms, where 'strong competition refers to potentially positive-sum attempts to improve the overall competitiveness of a locality through innovation, [while] weak competition refers to essentially zero-sum attempts to secure the reallocation of existing resources at the expense of other localities'. Through innovation, then, localities have the opportunity to step beyond the primordial zero-sum struggle towards a more mature economic strategy.

The guarded optimism of Borja and Castells is shared by other commentators such as Amin and Graham (1997), who note that, contrary to 1980s doom-laden predictions of the death of the city, there has been a process of 'urban rediscovery' due to the peculiar characteristics of the informational economy. Far from leading to the decline of the city through homeworking from networked barn conversions, it has been given a new lease of life due to a growing need for face-to-face contact in the new industries:

> like all creative industries, they depend on interaction, on networking, on a certain amount of buzz and fizz . . . in other words, there was a paradox: these new activities were supposed to substitute for face-to-face communication, but actually they depended on it, and in turn fortified the need for it . . . So the evidence . . . suggested that the city as a place of congregation and interaction was far from dead . . . (Hall, 1996: 407–8)

However, the embrace of technology still leaves questions of equity unanswered. Graham (1995) details the expansion of the telematics movement as a means of increasing collaborative policy-making, yet stresses the possibility that councils with this technology will actually be enhancing their competitive status. He cites the case of the World Teleport Association, 'a loose collection of consultants, teleport operators, property developers, and telecommunications specialists' which, in its lobbying efforts, 'carefully constructs the pervasive argument that cities without teleports face inevitable exclusion and marginalization from future dynamics of global economic change' (1995: 507). Similarly, while councils have been welcoming high technology employment as a means of belonging to the 'sunrise' economy, Massey et al. (1992: 189) argue that science park policies are inherently elitist and polarize local labour markets, legitimating a division of labour between 'those who think and those who do, between conception and execution'. This is the downside of vibrant 'edge cities': the spatial splitting of local labour markets covered by a high-tech gloss. And, of course, such new sectors have to be housed, leading to booms in speculative office building and further pressure on communities in globalizing cities (Fainstein, 1994; and see McNeill, 1999 on how this has occurred in Barcelona). As Peter Hall notes, the booming central city areas – downtown San Francisco, Tribeca and SoHo in New York, Fitzrovia and Soho in London – still sit close to areas of poverty unlikely to enjoy the undoubted benefits of the informational city, a juxtaposition of 'infocities and informationless ghettos' (Hall, 1996: ch. 13).

The Just City

The impact of neo-liberalism on urban debate has been associated with the view that 'cities and metropolitan life are an economic liability – pits into which public subsidy and social support must go to prop up ailing and anachronistic urban areas' (Amin and Graham, 1997: 414). However, this view

> seriously underestimates the economic costs of unemployment, crime, fear of crime, depressed demand and a declining urban fabric. These include the cost of containing such problems, lost potential revenue, high costs of urban circulation and distribution and, above all, lost skills, know-how and creativity. Ultimately, the state of the entire urban collectivity feeds back into the circuit of economic activity. But our case may be strengthened by the argument that a sense of place and belonging taps into hidden potential and the sources of social confidence that lie at the core of risk-taking entrepreneurial activity. The just city, therefore, also makes economic sense. (Amin and Graham, 1997: 427)

It has been suggested, then, that social cohesion is an essential element of economic competitiveness. Could the growth of the local and the use of competitive strategies, by harnessing local resources and decentralizing decision-making, be used as a way forward in developing non-boosterist strategies? Several suggestions have been made on this question: that inter-locality collaboration in the form of networks and city 'clubs' can help to mitigate zero-sum competition between places; that there is a possibility of counteracting global capital mobility through creating an 'institutional thickness' or networked local

economy (Amin and Thrift, 1995); that a 'third sector' could be encouraged which helps solve the crisis of the welfare state (Catterall et al., 1996); that there are potential gains to be made in the provision of public goods through a recreated urban realm, which is often tied in with arguments for planning gain and developer contributions.

Accepting the need for and possibility of such a new settlement still leaves open the question of whether the lessons of local economic activity can be incorporated in a more progressive, less cut-throat style of behaviour on the part of urban governments. The frustrations of urban areas having to compete with each other for scarce resources have already led many councils to develop strategies of collaboration. The possibility of creating synergies (making the whole greater than the sum of the parts) has been given particular emphasis in recent European Union attempts to promote integration across national state boundaries. INTERREG, which encourages cross-border cooperation, and RECITE, which funds pilot projects in areas with particular sectoral or social issues in common, have demonstrated the willingness of local authorities to cooperate in areas of mutual interest. Similarly, the Eurocities movement, acting as a forum for Europe's second cities, has encouraged collaboration on two fronts. First, on a very practical level, it has aided the development of joint strategies on various issues such as sharing best practice or developing solutions to shared urban problems. Secondly, it and similar bodies have strengthened the lobbying power of local states in Brussels, and this increases the possibilities of making gains from redistribution through policy innovation (McAleavey and Mitchell, 1994).

However, these initiatives can be seen as being merely another way of *enhancing* a competitive edge. Through collaboration, select councils are able to enhance their positions relative to others through pooling resources in technology development. The Four Motors network involving Catalunya, Rhône-Alpes, Lombardy and Baden-Wüttemburg, which collaborates on innovation, has been criticized as being a self-selecting group of some of the strongest European regional economies. As we noted above, Graham (1995) details the expansion of the telematics movement as a means of increasing collaborative policy-making, yet stresses the possibility that councils with this technology will actually be enhancing their (advantaged) competitive status, where rich councils with capital to invest are able to get richer, leaving the poorer local states behind.

It is possible, however, to foresee a time when collaboration can be extended into a more formal political arrangement, with the EU's Committee of the Regions providing a possible clue as to whether sub-national cooperation could have a future, albeit with limited goals. The former Mayor of Barcelona, Pasqual Maragall, used his position as president of this body to argue strongly for the need for an 'urban solidarity' between cities, recognizing their specific problems and shared interests. Borja and Castells (1997) set out in some detail the potential contribution of the United Nations in moving cities toward a common understanding, promoting issues such as sustainable development and multi-culturalism. The UN-sponsored Habitat II conference, held in Istanbul in 1996, was instrumental in formulating a common set of policy proposals for future urban development (Rogers, 1996).

There are practical problems with pursuing such strategies, which run through the entire history of global uneven development and resource inequality. City governments are torn between the perceived 'need to compete' and a variably expressed concern about social inequality. However, the weakest aspect of such visions of the 'just' city – however appealingly presented – is an unwillingness to engage with issues of power. As McNeill (1999) shows in relation to urban change in Barcelona, several of the public spaces which helped establish the city's reputation as a possible model of urban development were gained through non-violent – but illegal – community protest. The shifting political climate in the city, now focused very clearly on a bid to become an international service centre, has meant that for all the talk of citizenship and actual steps towards decentralization of service provision, the onus has moved back towards development, to the victory of exchange value over use value. Despite well-meaning rhetoric about 'civic empowerment' (Amin and Graham, 1997: 425), it is quite clear that difficult questions about the role of developers and business in the local state have to be addressed before such empowerment is achieved.

Urban Renaissance

Urban renaissance was an important motif of the 1980s. Old industrial cities shed their grim facades as they were endowed with bright new service-oriented structures both in their grimey hearts and their green peripheries. Yet these new structures – shopping malls, waterfront housing, warehouse conversions, art galleries and restaurants – have been portrayed as the logical extension of neoliberal social engineering. There are several strands to this critique. First, renaissance is often linked to the privatization of actual or potential public space, such as with waterfront developments or sports and entertainment complexes (see Pred, 1995 on Stockholm). Secondly, they contribute to a commodification of local cultures, a 'McDonaldization' or Disneyfication (Sorkin, 1992) of cities. This may include a denial, trivialization or depoliticization of labour history, or a general folklorization of working class cultures (Boyle and Hughes, 1991) as local history becomes commodified heritage (Philo and Kearns, 1993). Thirdly, the huge public investments in art galleries or opera houses often arouses controversy given their image of elite, establishment culture, drawing resources away from community-based arts programmes.

Furthermore, cultural geographers have suggested that a 'semiotics of growth' prevails, where shiny new developments help to create the impression that local states are effecting radical changes in the local economy. Growth coalitions have the ability to influence local opinion through advertising or the provision of large, attention-grabbing events or spectaculars (Fretter, 1993; Gold and Ward, 1994). As Fainstein (1994: 2) puts it, '[g]overnments have promoted physical change with the expectations that better-looking cities are also better cities . . . and that property development equals economic development'. Crilley (1993) argues that developments such as Canary Wharf and Battery Park City use 'architecture as advertising', employing urban design strategies to increase their acceptability to the public. While the 'creative city' has been welcomed by centre–left commentators as a base for 'new, reflexive forms of consumption and cultural production, and sites for intense webs of information and communication flows oriented around their night-time economies' (Amin and Graham, 1997: 415; Landry and Bianchini, 1995), doubts remain over how far this escapes the commodification of city life detected by less optimistic commentators (Sorkin, 1992).

The logical extension of all this is that certain social groups – the poor, the elderly, unintegrated ethnic minorities – are being excluded from renaissance. Loftman and Nevin (1996) note that major flagship developments such as Birmingham's ICC have little effect on providing employment for marginalized groups, the benefits rather being spread to the white collar commuter belt. Christopherson (1994) argues that the 'consumer citizenship' of the new leisure economies, with their managed shopping environments and defensive design strategies, involves a privatization of public space, which is exclusionary to those not possessing the requisite disposable income. This is closely related to gentrification (Smith, 1996), and the increasingly harsh treatment given to the homeless in US cities (Davis, 1990). While Neill's (1995) account of Detroit's reimaging provides an interesting twist on racial aspects of such a renaissance, there is a depressing similarity to newspaper tales of the 'dual city' and the rise of an 'urban underclass' (these terms in themselves often seen as being disempowering).

CONCLUSIONS

In this chapter we have tried to pull out the political implications of the 'new urban economies'. Our main concern has been to pick out some of the discourses by which these new economies have been presented, and to stress their stratified impact on labour markets, social groups and local identities. Whether these new economies are the only way forward is open to debate, but there is a need to establish their political dimensions, and to dispel the notion that they are some kind of panacea to structural economic problems. The appeals to civic identity, the need to compete and innovate, the benefits of networking or local proactivity made through these discourses serve to obfuscate the socially and spatially uneven effects of these new economic strategies, and head-off the need to make difficult political choices.

NOTES

1. For a further interpretation of NUP see Chapter 26.

BIBLIOGRAPHY

Amin, Ash (1994) 'Post-fordism: models, fantasies and phantoms of transition', in Ash Amin (ed.), *Post-Fordism: A Reader*. Oxford, Blackwell. pp. 1–39.

Amin, Ash and Graham, Stephen (1997) 'The ordinary city', *Transactions of the Institute of British Geographers*, 22(4): 412–29.

Amin, Ash and Thrift, Nigel (1995) 'Globalisation, institutional "thickness" and the local economy', in Patsy Healey et al. (eds), *Managing Cities: the New Urban Context*. Chichester: John Wiley, pp. 91–108.

Barnes, T.J. (1995) 'Political economy I: "the culture, stupid"', *Progress in Human Geography*, 19(3): 423–31.

Barnes, T.J. (1996) *Logics of Dislocation: Models, Metaphors, and Meanings of Economic Space*. New York: Guilford.

Bassett, K. (1996) 'Partnerships, business elites and urban politics: new forms of governance in an English city?', *Urban Studies*, 33(3): 539–5.

Beauregard, Robert A. (1993) *Voices of Decline: the Postwar Fate of US Cities*. Cambridge, MA: Blackwell.

Berman, Marshall (1983) *All That Is Solid Melts Into Air: the Experience of Modernity*. London: Verso.

Bianchini, Franco and Parkinson, Michael (1993) *Cultural Policy and Urban Regeneration: the West European Experience*. Manchester: Manchester University Press.

Bianchini, Franco and Schwengel, Hermann (1991) 'Re-imaging the city', in J. Corner and S. Harvey (eds), *Enterprise and Heritage: Crosscurrents of National Culture*. London: Routledge. pp. 212–34.

Borja, Jordi and Castells, Manuel (1997) *Local and Global: the Management of Cities in the Information Age*. London: Earthscan.

Boyle, Mark and Hughes, George (1991) 'The politics of the representation of "the real": discourses from the Left on Glasgow's role as European City of Culture, 1990', *Area*, 23(3): 217–28.

Buxton, James (1994) 'Cultural tug-of-war divides Scots', *Financial Times*, 22 January.

Castells, Manuel and Hall, Peter (1994) *Technopoles of the World: the Making of Twenty-First-Century Industrial Complexes*. London: Routledge.

Catterall, Bob et al. (1996) 'The third sector and the stakeholder', Conference extract (with Alain Lipietz, Will Hutton and Herbie Girardet). *City*, 5–6: 86–97.

Christopherson, Susan (1994) 'The fortress city: privatized spaces, consumer citizenship', in Ash Amin (ed.), *Post-Fordism: A Reader*. Oxford: Blackwell. pp. 409–27.

Cooke, P.N. (ed.) (1989) *Localities: the Changing Face of Urban Britain*. London: Unwin Hyman.

Cox, K. (1993) 'The local and the global in the new urban politics: a critical view', *Environment and Planning D: Society and Space*, 11: 433–48.

Cox, K. and Mair, K. (1988) 'Locality and community in the politics of local economic development', *Annals of the Association of American Geographers*, 78(2): 137–46.

Crilley, Darrel (1993) 'Architecture as advertising: constructing the image of redevelopment', in Gerry Kearns and Chris Philo (eds), *Selling Places: the City as Cultural Capital, Past and Present*. Oxford: Pergamon. pp. 231–52.

Davis, Mike (1990) *City of Quartz: Excavating the Future in Los Angeles*. London: Verso.

Dicken, Peter (1992), *Global Shift: the Internationalization of Economic Activity*, 2nd edn. London: Paul Chapman.

Dunford, Mick and Kafkalas, Grigoris (1992) 'The global–local interplay, corporate geographies and spatial development strategies in Europe', in Mick Dunford and Grigoris Kafkalas (eds), *Cities and Regions in the New Europe*. London: Belhaven Press. pp. 3–38.

Elkin, S.J. (1987) *City and Regime in the American Republic*. Chicago: University of Chicago Press.

Fainstein, Susan S. (1994) *The City Builders: Property, Politics and Planning in London and New York*. Oxford: Blackwell.

Fretter, David Andrew (1993) 'Place-marketing: a local authority perspective', in Gerry Kearns and Chris

Philo (eds), *Selling Places: the City as Cultural Capital, Past and Present*. Oxford: Pergamon. pp. 163–74.

Gold, John R. and Ward, Stephen V. (eds) (1994) *Place Promotion: the Use of Publicity and Marketing to Sell Towns and Regions*. Chichester: John Wiley.

Graham, Stephen (1995) 'From urban competition to urban collaboration? The development of interurban telematics networks', *Environment and Planning C: Government and Policy*, 13: 503–24.

Graham, Stephen and Marvin, Simon (1996) *Telecommunications and the City*. London: Routledge.

Hall, Peter (1996) *Cities of Tomorrow: an Intellectual History of Urban Planning and Design in the Twentieth Century*, 2nd edn. Oxford: Blackwell.

Hall, Tim and Hubbard, Phil (1996) 'The entrepreneurial city: new urban politics, new urban geographies?', *Progress in Human Geography*, 20(2): 153–74.

Harding, A. (1994) 'Urban regimes and growth machines: towards a cross-national research agenda', *Urban Affairs Quarterly*, 29(3): 356–82.

Harvey, David (1989) 'From managerialism to entrepreneuralism: the transformation in urban governance in late capitalism', *Geografiska Annaler*, 71B(1): 3–17.

Häussermann, Hartmut and Strom, Elizabeth (1994) 'Berlin: the once and future capital', *International Journal of Urban and Regional Research*, 18(2): 335–46.

Hay, C. (1994) 'Moving and shaking to the rhythm of local economic development: towards a Schumpeterian workfare state?', *Lancaster Working Papers in Political Economy*, no. 49.

Jauhiainen, Jussi S. (1995) 'Waterfront redevelopment and urban policy: the case of Barcelona, Cardiff and Genoa', *European Planning Studies*, 3(1): 3–23.

Jessop, Bob (1998) 'The narrative of enterprise and the enterprise of narrative: place marketing and the entrepreneurial city', in T. Hall and P. Hubbard (eds), *The Entrepreneurial City: Geographies of Politics, Regime and Representation*. Chichester: John Wiley. pp. 77–99.

Keating, M. (1993) 'The politics of local economic development: political change and local economic development policies in the US, Britain and France', *Urban Affairs Quarterly*, 28(3): 373–96.

Landry, Charles and Bianchini, Franco (1995) *The Creative City*. London: Demos/Comedia.

Ley, D. and Olds, K. (1988) 'Landscape as spectacle: world's fairs and the culture of heroic consumption', *Environment and Planning D: Society and Space*, 6: 191–212.

Leyshon, A. (1997) 'True stories? Global dreams, global nightmares, and writing globalization', in R. Lee and J. Wills (eds), *Geographies of Economies*. London: Arnold. pp. 133–46.

Loftman, Patrick and Nevin, Brendan (1996) 'Going for growth: prestige projects in three British cities', *Urban Studies*, 33(6): 991–1019.

Logan, John R. and Molotch, Harvey L. (1987) *Urban Fortunes: the Political Economy of Place*. Berkeley, CA: University of California Press.

Lovering, John (1995) 'Creating discourses rather than jobs: the crisis in the cities and the transition fantasies of intellectuals and policy makers', in Patsy Healey et al. (eds), *Managing Cities: the New Urban Context*. Chichester: John Wiley. pp. 109–26.

Lovering, John and Boddy, Martin (1988) 'The geography of military industry in Britain', *Area*, 20(1): 41–51.

McAleavey, P. and Mitchell, J. (1994) 'Industrial regions and lobbying in the structural funds reform process', *Journal of Common Market Studies*, 32(2): 237–48.

McNeill, D. (1999) *Urban Change and the European Left: Tales from the New Barcelona*. London: Routledge.

McNeill, D. (2000) 'McGuggenisation: National Identity and Globalisation in the Basque Country', *Political Geography*, 19(4).

Marshall, Tim (1996) 'Barcelona – fast forward? City entrepreneurialism in the 1980s and 1990s', *European Planning Studies*, 4(2): 147–65.

Martin, S. and Pearce, G. (1992) 'The internationalization of local authority economic development strategies: Birmingham in the 1980s', *Regional Studies*, 26: 499–509.

Massey, Doreen (1995) *Spatial Divisions of Labour: Social Structures and the Geography of Production*, 2nd edn. London: Macmillan.

Massey, Doreen et al. (1992) *High-Tech Fantasies: Science Parks in Society, Science and Space*. London: Routledge.

Mollenkopf, J.H. (1983) *The Contested City*. Princeton, NJ: Princeton University Press.

Morgan, Kevin (1992) 'Innovating by networking: new models of corporate and regional development', in Mick Dunford and Grigoris Kafkalas (eds), *Cities and Regions in the New Europe*. London: Belhaven Press. pp. 150–69.

Neill, William J.V. (1995) 'Lipstick on the gorilla: the failure of image-led planning in Coleman Young's Detroit', *International Journal of Urban and Regional Research*, 19(4): 639–53.

Peck, Jamie (1995) 'Moving and shaking: business elites, state localism and urban privatism', *Progress in Human Geography*, 19(1): 16–46.

Peck, Jamie and Tickell, Adam (1994a) 'Jungle law breaks out: neo-liberalism and the global–local disorder', *Area*, 26(4): 317–26.

Peck, Jamie and Tickell, Adam (1994b) 'Searching for a new institutional fix: the *after*-fordist crisis and the global–local disorder', in Ash Amin (ed.), *Post-Fordism: A Reader*. Oxford: Blackwell. pp. 280–315.

Peck, Jamie and Tickell, Adam (1995) 'The social regulation of uneven development: "regulatory deficit", England's South East, and the collapse of Thatcherism', *Environment and Planning A*, 27: 15–40.

Philo, Chris and Kearns, Gerry (1993) 'Culture, history, capital: a critical introduction to the selling of places', in Gerry Kearns and Chris Philo (eds),

Selling Places: the City as Cultural Capital, Past and Present. Oxford: Pergamon. pp. 1–32.

Pred, Allan (1995) *Recognising European Modernities: a Montage of the Present*. London: Routledge.

Rogers, Alisdair (1996) 'Prospects for an urbanizing world', *City*, 5–6: 144–6.

Rustin, Michael (1989) 'The politics of post-fordism: or, the trouble with "new times"', *New Left Review*, 175: 54–77.

Sabel, Charles F. (1994) 'Flexible specialisation and the re-emergence of regional economies', in Ash Amin (ed.), *Post-Fordism: A Reader*. Oxford: Blackwell. pp. 101–56.

Sadler, David (1993) 'Place-marketing, competitive places, and the construction of hegemony in Britain in the 1980s', in Gerry Kearns and Chris Philo (eds), *The City as Cultural Capital, Past and Present*. Oxford: Pergamon. pp. 175–92.

Sassen, Saskia (1991) *The Global City: New York, London, Tokyo*. Princeton, NJ: Princeton University Press.

Scott, A.J. (1988) *New Industrial Spaces: Flexible Production Organization and Regional Development in North America and Western Europe*. London: Pion.

Smith, N. (1996) *The New Urban Frontier: Gentrification and the Revanchist City*. London: Routledge.

Sorkin, Michael (ed.) (1992) *Variations on a Theme Park: the New American City and the End of Public Space*. New York: Hill and Wang.

Stoker, G. and Mossberger, K. (1994) 'Urban regime theory in comparative perspective', *Environment and Planning C: Government and Policy*, 12: 195–212.

Stone, C.N. (1989) *Regime Politics: Governing Atlanta, 1946–1988*. Lawrence: University Press of Kansas.

Sudjic, Deyan (1992) *The 100 Mile City*. London: Andre Deutsch.

Swyngedouw, Eric (1992) 'The Mammon quest, "Glocalization", interspatial competition and the monetary order: the construction of new scales', in Mick Dunford and Grigoris Kafkalas (eds), *Cities and Regions in the New Europe*. London: Belhaven Press. pp. 39–67.

Taylor, I., Evans, K. and Fraser, P. (1996) *A Tale of Two Cities: Global Change, Local Feeling and Everyday Life in the North of England: A Study in Manchester and Sheffield*. London: Routledge.

Thomas, Huw (1994) 'The local press and urban renewal: a South Wales case study', *International Journal of Urban and Regional Research*, 189(2): 315–33.

Thrift, Nigel (1989) 'New Times and spaces? The perils of transition models', *Environment and Planning D: Society and Space*, 7: 127–9.

Thrift, Nigel (1994) 'On the social and cultural determinants of international financial centres: the Case of the City of London', in Stuart Corbridge et al. (eds), *Money, Power and Space*. Oxford: Blackwell. pp. 327–55.

Ward, K. (1996) 'Rereading urban regime theory: a sympathetic critique', *Geoforum*, 27: 427–38.

Zukin, Sharon (1995) *The Cultures of Cities*. Cambridge, MA: Blackwell.

20

The Growth of Urban Informal Economies

COLIN C. WILLIAMS AND JAN WINDEBANK

In recent years, the emerging orthodoxy in urban studies is that our cities are witnessing the growth of informal employment as it becomes an increasingly important survival strategy for marginalized populations (for example, the unemployed, women, ethnic minorities and immigrants). The aim of this chapter is to evaluate critically this conceptualization concerning both the development and character of urban informal employment. Synthesizing the findings of the vast array of studies of such work, first, whether or not urban informal employment is growing or declining will be analysed and, secondly, the character of this work will be evaluated. On the one hand, this will reveal that rather than a universal process of either formalization or informalization, the trajectory of urban economic development is much more heterogeneous than has been so far considered. On the other hand, it will show that urban informal employment is not solely a survival strategy for the marginalized, but rather, that there is a heterogeneous informal labour market with a hierarchy of its own which reproduces the socio-spatial divisions prevalent in the urban formal labour market. To explain the heterogeneous magnitude, direction and character of different informal labour markets, the third section argues that this is due to the way in which local economic, social, institutional and environmental conditions combine in multifarious 'cocktails' in different places to produce specific local outcomes. In the concluding section, the policy opinions and implications are assessed. Evaluating critically the three dominant policy approaches towards urban informal employment, we conclude that the option chosen is inextricably related to broader prescriptions for work and welfare.

Before commencing, however, it is first necessary to define what we mean here by urban informal employment. In this chapter, we are discussing the paid production and sale of goods and services that are unregistered by, or hidden from, the state for tax, social security or labour law purposes, but which are legal in all other respects. As such, unpaid informal work is outside the scope of this chapter, as is criminal activity more widely defined. Instead, we are concerned here with activity which is illicit only because it involves social security fraud, tax evasion and/or the avoidance of labour legislation.

THE DEVELOPMENT PATH OF URBAN AREAS IN THE ADVANCED ECONOMIES: FORMALIZATION, INFORMALIZATION OR HETEROGENEOUS DEVELOPMENT?

The fact that there is a substantial amount of paid work in urban areas of the advanced economies which does not appear in the official statistics is without contention. However, the precise amount of such activity and whether it is increasing or decreasing faster than the measured economic activity is a matter of heated debate. Here, we evaluate critically both the popular belief that there is a formalization of urban economies as well as the theory that they are undergoing a process of informalization.

The Theory of Formalization of Urban Economies

One of the most widely held beliefs about urban economic development is that as urban economies become more 'advanced', there is a natural and inevitable shift of economic activity from the informal to the formal sphere (that is, the theory of formalization). Indeed, this is often the 'measuring rod' which defines Third World cities as 'backward' and First World cities as supposedly 'advanced'. However, there are good reasons for believing that this trajectory of urban economic development is neither natural nor inevitable. In the advanced economies, the natural culmination of formalization – full-employment and a comprehensive welfare state – can no longer be accepted as the end-state of economic development. Between 1965 and 1995 in the European Union (EU), for example, the share of the working age population with a job fell from 65.2 per cent to 60.4 per cent (European Commission, 1996a). When this is coupled with the shift away from permanent full-time jobs towards temporary and part-time work (Nicaise, 1996), full- or even fuller-employment seems to be receding ever further into the distant past. So too is a comprehensive welfare 'safety net'. Even in those EU nations wishing to retain it, universal welfare provision is in demise, not least due to the pressure from other trading blocs to keep social costs low (Conroy, 1996; Ferrera, 1996; Williams and Windebank, 1995a). As the European Commission (1995) reports, expenditure on social protection benefits per capita fell or remained static in six out of 15 member states during the 1990s compared with the late 1980s.

It is not only in the advanced economies, however, that formalization appears to be neither natural nor inevitable. Across the Third World, formal employment is by no means universally growing. Although employment has risen at a rate well in excess of the increase in the labour force in many East and South East Asian nations, there are whole swathes of the Third World not only at a very low base-level so far as formal jobs are concerned but which are either standing still or are undergoing a process of informalization. In low-income nations, for example, regular waged or salaried employment accounts for only 5–10 per cent of total employment, and in many regions such as Latin America, sub-Saharan Africa, North Africa and the Middle East, employment growth has been either static or decreasing relative to the growth in the labour force (International Labour Office, 1996).

Therefore, the theory of formalization, upon which much of our thinking about both the future of work and welfare in our cities is founded, is cast into doubt by the above findings. In consequence, it might be assumed that the opposite is occurring. If people are so widely and increasingly excluded from both employment and formal welfare provision, then they must be turning towards informal employment as a means of survival. Are we witnessing, therefore, the renewed informalization of urban life?

The Theory of Urban Informalization

Based on the assumption that informal employment is a new form of advanced capitalist exploitation (e.g., Castells and Portes, 1989; Sassen, 1991) or a response to over-regulation by the market (de Soto, 1989), an increasingly popular view is that as the advanced economies fall into economic crisis and the neo-liberal project of deregulation takes hold, we are witnessing the absolute and/or relative growth of informal employment in our urban economies. The principal problem, however, is that there is little solid evidence that this process is taking place. The indirect macro-economic methods which have generated data to support such an assertion suffer from fatal flaws in their design which render the results of dubious quality and validity (see, for example, Smith, 1986; Thomas, 1988). For the same country in the same year, for example, it is normal to find major variations ranging from as little as 2–3 per cent to 30–35 per cent of GNP (Barthelemy, 1991). As Thomas (1988) displays, the size of informal employment has been estimated to range from 1.5 to 27 per cent of GDP for the US and 2 to 22 per cent of GDP for the UK. In consequence, to rely on any of these indirect measures as evidence for the informalization of the advanced economies would be to close one's eyes to the inherent problems with the methodologies from which they are derived.

The only other source of potential evidence, the micro-social studies of informal employment, although more direct and accurate in their portrayals of such work, tend to be snapshots at one point in time, usually in specific localities. Even if they are an accurate snapshot of informal employment in a particular locality, therefore, they do not show whether there is an informalization of urban economic life taking place. Indeed, the only two longitudinal micro-economic studies of informal employment both refute the informalization thesis. Mogensen et al. (1995) in Denmark find that the proportion of people engaged in informal employment has remained level, at 13–15 per cent, throughout the period from 1980 to 1995, whilst Fortin et al. (1996) in Quebec reveal that between 1985 and 1993, informal employment as a percentage of GNP has stayed constant at 0.65 per cent of GDP. Micro-economic studies, therefore, provide little evidence of informalization.

For some, it might be assumed that the demise of formal work and welfare indicated above is sufficient proof of informalization. This, however, assumes that formal and informal employment are substitutes for each other (e.g. Castells and Portes, 1989; Gutmann, 1978) and that the rise of one leads to the fall of the other. The problem is that an increasing number of micro-social studies suggest that the relationship is not one of substitution but rather one of complementarity (e.g. Cappechi, 1989; Cornuel and Duriez, 1985; Leonard, 1994; Morris, 1995; Pahl, 1984). Given this, it would be misleading to take the de-formalization of urban economic life as a surrogate measure of the growth of informalization. The picture is far more complex. Neither can the supposed recent growth of informal employment simply be viewed as a new form of capitalist exploitation or a contemporary response to over-regulation by the market. Informal employment, after all, is not a recent phenomenon, having been in existence as long as there have been rules and regulations with regard to employment (Henry, 1978; Smithies, 1984) and its recent growth is by no means obvious.

Transcending Informalization/ Formalization: the Heterogeneous Development Paths Approach

Here, therefore, precisely because informal employment is not everywhere a substitute for formal jobs, we do not assert that the demise of formal work and welfare is necessarily leading to a relative informalization of the advanced economies. Instead, we introduce the idea that both history and geography matters. Examining the findings of the vast array of direct studies, it is evident that there are different processes in different places at varying times. As such, it is more accurate to describe the development trajectory of the advanced economies as one of heterogeneous development, with different places undergoing either informalization or formalization at varying times and at different rates. The rationale for such a theorization of the trajectory of urban economic development will be more fully explained below. Here, we simply state this as an assertion arising out of our reading of these studies of informal employment. Before more fully theorizing this temporally and spatially refined theorization of the level of urban informal employment in the advanced economies, it is first necessary to explore the nature of such work.

CHARACTERIZING THE NATURE OF URBAN INFORMAL EMPLOYMENT: PERIPHERAL LABOUR OR HETEROGENEOUS INFORMAL LABOUR MARKETS?

Based on the assumption that it is marginalized groups who engage in urban informal employment as a means of survival, the debate concerning how to explain the existence of such work has thus revolved around whether it is a leftover of classical capitalism (that is, the theory of formalization) or a new form of advanced capitalism (that is, the theory of informalization). This section, however, reveals that whichever position is adopted (Castells and Portes, 1989; Sassen 1989), both views are founded upon an erroneous premise.

Urban Informal Employment: Peripheral Labour for Marginalized Groups?

The 'marginality thesis' which views informal employment as a form of peripheral labour for marginal groups (for example, women, migrants, the unemployed, poor and children) has a long historical antecedent and can be traced back to the contemporary origins of the study of such activity in the work of Hart (1973) in Ghana. Since then, it has been taken up widely by writers in both the developing and advanced economies (for example, Button, 1984; Lagos, 1995; Maldonado, 1995). In this view, such work is seen as an exploitative peripheral form of employment which is at the bottom of a hierarchy of flexible types of formal employment and is undertaken mostly by marginalized population groups as a survival strategy. However, such a view of informal employment covers but one type of informal employment and one section of the informal workforce.

Informal employment, that is, is not a homogeneous category but is composed of a heterogeneous spectrum of paid economic activities (for example, Fortin et al., 1996; Jensen et al., 1995; Leonard, 1994; Mingione, 1991; Pahl, 1984; Renooy, 1990). It ranges from 'organized' informal work undertaken as an employee for a business which conducts some or all of its activity informally, to more 'individual' forms of such work. These cover not only self-employed activities which may involve a large proportion, if not all, of their earnings, but also casual one-off jobs undertaken on a cash-in-hand basis, such as for a neighbour, friend or relative. Not all of informal employment, moreover, is low-paid, exploitative and undertaken by marginal groups. Although there are forms of organized informal employment, such as in labour-intensive small firms with low levels of capitalization dependent on larger companies where marginal populations engage in low-paid exploitative activity (for example, Lim, 1995; Sassen 1991), there are also forms of organized informal employment in independent, modern, more capitalized, high technology firms producing higher priced goods and services and whose informal employees are well paid, use higher skills and have more autonomy and control over their work, with relations between employers and employees based more upon cooperation than domination (e.g.

Benton, 1990; Cappechi, 1989; Warren, 1994). There are also forms of individual informal employment which are not only deeply socially embedded in urban life, but have more in common with self-employment and/or unpaid community exchange than they do with exploitative forms of informal or formal employment. So, despite the emphasis in much of the literature on informal employment as an exploitative phenomenon, the reality is that this is just one, albeit important, part of the spectrum of activities which constitute this form of work. If informal employment is not solely an exploitative form of peripheral labour for marginalized populations, then how should it be conceptualized?

Retheorizing Urban Informal Employment as a Heterogeneous Informal Labour Market

Based on a detailed review of the numerous one-off locality studies of informal employment in the advanced economies, it is here suggested that existing alongside the urban formal labour market and often segmented along the same lines, is a segmented urban informal labour market. This informal labour market, to adopt a simplistic dual labour market model, is composed of 'core' informal employment which is relatively well paid, autonomous and non-routine and where the worker often benefits just as much from the work

as the employer, and 'peripheral' informal employment which is poorly paid, exploitative and routine and where the employee does not benefit as much as the employer. Just as in the urban formal labour market, moreover, there are also those excluded from even the most exploitative peripheral informal employment. Who, however, receives these high income informal jobs? And who is involved in the lower paid more exploitative informal employment?

To start to answer such questions, we here take a case study of the urban informal labour market in Quebec city (Lemieux et al., 1994). In 1986, 2,134 adults aged 18 years or over were surveyed in the Census Metropolitan Area of Quebec City, with a 63.8 per cent response rate. As Table 20.1 reveals, although 8.5 per cent of the total sample report working in informal employment, participation in this form of work is not evenly distributed across all social groups. Men, students, the unemployed and lower income groups are all more likely to conduct such work. Superficially, therefore, these results appear to reinforce the marginality thesis. However, closer analysis reveals that such a view does not fully describe this urban informal labour market.

Although the unemployed, for example, are more likely to engage in informal employment, they do not constitute the vast bulk of this informal workforce. Indeed, just 12.8 per cent of all informal workers are unemployed. The formally employed, meanwhile, constitute 35.2

Table 20.1   *Participation in informal employment in Quebec city, 1986*

Characteristics	% of sample	% doing inf. wk	% of all inf. wkrs	Hours	Total	Average/ informal worker	Av.$/h	Hours	Value
					Earnings			% of total inf. wk by	
All	100.0	8.5	100.0	357	2006	171	5.61	100.0	100.0
Gender									
Men	48.7	9.9	56.7	331	2294	227	6.93	52.6	64.9
Women	51.3	7.1	43.3	391	1628	116	4.16	47.4	35.1
Employment status									
Student	11.4	28.2	38.0	332	1976	557	5.45	35.6	37.4
Retired	5.1	1.9	1.1	120	490	9	4.08	0.4	0.3
Housekeeper	17.6	6.2	12.8	581	2251	140	3.87	21.1	14.4
Unemployed	4.0	27.4	12.8	369	1904	522	5.16	13.4	12.2
Employed	61.9	4.8	35.2	297	2034	98	6.85	29.5	35.7
Official income (C$)									
0–10,000	51.4	12.9	74.7	400	1984	256	4.96	84.3	73.9
10,000–20,000	17.2	7.0	13.4	286	2302	160	8.05	10.8	15.4
20,000–30,000	15.5	3.9	7.0	190	1943	75	10.23	3.7	6.8
30,000–40,000	9.9	2.0	2.2	58	1431	29	24.67	0.4	1.5
40,000+	11.0	2.3	2.7	104	1790	41	17.21	0.8	2.4

Source; Derived from Lemieux et al. (1994: Table 1)

per cent of all informal workers and earn more per hour for working informally than their more marginalized counterparts. So too do those who earn a higher formal income. Whilst those earning C$0–10,000 p.a. receive an average of C$4.96 per hour from their informal employment, the hourly wage rate rapidly rises, so that those earning C$30,000–40,000 p.a. from formal employment command informal wage rates of C$24.67 per hour. Informal employment, therefore, is clearly segmented along the lines of formal income and employment status. It is a similar story when gender is examined. Not only do a slightly higher proportion of men than women participate in informal employment (9.9 per cent compared with 7.1 per cent), but men receive higher hourly informal wage rates than women (C$6.93 compared with C$4.16). The informal labour market, therefore, is clearly segmented along similar lines to the formal labour market.

Contrary to popular prejudice, furthermore, the informal labour market does not necessarily have lower wage rates than the formal labour market. Lemieux et al. (1994) find that although the average gross wage per hour is 13 per cent higher in formal than informal employment (C$7.98 compared with C$6.99), the average net wage per hour in formal jobs, which takes account of marginal tax rates and tax-back rates associated with social transfers, is lower than the average informal wage. Informal employment, therefore, is not merely an exploitative low-paid sector of the economy. It has a higher average net wage rate than formal employment. Nevertheless, the standard deviation of the informal wage rate is larger than the standard deviation of the formal wage rate, suggesting that wage differentials are greater in this sector, adding yet further weight to the idea of a heterogeneous informal labour market. Moreover, whilst hours and wages are positively correlated in formal employment, they are negatively correlated in informal employment (Lemieux et al., 1994), suggesting that many more in informal than formal employment work long hours for little pay and short hours for high pay.

Are these findings from the Quebec City informal labour market replicated elsewhere? Do the unemployed, for instance, always constitute a smaller proportion of informal workers than the employed? Do they everywhere earn less than the employed? Does gender always relate in the same way to the informal labour market? And do ethnic minority and immigrant groups, not considered in the Quebec study, relate to the informal labour market in similar ways to these other marginalized groups?

Starting with the relationship between employment status and informal employment, the over-whelming finding of the vast majority of locality studies is that informal employment is primarily a means of accumulating advantage for those already in employment and that the number of regular 'working claimants' accounts for a very small proportion of all informal employees. In Spain, for example, the Ministry of Economy esti-mates that 29 per cent of those in employment also have an informal job (in Hadjimichalis and Vaiou, 1989), whilst Lobo (1990a) finds that just 12 per cent of those claiming benefit perform such work. Benton (1990) thus reveals that 65.7 per cent of all informal employees have a formal job, whilst just 5.2 per cent are working informally and receiving social security benefit at the same time. The remaining 29 per cent of informal workers, we must assume, are those unemployed not enti-tled to benefit. In Greece similarly, the Ministry of Planning recognizes the fact that it is more likely to be the employed than the unemployed who have informal jobs when it states that 40 per cent of those employed in the private sector also have non-declared jobs, as do 20 per cent of those in public sector jobs (in Hadjimichalis and Vaiou, 1989). This finding that informal employment chiefly benefits those already in employment is also identified in the Netherlands (Koopmans, 1989; Van Eck and Kazeier, 1985; Van Geuns et al., 1987, France (Barthe, 1988; Cornuel and Duriez, 1985; Foudi et al., 1982; Tievant, 1982), Germany (Glatzer and Berger, 1988; Hellberger and Schwarze, 1987), Britain (Economist Intelli-gence Unit, 1982; Howe, 1990; Morris, 1994; Pahl, 1984; Warde, 1990), Italy (Cappechi, 1989; Mingione, 1991; Mingione and Morlicchio, 1993; Warren, 1994) and North America (Fortin et al., 1996; Jensen et al., 1995; Lemieux et al., 1994; Lozano, 1989).

So far as the character of their work is concerned, the vast majority of studies find that the employed tend to engage in more autonomous, non-routine and rewarding informal jobs than the unemployed, who conduct more routine, lower paid, exploitative and monotonous informal employment (Fortin et al., 1996; Howe, 1990; Lemieux et al., 1994; Lobo, 1990a, 1990b; Pahl, 1987; MacDonald, 1994; Williams and Winde-bank, 1995a; Windebank and Williams, 1997). A plumber employed by one of the large utilities, for example, may put in central heating for cash-in-hand during the weekend for a customer she/he has met during the course of his/her formal employment, whilst an unemployed person may only have access to low-paid homeworking activity such as assembling Christmas crackers which she/he has learned about through an adver-tisement in a shop window or newspaper or through a friend who already does such activity. The result, therefore, is a segmented informal labour market in which many of those who already have a formal occupation find relatively well-paid informal employment, often conducted on a self-employed basis, whilst the unemployed generally engage in relatively low-paid organized informal employment which tends to be more exploitative in nature. As Van Eck and Kazeier (1985) identify, informal employment found through employers or colleagues is by far the best paid, namely two or three times higher than informal employment found through other channels. Given that the unemployed have neither employers nor colleagues, the outcome is limited access to these better paid informal employment opportunities. Instead, their networks confine them to seeking work via family or acquaintances. The result is lower wage levels since the average wage for work conducted for one's family is Dfl10, for acquaintances it is Dfl14, for colleagues Dfl20 and for employers Dfl33 (Van Eck and Kazeier, 1985). Indeed, it is the variation in informal wage rates which perhaps highlights better than any other variable the segmented nature of the informal labour market by employment status.

It is not only the Quebec findings on the relationship between employment status and informal employment which are replicated in many other urban areas. When the gendered configuration of the informal labour market is analysed, the overarching finding of the vast majority of studies is that the extent of women's participation is less than that of men and that men constitute the majority of the informal labour force. This is identitied to be the case in the Netherlands (Van Eck and Kazeier, 1985; Renooy, 1990), the UK (Macdonald, 1994; Pahl, 1984), Italy (Mingione, 1991; Vinay, 1987), Denmark (Mogensen, 1985) and the United States (McInnis-Dittrich, 1995). It is important to state, nevertheless, that the relative gap in the participation rates of men and women is not great and is perhaps not so surprising as it first appears when it is realized that men more frequently engage in paid employment than women in nearly every context.

Examining the types of informal employment men and women perform, it is striking that women undertake a very different set of tasks than men. Just as there is a clear gender segmentation by sector in the formal labour market, with women heavily concentrated in service sector jobs (e.g. Townsend, 1997), the same is true of the informal labour market. As Hellberger and Schwarze (1986) identify in Germany, whilst 12.3 per cent of informal workers are in the primary sector, 35.8 per cent in manufacturing and 51.9 per cent in services, these figures are 6.2 per cent, 11.8 percent and 82.0 per cent respectively for women (and 15.7 per cent, 49.3 per cent and 35.0 per cent for men). It is similar elsewhere. Women tend to engage in service activities such as commercial cleaning, domestic help, child-care and cooking when they undertake informal

employment. Men, on the other hand, tend to undertake what are conventionally seen as 'masculine tasks' such as building and repair work (Fortin et al., 1996; Jensen et al., 1995; Leonard, 1994; Mingione, 1991; Pahl, 1984). There is also tentative evidence that in some particular contexts men's participation in informal employment is more infrequent but full-time than women's, which although continuous, tends to be part-time (Leonard, 1994; McInnis-Dittrich, 1995). So, not only is informal employment sectorally divided in similar ways to formal employment, but so does the part-time/full-time dichotomy which is often prevalent in formal employment appear to be replicated in the informal labour market. Women, moreover, generally earn lower informal wages than men (Fortin et al., 1996; Hellberger and Schwarze, 1986; Lemieux et al., 1994; McInnis-Dittrich, 1995). This is because women have to engage in informal employment which reflects their domestic roles (resulting in sectoral division) or which fits in with their domestic duties (resulting in the tendency for regular but part-time informal employment so that family needs can be met).

Finally, there is the issue of whether ethnic minorities and immigrants relate to the informal labour market in the same way as these other marginalized groups. Do ethnic minority and immigrant groups more frequently participate in informal employment than their white or indigenous counterparts? Is the informal labour market heavily composed of these groups? To answer these questions, two different groups need to be distinguished, each likely to have contrasting experiences with regard to the informal labour market. On the one hand, there are ethnic minority populations, and on the other hand, there are illegal immigrants. Given that different types of legal immigrant ultimately witness the same experiences as one of these two groups, they are here dealt with under these two headings (see Williams and Windebank, 1998).

Most of the research on this topic has been conducted in the USA and focuses upon sectors and/or occupations in urban neighbourhoods with high concentrations of ethnic minority populations, mostly in global cities. These identify that informal employment is closely associated with such groups (Fernandez-Kelly and Garcia, 1989; Lim, 1995; Portes, 1994; Sassen, 1989; Stepick, 1989). Analysing ethnic minorities, the problem is that even if they are more likely to work informally in some parts of the global cities, this does not mean that the vast bulk of the informal labour force in the advanced economies as a whole are ethnic minorities, as reading the US literature outside its context might imply. Indeed, where ethnicity has been considered as a variable alongside others in alternative geographical contexts, such as in rural Pennsylvania (Jensen et al., 1995), no strong correlation has been identified between ethnic minorities and participation in informal employment. The informal labour market, therefore, appears to be configured differently in different areas so far as ethnicity is concerned.

Similarly, it might be assumed that the participation of illegal immigrants in informal employment is very high, possibly the highest of any socio-economic group in the advanced economies. The reason is supposedly simple: excluded from formal employment as well as social security benefits due to their illegal status, such immigrants have little choice but to engage in informal employment as a means of survival. However, in reality, they do have other options. They can use falsified or other people's documents (for example, national insurance or social security numbers) to gain access to either welfare benefits or employment without their employers' collusion. Alternatively, they can work as an employee with the consent of the employer who pays taxes for the worker but the worker remains undetected as an illegal immigrant due to the imperfections in, and lack of coordination between, the various regulatory institutions of the state. Indeed, the Internal Revenue Service (IRS) in the USA has estimated that as many as 88 per cent of illegal immigrants may be paying taxes (cited in Mattera, 1985). It is therefore not the case that all illegal immigrants who work are engaged in informal employment. Although they obviously engage to a greater extent in informal activity than other groups, the picture is much more complex than is sometimes thought. The important point, however, is that even if the vast majority of illegal immigrants do engage in informal employment, this does not mean that all informal workers are illegal immigrants. As Wuddalamy (1991: 91) argues in France: 'Immigrant clandestinity . . . only covers fragments of the underground economy.'

However, when ethnic minorities and immigrants (whether legal or illegal) do engage in informal employment, the overwhelming finding is that they engage in peripheral exploitative forms of organized informal employment rather than in core, more rewarding autonomous types of work (Lim, 1995; Mingione, 1991; Moulier-Boutang, 1991; Portes, 1994; Portes and Stepick, 1993; Sassen, 1989; Stepick, 1989). Therefore, the informal labour market mirrors many of the racial inequalities prevalent in the formal labour market. As Moulier-Boutang (1991: 119) puts it, 'all large world labour markets are hierarchical along ethnic lines: foreign workers and different nationalities in different legal positions occupy the posts at the bottom of the scale'.

Indeed, this is the principal finding so far as all marginalized groups are concerned. Nevertheless, there are localities where important exceptions to this rule can be identified. Other locality studies reveal that the unemployed engage in more informal employment than the employed. This is identified to be the case in Belfast (Howe, 1988; Leonard, 1994) and Belgium (Kesteloot and Meert, 1994; Pestieau, 1984). Pestieau (1984), for example, in the Liège area of Belgium, identifies that suppliers of informal labour were especially likely to be found among manual workers, unemployed or unoccupied persons, low income and uneducated groups and younger people. A similar configuration is also identified in Spain (Lobo, 1990a) and Greece (Hadjimichalis and Vaiou, 1980), albeit only so far as organized informal employment is concerned. Ethnic minorities and immigrants (whether legal or illegal), moreover, are not always found to occupy the most peripheral positions in the informal labour market. Rather, different ethnic minority and immigrant groups are to be found in different positions in the hierarchy of particular local informal labour markets and these positions can change over time as some groups progress up the hierarchy of paid informal activities and even formalize them (e.g. Sassen, 1989; Stepick, 1989). Informal employment, therefore, has a hierarchy of its own which mirrors the formal labour market so far as ethnicity and immigrant populations are concerned.

However, such exceptions should not hide the fact that the dominant way in which the informal labour market is segmented is very similar to the formal labour market. If marginal groups find themselves in exploitative informal employment, it is because they find themselves in this kind of work in formal employment, not because informal employment is comprised of this sort of work alone and these sorts of workers only.

EXPLAINING THE MAGNITUDE AND CHARACTER OF URBAN INFORMAL EMPLOYMENT

To explain why the growth, magnitude and character of urban informal employment varies temporally and spatially, it is here argued that its configuration in any place is the result of the way in which economic, social, institutional and environmental conditions combine in multifarious 'cocktails' in different urban economies to produce specific local outcomes. The consequence, therefore, is that at any time, there is neither universal formalization nor informalization but, rather, different processes in varying urban areas. Neither is there one unique composition of the urban informal labour market which is homogenous across all places. Rather, there are different configurations in different areas. Here, therefore, we outline the various conditions and combination factors which regulate the configuration of informal employment in any urban economy through an analysis of the general ways in which each condition normally influences the nature and level of informal employment.

Economic Regulators

Many explanations of the magnitude and character of urban informal employment heavily rely on economic explanations (e.g., Benton, 1990; Portes and Sassen-Koob, 1987). Here, we outline the key regulatory variables considered to be influential in configuring such work. On their own, however, such explanations are insufficient. They fail to address the social, institutional or environmental context in which economic processes are embedded.

Level of affluence and employment

For many, there is a strong correlation between the magnitude of informal employment and the level of affluence and employment in a locality. Some, however, have found that the higher the level of affluence and employment in a place, the greater the amount of such work and the more autonomous and well paid is the informal employment (e.g., Leonard, 1994; Pahl, 1984; Renooy, 1990), whilst others have discovered the opposite (e.g., Frey and Weck, 1983; Matthews, 1983). On its own, therefore, this condition is insufficient to produce a particular configuration of informal employment (see Williams and Windebank, 1995b; Windebank and Williams, 1997). It is the way in which it combines with a range of additional economic, social, institutional and environmental regulators in a place to produce a particular configuration of informal employment which is important.

Industrial structure

The vast majority of studies find that relatively little informal employment takes place in areas dominated by a few large companies because the skills acquired in such industries are less transferable to individual informal work and they less often use organized informal work than smaller firms (Barthelemy, 1991; Howe, 1988; Van Geuns et al., 1987). Informal employment usually flourishes, meanwhile, in local economies with a plethora of small firms (Barthelemy, 1991; Blair and Endres, 1994; Capecchi, 1989; Pahl, 1987; Sassen, 1996; Van Geuns et al., 1987) since trade unions are less active and organized, there is greater scope for employing organized informal work either directly or through sub-contracting

arrangements and skills acquired are more likely to be useful in individual informal employment. Similarly, many have drawn a strong positive correlation between the level of self-employment and the extent of informal employment (e.g. O'Higgins, 1981; Pahl, 1990). As Mingione (1991) reveals, however, although tax evasion may be higher amongst the self-employed and small firms than the employed and large firms, not only is this just one part of all informal employment but the magnitude of informality amongst small firms and the self-employed varies considerably between areas due to a host of other factors, including the structure of taxation, level of affluence and employment. Thus, one cannot automatically read-off from the level of self-employment or small firms in a locality that informal employment is necessarily higher.

Sub-contracting

In some localities, a link has been established between sub-contracting and the growth of informal employment (Benton, 1990; Thomas and Thomas, 1994). As workers previously employed in large firms have been forced to seek jobs in smaller marginal firms or to do piecework in their own homes, the result in economically vulnerable areas has been a weakening of the workers' bargaining position, decreased wages, reduced working conditions and a loss of state benefits (Benton, 1991; Thomas and Thomas, 1994). Such a deterioration in working conditions is not inevitable. In other areas, the outcome is thriving enclaves of small informal enterprises that have adapted to changing market demands for specialized products while sustaining relatively high wages and good working conditions (Benton, 1990; Capecchi, 1989). This is because the effect of increased sub-contracting depends on the local economy and the local labour market conditions as well as the local political circumstances.

Tax and social contributions

Although some argue that this condition alone is sufficient to explain the growth of informal employment (e.g. Matthews, 1983), it is but one, albeit important, part of the 'cocktail' of explanatory factors which cannot be ruled out. As taxes rise in a nation or region, so the differential cost of engaging in and using formal rather than informal employment increases for both consumers and firms (Frey and Weck, 1983; Gutmann, 1978; Renooy, 1990). However, not all individuals will avoid such burdens by employing or offering themselves as informal workers. Instead, some consumers will engage in self-provisioning. Neither do all companies automatically switch some or all production to informal modes. It depends on what other options are open to them. Indeed, it is not just the level but also the structure of tax and social contributions which influences the configuration of informal employment. In places where taxes and social contributions are raised more from companies than individuals, organized informal employment might well be more prevalent, whilst if contributions are raised more from individual workers, then individual informal employment might be more likely (Barthelemy, 1991). The way in which the level and structure of taxation influences the configuration of informal employment, nevertheless, is dependent upon a range of other factors such as attitudes towards tax evasion, labour and welfare laws and the interpretation and enforcement of rules and regulations.

Social Regulators

Local and regional traditions, social norms and moralities

Economic processes are often mediated by cultural norms, traditions and moralities. In some areas, informal employment is more acceptable than in others, such as when feelings of resentment and of being let down by the state lead to less acquiescence to the laws of the state (Howe, 1988; Legrain, 1982; Leonard, 1994; Van Geuns et al., 1987; Weber, 1989) or when informal employment is undertaken more for social than economic reasons (Barthelemy, 1991; Cornuel and Duriez, 1985; Komter, 1996). Moreover, tax morality (that is, the willingness to cheat on taxes) varies not only cross-nationally (Frey and Weck, 1983) but also intra-nationally, where tax morality is lowest in big cities (Chavdarova, 1995). The precise ways in which these cultural processes mediate economic activity, however, are shaped by the economic and institutional conditions of the locality as well as by other social regulators.

The nature of social networks

Areas characterized by populations with dense social networks generally undertake a greater proportion of the total work using informal employment than areas with sparse or dissipated social networks, who use relatively more formal purchases (Legrain, 1982; Mingione, 1991; Morris, 1994; Van Geuns et al., 1987). Similarly, individuals with denser social networks tend to have more opportunity for informal employment than those without (e.g., Leonard, 1994; Warde, 1990). Dense social networks alone, however, are insufficient to lead to informal employment if their members are unable to pay for such goods and services (Lysestol, 1995). As Morris (1993) discovers in the UK, the long-term unemployed often only mix with other long-term unemployed and thus are less likely to be offered opportunities for engaging in informal employment. However, it must be remembered that she was studying an area characterized by an industrial structure undergoing contraction and dominated by large enterprises. Dense social networks in conjunction with other factors (for example, in areas with a wide socio-economic mix or rural areas) can lead to informal employment alleviating poverty, if not providing a complete survival strategy (e.g., Jessen et al., 1987; Weber, 1989).

Socio-economic mix

An area that combines a population with high incomes but little free time with a population who have low incomes but much free time will usually witness both high absolute and relative levels of informal employment (Barthelemy, 1991; Portes, 1994; Renooy, 1990). This, however, need not always be the case. Strict enforcement of social security regulations coupled with a low tax regime would not necessarily lead to such a configuration, since suppliers would feel restricted from participating and customers would receive lower marginal benefits from using informal rather than formal employment.

Educational levels

The more educated the population, the more likely they are to supply informal labour and to perform autonomous well-paid informal employment, whilst those with fewer qualifications engage in more exploitative poorly paid informal employment (e.g. Fortin et al., 1996; Lemieux et al., 1994; Renooy, 1990). Again, however, there will be exceptions depending on the way in which this human capital factor is mediated by other economic, institutional, social and environmental conditions in any area (for example, the structure of tax and welfare benefits and the degree to which they are enforced).

Institutional Regulators

Besides economic and social regulators of the level and character of informal employment, there are also a range of institutional conditions that regulate such work.

Labour law

By definition, informal employment is created by government rules and restrictions (e.g. Gutmann,

1978). As labour law differs between nations, so too does the level and nature of informal employment. In some nations, for example, industrial homeworking is illegal whilst in others it is not (Phizacklea and Wolkowitz, 1995), which influences the type of work undertaken informally and who participates. Such legal differences thus have a heavy influence on the level and nature of informal employment.

Welfare benefit regulations

The unemployed in countries with poor access to permanent state benefits are more likely to engage in informal employment as a survival strategy because they have little to lose if caught (Del Boca and Forte, 1982), thus helping to explain the higher levels of informal employment in southern EU nations and the USA. In contrast, those eligible to claim more generous benefits will have the major disincentive of losing them if caught engaging in informal employment, as Wenig (1990) highlights in Germany. Again, however, this general tendency will be mediated by other economic, institutional, social and environmental factors such as local cultural traditions, social networks and the buoyancy of the local labour market as well as by the extent to which such regulations are enforced.

State interpretation and enforcement of regulations

In addition to actual differences in rules and regulations, how they are interpreted and enforced will influence the level and nature of informal employment. Some authorities, both national and local, not only actively support informal employment so as to enable firms to compete in international markets and/or to help individuals and families raise adequate incomes, which would be difficult if 'regular' regulations were strictly enforced (Cappechi, 1989; Lobo, 1990a; Portes and Sassen-Koob, 1987; Van Geuns et al., 1987; Warren, 1994), but also passively tolerated its existence through lax enforcement (e.g. Lobo, 1990b; Van Geuns et al., 1987). This again is insufficient alone to facilitate the growth of informal employment.

Corporatist agreements

Labour is not only regulated by the state but by industry, company and individual agreements which can limit the rights of employees to take on outside jobs, do overtime and so on. Informal employment can be an attempt to circumvent these rules, which thus become a factor in local and national variations in the level and character of informal employment (e.g., Lobo, 1990b).

Environmental Regulators

Settlement size

Although little is known about how urban settlement size influences the magnitude and character of informal employment, some studies show that rural areas undertake more informal employment than urban areas (e.g. Jessen et al., 1987; Legrain, 1982), whilst other show the opposite (e.g., Fortin et al., 1996, Mogensen, 1985). The implication, therefore, is that urbanity alone is an insufficient determinant of the level of informal employment. It depends upon the presence or absence of the range of additional factors discussed above. Furthermore, it appears to be the case that rural areas are characterized more by autonomous forms of informal employment (e.g. Renooy, 1990), whilst organized informal employment seems to be more concentrated in urban areas (e.g., Portes and Stepick, 1993; Sassen, 1991). Again, however, this depends upon a range of additional economic, social and institutional characteristics of the area and how these combine to produce particular configurations of informal employment.

Type and availability of housing

Housing tenure seems to be an important determinant of the quantity and quality of informal employment in a locality. Areas with high levels of privately owned households undertake a wider range of informal work than areas dominated by rented accommodation (Morris, 1994; Pahl, 1984; Renooy, 1990). Moreover, owner–occupiers have been shown to be more likely to use informal employment to get a job done than tenants (Mogensen et al., 1985; Renooy, 1990). Again, whether this always holds depends upon the presence or absence of the other factors discussed above.

In sum, the level and character of informal employment in a place is the result of the way in which economic, social, institutional and environmental conditions combine in multifarious 'cocktails' in different places to produce specific local outcomes. As Mingione (1990: 42) asserts in relation to the spatial variations in Greece, 'Everywhere, the mixes appear to be very complex and fundamentally based on a range of combinations of local opportunities.' The temptation, therefore, is to compensate for the previously overgeneralized and simplistic typologies concerning the urban geography of informal employment by concluding with such an idio-graphic approach. We could conclude, similar to Smithies (1984: 89), that 'Black economies are as different from one another as "legitimate" economies'. However, to leave it at this would be to go too far to the other extreme, ignoring some important parallels and commonalities in the extent and character of informal employment between and within urban areas.

Take, for example, deprived urban areas in the advanced economies. These localities are usually characterized by high unemployment, economic vulnerability, little support or sympathy for people (especially the unemployed) working informally, a poor socio-economic mix, below-average educational qualifications, authorities stringently implementing rules and regulations concerning benefit fraud, tax evasion and labour laws, and low levels of private ownership of housing. The result of this cocktail of factors is low levels of informal employment and that informal employment which does exist tends to be mostly of the exploitative variety (Barthe, 1988; Foudi et al., 1982; Howe, 1988; Jordan et al., 1992; Morris, 1995). This is because the popula-tion possesses the free time but lacks the addi-tional resources and opportunities necessary to engage in a wide array of informal employment. They lack the skills, tools and materials to participate in such work, the social networks to hear about informal employment opportunities, fear being reported to the authorities and have less scope for undertaking such work in the areas in which they live. The few opportunities for informal employment which do exist, moreover, tend to be in the form of exploitative organized informal employment (for example, contract cleaning, fruit picking, unlicensed taxi-driving, labouring, kitchen work). This would not perhaps be problematic so far as the urban geography of social inequality is concerned if these areas occu-pied only a minor space in the urban landscape of the advanced economies. The serious issue for socio-spatial disparities, however, is that this is the dominant way in which informal employment is configured in the majority of deprived urban areas in the advanced economies. This has major issues for the socio-spatial divisions of not only informal employment but standards of living in general.

Given the dominance of this particular cocktail in the deprived urban areas of our advanced economies, it can be argued that the principal outcome of integrating informal employment into the analysis of the growth/decline of urban economies is that it further exacerbates, rather than reduces, the overarching socio-spatial inequalities. On the whole, those urban areas witnessing contraction of their formal economies are also witnessing a decline of their informal work and vice versa. This applies both on an intra- and inter-urban level. Although, in specific situations, there are exceptions to this rule, the dominant way in which the relationship between formal and informal employment manifests itself in the contemporary urban landscape is one which reinforces rather than mitigates the socio-spatial inequalities produced by formal employ-ment. What, therefore, are the available policy options towards informal employment and what are the implications of pursuing each option?

POLICY OPTIONS AND IMPLICATIONS

To examine the policy options and their implica-tions, we here examine the three most popular policy approaches in the literature: the regulatory approach, which seeks to eradicate informal employment through tougher rules and regula-tions; the neo-liberal approach, which seeks to integrate formal and informal employment by de-regulating formal employment; and the new economics approach, which seeks a radical restructuring of work and income and, in so doing, to abolish the necessity for people to undertake exploitative informal employment whilst liberating them to more fully participate in other forms of work. Each is here evaluated in turn in terms of, first, their vision for work and welfare and, secondly, the practicability and desirability of their approach towards informal employment. Before commencing, however, it is necessary to note that few, if any, analysts advo-cate the development of informal employment. That is a common misconception. Instead, and in all approaches, the intention is to formalize informal employment as a means of harnessing its productive energies.

Given the way in which the informal labour market generally tends to reinforce rather than reduce the socio-spatial disparities produced by the formal labour market, it might seem superfi-cially that the overwhelming popularity of the regulatory approach with its desire to eradicate informal employment through tougher regulatory conditions is the most appropriate response. Seeking the regulation of informal employment due to its fraudulent (e.g., Feige, 1990; ILO, 1996) and/or exploitative nature (e.g., Portes, 1994), the goal of this approach is to replace such work with full employment and, failing this, a formal welfare 'safety net'. The principal problem, however, is that little consideration is given to the notion that full-employment and/or a comprehensive welfare state might be unachievable and thus, that people need to be given alternative ways of engaging in meaningful and productive activity both as a social entitlement and as part of a policy of active citizenship. There are also major problems with the feasibility of seeking the eradication of informal employment. Not only are there 'resistance cultures' to this policy in governments, such as in southern Europe (Benton, 1990; Cappechi, 1989; Mingione, 1991), but there is also the problem that because informal employment is deeply entrenched in everyday urban life and frequently undertaken as much for social as for economic reasons (e.g., Fortin et al., 1996; Komter, 1996), it will be near impossible to eliminate. Moreover, given that it is simpler to eradicate organized than autonomous informal employment, and that the former is more likely to be undertaken by marginalized populations, the unintended consequence of such an approach is likely to be the further exacerbation of socio-spatial inequalities.

The implication, therefore, is that any attempt to eradicate informal employment through more stringent regulations will not only fail but will extenuate current socio-spatial inequalities in our urban economies. Take, for example, current attempts to introduce tougher regulations through 'welfare-to-work' programmes in the USA and UK. Although little, if any, research has been conducted in either nation in terms of its impacts on informal employment, it seems likely that the principal consequence will be merely to exacerbate socio-spatial disparities by constraining the ability of the unemployed to supplement their benefit through informal employment whilst leaving the already employed free to continue with their informal working. Research, nevertheless, is badly required on the implications of 'welfare-to-work' programmes before any firm conclusions can be reached. What, for example, happens to people who fall out of welfare-to-work programmes? Do they depend on kin for support, find formal jobs, get by in informal employment, turn to crime or end up in prison? Currently, nobody knows the full consequences of the introduction of these tougher welfare approaches.

The second approach, similar to the regulatory approach, also wants to see a return to full employment but has a different means of achieving it. Rather than bolstering the flagging edifices of welfare capitalism by regulating informal employment, this neo-liberal approach advocates de-regulating formal work and welfare so as to close the gap between formal and informal employment (e.g., Matthews, 1983; Sauvy, 1984; de Soto, 1989). Although stripping away formal regulations would greatly reduce informal employment since such a form of work is by definition a product of such regulations, there is little reason to believe that this would do much to relieve socio-spatial inequalities. Indeed, the only possible reason that greater economic competitiveness and socio-economic equity would be achieved through de-regulation is because there will be a levelling down rather than up of material and social circumstances as urban economies engage in a race to the bottom.

The final policy stance towards informal employment is the 'new economics' approach. Similar to the de-regulationists, the existence of informal employment is seen as an indictment of the present system, but not of the over-regulation of the economy and welfare provision by the state. Instead, it is an indictment of the present system of distributing income and work in society, of which exploitative informal employment is a symptom. This approach, therefore, advocates a radical restructuring of work and income and, in so doing, seeks to abolish the necessity for people to undertake exploitative informal employment whilst liberating them to more fully participate in other forms of work. Assuming that full-employment and comprehensive welfare provision are impractical and undesirable goals, this perspective seeks to replace the employment-focused mentality of the above approaches with a 'whole economy' approach based on 'full-engagement' (e.g., Lipietz, 1992; Mayo, 1996) through the introduction of a citizen's income and greater assistance to help people help themselves. However, there are major difficulties over the practicality and desirability of implementing this approach (see Williams and Windebank, 1998). So far as the basic income scheme is concerned, there are many imponderables concerning not only both the cost of implementing it and whether there is the political will to do so, but also whether it will significantly reduce either the level of informal employment in general or exploitative informal employment more particularly. Moreover, there is the issue that the new economists are racing ahead with implementing self-help initiatives before they have a clear conception of the barriers to participation of different groups.

In sum, and having reviewed three different approaches towards urban informal employment, we can see that the option chosen is very much dependent upon the wider issue of the future for work and welfare desired. Should we try to return to the commodification of both work and welfare in the form of full employment and the welfare state, despite all of the evidence suggesting that this is not practical or desirable? Or, alternatively, should we pursue a future of work and welfare in which the formal labour market is de-regulated and the welfare safety net taken away so as to try to recapture full employment? Or should we accept that full employment is gone and seek to restructure work and welfare using the whole economy and do so in a manner which helps people to help themselves and pursue greater self-reliance? These are ultimately the questions that need to be answered when deciding what can and should be done about informal employment in the advanced economies. What is certain, however, is that the *laissez-faire* option is not available. To do this will only consolidate through the informal labour market the socio-spatial disparities and injustices perpetrated by formal employment.

REFERENCES

Barthe, M.A. (1988) *L'Economie cachée*. Paris: Syros Alternatives.

Barthelemy, P. (1991) 'La croissance de l'economie souterraine dans les pays occidentaux: un essai d'interpretation', in J-L. Lespes (ed.), *Les pratiques juridiques, economiques et sociales informelles*. Paris: Presses Universitaires de France.

Benton, L.A. (1990 *Invisible Factories: the Informal Economy and Industrial Development in Spain*. Binghampton: State University of New York Press.

Blair, J.P. and Endres, C.R. (1994) 'Hidden economic development assets', *Economic Development Quarterly*, 8(3): 286–91.

Button, K. (1984) 'Regional variations in the irregular economy: a study of possible trends', *Regional Studies*, 18: 385–92.

Cappechi, V. (1989) 'The informal economy and the development of flexible specialisation in Emilia Romagna', in A. Portes, M. Castells and L.A. Benton (eds), *The Informal Economy: Studies in Advanced and Less Developing Countries*, Baltimore, MD: Johns Hopkins University Press.

Castells, M. and Portes, A. (1989) 'World underneath: the origins, dynamics and affects of the informal economy', in A. Portes, M. Castells and L.A. Benton (eds), *The Informal Economy: Studies in Advanced and Less Developing Countries*. Baltimore, MD: Johns Hopkins University Press.

Chavdarova, T. (1995) 'Tax morality and households' strategies of economic activity: challenges for social policy in Bulgaria'. Paper presented at Euroconference in Social Policy in an Environment of Insecurity: Contemporary Dilemmas and Challenges for Social Policy. Lisbon, November.

Conroy, P. (1996) *Equal Opportunities for All*. European Social Policy Forum Working Paper I, DG-V. Brussels: European Commission.

Cornuel, D. and Duriez, B. (1985) 'Local exchange and state intervention', in N. Redclift and E. Mingione (eds), *Beyond Employment: Household, Gender and Subsistence*. Oxford: Basil Blackwell.

Del Boca, D. and Forte, F. (1982) 'Recent empirical surveys and theoretical interpretations of the parallel economy', in V. Tanzi (ed.), *The Underground Economy in the United States and Abroad*. Lexington, MA: Lexington Books.

Economist Intelligence Unit (1982) *Coping with Unemployment: the Effects on the Unemployed Themselves*. London: Economist Intelligence Unit.

European Commission (1995) *Social Protection in Europe 1995*. Luxembourg: Office for Official Publications of the European Communities.

European Commission (1996a) *Employment in Europe 1996*. Luxembourg: Office for Official Publications of the European Communities.

European Commission (1996b) *For a Europe of Civic and Social Rights: Report by the Comité des Sages*. Brussels: European Commission.

Feige, E.L. (1990) 'Defining and estimating underground and informal economies: the new institutional economics approach', *World Development*, 18(7): 989–1002.

Fernandez-Kelly, M.P. and Garcia, A.M. (1989) 'Informalisation at the core: Hispanic women, homework, and the advanced capitalist state', *Environment and Planning D: Society and Space*, 8: 459–83.

Ferrera, M. (1996) *New Problems, Old Solutions? Recasting Social Protection for the Future of Europe*. European Social Policy Forum Working Paper III, DG-V. Brussels: European Commission.

Fortin, B., Garneau, G., Lacroix, G., Lemieux, T. and Montmarquette, C. (1996) *L'economie souterraine au Quebec: mythes et realités*. Sainte-Foy: Presses de l'Université Laval.

Foudi, R., Stankiewicz, F. and Vanecloo, N. (1982) 'Chomeurs et economie informelle', *Cahiers de l'observation du changement social et culturel*, no. 17. Paris: CNRS.

Frey, B.S. and Weck, H. (1983) 'What produces a hidden economy? An international cross-section analysis', *Southern Economic Journal*, 49: 822–32.

Glatzer, W. and Berger, R. (1988) 'Household compositions, social networks and household production in Germany', in R.E. Pahl (ed.), *On Work: Historical Comparative and Theoretical Approaches*. Oxford: Basil Blackwell.

Gutmann, P. (1978) 'Are the unemployed, unemployed?', *Financial Analysts Journal*, 34(1): 26–9.

Hadjimichalis, C. and Vaiou, D. (1980) 'Whose flexibility?: the politics of informalization in Southern Europe'. Paper presented to the IAAD/SCG Study Groups of the IBG Conference on Industrial Restructuring and Social Change: the Dawning of a New Era of Flexible Accumulation?, Durham.

Hart, K. (1973) Informal income opportunities and urban employment in Ghana, *Journal of Modern African Studies*, 11: 61–89.

Hellberger, C. and Schwarze, J. (1986) *Umfang und stuktur der nebenerwerbstatigkeit in der Bundesrepublik Deutschland*, Berlin: Mitteilungen aus der Arbeits-market- und Berufsforschung.

Hellberger, C. and Schwarze, J. (1987) 'Nebenerwerbstatigkeit: ein indikator fur arbeitsmarkt-flexibilitat oder schattenwirtschaft?', *Wirtschaftsdienst*, 2: 83–90.

Henry, S. (1978) *The Hidden Economy*. London: Martin Robertson.

Howe, L.E.A. (1988) 'Unemployment, doing the double and local labour markets in Belfast', in C. Cartin and T. Wilson (eds), *Ireland from Below: Social Change and Local Communities in Modern Ireland*. Dublin: Gill and Macmillan.

Howe, L. (1990) *Being Unemployed in Northern Ireland: an Ethnograhic Study*. Cambridge: Cambridge University Press.

ILO (International Labour Office) (1996) *World Employment 1996/97: National Policies in a Global Context*. Geneva: International Labour Organization.

Jensen, L., Çornwell, G.T. and Findeis, J.L. (1995) 'Informal work in nonmetropolitan Pennsylvania', *Rural Sociology*, 60(1): 91–107.

Jessen, J., Siebel, W., Sieble-Rebell, C., Walther, U. and Weyrather, I. (1987) 'The informal work of industrial workers'. Paper presented at 6th Urban Change and Conflict Conference, University of Kent at Canterbury, September.

Jordan, B., James, S., Kay, H. and Redley, M. (1992) *Trapped in Poverty*. London: Routledge.

Kesteloot, C. and Meert, H. (1994) 'Les fonctions socio-economiques de l'economie informelle et son implantation spatiale dans les villes belges'. Paper presented to International Conference on Cities, Enterprises and Society at the Eve of the XXIst Century, Lille.

Komter, A.E. (1996) 'Reciprocity as a principle of exclusion: gift giving in the Netherlands', *Sociology*, 30(2): 299–316.

Koopmans, C.C. (1989) *Informele Arbeid: vraag, aanbod, participanten, prijzen*. Amsterdam: Proefschrift Universitiet van Amsterdam.

Lagos, R.A. (1995) 'Formalising the informal sector: barriers and costs', *Development and Change*, 26: 11–131.

Legrain, C. (1982) 'L'economie informelle à Grand Failly', *Cahiers de l'OCS*, no. 7. Paris: CNRS.

Lemieux, T., Fortin, B. and Frechette, P. (1994) 'The effect of taxes on labor supply in the underground economy', *American Economic Review*, 84(1): 231–54.

Leonard, M. (1994) *Informal Economic Activity in Belfast*. Aldershot: Avebury.

Lim, J. (1995) 'Polarized development and urban change in New York's Chinatown', *Urban Affairs Review*, 30(3): 332–54.

Lipietz, A. (1992) *Towards a New Economic Order: Post-Fordism, Ecology and Democracy*. Cambridge: Polity.

Lobo, F.M. (1990a) 'Irregular work in Spain', in *Underground Economy and Irregular Forms of Employment: Final Synthesis Report*. Luxembourg: Office for Official Publications of the European Communities.

Lobo, F.M. (1990b) 'Irregular work in Portugal', in *Underground Economy and Irregular Forms of Employment: Final Synthesis Report*. Luxembourg: Office for Official Publications of the European Communities.

Lozano, B. (1989) *The Invisible Workforce: Transforming American Business with Outside and Home-based Workers*. New York: The Free Press.

Lysestol, P.M. (1995) '"The other economy" and its influences on job-seeking behaviour for the long term unemployed'. Paper presented at the Euroconference on Social Policy in an Environment of Insecurity. Lisbon: November.

MacDonald, R. (1994) 'Fiddly jobs, undeclared working and the something for nothing society', *Work, Employment and Society*. 8(4): 507–30.

Maldonado, C. (1995) 'The informal sector; legalization or laissez-faire?', *International Labour Review*, 134(6): 70a5–28.

Mattera, P. (1985) *Off the Books: the Rise of the Underground Economy*. New York: St Martin's Press.

Matthews, K. (1983) 'National income and the black economy', *Journal of Economic Affairs*, 3(4): 261–7.

Mayo, E. (1996) 'Dreaming of work', in P. Meadows (ed.), *Work Out or Work In? Contributions to the Debate on the Future of Work*. York: Joseph Rowntree Foundation.

McInnis-Dittrich, K. (1995) 'Women of the shadows: Appalachian women's participation in the informal economy', *Affilia: Journal of Women and Social Work*, 10(4): 398–412.

Mingione, E. (1990) 'The case of Greece', in *Underground Economy and Irregular Forms of Employment: Final Synthesis Report*. Brussels: Office for Official Publications of the European Commission.

Mingione, E. (1991) *Fragmented Societies: a Sociology of Economic Life Beyond the Market Paradigm*. Oxford: Blackwell.

Mingione, E. and Morlicchio, E. (1993) 'New forms of urban poverty in Italy: risk path models in the North and South', *International Journal of Urban and Regional Research*, 17(3): 413–27.

Mogensen, G. V. (1985) *Sort arbejde i Danmark*. Copenhagen: Institut for Nationalokonomi.

Mogensen, G., Kvist, H.K., Kormendi, E. and Pedersen, S. (1995) *The Shadow Economy in Denmark 1994: Measurement and Results*. Copenhagen: The Rockwool Foundation Research Unit.

Morris, L. (1993) 'Is there a British underclass?', *International Journal of Urban and Regional Research*, 17(3): 404–12.

Morris, L. (1994) 'Informal aspects of social divisions', *International Journal of Urban and Regional Research*, 18: 112–26.

Morris, L. (1995) *Social Divisions: Economic Decline and Social Structural Change*. London: UCL Press.

Moulier-Boutang, Y. (1991) 'Dynamique des migrations internationales et economie souterraine: comparaison internationale et perspectives européenes', in S. Montagne-Villette (ed.), *Espaces et Travail Clandestins*. Paris: Masson.

Nicaise, I. (1996) 'Which partnership for employment? Social partners, NGOs and public authorities'. European Social Policy Forum Working paper II, DG-V. Brussels: European Commission.

O'Higgins, M. (1981) 'Tax evasion and the self-employed: an examination of the evidence–II', *British Tax Review*, 26: 367–78.

Pahl, R.E. (1984) *Divisions of Labour*. Oxford: Blackwell.

Pahl, R.E. (1987) 'Does jobless mean workless? Unemployment and informal work', *Annals of the American Academy of Political and Social Science*, 493: 36–46.

Pahl, R.E. (1990) 'The black economy in the United Kingdom', in Underground Economy and Irregular Forms of Employment: Final Synthesis Report. Brussels: Office for Official Publications of the European Communities.

Pestieau, P. (1984) *Belgium's Irregular Economy*. Brussels: Université Libre de Bruxelles.

Phizacklea, A. and Wolkowitz, C. (1995) *Homeworking Women: Gender, Racism and Class at Work*. London: Sage.

Portes, A. (1994) 'The informal economy and its paradoxes', in N.J. Smelser and R. Sedberg (eds), *The Handbook of Economic Sociology*. Princeton, NJ: Princeton University Press.

Portes, A. and Sassen-Koob, S. (1987) 'Making it underground: comparative material of the informal sector in Western market economies', *American Journal of Sociology*, 93: 30–61.

Portes, A. and Stepick, A. (1993) *City on the Edge: the Transformation of Miami*. Los Angeles: University of California Press.

Renooy, P. (1990) *The Informal Economy: Meaning, Measurement and Social Significance*. Netherlands Geographical Studies no. 115, Amsterdam.

Sassen, S. (1989) 'New York city's informal economy', in A. Portes, M. Castells and L. Benton (eds), *The Informal Economy: Studies in Advanced and Less Developed Countries*. Baltimore, MD: Johns Hopkins University Press.

Sassen, S. (1991) *The Global City*. Princeton, NJ: Princeton University Press.

Sassen, S. (1996) 'Service employment regimes and the new inequality', in E. Mingione (ed.), *Urban Poverty and the Underclass*. Oxford: Blackwell.

Sauvy, A. (1984) *Le travail noir et l'economie de demain*. Paris: Calmann-Levy.

Scott, W.J. and Grasmick, H.G. (1981) 'Deterrence and income tax cheating: testing interaction hypotheses in utilitarian theories', *Journal of Applied Behavioural Sciences*, 17(3): 395–408.

Smith, S. (1986) *Britain's Shadow Economy*. Oxford: Clarendon.

Smithies, E. (1984) *The Black Economy in England since 1914*. Dublin: Gill and Macmillan.

de Soto, H. (1989) *The Other Path*. London: IB Taurus.

Stepick, A. (1989) 'Miami's two informal sectors', in A. Portes, M. Castells and L. Benton (eds), *The Informal Economy: Studies in Advanced and Less Developed Countries*. Baltimore, MD: Johns Hopkins University Press.

Thomas, J.J. (1988) 'The politics of the black economy', *Work, Employment and Society*, 2(2): 169–90.

Thomas, J.J. (1992) *Informal Economic Activity*. Hemel Hempstead: Harvester Wheatsheaf.

Thomas, R. and Thomas, H. (1994) 'The informal economy and local economic development policy', *Local Government Studies*, 20(3): 486–501.

Tievant, S. (1982) 'Vivre autrement: échanges et sociabilité en ville nouvelle', *Cahiers de l'OCS*, vol. 6. (Paris: CNRS).

Townsend, A. (1997) *Making a Living in Europe: Human Geographies of Economic Change*. London: Routledge

Van Eck, R. and Kazeier, B. (1985) *Swarte inkomsten uit arbeid: resultaten van in 1983 gehouden experimentele enquetes*. CBS- Statistische Katernen no. 3. The Hague: Central Bureau of Statistics.

Van Geuns, R., Mevissen, J. and Renooy, P. (1987) 'The spatial and sectoral diversity of the informal economy', *Tijdschrift vor economische en sociale geografie*, 778(5): 389–98.

Vinay, P. (1987) 'Women, family and work: symptoms of crisis in the informal economy of Central Italy', *Sames 3rd International Seminar Proceedings*, University of Thessaloniki, Thessaloniki.

Warde, A. (1990) 'Household work strategies and forms of labour: conceptual and empirical issues', *Work, Employment and Society*, 4(4): 495–515.

Warren, M.R. (1994) 'Exploitation or co-operation? The political basis of regional variation in the Italian informal economy', *Politics and Society*, 22(1): 89–115.

Weber, F. (1989) *Le Travail à Côté: étude d'ethnographie ouvrière*. Paris: Institut National de la Recherche Agronomiques.

Wenig, A. (1990) 'The shadow economy in the Federal Republic of Germany', in Underground Economy and Irregular Forms of Employment: Final Synthesis Report. Brussels: Office for Official Publications of the European Communities.

Williams, C.C. and Windebank, J. (1995a) 'The implications for the informal sector of European integration', *European Spatial Research and Policy*, 2(1): 17–34.

Williams, C.C. and Windebank, J. (1995b) 'Black market work in the European Community: peripheral work for peripheral areas?', *International Journal of Urban and Regional Research*, 19(1): 23–39.

Williams, C.C. and Windebank, J. (1998) *Informal Employment in the Advanced Economies: Implications for Work and Welfare*. London: Routledge.

Windebank, J. and Williams, C.C. (1997) 'What is to be done about the paid informal sector in the European Union? A review of policy options', *International Planning Studies*, 2(3): 315–27.

Wuddalamy, V. (1991) 'Les Mauriciens en France: une insertion socio-professionelle caracterisée par l'immigration spontanée', in S. Montagne-Villette (ed.), *Espaces et travail clandestins*. Paris: Masson.

Part V

THE CITY AS ORGANIZED POLITY

Questions of how cities are, and should be, governed have been the subject of ongoing debate. Frequently the debate has been expressed in normative terms: accounts proffering how metropolitan governmental structures should be recast to meet strategic and local needs, how the allocation of functional responsibilities should be divided between metropolitan-wide and more local authorities, or how modes of citizen participation should be bolstered so as to make urban local democracy more responsive and accountable. Alongside this prescriptive analysis, and largely dating from work undertaken in the post-1945 period, has been the study of urban government/governance in terms of explaining the nature of local politics, where power resides and what are the outcomes of local political processes in conferring advantage and disadvantage for particular groups or neighbourhoods. Much of this analysis has been theoretically positioned, in terms of the distribution of power (elite and neo-pluralist analyses), or in the role of institutions or within theories of the state.

Contemporary analysis of the city distinguishes between government and governance, even though such differences – say between conditions in metropolitan areas a generation ago and today – can be exaggerated. Earlier studies talked more in terms of urban government, more recent analyses of governance, the distinction marking the changing roles of the state, market and civil society. How these relations have been recast varies between the advanced economies. In Britain the shifts are linked with the restructuring, typically the downsizing, of the state, the spread of privatization of service provision previously provided by the local public sector, the recasting of the role of urban local government from a producer to a facilitator, the drawing in and together of the private and voluntary sector into the processes of restructuring with the public within partnerships. In the developing economies the meaning of governance has taken somewhat different meanings attached to structural adjustment programmes in which governments are urged to adopt participatory and other practices which aim at ensuring the greater transparency and accountability of the practice of government.

While the shift from government to governance reflects the changing relationships between the local public sector and the market, the adoption of private sector practices through more entrepreneurial forms of local public sector intervention, and often a tilting of the balance of power towards private sector interests, the significance of urban governments continues to be acknowledged. There is renewed recognition that cities require a governmental structure which can facilitate the strategic development of the metropolitan area – this in turn has seen the re-emergence of the debate, last visited several decades ago, surrounding the restructuring of metropolitan government/governance. Much of this renewed interest stems from the deepening effects of globalization, the need to ensure the competitive position of the city, to bolster the legitimacy of processes of urban restructuring and to enhance the capacity of local governance to manage the strategic development of the metropolitan region. The logic of replacing a local government for London, dismantled in 1985, was premised in part on the need to give the city voice within the global arena and to enhance the position of key services, notably transport, essential to maintaining its position as a world city.

The effects of the adoption of neo-liberal policies may not have undermined the need for urban government, but it has affected its role as a producer of services. This has become particularly apparent in those countries, such as Britain, where local (urban) governments had become deeply involved in the delivery of the welfare state. As the welfare role has been redefined, and with the growth of quasi-markets, so the role of urban governments has altered, as has the relationship between the citizen and service producers. In education there have been radical reforms in many advanced economies in the way the service is structured and provided based on the adoption of market-led ideas. Citizens as consumer-parents have been given more choice while, as in the reforms in New Zealand and other countries, there is more local (parental) control over how schools are run.

These changes have re-kindled old questions of equity in service provision. Debate over the equity implications of the balkanization of the urban school system in metropolitan areas, as in much of the United States, have re-emerged as market-led ideas have led to the introduction of innovations in urban education – the spread of Charter schools in the US for instance – whose effects have been to sharpen the debate on social equity. Nor are such effects limited to cities in the core economies. In African cities the effects of structural adjustment programmes have been to raise the costs of (urban) schooling, the burdens of which fall unequally. Elsewhere in cities in the less developed nations other examples of the effects of marketization of services, such as the provision of potable water supplies, demonstrate how the urban poor may be further disadvantaged.

The paradox is that while the contours between public and private are shifting, the need for (local) state intervention has not diminished as cities become more unequal. Social policy has become a vital area of concern in the city as social exclusion has moved up the political agenda. In Britain, the United States and other countries there is a burgeoning suite of programmes designed to combat social exclusion. Following the election of the Labour Government in Britain in 1997 there has been a renewal of interest in area-based programmes aimed at countering social disadvantage and exclusion covering health, education, physical regeneration, social services, employment, training and others. The proliferation of initiatives has in turn raised problems of co-ordinating effort on the ground. Increasing emphasis has been given to the need to draw local participation into neighbourhood change through partnership and empowerment. Within the discourses of neighbourhood renewal and social exclusion lessons from the past have been learnt; 'top-down' modes of implementation have been replaced by more inclusive mechanisms mirroring the changing but nevertheless fundamental role played by urban governments.

21

Urban Governance

MICHAEL GOLDSMITH

For much of the past 30 years, the question of metropolitan government has hardly been on the agenda. Since the extensive reforms of metropolitan government in such cities as Toronto and London in the mid-1960s, through the extension of the principle of a single- or two-tier system in other metropolitan areas such as Winnipeg and conurbations such as Manchester, Liverpool and Birmingham in the UK, there has been a general retreat from the subject, both amongst practitioners and academics (Sharpe, 1995: 20). At least until recently that is, when both new reforms in places such as Rotterdam, Stuttgart, or re-structuring as seen in Toronto, as well as a concern for what might properly be called the governance of metropolitan regions, have all occasioned a re-awakening of interest.

This chapter will review both the experience with metropolitan governments – by which is meant some institutional arrangement (either single-tier but more generally a two-tier system) whereby the metropolitan area as a whole is governed – and the more general practice whereby different types of arrangement come into place to secure some control and regulation of public and private activities within a metropolis – in other words, forms of metropolitan governance.

At the outset, it is as well to be aware that despite the current fashion for the word 'governance', most metropolitan areas have and continue to operate under such systems, whereby governmental responsibilities in such areas are shared between a wide range of public bodies, some elected, many appointed, and where cooperation with the private sector is often a requirement of successful policy implementation.

CONTEXT

Given the continued growth of world population and its increasing concentration in urban areas, it is hardly surprising that both long-standing metropolitan areas (through the extension of suburbs) continue to grow and that new ones continue to emerge. Clearly such change – which is often rapid – gives rise to definitional problems as to what we mean by a metropolitan area. For example, the local government reforms in Britain in the early 1970s included the designation of six metropolitan areas, in addition to London, which had been reformed almost 10 years earlier in 1963. In effect such a designation was a reflection of the major conurbations (contiguous urban areas) which had first emerged during the Industrial Revolution and had come together as successive processes of suburbanization brought the original industrial towns in these areas closer to each other – physically if not culturally and politically. Most of these areas had populations well in excess of a million inhabitants. By contrast, in the 1960s the US Census adopted the term Standard Metropolitan Statistical Area (SMSA), which was used to refer to any urban area with a population in excess of 100,000 inhabitants – a scale far smaller than matched the case in Britain, but which was perhaps appropriate in the USA of that time.

Since the 1960s, population and urban growth has continued. In particular many new metropolitan areas have emerged, particularly in Third World or developing countries, and many world cities have populations numbered in millions. Places such as Mexico City and San Paulo now lead in terms of population, and many Third World countries have populations concentrated in a few large cities. In the USA, with the move from rustbelt to sunbelt, many old metropolitan areas, such as Detroit, have declined, whilst new ones, such as Dallas, Houston and Atlanta, have grown substantially. A recent Franco-Swiss text discusses metropolitan areas in terms of populations running from 200,000 to around a million, but acknowledging the possibility of even larger areas, such as the Randstad in the Netherlands (Saez et al., 1997: 18–19; see also Leresche et al., 1995). In Britain, most population growth has taken place in small and medium-sized towns, with the result that no new metropolis has emerged over the past 30 years. Nevertheless, most British towns would come close to both the old US SMSA definition of a metropolitan area and to that discussed by Saez et al. In this context, one cannot help but agree with the view of Young and Garside (1982: 338): 'The irony of metropolitan life is that its dynamic nature constantly draws the sought-for solution to the problems of urban form and urban institutions out of the grasp of government, just as it seems to come within reach.'

Clearly then there is a problem for us in what we mean by a metropolitan area, since the form of government and governance may vary accordingly. For our present purposes, however, it is perhaps less of a problem if the formal definition employed in a country or generally in the literature is adopted, and if we recognize some of the key characteristics of metropolitan areas. These we can take as some or all of the following: relatively large land area; relatively large populations; large range of economic activities; extensive number of governmental bodies or agencies and/or some reformed structure; probably considerable social segregation and possible ethnic diversity reflected in a range of social problems (crime, drugs, homelessness, unemployment); poor physical infrastructure – existence of shantytowns, poor physical communications and possible transport crisis – and a range of environmental problems, particularly poor air quality. But they are also honeypots, continuing to attract people from outside to their centre. They are the places in which the process of globalization effectively occurs, acting as centres of innovation, creators of new markets and trends. Finally, such areas may also face a public resource crisis, particularly financial, in terms of their capacity to deal with these various problems and trends, as exemplified by the fiscal crisis faced by New York in the 1970s.[1]

Despite these difficulties and characteristics, cities and specifically metropolitan ones, continue to perform a honeypot function – attracting people and activities from outside the area for all kinds of reasons. London, Paris, Copenhagen, New York and Tokyo all continue to be major centres of commerce, culture and social life, whilst cities like San Francisco, Sydney, Seoul and Singapore attract tourists on a worldwide scale. Savitch (1997) has usefully distinguished between world cities, of which London, Paris, New York and Tokyo are all examples, primate cities, of which places such as Berlin and Karachi would be examples, and regional cities, such as Marseilles

and Osaka, which act as key gateways to smaller markets. One French study (Sallez, 1993: 58–9) has produced a useful classification of European metropolitan areas according to their international status and their economic base. Thus, under this classification only London and Paris are seen as truly dominant international metropolitan areas, whereas places like Rotterdam, Stuttgart and Barcelona are seen as emerging on to the international stage as key players.

Such developments reflect the process of globalization and internationalization which has brought about a system of 'world cities' in which practically every corner of the world is in effect part of such cities' hinterland. And arguing from a postmodernist perspective, Soja (1995: 130) contends that it is the 'rise of the postfordist metropolis [which] has been an integral part of the development of the postmodern city', in which a process of industrial restructuring 'has had a dramatic impact on the urban economy and the residential as well as employment geography of the city', giving rise to a new geography of the metropolis in which terms such as 'Edge City; technoburbs, techno/metropoles, exopolis' are used to describe the new shape of the metropolitan area. Such areas, argues Soja, are increasingly 'ungovernable, at least within the confines of . . . traditional local government structures' (Soja: 1995: 133; see also Davis, 1990), giving rise to private forms of regulation and control – private police forces, walled estates – which seek to reinforce the exclusionary and exclusive nature of many activities in the metropolitan area. One does not have to follow Soja all the way in his argument to accept that the modern metropolis has become a far different place from its counterpart of even 30 years ago.

FORMS OF METROPOLITAN GOVERNANCE

Although it may have been 30 years ago when forms of metropolitan government were last high on the political agenda, most metropolitan areas operated then, as they do now, under a system of *governance* rather than *government*. The distinction is an important one – it symbolizes the wide variety of interests incorporated in a metropolitan area and the range of governmental arrangements employed both to accommodate those interests whilst at the same time ensuring the minimum amount of governmental regulation necessary to ensure the continued existence of the metropolis itself. Prud'homme (1996: 174) argues that it is not the form of government which matters, nor the size of the city which is important, but rather the ability to manage the metropolitan area. In economic terms, he sees the task of urban management to be one of 'moving the benefit curves upwards and the cost curves downwards'. As such 'good management requires good coordination', so that mega-cities are well managed. As Prud'homme argues, 'good management is a matter of people, but also of institutions'. Large cities contain large numbers of actors, concerns, parties, institutions, interests etc. – they all require good leadership and coordination, the need for which increases rapidly as city size increases. In terms of governance, managing these networks is a key task in metropolitan areas, a topic to which we shall return later.

Institutional Forms

The range of institutional arrangements for governing metropolitan areas is extensive, but they can be classified along the following lines (Bollens and Schmandt, 1970) – the one-government approach, the two-level approach and the cooperative approach. In the USA, the single-government approach was largely achieved through two means – either by annexation by the central city of unincorporated lands outside its boundaries or through consolidation of the central city with other municipalities lying in the suburbs. Similar approaches were used in Canada until after the Second World War. By contrast, European metropolitan areas have not adopted either approach, preferring instead to rely either on reform of the structure by higher-level governments (the British case) or depending on some form of horizontal intergovernmental cooperation across municipal boundaries in other cases.

Single-tier government: annexation and consolidation

The adoption of a single tier of government to cover the needs of populations in excess of one million would indeed be rare. As has been noted, single-tier governments covering metropolitan areas largely resulted from processes of annexation, consolidation and incorporation used in countries like the USA and Canada, largely in the inter-war period. (Earlier, St Louis, for example, more than tripled its area through a process of annexation in 1876.) Such a process was deemed logical and seen as a natural extension of the original city. Annexation was the means whereby the expansion of municipal boundaries kept pace with population growth. Without annexation, it is likely that few large US cities could have developed in the late nineteenth and early twentieth centuries. However, by the end of the First World War, annexation as a means of establishing area-wide metropolitan government had declined in

popularity. In the 1920s, for example, the total amount of territory absorbed through annexations in the USA was substantially less than in the 1890s. Only a few cities – notably Detroit and Los Angeles – increased substantially through annexations between 1900 and 1945.

Some cities made spectacular use of annexation: Oklahoma City, Dallas, Houston, Phoenix, San Diego and Kansas City all achieved large areas through annexation, in some cases continuing to add territory into the 1960s. But in most cases these increases were made under conditions favourable to the annexation route for establishing a single-tier metropolitan government. Annexation has to be legally possible and relatively easy to achieve. Substantial unincorporated land has to be available for annexation – and these factors applied in the cases already quoted. Again Bollens and Schmandt (1970: 289) note that only two (Milwaukee and Chicago) of the ten most populous US metropolises of the 1960s had gone through a process of annexation after 1945.

There is a simple reason why annexation generally proved a less than useful way of establishing area-wide metropolitan government. Unless annexation takes place at a time before substantial population growth occurs, it is likely to leave considerable urbanized areas outside the boundaries of the expanded area – housing opponents to the process of what is usually seen as central city expansion. Such areas are often large in population, predominantly residential, provide minimal services, and may be seriously deficient in basic services – water supply, refuse collection and sewerage, and lacking in physical protection services. Despite such problems, the reason for the decline in annexations is clear – technological improvement opened up suburbs, which themselves incorporated as a municipality. As such they opposed takeovers by the unfriendly central city, successfully forming coalitions with other municipalities, rural areas and state legislators to resist both annexation and consolidation. At the same time as annexation and consolidation became more difficult, the process of municipal incorporation, whereby new growing suburbs achieved local government status, remained simple and easy to achieve (Bollens and Schmandt, 1970: 284).[2]

City–county consolidation is a broader one-government approach than municipal annexation, involving a complete or substantial merger of the county with the central city and other municipalities. Despite its attractions, it has been relatively little used in the USA and Canada, for example. In the nineteenth century it was the means whereby area-wide government was achieved in places such as Boston, New York and Philadelphia – but it has been largely ignored during this century. More recent examples – achieved generally after considerable political struggle – include Miami–Dade County (Florida: 1953) and Nashville–Davidson County (Tennessee: 1962). In both cases the merger involved only one large city and a single county – and there was initial opposition to both.

The result has been that in the United States and elsewhere the single-tier or one-government approach to metropolitan areas has generally been rejected over the past 30 years. Metropolitan areas have simply become too large and too complex in economic and social terms to be governed by a single tier; even with some form of area decentralization, such as the establishment of neighbourhood governments. The response has been to do nothing, adopt a two-tier approach, or else rely on some form of cooperative arrangement between existing municipal institutions.

Two-tier approaches: federations

The governments of Toronto and London in the 1960s and 1970s remain the best-known examples of two-tier metropolitan government, generally the approach preferred by reformers seeking an integrated governmental system for metropolitan areas. What it involves is an area-wide – county – level of government, responsible for what would be regarded as area-wide services such as physical infrastructure, police, fire and ambulance, as well as area-wide planning. Below this tier a number of metropolitan districts are put in place to deal with those services which are closer to the consumer or citizen – housing, education, personal social services, etc. – whilst some functions might be shared between the two tiers.

The introduction of such two-tier metropolitan government, also found in the USA in such places as Miami–Dade County, has usually come about because of some service crisis, generally in terms of the physical infrastructure. Such was the case with London, where the fear of gridlock in the 1960s and poor housing conditions made reform possible, while in Toronto, lack of infrastructure and school provision in the face of a rapidly expanding metropolitan area were key factors assisting the reform process. Additionally, many Toronto municipalities faced a fiscal crisis. Such factors came together at a propitious time for reform in both locations, and the use of an independent review procedure to assess the need for reform (the Herbert Commission in London and the deliberations of the Ontario Municipal Board in the Toronto case) helped the process. Further, the existence of upper-tier governments (the Ontario provincial legislature and the Conservative government of Harold Macmillan) willing to consider and support reform was also crucial.

During the same period as the reforms of Toronto, other Canadian areas followed suit – Montreal adopted a two-tier system , and the idea was designed to be spread more generally to large cities throughout Ontario in the 1970s.[3] Winnipeg went one stage further, going for a single-tier system that quickly proved unsatisfactory (Kiernan and Walker, 1983: 222–46). Elsewhere, the London reforms were extended to other English conurbations at the time of local government reform in the early 1970s, with a division of functions and powers which generally followed the London pattern.

Cooperative forms

If the introduction of some formal one- or two-tier structure has been increasingly ignored, then one result has been a general flourishing of varying degrees of cooperative arrangement between the different municipal institutions to be found in metropolitan areas, whether these be some formal Council of Governments (CoGs) in the USA, Communautés Urbaines (CUs) in France or some other arrangement to identify and solve policy problems, as with London and the former metropolitan counties in Britain after abolition of the upper tier in 1986.

By far the most common form of cooperative arrangement is, of course, the agreement to establish a single purpose body to oversee a particular function or to solve a particular problem. Such bodies are all too easy to create, as the American experience reveals. More recently, such bodies have become more common in the UK. Often appointed and given extensive powers and resources, such bodies can be difficult to make accountable, as Caro's account of the activities of Robert Moses in New York makes clear (Caro, 1974), despite the fact that they are also often successful in terms of the problems or functions they are expected to undertake. Accountability is especially a problem where the fragmentation of institutions is extensive. With large numbers of institutions, it becomes very difficult for people to know which unit is responsible for what: 'citizens neither have perfect knowledge nor could they be expected to do so' (Jones, 1983: 215). The fact that many such bodies are appointed means there is no way the citizens can 'kick the rascals out', even if they know who is responsible for any mess that may arise.

At the root of the rise of cooperative arrangements lies the principle that such arrangements are entered into on a voluntary basis. In the USA, such arrangements are often negotiated by paid officials. They are then put in place solely on the basis of municipal agreement through resolution in the appropriate council, thus avoiding the need to 'trouble the electorate' who might be expected to reject such propositions if they were to appear on a municipal ballot. In the 1950s and early 1960s, the national Advisory Council on Inter-governmental Relations often endorsed such an approach, as did the National League of Cities and National Association of Councils in the USA.

Most such arrangements pertain to services rather than facilities, notwithstanding the role played by Moses in supplying much of the infra-structure for New York City. Service agreements are far more common than facility provision agreements. Such agreements are also not neces-sarily permanent, often having a limited time period written into the agreement. Furthermore, many are of a standby nature, coming into oper-ation only when certain conditions have been met – as with fire, public disturbance or some other local emergency agreements.

A very early example of such a cooperative agreement was the Lakewood Plan under which the municipalities agreed to take services provided by Los Angeles County. Although the provision of services was central to the Lakewood Plan, not all cities operating under the system took the same services. Services provided include fire and ambulance, law enforcement, planning and zoning, street construction and mainte-nance, as well as such things as tree trimming and running elections.

In the USA, cooperative metropolitan or regional councils of government (CoGs) date from the 1950s. It may be regarded as a logical device for securing some degree of integrated metropolitan government in a politically frag-mented system, even if some people may consider such CoGs as little more than paper tigers. What such arrangements require is a process of bargaining and negotiation, compromise and conciliation amongst a (often large) number of sovereign municipalities. CoGs essentially began as metropolitan planning commissions (Jones, 1983: 218) and there were few in place by as late as the mid-1960s. This was due largely to the absence of finance and because there was no mechanism to enforce cooperation. In 1965, the US federal government, through the Housing and Urban Development Act, provided both the finance and the means of ensuring cooperation. Even today such CoGs remain heavily dependent on federal finance for their existence.

Metro Washington provides a present day example of the CoG approach. Savitch and Vogel (1996: 281) indicate that its power and authority are quite limited, suggesting that it acts as a clearing house and as a recipient of grants which would otherwise not come to the area, which covers Washington as well as 15 cities and coun-ties in Virginia and Maryland. They suggest that whilst this example supports the public choice

position that coordination and cooperation can take place without formal arrangements, the CoG is not a replacement for metropolitan govern-ment, even being unable to serve as a forum for regional issues (Henig et al., 1996).

Savitch and Vogel (1996: 297) suggest that such bodies are in an awkward position, acting at the behest of others and having to 'work tactfully at the margins and splice together the pieces of authority . . . regional bodies seek a more tenable role by preferring to offer benefits rather than exact costs'. In effect, such organizations pursue strategies of mutual adjustment, giving strength to Holden's description of the relationships between the different members as being similar to those amongst nations in the international arena (Holden, 1964).

With the abolition of the two-tier metropolitan system in Britain, London and the other metro-politan areas have had to follow a similar path. France, with its *Communautés Urbaines*, in cities such as Lyons, Lille and Bordeaux, provides another European example. In both countries, metropolitan areas operate under a system which (following Biarez, 1997: 140), might be considered as 'flexible' or with 'variable geometry'. But Biarez also argues that the three examples are not similar, and that unlike Lyons and Lille, Bordeaux, as well as Marseilles, does not function well.

The Urban Communities of Lille and Lyons were created in 1967 and 1969 respectively. Lille contains 86 communes, and Lyons covers 55; both have about a million inhabitants. The governing bodies of both communities are indi-rectly elected and both have had strong leaders in recent years – Noir in Lyons and Mauroy in Lille. Both urban communities have extensive functions and resources, and both have adopted strategic development plans – in the case of Lille built around the TGV and other big projects in telecommunications, transport and the develop-ment of a technopole, linked to the community's ambition to be at the centre of Europe.

A similar objective has driven Lyons in recent years, where the *Communauté Urbaine* has been extended to the regional level. The result has been to bring together the original *Communauté Urbaine*, the region with four *départements* and other state offices. Again physical development and big projects have characterized the work of these cities, seeking to create 'an international city' or one of the 'motors' of Europe.

What characterizes the Lyons and Lille exam-ples is the relatively closed nature of decision-making and the lack of democratic control (deci-sions being taken by the key mayor or president of the urban community together with a number of other elected officials supported by a strong administration) (Biarez, 1997: 145), with these leaders supported by key elites from other sectors.

Another European example from this period is Stuttgart, which again has a form of metropolitan government. The Stuttgart regional community covers 179 localities, five counties and has almost two and a half million inhabitants. It is the centre for four major economic activities – cars (Mercedes Benz), aerospace (Deutsche Aerospace), electronics (AEG) and financial services. Again, political leadership, in the form of Manfred Rommel, mayor of Stuttgart, together with support from the *Land*, Baden–Württemberg, was important. Faced with the problems of economic recession and financial crisis in the early 1990s, it was Rommel who proposed the idea of a metropolitan form of government for Stuttgart.[4] After a number of years of negotiation and discussion, the Stuttgart Regional Community came into existence in 1994. Hoffman-Martinot, (1994) describes the reform as a 'substantial innovation' involving a directly elected 90 member Assembly, with substantial powers in five main areas: regional planning; transport; economic development and tourism; waste management and treatment; and regional sporting and cultural activities. No special financial arrangements were made, finance being expected to be supplied from within the region itself.

A number of features underlie this reform, one of the most recent in Europe. First, there is its highly pragmatic nature – at least in the early stages of its life. None of the functions involves much in the way of direct service provision, for example, and most of the services remain the responsibility of existing municipalities. Secondly, the key feature of the reform – the directly elected assembly – provides a mechanism designed to give a new institution extra political legitimacy whilst at the same time freeing it from the kind of municipal interests which indirectly elected systems often reflect, making the new institution somewhat stronger than the more common indirectly elected bodies. Third, primacy is given to planning and coordination of the work of 179 municipalities and five counties, even if involvement in direct service provision remains limited. Finally, at all stages of the reform there has been the clear involvement of leading private sector interests, such as chambers of commerce, and major entrepreneurs, such as Mercedes Benz, again perhaps reflecting the type of regime suggested by Stone (1993). Yet one cannot help but agree with Biarez (1997: 147) that the establishment of the regional community is much more a question of urban management designed to minimize conflicts between municipalities still jealous of their autonomy rather than a body for developing strategic policy on future development and service provision in the wider Stuttgart area.

By contrast, we can consider the case of London and the other British metropolitan areas who first gained two-tier metropolitan government in 1963 and 1974 respectively, only to lose it again under Margaret Thatcher in 1986 although it was reintroduced in the case of London by the end of the century. Notwithstanding their lack of popularity, it is difficult to understand exactly the logic that lay behind the abolition of the GLC (Greater London Council) and the metropolitan counties in 1986, other than partisan arguments. For example, it must have been particularly galling for Margaret Thatcher to be greeted by a sign reflecting London's unemployment totals in the mid-1980s as she passed the GLC headquarters on her way to the Houses of Parliament. Certainly the attack was part of a wider assault on local government at the time – as the largest part of the public sector, producing rather than enabling services, and a bastion of trade unionism. In the context of a government concerned to roll back the state and give vent to market forces, local government was a ready-made target.[5]

Abolition resulted in previously integrated metropolitan areas now being governed under a more fragmented system as the bottom-tier metropolitan districts took on a wider range of responsibilities. As Leach et al. (1991) make clear, the reform was not a minor tidying up operation but had wider consequences. Nor did it result in less government or less expenditure – both reasons given for abolition at the time. Nor did all the functions simply revert to the districts. The INLOGOV team's research shows that, in the former metropolitan counties at least, expenditure on both economic development and leisure/cultural services increased after abolition (both given as a justification for abolition), whilst services like fire, police and public transport were still being run on an area-wide basis by joint boards as late as 1990. In some areas other services also continued to be run on a county-wide basis by a joint board, and the team estimates that only about 25 per cent of former metropolitan county expenditure actually ended up at the district level – the rest went to county-wide bodies of one form or another (Leach et al., 1991: 30.

This limited change leads Leach et al. to consider the reform as a move from a two-tier system based on direct elections to one based on indirect elections – in effect, however, many of the joint boards were appointed (albeit from within the elected ranks of the member authorities), whilst other special purpose bodies also came into operation in the metropolitan areas – Training and Enterprise Councils (TECs) and Development Agencies being two examples.[6] And as with other countries with fragmented metropolitan systems of government, the recent British experience again raises the issue of accountability for action as well as that of public understanding of the system and its workings. It also raises questions about service efficiency and quality, with Leach et al. claiming that both coordination of activities in areas such as strategic planning and the quality of such services as waste management and trading standards decreased after abolition.

The experience of London was a little different in this respect. Admirers of the GLC at the time of its reform and during its life saw it as an example *par excellence* in the reform tradition: an area-wide body (almost), combining the strategic functions – planning, transport, traffic housing and others, and serving a large population (almost 7 million). As such, the London two-tier system served as a model for others.

Critics on the other hand complained of its excessive bureaucracy (Flynn et al., 1985), whilst others felt the allocation of functions between the different levels did not work and that politically the GLC suffered from political hostility from the lower tier authorities (Young and Garside, 1982). Reformers believed that the lower tier authorities were large enough to take on some of the GLC's functions and few enough to be able to run others through joint bodies. But though ministers often thought the GLC's strategic functions were overstated, Hebbert (1995: 348) is right when he characterizes abolition as a 'leap in the dark'.

In the first place it ended a one-hundred-year history of metropolitan government in one of the world's leading cities. Though the old London County Council had been a Labour stronghold, its history was largely one of success. The GLC – at least until abolition – failed to build up a sense of shared London-wide identity between itself and the lower-tier authorities in a way which the old LCC avoided. In this sense the GLC shared a similar position with the metropolitan counties elsewhere in England. Generally unloved, it and they suffered crucially from a lack of political support, and even some of the Labour boroughs in London (as elsewhere) were not too unhappy to see the GLC (and the metropolitan counties) disappear. Despite the fact that the reform proposals appeared to elevate the (Labour) leader of the GLC to the status of a martyr, the GLC was always seen as imperialistic and concerned with territorial domination. However, abolition meant that no clear political leadership for London existed, although the continuation of services was generally satisfactory and their transfer worked smoothly (Hebbert, 1995: 351–3). Much of this was due to the determination of central government to see that London's services did not suffer unduly from abolition and to the professionalism of the many staff involved

in handling the transfers through the London Residuary Body (LRB). So successful was this body that not only did it raise considerable financial resources from this operation, it was also seen as a suitable vehicle for handling the arrangements for the break-up of the one remaining London body – the Inner London Education Authority – in 1990.

As the millennium approached, the need for some *political* voice for London became clearer. Experience of working together amongst the boroughs helped reduce competitiveness between them – the greater majority of London residents identify with London (unlike their counterparts in other English metropolitan areas) – to the extent where calls for a stronger voice resulted in the introducction of an elected mayor and a new assembly for London. Abolition may 'have successfully dismembered the machine of London government, [but it has failed] to lay the soul to rest' (Hebbert, 1995: 357).

THE POLITICS OF METROPOLITAN GOVERNANCE

This review of the structures and workings of metropolitan governance raises issues about the nature of politics under these systems. Which interests come to the fore and which tend to be under-represented or ignored? How best can one describe the political task in relation to the fragmented systems of governance which predominate in metropolitan areas? What policies and issues are favoured and which prove less tractable to initiatives?

In his seminal and influential work on Atlanta, Stone (1989) uses what he later calls regime analysis as a mode of explanation for the operation of Atlanta politics. As such his work represents the latest in the 40-year tradition of community power studies in the USA from Hunter and Dahl onwards. In other cases, the importance of political leadership (Toronto: Colton, 1980; Kaplan, 1967, 1982; French cities such as Lille, Rennes and Lyons: Biarez, 1997; and in other British and American cities: Judd and Parkinson, 1990) is seen as the key factor enabling systems of metropolitan governance to adapt satisfactorily to their ever-changing environment.

What is clear is that the ever-increasing fragmentation of political institutions in the metropolis leads to an increasing number of political networks. Following Prud-homme (1996) and Rhodes (1997), the key political task is the *management* of these networks – which for Rhodes is the essence of governance. At its heart is the working together of local governments and other local institutions to address the challenges posed by post-industrialism. As such the boundaries between public and private spheres and sectors become blurred, and flexibility of approach in a variety of mechanisms is essential if policies and problems are to be tackled successfully. For Stone, it is the regime – a public/private partnership of key elites, political, economic and social – which achieves this task, more or less successfully depending on the nature of the regime and its longevity and legitimacy. For others, it is the skill of political leaders in successfully building and maintaining coalitions – the popular phrase is partnerships – covering different policy areas. However one puts it, it is the supply of management and leadership skills which is likely to determine overall success.

Reviewing the experience of metropolitan governance in a number of US cities, Savitch and Vogel conclude that the pattern of governance is:

> made out of politics . . . Coalitions and splits between groups make certain kinds of regionalism [governance] possible and preclude others. The process of eliciting cooperation creeps along slowly; it is incremental, and it is based on trial and error. Generally, solutions are negotiated around obstacles, so that thorny problems are avoided. (1996: 292)

They go on to suggest that the forms of governance differ according to the different problems which different metropolitan areas face, and that the cooperation engendered, though it is managerially viable, can also be politically fragile. The reason is simple – governance networks lack 'a loyal and dedicated constituent base' (Savitch and Vogel, 1996: 295). Metropolitan areas contain a large number of differentiated areas – some rich suburbs, some poor; as well as different classes and ethnic groups, each with a loyalty of their own. Diversity is the chief characteristic of metropolitan areas – and their governance is likely to reflect it. Reflecting multiple adversaries and narrow political turf, the political logic of regional/metropolitan cooperation is weak 'because it offers little direct, tangible and immediate payoffs to enough constituents' (Savitch and Vogel, 1996: 297). It is hard not to agree with their conclusion that the result is 'metropolitan governance without metropolitan government' – an appealing phrase which 'obfuscates the meaning of government' through a process of mutual adjustment (Savitch and Vogel, 1996: 298).

But to do so is to ignore the examples where the process of coalition or partnership building has met with success and where skilled political leadership has secured significant benefits for much of a particular metropolitan area. Such would be the view of those following in the footsteps of Stone and his analysis of Atlanta. Partnerships and skilled political leadership have helped bring about the successful and continuing regeneration of cities like Pittsburgh, Lille, Barcelona, Glasgow and Manchester. Key politicians have managed to put together a partnership and network of key actors behind a policy designed to secure economic growth. What Logan and Molotch (1987) call growth machines, or Elkin (1987) also refers to as regimes, have secured significant benefits in the area of economic development for many metropolitan areas, which in turn become the role models to be imitated by other cities.

Much research has been conducted in this area, hardly surprising in the face of the fact that economic regeneration and economic development have become major policy foci for all metropolitan and local governments in recent years. But such a research focus obscures or ignores two other features. First, there is the continuing success of 'world cities' such as London, Tokyo, New York and Paris, where political leadership may be weak or even lacking; where the system of government is highly fragmented and governance is fragile; and where the policy problems continue to grow. How is it that these cities continue to be successful in adapting to their changing global environment? As Fainstein et al. (1992: 4) note, cities such as London and New York:

> have been affected by the internationalization of capital and the rise of new technologies that have caused the outflow of old manufacturing from old urban centres to peripheral areas of home and abroad. They have at the same time profited from the increasing importance of financial and coordinating functions, as well as tourism and cultural production.

Putting it in simple terms, such cities have continued to innovate and to develop markets for new activities. All four cities are, in Sassen's (1991: 5) terms:

> Sites for (a) the production of specialized services needed by complex organizations for running a spatially dispersed network of factories, offices. and service outlets; and (b) the production of financial innovations and the making of markets . . .

Secondly, the focus on economic development, strategic planning, transportation and land use ignores other important policy areas such as education and welfare. Yet world cities are also, in a term used by Mollenkopf and Castells (1991), 'dual cities', containing within the same urban space increasingly separate economies, segregated neighbourhoods and social systems, highlighting once again the continued existence of an 'underclass' – a feature of such cities from the nineteenth century onwards. As Harloe and Fainstein (1992: 264–7) argue, this social marginalization is the major policy problem these cities and others face in future years, yet they are not optimistic about the likely outcome.

The characteristic diversity of metropolitan areas inevitably means that such areas contain widespread social differences and inequalities, for which governance, cooperation and fragmentation are ill suited. There may be some trickle-down benefits from successful economic development, but the evidence is that most major metropolitan areas contain significant social segregation; that disadvantaged groups continue to remain disadvantaged, and that the problems associated with such groups are growing rather than diminishing.

Whilst such a view holds true for most Western metropolises, it is even truer for the rapidly growing cities of the developing world, where government and governance are even weaker. Third World metropolises have a number of characteristics. First, they continue to demonstrate rapid growth of population and area, and Third World metropolises are amongst the largest concentrations of population in the world. Thus cities such as Mexico City (25 m), Sao Paulo (24 m), Calcutta and Bombay (both around 16 m) are amongst the largest in the world. Only Tokyo (20 m) and New York (16 m) are of a similar size amongst developed countries. Moscow (11 m) is the largest European city in this context (Colton, 1995). Third World cities lack infrastructure, public services, housing – especially for low income households – and generally face environmental problems. Thirdly, they often have poor quality administrative and political leadership, and lack the financial resources to deal with their problems (Bolay, 1995: 89–90). Yet such cities are also often the dominant economic centre of their country: Bangkok in Thailand, for example, accounts for over 37 per cent of the country's economic activity. Indian metropolises account for nearly 40 per cent. Yet in Bombay, riots in early 1993 resulted in the deaths of over 550 people and damaged property in excess of £1 billion. Massolos (1995: 202–3) describes how in

> most suburbs, and especially the middle class ones, whole compounds and blocks of buildings were converted into armed fortresses . . . it was clear that no one could be trusted and that there was no guarantee of security . . . people found themselves being attacked by people who had been their friends for a number of years . . . the norms on which the city had functioned seemed to dissolve.

Bombay has India's largest urban population (circa 16 m), is the country's major financial centre, has high-tech industry as well as cottage-style production centres, is a main attraction for immigrants, many of whom live in shanties or on the streets (Massolos, 1995: 210), and all lacking any real centre of control, i.e. governance or government. In South Africa, following the emancipation of black citizens, 'confusion and uncertainty about urban management necessarily follows in a period in which the ground has shifted so profoundly' (Mabin, 1995: 191), so that cities such as Johannesburg, Durban and Cape Town face difficulty in adapting to the new circumstances in which practically everything changes. Again, South African cities are characterized by levels of violence 'as high as any in the world: Bogota and some Brazilian cities come to mind as parallels' (Mabin, 1995: 193). Riots in other cities like Seoul, the pollution problems of some Chinese metropolises or in places like Mexico City and Buenos Aires, all remind us that by contrast most Western cities are largely still well ordered and well governed. Yet if such centres are to continue to adapt and grow, some solution to the problem of governance will have to be found.

The problem is not new. More than 25 years ago, Oliver Williams wrote perceptively:

> Present practices place in private hands the socially important task of city building. Economic market power is translated too directly and too immediately into the priorities for urban change. The result is that we have no concept of a basic minimum or floor for urban utilities . . . (Williams, 1971: 110)

Little appears to have changed in many metropolitan areas. If governance is the best we can do to manage the complex networks and interactions of interest at play in the metropolitan area, then the mobilization of bias in favour of major economic interests will continue to dominate public policy. Whilst such a strategy may well ensure that some metropolitan areas continue to compete in the face of the forces of economic globalization, they are likely to do so at the expense of significant numbers of city dwellers. One reason why there was both a strong interest in reform in the 1960s and 1970s and why some reforms took place, was the belief that such reforms could be more encompassing of all social groups and problems. Perhaps in the millennium attention will turn again to such issues, and the question of the forms of government for metropolitan areas will return to the agenda once more.

NOTES

1. The literature on the fiscal stress of cities is extensive. See, for example, Clark and Ferguson (1973) and Mouritzen (1992).

2. Annexation did undergo something of a renaissance in the post-war period until the mid-1950s, but was largely used by relatively small cities (under 5,000 population) to secure additional territory.

3. For a fuller discussion of metropolitan reform in Canada, see Kaplan (1982) and Magnusson and Sancton (eds) (1983). In Ontario the government

adopted a new Act in April 1997 which amalgamated the six metro municipalities into one area-wide authority, but less than the whole urban region. It came into effect, in April 1998.

4. For a fuller discussion of the emergence of the Stuttgart metropolitan authority, see Hoffman-Martinot (1994).

5. See *inter alia* Cochrane (1993); Stoker (1992).

6. Leach et al. (1991) are, however, correct in their assessment that abolition de-stablized the local government system still further, strengthening the claims of the former large shire districts (cities such as Nottingham and Bristol) for unitary status, and paving the way for the further reform of British local government in the mid-1990s.

REFERENCES

Biarez, S. (1997) 'Metropolisation against metropolitan government', in G. Saez, J.-P. Leresche and M. Bassand (1997) (eds), *Metropolitan and Cross Frontier Governance*. Paris: L'Harmattan. pp. 133–53.

Bolay, J. (1995) 'Third World metropoles: what environment for tomorrow?' in J.P. Leresche, D. Joye and M. Bassand (eds), *Metropolisations*. Geneva: Editions Georg.

Bollens, J. and Schmandt, H. (1970) *The Metropolis: Its People, Politics and Economic Life*. New York: Harper & Row.

Caro, R. (1974) *The Power Broker: Robert Moses and the Fall of New York*. New York: Vintage.

Clark, T. and Ferguson, L. (1973) *City Money*. New York: Columbia University Press.

Cochrane, A. (1993) *Whatever Happened to Local Government?* Milton Keynes: Open University Press.

Colton, T.J. (1980) *Big Daddy*. Toronto: University of Toronto Press.

Colton, T.J. (1995) *Moscow: Governing the Socialist Metropolis*. Cambridge, MA: Harvard University Press.

Davis, M. (1990) *City of Quartz*. London: Verso.

Elkin, S. (1987) *City and Regime in the American Republic*. Chicago: University of Chicago Press.

Fainstein, S.S., Gordon, I. and Harloe, M. (eds) (1992) *Divided Cities*. Oxford: Basil Blackwell.

Flynn, N., Leach, S. and Vielba, C. (1985) *Abolition or Reform? The GLC and the Metropolitan Councils*. Birmingham: INLOGOV.

Harloe, M. and Fainstein, S. (1992) 'Conclusion', in S. Fainstein, I. Gordon and M. Harloe (eds), *Divided Cities*. Oxford: Blackwell. pp. 236–68.

Hebbert, M. (1995) 'Unfinished business: the remaking of London government, 1985–1995', *Policy and Politics*, 23(4): 347–58.

Henig, J., Brunori, D. and Eberrt, M. (1996) 'Washington, DC: cautious and constrained cooperation', in H. Savitch and R. Vogel (eds), *Regional Politics*. Thousand Oaks, CA: Sage. pp. 101–29.

Hoffman-Martinot, V. (1994) 'Le Relance du gouvernment metropolitaine en Europe: le prototype de Stuttgart', *Revue Française d'Administration Publique*, no. 1 71.

Holden, M. (1964) 'The governance of the metropolis as a problem in diplomacy', *Journal of Politics*, 26: 627–47.

Jones, B. (1983) *Governing Urban America*. New York: Little Brown.

Judd, D. and Parkinson, M. (eds) (1990) *Leadership and Urban Regeneration*. London: Sage.

Kaplan, H. (1967) *Urban Political Systems: A Functional Analysis of Metropolitan Toronto*. New York: Columbia University Press.

Kaplan, H. (1982) *Reform Planning and City Politics*. Toronto: University of Toronto Press.

Kiernan, M. and Walker, D. (1983) 'Winnipeg', in W. Magnusson and A. Sancton (eds), *City Politics in Canada*. Toronto: University of Toronto Press. pp. 222–5.

Leach, S., Davis, H., Game, C. and Skelcher, C. (1991) *After Abolition*. Birmingham: INLOGOV.

Leresche, J-P., Joye, D. and Bassand, M. (eds) (1995) *Metropolisations*. Geneva: Editions Georg.

Logan, J. and Molotch, H. (1987) *Urban Fortunes*. Berkeley, CA: University of California Press.

Mabin, A. (1995) 'On the problems and prospects of overcoming segregation and fragmentation in Southern Africa's cities in the postmodern era', in S. Watson and K. Gibson (eds), *Postmodern Cities and Spaces*. Oxford: Basil Blackwell. pp. 187–98.

Magnusson, W. and Sancton, A. (eds) (1983) *City Politics in Canada*. Toronto: University of Toronto Press.

Massolos, J. (1995) 'Postmodern Bombay: fractured discourses', in S. Watson and K. Gibson (eds), *Postmodern Cities and Spaces*. Oxford: Basil Blackwell.

Mollenkopf, J. and Castells, M. (eds) (1991) *Dual City*. New York: Russell Sage Foundation.

Mouritzen, P.-E. (1992) *Managing Cities in Austerity*. London: Sage.

Prud'homme, R. (1996) 'Managing megacities', in *Cities*. Paris: CNRS for Habitat 2, Istanbul. pp. 170–76.

Rhodes, R.A.W. (1997) Preface in W. Kickert, E.-J. Klijnn and J.F. Koppenjan (eds), *Managing Complex Networks*. London: Sage. pp. xi–xvi.

Saez, G., Lersche, L.-P. and Bassand, M. (eds) (1997) *Metropolitan and Cross Frontier Governance*. Paris: L'Harmattan.

Sallez, A. (ed.) (1993) *Les Villes de l'Europe*. Paris: DATAR/I-éditions de l'aube.

Sassen, S. (1991) *Global City: New York, London and Tokyo*. Cambridge: Cambridge University Press.

Savitch, H. (1997) 'Globalisation and local politics'. Paper presented at the FASP Colloque on Local Politics and the Changing Public Sphere, Lyons, September.

Savitch, H. and Vogel, R. (eds) (1996) *Regional Politics*. Thousand Oaks, CA: Sage. pp. 101–29.

Sharpe, L.J. (ed.) (1995) *The Government of World Cities*. London: Wiley.

Soja, E. (1995) 'Postmodern urbanization: the six restructurings of Los Angeles', in S. Watson and K. Gibson (eds), *Postmodern Cities and Spaces*. Oxford: Basil Blackwell. pp. 135–7.

Stoker, G. (1992) *The Politics of Local Government* (2nd edn). London: Macmillan.

Stone, C. (1989) *Regime Politics: Governing Atlanta, 1946–1988*. Lawrence, KS: University of Kansas Press.

Stone, C. (1993) 'Urban regimes and the capacity to govern: a political economy approach', *Journal of Urban Affairs*, 15(1): 1–28.

Williams, O.P. (1971) *Metropolitan Political Analysis*. New York: Free Press.

Young, K. and Garside, P. (1982) *Metropolitan London*. London: Arnold.

22

Cities and Services: a Post-Welfarist Analysis

STEPHEN J. BAILEY

Combined with increasing urbanization, the development of the welfare state led to substantial increases in the size of local governments in developed countries in terms of population, employment, expenditure, administrative powers and duties, geographic area, etc. Local public services in cities have typically been directly provided by municipal governments themselves. However, the local public sector is becoming increasingly complex in many countries. Market mechanisms are being progressively introduced in the management of public services in Western Europe, Australasia, North America, Central and Eastern Europe and in developing countries.

These changes have been analysed through a number of analytical perspectives, including a post-welfare agenda (Bennett, 1990), a post-bureaucratic form (Hoggett, 1996), post-fordism (Burrows and Loader, 1994), the hollowing out of the state (Rhodes, 1994), and as a shift from government to governance (Bailey, 1993). Such categorizations are not necessarily mutually exclusive. They all emphasize a shift away from monolithic, hierarchical, highly standardized bureaucratic production technologies to micro-corporatist networked organizations dominated by meeting the needs of consumption rather than of production. The shift is driven as much by cultural and philosophical considerations regarding the role of the state in society as by changing technological parameters. Changes in management practice are influenced by changing philosophies for the welfare state, and both reflect relatively more emphasis on efficiency issues, serving to qualify earlier prioritization of addressing welfare needs.

Whilst addressing the same general areas, the prespectives and focus of these more recent themes differ quite significantly. For example, the post-bureaucratic theme pays particular attention to management issues, whilst the post-welfarism agenda is concerned with the impacts (through marketization and privatization) of the New Right project on the so-called *three Es* of service delivery: economy, efficiency and effectiveness.

The characteristics of the post-welfare agenda vary between the national and local contexts. In particular, it has to be recognized that welfare and post-welfare capitalism have different outcomes at the local, regional and national scale within any one country. Experience may differ not only according to the level of sub-national government (for example, federal, state and local) but also according to urban scale (if only because of different service responsibilities). Similarly, there are subtle differences between countries in the agenda, the danger being of overgeneralizing the experience of the USA by assuming that it is applicable throughout the developed world. There may, in fact, be fairly radical differences in the post-welfare agenda even within Western Europe, for example, comparing Sweden and the UK.

It is not possible to take account of such different outcomes in a chapter of this size and, besides, individual country case studies are available elsewhere (Batley and Stoker, 1991; Bennett, 1990; Bird et al., 1995). Instead, the following analysis concentrates on providing an overview of the broad themes, principles and initiatives of post-welfarism, subsequently integrated within a simplified model. Besides considering the delivery of city services, it also examines relevant aspects of their financing since the revenue consequences of welfarism have featured so strongly in the debate about the role of the state.

METHODOLOGICAL CAVEATS

No attempt is made to define precisely cities and their services since, besides being problematic, it is not the purpose of this chapter to make detailed international comparisons. A general conception of city government and the services typically provided will suffice.

Local governments can be thought of as democratically elected bodies whose jurisdiction is of a local (rather than regional or national) scale, backed by powers to levy local taxes by which to exercise genuine discretion over service provision (Cole and Boyne, 1995). Cities can be thought of as dense urban agglomerations generating economic growth from their own local economies (Jacobs, 1969). They sometimes develop into metropolitan areas or conurbations, often encompassing more than one municipality within their boundaries.

It is generally accepted that local and regional governments should provide services whose benefits are restricted to a single jurisdiction, for example local roads, local public transport, local parks and other leisure and recreational facilities, local and regional water and sewerage services. However, the situation is less clear-cut where there are significant benefits extending beyond administrative boundaries, for example education and health care. Such services may be provided by regional rather than local government. Alternatively, there may be some sharing of responsibilities between central and local government.

Ultimately, however, the designation of city titles, as well as the division of service responsibilities between central, regional and local governments, reflects past practice and historical factors more than theoretical prescriptions. Hence, there are substantial differences in the characteristics of city governments and in the array of powers attributed to them in different countries. The relevance of these methodological caveats will become apparent later.

WELFARISM AND CITIES

Welfarism refers to the development of the post-1945 welfare state within developed countries. The main features were the development of direct provision by the state of major services, notably education, health, public sector housing and personal social services. Service development went hand-in-hand with a commitment to full employment, the achievement of which was thought to be possible by adoption of Keynesian-style economic policies and programmes, themselves compatible with increasingly elaborate social welfare programmes. The state adopted an increasingly paternalistic, protective stance, supporting economically deprived groups whilst managing aggregate demand in the economy so as to promote economic growth and general prosperity.

A commitment to support people and families, effectively from the cradle to the grave, led to a general perception of entitlement and rights to welfare services. Rising expectations of increased welfare provision led to substantially increased levels of state intervention, further fuelling rising expectations, such that state involvement in society and the economy rose inexorably. The mechanisms by which such involvement occurred inevitably involved local government in general and, given increasing urbanization, city governments in particular. The inevitable result was the increased provision by municipalities of public sector services.

Despite the large number of local government systems and structures, local governments in different countries have provided many of the same services, the typical arrangement being one of multipurpose local governments (Karran, 1988). Throughout Western Europe, municipal provision has been the norm for refuse collection and disposal, theatres and concerts, museums, art galleries and libraries, sports and leisure pursuits, urban road transport, district heating, electricity, water supply, police, fire protection, pre-school and school education, family welfare services, welfare homes and housing. Other services such as parks and open spaces, roads, tourism, vocational and technical education, hospitals and health have typically been either a local or regional government function.

This emphasis on local governments as direct service providers, implementing the local welfare state, has arguably led to the democratic role of local government being underplayed. Local governments are not simply local providers of welfare services (perhaps in accordance with local, rather than central, wishes). They are also integral parts of the democratic system, promoting pluralism and participation, as well as public choice (Young, 1988). The justification of local government is ultimately to be found in political and constitutional theory rather than in administrative convenience.

THE BASIS OF POST-WELFARISM

The key features of post-welfarism are decentralization to markets (typified by the UK) and to sub-national levels of government (typified by 'new federalism' in the USA), the balance between them varying from country to country. The shift from welfarism to post-welfarism is the outcome of a complex amalgam of influences, not simply rolling back the frontiers of the state as a result of economic necessity.

Bennett (1990) argues that, consequent upon global changes which have served to restrict governmental actions, there has been a general reappraisal of the role of government throughout the member states of the Organization for Economic Co-operation and Development (OECD). The increasing economic interdependence of developed economies has progressively restricted the degree to which countries can follow their own tax, spending and other policies at odds with the rest of their trade partners.

Certainly, the efficacy of macro-economic (demand-side) policy as a means of maintaining the welfarist commitment to full employment came increasingly to be questioned as citizens' faith in governments' abilities to spend their way out of recession diminished. Increasing use was made of supply-side policies designed to help markets work better, namely those for labour, land, capital, and goods and services (Bailey, 1995a: 113–45).

The development of public choice theory also questioned acceptance of the despotic, benevolent model of government underpinning welfarism. Instead, it can be argued that the state grows monstrously, serving the interests of those who work in the public sector (bureaucrats, for example) and of particular pressure groups upon whose votes politicians depend: the Leviathan model of government (Bailey, 1995a: 99–112). This inhospitality thesis, analysed in more detail elsewhere (Bailey, 1993), argues that it is idealistically naive to regard governments as protectors and promoters of the 'public interest', if such a paternalistic concept could be defined. Voting systems do not gauge the intensity of preferences and voters are known to be poorly informed about alternative policies on offer. Furthermore, the voter's knowledge may be heavily biased in favour of his or her special interests, the imbalance between focused benefits and diffused costs arguably leading to excessive public spending on voted-for services. Moreover, simple forms of representative government are becoming increasingly inadequate, both in representing an increasingly diverse multicultural society with increasingly divergent socio-economic conditions and in ensuring public choice and accountability.

Deficiencies in demand articulation, representation and accountability may allow both local politicians and bureaucrats to act opportunistically rather than promote the public interest. Politicians may seek to increase their chances of re-election and political survival by serving the needs of special interest groups which seek to obtain the largest possible share of output for their members. Bureaucracies tend to become larger (empire building) and, since bureaucrats themselves have votes and are perhaps more likely to use them than other groups, they can acquire sufficient political power to guarantee self-preservation and self-interest.

These tendencies towards suboptimal outcomes for the public interest are not necessarily resolved by attempting to improve local democratic accountability. Democracy is essentially predicated by competitive processes within electoral politics, voters being able to replace an underperforming political party with another which seems more able to meet voters' wants. There is, however, only periodic contestability and often limited choice between alternative parties because of high entry costs. Combined with a general disinterest in local politics (possibly reflected in low rates of voter turnout at local elections), the conditions described above lead to a mismatch between service needs/demands and service supply (allocative inefficiency). They also create 'organizational slack', leading to unnecessarily high costs of service provision (X-inefficiency).

Put more simply, the ability of local governments to adapt to changing socio-economic conditions is severely constrained because their services tend to be targeted at particular groups or localities. Any reduction of services leads to a reduction in the benefits accruing to them but no change in tax liability. Hence, local government services are difficult to restructure in response to changing service needs, this characteristic reinforcing self-serving behaviour by local bureaucrats and politicians.

There are two broad approaches to such inefficiencies. *Demand-side reforms* could make voters better informed, for example through larger electoral assemblies and annual elections to single-tier local governments (to strengthen the connection between voters and their political representatives) or by single-issue voting (referenda). Whilst such measures have the potential to improve allocative efficiency, it is simply too expensive (in terms of time and effort) for voters to become knowledgeable on all issues. They may prefer to trust politicians to make the right decisions or simply not be concerned with local democratic processes at all.

Supply-side measures are the basis of attempts to secure improved economy, efficiency and effectiveness. They have the potential to improve X-efficiency by constraining the pursuit of self-interest by bureaucrats. They include value-for-money audits, performance review, efficiency studies and the opening up of in-house provision to competitive forces. However, there is no obvious reason why the replacement of monolithic by multiple solutions should be contained within the existing institutional framework of local government itself, nor even within the public sector. The lack of clear answers to socio-economic problems is increasingly seen as requiring more variety of response and multiple solutions. It may be regarded as unreasonable that individuals should have to uproot themselves and move between individual local governments to secure preferred service packages, assuming local governments compete in such ways (Dowding et al., 1994). Local citizens should perhaps be in a position to choose between alternative providers of local public services in their present locality. Monolithic public sector solutions are increasingly being seen as both inappropriate and inimical to promoting choice. The perceived solution is to promote *consumerism* rather than *producerism*. This increasingly consumerist orientation for public sector services has been occurring throughout the OECD since the early 1980s (OECD, 1987a).

PRINCIPLES OF POST-WELFARISM

Besides promoting democratic decision-making by increasing opportunities for participation by citizens (Council of Europe, 1995), the creation of stronger local democratic institutions is increasingly being seen as a supply-side policy in the form of national capacity-building and institutional strengthening. These are increasingly being regarded as prerequisites of sustainable economic growth, most notably in the Central and Eastern European 'economies in transition'. It is increasingly believed that decentralization reduces the inherent inefficiencies of the previously centrally controlled welfarist system and facilitates privatization.

Central decisions impose uniformity of service provision on localities with divergent preferences, some localities being forced to consume more than they wish of certain services and less than they wish of others (Oates, 1972). This mismatching of supply of service with the demands expressed by their users results in a loss of economic welfare (allocative inefficiency). Hence, where administratively feasible, there should be a general presumption in favour of *political decentralization*, namely devolution to local and regional governments of political decision-making so as to better reflect needs (OECD, 1987b). Besides accepting the principle of political decentralization, the OECD also accepted the need for *administrative decentralization* to managers for improved accountability for responsibilities (OECD, 1994a, 1994b).

The ultimate form of decentralization, however, is *economic decentralization*, making economic decisions more fully reflect consumer sovereignty by increasing the role of private markets (through privatization, for example). Where private market provision is not feasible on equity grounds, in school education for example, internal or quasi-markets are a means of facilitating both economic and administrative decentralization. Put simply, government can be made much more responsive to consumers' preferences and wants by not imposing centrally assessed needs on individuals. Facilitating user choices in such ways also has the potential to improve the efficiency with which services are provided, competition reducing organizational slack and so increasing X-efficiency.

The consumerist principle relates not just to *levels* of service provision but also to the *quality* of service. Like that relating to the private sector, the literature on public service quality is now similarly permeated with terms such as 'choice', 'customer', 'responsiveness', 'competition', 'market', 'quality', 'service options' and 'consumerism' (Younis et al. ,1996).

In the UK, for example, concern for quality underpins the six principles of public service set down within the 1991 Citizen's Charter (Office of Public Service and Science, 1991). Despite its official title, the Charter is more a consumer's charter than a citizen's charter, supposedly making services more responsive to the wishes and needs of their users by the setting, monitoring and publication of explicit standards which individual service users can reasonably expect. Performance against the Charter's standards is periodically reviewed. In addition, the UK's Local Government Act 1992 requires each individual local government to publish information in local newspapers about its performance in providing its major services. Performance data is also published nationally (Accounts Commission, 1995; Audit Commission, 1995). UK local governments are also acting independently of the Citizen's Charter in paying increasing attention to corporate quality programmes (LAA, 1993) and issuing service or charter guarantees, for example target response times for dealing with the public (Office of Public Service and Science, 1995).

However, a distinction has to be drawn between meeting the direct needs of users in accordance with total quality management (TQM) systems (for example, timeliness, accessibility and courtesy) and programme effectiveness (namely, outputs and outcomes). In that TQM relates to quality of service *delivery* rather than to service *outcomes*, a customer orientation is not the only yardstick by which to evaluate public sector services. Nevertheless, it represents a move away from quality as a collectivist, paternalistic and inward-looking, institution-based, concept to outward-looking responsive relationships with individual customers. The intended shift is from a *despotic, benevolent model* of government to a *service exchange model* (where service users make specific tax payments in exchange for specific agreed services).

These changes are broadly in accordance with the 10 principles of 'entrepreneurial government' (Osborne and Gaebler, 1993). They are:

- *competition* between service providers;
- *empower citizens* by transferring control from the bureaucracy to the community;
- emphasize *outcomes*;
- replace rules and regulations by *missions*;
- redefine clients as *customers* who are offered choices between schools, training programmes, housing and other public services;
- *prevent* problems before they occur;
- increase emphasis on *earning* money rather than simply spending it;
- *decentralize* authority;
- replace bureaucratic mechanisms with *market mechanisms*;
- replace direct provision by *catalysing all sectors* (public, private and voluntary) into action to address community problems.

These principles are clearly different in kind from the conventional arguments about whether more or less government intervention is required. They are concerned with the quality rather than the quantity of services, which in turn depends on the kind of government. 'To be more precise, we need better *governance*' (Osborne and Gaebler, 1993: 24).

Defined in terms of these 10 principles, governance is clearly not the same as privatization. Traditional approaches based on legal rules and sanctions, regulation/deregulation, licensing, grants, loans, subsidies and contracting give way to innovative methods such as public–private partnerships and corporations, rewards, assistance and information, voluntarism, charges and vouchers (OECD, 1987b). Entrepreneurial governments separate policy decisions from service delivery. Put more simply, they *steer rather than row*, stimulating competition between alternative service providers to maximize flexibility and accountability.

Whilst believing deeply both in the need for government and in the integrity of those who work in it (unlike the inhospitality thesis of public choice theory), Osborne and Gaebler argue that the *system* of government creates inefficiencies. 'We believe that industrial-era governments, with their large, centralized bureaucracies and standardized, "one size fits all" services, are not up to the challenges of a rapidly changing information society and knowledge-based economy' (Osborne and Gaebler, 1993: xviii). Hence, they are not concerned with what government should do but rather how it operates, at all levels. 'The central failure of government today is one of means, not ends', (Osborne and Gaebler, 1993: xxi). Hence, they argue that government needs to be reinvented.

A preoccupation with fiduciary duty led to stifling bureaucratic controls over inputs and processes, outputs and outcomes being ignored as a consequence. They argue that the bureaucratic model may have been appropriate in a past slowly changing age of hierarchy and mass markets but it is no longer appropriate in the fast-moving global market place and information society. In the post-industrial era institutions need to be more flexible and more responsive to consumer demands, leading by persuasion and incentives rather than by command, empowering citizens rather than simply serving them.

The UK government also has in mind similar objectives. It argues that local governments need to operate in a pluralistic way, working alongside other public, private and voluntary agencies to *enable* provision of services, rather than provide them directly. Local governments need not necessarily be the direct provider of services for which they are statutorily responsible and have a policy-making role. They should focus on developing strategic policies to address needs and concentrate on setting standards, awarding contracts, determining and enforcing regulations and monitoring performance. The belief is that the need to compete for service users motivates producer organizations to search for efficiency improvements, leading not just to expenditure savings and improved cost control but also to increased programme effectiveness.

The boundary between the public and private sectors is likewise becoming increasingly blurred in many other countries by the development of pluralistic public sector *governance* structures seeking greater efficiency in the provision of public sector services (OECD, 1993: 32–42).

THE SHIFT TOWARDS WELFARISM

In addition to the USA and UK, these principles of post-welfarism have been adopted in many other OECD member countries over the past decade or so, notably Australia, Canada, Denmark, Finland, France, New Zealand, Norway and Sweden. They display differing balances on the continuum between the two polar extremes of decentralization to sub-national levels of government and decentralization to markets. However, almost all of the 27 member states of the Council of Europe have ratified the 1985 European Charter of Local Self-Government and the European Union has adopted the decentralization principle under the name of 'subsidiarity', namely that government powers should be exercised at the lowest level of government possible (Council of Europe, 1994). Similarly, there has been an almost universal tendency to reform administrative boundaries and functional responsibilities, and a growing tendency to create special bodies for the delivery of local public services (Martins, 1995: Oates, 1990).

Reform of Politico-Administrative Boundaries

Introduced in many European countries, boundary reforms have been intended to secure the most efficient, most democratic size of municipalities within a general context of the urban–rural shift of manufacturing, de-urbanization and increasing mobility of people and economic activity. Whilst this led to dramatic reductions in the total number of local governments in many of the 27 member states of the Council of Europe between 1950 and 1992, many countries still have local governments with populations both less than 1,000 and more than 100,000 (Martins, 1995). More than half the local governments (municipalities) in 15 European countries have fewer than 5,000 citizens. At the other extreme, no country has more than a tenth of its municipalities with more than 100,000 population.

Elsewhere, separatist and decentralization issues have been prominent, for example Quebec in Canada and Catelonia in Spain. Countries in Central and Eastern Europe have tended to move towards increasing decentralization following the collapse of the former Soviet Union and the transition from a command to a market economy (Bird et al., 1995). Such decentralization was often aided by development agencies and bilateral assistance programmes, including the World Bank's Fiscal Decentralization Initiative, the European Union's PHARE programme (which provides finance in support of economic and democratic reform in Central and Eastern Europe), the United Nations' Centre for Human Settlement (UNCHS) and the Council of Europe's LODE (an acronym for LOcal DEmocracy) Programme.

Reform of Functional Responsibilities

Redistribution of service responsibilities between different levels of government has taken place within the context of the European Charter of Local Self-Government and the EU's subsidiarity principle, both of which require public responsibilities to be exercised by those authorities closest to the citizen.

However, there is a potential trade-off between larger local governments to achieve economies of scale (that is, reduce average costs of service provision) and smaller local governments to improve citizen representation. Where maximum size is not sufficient to minimize the costs of providing particular services, intermunicipal cooperation may achieve economies of scale. Intermunicipal enterprises are sometimes used for provision of public utility services such as public transport, energy, water and sanitation and waste removal. For example, single function 'special districts' are common for water supply, sewers, highways, hospitals, libraries, housing and fire protection in the USA, as are 'secondary communes' in the Nordic countries.

Whilst technical services such as domestic waste collection and water and sewerage services can achieve scale effects, other services cannot since they are not standardized (for example, personal social services). Moreover, it is difficult to measure the extent of any economies of scale for non-technical services since, in many cases, neither costs nor output are clearly definable and measurable (Boyne, 1995). Single-purpose special district authorities are less suitable for such services. Where local governments provide an array of services, the size that achieves economies of scale for some services may lead to diseconomies of scale for others. It may also encompass diverse groups of residents with discrete preferences, provision of standardized services resulting in allocative inefficiency. In these cases, decentralization of service provision may be necessary, for example in the management of municipal housing. Such initiatives will inevitably have different implications for different sizes of city governments, in part reflecting their different service responsibilities and different national contexts.

Changes in the Ways Local Governments Provide Services

Changes in the ways local governments undertake their given functions have involved adoption of new models of public service delivery (OECD, 1987b; Walsh, 1995). Except for privatization, the following initiatives involve the centralization of strategic commands simultaneous with the decentralization of operations, introduction of (market and non-market) competition to coordinate activities of those decentralized units, and further development of performance management techniques by which to appraise them. Competition is being promoted in producer markets via market testing (the state usually retaining sole purchaser powers), in consumer markets via increased scope for user choice, and in internal markets via competition between public sector service units.

Privatization

Privatization has been used to describe so many different courses of action that its use has become highly misleading (Butcher, 1995: 107–20). If, however, it is used only to refer to the replacement of collective decisions regarding *both* production and consumption of services by private market decisions, then it strictly refers only to the sale of public assets to the private sector, for example municipal airports and seaports, bus companies and municipal housing. Such economic decentralization is only feasible in a minority of cases so that other measures are necessary if core functions are to be subject to increased competition.

Market-testing the provision of public services

This focuses on liberalization (that is, the removal of statutory monopolies), allowing private firms to compete with public providers. Ultimately, competitiveness in the provision of services is seen as the key principle which should underpin provision of local government services in order to ensure economy, efficiency and effectiveness (Audit Commission, 1987). Conventional economic theory regarding the degree of competitiveness in any industrial or service sector suggests that the greater the market share of the dominant firm the greater the degree of monopoly power and the greater the risk of exploitation of the consumer through high prices. More recent theoretical developments show that the greater the *threat* (as opposed to the fact) of market entry by potential competitors, the greater the efficiency of incumbent firms, irrespective of their actual number or proportion of the market for which they account (Baumol et al., 1982). The lower the sunk costs of entry and irrecoverable costs of exit, the greater the contestability of an individual market and so the greater the pressures on incumbent firms to achieve maximum economy, efficiency and effectiveness. Hence, public policy should attempt to maximize wherever possible the scope for *contestability* rather than necessarily attempt to increase the number of companies producing a particular good or service.

Market testing requires bids for service contracts to be invited from private as well as public sector suppliers and public procurement practices to be opened up to greater international competition. Contracting-out occurs where private provision is cheaper than public provision and contracting-in in the reverse case. Even contracting-in leads to efficiency improvements because creating a market for provision of the service introduces the threat of competition.

Although there has not been a major shift to contracting-out throughout the OECD over the previous decade, formal contracting procedures have been applied in UK local government since 1980. Referred to as the compulsory competitive tendering (CCT) regime, it applied to both technical services and non-technical services including construction, office cleaning, refuse collection, street cleaning, school and welfare catering, vehicle maintenance, management of sports and leisure facilities and management of municipal housing (Bailey, 1995a: 367–401).

The UK government subsequently replaced the CCT regime with a more comprehensive system of 'Best Value', the latter being not just concerned with efficiency but also with effectiveness and the quality of local government services (DOE, 1997). The new regime therefore addresses the three Es more effectively and covers all services, not just those previously subject to CCT. Whilst competition continues to be an important part of the new regime, it is no longer the only means of securing value for money. The rationale for the new regime is that market testing does not, in itself, take account of the needs and preferences of service users. The same can be said of quasi-market mechanisms (see below). In comparison, placing a statutory duty on local authorities to provide Best Value reaffirms that they exist to serve their local citizens. Local authorities are required to demonstrate to service users, auditors and central government officials that they are achieving best value. In that they are required to make comparisons with alternative means of service provision, local choices and local accountability are supposedly promoted.

Quasi-market mechanisms

Quasi-markets in health and local government are bureaucratically managed by institutional purchasers and outcomes are necessarily constrained by the public purpose considerations underpinning the provision of services. Provision of these services remains within the public sector when it is thought that attempts to increase private market contestability are likely to lead to suboptimal outcomes. For example, competitive pressures may lead potential entrant companies to submit unrealistically low tender prices for public sector service contracts in order to win contracts (the 'winners curse'), ultimately leading to bankruptcy and so collapse of service provision.

Furthermore, community-wide social objectives for such activities as sports and leisure are broad and vague and performance measurement against objectives is often haphazard. Contractors may act opportunistically to increase their profits by reducing quality of service in order to cut costs. Combined with limited information about the benefits of service and lack of alternative supply (because of highly specific dedicated physical capital), the result is high transaction costs, namely the costs of using contracts to secure the intended levels and qualities of services (Williamson, 1986). Either the service outcomes intended by the local government will not be achieved or the costs of ensuring that they are will be higher than the costs of provision by the local government itself.

Although rather limited in scope, the evidence in respect of UK local government services subject to contestability is that quality of service has been maintained or improved simultaneously with a reduction in costs (Bailey and Davidson, 1997). However, Kirkpatrick and Martinez Lucio (1995) argue that the micro-political uses of the emphasis on service quality have not simply been to priortize the interests of users. They argue that quality and 'value for money' issues have been used to legitimize increased managerial control over professional and other 'street level' staffs in the provision of public sector services, with little or no real effort to democratize and open producer interests to new external considerations by extending the involvement of users in the design, delivery and assessment of public services. According to this analytical framework, the introduction of competition, operational decentralization and increased use of performance management systems may become 'technologies of control' used to strengthen managerial positions (Miller and Rose, 1990).

Nevertheless, where it is expected that transaction costs will be too great, quasi-markets can be created to stimulate competition between public sector providers and so, in theory, achieve many of the benefits of competitive market provision by the private sector. The structural constraints of quasi-markets necessarily restrict both user choice (in terms of exit to alternative providers) and user voice (in terms of being better represented in management and decision-making processes). Ideally, however, service users can 'shop around' amongst alternative public sector suppliers with the result that providers face incentives to be more responsive to the requirements of their 'customers'.

Whilst sharing the common concept of quasi-markets, countries have differing institutional arrangements in both the public and private sectors for the provision of health, education and other such services. They may have differing provisions regarding the degree of schools' autonomy, for example, the extent to which individual schools can determine the subjects taught and their syllabuses, educational methods employed and the size of the pupil intake. They may also differ in terms of the degree of pupil and parent choice regarding school attended, the role of parents on any school governing bodies, the use of educational vouchers etc. Similar comments can be made about health services and the degree of patient choice. Ultimately, however,

the main principle is that money should follow the patient or other service user so as to reward the most effective service providers. In its most radical form, ineffective service providers are either allowed to become bankrupt or are closed down.

Reform of public sector labour markets

This has typically sought to relax highly restrictive centralized conditions of employment relating to the work to be undertaken and the method of payment. Standardized national pay and conditions do not reflect local labour market conditions, creating difficulties recruiting some categories of staff due to below-market rates of pay (and the reverse case). Staff have typically been difficult to redeploy in response to changing service requirements. Throughout the OECD, there has been fairly extensive and increasing use of decentralized payment systems (usually within budgetary ceilings), performance-related pay (particularly for higher levels of management), employment of workers under contract, and promotions based on merit (rather than on length of service).

Increased use of user-charges

In almost all OECD countries for which consistent time-series data is available, there is a clear trend for user-charge revenues to grow substantially faster than other income sources (Bailey, 1994). They are not, however, being used simply to replace income from local taxes and intergovernmental grants. Countries with the highest percentage increases in user-charge revenues also have the largest proportionate increases in grants and local taxes. Charges have therefore not been used as a means by which to reduce the (national and local) tax implications of public expenditure. Their growth reflects a myriad of factors including changing service responsibilities, faster growth of services for which charges are already major sources of revenue (municipal energy and water supply, for example), new charging technologies (such as for road use), changing patterns of urbanization and so on. Otherwise, within these parameters, user-charges seem to be increasingly used as a means by which local government services can be made more sensitive to user preferences, for example equating supply of particular sports and leisure facilities with willingness to pay. Allowing individual service users greater discretion over their take-up of services, and relating charges to it, is simply a variant of subsidiarity. This has the potential to improve both allocative and X-efficiency if both the quantity and quality of services increasingly match users' preferences and producers face incentives to reduce service costs, particularly when in competition with alternative service providers.

Other forms of private finance

User-charges are obviously a demand-side form of private finance. Other forms of private finance have the potential to reduce the cost of supply (a supply-side instrument). Private finance has been used for highways, bridges and tunnels in Norway, the Netherlands and the UK, the last being the most radical in having a Private Finance Initiative (Bailey, 1995a: 136–42; HM Treasury, 1996). This initiative is being used to enable provision of a diversity of capital and infrastructure projects relating to transport, health, defence, accommodation, urban regeneration, information systems, prisons, higher and further education, and water and sewerage. The public sector authority decides in broad terms the service it requires and then invites bids for the work from the private sector. The latter is expected to share total project risks with the public sector in bearing the risks of projected design, finance, construction and subsequent management. The private sector provides management expertise and access to capital markets whilst the public sector provides policy-making and strategy. Private operators recover their costs through payments, for example actual or shadow tolls for use of transport infrastructure, rents for accommodation and volumetric charges for water. The project is still a public service and there is a clear presumption that the better project management will secure increased value for money (economy, efficiency and effectiveness).

Competitive bidding systems

These are introduced in accordance with the post-welfarist principle of earning money rather than simply spending it. For example, the UK has introduced competitive bidding systems for the distribution of its National Lottery funds in support of local arts, sports, national heritage, charity and millennium projects (Bailey, 1995b), and for Single Regeneration Budget and Rural Challenge funds (CIO, 1995; Scottish Office, 1996). Similar systems are being employed for distribution of European Union funding.

MODELS OF WELFARISM AND POST-WELFARISM

The main features of welfarism and post-welfarism can now be built into two very simplified models by means of which to summarize the

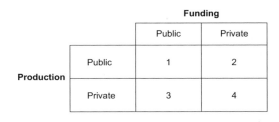

Figure 22.1 A general model of welfare provision and finance

foregoing discussion. The first model, taken from Klein (1984), relates to government in general whilst the second model relates to local government in particular.

Cell 1 in Figure 22.1 depicts welfarism. This is the 'pure' model of the welfare state as universal provider, the state directly producing as well as financing (through taxes and other public funds) public sector services free at the point of use. The state still provides services in cell 2 but now they are financed by user-charges and other sources of private finance. Cell 3 represents the scenario where the role of the state changes from being universal provider to the regulatory state. Cell 4 is the mandating welfare state (a welfare society without a welfare state), where the state passes legislation to make compulsory the private provision and financing of welfare services, for example compulsory health insurance. Whilst welfarism is represented only by cell 1, post-welfarism contains elements of all of cells 2, 3 and 4, illustrating its diverse nature.

Cell 3 in Figure 22.1 can refer to two cases. First, where the state makes transfer payments to individuals by means of which they can buy services from private producers. Second, where services are contracted-out to private sector producers. Ignoring any constraints arising from wider social benefits, the first necessarily improves allocative efficiency since purchases reflect individual wants and willingness to pay. On the same basis, the second does not improve allocative efficiency since decisions regarding the quantity and quality of service remain within the public domain: in other words consumers are not sovereign. Increased consumer sovereignty would be facilitated by a cell 3/4 hybrid, user-charges being levied at less than full cost (so that publicly financed, partial subsidies continue) combined with competition for customers between alternative providers.

Cell 2 is potentially unstable since if user-charges can fully finance production costs and payment really is voluntary, there is not necessarily any reason why production of the service should remain within the public sector. Economists would recommend privatization in such cases, namely sale to the private sector of public sector trading bodies and removal of statutory monopolies (liberalization) in order to allow private firms to provide the service. A shift from cell 2 to cell 4 would then occur. Activities would only remain in cell 2 if they were in joint supply with other activities in cell 1 and if their separation were technically impossible or would incur inordinately high costs, or a decision were taken by the public sector provider to continue to make public supply available even though private sector provision existed. An example would be information brokerage services supplied by local government libraries, even though similar services are available from the private sector, because public libraries have the necessary staff and expertise and infrastructure (that is, computerized network services) and also wish to widen access to such services.

Cell 4 is typified by the growth of private policing in the UK (Johnstone, 1992) and, more radically, by the dramatic growth of private government in the USA, in the form of 150,000 or so residential community associations (RCAs) and 1,000 or so business improvement districts (BIDs) (Ashworth and Clark, 1997; Broom and Wild, 1996; Dilger, 1993; Houston, 1996; ICMA, 1997; Lavery, 1995; Travers and Weimer, 1996). RCAs and BIDs are groups of private property owners agreeing by majority voting for additional municipal-style services to be provided in their own areas (neighbourhood, commercial centre, office complex or town centre) and for which charges or property-based taxes are levied on their members. These broadly democratic but private organizations take collective decisions for provision of services such as security, leisure and recreation, street cleaning, refuse collection, grounds and roads maintenance and local economic development in residential areas (RCAs) and in city or town centres (BIDs). They have developed out of a context of significant involvement of non-profit organizations and voluntary action. Payments to RCAs and BIDs are in addition to those paid to their true (constitutional) local governments.

A move to call 4 could be progressively self-reinforcing. As people make their own insured provision of private health care, for example, this could weaken the political voice for more spending within the public sector. The result could be that public sector provision of the service becomes residualized, making private provision more attractive. In such cases, the shift from welfarism to post-welfarism increases, the so-called 'overspill model' of privatization (Klein, 1984).

Hence, the shift from welfarism to post-welfarism can be thought of as a continuum rather than as a discrete categorical shift, and one which has several strands. The question is whether the shift reaches a limit short of complete post-welfarism. That depends on the nature of individual services and, in particular, whether private provision is complementary with, or substitutes for, public provision. For example, it can be argued that private sector provision of health care services would not find it profitable to treat all medical conditions (particularly the chronically sick and terminally ill). In this case the shift will reach a natural limit. Other services such as private company refuse collection can be considered as full substitutes for public provision, typified by the community associations in the USA, such that the shift from welfarism to post-welfarism could be complete.

The shift from welfarism to post-welfarism may also be encouraged if restraints on government expenditure are such that the informal sector of the economy takes on a growing absolute and relative responsibility for service provision. For example, local government provision of residential care places for the elderly may be limited by fiscal stress. This occurs when there is a structural gap between revenues and expenditures such that the costs of public service provision tend to rise faster than the revenue needed to finance them (Bailey, 1991). Coinciding with an ageing demographic structure, the result may be that families and local community voluntary organizations take on more responsibility for care of the elderly. This may be encouraged by tax reliefs and other fiscal incentives, leading to a progressive shift from cell 1 (or cell 3) to cell 4 in Figure 22.1 as demonetization and/or redomestication of service provision takes place.

Whilst highly illustrative of the differences between welfarism and post-welfarism, Figure 22.1 fails to differentiate between the different forms of contracting in the provision of public services. There is necessarily a split between the state as purchaser and the private company as provider in cell 3 of Figure 22.1 above. However, cell 1 can include both a formal split between purchaser and (public sector) provider or no such split. Figure 22.2 is therefore required to further illustrate the purchaser–provider split within local government.

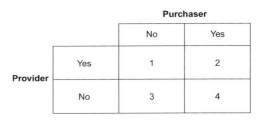

Figure 22.2 A local government model of the purchaser–provider split

In cell 1 of Figure 22.2 the local government directly provides the service with no purchaser–provider split. In cell 4 provision of the service is contracted-out to a private producer (ideally by competitive bidding) and so purchased by the local government. Cell 2 denotes contracting-in where local government employees win service contracts in competition with local companies. Both cells 2 and 4 encompass the purchaser–provider split.

The purchaser–provider split creates contestability for providers but not for purchasers. Hence, whilst it has the potential to improve X-efficiency (through the threat of competition), it will not improve allocative efficiency because it does not make purchasers' decisions more fully reflect the preferences of service users. In effect, market features are combined with a modified hierarchy of control to create managed markets. Increased centralization (of strategy) and decentralization (of self-managed functional units) are achieved simultaneously, a distinctive feature of post-bureaucratic control. In most cases, decentralized functional units have regulated autonomy rather than real autonomy since their remits and budgets are strictly controlled by the centre.

Cell 3 represents transfer of responsibility for service provision outwith local government. The cell 3 scenario refers to provision of services by the non-local government public sector (as defined in the methodological caveats section). Central government (or state governments in a federation) could decide that a former local government service should become the responsibility of a public sector body outwith local government. These quasi-autonomous non-governmental organizations (quangos) operate at the national, regional or local level. In respect of the last, they are 'local bodies which are independent or self-governing, but which spend public money and perform public functions. In many cases these functions were previously provided by national or local government. In Britain this includes bodies such as Training and Enterprise Councils (TECs), Local Enterprise Companies (LECS) in Scotland, Housing Associations, Further Education Corporations and the Boards of Grant Maintained Schools' (Nolan, 1995: 67).

It has been estimated that UK quangos account for about a fifth of public expenditure. Not surprisingly, concerns have been expressed as to whether 'quangocracy' is the right way to govern a country, the main issue being lack of democratic accountability (LGIU, 1994). In sharp contrast with local government, the meetings of quangos are usually not open to the public, the basis on which decisions are made is therefore unknown and there is usually no requirement for members of quangos to declare potential conflicts of interest, for example in the award of contracts.

The power of patronage is therefore increased. Hence, rather than enabling individuals, critics argue that quangos reduce the influence that individuals have on decisions that affect their lives, a non-elected elite assuming responsibility for a large part of local governance (Mulgan, 1994). This is clearly contrary to the 'empowering citizens' principle of post-welfarism.

The counter-argument is that democratic accountability does not, in itself, ensure that services respond to users' needs. The usual assumption of economic theory is that politicians operating within competitive electoral systems respond to (median) voter needs in order to ensure that they are re-elected to political office. In contrast, some political scientists argue that politicians adopt policies on the basis of party ideology rather than in accordance with median voter preferences, there being some empirical evidence in support of this hypothesis that politics and political parties do matter (Boyne, 1996a; Castles, 1982; Sharpe and Newton, 1984).

However, other political scientists argue that even party ideology does not necessarily determine service provision and its financing, ideology being tempered by the institutions of government, the inheritance of past programmes and pragmatism (Rose, 1984, 1990). There are, of course, other potential explanations of the spectrum of service provision, socio-economic factors, for example traditionally being used to derive the presumed wants of local electorates (Davis et al., 1966).

In addition to these more general factors, service provision at the local government level may diverge from local voter preferences because of conflicts between central and local governments. In the UK, for example, increasing central control over local authority spending during the 1980s and 1990s led to a progressively closer correlation between central government's assessment of what each local authority needed to spend and the level of spending actually undertaken. The statistical relationship (r^2) between local government spending (by type of authority) in England and Wales and central government's spending guidelines rose from around 50 in the early 1980s to well over 90 in the early 1990s (Boyne, 1996b).

In short, democratic accountability does not necessarily transform a producer-oriented service into one which is consumer-responsive. The latter requires service users to be given effective choices between providers whose standards of service have been publicly approved and which are subject to right of redress when they are not attained. This is market accountability rather than political accountability. The former is justified on the grounds that some consumerist rights are more effective than democratic rights in securing accountability.

As already noted, single party political dominance may combine with low voter turnout at local elections to create limited accountability for local governments in providing services. Much the same could be said of central government, however. Moreover, where quangos face little competition there is every reason to suppose that they are more likely than local government to be inefficient, self-serving and unresponsive, especially given their hidden decision-making processes.

IMPLICATIONS OF THE POST-WELFARIST MODEL

Competition is a powerful force shaping the behaviour of all public sector employees, even those with traditional notions of public service as distinct from the new values of public enterprise. This creates an immediate dissonance between the values of workers and the ways in which they act. If attitudes follow behaviour rather than vice versa, then cultural values change, with the result that competition introduces an entrepreneurial culture which, in turn, should transform local governments from a producer-oriented universal provider to a consumption-oriented regulatory body (steering rather than rowing). This is what is referred to by *governance* as distinct from *government*, the latter implying a standardized form of polity, a highly organized and coordinated form of civil government. Governance suggests a multiplicity of ways in which representation is achieved and in which services are delivered.

A shift from government to governance will have profound implications for the way in which services are delivered. In theory at least, the purchaser–provider split model depicted in Figure 22.2 focuses local government's attention on securing the desired service outputs and outcomes, since inputs and processes are now the responsibility of the provider. There will also be implications for the way in which local political processes operate, the respective roles of local politicians, administrators, service providers and service users, the management function, budgeting, the contribution of voluntary action and 'voluntarism', the interface between the public and private sectors and the emphasis on consumerism as distinct from producerism.

Vertically integrated forms of organization, which have traditionally provided services in-house, are broken down into out-house, horizontally integrated devolved service units. These service units or cost (or profit) centres effectively become small and medium-sized public enterprises providing education, health, local economic development and other services. This is

what is characterized by *entrepreneurial* government.

The development of local public service networks requires a shift away from the relatively simple vertically integrated collegial hierarchy within local government to a form of contractual relationships between individual departments operating as competitive internal consultancies charging for professional services. External relationships already exist between local governments and the private and voluntary sectors but they will become more comprehensive and encompass the new agencies providing education and other services, part of a complex, disaggregated, horizontally linked system of service-level agreements. Local democratic public accountability remains as part of the enabling function but local government will increasingly be defined by new forms of accountability relating to the fulfilment of contractual and quasi-commercial arrangements with contractors, residents, partners in the voluntary and commercial sectors and even with central government.

This wider form of accountability will permeate not just the local government purchaser–provider relationship but also the intra- and inter-departmental relationships within local authorities. Political power and authority will increasingly give way to negotiation and persuasion. Local politicians' internal committee responsibilities will have to be supplemented by increased representation on the boards of external agencies. Increased emphasis on negotiation and persuasion will require behavioural changes on the part of officers and politicians alike. Power relationships will be transformed as functional representation of different interest groups increasingly takes place through a variety of elected and non-elected organizations of which local government is only one, even if the main, constitutional form.

CONCLUSIONS

Post-welfarism does *not* mean the end of the welfare state. Instead, it denotes an increasing emphasis on securing improved economy, efficiency and effectiveness of public services. Post-welfarist analysis does *not* assume that the private sector is necessarily more efficient than the public sector. Instead, it is more concerned with making the public sector more efficient and, where public sector provision is more effective than private sector provision, the activity can legitimately be retained within the public sector. The move from welfarism to post-welfarism represents a shift from a producer state to a consumer-oriented state; from a 'nanny' state dispensing general well-being, but supposedly fostering a dependency culture based on collective rights and entitlement, to one based on the primacy of individual responsibility for personal welfare.

This is not a new political philosophy. It dates from as far back as Aristotle's view of government as the art of governing free men, resurfacing periodically in the writings of St Thomas Aquinas and John Locke. It does not question the ends of society but, instead, addresses the principles governing its organization. Post-welfarism is consistent with (and, indeed, still requires) the existence of social groups and collective action. Hence, post-welfarism is *not* privatization in the narrower sense of the term used in this chapter.

The issue essentially boils down to how to reconcile efficiency, responsiveness and accountability in providing collective services. On the one hand, there is little virtue in highly democratically accountable local government if it is at the cost of service inefficiency. On the other hand, there is little virtue in narrowly defined efficiency if decisions are not democratically legitimated. Whilst the efficiency and effectiveness of conventional models of central and local government control is increasingly brought into question, pure consumer markets for education, health and such like services are likely to yield unacceptable outcomes in terms of equity. The real question is one of balance, drawing the best from each approach whilst avoiding the major pitfalls.

Local governments clearly draw on both welfarist and post-welfarist approaches in the provision of services. There is a danger that, in emphasizing what is new, attention will be drawn away from what is not new, so giving a misleading impression of radicalism. Certainly there is much innovation, but many mechanisms are common to both welfarism and post-welfarism. The fact is that user-charges, public–private sector partnerships, contracting-out and elements of internal markets have long been used by city governments in providing their services. Whilst it is undoubtedly the case that much more extensive use is being made of these and other initiatives, the main differences are the perception of the role of the state and the culture of its management.

The states of many countries are seeking more diverse approaches to long-standing problems and issues, less sure in themselves of their capacity to know how best to achieve social and economic goals. The role of the state is increasingly seen in terms of its core functions of strategic policy-making, thereafter encouraging the private and voluntary sectors to make greater contributions in the delivery of welfare services.

The scope for such intervention by these other sectors is probably a direct function of the size of municipalities. Larger city authorities constitute larger potential markets for private sector providers of public services, especially in affluent areas. In contrast, smaller (particularly non-urban) authorities offer less profit potential because of the smaller potential markets and higher costs associated with the delivery of services in sparsely populated areas. This seems to be the case from the UK's experience of compulsory competitive tendering for sports and leisure services, refuse collection, etc. Similarly, the voluntary sector tends to be more pro-active in urban areas, if only because of the greater concentrations there of particularly needy groups. Moreover, affluent middle-class areas also seem to be more adept at 'earning' money through systems of competitive bidding for national funds spent at the local government level.

To the extent that this analysis is valid, the scope for post-welfarism is greater in city governments than elsewhere and this, in turn, may lead to even greater disparities in the provision of services between metropolitan areas and their hinterlands. Such disparities may be exacerbated by the implications for cities as generators of economic growth. Post-welfarism has potentially profound implications for local economic development initiatives and the strategic role of city governments as facilitators and enablers: as catalysts of change rather than as direct bringers of it. Their potential as initiators of supply-side measures is substantially increased, as is their strategic role in networking with local economic development agencies in the non-governmental public sector, the private sector and the voluntary sector. An appreciation of the urban condition, and the means by which urban problems can be addressed, become ever-more complex as monolithic 'remedies' increasingly give way to multiple post-welfarist solutions. It is now well recognized that the experience of the UK and USA is that deprived groups tend not to benefit significantly from such a reorientation. Hence, it is arguable that the progressive adoption of post-welfarist principles will heighten both the distinction between metropolitan and non-metropolitan areas and the differentiation between deprived and affluent neighbourhoods within any one municipality.

REFERENCES

Accounts Commission (1995) *Performance Information for Scottish Councils 1993/94*. Edinburgh: Accounts Commission.

Ashworth, S. and Clark, G. (1997) 'A bid to revive town centres', *Housing Agenda*, October: 22–23.

Audit Commission (1987) *Competitiveness and Contracting Out of Local Authorities' Services*. London: HMSO.

Audit Commission (1995) *Local Authority Performance Indicators*, Volumes 1, 2 and 3. London: HMSO.

Bailey, S.J. (1991) 'Fiscal stress: the new system of local government finance in England', *Urban Studies*, 28(6): 889–907.

Bailey, S.J. (1993) 'Public choice theory and the reform of local government in Britain: from government to governance', *Public Policy and Administration*, 8(2): 7–24.

Bailey, S.J. (1994) 'User charges for urban services', *Urban Studies*, 31(4/5): 745–65.

Bailey, S.J. (1995a) *Public Sector Economics: Theory , Policy and Practice*. Basingstoke: Macmillan.

Bailey, S.J. (1995b) 'The National Lottery: public expenditure control and accountability', *Public Money and Management*, 15(4): 43–8.

Bailey, S.J. and Davidson, C. (1997) 'Did quality really increase under local government CCT?', in L. Montanheiro, B. Haigh, D. Morris and Z. Fabjančič (eds), *Public and Private Sector Partnerships: Learning for Growth*. Sheffield: SHU Press.

Batley, R. and Stoker, G. (eds) (1991) *Local Government in Europe: Trends and Developments*. Basingstoke: Macmillan.

Baumol, W.J., Panzer, R. and Willig, R.D. (1982) *Contestable Markets and the Theory of Industry Structures*. New York: Harcourt Brace Jovanovich.

Bennett, R.J. (ed.) (1990) *Decentralization, Local Governments and Markets: Towards a Post Welfare Agenda*. Oxford: Clarendon Press.

Bird, R.M., Ebel, R.D. and Wallich, C.I. (1995) *Decentralization and the Socialist State: Intergovernmental Finance in Transition Economies*. Washington, DC: IBRD/World Bank.

Boyne, G.A. (1995) 'Population size and economies of scale in local government', *Policy and Politics*, 23(3): 213–22.

Boyne, G.A. (1996a) 'Assessing party effects on local policies: a quarter century of progress or eternal recurrence?', *Political Studies*, 44: 232–52.

Boyne, G.A. (1996b) 'Competition and local government: a public choice perspective', *Urban Studies*, 33(4–5): 703–21.

Broom, D. and Wild, D. (1996) 'Manhattan Transfer', *Public Finance*, 23, August: 10–13.

Burrows, R. and Loader, B. (eds) (1994) *Towards a Post Fordist Welfare State*. London: Routledge.

Butcher, T. (1995) *Delivering Welfare: The Governance of the Social Services in the 1990s*. Buckingham: Open University Press.

Castles, F.G. (ed.) (1982) *The Impact of Parties: Politics and Policies in Democratic Capitalist States*. London: Sage.

CIO (1995) *Urban Regeneration*. London: Central Office of Information. HMSO.

Cole, M. and Boyne, G. (1995) 'So you think you know what local government is?', *Local Government Studies*, 21`(2): 191–202.

Council of Europe (1994) 'Definition and limits of the principle of subsidiarity', *Local and Regional Authorities in Europe No. 55*. Strasbourg: Council of Europe.

Council of Europe (1995) 'The size of municipalities, efficiency and citizen participation', *Local and Regional Authorities in Europe No. 56*. Strasbourg: Council of Europe.

Davis, O.A., Dempster, M.A.H. and Wildavsky, A. (1966) 'A theory of the budgetary process', *American Political Science Review*, 60: 529–48.

Dilger, R.J. (1993) 'Residential community associations: their impact on local government finance and politics', *Government Finance Review*, 9: 710.

DOE (1997) *Compulsory Competitive Tendering: Changes to Regulations and Guidance*. London: Department of the Environment, Transport and the Regions/Welsh Office. Consultation Paper, 25 July 1997.

Dowding, K., John, P. and Biggs, S. (1994) 'Tiebout: a survey of the empirical literature', *Urban Studies*, 31(4/5): 767–978.

HM Treasury (1996) 'The Private Finance Initiative', *Economic Briefing*, No. 9, April: 3–5. (London: HMSO.)

Hoggett, P. (1996) 'New modes of control in the public service', *Public Administration*, 74 (Spring): 9–32.

Houston, L.O. (1996) 'Business Improvement Districts', *Development Commentary*, 20: 4–9.

ICMA (1997) 'Business Improvement Districts: Tool for Economic Development', *MIS Report*, 29(3). Washington: International City/County Management Association.

Jacobs, J. (1969) *The Economy of Cities*. London: Jonathan Cape.

Johnstone, L. (1992) *The Rebirth of Private Policing*. London: Routledge.

Karran, T. (1988) 'Local taxing and local spending', in R. Paddison and S.J. Bailey (eds), *Local Government Finance: International Perspectives*. London: Routledge.

Kirkpatrick, I. and Martinez Lucio, M. (eds) (1995), *The Politics of Quality in the Public Sector: The Management of Change*. London: Routledge.

Klein, R. (1984) 'Privatisation and the welfare state', *Lloyds Bank Review*, no. 151, January: 12–29.

LAA (1993) 'Quality initiatives: 1993 Directory of Local Authority Activity', *Local Authority Associations Quality Group*. London: Association of Districts Councils.

Lavery, K. (1995) 'Privatisation by the back door: the rise of private government in the USA', *Public Money and Management*, 15(4): 49–53.

LGIU (1994) 'Quangos under examination: the rise and rise of the unaccountable state', *LGIU Briefing*, July: 7–13. (London: Local Government Information Unit.)

Martins, M. (1995) 'Size of municipalities, efficiency, and citizen participation: a cross-European perspective', *Environment and Planning C: Government and Policy*, 13: 441–58.

Miller, P. and Rose, N. (1990) 'Governing economic life', *Economy and Society*, 19(1): 1–31.

Mulgan, G. (1994) 'Democratic dismissal, competition, and contestability among the quangos', *Oxford Review of Economic Policy*, 10(3): 51–60.

Nolan, Lord (1995) *Standards in Public Life: First Report of the Committee of Standards in Public Life*, Volume 1. London: HMSO. Cm 2850–I.

Oates, W.E. (1972) *Fiscal Federalism*. New York: Harcourt Brace Jovanovich.

Oates, W.E. (1990) 'Decentralization of the public sector: an overview', in R.J. Bennett (ed.), *Decentralization, Local Governments and Markets: Towards a Post Welfare Agenda*. Oxford: Clarendon Press.

OECD (1987a) *Managing and Financing Urban Services*. Paris: Organization for Economic Co-operation and Development.

OECD (1987b) *Administration as Service: The Public as Client*. Paris: Organization for Economic Co-operation and Development.

OECD (1993) *OECD Economic Outlook*, No. 54, December. Paris: Organization for Economic Co-operation and Development.

OECD (1994a) 'Performance measurement in government: issues and illustrations', *Public Management Occasional Papers*, No. 5. Paris: Organization for Economic Co-operation and Development.

OECD (1994b) 'Performance management in government: performance measurement and results-oriented management', *Public Management Occasional Papers*, No. 3. Paris: Organization for Economic Co-operation and Development.

Office of Public Service and Science (1991) *The Citizen's Charter: Raising the Standard*. London: HMSO. Cm 1599.

Office of Public Service and Science (1995) *The Citizen's Charter: The Facts and Figures – A Report to Mark Four Years of the Charter Programme*. London: HMSO. Cm 2970.

Osborne, D. and Gaebler, T. (1993) *Reinventing Government: How the Entrepreneurial Spirit is Transforming the Public Sector*. New York: Plume.

Rhodes, R.A.W. (1994) 'The hollowing out of the state: the changing nature of the public services in Britain', *Public Administration*, 65(2): 138–51.

Rose, R. (1984) *Do Parties Make a Difference?* London: Macmillan.

Rose, R. (1990) 'Inheritance before choice in public policy', *Journal of Theoretical Politics*, 2(3): 263–91.

Scottish Office (1996) *The Rural Challenge Fund: Bidding Guidance 1996*. Edinburgh: Scottish Office.

Sharpe, L.J. and Newton, K. (1984) *Does Politics Matter? The Determinants of Public Policy*. Oxford: Oxford University Press.

Travers, T. and Weimar, J. (1996) *Business Improvement Districts: New York and London*. London: London School of Economics and Political Science.

Walsh, K. (1995) *Public Services and Market Mechanisms: Competition, Contracting and the New Public Management*. Basingstoke: Macmillan.

Williamson, O.E. (1986) *Firms, Markets and Policy Controls*. Hemel Hempstead: Wheatsheaf.

Young, K. (1988) 'Local government in Britain: rationale, structure and finance', in S.J. Bailey and R. Paddison (eds), *The Reform of Local Government Finance in Britain*. London: Routledge.

Younis, T., Bailey, S.J. and Davidson, C. (1996) 'The application of total quality management to the public sector', *International Review of Administrative Studies*, 62(3): 369–82.

23

Social Policy and the City

SUSANNE MACGREGOR

Drug-dealing, inner-city crime, family disintegration, mass unemployment, are now all aspects of everyday experience.

Watkins, 1995: 5

POVERTY AT THE END OF THE TWENTIETH CENTURY

The economic policies that have swept through the world since the 1980s, and now exercise almost complete dominance over politics and minds, have brought in their wake a series of social problems, most evident in cities, which are characterized as social exclusion, rising crime and violence. The irony is that these problems have grown in a period of unparalleled affluence.

Since the Second World War, global economic wealth has increased sevenfold and average incomes have tripled (Watkins, 1995: 1). Improvements have been evident in increased life expectancy, falling infant mortality, improved nutrition and educational attainment. Yet, '[t]oday, one-in-four of the world's people live in a state of absolute want, unable to meet their basic needs' (1995: 2).

A UN report in 1994 saw the situation thus: 'the incidence of poverty in the OECD countries is distributed as follows: the number of single-parent families in poverty has grown; unemployment and insecure employment are increasing in significance as causes of poverty; and the elderly form a declining proportion of the poverty population' (UN, 1994: 86). 'In both Europe and North America, sluggish and uneven growth has left enclaves of unemployed ethnic minorities in declining industrial centres' (McFate, 1995: 1).

In the USA, an additional four million children fell into poverty during the 1980s (UNICEF,

1994). 'By 1992, child poverty affected 22 per cent of all [American] children, and infant mortality rates for black children were more than double those for white children' (Watkins, 1995: 5). There were estimated to be 36.9 million poor people in the USA in 1992 – 14.5 per cent of the population. In 1971, this proportion had been 12.5 per cent. In the European Union, the number of people living in poverty grew from 38 million to 52 million between 1975 and 1988 (UN, 1994: 84). In the same period in the UK, the number of individuals living below the poverty line increased from five million to almost 14 million (Joseph Rowntree Foundation, 1995).

The phenomenon of a sense of reduced quality of life alongside rising overall income is a key feature of contemporary northern cities: 'substantial improvements in living standards and material well-being have been accompanied by a weakening of the social fabric reflected in the rise of homicide, crime, drug abuse, suicide, divorce and illegitimate births' (UN, 1994: 228–9).

CONCEPTS OF POVERTY

Some commentators question whether poverty can be measured and compared across continents and time periods. Increasingly, it is argued, not only can this be done but it must be so if progress is to be made (Townsend, 1993). 'Poverty is at the same time culture-bound and universal' (Oyen, 1996: 4). Key ideas in concepts of poverty are subsistence, basic needs and relative deprivation. The major part of poverty research, since its beginnings one hundred years ago, has focused on counting and measuring the extent and nature of poverty. But such statistics do not lead automatically to clear-cut solutions. 'Unacceptable numbers of poor people can be portrayed both as a demonstration of unworthy poverty conditions and as a demonstration of a spreading moral ill' (Oyen, 1996: 9). Constantly, distinctions have been made between the deserving and the undeserving poor: '[t]he language of poverty is a vocabulary of invidious distinctions . . . we use our language to exclude, to distinguish, to discriminate . . . by mistaking socially constructed categories for natural distinctions, we reinforce inequality and stigmatize even those we set out to help' (Katz, 1989: 5–6).

The poor and poverty cannot be understood in isolation from an understanding of their society and city as a whole. At the centre of any explanation are the relations between these groups, these individuals, and the rest of their city and society. 'The non-poor and their role in creating and sustaining poverty are as interesting an object for research on poverty as are the poor' (Oyen, 1996: 11). This is why the study of policy and policy responses is so important. Through an analysis of policies and practices, we can discern the crucial relations between groups in society, in the cities, which shape the size and nature of the social problem of poverty.

It is worth reminding ourselves of what Gans has described as the functions of poverty (Gans, 1991). The existence of poverty may be of benefit to others in the city: it may encourage discipline, hard work and proper behaviour patterns – serving as a warning of what befalls the deviant and miscreant. The poor have even been seen more recently as crucial to sustainable living, by ecologists who applaud their role in recycling waste products, like the rag-pickers of Bombay and the refuse-collecting Zabbaleen of Cairo. Being willing to work for low wages, the poor also contribute to the control of labour demands and inflationary pressures. The poor may serve as the scapegoats for deep fears about the city in a late modern climate, carrying on their backs fears of violence, crime and of the stranger.

Discussions of poverty in the particular setting of the city have often been concerned to point to quality of life issues as much or more than measures of absolute or simply material poverty. The components of quality of life indexes are open to debate, ranging as they do from years of schooling to measures of political freedom and civil rights, but they are valuable attempts to include aspects of human dignity as well as mere survival in the account. Most commonly, life expectancy at birth has been taken as a proxy measure of human welfare. On this indicator, it has been shown that for some groups in some parts of First World cities, their quality of life is lower than that of some groups in the developing world. Urban poverty in both First World and Third World contexts appears to have some distinctive characteristics: the influence of the informal labour market (both illegal and semi-legal or unregulated); the notable presence of female-headed households; the impact of environmental hazards; and the stress of social and personal isolation. Other issues that figure in the context of discussion of the cities concern matters like personal suffering, fear, vulnerability, injustice and a sense of dignity. Granted that the miseries of the city and those of rural areas may simply not be comparable, that miseries exist in the cities, even in countries and cities which are overall wealthier than they have ever been is the stark fact which characterizes concerns about social problems in the contemporary city.

These problems have posed challenges to traditional approaches to social policy: 'in the midst of the affluence and excess of the 1980s, most Western nations experienced an increase in unemployment and poverty that has left traditional safety-nets sagging and policymakers scrambling to adjust programs to fit a changing economic environment' (UN, 1994: 229). Slower growth, more employment insecurity, restructured families and more diverse communities pose questions for social policy in the cities of the advanced industrial countries.

EXPLANATIONS AND CHANGING PERCEPTIONS

What is now evident is that we are living through a period of profound social and economic transformation, one with political and cultural ramifications. The changes in this late industrial period are equal in scope to those of that earlier great transformation, brilliantly analysed by Karl Polanyi (1944). Polanyi argued that in all previous human societies, the economy was submerged or embedded in social relationships. The distinctiveness of modern industrial capitalism lies in the way the market is seen as separate and dominant. The central argument of *The Great Transformation* is that the liberal utopia of a generalized self-regulating market is a prescription for disaster.

Polanyi wrote in the 1940s after a period which had, like that of the 1980s, attempted to restore the nineteenth-century liberal economic order. The results then included the rise of fascism and Nazism in Europe. In the USA, the New Deal demonstrated a rather different kind of response. Now some argue that the lessons drawn by Polanyi are relevant once again. Prioritizing the profit motive above all else is a recipe for social and cultural degradation. This has been identified (perhaps unfairly) as a specifically Anglo-Saxon cultural trait, one which elevates 'institutions of private property, private enterprise and private profit to normative ends in and of themselves and the fiction of "economic man" to a pseudo-scientific shibboleth' (Levitt, 1995: 8–9).

This sense of a clash or contradiction of interests and values is encapsulated in the concept of the dual city, which has been popular in explaining social polarization. The aggravation of social exclusion and structural unemployment is thought to have resulted in the rise of dual societies almost everywhere (Bessis, 1995: 7). The dominance of the market and deregulation approaches in recent years has produced the two-speed society, most visible in the large cities.

This concept has, however, been criticized for resting on a vague and confused image of the post-industrial city, not standing up to evidence and glossing over the complexity of social processes and social relations (Mooney and Danson, 1997; Vankempen, 1994).

Undoubtedly, however, urban inequalities have increased in all advanced countries in recent years. 'Persistent joblessness, material deprivation, and ethnoracial tensions are on the upswing throughout much of Western Europe and North America as segments of the urban population of these countries appear to have become increasingly marginalized and segregated economically, socially and spatially' (Wacquant, 1995: 543). Fears about a black underclass trapped and cut off in the urban core in America are mirrored by fears about the new poverty of groups on outer estates or in the banlieues of Europe.

So as the 1980s moved on into the 1990s, there grew an increasing awareness that all was not well in the cities. The most common explanation offered for these social ills was that an urban underclass had developed, composed of non-working poor whose moral and individual shortcomings accounted for what was happening. These should be the target of new social policies. These underclasses were usually seen as being primarily composed of particular ethnic groups, in the USA of African-Americans, in other countries of 'immigrants'. Racism and reaction found an outlet in the rising political impact of far-right groups, whose activities tended to violence.

The alternative structural explanation, stressing factors other than individual failings, pointed to increasing bifurcation of the job market between, on the one hand, highly skilled and well-paid jobs, and on the other, low-skilled, low-paid, insecure and part-time employment. Unemployment had become concentrated among the least skilled and long-term unemployment had increased. Such structural unemployment was seen to pose challenges to conventional social protection programmes.

At the same time, the share of service jobs in the overall economy of all these countries was rising. Wage inequality increased, with the least educated suffering most decline. Part-time work increased, as did other forms of temporary, contingent, insecure employment (contract, short-term, self-employment) and women were more likely to be found in the category of part-time employment. These non-standard employees are least likely to have robust social insurance or other protection, such as access to health, disability or accident insurance and pension entitlements. 'There has been a new "flexibility" in private sector work arrangements, but it has not been matched by a new flexibility in social protection programs' (McFate, 1995: 9). Households with young children, especially where the parents are unemployed or unskilled, or where a mother is the head of the household, are at risk of being in poverty and these risks have increased in recent years.

All explanations, behavioural and structural, concluded that welfare reform was required (Atkinson, 1995).

FROM POVERTY TO EXCLUSION

In attempting to grasp these changes, and develop appropriate policy responses, social analysts developed a new language. The social question is now one of *exclusion* rather than of *poverty* (cf. Social Summit, UN World Summit on Social Development, Copenhagen, 6–12 March 1995). The replacement of poverty with exclusion is primarily a switch from static to dynamic analyses, focusing not merely on counting the poor but highlighting the processes which lead to exclusion. These debates also link into analyses of social disintegration and pauperization. Of course, academic discourse is never wholly innocent and rarely less so than in the field of poverty research. Discourses of poverty are part of a political contest where those involved aim 'to act upon reality by acting upon its representation' (Wacquant, 1995: 546). The concepts used create, and are themselves permeated by, commonsense theories of poverty and social relations. They are shot through with assumptions about causes and solutions to the social problems so constructed.

One issue on which this chapter will try to cast light is whether or not there are fundamental differences in approach between the 'Anglo-Saxon' liberal regimes and the conservative, continental European regimes (Esping-Andersen, 1996). It is interesting to note that in many ways these approaches appear to be converging. Or, it may be concluded, they are being superseded by a new consensus on social policy, one which focuses on partnership and places greater stress on both individual responsibility and state compulsion at one and the same time.

Wacquant argues that there are fundamental differences between American and European discourse, with Europeans emphasizing a language of class, labour and citizenship and Americans being anchored in vocabularies of family, race and individual moral and behavioural deficiency (1995). Whereas in the USA the central concepts are those of underclass and the new poverty, in Europe, the key term employed is that of social exclusion. It is in France that the language of exclusion has been used most explicitly in policy design and from there that ideas have spread into the discourses of the European Union:

> The emphasis is now on the structural nature of a process which excludes part of the population from economic and social opportunities. The problem is not only one of disparities between the top and bottom of the social scale but also between those who have a place in society and those who are excluded. (European Green Paper on Social Policy, CE–81–93–292–EN–C)

Those involved in consciously constructing the European Union paid a great deal of attention to this issue. The final report of the Second European Poverty Programme commented that 'the completion of the single market . . . will not be seen as an unqualified success if it does not take the phenomenon of social exclusion into account (CEC, 1991: 27). The report warned:

> there is a considerable risk of two different societies developing within member states, one of them active, well-paid, well-protected socially and with an employment conditioned structure, the other poor, deprived of rights and devalued by inactivity. (CEC, 1991: 8)

POLICY OPTIONS AND DEBATES

Welfare state protections have been under attack in many countries in campaigns to make labour markets more flexible. New forms of social protection are thought now to be required to respond to the growth of the informal economy and the decline of traditional work careers.

In both the USA and Europe, there are urban enclaves of entrenched poverty where housing blight, joblessness, educational failure and ethnic/racial discrimination cumulate into constellations equally impervious to policy intervention (Wacquant, 1995: 543). Inner-city areas in the USA are on the whole more degraded than those found in European cities. Yet the lived experience of the inhabitants of poor neighbourhoods 'exhibit striking similarities, owing to their structural position at the bottom of their respective

social orders' (Wacquant, 1995: 546). American ghettos share the ubiquity of joblessness, welfare receipt and poverty, the destruction of housing and the waning of the economic fabric. In Europe, outer estates are stigmatized public housing projects, on the peripheries of cities, decimated by deindustrialization, dependent on public assistance and characterized by multiple deprivation.

But while there are similarities, there are also differences between countries. Wacquant sees the core of the American situation as racial exclusion; the core of the European as class fragmentation. France and the USA, for example, differ in the extent to which they 'regulate the lower segments of the labor market and have elaborated policies to absorb and redistribute the social costs of the restructuring of their economy . . . [and in the extent to which] they provide their citizens and residents with minimum standards of living and segregate social provision from "welfare"' (Wacquant, 1995: 544).

The French state has an extensive array of industrial reconversion, unemployment and training policies and retains an active, interventionist labour policy (in spite of relatively low rates of union membership). There is universal health coverage and also family allowances and a national guaranteed minimum income plan. The USA lacks most of these, although having extensive educational provision and social security for older people.

The US Welfare Regime

American ghetto residents traditionally relied on public assistance benefits: Aid to the Aged, Blind and Disabled; Aid to Families with Dependent Children (AFDC); General Assistance; and Food Stamps. These varied in different states. In most, the standard package was insufficient to lift recipients close to the poverty line. 'Receiving welfare . . . is tantamount to enforced destitution' (Susser and Kresnicke, 1987). Escape was difficult because there is little public transport and most do not own a car.

The conditions so eloquently described over many years by W.J. Wilson and his colleagues at the University of Chicago led to a disconnection between these areas and the rest of the city. In these conditions, drug-peddling, prostitution and thieving became common. The invasion of crack towards the end of the 1980s added to this plight and increased the prevalence of violence. Soup kitchens tried to meet the needs of the hungry, malnourished and homeless (DiFazio, 1995).

Medicaid was introduced in the USA to provide better access to health care for those with low incomes in 1965. But inadequate primary care is a key problem in low income areas. One survey in 1988 of community-based physicians in nine of New York City's low-income communities found that there was a shortage of nearly 500 primary care physicians. Use of hospital out-patient departments and emergency rooms was two to four times higher than in more affluent communities in the city. More than 70 per cent of the practitioners were graduates of foreign medical schools – that is, very few graduates of New York State or City medical schools practised in these poor communities. The Medicaid reim-bursements were inadequate to sustain a practice in these areas (Brellochs et al., 1990).

Under President Clinton, the USA began to wage war on welfare. This reform of welfare involved a change from entitlement to one where cash assistance would last only two years. After this time transitional work would be required.

There are two underlying problems with work-fare schemes. First, if the jobs are of any value, the system will discriminate against unemployed people who are not on welfare or have not exhausted their entitlement, to whom such jobs will not be available. Secondly, if such discrimi-nation is avoided, they are likely to be filled by workers displaced by restructuring, who are not likely to be long-term welfare dependents in any case. The logic is therefore that welfare jobs will be poor jobs.

Nat Glazer argued in 1995 that Clinton needed a welfare reform plan mainly for symbolic reasons not because there really was a welfare crisis. 'Welfare', he said, 'has come to stand for the rise of a permanent dependent population cut off from the mainstream of American life and expec-tations, for the decay of inner cities, for the problem of homelessness, for the increase of crime and disorder, for the problems of the inner-city black poor . . . Ending "welfare as we know it" seems to promise some relief from these social disorders' (Glazer, 1995: 21).

The policy debate was fuelled by stereotypes about poor city residents. Research has shown, however, that high poverty areas contain a good deal of social and economic diversity. 'Although some residents clearly engage in "underclass" lifestyles, many of their neighbours are not public-assistance recipients and do participate in the labor market, albeit in lower-skill occupations and for fewer hours and lower wages' (Jargowsky, 1996: 579).

Two changes have occurred in social values in the USA which encouraged moves to welfare reform. One is the acceptance, even the require-ment, that mothers with children should take paid employment: bringing up children is not seen as sufficient employment for mothers who are not supported by a male breadwinner. And there is emphasis on responsibilities in return for oppor-tunities – people who receive income support should pay back in some way to the society that is providing their income. People are not seen as being poor because of factors outside their control – they are seen as being in some way responsible for the condition they are in. The poor are no longer seen as respectable and deserving but as mainly disreputable and unde-serving. Education and training and work experi-ence are expected to change the character of the poor as much as provide them with new technical skills.

It was the rise in the poverty population after 1973 that caused alarm and led to the conclusion that previous anti-poverty programmes were not working. The main programmes were AFDC, Food Stamps and Supplemental Security Income (SSI). Through these programmes, the USA was spending around 1 per cent of its gross domestic product on families and individuals with incomes below the poverty line. It was AFDC which was equated with 'welfare'. The overwhelming bulk of recipients were women and children. The real value of these benefits had been declining since 1973 but the numbers relying on it had grown. AFDC spending accounted for about 1 per cent of the federal budget and about 2–3 per cent of the budgets of most states. The decline in AFDC benefits' value had been offset by a rapid increase in expenditures on the Food Stamp programme, the only programme available to all the poor. SSI, payable to the elderly, blind and disabled, had also been growing steadily. Medicaid and public housing supplements were the other important elements in the American system of support for the poor.

Innovations aimed to encourage return to work included the Earned Income Tax Credit (EITC), a refundable subsidy to earned income directed primarily at low-income workers with children. It provided the model for the UK's Working Fami-lies Tax Credit (WFTC), introduced by the new labour government in the late 1990s. Such targeted programmes are generally shown to have clear effects in removing considerable numbers of people from living in poverty or below conven-tionally determined poverty lines (whatever debate there might be about the adequacy of these poverty lines). Without these programmes, many more people would live in poverty. 'The erosion of labor market opportunities for people with low levels of education has placed an enor-mous strain on the nation's anti-poverty programs' (Haveman, 1995: 190).

It is this situation which led in all Western countries to the move to reform welfare programmes, to make work pay. The aim was to add into EITC or WFTC schemes forms of childcare assistance and job training, and to encourage more parental responsibility, including stressing the responsibility of biological fathers for their offspring (as through the Child Support Agency in Britain), and utilize workfare measures, including education and training and signed contracts (as in the French RMI). Teenage mothers are encouraged to live with their parents rather than form independent households: more case work by social welfare officers is part of the new systems (as it has been in Sweden for some time and is now in Australia). The aim is to 'change the expectations of the poor and establish a new norm' (Haveman, 1995: 191). There is less use of incentives and more use of compulsion in these new welfare policies, which stretch across industrialized countries from Australia to Europe and the USA – countries which follow each other's experiences and deliberations closely and borrow ideas and plans at remarkable speed.

Many of the ideas originated in the USA and were transplanted to other countries. How generally appropriate are they? A central continuity in American society is its racism – what Myrdal called the American Dilemma and what Gans says could now be called the American Impasse – 'for racial segregation and discrimination have proven more stubborn than Myrdal believed' (1991: 273). Race and poverty are tightly interconnected in the USA. In recent decades, blacks, both male and female, have been 'driven further to the margins and out of the economy' (Gans, 1991: 295). The decline in the labour force participation of young black men is of staggering proportions. This is explained by the lack of jobs and also by the fact that a high proportion are in prison or have dropped out altogether and committed themselves to the underground economy. Others have died – to such an extent that some have talked of young black men as an endangered species – because violence is ever present in these poor neighbourhoods, making death from homicide a leading cause of death in young black men. This has produced what W.J. Wilson has termed a decline in the pool of marriageable men – explaining the increasing likelihood that young women will remain unmarried even if they have children. Conservative researchers see this family form as the main cause rather than a result of black poverty. The difference is one of emphasis and of attitude – whether to condemn or to understand. The conservative approach is that marriage solves all problems.

A key point made by Gans is that the problems for African-American city residents did not begin with the de-industrialization of the 1980s. Feminization of poverty has a longer history in these communities too. What changed, according to Gans in the 1980s was that the numbers Moynihan reported rose and heroin and crack cocaine literally destroyed a number of poor black families (1991: 276). While these differences in explanations reflect old divisions, what was new, and added urgency to the policy debate, was that the economic problems of poor people became so much worse, and the disorder in the cities much greater. People began to describe their local areas as war zones and those working in accident and emergency rooms in big city hospitals became experts in dealing with injuries otherwise usually only seen on the battlefield. As these conditions got worse, the attitude of the better-off became less caring, more rejecting, more willing to blame the victim, and more willing to support intensely punitive policies. 'Labelling needy people as undeserving as a way of depriving them of help is an old American solution; and it is also a cheap one because then anti-poverty programs do not need to be considered' (Gans, 1991: 277).

Gans pointed out that this is a short-sighted view, because poverty produces costs for society as crime and other forms of resistance and rebellion increase, and the drug economy grows: 'once the price of extra police, courts, prisons, mental hospitals and the like is factored in . . . the total price is not much less than a set of effective anti-poverty programs which would begin to reintegrate the poor into the economy and incidentally turn them back into full taxpayers' (1991: 277).

The problem is that such policies rely on the support of the majority for implementation and such support is no longer present.

Specific Urban Initiatives in the USA

Over the past two decades, a number of policy changes in the USA have constrained the role of local governments. Fiscal straitjackets, fiscal stress and recession forced tax increases and service cuts at state and local levels. In the 1960s and 1970s, American cities had seen dramatic and destructive riots and social violence. The poverty programme resulted. It was seen that there was an urgent need for action in the ghettos. Public interventionist approaches were adopted, which relied on federal assistance to localities through various anti-poverty measures. This approach was pulled back by the Nixon administration. Later President Carter introduced a new urban policy, which led to an expansion of several major programmes of community and human development. Increased financial support for urban housing and social services was provided. The aim was to create liveable cities, utilizing in some areas joint public–private partnerships. Still the infrastructure deteriorated. Under President Reagan there were further budget cuts in public programmes. The cities were singled out for budget reductions as part of general policy to reduce the inflationary effects of public expenditure, shifting responsibility to states and reducing the role of federal government. Paradoxically, this led to renewed dynamism at some city and local levels as local leaders responded energetically. Even tighter controls made things more difficult for cities, however, with the impact of the Gramm–Rudman law aiming to move to balanced budgets. More budget cuts followed, plunging some cities into fiscal crises and leading to cuts in social provision, from higher education to soup kitchens and night shelters.

These developments were paralleled by an intellectual assault on the poor from the right, wherein welfare programmes were seen as the cause of urban poverty. On the contrary, argued their opponents, the main cause was unemployment. Herb Gans, for example, has consistently argued that 'America's major urban problems [are] poverty, joblessness, the scarcity of decent jobs . . . racism and other kinds of inequality' (1991: ix) and that 'the central problem of our cities [is] decent and secure jobs for everyone and income grants for those unable to work or unable to find it. . . . [M]any other urban problems would almost solve themselves if such jobs and grants could be supplied' (1991: 189). Private business will never be willing or able to provide the necessary resources for this – only government can do so – but this is not the conventional wisdom at this time. But, states Gans, conventional wisdom has often been wrong, so those who think otherwise must continue to put their case.

The booming US economy in the 1990s appeared to have solved the problem with all but the very poor feeling as well or better off. Some argued however that this was a mirage with most new jobs being part-time or insecure and inequality increasing. Those who saw a jobless future (Aronowitz and DiFazio, 1994) stressed the need for policies like job-sharing or job-creation in socially useful areas. A central dilemma for all employment policies is whether to create decent jobs for the unskilled and semi-skilled, who are those most often in need of work, or whether to train the low-skilled to take up more skilled jobs, which are those most likely to become available. Gans sees a place for both approaches. And he stresses there is a clear need for public works projects. Schemes to reduce the length of the working week and share available work around more fairly are also needed. 'Despite its drawbacks, work-sharing is far superior to condemning a quarter or more of the population

to permanent or near-permanent joblessness' (Gans, 1991: 191).

European Welfare Regimes

Compared to the USA, the European countries confront the problems of economic change from a different base, having all established universalistic welfare states in the post-war years. The issue for them lies in the extent to which these systems can cope with the new situation. 'Long-term unemployment is at the centre of the debate over exclusion in Europe . . . The problem is the risk that those who lose their jobs enter a spiral of cumulative exclusion' (Rodgers, 1995: 254). The best way to tackle unemployment is to create jobs but on the whole this attempt has been unsuccessful. Where jobs have appeared, they have been increasingly part-time, insecure work, often taken by women. Special programmes for the unemployed exist but they do not show good results in terms of long-term reinsertion of the unemployed. Yet this 'churning' of the unemployed may have some value in preventing a proportion sinking to the bottom and becoming the long-term unemployed with a range of secondary problems; and for the individuals concerned, the provision of advice, training and support is appreciated. Nevertheless, it is obvious that policies of reintegration are of limited value without related, macro-level, effective employment policies.

There is a general tendency in Europe (as in America and Australia) to place more stress on welfare-to-work programmes of varying degrees of punitiveness, levels of remuneration, quality of experience and training and general support, especially with regard to childcare arrangements. Extensions of these include development of environmentally useful schemes, community schemes, financial incentives to employers to take on long-term unemployed people, temporary employment schemes with an element of restoring the work habit and training in social skills and work-seeking skills. Self-employment is also encouraged.

Much hope is now being invested in training as the ultimate panacea, with real benefits in some cases and in others mere accreditation – the effect being one of an inflation of qualifications. Training which is make-work and has the purpose simply of keeping the unemployed off the count has bred much cynicism and may increase exclusion, as some people decide to opt out altogether from the formal society with its compulsions and poor rewards. Longer-term policies argue for improving the general level of education, from primary school onwards, which may have wider benefits than mere increase in qualifications and individual knowledge and skills, by aiding the socialization process, producing a better integrated citizenry.

In France, the *revenu minimum d'insertion* (RMI) adopted in 1988 has led to a great deal of discussion. 'The fundamental goal of the law was to link the right to the satisfaction of basic needs with the aspiration for social and professional reinsertion' (Rodgers, 1995: 258). This was a particular response to the new poverty, that is, poverty among that old category of the able-bodied adult population. These groups were not covered by the existing insurance-based social security systems (a common problem in welfare systems, consequent on long-term structural unemployment, especially among younger people). The RMI combines an income allowance with an insertion contract. The social insertion is based on a social psychology of behaviour. The contract (a familiar tool in social work) is signed by the individual and the administration. It is claimed by the French as revolutionary, although some observers are more sceptical and see it as not very different from older principles of managing the able-bodied poor. The methods employed to establish reinsertion include actions to aid individual autonomy – providing shelter, health care, education and cultural assets; training relevant to the areas of future employment; instruction in enterprises, especially in conjunction with employers; and participation in NGOs or public administration. A key element is the location of the training and instruction – not being in separate locations but involving direct work experience in normal settings. It is important as a device to legitimate payments of income assistance but its effects in creating permanent employment in the long term are inevitably less impressive. These principles have also informed New Deal initiatives under Labour in the UK.

In all countries, the core concept of social insurance is under strain, not least in Germany, coming to terms as it has been with the effects of reunification and the constraints of a single European currency. Similarly

[i]n the UK which has created one of the most comprehensive (if not generous) social insurance schemes, the proportion of unemployed men receiving insurance benefits [fell] from 55 per cent in 1970 to 25 per cent in 1987 . . . In Germany, the proportion of unemployed with a claim to unemployment compensation or unemployment assistance fell from 86 per cent in 1975 to 67 per cent in 1987 . . . In France, about one third of the unemployed had no entitlement in the mid-1980s while another quarter had exhausted whatever entitlement they had. (Room et al., 1990: 3)

In response, a number of countries have emphasized targeting and the area approach. France and the UK have been strong on area compensation policies. Spanish interventions have also adopted an area and client-targeted approach.

Underlying these developments was the fact that the racial and ethnic differences between the poor and the middle classes challenged the basis of welfare state solidarities. If the welfare state assumed 'I am my brother's keeper', some began to raise the issue of 'Who is my brother?' How far can a sense of shared responsibility be extended? 'An estimated 4–4.4 million foreign immigrants now live in France; a little over a third are Muslims from North Africa. In Germany, there are currently 1.8 million Turkish guestworkers' (McFate, 1995: 13). 'In fact, multi-ethnic large cities have become the norm in most Western nations . . . such residential patterns often reinforce discrimination and separation from majority institutions and cultures' (1995: 13–14).

Concentration increases visibility. Prejudices are exacerbated. The 'creation of such "islands of isolation" defers integration . . . heavy ethnic concentration increases the awareness of the presence of a particular group and becomes the vehicle for social and political movements against further integration' (OECD, 1987). 'If poor market conditions and discrimination force minorities into informal or illegal sectors of the economy, a damaging dynamic may be set in motion' (McFate, 1995: 14).

European Urban Initiatives – Some Country Examples

As central governments became increasingly restricted in the range of options open to them, more attention was directed to lower levels of city and local governments as providing a space for interventions. Responsibility for a number of functions was passed to these lower levels, not necessarily with any transfer of funds. At this level, business can have more influence: local governments compete with each other to lure business to invest in their area. Public–private partnerships are now the order of the day.

Specific policies focus at an area level. A common approach in Europe, as in the USA, stresses a strategically linked, multi-agency, multi-disciplinary approach targeted on poor areas. A geographical focus is helpful for cross-sectoral working. The components of these approaches include local labour market policies, the provision or improvement of schools, roads and community services, incentives for enterprises, shops and banks to locate (or stay) in the area and housing improvements. 'What is important is the institutional base of such initiatives, involving partnership and co-operation' (Rodgers, 1995: 259). These partners will almost always include central government and local administrations, local associations, trade unions (to varying degrees), private business and NGOs.

The general conclusion is that these area-based policies have had an uneven impact. Interesting innovations have been explored, especially in the areas of cultural, environmentally focused and youth and sporting initiatives (Bianchini, 1990). The main problem has been the relatively small scale of the interventions in the context of the very large macro-level problems facing the countries concerned. Obstacles remain in implementing the idealized multi-agency, flexible approach. The cultures of different agencies and authorities are very different, the pace at which they work, whether there is a 'can-do' approach or whether caution and conservatism dominates, and issues of accountability and coordination, are not easily solved, certainly not merely by the rhetoric of partnership and working together. Expectations raised to be then dashed may produce cynicism and alienation, worse than what went before. NGOs and community organizations resent the contract culture, fear being taken over and having their goals and priorities distorted through involvement with state agencies: they fear they may lose their ability to innovate. The partnership approach may end up producing lowest common denominator results, catering only for the deserving needy, under the pressure to reach agreement before action can be taken. Ironically, this may produce standardized responses, rather than the diversity aimed at and necessary in the pluralistic, multi-factored situation in contemporary cities.

In France, socialist administrations in the 1980s launched comprehensive programmes for the social redevelopment of neighbourhoods, initially in peripheral social housing estates but later in inner-city areas also. These interventions included improvement of public service infrastructures, job creation and citizen participation in urban planning and they built on the concept of a contract between the state and local authorities. Urban problems rose high on the agenda in the 1980s, especially visible in the violent incidents which flared up on peripheral social housing estates. In response, the Rocard administration announced the creation of a Ministry of Urban Affairs to expand the Neighbourhood Social Development Scheme, a comprehensive package of housing, educational, crime prevention, job training and search assistance programmes, youth activities, and improved public services, in 400 officially designated 'sensitive neighbourhoods' across the country. These urban rejuvenation programmes incorporated explicit social objectives. Some accounts conclude that these interventions have proved effective (Donzel, 1993).

From an analysis of projects in Marseilles, Donzel concluded that the greatest gains were in

education, vocational training and social reinsertion measures. A richer collective life has developed and a more robust community self-confidence. French urban policy at this time, according to Donzel, was relatively successful, largely because of the direct involvement and control of the central state, working in partnership with local authorities. The French Ministry for Cities was later abolished, it is said partly because it was too much tainted with the rhetoric of socialist administrations.

While, perhaps reasonably, there is a tendency to discuss European policies as one bloc, especially when making comparisons with the USA, it is important to note the substantial differences between member-states of the EU. Just as there are differences between states and cities in the USA, there are differences between the different countries and regions of the EU.

Greece, for example, shows an absence of a well-organized welfare state but, on the other hand, there are strong family support networks. The informal sector and religious associations play a large part. For this reason, urban inequalities and social exclusion in Greek cities have tended to be overlooked by policy-makers and also by researchers. But Greece has started to exhibit similar negative trends to those found elsewhere and is also similarly influenced by pressure from the EU for policy reforms.

In the UK, the 1977 White Paper, *Policy for the Inner Cities*, marked the first explicit recognition by government of the economic causes of many urban problems. It discussed economic decline, physical decay and social disadvantage. Mismatch between the skills of the local labour force and the demand for labour, as well as a lack of demand for labour in general were noted as important causes of unemployment (Department of Environment, 1977). But the British learnt quite early on after this White Paper that economic benefits of interventions do not substantially accrue to inner-city populations themselves. Trickle-down has hardly been credible to many involved in regeneration schemes, however loudly proclaimed by politicians in the 1980s.

Many of the most difficult poor housing estates are those which were used by local or city authorities as dumps for their problem families or as places to house those affected by urban renewal. In the UK, the sale of council property sliced away the more desirable residences and left much of the public or social housing sector with only those properties which no one wants to live in. Many policies have been concerned in recent years to improve these residual estates through better management, involving residents, sale to housing associations and special injections of funds. Rent arrears have been a particular problem. Where this is dealt with by housing benefits, soaring bills for public expenditure present a different problem (Stewart and Taylor, 1995).

Throughout the 1980s, a whole series of initiatives were launched targeted at inner cities (MacGregor and Pimlott, 1990), partly in response to the riots and disorder that characterized that decade. A review of the effectiveness of some of these policies by the National Audit Office concluded that City Action Teams promoted coordination at local level and provided a focus for action in some urban areas but their effectiveness needed further improvement. Urban Programme partnerships worked well in some areas but not in all. Enterprise Zones provided some benefits but proved to be an expensive means of regenerating run-down areas. The National Audit Office concluded that performance varied and there was scope for better communication (NAO, 1990).

Other reviews of national policy development have concluded that surprisingly little has been achieved. The gap between the deprived areas and the rest remains as wide as before and in some respects has widened.

THE MOVE TO PARTNERSHIP

In the UK there is now a political consensus that a multi-sectoral partnership approach is essential to achieve urban regeneration. The Single Regeneration Budget and the establishment of integrated regional offices brought together the Departments of Environment and Trade and Industry to try to work out a common approach to the problems of run-down areas. 'As a term however "partnership" is overused, ambiguous and politicized', (Hastings, 1996: 253). There are at least two interpretations of partnership. The Thatcherite approach stressed privatization and the centralization of urban policy. The centre–left are now more prepared to see a place for local involvement, not necessarily, however, organized through elected local authorities.

Michael Ward, now leading development in London in the new governance structures has argued that to deal with disadvantage and to exploit opportunity, city-wide strategies are necessary to overcome economic, social and environmental problems (Ward, 1994). Such policies should build a consensus of partnership involving civil society as a whole, seen as consisting of the private sector, voluntary and community groups. These approaches have been shown to be successful in the management of health care in Canada and in the European Healthy Cities network. The main aim is to replace a *sectoral* approach with a *wider vision* of the context in which problems are situated. In Canada, a community-type approach, based on local action, led to the creation of health care councils in a number of Canadian cities. The activities of these councils included the discussion of problems, development of solutions and the collecting and dissemination of ideas (Bessis, 1995: 47). Similarly, the European Healthy Cities initiative, implemented in 30 European cities, aimed to act in all domains having an impact on health care, such as housing, transport and the environment and to encourage decision-makers to take into account the consequences of their policies on health care as a whole, not merely thinking of the interests of their own sector.

In urban policy, Ward (1994) argues for a people-led not a property-led approach. City government should have a leading role but should be driven by wide participation and as much devolution of control to local neighbourhood level as possible. For this to be effective, however, new sources of finance will be required, such as local business rates; a national Urban Programme; and European investment.

THE ROLE OF LOCAL GOVERNMENT

Local consultation, community involvement and partnership are terms that are assuming increasing prominence in the discourse on urban regeneration, not least in the British New Deal for Communities. There are differing views however on the role of local authorities. Some argue that local authorities should be at the centre of efforts aiming to improve the environment of cities. In housing and employment schemes and in provision of public transport, local authorities can partner with groups in their own communities and striking improvements can be made. Others are more sceptical, seeing local authorities as corrupt or incompetent. The World Assembly of Cities and Local Authorities' declaration at the Habitat II conference in Istanbul, May 1996, included the principle that 'towns and cities must give more attention to social integration and the struggle against social exclusion'. This could be achieved through greater decentralization and acknowledgement of the strategic role of local authorities. In this new era, it is thought that mayors can play a particularly important role as agents of change.

The key question is whether or not local politicians closely link to and represent the poor in poor communities. Traditionally, the powers of exit and voice have been few in poor areas. Greater stress is now placed on community involvement and more open structures of local government. In practice, community involvement can range from a minimal consultation, largely designed to fend off dissatisfaction (even riot), to more robust approaches which rest on a pooling of resources and the development of autonomous organizations. More attention is now being given to fostering an older tradition of mutual help through encouragement of self-help, skills development and participation. Methods adopted have included localized estate-based housing management. Many of these community initiatives are, however, small and fragile and lacking in power compared to the power of city government, large employers, and the like. And it must be remembered that the central cause of the disempowerment of the poor is the contrast of conditions and opportunities in rich and poor areas.

In many of these developments, the idea of networking has assumed greater importance – networks within localities and between localities. As poverty and inequality rose in the 1980s and 1990s, a number of cities in Europe established committees to examine the problem. The Association of Metropolitan Authorities in Britain created an Anti-Poverty Network (Balloch and Jones, 1990), which spread out to link with similar initiatives across Europe. Sharing experiences has been a key aim of these proliferating networks. For example, a global Inter-City Network launched by the Prince of Wales Business Leaders Forum aimed to share experience of corporate–community partnerships to regenerate urban areas. Innovative inner-city projects in New York, Atlanta and Los Angeles were taken as models from which other cities as far apart as Russia, Asia, Latin America and South Africa might learn. In the USA, the Coalition of Mayors puts pressure on federal government and raises the profile of cities and urban issues. And the 'Neighbourhoods in Crisis' programme, sponsored by the European Commission, 'brought together disadvantaged communities from a number of European countries to examine the basis of renewal strategies' (Ward, 1994: 40). The European Union's Urban Initiative was designed to build on these networking experiences in order to spread innovative approaches to urban policy.

COMMUNITY OR CITIZENSHIP?

In the First World cities, policy in post-war years moved through a number of phases: encouragement of new towns and/or migration to the suburbs; responding to racial tensions; developing social and welfare programmes; increasing stress on inner-city areas; property-led regeneration; and then a return to more social approaches to renewal. These more social approaches can include: training and education; city economic strategies; community renewal; urban industrial policy; institutions for regeneration; and job creation programmes (Ward, 1994: 14).

A wide range of options are available to city governments to alleviate or reduce poverty, if these governments have the will and the capacity to implement such policies. Options include employment creation, improving access to justice, protection from crime, improvements to services such as water supply, sanitation, solid waste management, public transport, health care and education. These needs are shared between cities of the developing and developed worlds. Notably also all proposals now stress the importance of encouraging and supporting the activities of community-based organizations, NGOs and the private sector – through various partnership arrangements. For these to work effectively, changes in local systems of regulation may often be required (Wegelin and Borgman, 1995).

Many schemes now aim to stimulate community involvement to improve poor or disturbed neighbourhoods. Malfunctioning is assumed in many of these approaches but others stress the strengths of communities, on which new initiatives can be built. Examples are community safety and community policing schemes, and community development approaches to drugs prevention.

Experience has shown that community involvement in local improvement projects does not always work. Especially where local residents are not home-owners or have no other investment or commitment to their property or neighbourhood, there is little in it for them should they become involved. Growing separation of areas by housing tenure is thus a problem, which, some have argued, can be altered by stressing housing mix more strongly in both public and private sector housing developments. Another route is to stress that the organizations promoted should be concerned with more than just housing. Others have advocated the sale of public housing to tenants as a means of giving them a stake in the community and thus the motive to prevent deterioration. However, there are others who argue that evidence does not support the claim that residents of inner-city public housing are less likely to become involved in community affairs than inner-city home-owners. Factors other than housing tenure status, such as church attendance, family structure and the number of families in the neighbourhood, explain the variation in rates of community participation (Reingold, 1995).

Philanthropists, such as the Kellogg Foundation, have provided funds to develop sustainable communities (for example, the Healthcare Forum supported by the W.K. Kellogg Foundation in the USA).

In order to address the problems of American urban communities, a new approach is proposed using the concept of 'collaborative empowerment'. Community and neighbourhood-based organizations are allowed to design, implement and assess problem-solving strategies, with advice being provided to them on organizational management and to strengthen community leadership (Johnson, 1996). The Kellogg Foundation has promoted and fostered the concept of urban and community leadership through support for projects in the USA and in other countries, such as South Africa.

But while *community* is the key term in US approaches, it is *citizenship* in Europe. The solution to urban poverty and exclusion, according to the Centre for Local Economic Strategies in Manchester, lies in renewed citizenship, based on rights 'to work, to housing, to an income, to public services, and above all to human dignity; a citizenship based on equality before the law, equal treatment, and equal access to public services. It is also a concept of active citizenship: participating in a restored and renewed local democracy with opportunities for direct citizen involvement at local level' (Ward, 1994: 2).

But, in truth, how much difference is there between the community and citizenship approaches when it comes down to the level of practical policies? There may be a different language and terminology in use but the practical proposals often look very similar.

There is an unfortunate tendency (perhaps understandable given the breadth of material to be covered) in discussions at international level to contrast the 'Anglo-Saxon' with the 'European' models with Britain seen as wavering between the two. These stereotypes dangerously ignore important differences *within* countries and the clashes of interest and the policy debates raging there. Crucially, the interests of business, multinational companies and banks, for example, cross-cut national differences, as do the linkages between trade unions, anti-poverty groups, churches, social democratic associations and social analysts.

For example, the package of proposals proposed by Ward and others in Europe is very similar to that put forward by people like W.J. Wilson and Elliott Currie for American cities (Currie, 1993; Wilson, 1996). They share a common *social* and *democratic* framework.

Wilson argues that the problems of the poor in the cities are so great that they 'require bold, comprehensive, and thoughtful solutions not simplistic and pious statements about the need for greater personal responsibility' (1996: 209). He calls for 'the integration and mobilization of resources from both the public and the private sectors . . . to generate a public/private partnership to fight social inequality' (1996: 210). The

USA has to catch up with other industrial democracies: 'the United States is alone in having no universal pre-school, child-support or parental leave programs' (1996: 215).

'The French *cité* and the American ghetto are both viewed as "dangerous places" in which violent crime and delinquency are commonplace and create a climate of fear and insecurity' (Wacquant, 1995: 559–60), but they are not comparable in reality. 'The incidence of homicides and street violence (shootings, muggings, rape and battery) in Chicago's ghetto has reached levels comparable with that of a civil war and caused the virtual disappearance of public space' (Wacquant, 1995: 560).

Wilson agrees with this analysis and states that 'none of the other industrialized democracies has allowed its city centres to deteriorate as has the United States' (1996: 218). What is needed is better public transportation, more effective urban renewal programmes and good public education. But it will be difficult to tackle any of these problems or the racial tension associated with them until the country faces up to the issue of the 'shrinking revenue and inadequate social services and the gradual disappearance of work in certain neighborhoods' (Wilson, 1996: 218). To do this would require restoring the federal contribution to the city budget and an increase in the employment base. The fiscal crisis of the cities could be alleviated if the employment base could be increased significantly. Wilson supports ideas for neighbourhood revitalization, such as community development banks, non-profit inner-city housing developments and enterprise zones but, without greater collaboration between the suburbs and the cities, these will prove inadequate to the problem. This will require some form of tax-sharing, collaborative planning (especially of public transit systems) and the creation of regional authorities.

A popular alternative view in the USA however is that zero-tolerance and high rates of imprisonment can be most effective in restoring city areas, as in New York's Manhatten.

European countries are facing similar problems to those of the USA and their welfare states are under strain. Resentment against immigration and other excluded groups is growing. The question is whether welfare states can evolve to adapt to changed conditions without abandoning their core principles.

FROM WELFARE TO WORK

In most social policy regimes, there is increasing emphasis on encouraging work rather than expanding welfare entitlements. The choice is between minimum wage policies, based on greater sharing out of the available work, and minimum income policies, which do not require a work contribution. European countries have developed policies to get unskilled workers into low-wage jobs. These initiatives should be 'buttressed by maintaining certain desirable aspects of the safety-net, such as universal health insurance, that prevent workers from slipping into the depths of poverty, as so often happens to their American counterparts' (Wilson, 1996: 221).

In the end, a major source of poverty alleviation must be the generation of employment. 'Employment creation remains the most powerful instrument for reducing both inequality and poverty. For people to be employable, they must be healthy, educated and skilled' (UN, 1994: 87). One approach is to act directly on the job market through voluntaristic policies involving a mix of sanctions and incentives on employers. An active job creation programme might involve the creation of social service jobs, which are generally undersupplied. Another related approach, much favoured at present in many countries, is to concentrate on improving the skills of the work-force through better education and training. Linked to this are equal opportunity policies. Day care and family planning policies are also relevant here. These policies, customized to local conditions, of course, are as applicable in the ghettos of the USA as in the outer estates or *banlieues* of Europe or the towns and villages of India or China.

Minimum wages set at reasonable levels clearly have a part to play. These wages, combined with medical insurance and child provision (a classic Beveridgean idea), should be adequate to pull a family out of poverty (albeit the old problem of housing costs is still around). Yet in the USA, the issue of moving to universal health care provision – a major lack in that country – is still one which has defeated politicians. Underlying all these discussions is the decline in unskilled work or work suitable for those with limited capacities and competencies. To solve this would require a policy of public sector employment of last resort. These jobs might take the form of public sector infrastructure maintenance jobs, or public service jobs. The need is clearly there for both, and society as a whole would benefit, but such solutions would not be cheap.

While leading figures, like Wilson, try to persuade Americans to adopt European solutions, there is the issue of whether social democracy is itself unravelling in Western Europe. Social democracy represented the third way between America and the Soviet Union during the period of the Cold War. The end of the Soviet Union has posed a threat to social democracy as the fulcrum has shifted. In addition, all parties take for granted middle-class resistance to paying taxes.

Given growing needs for human services, there is a need to find new ways to get money into the system, perhaps through forms of hypothecation or independent social insurance funds – various proposals have been suggested. The key problem is the underlying lack of social solidarity, as the base for a sense of common interest, common identity and common threat appears to have eroded. Perhaps in future the locality may provide a more hospitable base for such commonalities to be identified, although this runs the risk of excluding from all localities those who are viewed as outsiders and undeserving – creating a new vagrant class.

CONCLUSION

The two-speed city on the American model is more than just a potential risk: it is becoming a reality. In most European cities, access to work, housing, education and training, to culture and to health, is a right from which certain social groups – young people, women, ethnic minorities, the long-term unemployed – are excluded. This fact is made even more unacceptable in that the excluded are concentrated in vulnerable areas of cities, often next to areas of conspicuous affluence. (Delegation Interministerielle á la Ville, 1990 report for the European Commission)

The policies which emerged in post-war years and were labelled welfare state policies were themselves responses to what were seen as the failures of the 1930s. At the international and at national levels, the motive was 'never again': governments took responsibility for the basic social and economic welfare of their citizens and poverty and mass unemployment were identified as evils to be outlawed. A post-war social settlement with similar features emerged in most countries, although to varying degrees. The key characteristic of this settlement was acceptance of the role of markets (capitalism), but its power was to be tempered by state regulation, in the public interest. Areas seen as crucial to effecting social justice were universal health care, universal education, secure employment and income support when employment was not possible (through old age, disability, childcare, maternity, etc.). These ambitions were pursued and achieved to varying degrees in different countries but the broad shape of what would be needed to abolish poverty was roughly seen in the same way. As this settlement crumbled in the late 1970s and policies of neo-liberalism and structural adjustment were applied extensively in the 1980s and 1990s, the social evils which some had thought conquered returned with a new force.

Arguments for a need for change in welfare systems point to the changed basis of poverty. Income support schemes based on insurance contributions and/or employment records are inadequate for young people, those with part-time work and cycles of employment/unemployment – because of their infrequent and unstable contribution records. As important here is the need felt by centrist politicians to keep taxpayers on board: it seems that to maintain their willingness to pay taxes to support the poor, an element of punishment must be built into systems, to satisfy the need for balance in the contractual relationship. The poor must be grateful and subservient or (if they are not) they must be seen to be punished to some degree. The alternative is an even greater tax backlash, which would leave the poor with no support at all. The price of maintaining a reduced or minimal welfare state appears to be some symbolic retribution.

The simple conclusion to our review of poverty and social problems in the city is that social policy makes a difference. Wider social policies provide the framework within which specific urban policies work. If urban policies are asked to do too much they will fail.

Over the past 20 years, there has been increasing attention to local economic initiatives as the way forward. As central government loosened its sense of responsibility for unemployment and social justice, so a devolution of responsibility to lower levels occurred. Broadly, the right wing wanted to devolve responsibility to family and neighbourhood, business and charities; more centre–left politicians saw a role for city governments but these are increasingly harnessed, tempered and controlled through partnerships with business.

There remain those who argue against targeting area programmes on the urban poor. Such programmes, they say, attack symptoms rather than causes and the causes must be dealt with at the macro-level. They point out that there are as many or more poor people in non-poor areas than in targeted areas and that targeted, selective programmes always stigmatize the recipients. In addition, there is always the problem of where to draw the line, posing problems for contiguous areas, which are left out of regeneration initiatives yet perhaps may be as deserving of attention.

ILO/UNDP conclude that action is needed at international/global level, national level and at regional and local level if social exclusion is to be effectively combatted (Rodgers et al., 1995). Capital may be difficult to attract to countries which impose social policy objectives (Rodgers, 1995: 268) in a context of increasing competition between countries. For national policy, the fundamental issue has to do with the extent to which, and the ways in which, the public attempts to restrain or regulate the market. The shape, extent and direction of social policy at the national level provides the crucial context within which city level policies have to work.

Partnership is now proposed as the way forward, involving the state, the market and the tertiary sector. In these new forms of social organization, a greater place is available for local initiatives and intermediate levels of decision-making. 'The danger of this approach', as Rodgers points out, 'is that problems of exclusion will then be given less priority in national policy, leaving local policies as a palliative within an exclusionary model of development' (1995: 272).

The 'social question' re-emerged as the twentieth century drew to a close. With the end of the Cold War, poverty, exclusion and inequality appear now to be the main reasons for instability in the world (Bessis, 1995: 11). The social crisis is increasingly urban in character and this is likely to increase in the decades of the twenty-first century.

> In the view of increasing numbers of researchers, the economistic drift, the supremacy of exchange value over the notion of use value, the running of the planet according to the sole criterion of the profitability of enterprises and the extension of the cash nexus to the ensemble of human activity are in the process of pushing humanity into an impasse. (Bessis, 1995: 34)

These trends have led to a disregard of collective needs not linked to the market, the waste of finite resources and the exclusion of increasing proportions of the world's population. Policies may be either an agent for integration or a force for exclusion. The challenge is to turn round the policies which have led to increasing poverty and disintegration in the cities and develop new ways of meeting human and social needs. For this, new political alliances and new political cultures will be required.

REFERENCES

Aronowitz, S. and DiFazio, W. (1994) *The Jobless Future*. Minneapolis, MN: University of Minnesota Press.

Atkinson, A.B. (1995) *Incomes and the Welfare State*. Cambridge: Cambridge University Press.

Balloch, S. and Jones, B. (1990) *Poverty and Anti-Poverty Strategy: The Local Government Response*. London: Association of Metropolitan Authorities.

Bessis, Sophie (1995) *From Social Exclusion to Social*

Cohesion: A Policy Agenda. MOST Policy Paper 2. Paris; UNESCO.

Bianchini, F. (1990) 'Urban renaissance? The Arts and the urban regeneration process', in S. MacGregor and B. Pimlott (eds), Tackling the Inner Cities. Oxford: Clarendon Press. pp. 215–50.

Brellochs, Christel, Carter, Anjean B., Caress, Barbara and Coldman, Amy (1990) Building Primary Health Care in New York City's Low-Income Communities. New York: Community Services Society of New York.

CEC (Commission of the European Community) (1991) Final Report on the Second European Community Programme to Combat Poverty. Brussels: CEC.

Currie, Elliott (1993) Reckoning: Drugs, the Cities, and the American Future. New York: Hill & Wang.

Department of Environment (1977) Policy for the Inner Cities. London: HMSO. Cmnd 6845.

Difazio, W. (1995) 'Soup kitchen blues: post-industrial poverty in Brooklyn', in S. MacGregor and A. Lipow (eds), The Other City: People and Politics in New York and London. Atlantic Highlands, NJ: Humanities Press. pp. 40–60.

Donzel, Andre (1993) 'Suburban development and policy-making in France: the case of Marseilles', in S. Mangen and L. Hantrais (eds), Polarisation and Urban Space. Cross-National Research Papers. Third Series: Concepts and Contexts in International Comparisons. pp. 44–53.

Esping-Andersen, Gosta (ed.) (1996) Welfare States in Transition: National Adaptations in Global Economies. London: Sage.

Gans, Herbert J. (1991) People, Plans, and Policies. New York: Columbia University Press, Russell Sage Foundation.

Glazer, Nathan (1995) 'Making work work: welfare reform in the 1990s', in D.S. Nightingale and R.H. Haveman (eds), The Work Alternative: Welfare Reform and the Realities of the Job Market. Washington, DC: Urban Institute Press.

Hastings, A. (1996) 'Unraveling the process of partnership in urban regeneration policy', Urban Studies, 33(2): 253–68.

Haveman, Robert H. (1995) 'The Clinton alternative to "Welfare as we know it": is it feasible?', in D.S. Nightingale and R.H. Haveman (eds), The Work Alternative: Welfare Reform and the Realities of the Job Market. Washington, DC: Urban Institute Press. pp. 185–202.

Jargowsky, P.A. (1996) 'Beyond the street corner: the hidden diversity of high-poverty neighbourhoods', Urban Geography, 17(7): 579–603.

Johnson, D. (1996) 'Building capacity through collaborative leadership', International Journal of Health Planning and Management, 11(4): 339–44.

Joseph Rowntree Foundation (1995) Inquiry into Income and Wealth, Vol. 1. York: Joseph Rowntree Foundation.

Katz, Michael (1989) The Undeserving Poor: From the War on Poverty to the War on Welfare. New York: Pantheon Books.

Levitt, Kari Polanyi (1995) 'Toward alternatives: re-reading The Great Transformation', Monthly Review, 47 (June): 1–16.

MacGregor, S. and Pimlott, B. (1990) 'Action and inaction in the cities', in S. MacGregor and B. Pimlott (eds), Tackling the Inner Cities: The 1980s Reviewed, Prospects for the 1990s. Oxford: Clarendon Press. pp. 1–21.

McFate, Katherine (1995) 'Introduction: Western states in the new world order', in K. McFate, R. Lawson and W.J. Wilson (eds), Poverty, Inequality and the Future of Social Policy. New York: Russell Sage Foundation. pp. 1–28.

Mooney, Gerry and Danson, Mike (1997) 'Beyond "culture city": Glasgow as a "dual city"', in N. Jewson and S. MacGregor (eds), Transforming Cities: Contested Governance and New Spatial Divisions. London: Routledge.

NAO (National Audit Office) (1990) Regenerating the Inner Cities. London: HMSO.

Organization for Economic Co-operation and Development (1987) The Future of Migration. Paris: OECD.

Oyen, Else (1996) 'Poverty research rethought', in Else Oyen, S.M. Miller and Syed Abdus Samad (eds), Poverty: A Global Review – Handbook on International Poverty Research. Oslo: Scandinavian University Press. pp. 1–17.

Polanyi, K. (1944) The Great Transformation. New York: Free Press.

Reingold, D.A. (1995) 'Public housing, home ownership and community participation in Chicago Inner City', Housing Studies, 10(4): 445–69.

Rodgers, Gerry (1995) 'The design of policy against exclusion', in G. Rodgers et al., Social Exclusion: Rhetoric, Reality, Responses. Geneva: International Labour Organization (Institute for Labour Studies). pp. 253–82.

Rodgers, Gerry, Gore, Charles and Figueiredo, Jose B. (eds) (1995) Social Exclusion: Rhetoric, Reality, Responses. Geneva: International Labour Organization (Institute for Labour Studies).

Room, G. (1990) 'New Poverty' in the European Community. London: Macmillan.

Stewart, Murray and Taylor, Marilyn (1995) Empowerment and Estate Regeneration. Bristol: Policy Press.

Susser, Ida and Kresnicke, John (1987) 'The welfare trap: a public policy for deprivation', in Keith Mullings (ed.), Cities of the United States: Studies in Urban Anthropology. New York: Columbia University Press. pp. 51–68.

Townsend, Peter (1993) The International Analysis of Poverty. New York: Harvester Wheatsheaf.

UN (United Nations) (1994) World Social Situation in the 1990s. New York: United Nations.

UNICEF (1994) The State of the World's Children. Oxford: Oxford University Press.

Vankempen, E.T. (1994) 'The dual city and the poor –

social polarization, social segregation and life-chances', *Urban Studies*, 31(7): 995–1015.

Wacquant, Loic (1995) 'The comparative structure and experience of urban exclusion: "race", class and space in Chicago and Paris', in K. McFate et al. (eds), *Poverty, Inequality and the Future of Social Policy*. New York: Russell Sage Foundation. pp. 543–70.

Ward, Michael (1994) *Rethinking Urban Policy: City Strategies for the Global Economy*. Manchester: Centre for Local Economic Strategies.

Watkins, Kevin (1995) *The Oxfam Poverty Report*. Oxford: Oxfam.

Wegelin, E.A. and Borgman, K.M. (1995) 'Options for municipal interventions in urban poverty', *Environment and Urbanisation*, 7(2): 131–51.

Wilson, William Julius (1996) *When Work Disappears: The World of the New Urban Poor*. New York: Alfred A. Knopf.

Part VI

POWER AND POLICY DISCOURSES IN POSTMODERN CITIES

Appreciation of the changing nature of the city reflects the 'cultural turn' as well as being representative of it. As much as postmodernism is a contested, and certainly multi-dimensional, term, shifts within the city at the closing decades of the twentieth century were to increasingly suggest that urban development was experiencing its own 'turn'. The role model used to develop a sense of understanding the 'new' city was Los Angeles, replacing the role Chicago had played in so much earlier urban analysis. Physically, postmodern urbanism as represented by Los Angeles was qualitatively different from earlier urban development, being simultaneously more fragmented and dispersed, yet focused around the needs of information capitalism, the ever expanding expressions of consumerism, and of the city as cosmopolitan place. Of course Los Angeles is no more representative of individual cities than was Chicago, but a greater sympathy to the uniqueness of place than was necessarily always apparent in charting the spread of influence of the Chicago role model is the more likely to ensure that history will not repeat itself. Yet elements of Los Angeles as the role model postmodern city are evident in the contemporary development of cities from Kuala Lumpur to Manchester. Globalization, cultural hybridity, social polarization, fragmentation and the rapidly developing influence of cyberspace are among the defining features of postmodern urbanism, each of which to varying degrees, and in different ways, are recasting the world's cities.

Nor is it just in terms of architectural shifts or the physical form of the city that postmodern influences are posited at play; how cities are to be read, the interplay of forces influencing their development, pose new epistemological questions. One of the benefits of the developing postmodern critique has been to question the nature of modernity and its influences, as on urban development and restructuring. Cities continue to carry a legacy of the optimism of rational planning with the urban planner as the professional change-agent confident in their ability to (re)engi-

neer cities to meet social and economic goals. After 1945 in the advanced economies urban planning became a powerful means by which to rationalize the meeting of desired ends, its dominance reflecting the power of the discourses through which it was mediated, and of the professionalism which permeated planning practice. Yet, discourses, once they are exposed to the scrutiny of deconstruction, are as revealing in what and who is excluded as they are in emphasizing the idea(l)s and voices which are represented. While there could be spontaneous opposition to urban reconstruction once its local impacts became known – to urban motorway proposals in British cities, for example, in the 1960s – it is perhaps only more recently that there has been an appreciation of the discursive practices through which planning operates, the ideologies on which it is based and its power in being able to suppress alternative voices.

Confronted by the postmodern turn planning and planners need to re-orientate the more dynamic world economy. Notions of collaborative planning are suggestive of the ways in which planners will need to be more responsive to the multiple voices of the city, to the need to be able to orchestrate and foster coalitions between private and public sector and community interests. Further, in responding to the competition for growth between cities, planners, it is argued, should become more entrepreneurial. In part these shifts reflect the trend towards governance, while in part they represent the response to the increasing role of the market. Their effect has tended to downplay the status of the planner as an urban change-agent, while market reform itself has simultaneously resulted in the increasing privatization of the profession. Yet, in countries such as Britain and the Netherlands, in which the planning profession has been a powerful agent of urban change in post-war society, it remains significant as a regulator of land-use development and change. Postmodern urbanism poses particular challenges to the inherited legacy of the planning profession, to the logic of rational planning, and to the need to become more flexible as well as entrepreneurial.

Within the postmodern city no less than in the modern city, power and power relations are central to understanding the processes underpinning urban change. Challenged by increasing competition and the need to establish market position within the world economy, the style of urban politics, together with the array of agencies involved in formulating and implementing the strategic development of the city have refashioned power structures and relations within cities. Yet old questions remain: whose interests are represented in urban change; to what extent does (and should) urban politics address redistributive justice? As ever urban change is a conflictual process, managing conflict without suppressing minority voices is a major challenge.

One of the paradoxes of the contemporary city is that in many ways the postmodern is accompanied by the modern. Indeed, glimpses of the postmodern were apparent in the modern city. Consumption and consumerism, taken as one of the defining features of postmodernism, were represented in the department stores and the shopping arcades of Paris of the Second Empire, and so painstakingly described by Zola in *Au Bonheur des Dames*. In the contemporary city planners work within postmodern and modern concepts simultaneously. Postmodern urbanism is not a development which in some simplistic sense has emerged after the modern city; it is a complex weave of trends and concepts which are superimposed over cities whose development has been, and in different ways, still is guided by modernist idea(l)s.

24

Communicative Planning, Emancipatory Politics and Postmodernism

LOUIS ALBRECHTS AND WILLIAM DENAYER

THE CHANGING POSITION OF PLANNERS

In the 1960s, planners believed in a future in which social problems could be tamed and humanity could be liberated from the constraints of scarcity and greed (see Albrechts, 1991). Planning was thought to be an adequate tool of the welfare states to ameliorate and to equalize the conditions of living. In the 1990s, however, an entirely new situation is faced. The state has become much more ideologically conservative and more subservient to the needs and demands of capital, turning away from the simultaneous pursuit of both economic growth and welfare (Beauregard, 1989). Moreover, everywhere in Europe the proponents of the welfare-state project seem to be ideologically in crisis, theoretically confused and internally divided.

Although in this chapter the focus is not on explaining the general causes leading to this situation, it has now become clear that the 1980s witnessed a general process of industrial restructuring throughout the world (Albrechts and Swyngedouw, 1989; Priore and Sabel, 1984; see also Amin, 1994). There are several versions explaining this round of restructuring. A dominant interpretation, however, is that the Fordist mode of development based on an international spatial division of labour in the industrial realm and on regulatory state intervention resulting in the building of the welfare state, has been superseded by a new, more geographically open and market-based mode of production founded on a growing and all-encompassing flexibility (see Albrechts and Swyngedouw, 1989). By the end of the 1960s, the organization and technical limitations of the Fordist mode of production, the countervailing practices of institutionalized regulation, and the inherent tendencies of a fall in the rate of profit and overaccumulation, threatened the existence of this mode of production. This restructuring radically changed the roles planning and planners could play in society.

During the 1980s and early 1990s, mainstream planning was basically aimed at smoothing the negative implications of uncontrolled economic development and is concerned with optimizing the environmental conditions for ever-widening economic expansion. Following this route, the scope of planning not only became severely limited but also led towards an inherently contradictory position of planners towards their own practice (Albrechts, 1991). The question to be asked was no longer how to minimize the socially negative consequences of economic development through redistributive policies, but how to maximize opportunities. Many planners became obedient to the 'imperative' of providing a good business climate and the 'construction' of all sorts of incentives to attract inward investment. As Fainstein pointed out, planners became more deal-makers rather than regulators (Fainstein, 1988, see also Harvey, 1982, 1989, on the rise of urban entrepreneurialism). Economic development, certainly, can be considered *the* most highly valued political aim of the 1980s and 1990s. The competition for investments, however, led to a divorce between development policies and democratic politics. Public–private partnerships, for example, were ideologically conceived as politically neutral decisions, leaving the deliberations and the politics inherent in it in the hand of 'experts' (see Fainstein, 1988; Harvey, 1989). More than ever before, decisions were made in private, with no eye to the common interest at all. Welfare corporate society depoliticizes public life by restricting discussion to distributive issues in a context of interest-group pluralism where each group competes for its share of public resources. The government authority is carved out in favour of institutionalized interest groups and private entrepreneurs (see also Young, 1990: 71ff; Denayer, 2000). The problem with this is that it institutionalizes and encourages an *egoist*, self-regarding view of the political. Interest-group pluralism, indeed, allows little space for claims that parties have a responsibility to attend to the claims of others because they are needy or oppressed (see Young, 1990). As an answer, it is argued that planning has to be normative in purpose, pro-active and political in attitude, also towards traditionally unchallenged power structures (Albrechts, 1993, 1996, 1999).

CHANGES IN SOCIETY AFFECTING PLANNING

All of this has to be placed within the more general framework of a changing view of how society should be organized. Under the spell of neo-liberalism, planning became increasingly associated with inefficiency, regulation and excessive cost, hindering individual freedom and the functioning of the free market economy. Neo-liberals and conservatives assume that the economic factors spontaneously develop towards an optimal state of affairs or, if this would prove to be only partially true, that only very limited state intervention is desirable. In this conception planning is considered to be largely superfluous. Moreover, many planners radicalized the view that planning was politically neutral. Planning became, still more than this had already been the case in the 1960s, more concerned with *how* to plan rather than with the *outcomes* of planning. Put otherwise, procedural rationality and formal efficiency gained in respect to the disadvantage of substantive and normative rationality, which deals with questions as to what planning is all about, who profits from it, what kind of society planners really help to plan and where the societal responsibilities of planners lie. As a consequence of this proceduralist approach, planning came more than ever under the spell of technical and positivistic reasoning (Harvey, 1982; Swyngedouw, 1987).

As during the high tide of Fordism, many people were able to ameliorate their position of living; now an 'underclass' (Gans, 1991) is emerging. The members of this heterogenic 'class' have in common that they do not have the opportunities, the financial means, the proper education or the organizational capacities needed to conquer for themselves a place in society. Nor does there seem to exist the solidarity needed to help these people (see Galbraith, 1992). Underclasses are emerging especially in poor regions and cities (see Wilson, 1991, on the underclass and the city), leading further to a stratification of poor and rich regions which, in some cases, threatens the regional–political compromises

made in the context of the nation-states. During the past decades, cities and regions have become direct players in the world economy. Some argue that the location of cities within the international division of labour under conditions of heightened inter-urban competition has strengthened the bargaining position of cities (see Mayer in Peck and Tickell, 1994: 303). Peck and Tickell, however, comment:

> The nation state *may* have been eroded from above and from below, but the nature of this erosion has been different in each case. Below the nation state, local regulatory systems . . . have been conferred *responsibility without power* . . . Above the nation state, supranational regulatory systems have inherited *power without responsibility*: remaining wedded to a neo-liberal agenda, they continue to fuel global economic instability with apparent disregard for its damaging effects on national and local economies and its pernicious ecological and social consequences. (Peck and Tickell, 1994: 311)

Another but related issue is that, in the future, policies aimed at increasing the productivity very likely will be propagated under the banner of sustainability, of which a technocratic conception, in which all relevant questions concerning the desired balance between the intergenerational and the intragenerational aspects of social and political justice are set aside is in the making (Sachs, 1995; Van Dieren, 1995). Another related problem concerns the rebirth of nationalism, an ideology which again gains respectability everywhere (see Gellner, 1983; Hobsbawm, 1992). This ideology only too well couples itself to the economic restructuring and changed ideological outlook mentioned. Castells echoes Gellner in arguing that 'The globalization of power flows and the tribalization of local communities are part of the same fundamental process of historical restructuring; the growing dissociation between the techno-economic development and the corresponding mechanisms of social control of such development' (Castells, 1989: 350). In fact, nationalism is aimed to spread grimness and to direct widespread discontentment towards the weakest parts of society, such as the 'ethnic' minorities, the disadvantaged, the unemployed and the excluded. There are still other difficulties. Political activities move away from the traditional political representational institutions to spheres which traditionally were not considered to be political, such as the mega-enterprise, high-tech research and the media (see Beck, 1988; Huyse, 1994). The ability of capital to circulate globally redefined and weakened the role of traditional state-based politics (see Castells, 1989: 349; Offe, 1984: 76). Speculators determine the rate of interest of money and, by consequence, the overall level of investment and employment. Although the question of technology is paramount and although technology meanwhile has successfully been analysed as a 'social process' (Bijker et al., 1993), no serious attempt has yet been made (in planning or in politics) to design a strategy aimed at politicizing democratically decision-making concerning technology (Kirsch, 1995; Van Dijk, 1995). In our societies, politicians and populations only become partners in a discussion about the desirability of a new technology when that progress is already under way. Transnational enterprises pursue a politics of *faits accomplis*. All these developments contributed to the destruction of the ideal of a society guided by the intersection of planners and democratic popular control, and participation aimed at democracy, normative rationality and social justice.

CHALLENGES FOR PLANNING

All of this makes abundantly clear that there is a need for a new role for planners if planning is still to be taken seriously in the future. The planner's tool-kit, which was designed during a period of uncontested belief in overall economic and social progress in an environment characterized by a seemingly relative abundance of resources, is outdated and must be redesigned and adjusted to the needs and challenges of today's society. This, however, will prove a difficult task. As Albrechts writes, 'the impact of incentives, motivation and social mobilization is rather primitive in our planning' (Albrechts, 1991: 125). The planning challenge seems to be either to adjust to changing conditions or to tackle the roots of the current deterioration in order to find a sounder basis for shaping new and better conditions in society. Accepting the latter view implies the re-emphasis of the *political role* of the planners. Planners have to present themselves openly as strong partisans for certain outcomes as opposed to others, for the interests of some groups over others, for some styles of governance, for some conceptions of justice, some patterns of future development, and so on (Albrechts, 1991, 1999; Forester, 1989). Planners have to operate in close collaboration with other actors and target groups in the decision-making process as well as to comprehend their interests and power relationships. A new concept for planning, which can be partly drawn from the literature of the 1960s, ideally, is aimed at inducing structural change. The planner's political role comprises a contribution not only to the substantiation of these changes, but also to the mobilization of the social forces necessary to realize proposed policies.

NEWLY EMERGING CONCEPTS

The planner's tool-kit is now being filled with highly sophisticated (and highly academic) concepts which, among others, refer to 'building identities' (Crow, 1994), 'seeking consensus', 'transferring critical knowledge to action' (Friedmann, 1987), 'letting unfold and confront situated knowledges of different communities' and 'planning through debate' (Healey, 1992; Hillier, 1993), providing the necessary infrastructure to make debate, and consequently consensus, possible, maximizing the potential of human capital, establishing mutually beneficial contacts between different actors such as firms, financial sources, knowledge centres and groups, getting in contact with target groups, providing society with a conception of 'the common good', mobilizing the social forces necessary to realize politics aimed at structural change. Reading the list, one remembers Wildavsky's complaint about the literature on planning theory made in 1973: 'The planner has become the victim of planning; his [*sic*] own creation has overwelmed him. Planning has become so large that the planner cannot encompass its dimensions. Planning has become so complex planners cannot keep up with it. Planning protrudes in so many directions, the planner can no longer discern its shape' (Wildavsky, 1973: 127). However, Wildavsky warned, if planning is everything, then maybe it is nothing (Wildavsky, 1973). Most planners have not left the modernist mode of thinking, but often seem to be engaged in an enterprise which, on the theoretical level, consists of a rather associative and intuitive made-up mix of both modernist and postmodernist insights, some of which even contradict each other. This mix obstructs conceptual clearness to the identification of planning key words and fundamentals. This state of affairs makes both the theoretical and the practical work vulnerable. It will be necessary to take postmodern objections seriously and to incorporate these views substantially and systematically – and not marginally or rhetorically – in planning theory. If the postmodernists are right, many actions now undertaken by planners will prove to be largely done in vain or even will prove to be counterproductive. Instead of regarding the admittedly many-faced and polymorphous postmodernist literature as a monster from a safe distance, planners have to confront themselves with these views and have to make up their minds about the fundamentals of planning. Such a confrontation, which eventually may lead to a mutually beneficial debate between postmodern theory and planning practice, cannot be selective and free-ended. In order to elucidate this claim, the focus is on consensus-building as a central concept in planning theory.

Comments are also made on the ideal of social justice communicative planners claim to pursue.

WHAT IS CONSENSUS-BUILDING?

Critical planning theory, communicative planning and radical planning are related (see Friedmann, 1987: 225). They are concerned with equity and with the distribution of power in society and with the extent to which planning reflects this distribution.

Critical planners do not consider planning to be solely a professional or technical activity. Instead, planners have a responsibility to act militantly and to challenge the status quo. If we want plans to work, planners must gain legitimacy and support in the society in which plans are meant to be implemented. Planners have to redefine problems and have to learn to work as political actors in partnerships with organizations that represent the interests of disenfranchised and unorganized people. This means that a double learning process is called for. Social learning is essential to radical planning. Since planners, like everyone else, draw on different rhetorics and discourses to construct different versions of the world, this would mean that, minimally, planners are able to reach a consensus between themselves on the one hand and the target groups for whom they are advocates on the other. However, the critical question here is whether this is possible.

The case for consensus-building has, of course, very often been made. As Hillier explains (1993: 107), practical reason for planning decisions should involve the Aristotelian notions of persuasion, reflection upon values, prudential judgement and free disclosure of ideas (see also Forester, 1989). Only when each community involved listens to the others and recognizes the legitimacy of the different perspectives, a level of shared understanding can be reached in which areas of congruence or overlapping qualities can be discovered. For Hillier, rational debate is possible between proponents of different truths, provided that their systems of thought intersect at some point and that the instrumentalism of 'expert culture' (Dryzek in Hillier, 1993: 108) is not overpowering.

Planning, instead of being dominated by technical approaches and panaceas which make the process relatively inaccessible to laypersons, would then have to be interpretative and interactive. Interaction means that different individuals and organizations be engaged with one another in debate and negotiation. Each group uses its own discourse, knowledge and meaning systems. The role of planning should be to engage in 'respectful discussion within and between discursive communities, respect implying recognizing, valuing, listening and searching of translative possibilities between different discourse communities' (Healey, 1992: 9–10).

Hillier (1993) presents an empirical example of the systematic distortion of information and the consequent impacts of it in a particular planning context. She explains that choices are made according to certain power structures and decision rules for preferring certain solutions over others. Planners engage in discourse, conversation, negotiation and persuasion with several groups in society which tell different stories, which see reality differently from each other. Hillier recognizes that in society implicit divergences exist on what reality 'really' is and, by consequence, on what is 'really' going on. Schwarz and Thompson explain that public discourse suffers from these implicit divergences, because societies like ours have political mechanisms only for resolving conflicting interests, not for conflicting views of reality. *Because* the mechanisms for dealing with conflicting world-views and different, discourse communities are lacking (and because, in discourse, we mainly stick to our own group and the language we 'understand'), we only seldom notice that perceptions and not only interests in society differ markedly (see Schwarz and Thompson, 1990: 30 on 'the' – for different actors in different and unequal ways real enough – reality of the energy debate). The word 'world-view' is not used here to suggest that different world-views are dependent from different positions that lead to the adhering of different and conflicting ideologies. On the contrary, different variations of the same story told to and by various interest groups, demonstrate the power of discourse not merely to describe but actually to *constitute* different realities (see Hillier, 1993: 89). By this is meant something like the social construction of reality, reality being the domain made up both by objects and facts, values and social relations leading to various perceptions (and conceptions) of 'what is' (see also Schwarz and Thompson, 1990 and Mandelbaum, 1992 on 'multiple realities'). This does not mean that audiences are not being manipulated in terms of their subject position. As Hillier also puts it, the reality of different realities creates an ideologically potent 'way of seeing' issues, which affects the way people act. Habermas's work on the distortion of communication is a case in point here. Planners, however, have to communicate and also let all those involved communicate. Planning, if not overpowered by technical reason, is an inherently rhetorical project for which there exists no rule books – a project that cannot be properly understood apart from 'the audiences to which it is directed and the styles in which it is communicated' (Throgmorton, 1993). This means that planners must have specialized skills for dealing with 'different truths', but also that planners cannot avoid asking questions about *their* values and the legitimation of *their* professional actions which, of course, intersect with their 'truth'. In evading such – admittedly complex – issues, many planners attempt to justify their own legitimacy by engaging in a complex set of distortions in communication, designed to obscure and to prevent understanding of their practice and the outcomes *they* propose.

As long as planning was understood as a solely technical activity, planners did not openly have to bother with the ethical implications of their work. In our society, the society of risks, the noun 'technical' functions as a synonym for not having to bother about ethical implications (Beck, 1988). Even in the choosing of goals or the choosing of the 'optimal' or most efficient means for achieving 'desired' objectives, planning was not considered to be political. However, as Healey and Beck remind us, even if one adopts this view, it does not eliminate the responsibility of planners, engineers and other scientists to deal with the ethical issues inherent in the actions they undertake. Responsibility can never be abdicated. As Hillier writes, 'planning is itself essentially political. Planners' actions help to determine who gets what, when and how. So, the question is not whether planning reflects politics, but whose politics it reflects. For whom does the planner speak? What does a planner want to do? Who does she or he serve? Who does she or he exclude? What is the position of the narrative the planner puts forward?' (Hillier, 1993: 90). What is the planner's conception of justice and of citizenship? Must he or she hide partiality or bias or be completely open on it? What sort of information is most consonant to the planner and which audiences or groups will he/she find the easiest to communicate with? (See Schwarz and Thompson, 1990 on 'different rationalities'). Is it, in practice, really possible for the planner to resist the temptation to engage unproportionally more in communication with those with whom she or he feels related, for example, with the higher educated or others who adhere to the same rationality or world-view? (See also Forester, 1989: 31ff.; Healey, 1992.)

Members of a community tend to agree because they see everything in relation to their assumed purpose and goals. Members of different communities disagree because, from their respective positions, they see the same issue in completely different manners. Each community constitutes within itself an ensemble of rules as to what constitutes truth within their discourse. These truths often vary and conflict, yet they are all, in themselves, legitimate. Since communication is made of language and narratives, the language and the narrative of planning are not neutral. Rather, they form 'an institutionalized structure of meanings which channel thought and action in certain directions' (Hillier, 1993: 92). Planners use different rhetorical styles for different audiences and different partners. This 'rhetoric of planning' shows a way in which a planner's work may become 'a creative process' (Dear, 1993), changing presentational information to become more convincing or responsive to several audiences in order to win acceptance for a particular explanation or interpretation.

In this interdisciplinary mix of modernist and postmodernist views, consensus-building does not mean (any more) to speak by means of a ('the') language common to all or to create such a language (through the elimination of structures and balances of power that distort free and rational communication) and to reach agreement. Consensus-building has to let unfold human plurality and to respect the different world-views and multiple truths. The planner (still) faces the task to speak in a language that is (equally) comprehensible to all. As Forester writes, mutual understanding depends on 'the satisfaction of these four criteria: comprehensibility, sincerity, legitimacy and accuracy or truth' (Forester, 1989: 144). Otherwise, the door is open for confusion, manipulation, deceit and abuse. When it is impossible to gauge the truth, Forester asserts, it is impossible to make a difference between fact and fantasy. The planner has to be able to act as a mediator between his or her own world-view and all the other world-views concerned as well as between the different world-views planners have to bring in communication with each other. The planner must be able to assemble the conflicts and inconsistencies and to alter the balance of power in the advantage of the people she or he speaks and acts for. This, however, is precisely the standpoint criticized and abandoned by postmodernism.

POSTMODERN CRITICISM

In contemporary literature on communicative planning, the planner appears as an understander and beholder of a meta-language, that, following postmodernism, cannot exist. There is no planner – or anyone else for that matter – who is able either to incorporate the different narratives and world-views and make the best of them, or, in a Rousseauian manner, to listen to all and to select the best option that then can serve as the basis for a consensus. Also noticeable is that no answer is provided to questions as to how planners should recognize a consensus that is *morally* right from a consensus that is morally wrong. What, for example, has to happen if mobilization stimulated by planners leads to *privativistic* forms of life which provide the psychological base for unleashing forces that are of their nature unpolitical (see also Young, 1990)? This is a distinct possibility if communication does not lead to a concept of truth that is also bestowed with moral meaning. Do planners in such cases have the moral duty to amend the agreement or must she or he strive towards agreement irrespective of the content and put their trust in the common sense? What must happen when a consensus or agreement – in the name of democracy and participation – leads to *bad* planning, because, for example, several goals, some explicit and others implicit, that contradict each other with tactics that are not compatible are pursued at the same time? What must happen if, through mobilization, target groups acquire a sense of identity, but, in doing so, become entrapped in a *counter-emancipatory* ideology propagated by a political party or social movement (see Crow, 1994; Friedmann, 1987: 276)? Communicative planning requires collective action, which means bottom-up organization. This requirement is often not properly understood by planners who want participation. How do we get the participation we want from the people we envisage? In practice, this is often very difficult or even impossible to achieve. Some groups never organize. Organization supposes a sense of common identity, but a common identity already supposes organization to a certain degree. One also may honestly believe that participation, of itself, will increase the power of poor and underprivileged people. As Benveniste (1989) writes, power equalization does not come automatically with participation (of course, Benveniste makes himself completely incredible when he states that participation, *per se*, reduces the power of the weak by making them more dependent on the expertise of those who have access to knowledge; see Benveniste, 1989: 67). The conclusion is that working out an emancipatory planning strategy is more difficult than should be expected at first sight.

Communicative planning is not possible without the explication of norms, values and world-views which lead to certain actions. But what has to happen when values and world-views differ so considerably that no consensus concerning goals can be reached? This is the crucial question. Since no homogeneity on values and world-views can be provided for, the traditional idea (and ideal) of a consensus has to be questioned.

MODERN AND POSTMODERN POLITICAL THEORY

Since the focus is on the discussion on the idea (and the ideal) of consensus to be pursued through communicative planning and initiatives of empowerment, it is necessary to explore the debate between modernists and postmodernists. First some general remarks are made. Thereafter, the focus is on the debate between Habermas and Lyotard. Other modernists and postmodernists, such as Arendt on the one side and Foucault and Baudrillard on the other, are used as well.

Modernist thinkers consider themselves to be the inheritors of the philosophy of the Enlightenment. Modernists consider it their task to develop theoretical knowledge that can be of practical use. Rationally, by way of *critique*, it can be proved that certain ruling conceptions, scientific statements or popular beliefs are one-sided or even fundamentally wrong, this is to say, counterproductive (see Boehm, 1977; Sachs, 1995). This claim is essential if one is to prove that certain emancipatory goals, such as the realization of 'sustainable development' (however defined, see Sachs, 1995) or social justice, cannot be realized given the framework of the present route of (capitalist) development and distribution of power (Harvey, 1996; Smart, 1993).

Postmodernists, on the other hand, take the position that the confidence of modern man in her or his thinking and judging faculties as adequate tools for understanding reality has to be relativized. They emphasize that scientific knowledge functions as a filter of its own. This filter lets through certain phenomena, but deliberately ignores others (Kunneman, 1986). Seen from this angle, the postmodernists claim modern science to be a peculiar sort of ideology of its own. To them, modern science cannot function without the propagation of certain myths. One of these is that the modern scientific project is all-encompassing, while, in reality, it is highly selective (see on this from a different angle, Latour, 1987) and to some highly irrational (Beck, 1988; van Reijen, 1988).

As van Reijen (1988) notices these antagonistic social-philosophical and epistemological options have, of course, very different consequences concerning *the types of actions*, such as planning, one sees fit to undertake in the world. Modernist thinkers consider it possible to formulate appropriate proposals and take actions which, if carried out well, can effectively fight undesirable developments. On the basis of communicative rationality, argumentative processes in society can lead to the popularization of rational insights. *Ideally*, a consensus, with the maximum level of legitimation, democratic support and approval, can be reached concerning actions to be undertaken.

Postmodernists, however, evaluate the adherence of the part of contemporary social scientists to modern theoretical presuppositions, such as the trust put in communicative rationality as an untenable, now outdated and *counter-emancipatory ideology* (see Kunneman, 1996). 'The position taken is theoretically untenable because it is assumed that reality is homogenic and knowable to man by way of objective criteria and methods. The position is the more outdated because the path since the Enlightenment does not point to a historic development towards more "humanity"' (van Reijen, 1988: 207, our translation).

HABERMAS AND LYOTARD ON RATIONALITY AND JUSTICE

The debate between Habermas and Lyotard concerns the crucial question whether we have our thinking and acting still under control. This obviously is an important question for planning.

Habermas

Habermas explains that in the philosophy of the Enlightenment a pure and emancipatory concept of rationality can be found, but, also, that from the eighteenth century onwards the West witnessed an explosion of technical knowledge (and hardware technology – a theme to which Habermas unfortunately pays too little attention, see Feenberg, 1995) and social engineering techniques, '*Humantechniken*', which invades through economy, law and the state into the life of the individuals, with the result that the potential of communicative rationality becomes severely threatened. To Habermas, the process of reflective modernization disconnected domains such as economy, politics, law and ethics from each other and from the lifeworld. This had the positive result that our capacity to solve problems in each of these domains increased dramatically (see for critique, Boehm, 1977; Achterhuis, 1988). The evolution, however, also had the severe negative effect that goals and norms were disconnected from the emancipatory vision of a *humane society* (see van Reijen, 1988; Kunneman, 1996). Economy, politics and ethics became relatively autonomous. However, for Habermas, *common* desirable societal goals, such as manageability, rationality and justice, can be reached intersubjectively through speech acts – debate, agreement, consensus – by equal, well-informed and honest participants. An idea of the good society can take form in practice through the spread of such acts. Instead of freeing people and increasing the capacity of individuals to speak and take care of

themselves for themselves, modern politics and economy 'parasite' on the lifeworld of individuals. Habermas uses the word 'parasitizing' because the subsystems always remain dependent – although they intrude on it violently – on the original reserve of common sense and solidarity. Eventually, if, due to all sorts of developments, the 'parasitizing' of the subsystems become counterproductive, one faces the situation of societal unmanageability ('*Unübersichtlichkeit*', see Denayer, 1994; Habermas, 1985; Kunneman, 1986; van Reijen, 1988;).

Put in a nutshell, this can be said to be the critique Habermas develops throughout his work. His aim is to restore the lost unity of technical (functionalist and instrumentalist) reason and moral responsibility. This becomes clear in his *magnum opus The Theory of Communicative Action*. Here Habermas partly repeats his diagnosis: Western culture is engaged in a process of modernization (see *The Theory*, Vol. I. § 2; see also Mannheim, 1951; Sass, 1992). This process leads to the differentiation of subsystems (Vol. I. § 3). This evolution in turn leads to the reign of instrumental reason ('*Zweckrationalität*') (Vol. I, § 4). As a consequence, men and women become cogs in a machine, since lifeworld and system decouple under a regime of exponential technical progress (Vol. 2, § 5). The critical questions as to the why and to the justness of things cannot be put anymore. The communicative reason ('*Wertrationalität*') becomes destroyed. However – and this is Habermas's main thesis – through action people can redeem communicative rationality. It is the task of critical theory to show that the disparities and contradictions between the subsystems can be united on a higher level of modernization (Vol. 2, §§ 6, 7 and 9). The historical development of modernity may lead to all sorts of pathological phenomena, but this does not prove the wrongness of the modern project. For Habermas, through rational critique and communicative practice, modernity can still be 'completed' in an emancipatory way (Kunneman, 1996; Smart, 1993; van Reijen, 1988).

Lyotard

Lyotard takes a totally different position. Lyotard argues that the 'grand narratives', the philosophy of the Enlightenment (liberalism) and Marxism, have failed. This is due to the fact that in these narratives a heterogenic, many-faced and antagonistic reality is analysed under one point of view only (see van Reijen, 1988; Kunneman, 1996). Both the philosophy of the Enlightenment and Marxism make the fault of forcing (!) everything under the yoke of a great theory which, under the pretence and the pretext of universal emancipation, denies other ('non-universal') views the right to exist (see also Latour, 1987). Lyotard explains postmodernism to be 'the state of our culture following the transformation which, since the end of the nineteenth century, has altered the game rules for science, literature and the arts' (Lyotard, 1979: 5). Postmodernism is also defined as 'incredulity towards meta-narratives' (Lyotard, 1979: 5). Lyotard explains that he uses the term modern to 'designate any science that legitimates itself with reference to a meta-discourse . . . making an explicit appeal to some grand narrative, such as . . . the emancipation of the rational, . . . or the rule of consensus between sender and addressee of a statement' (Lyotard, 1979: 16). As van Reijen (1988) explains, for Lyotard, the status of knowledge altered as societies entered the post-industrial age. Lyotard dates this development at the end of the 1950s. Lyotard concentrates on scientific knowledge understood as discourse. This allows him to treat knowledge as something which stands or falls depending on its relationship to other discourses or meta-narratives. This in turn makes it possible to treat science as something which can be commodified and alienated from its producers (see also Smart, 1993; Kunneman, 1996). The knowledge–society is an exploitative and alienating one, because it is built on capitalist lines with an elite class of decision-makers having access to knowledge and the distribution of it. The postmodern condition, however, can alter these circumstances – and thus can be *emancipatory* – by way of an atomization of the social into flexible networks of language games and by the critique of games which attempt to increase performance and to appropriate power unto themselves. Obviously one cannot be further removed from the position taken by Habermas. Lyotard sharply attacks Habermas and the possibility of consensus:

> [C]onsensus is a component of the system, which manipulates it in order to maintain and improve its performance [the legitimation of the system–power]. Consensus is a horizon that is never reached . . . For this reason, it seems neither possible, nor even prudent, to follow Habermas in orienting our treatment of the problem of legitimation in the direction of a search for universal consensus through what he calls *Diskurs* . . . (Lyotard, 1979: 36)

Modern life was organized around grand narratives, the Ideas of Progress, Emancipation, Enlightenment. They convey a sense that the narratives of the separate value-spheres are moving in the same direction. But these narratives have lost their credibility. Post-industrial developments led to the commodification of knowledge (see also Beck, 1988). As van Reijen (1988: 212ff.) explains, for Lyotard the performance of the total system replaces truth as the yardstick of knowledge and justice as the yardstick of a humane society. The cultural production of society has dissolved into a series of localized and flexible, but incommensurable networks of language games (see also Smart, 1993; see for a critique Callinicos, 1989). Since the cultural is the source of legitimation for the social, traditional social structures, such as nation-states, political parties, professions (such as planning) and institutions, historical traditions lose their capacity to attract engagement and support. Lyotard envisions a society in which, at best, flexible networks of language games provide localized capacity for resistance to total (becoming totalitarian?) performativity (Smart, 1993; van Reijen, 1988).

To substantiate this, Lyotard cites the contradiction of labour and capital. In Western democracies, there exist institutions that mediate between labour and capital. Individual conflicts eventually can get resolved by courts, specialized in these matters. But this constitutes no proof of homogeneity. Courts can take decisions between parties, but leave intact the antagonistic relation between the two systems. A labourer cannot prove before a court of law that his or her labour creates surplus value which the employer creams off. He or she only can make use of the dominant capitalistic idiom to say what labour is – and by doing so labour becomes encapsulated as a value (see also Baudrillard, 1988: 98ff.). The labourer cannot use expressions he or she finds more appropriate. There simply is no institution in society through which men and women can speak truthfully and authentically to each other (see van Reijen, 1988: 216ff).

As van Reijen (1988: 218ff.) writes, Lyotard also gives a few clues which may be relevant for planning, politics and justice. To begin with, the attempt to realize a certain 'blueprint' or utopian ideal, such as the realization of a more humane and free society, cannot function as the guiding principle for political actions. Instead, men and women can only try to treat others 'as if' such a society already existed. Justice and rightness are thus not 'great' ideals to be pursued as an apart programme, with a lot of specialized machinery, 'great' ideology and fastidious fuss, but something one has to practise in everyday life towards the others one is or feels connected with (Lyotard, 1983). As van Reijen also notes, since language games and world-views differ and since justice and rightness remain somehow vague, such an undertaking necessarily remains always fragile (see Nussbaum, 1989). The argument on what, for example, justice is, cannot be closed, communicatively or otherwise, not only because there is no discursive bridge between the ideal and the concrete, but because actors are engaged in different incommensurable language games. For

Lyotard, the idea of a free, egalitarian society is unimaginable because no empirical reality corresponds with it (Lyotard, 1983). In consequence, *every* theory in which statements and proposals in favour of the realization of such an utopian society are made in fact *abuses* people by forcing them into an *instrumental* way of thinking and acting, as if the realization of a humane society could ever be achieved by either decision-making (*poiesis*) (such as planning) or/and politics (*praxis*).[1] For Lyotard, freedom, and all other political ideals (normative rationality, justice, proper steering), are contrafactual. They cannot be seen as goals that can be implemented technically (Lyotard, 1983).

POSTMODERNS ON POWER AND REALITY

The postmodernist critique on the modern project is not exhausted by reproducing Lyotard's forceful claims. In the next section the implications of postmodern criticism for communicative planning practices and planning theory will be looked at.

Villa (1992: 713) notes three postmodern objections to the modern project of the 'recovery of the public realm'. This motto indeed adequately captures the goal of Arendt's political theory and of Habermas's critical theory (see Villa, 1992). The postmodern objections point to fundamental difficulties and contradictions concerning the presuppositions of the *type* of project theorists such as Arendt and Habermas – and, by consequence, communicative planners – engage in. The first objection, originated in the work of Foucault, is called the *power* objection. Foucault radically questioned the idea of a coercion-free space by retheorizing the nature of power in the modern age. Villa's second objection is the *epistemological* one. This has already been touched upon in the discussion of the possibility of a unified, consensus-based public realm in an era that has witnessed the death of legitimating meta-narratives and the fragmentation of discursive practice into heterogeneous language games. Villa's third objection is called the *ontological* one. This objection perhaps is even the most powerful because in it, following Baudrillard, the peculiar reality attributed to a space, called the public realm, is questioned.

To begin with the first objection, both Arendt and Habermas make a distinction between power on the one hand and violence, force, coercion, repression and so on on the other (see Arendt, 1972; Habermas, 1977). For both Arendt and Habermas, power is the capacity of people to act freely and equally in concert, to let unfold human plurality. Both Arendt and Habermas refer to the possibility of power, which grows out of discursive and agonistic interactions, as opposed to coercion and violence, and, for both Arendt and Habermas, hierarchy has no place in politics properly understood. However, a good deal – but not all (see Denayer, 1994; Zwart, 1995) – of Foucault's research into the internalized microtechniques and panoptical mechanisms at work in society, leads to the view that the Arendtian/Habermasian distinction between power and violence has become outdated in the modern age. If (repressive) 'normalization–power' works through each of us, indeed, if, as Foucault explains, the (modern invention of the) individual only comes into being *through* this repressive but intensely productive power, then there is little or no point in pursuing an empowerment programme aimed at liberation or dissociation from, to borrow Gramscian/Althusserian language, openly recognizable repressive 'ideological state apparata', or in communicative planning that is aimed at letting people unfold their plurality in the hope they will agree on a consensus or on the need for social justice and so on (see Villa, 1992: 715ff.). Such undertakings risk being counterproductive for the reason that they bring with them the illusion that people can accumulate power without recognizing the effect that power can accumulate people. From Foucault's point of view, Arendt and Habermas deal with a 'power-free zone' in society that nowhere exists. As Villa writes, the positive model of power advocated by public theory and the negative one upheld by liberalism both blind one to the constitutive workings of modern power and its role in the *fabrication of subjects* (Villa, 1992: 715). The limitations of public realm theory are most apparent in its naive reliance upon and belief in conditions of symmetry, nonhierarchy, plurality and reciprocity as adequate safeguarders of a non-violent space and of undistorted communication. Following Foucault, the elimination of what constitutes the antipolitical foes according to the public realm theory – hierarchy and asymmetry in the political sphere – is not essential to the functioning of disciplinary power because, in his ontology, subjugation and subjectification go hand in hand. Foucault cannot imagine ideal speech situations. For him, the realization of an ideal (or even good) speech situation would mean no more than the achievement of 'pseudoautonomy in the conditions of pseudosymmetry' (Fraser in Villa, 1992: 715).

Leaving the second objection to later, the third objection made by postmodernists, the ontological one, is directed against the specific reality claimed for the public realm by public realm theorists. Here, Arendt is a case in point. Arendt describes the public sphere phenomenologically as a space common to all: the public space is the realm of appearances made up by action (*praxis*). Only action, which potentializes the plurality of men and women and leads to a diversity of perspectives, creates a distinctive, truly human world. This 'in-between' is the proper domain of freedom and intersubjectivity. Arendt believes that the reality of a space of appearances common to all is guaranteed as long as the in-between stays in function. Where the public spirit animates a community, Arendt believes, the *sensus communis* remains strong, providing a non-transcendental ground for action and judgement (Arendt, 1982). This common feeling for the world, in which all the others of the community are taken into account and which is grounded in the Aristotelian's prudent judgement and Kant's practical judgement and in which no claim whatsoever to extrapolitical forces is made, seems congruent with and sympathetic to communicative planning (especially also if one considers Arendt's localism). However, Arendt has come under attack from especially Baudrillard.

For Baudrillard, man really is 'a child gone astray in the forests of symbols' (Kundera in Beck, 1988: 62). Baudrillard does not only speak of the production of culture, but also of its consumption in which knowledge is centrally organized and controlled. The effectivity of culture is so strongly increased that only the relations between its symbolic contents, its 'signs', have force. The condition of postmodernity exists in 'hypersimulation', by which is meant a double counter-reflection in which life simulates the already simulated contents of the mass media (see Villa, 1992). The social does not mediate the signs since they have neither labour value nor use value, but only consumption value. Everything becomes absorbed by signs. They become a 'hyper-reality' in which only simulations exist. Baudrillard also describes the social as a mass, something without any meaning, which makes politics, intelligence, empathy, initiative and planning all quite impossible: 'With the masses, the logic of the social is at its extreme; the point where its finalities are inverted and where it reaches its point of inertia and extermination . . . The masses are the ecstasy of the social . . . the mirror where the social is reflected in its total immanence' (Baudrillard, 1988: 188; cf. Beck, 1988: 184).

Baudrillard and Lyotard now make two objections against Arendt (see also Villa, 1992: 718). Arendt's description of a public space adds up to a dichotomy between an authentic, real in-between and its inauthentic, manipulated counterpart, the former both condition and result of action, the latter being nothing more than the proliferation of private and idiosyncratic images that only too well serve to destroy our feeling in the world (see also Harvey, 1989). Baudrillard makes the point that this way of looking at things does not mark a final break with the modern representational *episteme*. Postmodernism has to leave the order of appearances/simulations behind. Only then one can observe that 'reality' – including Arendt's space of appearances – is presently generated as a *simulation effect*. From this point of view, Arendt's ontology prevents us from seeing just how truly lost the world is, or, in Arendt's language, the feeling of 'wordliness' and in consequence the possibilities for politics, communicative planning and empowerment (Villa, 1992: 718).

Returning to the second objection to Arendt's modernist ontology, this has been formulated by Lyotard. For Lyotard, politics has to exist without (tyrannical) ideas of, for example, justice (for, what is just if no consensus can be brought about the just or about justice and if consensus is a suspect value?). Arendt also bans all non-political, transcendental and other standards out of political life properly speaking, but she operationalizes the Kantian notion of *sensus communis* in an attempt to save politics from arbitrariness. For Lyotard, this appeal to the notion of *sensus communis* cannot be taken seriously: the withering away of common sense in postmodernity simply means that 'there cannot be a *sensus communis*' (Lyotard in Villa, 1992: 719). Social identities have exploded, with the consequence that what is acceptable to some remains illegitimate to others, while there is no way to settle this dispute.

POSTMODERN PLANNING, COLLECTIVE IDENTITY AND SOCIAL JUSTICE

Let us now see how the above-mentioned views can be incorporated in planning theory. Which consequences are involved? Healey (1992: 101) puts much focus on how democratic values are reflected in plans. The plan is a product of interaction between several parties. The plan may become a point of reference for continuing interaction within which discourses may evolve. One of the objectives is to offer rhetorical strategies. Recognizing the postmodern culture, this means 'recognition of the plurality of discourse communities . . . and means explicitly addressing the challenge of "interdiscursive communication"' (Healey, 1992: 103). Rhetoric is constitutive for planning. As Throgmorton (1993: 334) argues, planners should strive towards 'a rhetoric that helps to create and sustain a public, democratic discourse. This should be a persuasive discourse that permits planners to talk coherently about contestable views of what is good, right, and feasible. Planners should strive to create arenas that facilitate and encourage just such a persuasive, public discourse' (Throgmorton, 1993: 336; see also Faludi and Korthals Altes, 1994). What are the difficulties involved in view of what was said earlier?

Massey has commented on Mouffe's explanation that 'political practice in a democratic society does not consist in defending the rights of preconstituted identities, but rather in constituting those identities themselves in a precarious and always vulnerable terrain' (Mouffe in Massey, 1995: 284). Massey emphasizes the crucial point that political actors do not bring already appropriately constituted identities into political arenas and planning settings. Indeed, as Arendt already pointed out, the political actor has no distinctive identity before he or she sets foot in the public domain. Moreover, when he or she does this, his or her identity always remains somewhat hidden to him or her (see Arendt, 1958: 221). Politics is a fundamentally rhetorical project and people gain identity by engaging in common rhetorical enterprises. This does not mean that the past is a blank sheet: 'Political subjects are . . . constituted in political practice, but they are not constructed out of nothing' (Bauman, 1992; Massey, 1995: 286). There always is a history, a biography of the person (the same can be said of the identity of place; see Massey, 1995). This is, of course, as Healey (1992: 106) has pointed out, already important for political and planning purposes. People obviously can be hurt, frustrated, not able to tell the difference between manipulation and planning in their advantage, unwilling to cooperate and so on. Massey's point, however, is still more relevant. Planners often work with communities that are already well defined and therefore highly organized (see, for example, Amdam, 1995, on rural municipalities in Norway). But what about communities that are in the making or different target groups that partly overlap? Does the planner in that case have to establish the boundaries of the community or target group which he or she wants to work for? Obviously, his or her work has to remain manageable. The planner, at one point or another, has to decide who will participate in and therefore will 'belong' to the community or target group and who will not. This also seems necessary for the reason that there cannot exist an 'us', a community or target group, without an outside 'them'. Political communities and social groups precisely define themselves by defining others as non-members.

However, several questions do arise here. The first problem concerns the criteria planners use to establish boundaries in a way that is both effective and just. A first way out of this perhaps is to plan and advocate for the benefit of the human species. A second (and more realistic) way out is to recognize that people *can* act together without

a strong sense of an 'us' in any traditional sense, such as the family, the neighbourhood, the *Gemeinschaft*, the party affiliation, the nation and so forth. In reality, the 'us' always exists, both of proponents of common political principles and goals to be agreed on and dissenters to these principles and goals. However, proponents and dissenters are not always the same individuals. They change position. Therefore, the 'us' can also include those who do not agree only about the common political principles or those who do not agree about the political framework, necessary for the division lines to become manifest. It must be possible to empower a 'community', seen as something without boundaries, *without* knowing beforehand what political forms and goals will be agreed upon. In effect, paradoxical as it may sound, such an empowerment programme comes close to planning unpredictability. However, such planning would mean empowerment, because it would create the conditions, especially worldliness and plurality, for political initiative, coalition building, and freedom.

Our viewpoint is not only normative (or strategic). If empowerment means anything real, it needs to proceed in this way. Freedom and dissociation obviously cannot be planned by the state or its advocates. Communicative planning cannot only be pursued from a top-down approach in which planners of city A decide to advocate marginalized target group 22. Group 22, however, only exists as an essentialist abstraction in the planner's mind or – still more often – only in the minds of the politicians who decide to raise funds for this 'entity'. In this way, politicians play an important part in the definition of groups. However, politicians know only too well how to touch changing, latent, formations and bring them to the surface as new political identities, for example along nationalist lines (see Massey, 1995: 286). Communicative planning then can become a morally doubtful and practical counterproductive enterprise. As Hillier writes, politicians and officials have the power to manipulate information and actors. The state can use professional mystique to obscure, can make superficial concessions, can simulate procedural instead of substantive gains. There is still another danger here. This is that accepted within this way of proceeding is the classic ethnographic notion that each individual can belong to one, and only one, discrete (unambiguous, non-overlapping) cultural entity. However, in the real world a plurality of partially disjunctive, partially overlapping communities exist that criss-cross between the people planners address and for whom they speak (see Rosaldo, 1995: 179). As Alarcón (1995: 10) writes, social reality is made up of 'selves' who occupy simultaneous social axes of gender, class, race and sexuality. There is, for example, Alarcón argues, no target group corresponding to the entity women. Feminists – and communicative planners pleading for women – must give up appeals to a unitary idea of 'women', because it inevitably excludes some of them and impedes coalition-building, but instead must favour the possibility of subjects who are aware of themselves as occupying multiply social junctures, leading to pragmatic politics: 'The multiple-voiced subjectivity is lived in *resistance* to competing notions for one's allegiance or self-identification . . . The choice of one (I am a woman) or many themes (I am a white working-class woman) is *both a theoretical and a political decision*' (Alarcón in Seidman, 1995: 11). This is crucial. Planners cannot take identities for granted, for this is precisely the way in which, as Foucault explains, subjugation and subjectification become activated. If this is not understood, attempts for getting empowerment on the rails may lead to the emancipation of (wo)man, the concept of (wo)man, however, being a fraud (see Foucault, 1973; see also Rorty, 1995; Young, 1990). The planner must strive towards groups-in-the-making, without a sense of an 'us'. People must have the opportunity *to gather*. Then they will make out of themselves 'who' they are and which goals they decide to pursue (see Barber, 1988). Communicative planners must question social theories in which a priori given, intentional egos as the ontological origins of social relations are presented. Identities – just as theories – never exist a priori, but are always the products of rhetorical and practical interactions. If one does not recognize this, one can never gain access to the hidden power relations which structure identities and social life and present things as so-called natural, unproblematic givens and facts. As mentioned earlier, communicative planners must be very careful not to emancipate groups of people as egoistic creatures who, in turn, want for themselves as great as possible a piece of the social product, but without engaging in larger considerations concerning the justice of this distribution, the conditions of all or the common interest of the society or the world we live in. Furthermore, as Young (1990: 71) also states, communicative planners must realize that programmes aimed at redistributing power make little sense because power is not a possession that can be abstracted from its owners and distributed democratically into society. Rather, following Foucault, power structures relationships and presents these interpretations (of, for example, democracy, groups and solidarity), though disciplinary techniques and mimetic eagerness (see Girard, 1972) as normal and given.

The exclusion paradigm is also clear on a more immediate and practical level. In planning theory it is accepted that the communicative planner has to offer justification towards the in-group only. Moreover, definition of decision-situations can be understood only by those who have participated in their formulation. As Faludi says 'understanding a plan is reserved for those who, in a sense, are *part of it*' (Faludi, 1986: 102; added emphasis). Here the inherently contradictory situation of the planner, who has to act as advocate but has to do this within the margins and definitions laid down by policy, is obvious. Furthermore, in putting it this way, Faludi does not reckon with the fact that the planner *has an interest* in the existence of the social order which *lets* him or her execute his or her profession, which is to act in favour of groups given certain margins.

For Healey, Forester, Albrechts and many others, planning is considered to be an intrinsically political enterprise. To state this will certainly function to begin a discussion, not to end one (see Forester, 1989: 156). Young explains (Young, 1990: 39ff) that justice should refer not only to distribution, but also to the *institutional conditions* necessary for the development and exercise of individual capacities and collective communication and cooperation. She cites five aspects in which oppression becomes factual. These aspects are exploitation, marginalization, powerlessness, cultural imperialism and violence. Of course, planners cannot engage with each of these aspects, but planners should always be aware of these aspects and fight against them wherever possible. Exploitation refers to the structural process of transfer of the results of the labour of one social group to the benefit of other groups. As Young (1990: 53) explains, the injustice of exploitation consists in social processes that bring about transfers of energies from one group to another to produce unequal distribution, and in the way in which social institutions enable a few to accumulate while they constrain many more. Bringing about justice where there is exploitation requires reorganization of institutions and practices of decision-making, alteration of the division of labour, and similar measures of institutional, structural and cultural change. Marginalization primarily refers to unemployment. Marginals are people the system of labour cannot or will not use. Whole categories of people nowadays are expelled from participation in social life and often thus subjected to severe material, cultural and intellectual deprivation. Marginalization, as Young (1990: 54ff.) writes, surely touches on basic structural issues of justice. Powerlessness refers to several injustices, such as the inhibition in the development of one's capacities, lack of decision-making power and exposure to disrespectful treatment because of the status one occupies (see Young, 1990: 55ff.). To experience cultural imperialism means to experience how the dominant meaning of a society renders the particular perspective of one's own group invisible at the same time as it stereotypes one's group and marks it out as the Other (see also Melucci, 1989). Violence refers to attacks on persons or property with the motive to damage, intimidate, humiliate or destroy (see Young, 1990: 61ff.).

Young defines justice as 'the institutional conditions that make it possible for all to participate in decision-making, and to express their feelings, experience, and perspective on social life in contexts where others can listen' (Young, 1990: 91). In this, democracy is seen both as an element and a condition of social justice. Justice, for Young, requires participation in public discussion and processes of democratic decision-making. All persons should have the opportunity to participate in the deliberation and decision-making of the institutions to which their actions contribute or which directly affect their actions. However, as postmodern criticism shows us, justice and democracy will be much more difficult to 'achieve' than many communicative planners nowadays still are willing to accept.

EPILOGUE

This chapter has offered some thoughts as to how emancipatory politics might be realized. Obviously much work in both planning theory and political philosophy still has to be done. Science has no other duty than to destroy mystifications. Hopefully, at the end of this theoretical journey, undertaken by many scholars initiating many discourses, ultimately *the mystification of harmony*, the biggest of all, will not be spared. For, as Harvey also has written, this really is paramount:

> [T]he most imposing and effective mystification of all lies in the presupposition of harmony at the still point of the turning capitalist world. Perhaps there lies at the fulcrum of capitalist history not harmony but a social relation of domination of capital over labor . . . Should we reach *that* conclusion, then we would surely witness a markedly different reconstruction of the planner's world-view than we are currently seeing. We might even begin to plan the reconstruction of society, instead of merely planning the ideology of planning. (Harvey, 1982: 231)

Empowerment means *dissociation*: its aim is to pursue a more autonomous development that will be guided by the needs internal to the place, and whose results will directly benefit the people (see Friedmann, 1987: 311; Lipietz, 1994: 352; Sachs, 1995: 11ff.). This is an agenda aimed at empowerment: people's living places have to be extended outward in a political sense to gain control over the surrounding economy. However, as Friedmann mentions, the struggle over time and place, or over greater access to any of the other bases of social power will not, in itself, lead to empowerment. What makes the situation so problematic is that the creation of new productive employment along the traditional 'modernization' route is completely inadequate for meeting the aggregate needs of the population. An empowerment programme has to be aimed at making people less dependent on global capital (see Castells, 1989: 353).

NOTE

1. Here, the affinities between Lyotard and Arendt become obvious. Commenting on the antipolitical ideal of traditional philosophy and historicism, Arendt notes that '[w]hile only fabrication . . . is capable of building a world, this same world becomes . . . worthless . . . , a mere means for further ends, if the standards which governed its coming into being are permitted to rule after its establishment' (Arendt, 1958: 156). In other words, mere technical reason leads to meaninglessness. As she writes: 'Meaning, which can never be the aim of action and yet . . . will rise out of human deeds . . . was . . . pursued with the same machinery . . . of organized means . . . with the result that is was as though meaning itself had departed from the world . . . and men were left nothing but an unending chain of purposes. . . . It was as though men were stricken . . . blind to fundamental distinctions such as the distinction between meaning and end . . .' (Arendt, 1963: 87). However, if making cannot create a permanent (and just) world, neither can politics. The belief that politics can resolve injustice and create a new, better world is a grand narrative: 'Nothing could be more obsolete than to attempt to liberate mankind from poverty by political means; nothing could be more futile and dangerous' (Arendt, 1963: 114; see Denayer, 1993).

BIBLIOGRAPHY

Achterhuis, H. (1988) *Het rijk van de schaarste: van Thomas Hobbes tot Michel Foucault*. Baarn: Ambo.

Alarcón, N. (1995) 'The theoretical subject(s) of "This Bridge Called My Back" and Anglo-American feminism', in S. Seidman (ed.), *The Postmodern Turn: New Perspectives on Social Theory*. Cambridge: Cambridge University Press. pp. 140–52.

Albrechts, L. (1991) 'Changing roles and position of planners', *Urban Studies*, 28 (1): 123–37.

Albrechts, L. (1993) 'Reflections on planning education in Europe', *Town Planning Review*, 64 (1): ix–xii.

Albrechts, L. (1994) 'National planning: relic or pioneer?', *Acta Geographica Lovaniensia*, 34: 543–51.

Albrechts, L. (1996) 'Sul futuro della pianificazione spaziale', in *CRU*, 6: 84–9.

Albrechts, L. (1999) 'Planners as catalysts and initiators of change. The new structure plan for Flanders', *European Planning Studies*, 7: 587–603.

Albrechts, L. and Swyngedouw, E. (1989) 'The challenges for regional policy under a flexible regime of accumulation', in L. Albrechts, F. Moulaert, P. Roberts and E. Swyngedouw (eds), *Regional Policy at the Crossroads: European Perspectives*, London: Jessica Kingsley. pp. 67–89.

Amdam, J. (1995) 'Mobilization, participation and partnership building in local development planning: experience from local planning on women's conditions in six Norwegian Communes', *European Planning Studies*, 3 (3): 305–34.

Amin, A. (ed.) (1994) *Post-Fordism: A Reader*. Oxford: Blackwell.

Arendt, H. (1958) *The Human Condition: A Study into the Central Dilemmas Facing Modern Man*. Chicago: Chicago University Press.

Arendt, H. (1963) *On Revolution*. New York: Harcourt Brace Jovanovich.

Arendt, H. (1972) *Crises of the Republic*. New York: Harcourt Brace Jovanovich.

Arendt, H. (1978) *The Life of the Mind*, Vol. I. *Thinking*; Vol. II *Willing*. New York: Harcourt Brace Jovanovich.

Arendt, H. (1982) *Hannah Arendt's Lectures on Kant's Political Philosophy* (ed. R. Beiner). Chicago: Chicago University Press.

Barber, B. (1988) *The Conquest f Politics, Liberal Philosophy in Democratic Times*. Princeton, NJ: Princeton University Press.

Baudrillard, J. (1988) *Selected Writings* (ed. M. Poster). Cambridge: Polity Press.

Bauman, Z. (1992) *Intimations of Postmodernity*. London: Routledge.

Beauregard, R.A. (1989) 'Between modernity and postmodernity: the ambiguous position of US planning', *Environment and Planning D Society and Space*, 7: 381–95.

Beck, U. (1988) *Gegengifte: Die organisierte Unverantwortlichkeit*. Frankfurt-on-Main: Suhrkamp.

Benveniste, G. (1989) *Mastering the Politics of Planning: Crafting Credible Plans and Policies That Make a Difference*. San Francisco: Jossey-Bass.

Bijker, W.E., Hughes, T.P. and Pinch, T.J. (1993) *The Social Construction of Technological Systems: New Directions in the Sociology and History of Technology*, Cambridge, MA: MIT Press.

Boehm, R. (1977) *Kritiek der grondslagen van onze tijd*. Baarn: Het Wereldvenster.

Brown, R.H. (1995) 'Rhetoric, textuality, and the postmodern turn in sociological theory', in S. Seidman (ed.), *The Postmodern Turn: New Perspectives on Social Theory*. Cambridge: Cambridge University Press. pp. 229–41.

Callinicos, A. (1989) *Against Postmodernism: A Marxist Critique*. Cambridge: Polity Press.

Castells, M. (1989) *The Informational City: Information Technology, Economic Restructuring, and the Urban-Regional Process*. Oxford: Blackwell.

Crow, D. (1994) 'My friends in low places: building identity for place and community', *Environment and Planning D: Society and Space*, 12: 403–19.

Dear, M. and Scott, A. (1981) 'Towards a framework for analysis', in M. Dear and A. Scott, *Urbanization and Urban Planning in Capitalist Society*. New York: Methuen. pp. 3–19.

Denayer, W. (1993) 'Handelen en denken bij Hannah Arendt. Een ontologie van het actieve leven', unpublished PhD, Leiden State University, Leiden.

Denayer, W. (1994) 'De onoverzichtelijkheid in de welvaartsstaat. Over symbolische politiek, onbestuurbaarheid en herbronde handelingsbereidheid', *Tijdschrift voor Sociale Wetenschappen*, 3: 237–62.

Denayer, W. (1996) 'Politiek en de noodzaak aan consideratie', *Beleid & Maatschappij*, 4: 184–94.

Denayer, W. (2000) 'Stedelÿke *governance* en *workfare*. De conservatieve revolutie tegen de welvaartsstaat', *Ethiek & Maatschappÿ*, 2: 59–80.

Faludi, A. (1986) *A Decision Centred View of Environmental Planning*. Oxford Pergamon.

Faludi, A. and Korthals Altes, W. (1994) 'Evaluating communicative planning: a revised design for performance research', *European Planning Studies*, 2 (4): 403–18.

Fainstein, S. (1988) 'Urban transformation and economic development policies'. Paper presented at the Annual Meeting of the Association of Collegiate Schools of Planners, Buffalo, NY.

Feenberg, A. (1996) *Alternative Modernity. The Technical Turn in Philosophy and Social Theory*. Los Angeles: University of California Press.

Forester, J. (1989) *Planning in the Face of Power*. Berkeley, CA: University of California Press.

Foucault, M. (1967) *Madness and Civilisation: An History of Insanity in the Age of Reason* (trans. R. Howard). London: Tavistock.

Foucault, M. (1975) *Surveillir et punir*. Paris: Gallimard.

Foucault, M. (1976, 1977, 1978) *Geschiedenis van de seksualiteit*. Vol. 1 *De wil tot weten*; Vol. 2 *Het gebruik van de lust*; Vol. 3 *De zorg voor zichzelf*. Nijmegen: SUN.

Fraser, N. and Nicholson, L. (1995) 'Social criticism without philosophy: An encounter between feminism and postmodernism', in S. Seidman (ed.), *The Postmodern Turn. New Perspectives on Social Theory*. Cambridge: Cambridge University Press. pp. 242–64.

Friedmann, J. (1987) *Planning in the Public Domain: From Knowledge to Action*. Princeton, NJ: Princeton University Press.

Galbraith, J.K. (1992) *The Culture of Contentment*. Boston: Houghton Miflin.

Gans, H. (1991) *People, Plans and Policies: Essays on Poverty, Racism and Other National Urban Problems*. New York: Columbia University Press.

Gellner, E. (1983) *Nations and Nationalism*. Oxford: Blackwell.

Girard, R. (1972) *La violence et le sacré*. Paris: Grasset.

Habermas, J. (1973) *Theory and Practice*. Boston, MA: Beacon Press.

Habermas, J. (1975) *Legitimation Crisis*. Boston, MA: Beacon Press.

Habermas, J. (1977) 'Hannah Arendt's communication concept of power', in *Social Research*, 44: 3–24.

Habermas, J. (1979) *Communication and the Evolution of Society*. Boston, MA: Beacon Press.

Habermas, J. (1985) *The Theory of Communicative Action*, Vol. I *Reason and the Rationalization of Society*, Vol. II *The Critique of Functionalist Reason* (transl. by T. McCarthy). Cambridge: Polity Press.

Harvey, D. (1982) 'On planning the ideology of planning', in R.W. Burchell and G. Sternlieb (eds), *Planning Theory in the 1980s*. New Brunswick, NJ: Rutgers University. pp. 213–33.

Harvey, D. (1989) *The Condition of Postmodernity: An Enquiry into the Origins of Cultural Change*. Oxford: Blackwell.

Harvey, D. (1996) *Justice, Nature and the Geography of Difference*. Oxford: Blackwell.

Healey, P. (1992) 'Alla ricerca della democrazia: nuovi modi di usare strumenti vecchi. Forma e contenuto dei piani di sviluppo', in F. Archibugi and P. Bisogno (eds), *Per una teoria della pianificazione*. Milan: Prometheus. pp. 90–108.

Hillier, J. (1993) 'To boldly go where no planners have ever . . .', *Environment and Planning D: Society and Space*, 11: 89–113.

Hobsbawm, E. (1992) *The Age of Extremes: The Short Twentieth Century, 1994–1991*. Oxford: Blackwell.

Howell, P. (1993) 'Public space and the public sphere: political theory and the historical geography of modernity', *Environment and Planning D: Society and Space*, 11: 303–22.

Huyse, L. (1994) *De politiek voorbij. Een blik op de jaren negentig*. Leuven: Kritak.

Kirsch, S. (1995) 'The incredible shrinking world? Technology and the production of space', *Environment and Planning D: Society and Space*, 13: 529–55.

Kunneman, H. (1986) *De waarheidstrechter*. Meppel: Boom.

Kunneman, H. (1996) *Van theemutscultuur naar walkman-ego. Contouren van postmoderne individualiteit*. Amsterdam: Boom.

Latour, B. (1987) *Science in Action: How to Follow Scientists and Engineers through Society*. Milton Keynes: Open University Press and Cambridge, MA: Harvard University Press.

Lipietz, A. (1994) 'Post-Fordism and democracy', in A. Amin (ed.), *Post-Fordism: A Reader*. Oxford: Blackwell. pp. 338–58.

Lyotard, J.F. (1979) *La condition postmoderne*. Paris: Minuit.

Lyotard, J.F. (1983) *Le differend*. Paris: Minuit.

Lyotard, J.F. (1986) *The Postmodern Condition: A Report on Knowledge*. Manchester: Manchester University Press.

Mandelbaum, S. (1992) 'Sensibilità communitaria e progettazione della communita', in F. Archibugi and P. Bisogno (eds), *Per una teoria della pianificazione*. Milan: Prometheus. pp. 109–25.

Mannheim, K. (1951) *Freedom, Power, and Democratic Planning*. New York: Harcourt Brace.

Massey, D. (1995) 'Thinking radical democracy spatially', *Environment and Planning D: Society and Space*, 12: 283–8.

McHarg, I. (1982) 'Ecological planning: the planner as catalyst', in R.W. Burchell and G. Sternlieb (eds), *Planning Theory in the 1980s*. New Brunswick, NJ: Rutgers University. pp. 13–15.

Melucci, A. (1989) *Nomads of the Present: Social Movements and Individual Needs in Contemporary Society*. Philadelphia: Temple University Press.

Nussbaum, M. (1989) *The Fragility of Goodness*. Berkeley, CA: University of California Press.

Offe C. (1984) *Contradictions of the Welfare State*. Cambridge, MA: MIT Press.

Peck, J. and Tickell, A. (1994) 'Searching for a new institutional fix: the after-Fordist crisis and the global–local disorder', in A. Amin (ed.), *Post-Fordism: A Reader*. Oxford: Blackwell. pp. 280–315.

Priore, M. and Sabel, C. (1984) *The Second Industrial Divide: Possibilities for Prosperity*. New York: Basic Books.

Rorty, R. (1985) 'Habermas and Lyotard on post-modernity', in R. Bernstein (ed.), *Habermas and Modernity*. Oxford: Blackwell. pp. 161–75.

Rorty, R. (1995) 'Method, social science, and social hope', in S. Seidman (ed.), *The Postmodern Turn: New Perspectives on Social Theory*. Cambridge; Cambridge University Press. pp. 46–64.

Rosaldo, R. (1995) 'Subjectivity in social analysis', in S. Seidman (ed.), *The Postmodern Turn: New Perspectives on Social Theory*. Cambridge: Cambridge University Press. pp. 172–86.

Sachs, W. (1995) 'Global ecology and the shadow of "development"', in W. Sachs (ed.), *Global Ecology: A New Arena of Political Conflict*. London: Zed Books.

Sass, L.A. (1992) *Madness and Modernism. Insanity in the Light of Modern Art, Literature, and Thought*. New York: Basic Books.

Schwarz, M. and Thompson, M. (1990) *Divided We Stand: Redefining Politics, Technology and Social Choice*. New York: Harvester Wheatsheaf.

Seidman, S. (ed.) (1995) *The Postmodern Turn: New Perspectives on Social Theory*. Cambridge: Cambridge University Press.

Smart, B. (1993) *Postmodernity*. Milton Keynes. The Open University.

Swyngedouw, E. (1987) 'Planning! . . . What Planning?', in J. Angenent and A. Bongenaar (eds), *Planning without a Passport*. Netherlands Geographical Studies, 44: 77–89.

Swyngedouw, E. (1992) 'The mammon quest: glocalization, interspatial competition and the monetary order', in M. Dunford and G. Kafkala (eds), *Cities and Regions in the New Europe*. London: Belhaven. pp. 39–67.

Throgmorton, J. (1993) 'Planning as a rhetorical activity: survey research as a trope in arguments about electric power planning in Chicago', *Journal of the American Planning Association*, 59: 334–46.

Van Dieren, W. (ed.) (1995) *De natuur telt ook mee: naar een duurzaam nationaal inkomen – een rapport aan de Club van Rome*. Utrecht: Het Spectrum.

Van Dijk, A. (1995) 'Beleidsinstrumenten in technologiebelied', in H. Achterhuis, R. Smits, J. Geurts, A. Rip, E. Roelofs (eds), *Technologie en Samenleving*, Leuven - Apeldoorn: Garant. pp. 37–402.

Van Reijen, W. (1988) 'Moderne versus postmoderne politieke filosofie. Een vergelijking tussen Habermas en Lyotard', *Acta Politica*, 2: 199–223.

Villa, D.R. (1992) 'Postmodernism and the public sphere', *American Political Science Review*, 86 (3): 712–23.

Wildavsky, A. (1973) 'If planning is everything, maybe it's nothing', *Policy Sciences*, 4: 121–36.

Wilson, W. (1991) 'Studying inner-city social dislocations: the challenge of public agenda research', *American Sociological Review*, 56: 1–14.

Young, I. (1990) *Justice and the Politics of Difference*. Princeton, NJ: Princeton University Press.

Zwart, H. (1995) *Technocratie en onbehagen: de plaats van de ethiek in het werk van Michel Foucault*. Nijmegen: SUN.

25

Planning, Power and Conflict

JAMES SIMMIE

Large-scale industrial and commercial cities were mainly invented in Britain during the early part of the nineteenth century. Before that time Rome had probably been the largest city in the world during its imperial era. As a result of the first Industrial Revolution in England, the development of new industries, such as cotton, combined with new forms of production, such as factories, led to rapid urbanization and the growth of the first large industrial and commercial cities. The economic changes underlying this urban growth also led to major social changes in phenomena such as social stratification and consequential political changes such as the rise of the Labour Party in Britain.

Urban growth and change therefore need to be analysed in the wider context of regional, national and, increasingly, global economic, social and political change. Cities are the geographic concentrations where these wider societal forces are played out. Although each city has its own historically specific economic, social and political characteristics, these specifics are composed of the particular elements of the wider societal forces operating in that particular geographic location combined with the actions and reactions of local institutions and actors.

One major element of these specifics in all cities is the use of land and buildings. These constitute one of the most important physically visible defining characteristics of urbanization. Great concentrations of public and private buildings, factories, offices, shops and houses, parks, libraries, museums and stadia, roads and railways are both symbolically and functionally how most people define and recognize cities. Behind the development of these physical artefacts, however, lie the working of elements of economy, society and politics, usually mediated through some kind of public land use planning and regulation system. This is true even in the case of Houston, Texas, which has been misleadingly characterized as arranging all its land uses through markets (Siegan, 1972).

Nevertheless, Siegan (1972) raises the important question of why any city actually needs or develops a land-use planning system at all. This is a question that needs to be asked on a regular basis whether the cities in question are located in North America, Europe, or Australasia. It needs to be asked even more often in some Third World countries which have developed mega-cities with populations in excess of 25 million people.

Four main arguments that address this question in different ways are analysed in this chapter. They focus on the conflicts that develop over urbanization and consequential debates around the nature and powers of planning systems in relation to issues such as:

1 Private interests versus the public interest.
2 Private individuals versus public institutions.
3 Private property rights versus public regulation.
4 Markets versus plans.

In the first case, in capitalist economic systems the interests of private property owners are hegemonic. But the unfettered pursuit of these private interests gives rise to a whole range of undesirable externality effects as far as the generality of the urban public are concerned. As a result, there have been conflicts between the absolute abilities of private property owners to propose, oppose and dispose of their properties exactly as they see fit, and collective interests expressed in terms of a public interest to influence and curtail these freedoms. In practice the outcome of such conflicts may reduce the extreme freedoms of private property owners to site such noxious developments as bone crushing and blood boiling factories next to middle-class housing. In less extreme cases, the conflicts are more often resolved by compromises that favour the interests of private property owners. This is true of both the British and Californian planning systems compared and contrasted in this chapter.

The second set of conflicts arise once public land-use planning institutions have been established. They revolve around the issue of who influences and controls the goals and objectives of these institutions. This depends on the balance of elitist or pluralist control in different political systems over their public agencies. This is often expressed in the degrees to which public participation in planning is permitted or demanded. The highly elitist and centralized planning system in Britain does not provide adequate democratic controls over the main goals and objectives of land-use planning. In contrast, in California, such institutions as ballots, recalls, initiatives and referenda are available to middle-class groups wishing to hold their elected and appointed planning representatives to account.

The third set of conflicts that establish the characteristics of particular land-use planning systems revolve around the degree to which political systems are moved to introduce systematic formal legislation which seeks to regulate private property rights. This depends very much on the power of property owners to enshrine their own individual freedoms in the ideology of the planning legislation. This power changes through time. In general, however, British town planning legislation is still based on the fundamentals of the central government 1947 Town and Country Planning Act. This effectively nationalized all new urban property rights from that time to the present. It is much more draconian than the more piecemeal state and local legislation enacted in California. Both systems are subject to fairly regular pressures to change in somewhat contradictory directions. Urban property owners and developers often seek legislative relaxations in their interests. Rural property owners often press for tougher legislation to prevent new developments in their own backyards.

The fourth conflict at the heart of the establishment and subsequent nature of land use-planning systems focuses on the degree to which they seek to replace free markets with plans. The control and regulation of land and property markets changes the reward structures of private owners. Those that can acquire permissions to develop reap larger rewards in restrictive circumstances than would otherwise have been the case. Those who are refused permission to develop lose virtually the entire potential development value of their property. Not surprisingly such decisions, and the basis of the planning system that gives rise to them, lead to conflicts. One of the most significant of these is stimulated by attempts to recoup for the community some or all of the new property values that it has created by giving public planning permission to develop. In Britain such attempts have generally been reduced in scope until they are now limited to

specific 'planning gain' agreements concerned with individual projects. In California some cities such as Los Angeles resisted planned market interventions right up to the 1980s. In such an ideological framework the idea that some private property benefits should be appropriated by the state is anathema.

All these conflicts involve, in one way or another, the distribution of scarce land uses and property rights to different groups, organizations and social classes producing, working or living in cities. They deal with the quintessential political questions of who gets what, when and how (Lasswell, 1958). As a result they are also the focus of attempts by different interest groups to exercise power over the goals, objectives and outcomes of land-use planning. This is the case in virtually all cities. Empirical evidence to this effect abounds in the community power study literature (for a summary see Simmie, 1981: ch. 1; LeGates and Stout, 1996: ch. 4), which mainly refers to the USA; in British studies, such as those by Brindley et al. (1989), Saunders (1979), and Simmie (1974, 1981) all show the significance of power with respect to planning from the Second World War onwards; it has been shown to be a key feature of planning in Australasia by Badcock (1984), and across Europe, North and South America by Castells (1983).

Two major case studies are used to illustrate the working of these conflicts in this chapter. They are the British and Californian land-use planning systems. These two are selected partly because the author has worked within both of them and, more importantly, because they represent relatively extreme cases of tough and relaxed planning regimes set in significantly different economic, social and political circumstances. Many planning regimes in other countries fall somewhere between these two extremes of tough and relaxed regulation of land uses and development.

The two regimes are compared and contrasted in terms of how the four major conflicts identified in this introduction are worked out. Despite significantly different starting points, it is shown – when it comes to regulating private property rights in capitalist societies – public interests, institutions, regulations and plans often meet with limited success. One of the few major differences between California and Britain is the more extensive institutional arrangements for calling elected representatives and planning bureaucracies to account in California. There at least the vocal and organized middle classes can affect directly the goals and objectives set in their local planning system.

In general it is argued that, at any point in time, the current characteristics of both planning systems will reflect the outcomes of the balance of forces between those ranged against each other over the various aspects of the major categories of conflict outlined above. The systems are therefore in a constant condition of tension and change.

A final section will summarize the main points of the argument and draw some conclusions. It will be argued that, although the British and Californian land-use planning systems are often seen as significantly different from one another, when attention is focused on key phenomena such as the furtherance of private property rights and markets, they are not so different as they are made out to be.

PRIVATE INTERESTS VERSUS THE PUBLIC INTEREST

Britain

Britain has no written constitution. Parliament is sovereign. Party discipline in Parliament is enforced through the carrot of prime ministerial patronage and the stick of the 'whipping' system. In practice this means that a party that can gain a majority of Members of Parliament can override opposition and can introduce or change any law. This system has been likened to an elected dictatorship by Lord Hailsham, a former Lord Chancellor. It partly accounts for the highly centralized and elitist nature of government in Britain. If the relatively small governing elite can be persuaded to initiate or support a measure, then it can be made law despite opposition. For this reason the embryonic, pre-war town planning movement made its appeal by elites outside Parliament to the governing elite inside Parliament.

The war-time coalition government drawn from all parties was receptive to the idea of planning for two main reasons. The first of these was the recent experience of the economic collapse and depression during the late 1920s and early 1930s. This combined with wartime production needs gave rise to ideas for and an acceptance of the need for economic planning. The second reason was the need to provide a morale-boosting vision of post-war society. Ideas for economic and land-use planning were developed, along with other aspects of a 'welfare state', as part of this vision.

The ideas for the land-use planning part of this vision were developed by elites both inside and outside Parliament. According to Hall et al. (1973) these elites followed a unitary model of society in which social stability and harmony were of prime concern. These views failed to recognize the inherent conflicts of interests both between different elites and between them and the rest of the population which have dogged land-use planning in Britain ever since the Second World War.

These basic conflicts rest on the differences of interest between manufacturing industry and agriculture, often expressed as a difference between town and country; and between these and the mass of the working population. During the war there was some cooperation between manufacturing and agriculture because of their importance to the war effort. There was some acceptance by their elites that better conditions should be promised to the working masses than they had experienced before the war. As a result, an elite view of the public interest was developed.

Hall et al. (1973) argue that the British elites proposed an 'organismic' vision of the public interest. By this they mean that society as a whole had some interests which were different from those of partial and separate private interests. This organismic vision of the public interest was, however, a lofty and imprecise definition of the public good in terms of ends which were ideologically derived. Because the elites involved in developing this lofty vision were influenced by different and conflicting ideologies, the view of the public interest that they arrived at was also inconsistent. To disguise the inconsistencies, it was also vague.

The post-war history of the conflicts between private interests and the public interest in British land-use planning is one of a shifting balance. During the post-war years the public interest in planning was defined by relatively small numbers of politicians, civil servants and experts. It was elitist and inconsistent. It contained conflicting elements of both the ideologies of private property and the public interest.

During the 1950s and 1960s there was relatively full employment and rising affluence among the masses. Increasing profits could be made from development. As a result, private property interests both in terms of home owners and developers began to re-assert themselves over the public interest in planning. Despite this, planning continued to claim that it was guided by the public interest even though that was defined by political, technical and property-owning elites. With the notable exception of the New Towns programme, planning had no means for implementing plans without negotiating with these elites.

The late 1960s and the 1970s saw a temporary attempt to redefine the public interest by non-elite groups through the ideology and mechanisms of public participation in planning. There were two main reasons why this movement arose at this time. The first was the ability of members of the public to monitor the effects of post-war planning for themselves. Many of the original schemes for comprehensive redevelopment and high-rise public housing had been completed by this time. Frequently they did not measure up to the elitist

visions of the public interest portrayed before
they were started. Those who lived in them or
could see their effects demanded a say in future
schemes.

The second reason for the emergence of
demands for public participation at this time was
economic decline. This set in during the 1970s. It
meant that it became increasingly difficult to
follow social objectives without treating different
groups or areas differently. Policy-making became
more of a zero-sum game in which one group's
gain became a lost opportunity for another group
or area. Politics deals with these kinds of issues
and so planning became more overtly politiczed
during this period.

Continued economic decline during the 1980s
meant that it became even more difficult to satisfy
rising social demands in conditions of falling or
static economic growth. To meet this conflict the
ideology of private property was reasserted in
Britain during this period.

California

In contrast with Britain, America has a written
constitution. Any instance of the assertion of a
public interest over private interests therefore has
to follow from some provision of the Constitution. The provisions which provide such a basis
for land-use planning are twofold. First, there is
the Tenth Amendment to the Constitution – the
reserved powers doctrine. This states that any
powers not specifically granted to the federal
government in the constitution are reserved by
the individual states. As land-use planning powers
have not been specifically granted to the federal
government they therefore rest with individual
states which wish to take them up.

Secondly, the general power resting at the state
level which forms the basis of planning is called
'police power'. This is the power of a governmental entity to restrict private activity in order to
achieve a broad public benefit. In California
police power is used to protect the health, safety,
morals and welfare of the public. For more than a
century it has been used to justify governmental
regulation over the use of land.

Police power does not rest inherently with city
hall. Local government in California exercises
police power over land uses only because the state
government has delegated that power to them.
Article XI, Section 7, of the California Constitution states 'A county or city may make and
enforce within its limits all local, police, sanitary,
and other ordinances and regulations not in
conflict with general laws.'

Local ordinances and regulations must not
contradict constitutional principles. Among the
most important of these for planning are the

concepts of due process, equal protection and the
unlawful taking of property without compensation. The last has proved to be a long-running
conflict between private interests and the public
interest as expressed in land-use planning. It will
be seen later that private interests have often
taken the view that any effective planning regulations constitute a taking of property. Conflicts
such as these mean that the actual breadth and
depth of police powers expressed in planning are
almost always in a state of change.

Planning commentators tend to look back on
earlier eras with some nostalgia. One common
theme is that *the* public interest was easier to
define in planning at the time it was first established
than it is now. It is claimed that 'Planning does
not have a strong sense of the public interest any
more. Each group claims to represent the public
interest while in fact representing a much
narrower interest' (Fulton, 1991: 13). But, to
European eyes, American planning systems never
seem to have been based on the kind of temporary ideological and political consensus that
prevailed in Britain after the Second World War.

The most famous early zoning ordinance in
America, for example, was passed in 1916 by New
York City. It was designed to stop the march of
the *hoi polloi* up Fifth Avenue towards the homes
of the rich. Its primary purpose was therefore to
assert private interests over the public interest.
The history of planning in California is replete
with examples of political and legal conflicts. The
golden era of consensus and a public interest
based on it seems to be even more of a myth in
California than it has been elsewhere.

Conclusions

The main constitutional difference between
Britain and the USA is the existence of a written
constitution in the latter. This provides an important ongoing framework within which the two
parties must operate. In Britain the absence of a
written constitution allows any party with a
majority in Parliament to legislate on the basis of
almost any strongly held set of ideological beliefs.

In these differing contexts many planning
regulations have been enacted. In both cases,
however, it has proved extremely difficult to
introduce any lasting concept of the public
interest which runs counter to private property
interests. In Britain the highly elitist nature of
government has meant that both the development of a land-use planning system and its subsequent operation was both instigated by elites and
appealed to private interests such as major rural
landowners, who are still mainly exempted from
its provisions.

In California the planning system has been

effected by the presumptions in the US Constitution which are strongly in favour of private property. The spirit of frontier individualism which pervades politics in the state has also viewed any intervention in the freedom of individuals to dispose of their private property as they wish with deep suspicion. In both Britain and California conflicts between private interests and the public interest are usually resolved on terms which are reasonably satisfactory to the former.

PRIVATE INDIVIDUALS VERSUS PUBLIC INSTITUTIONS

Britain

Because Britain has no written constitution, Parliament is sovereign and the heads of the legislature, executive and judiciary all merge there. Power is thus highly centralized and private individuals have little control over public institutions. The administrative structure and procedures of the planning system are therefore largely determined by central government elites and those with access to them.

Parliament determines both the structure of government and the allocation of different planning functions to different parts of that structure. Both have been subject to periodic major changes since the Second World War. Throughout the entire post-war period, however, the relevant central government department has supervised both local government and the planning system. Until 1970 this was called the Ministry of Housing and Local Government (MHLG). In 1970 it was amalgamated with the Ministries of Public Building and Works, and Transport and its name was changed to the Department of the Environment (DoE). After the Labour Party election victory in 1997 its functions and name were changed once again to the Department of Environment, Transport and the Regions (DETR).

Both the old MHLG and DoE and the more recent DETR have the function of supervising local government and planning according to the legislation on these two subjects laid down by Parliament. Since 1997 the DETR has supervised the planning system, interpreted and made routine government planning policy and acted as the administrative court of appeal in disputed planning decisions. It is a powerful central ministry.

The original elitist, idealized planning system was intended to be centralized and unitary. In other words, it was envisaged that the old MHLG would perform most of the major planning functions. In the event, the day-to-day administration of the planning system was delegated to the largest local authorities after the war. These were the 58 counties, administering mainly rural areas, and the 83 county borough councils in urban areas.

In 1974 local land-use planning functions were split between two tiers of government. Strategic planning, in the form of structure plans, was allocated to the counties in England and Wales, and regions in Scotland. Local planning, in the form of local plans, was allocated to the district councils. This was a recipe for continual conflict between the two tiers of local government over planning matters. Many of these conflicts did not involve the general public directly. They were more often between property-owning county elites, seeking to protect their property values and amenities, and urban-based developers seeking to erect housing, industrial and commercial property in the counties.

The general public had least influence over the most important strategic, structure plans. The Secretary of State had a much greater degree of control here. These plans had to be approved by him and also to include any modifications which he demanded. Local plans then had to be certified as being in accordance with the relevant structure plan. Even then, the Secretary of State could 'call in' any politically controversial proposals or decisions made by the second tier district councils and decide the matter for himself.

Paradoxically, it was the rise of popular, participatory planning in the conurbations during the 1970s that led the government to propose changes to both the upper tier of local government and structure plans. Conflicts between a Labour-controlled Greater London Council (GLC) and central government over the non-private oriented aspects of the Greater London Development Plan (GLDP), the structure plan for London, contributed to the central government abolition of all the metropolitan counties in 1986. Their strategic planning was conducted within the DOE, whose dictates were then handed down to the boroughs in the case of London. They were bound to follow this 'advice' in drawing up what are now called 'unitary development plans' (UDPs).

The evolving system of public planning institutions in Britain is one that normally excludes the general public from any significant involvement in major and strategic decisions. Instead, organized elite private interests focus their lobbying and influence on Parliament in order to have their interests included in parliamentary legislation and on the DoE in order to influence major administrative decisions. Examples of such private interests operating in this way, and often in conflict with one another, are the Council for the Preservation of Rural England (CPRE) and the House Builders' Federation (HBF).

A major new development in this structure of

government is the growing importance of European Union (EU) institutions. At the moment two elements of these are of particular relevance for planning in Britain. The first is the Directorate General XVI, which deals with regional planning. This has the second largest budget of all the European Commission Directorates after agriculture. Already it has forced a Conservative government to spend funds on declining coalfields before receiving additional EU funds.

The second EU measure which is of significance for British planning is the 1986 Directive on Environmental Impact Assessment (EIA). This says that member states are required to assess the effects of both public and private projects which are likely to have significant impacts on the environment as a consequence of their nature, size or location. Despite some recalcitrance on the part of the British government to comply with the spirit of this Directive, most notably by avoiding its use on the Channel Tunnel project, they have had to comply with the letter of the Directive since 1988. The full implementation of this Directive should make it much easier to monitor the effects of major planning decisions in the future if only because of the information that EIAs will make public.

California

In California on the one side are public institutions with varying planning responsibilities. These include the federal government, the California state government, different regional agencies, numerous special districts, counties and cities. On the other side are private individuals who can seek to determine the planning objectives of these institutions using lobbying, political participation, ballots, recall, initiatives and referenda.

The power of Congress over planning is both ubiquitous and indirect. It is ubiquitous in the sense that the federal government can touch on all public policy-making. Thus, for example, federal environmental laws such as the Endangered Species Act and the wetlands provisions of the Clean Water Act have the power virtually to stop local development if the environment is threatened. It is indirect in the sense that the police powers, such as those used as the basis of local planning, are exercised by individual states.

In practice it is the Californian state legislature which has established the framework and roles of local planning in that state over the past 80 years. At the state level itself, the Governor can influence transportation planning. The State Department of Transportation, usually known as Caltrans, has played a crucial role in establishing the state's pattern of growth. But, as state transportation funds have dwindled so has the ability of Caltrans to determine the patterns of urban growth.

The state also has a major effect on what planning is required at the level of cities and counties. It passed the first Subdivision Map Act in 1907. The first General Plan law was passed in 1927. It first required that specific elements, such as housing and circulation, should be included in general plans in the 1950s. Finally, it required that zoning ordinances should be consistent with general plans in 1971. All these activities help to determine the nature and scope of local planning in cities and counties.

Within the state of California there are also a number of different kinds of regional agencies which also affect land-use planning directly or indirectly. The most important of these are the specifically regional planning agencies such as the San Francisco Bay Conservation and Development Commission, the Coastal Commission and the Tahoe Regional Planning Commission; and the agencies that build infrastructure on a regional level such as Caltrans and the Metropolitan Water District of Southern California.

In some parts of California there are also nominal regional governments partly concerned with land-use planning. These are known as Councils of Government (COGs). Examples are the Southern California Association of Governments (SCAG) in Los Angeles, the Association of Bay Area Governments (ABAG) in San Francisco and the San Diego Association of Governments (SANDAG). These are groupings of local governments and business interests rather like the London and South East Regional Planning Conference (SERPLAN) in Britain. Their only real power used to come from administering federal and state grants. As these were cut the power of the COGs has declined.

A more important group of state-established regional agencies that affect planning are special districts. Two in particular are important from a planning point of view. These are air pollution control districts and local agency formation commissions (LAFCOs). Air pollution districts have traditionally only dealt with stationary sources of pollution. Even so, this has important implications for planning. They also look set to become even more important by trying to cut down on the use of moving vehicles. LAFCOs rule on all incorporations of new cities, annexations and boundary changes. Boundary decisions, in particular, contribute to the regulation of the conversion of rural and agricultural land into urban development within cities and are therefore particularly significant for land-use planning.

Despite this plethora of different institutions concerned with planning, counties and cities are the two types of local government with primary responsibility for the production of general plans, zoning and the regulation of land-uses according to the criteria contained in them. Counties are created by the state. Every square inch of California falls into one of its 58 counties. Cities are created by local citizens to serve their own purposes. These usually include the provision of urban services such as water, sewers and police. There are about 450 cities in California.

Both counties and cities conduct land-use planning. Planning powers within city boundaries rest with those cities. Outside the cities and in unincorporated areas they rest with the counties. The form of local government in both is non-partisan. That is to say, individuals do not stand for election to them on the basis of party political divisions. Once elected they do not divide or vote along party lines. Counties are run by a board of supervisors. Cities are run by a city council. Unlike in Britain, both these bodies usually consist of only five members. The same is true of the separate planning commission which is appointed by the supervisors or council.

Both counties and cities usually have a department of officials to carry out planning on their behalf. In cities this may be part of a much larger community development department which encompasses other related activities such as housing, transportation and building code enforcement. These public institutions come into conflict with private individuals seeking to use their activities for their own private purposes.

American government in general is subjected to the activities of many full-time professional lobbyists. At the state level any major land-use planning bill is likely to attract the attentions of the League of California Cities, the County Supervisors of the individual counties and cities of California, the California Building Industry Association, the California Association of Realtors and environmental groups such as the Sierra Club. As a result, planning legislation comes to reflect compromises between the kinds of interests that can afford the time and resources to lobby in this way.

Individuals, on the other hand, can determine the content of state planning policies by the use of the ballot. Any issue can appear on the state wide ballot as an initiative or referendum if enough signatures are gathered to place it on the ballot. Major examples of this procedure in practice are the Coastal Act, 1972, which established the Coastal Commission to protect the coast from overdevelopment; and Proposition 13, the 1978 initiative that cut property taxes dramatically.

Similar procedures can be used at the local level. Citizens can also use the procedure known as 'recall' to bring an elected official back to the ballot box before their normal term of office has expired. Because most day-to-day land-use planning powers rest with local elected officials, the attempts of private individuals to use the public institution of planning to satisfy their private interests results in a high degree of politicization of planning. Planning in California is inseparable from politics.

The conflict between the interests of private individuals and those of public planning institutions has been seen at its sharpest following the passage of Proposition 13. That cut property taxes as a source of public authority revenue. It left sales tax as the remaining source of public revenue that was not constrained. Local authorities have reacted by aiming for fiscal growth by attracting large new sources of sales tax revenues into their jurisdictions. On the other hand, many of the same individuals who voted for tax cuts are also interested in the new planning growth management controls. In many cases fiscal growth by local governments is in direct conflict with growth management.

Conclusions

In Britain private individuals have little influence over the public institutions that make strategic decisions in land-use planning. As a result they have sometimes resorted to direct action as the only means of insisting that their voices are heard. In the past much of this form of protest has involved major infrastructure projects such as new motorways or airport runways. It has been met as much by administrative changes that make it even more difficult to object to such proposals legally as it has with success in stopping them.

In California, in contrast, there are constitutional methods for introducing the interests of private individuals into the decisions of public planning institutions. Individuals in California have several ways of holding their elected representatives directly to account, as we have seen above. Although by no means perfect, they offer democratic opportunities to citizens not afforded to their British counterpart.

PRIVATE PROPERTY RIGHTS VERSUS PUBLIC REGULATION

Britain

The 1947 Town and Country Planning Act expropriated all the future urban private property rights of owners to change the existing uses of their buildings or to construct new ones at will. The major, and often elite, land users, agriculture, forestry, statutory undertakers, the Crown and the military, were exempt from these provisions. From that time, private property owners wishing to carry out urban development had to acquire planning permission from the relevant local planning authority (LPA).

Development was defined broadly as 'the carrying out of building, engineering, mining or other operations in, on, over or under land, or the making of any material change in the use of any buildings or other land'. A series of subsequent Town and Country Planning and closely related Acts in 1968, 1971, 1980 and 1990 have modified the details but maintained these basic principles of the 1947 Act.

These public regulations have created sharp distinctions between the private property rights of those who can acquire planning permission and those who cannot. The underlying justifications for public distinction between them rest on two quite different principles. On the one hand, the publicly created oligopolistic profits enjoyed by those who can acquire planning permission are justified as legitimate rewards for enterprise. On the other hand, those who cannot acquire the right to develop their property are prevented from doing so on the grounds that such development would cause nuisance to others (Reade, 1987: 22).

Local planning decisions which separate these two sets of private property rights are based on two main elements. The first of these elements is development plans. The second is development control. Like the structure of local government, both have been subject to change since their statutory arrival in the 1947 Town and Country Planning Act.

During the 1960s development plans were separated into general, strategic 'structure' plans and detailed 'local' plans. It was wrongly assumed that, with the exception of London, these would be produced and operated within a single local planning authority. In the event, structure plans were made the responsibility of counties and local plans were given to the districts after the reorganization of local government in England and Wales in 1974.

Initially, structure plans were broad strategic plans which included some economic and social considerations. They had to be approved by the Secretary of State who, by this time, was ensconced in the DoE which had superseded the MHLG. Once approved, they became statutory documents. Local plans produced by the districts had to be certified as being in accordance with their relevant county's structure plan.

The history of structure plans is one of the progressive reassertion of private property rights as their scope has been whittled away by the Secretary of State. First to go was their economic and social content. The DoE soon insisted that they confine themselves exclusively to matters of land use. The next nail in their coffin was the conservative government's 1980 Local Government, Planning and Land Act. This effectively emasculated structure plans by allowing districts to prepare local plans without waiting for an approved structure plan, and placing all control over development, apart from mineral extraction and waste disposal, in the hands of the districts. This was followed in 1985 by the Local Government Act which abolished the metropolitan counties and placed their strategic planning functions in the hands of the Secretary of State.

Local plans were introduced at the same time as structure plans. They were prepared by the district councils. They could take the form of general plans covering the whole range of local planning issues, 'subject plans' concerned with specific matters and 'action area plans' which were related to specific localities. They have always been concerned with the more traditional aspects of land-use planning and especially with questions of layout and design. Even so, a circular from the Secretary of State (circular 22/80) required them to devote less concern to layout, design, non-conforming uses and public participation. During the 1980s districts were under pressure from central government to pay more attention to private property rights and less to their public regulation.

Under the latest Unitary Development Plan (UDP) system, such strategic planning as will take place will be laid down by the Secretary of State in the form of regional planning guidelines. These will determine the framework within which local plans will have to be produced. It will depend very much on the attitude of central government to private property rights as to how far their interests will be written into regional planning guidelines. The evidence so far suggests that the rights of certain kinds of private property will carry more weight against their regulation than at any time since the Second World War.

The second main element of the British planning system is development control. Development control is the power to decide whether or not a specific development can take place on a specific site, to control the intensity of the development permitted and to control its layout and design. It is at the heart of the British land-use planning system. It is the point at which individual property rights come into direct conflict with public regulation.

Development control has traditionally been operated mainly by the lowest tier of government. At first this was the counties and county borough councils. Since 1974 the districts have had the main responsibility for operating the development control system. Decisions are expected to be in accordance with the contents of any relevant public and formal plans. They are also expected to be consistent with the provisions of two central government orders. These are the General Development Order (GDO) and the Use Classes Order (UCO). The GDO specifies which limited types of development can take place without planning permission. The UCO categorizes land uses. Planning permission is not required for changes of use within certain defined categories.

Until the 1980s the development control system regulated strongly urban types of private property rights. Since then its operation has been relaxed and speeded up by central government. The exceptions to these general principles are in what are called 'designated areas'. These are green belts, conservation areas, national parks and areas of outstanding natural beauty.

In addition to the statutory framework, development plans and development control, planning decisions in Britain are also influenced strongly by circulars from the Secretary of State, dispatched to LPAs at regular intervals. One of the more important of these was 22/80. It recommended to LPAs various ways in which the operation of the planning system should be speeded up and made more responsive to the private property rights of large-scale developers.

Circulars do not have any statutory force, but LPAs are advised to follow their requirements because, should they refuse planning permission and that decision be taken to appeal, the Secretary of State is the ultimate judge and jury, through the ministerial inspectorate, of that appeal. The Secretary of State can therefore reverse any decision taken at the local level to accord with his or her wishes as expressed in circulars.

The history of conflicts between private property rights and their public regulation can be summed up as one of strict elitist regulation of urban rights after the Second World War, an element of popular public regulation during the 1970s, and the steady reassertion of urban private property rights through the 1980s. The system is still changing.

California

The third conflict at the heart of the Californian planning system is also between private property rights and their public regulation. The results of this conflict have swung the nature of planning in California from a primary concern with the rights of private property owners before and after the Second World War, through growing regulation during the 1960s and 1970s, and back again to private property rights during the 1980s and 1990s. This reflects the continuing conflict between private property rights on the one hand and public regulation on the other.

The rights of private property owners are a fundamental constitutional principle in California. The degree of significance attached to them there is a major distinguishing feature between the political economy of California and Britain. In California private property rights are respected by the courts and guarded by civil libertarians. Planning, on the other hand, is a classic regulatory system which, in principle, is designed to restrain the exercise of private rights in order to achieve a public good. The technical basis of planning regulation is a series of laws which have been passed by the state legislature and interpreted by the courts right up to the Federal Supreme Court.

The characteristic American and Californian planning regulation of private property rights is zoning. This is a system which allocates all urban land to zones in which only certain types of building are permitted. Early zoning was cumulative. This placed single family houses at the top of a pyramid of desirable land uses and industrial property at the bottom. In this system any land-use zone was permitted to accommodate any uses below it but not above it. Thus, industrial zones could contain a mixture of all uses because they formed the base of the pyramid. Single family zones could only contain single family houses because they were at the top of the pyramid. Cumulative zoning was replaced during the building boom after the Second World War by exclusive zoning. This allowed only one type of use in each zone.

Although zoning regulates the uses of private property this has not always been against the interests of private property owners. Initially it was used in New York, in 1916, to prevent the continued encroachment of rented tenements up Fifth Avenue towards the privately owned mansions of the bourgeoisie. The constitutionality of such zoning was upheld by the Supreme Court in 1926 in the case of the Village of Euclid, Ohio versus Ambler Realty Company. The village in this case lent its name to the subsequent Euclidean zoning which divided all a municipality's land into districts and aimed to treat all property owners within those districts equally.

Although property owners within districts have been treated more or less equally, residents within different districts have not. Post-war exclusionary zoning practices tended to operate in the interests of property owners and against those of non-owners. Single-family house zoning on quarter-acre plots, for example, was covertly designed to prevent reductions in property values by not allowing poor urban renters to escape to the privately owned suburbs. Prices were maintained at too high a level for them to be able to do so.

During the 1960s and 1970s planning regulation of private property rights was increased by the growing use of existing or new planning laws. These included the planning, zoning and development laws discussed above. In addition to these, there are six other major laws which are significant for planning in California.

First there is the General Plan Law. This requires that all local governments prepare a general plan for the future development of their city or county, and lay out all the state's requirements governing what a general plan should contain. The general plan is the same as the document that most planners outside California refer to as the Comprehensive or Master Plan. It is a comprehensive document that establishes the city or county's land-use policies and also details the likely future development patterns.

Second, the Subdivision Map Act governs all subdivisions of land. It requires that local governments establish regulations to guide subdivisions. It grants powers to local governments to ensure that the subdivision occurs in an orderly and responsible manner.

Third, the California Environmental Quality Act (CEQA) requires local governments to conduct some form of environmental review on virtually all public and private development projects. CEQA's requirements, which are mostly procedural, sometimes cause local governments to prepare environmental impact reports on specific development projects, detailing the likely environmental damage the projects would cause.

Fourth, the Coastal Act, which was originally passed as a ballot initiative in 1972, establishes special planning requirements for coastal areas and creates a powerful state agency, the Coastal Commission, to oversee coastal planning.

Fifth, the Community Redevelopment Law gives cities and counties great power to redevelop blighted areas. It is perhaps the most powerful single tool local government possesses, other than the basic laws permitting them to engage in planning at all.

Finally, there is the Cortese–Knox Local Government Reorganization Act, which is not strictly speaking a planning law. Its provisions play an important role in local planning because it governs procedures by which local government boundaries may be changed. Under the Act, all annexations, incorporations and other boundary changes must be processed through a special countrywide agency called the Local Agency Formation Commission (LAFCO) (Fulton, 1991: 223).

During the 1960s and through until the mid-1980s the Californian courts gave planning agencies a relatively free hand to apply these laws in regulating land uses. But, towards the end of the 1970s, the past 15 years of the expansion of these regulatory powers stimulated a counter-reaction in the form of the property rights movement. When this coincided with the political shift to the right under Ronald Reagan, the property movement began to have some successes. The most notable of these were, first, in 1981, when the legislature passed the development agreement statute. This permitted developers to enter into long-term contracts with local governments to secure their vested property rights irrespective of future conditions.

In 1987 two US Supreme Court rulings known as the First English and the Nollan cases also upheld the constitutional rights of property owners against governmental agencies seeking to regulate the use of their land. In the First Evengelical Lutheran Church versus the County of Los Angeles, it ruled that a property owner whose land was taken by regulation is entitled to just compensation even if the taking is only temporary. Thirdly, in the case of Nollan versus the Californian Coastal Commission, the Supreme Court ruled that exactions and other conditions of approval placed on a development permit must be directly related to the project under consideration (Fulton, 1991: 60).

These three decisions swung the conflict between private property rights and their public regulation by planning agencies firmly back in favour of the former. That such conflicts are nearly always in a state of change, however, is illustrated by the fact that, at the same time as the rights of individual private property owners were being re-asserted in California, the rights of individual property developers were being restricted in some parts of California by the political pressure exerted by some of those very same individual owners. This pressure arose in the mid-1980s from concerns about traffic congestion and urban sprawl. This led to state action in Florida and New Jersey to manage growth. Individual cities have also begun to introduce growth management schemes in California.

Conclusions

The public regulation of private property rights is conducted in quite different ways in Britain and California. The 1947 Town and Country Planning Act which expropriated all new urban development rights in Britain was a draconian measure by Californian standards. There a much more piecemeal series of legislative activities have sought to exercise rather weaker powers over private property rights.

The system in Britain has been more uniform than in California. It has rested on the production of various kinds of development plans. Their provisions have then been enforced on private development proposals through a uniform system of development control which in turn rests on an extensive body of specialized planning law.

In California the rights of private property are enshrined and generally protected by the US Constitution. There has always been a major conflict between these rights and their legal regulation through a land-use planning system. The main vehicle for achieving regulation has been zoning. This is a much more general provision than the development control system in Britain. Given a particular zoning provision, private property owners may assume that they can exercise their development rights within that framework. In Britain there is much more detailed regulation on a case-by-case basis.

MARKETS VERSUS PLANS

Britain

One of the key reports which inspired the post-war British planning system was the Uthwatt (1942) Final Report of the Expert Committee on Compensation and Betterment. It dealt with the significance of introducing effective plans on land markets. Had its advice been followed in its entirety, markets in land would have been abolished altogether and replaced by plans. This was not done, however, and the post-war system was established on the basis of a combination of both markets and plans.

The initial intention was to alter the ways in which land markets had functioned before the war not only by the introduction of plans but also by the use of taxes on the oligopolistic profits made as a result of those plans. Three abortive attempts were made to alter the operation of the market and tax 'betterment' as it is called. The first, introduced in 1947, levied a 100 per cent tax on the difference in price of a piece of land before and after planning permission was granted. This did not stop land owners adding this and more to the price of their land. It was abolished by the Conservative government in 1953. A Labour government tried to reintroduce the tax on betterment in 1967 in the Land Commission Act. This time the tax was set at 40 per cent. This was again repealed by the incoming Conservative government in 1971. Labour tried again in 1976 with the Development Land Tax. This varied between two-thirds and four-fifths of the assessed betterment gains. It was modified and then abolished altogether by the Conservative government in 1985.

These attempts to replace markets in land by taxing the gains made as a result of planning were too radical to be willingly accepted by the dominant private property interests. As a result they were a failure. Land owners did their best to avoid them or pass on the extra costs to developers. No Conservative government would accept them on a permanent basis and so they became an ineffective political football. Without them there is an inherent contradiction in the attempt to operate a planning system. This is that without an effective tax on betterment, it will usually be the markets rather than plans that make the significant land-use planning decisions (Reade, 1987: 23).

An alternative method used to acquire an element of betterment arising from planning decisions is planning agreements. These involve bargains struck in secret between planners and developers over payments in kind or cash for planning permission. They were not much used before the 1971 Town and Country Planning Act and the 1974 Housing Act. They used to take their jargon name from Section 52 of the 1971 Act, and are now known as Section 106 Agreements after their place in the 1990 Town and Country Planning Act.

Planning agreements have been used to specify additional land uses, provide public rights of way and public open space, to extinguish existing use rights, to provide community buildings, to rehabilitate property, to provide infrastructure, to acquire free public buildings and to pay for car parking (Jowell, 1977). They are only common in conditions of economic growth. They do not cost developers as much as the various taxes on betterment were intended to.

During the economic decline of the 1980s, market forces were reasserted over plans by successive Conservative governments. Their general intention was to free up the restrictions contained in plans and to allow market forces to operate both in terms of the development of housing and national economic 'growth'. The 1980 Local Government, Planning and Land Act and a whole string of subsequent circulars were all designed to achieve these ends. The result is that plans are now more market-led than they have ever been since 1947.

California

In California there is a general conflict between those who assert that decisions about the use of land and buildings should be left to the operation of relatively unfettered markets, and those who argue that some desirable outcomes do not arise if markets are left to decide and therefore plans should be made to ensure that they do arise. This conflict is expressed in California's zoning, planning and development laws. These say that the purpose of planning is to ensure the preservation and use of land in ways which are *economically* and *socially* (emphasis added) desirable in an attempt to improve the quality of life in California (Fulton, 1991: 3).

Planning, as opposed to markets, is the process by which government agencies determine the intensity and geographical arrangement of various land uses in a community. These include residential projects, shopping centres, office and industrial employment centres, transportation facilities, agricultural land and parks. Most of these land uses are established or built by private developers. What planning does is to regulate where and what these uses are. In California this means that planning is mostly a regulatory and reactive activity. It seeks to regulate what private property owners do in order to achieve some public good that unrestrained market actions do not provide. In most cases the only way that it can seek to obtain a public good is by reacting to proposals for land-use changes and development from the private owners of property. Planners do not realize their own plans themselves.

The prerequisite for the realization of the provisions of regulatory and reactive planning is economic growth. Only with growth do developers come forward with proposals that such a planning system can regulate and react to. It is economic growth that has raised the concern with and significance of planning.

The state of California required every city and county to produce a general plan as long ago as 1937. They were generally based on ideas derived from architecture and urban design. In most cases they remained ineffectual pieces of advisory paper until rapid economic and consequential physical growth were stimulated in California by the Second World War. In 1953 the state made the requirement to produce a general plan a central legal condition. Two years later it specified that all plans must contain at least land-use and circulation elements. Such specific requirements have now been extended to include not only the original land-use and circulation elements but also housing, conservation, open space, noise and safety.

The continuing conflict between plans and markets can be seen in the fact that, until 1971, general plans had no legal force in California. Until that time they were rarely used in conjunction with zoning, which did have legal force. In 1971 state law was changed to require consistency between general plans and zoning. Even so, Los Angeles did not reconcile its planning and zoning until forced to do so by the courts in the late 1980s.

In some senses it can be seen that planning, as opposed to zoning, did not have much overall effect on land and property markets in Californian cities in general until after 1971, and until after the late 1980s in Los Angeles in particular. Indeed, some of the state's most handsome cities, such as Carmel, Pasadena and Santa Barbara, were developed as the result of private oligarchical rather than public planning.

Such effects as planning *per se* has had on Californian markets have also been undermined since the passage of Proposition 13 in 1978. This cut property taxes and left local governments increasingly reliant on local sales taxes for revenue and local deals with private developers for the provision of infrastructure. As a result of Proposition 13, planning in California has become 'fiscalized'. Cities and counties now compete with each other in market fashion to obtain physical sources of big sales tax revenues. These include car dealerships, hotels and shopping malls. They do not include housing. Bargains are also struck with developers for the construction of highways and sewers as quid pro quo for development permits. The net result of these new forms of market is that huge regional imbalances have been created between those cities that can capture economic growth in the form of high sales tax developments, and those areas in which their employees can find housing. The former gain the most public revenues while the latter actually need them the most in order to provide for the collective needs of their populations.

Conclusions

Attempts to replace markets with plans in both Britain and California have met with effective opposition. This is particularly the case when they have sought to intervene in the financial rewards gained by development in market conditions. Although it is clear that, once land-use regulations are introduced, they both create and reduce land values, governments have seldom been allowed to collect much of the publicly created betterment.

Attempts to regulate land and development markets have been much weaker and more recent in California than in Britain. Some cities such as Los Angeles resisted effective interventions right into the 1980s. In Britain various attempts have been made to claw back for the public authority gains made as a direct result of planning decisions. These are now confined to planning agreements associated with specific developments. They can only be enforced if the benefits required also relate directly to those particular projects.

SUMMARY AND CONCLUSIONS

The British planning system has to be seen as being based on conflicts. At the core of these are conflicts between private property interests, which favour the use of markets to decide the uses to which they can put their property, and public interests, which look to public institutions, regulation and plans to provide goods and services which markets characteristically do not. There are also conflicts between different groups within these two large and general categories.

The outcomes of these conflicts rarely conform to the original intentions of all developers or public interests. One of the main divergences from the formal intentions of planning is the partial replacement of competitive markets in land and development with oligopolistic markets. In these modified circumstances a small number of large scale developers come to dominate the markets in land and development where development is permitted by the planning system. This reduces competition in these local markets and is therefore unlikely to benefit the eventual consumers of development.

It is possible that this result suits both developers and the supporters of planning. This may be seen in the fact that the history of British planning is one of elite and paternalistic decisions to introduce effective public regulation over land uses in the name of the public interest. When other interests were introduced into policy-making for a relatively brief period during the 1970s, as a result of increased public participation, they were soon reduced in the face of economic problems. The 1980s saw the reassertion of oligopolistic market forces over plans.

The reassertion of oligopolistic market forces in planning brought to the fore the basic contradiction on which the whole enterprise rests. This is the conflict between the drive to assist large enterprises to maintain or increase their profits in the name of market forces by providing infrastructure, information and other enabling functions; and the need to legitimate planning to the general public by demonstrating that it also provides some benefits to them in the form of public participation, establishing social objectives and generally assisting to create social harmony.

It was possible to follow both these contradictory objectives during times of economic growth. But the first objective has taken precedence since the 1980s. This is partly because of economic limitations and partly because of the ideological predisposition of the Conservative central government during that period. Nevertheless, the creation of oligopolistic markets in land and development by public action does not produce universal, public or mass benefits. This is shown by the analysis of the actual effects of the British planning system.

The background to the effects which can be directly attributed to planning is that social change and polarization have been increasing in Britain. This is marked by increases in the service class, decreases in manual workers and increases in unemployment. This has been combined with increases in the residential segregation of these groups, particularly between prosperous and less prosperous areas. Those groups, such as ethnic minorities, have been trapped in the inner cities where they form a growing and disenchanted 'underclass'.

The introduction into these ongoing changes of containment, suburbanization and rising land prices by planning has increased residential segregation and allocated better living environments to those who are already better off. This is particularly true of those suburbanites who live beyond the major cities in their green belts and the smaller towns and villages both in and beyond them. They have acquired the best housing, the highest capital gains and have been able to protect those gains for themselves by using the planning system.

The planning system has also influenced access to public goods and services. On the whole, planning decisions have improved the life chances of those for whom they are already highest. In contrast, those without cars and those who cannot afford house prices in the prosperous rural South East are excluded from access to the publicly provided goods and services there.

Many of the effects of British planning were not formally intended by the planners themselves. They have followed as the unintended consequences of policies such as urban containment. The paradox is that few of the major post-war reasons for constraining the supply of housing land are valid today.

The first of these was that agricultural land should be preserved for defence purposes. This is no longer valid as no one really expects a modern war to last long enough to make the growing of crops a worthwhile or un-radioactive proposition. In addition, the Common Agricultural Policy (CAP) of the European Union generates food surpluses on the one hand and pays farmers to 'set aside' land from production altogether on the other. In such conditions there is a large surplus of land in Britain which is not needed for agricultural production.

Secondly, many of the objections to pre-war urban sprawl were often voiced by design professionals in terms of its visual appearance rather than its functional provision of inexpensive housing for the masses. If anything, the alternative restricted estate developments that have been located away from the major cities are even worse in design terms, lacking in urban facilities and less affordable than their pre-war counterparts. In Britain, at least, where 4.4 to five million new homes are needed by the early twenty-first century, some structured and sustainable release of building land is essential if housing is not to become even less affordable than it is now and slum densities are not to increase considerably.

Thirdly the decentralization of industry and commerce has taken place even with restrictive planning. It is no longer the dirty smoke-stack nuisance that was familiar before the war. It is increasingly high-technology based. The professionals who work in such new industry look for working environments which, among other things, improve the locality. This combined with the desperate need for economic growth of any kind argues for locational freedoms which are not provided within current policies of land-use constraint.

While this much has become increasingly obvious even in DoE sponsored research under the previous Conservative government, there remains a political reluctance to alter the fundamentals of the British planning system as it has operated since 1947. This is true even after the detailed changes made to the system by the Conservative governments of the 1980s.

The question therefore arises as to what holds these basic fundamentals of the land-use planning system in place. The answer to this question lies in who makes the significant planning decisions and who benefits most from the operation of the system.

In highly centralized Britain, it is powerful individuals and groups who are able to use the existing social and political institutions, like the planning system, to influence urban structure and the ways in which it is developed. Reade (1987) makes the pessimistic point that in any society in which land-use controls exist they will be 'misused'. This is because it is always difficult to say what the public interest in land uses is and even more difficult to determine what combination or arrangement of them would actually be 'best' in any circumstances. On the other hand, it is much more possible for powerful groups such as the House Builders' Federation to discern what kinds of planning decisions would be in their own best interests. It is also possible for the high-income and well-educated members of the service class to define and follow their interests in planning. It is much more difficult for the urban underclass to express 'acceptable' demands and to negotiate them successfully with the British land-use planning system.

The guiding principles of the Californian planning system are the protection and enhancement of the rights of private property owners and limited intervention in development markets. This is not to say that every individual planning action takes one or both of these forms. This is because, like most public policies, it is a system responding to mutually incompatible, external conflicts of interest. In the case of planning these conflicts are between private and public interests; between private individuals and public institutions; private property rights and public regulation; and between markets and plans. These conflicts pull the system in different directions. The outcomes at any particular point in time depend upon the balance of forces in contention over particular issues.

The protection and enhancement of the rights of private property owners, and limited intervention in development markets come to be the most common guiding principles of the Californian planning system, because private property owners and large scale developers are the best equipped and organized to obtain these outcomes from the conflicts surrounding planning. They do not always win all the individual conflicts and this is why not every individual planning action results in the protection and enhancement of the rights of private property owners and limited intervention in development markets. The majority, however, do take one or other of these forms.

One major distinguishing feature between British and Federal and Californian planning policy is that there are no national or state policies for the containment of urban growth. In California there were no effective planning mechanisms which could have been used to contain urban growth before 1971 when state planning law was changed to require the coordination of general plans and zoning ordinances.

In California, planning and zoning have combined with a discriminatory housing finance system and the lack of a significant public housing programme to spread out different social and racial groups over space. This has produced residential and social segregation. Its most extreme forms are racial segregation and the urban ghettos.

As far as planning is concerned, exclusionary zoning and growth management policies have served to segregate populations according to their different abilities to pay for housing in the market place. Growth Management Policies (GMPs) have tended to make housing relatively scarcer, more expensive and larger. All these factors make for increased residential and social segregation.

Residential segregation is important because of the differences in access to a wide range of both public and private goods and services that follows where families are able to live. On the one hand some locations give access to clean environments, good schools, parks and other public and private benefits. Other locations trap those who cannot escape from them in unhealthy surroundings with minimal or non-existent public facilities. The arrangement of these locational differences and opportunities on the ground in terms of new land uses and the built environment, has, in the past, been a prime official concern of planning.

Underlying planning, financial and government programmes in California, however, is the objective of protecting existing land and property values. Financial appraisal systems, zoning ordinances, GMPs and other policies have all been supported by existing and potential private property owners. They do not see residential integration as a way of maintaining and enhancing their property values. For this reason and because of their political power in the conflicts over planning policies, residential integration is unlikely to become a major planning goal in California.

Both the British and the Californian land-use planning systems have therefore reflected the distribution of power in their respective societies. Even the radical post-war Labour government in Britain was unable to establish a planning system with significant progressive redistributions of the costs and benefits derived from the uses of land and buildings. It is hardly likely therefore that the more conservative parties in California would even seek to establish such a system.

In practice, collective interventions in the rights of private property have been limited in both places. Where they have had some impact this has usually been where not to intervene would produce costs for all social classes. Key examples of this type include the need to provide public health infrastructure for all. Many regard the introduction of clean water and sewerage regulations, in the late nineteenth century, as the foundation of town planning in Britain. In California the more recent concern with the environment has stemmed from similar concerns but now on a wider scale. The growing concern with the whole issue of sustainability on both sides of the Atlantic may also eventually give rise to effective regulation.

In the meantime, it is no good expecting powerful property owners to establish a socially progressive system of land-use planning. It would not be in their interests. They have most access to the institutions that take such decisions. Therefore, other things remaining equal, the land-use planning systems in Britain and California will continue to reflect the distributions of power in their respective societies.

REFERENCES

Badcock, B. (1984) *Unfairly Structured Cities*. Oxford: Basil Blackwell.

Brindley, T., Rydin, Y. and Stoker, G. (1989) *Remaking Planning: The Politics of Urban Change in the Thatcher Years*. London: Unwin Hyman.

Castells, M. (1984) *The City and the Grassroots*. London: Edward Arnold.

Fulton, W. (1991) *Guide to California Planning*. Point Area, CA: Solano Press Books.

Hall, P., Thomas, R., Gracey, H. and Drewett, R. (1973) *The Containment of Urban England*, 2 volumes. London: George Allen and Unwin.

Jowell, J. (1977) 'Bargaining in development control', *Journal of Planning and Environmental Law*, July: 414–33.

Lasswell, H.D. (1958) *Politics: Who Gets What, When, How?*, New York: World Publishing.

LeGates, R.T. and Stout, F. (1996) *The City Reader*. London: Routledge.

Reade, E. (1987) *British Town and Country Planning*. Milton Keynes: Open University Press.

Saunders, P. (1979) *Urban Politics: A Sociological Interpretation*. London: Hutchinson.

Siegan, B. (1972) *Land-use without Zoning*. Lexington, MA: Lexington Books.

Simmie, J.M. (1974) *Citizens in Conflict: The Sociology of Town Planning*. London: Hutchinson.

Simmie, J.M. (1981) *Power, Property and Corporatism: The Political Sociology of Planning*. London: Macmillan.

Uthwatt Report (1942) *Final Report of the Expert Committee on Compensation and Betterment*. London: HMSO.

FURTHER READING

Britain

Ambrose, P. (1986) *Whatever Happened to Planning?* London: Methuen.

Best, R.H. (1981) *Land Use and Living Space*. London: Metheun.

Best, R.H. and Anderson, M. (1984) 'Land use structure and change in Britain, 1971 to 1981', *The Planner*, 70 (11): 21–4.

Beveridge Report (1942) *Social Insurance and Allied Services*. London: HMSO.

Bowers, J.K. and Cheshire, P. (1983) *Agriculture, the Countryside and Land-Use: An Economic Critique*. London: Methuen.

Champion, A.G., Green, A.E., Owen, D.W., Ellin, D.J. and Coombes, M.G. (1987) *Changing Places: Britain's Economic and Social Complexion*. London: Edward Arnold.

Cheshire, P. and Sheppard, S. (1989) 'British planning policy and access to housing: some empirical estimates', *Urban Studies*, 26: 469–85.

Coleman, A. (1977) 'Land use planning: success or failure?', *Architects Journal*, 16 (3): 94–134.

Curtis, S. and Mohan, J. (1989) 'The geography of ill-health and health care', in J. Lewis and A. Townsend, (eds), *The North–South Divide: Regional Change in Britain in the 1980s*. London: Paul Chapman.

Davies, J.G. (1972) *The Evangelistic Bureaucrat: A Study of a Planning Exercise in Newcastle upon Tyne*. London: Tavistock.

Dennis, N. (1972) *Public Participation and Planners' Blight*. London: Faber.

Department of the Environment (1976) *The Recent Course of Land and Property Prices and the Factors Underlying It*. London: HMSO.

Department of Environment (1983) *Town and Country Planning Act 1971: Planning Gain*. Circular 22/83. London: HMSO.

Department of Environment (1985) *Lifting the Burden*. White Paper. London: HMSO.

Economist Intelligence Unit (1975) Housing Land Availability in the South East. Report to the DoE. London: EIU.

Elson, M.J. (1986) *Green Belts: Conflict Mediation in the Urban Fringe*. London: Heinemann.

Evans, A.W. (1988) *No Room! No Room! The Costs of the British Town and Country Planning System*. London: Institute of Economic Affairs.

Eve, G. and Department of Land Economy (1992) *The Relationship between House Prices and Land Supply*. DoE Planning Research Programme. London: HMSO.

Grigson, W.S. (1986) *House Prices in Perspective: A Review of South East Evidence*. London: SERPLAN.

Halcrow Fox and Associates/Birkbeck College, University of London (1986) *Investigating Population Change in Small to Medium Sized Areas*. London: DoE.

Hooper, A., Pinch, P. and Rogers, S. (1989) 'Land availability in the South East', in M. Breheny and P. Congdon (eds), *Growth and Change in a Core Region*. London Papers in Regional Science. London: Pion.

McAuslan, P. (1980) *the Idealogies of Planning Law*. Oxford: Pergamon.

Moore, B., Rhodes, J. and Tyler, P. (1986) *The Effects of Government Regional Economic Policy*. London: HMSO.

Nuffield Foundation (1986) *Town and Country Planning*. London: The Nuffield Foundation.

Regional Studies Association (1990) *Beyond Green Belts: Managing Urban Growth in the 21st Century*. London: Jessica Kingsley Publishers and the Regional Studies Association.

Simmie, J.M. (1971) 'Ideology and physical planning in Britain', Town Planning Discussion Papers, No. 16. London: Bartlett School of Architecture and Planning.

Simmie, J.M. (1985) 'Corporatism and planning', in W. Grant (ed.), *The Political Economy of Corporatism*. London: Macmillan.

Simmie, J.M. (1986) 'Structure plans, housing and political choice in the South East', *Catalyst*, 2 (2): 71–83.

Simmie, J.M. (1987) 'Planning theory and planning practice: an analysis of the San Francisco downtown plan', *Cities*, November: 304–24.

Simmie, J.M. (1991) 'Corporatism and planning in London in the 1980s', *Governance*, 4 (1): 19–41.

Simmie, J.M. (with S. French) (1989) 'Corporatism, participation and planning: the case of London', *Progress in Planning*, 31 (1): 1–57.

Simmie, J.M. (with D.J. Hale) (1978) 'The distributional effects of the ownership and control of land use in Oxford', *Urban Studies*, February: 9–21.

Simmie, J.M. (with S. Olsberg and C. Tunnell) (1992) 'Urban containment and land use planning', *Land Use Policy*, 9 (1): 36–46.

Skeffington, A.M. (1969) *People and Planning*. London: HMSO.

Thornley, A. (1991) *Urban Planning under Thatcherism: The Challenge of the Market*. London: Routledge.

USA and California

Adams, John S. (1987) *Housing America in the 1980s*. New York: Russell Sage Foundation.

Babcock, R. (1969) *The Zoning Game*. Madison, WI: University of Wisconsin Press.

Babcock, R. and Siemon, C. (1980) *The Zoning Game Revisited*. Madison, WI: University of Wisconsin Press.

Cervero, R. (1986) *Suburban Gridlock*. New Brunswick, NJ: Center for Urban Policy Studies.

Coyle, S. (1983) 'Palo Alto: a Far Cry from Euclid', *Stanford Environmental Law Annual*, 4: 83–103.

Deakin, E. (1989) 'Growth controls and growth management: a summary and review of empirical research', in D. Brower et al. (eds), *Understanding Growth Management*. Washington, DC: Urban Land Institute.

Dowall, D.E. (1984) *The Suburban Squeeze: Land Conversion and Regulation in the San Francisco Bay Area*. Berkeley, CA: University of California Press.

Fischel, W.A. (1985) *The Economics of Zoning Laws: A Property Rights Approach to American Land Use Controls*. Baltimore, MD: Johns Hopkins University Press.

Frieden, B.J. (1979) *The Environmental Protection Hustle*. Cambridge, MA: MIT Press.

Fulton, W. (1989) 'Wheeling and dealing in California', *Planning*, November: 4–9.

Haar, C. and Kayden, J. (1989) *Zoning and the American Dream*. Chicago: American Planning Association.

Hayden, D. (1984) *Redesigning the American Dream: The Future of Housing, Work, and Family Life*. New York: W.W. Norton.

Jackson, K.T. (1985) *Crabgrass Frontier: The Suburbanization of the United States*. New York: Oxford University Press.

Juelsgaard, J. (1983) 'Atherton: applying the state's fair share housing requirements', *Stanford Environmental Law Annual*, 4: 130–42.

Landis, J.D. (1986) 'Land regulation and the price of new housing', *Journal of the American Planning Association*. Winter: 9–21.

Miller, T.I. (1986) 'Must growth restrictions eliminate moderate-priced housing?', *Journal of the American Planning Association*, Summer: 319–25.

Mumford, L. (1938) *The Culture of Cities*. London: Secker & Warburg.

Niebanck, P.L. (1989) 'Growth controls and the production of inequality', in D. Brower, et al. (eds), *Under-*

standing Growth Management. Washington, DC: The Urban Land Institute.

Peterson, A.L. (1989) 'The takings clause: in search of underlying principles', *California Law Review,* 77: 1299–363.

Schwartz, S.I., Hansen, D.E., Green, R. (1984) 'The effect of growth control on the production of moderate-priced housing', *Land Economics,* 60: 110–14.

Siegan, B. (1972) *Land-Use Without Zoning.* Lexington, MA: Lexington Books.

Sigg, E. (1985) 'California's development agreement statute', *Southwestern University Law Review,* 15: 695–727.

van Vleit, W. (ed.) (1990) *Government and Housing.* Beverly Hills, CA: Sage.

International Comparisons

Ball, M. Harloe, M. and Martens, M. (1989) *Housing and Social Change in Europe and USA.* London: Routledge.

Clawson, M. and Hall, P. (1973) *Planning and Urban Growth: An Anglo-American Comparison.* Baltimore, MD: Johns Hopkins University Press.

Simmie, J.M. (1993) *Planning at the Crossroads.* London: UCL Press.

Smith, B. (1988) 'California development agreements and British planning agreements: the struggle of the public land use planner', *Town Planning Review,* 59: 277–87.

26

Power, Discourses and City Trajectories

MARK BOYLE AND ROBERT J. ROGERSON

In 1997, the Committee for Sydney, a newly formed partnership of private sector interests within the city, invited tenders to assist in the development and enhancement of Sydney's role as a 'world city'. The outline argued that most of the other major cities of the world – such as Barcelona, Paris, London, Rome and Venice (the cities mentioned by the Sydney Committee) – had already developed long-term strategic plans designed to ensure their status in the global economy. Recognizing the absence of just such a strategic vision and aware of the impending opportunities associated with the hosting of the 2000 Olympic Games, the Committee called for research to be undertaken: to benchmark Sydney against other world cities; to examine its capacity to grow as a centre for global capital; and to suggest strategies through which the city could be repositioned as a world city.

The research programme defined by the Sydney brief serves to draw attention to several issues which are at the forefront of contemporary urban studies and around which we wish to construct this chapter. First, it highlights the manner in which contemporary cities are defining their futures relative to global capital and emerging international divisions of labour. Second, it points to the consciousness within cities of the need to compete with other rivals to secure a future role and, therefore, of the importance of formulating coherent city trajectories to gain competitive advantage. Third, and perhaps most importantly for the discussion in this chapter, the brief illustrates the way in which, although recognizing the possibility of marginally different strategies, the Sydney partnership provides no opportunity to question the underlying language or logic of the need to be positioned relative to global capital.

Indeed, in the policy literature, it is now becoming impossible to read accounts of the fortunes of most Western cities without encountering such phrases as 'city competition', 'city imaging', 'benchmarking', 'market niche' and 'hallmark events'. So pervasive has been the adoption of this vocabulary that it has given birth to a whole new area of what might be termed 'instrumental knowledge'. For instance, urban 'handbooks' now adorn book shelves providing lessons and practical instruction on how to organize local economic development strategies (Blair, 1995; Blakely, 1989; Duffy, 1995) whilst business schools now run courses on the techniques of urban regeneration. The rise of such terminology has given birth to new areas of expertise within city councils and universities. The discourse of globalization and localization, therefore, sits at the heart of contemporary ways of thinking about the city.

Likewise, beyond the policy literature it has been argued that in the past two decades inter-city competition for economic development has become 'the' major activity of urban governance (Cox, 1995). Across advanced capitalist cities, the net effect has been the creation of what Cox (1993) terms a whole New Urban Politics (NUP). Written around the NUP as a building block, and breaking down the title of this chapter – power, discourses and city trajectories – into its constituent concepts, we wish to rework literature on the politics of local economic development from the vantage of a new perspective. Referred to herein as the discourse perspective, we seek to examine the multiple ways in which the notions of 'power' and 'discourse' interweave with the production and legitimation of city development trajectories. In so doing, we first explore the way orthodox policy literature has used the NUP to understand the formulation of city trajectories. We then seek to move beyond existing accounts by approaching the NUP as a language or vocabulary, or what we will term a discourse. In charting a path from the orthodox perspective of the NUP outlined in the first section to our alternative, discourse analysis of the NUP in the third section of this chapter, we do so by critically evaluating one of the more seminal examples of the discourse approach in urban studies, that provided by Beauregard (1993). We end by offering some tentative conclusions regarding issues requiring future development within the discourse perspective.

DISCOURSES AND URBAN GOVERNANCE

The New Urban Politics: The Orthodox Policy Perspective

According to Kevin Cox (Cox 1993, 1995; Cox and Mair, 1988; Cox and Wood, 1994), there has been something of a convergence in *orthodox policy* studies of the city during the past decade. Cast initially in terms of the impact of the new international division of labour on the so-called 'world cities' (Friedmann, 1986; Friedmann and Wolff, 1982), this literature has expanded to argue that the fundamental context in which all cities are currently acting is the heightened mobility of capital. The capacity of transnational corporations to switch their operations around the world has created a truly global economy, in which the world has become a single location for production outlets. Whilst there remains considerable debate over the definition and meaning of the phrase, this *globalization of capital* is argued to lie at the heart of contemporary urban restructuring.

Catapulted into such a context, many cities find their bargaining position in relation to capital substantially weaker. The capacity of capital to switch locations engenders a competition between places to secure investment. As such, the task of urban governance has increasingly become the creation of urban conditions sufficiently attractive to lure prospective firms. Whether this entails alterations to the city's image through manipulation of its soft infrastructure (cultural and leisure amenities, for instance) or a refashioning of the city's economic attractiveness (through provision of grants, property, transport facilities, or tax abatements, for instance) localities are now having to offer ever more inducements to capital to secure development and growth.

It is this self-perpetuating competition between localities that Cox refers to as constituting a New Urban Politics (NUP). Cox (1993: 45) summarizes conventional wisdom thus:

1   The economic space within which cities are situated is subject to change.
2   This change is a result of an increased footlooseness of capital with respect to cities as possible sites.

3   Within cities, there are a variety of economic interests which, as a result of immobility, are dependent upon the health of the urban economy. These include property owners, some businesses – such as banks and newspapers – local governments, and local residents. Taken together, Cox argues, these agents constitute 'cities' or 'communities'.
4   Changes in the space economy as a whole provide threats and opportunities to these economic interests.
5   These interests work through city governments in order to channel investment into their particular city through appropriate infrastructure, taxation, and regulatory practices.
6   Policies of this sort bring them into competition with corresponding 'cities' or 'communities'.

As an extension of the NUP thesis, it is worthwhile noting that faced with an uncertain future in the global division of labour, places are not only competing with each other to catch the eye of global capital. Inter-city competition in this context is also extended to include competition for *consumer expenditure* (or what Harvey (1989a) calls the 'tourist dollar') and for a role in the *spatial division of labour created by the state* (the allocation and reallocation of different state functions to different areas). So precarious has the economic health of cities become and so pervasive the competition ethos, that whilst there might be an emphasis upon one, every city is likely to be active on all three of these fronts. And, in yet another twist to this evolving story, it is possible to identify a different and fourth form of competition; that involving efforts to secure what are often termed 'prestige' or 'flagship projects'. Whether funded by the state or more likely by international bodies which organize events such as the Olympic Games, Commonwealth Heads of Government meetings, or jazz festivals, the importance of hosting a major leisure, cultural, sporting or political event to a city's profile has generated in itself an intense competition between localities to win the right to hold such events. What makes this form of competition different from the other three forms is that for such prestige events, it would appear that places are competing for the 'raw materials' themselves needed to facilitate a competitive edge in their efforts to lure other forms of capital and expenditure. Whilst recognizing these different types of inter-city competition, in turning to consider the question of city visions and trajectories, we focus more narrowly on global capital – the first of the four forms outlined above.

The NUP, City Visions and Trajectories

A key tenet of the NUP agenda is that there has been a shift in the position of cities away from local, regional or national roles towards roles ascribed by the global economy. As the Sydney tender with which we started this chapter illustrates, within the NUP there is a perceived acceptance that a position of marginality relative to the global economy is inappropriate for a thriving, dynamic and visionary city. In other words, city visions are fundamentally concerned with the degree of marginality/centrality of the city in relation to global capital.

Arising from both the engagement with the wider global setting and from the need to respond to the demands of capital, city visions are being shaped by the requirement not merely to attract capital investment in its many forms but also to adopt some of the characteristics of this capital. Not least, as Harvey (1989b) argues and Sassen (1991) exemplifies in relation to London, New York and Tokyo, cities have had to struggle to find a response to the values and virtues of flexibility and volatility, of instantaneity and disposability, which have become hallmarks of postmodern society and economy. The accelerating rate of change in production and consumption needs, aided by technological advances, has meant that city agencies attempting to engage with capitalist systems have, too, to respond to their volatility and dynamism. The extent to which capital is indeed so spatially mobile (Cox, 1995; Gertler, 1988; Schoenberger, 1988) and production opportunities so flexible (Allen and Massey, 1988) cannot be assumed unproblematically. Nevertheless, few observers would doubt that capital is more dynamic and volatile than in past decades. Equally, other dimensions of the postmodern condition – such as disposability and instant obsolescence of commodities – also reinforce the instability of capitalist systems.

As Harvey (1989b: 286) notes in relation to the corporate sector's response to market conditions, 'volatility, of course, makes it extremely difficult to engage in any long-term planning'. And as he continues for those attempting to undertake such 'designing' – acknowledging that in the postmodern condition there is no planning only designing – 'learning to play the volatility right' is now crucial for success within the market place. Harvey indicates that 'this means either being highly adaptable and fast-moving in response to market shifts, or masterminding the volatility' (Harvey, 1989b: 287). The former position encourages the taking of short-term gains – for example, in the form of acquisitions, or mergers – with the acceptance that visions and strategies will have to adjust rapidly and frequently to make most benefit of the volatility of market needs and

tastes. The latter, masterminding change, is more long-term and strategic for here the intention is to manipulate the market's tastes, opinions and needs to fit into the shape which the organization has designed. Such a strategy involves the construction of symbols and images as well as the redevelopment of the 'product' itself (see also Zukin, 1995).

Just as such ideas have an immediacy to companies developing a blend of long-term and short-term opportunistic strategies in the market place, so too are there parallels in the form of trajectories adopted by city designers in their attempts to position their location relative to capital opportunities. Traditionally, city 'planners' have been more comfortable with the adoption of long-term visions. With long-lead times for the development and implementation of infrastructural projects and the need to gain and hold local political support, most city planning has been seen to require a long-term vision. However, with the growing volatility of capital, and the rise of city competition, the adoption of long-term planning can appear to be at odds with the flexibility and mobility which capital demands.

Significantly, therefore, city agencies in an increasing attempt to offer the characteristics of place which capital desires are adopting characteristics of this capital – such as its volatility. Employing Harvey's distinction between the short-term strategy of responding to volatility and the long-term mastering of volatility, there are – like the limited number of global corporates able to master the tastes and opinions of capital – likely to be only a few cities able to have sufficient 'power' to fashion or manipulate the desires of capital over the long term. With command and control functions, as well as a concentration of key producer services, research and development functions, innovative centres of culture and crucially the capability to control and shape the informational economy (Castells, 1989), global cities are one clear example of locations which *may* have sufficient power to 'mastermind' volatility.

On the other hand, for most other cities, the lack of a pivotal role in mastering capital may result in each having to adopt a more short-term and less ambitious vision. For these cities, there may indeed be no coherent vision as the timescale involved in responding to capital is so short and the need for flexibility so great that the very notion of a vision or sense of direction becomes an anathema. Instead, a development path might be taken which is constructed around an ability to take short-term gains whenever and wherever they are available. As such the city's trajectory is shaped more by the imprint of each opportunity than by an overarching blueprint. In the following section, we explore further some of the possible consequences of adopting the distinction between visions based on masterminding and those based on opportunism via two selected case studies: Atlanta and Glasgow.

Atlanta

Atlanta, with its significant role as a centre for global capital – 'Atlanta International' as the place marketeers would like all to believe – has characteristics more typical of a global city than Glasgow. As befits the NUP agenda, the city has many of the appearances typifying a place successfully competing with other cities: a long history of public–private partnerships; entrepreneurial form of governance; and an inward investment strategy led by 'urban rentiers' (Logan and Molotch, 1987). Encouraged by the pro-growth leadership of the mayor's office and by significant national financial investment, Atlanta's international credentials have blossomed in the past decade, pulling significant investment from Canadian, European, Japanese and Asian firms into the city. The city, too, has long adopted the need for civic boosterism and place marketing, contentedly crafting together the notion of 'the city of dreams' with an ability to envision the future. Over the decades, Atlanta's history has been one of a series of successive visions of organized promotion and redevelopment, using its own distinctive history and location in the 'South' of the USA to bring the city to the threshold of 'the next great international city'. As the most recent academic interest in Atlanta testifies (Allen, 1996; Bayor, 1996; Rutheiser, 1996), the spectacle of the Centenary Olympic Games in 1996 represented the 'outing' of the city in terms of its goal of attaining international status.

In many respects, thus, Atlanta portrays the appearance of a city with a long-term vision and a successful engagement with global capital. Rutheiser's detailed chronicling of Atlanta's development and place marketing history up to and including the Olympic Games offers, however, an alternative reading. In *Imagineering Atlanta* he explodes what he sees as a myth that there has been a unified, collective vision and carefully crafted trajectory for the city. In its place, Rutheiser places 'emphasis on the conflicts and contradictions of the different mythologies that have been elaborated over the years through a multiplicity of media' (1996: 14). Drawing on local accounts and interviews, council records, private papers and media reports, he portrays a more speculative, opportunistic form of urban regeneration, sparked at times by the public sector and at other times by individuals in the private sector, and lacking a coherent vision of the future.

Whilst it may be debatable that all elements of the redevelopment of Atlanta can be seen as contested in the way that Rutheiser suggests, his account undoubtedly makes the notion of a coherent, long-term trajectory and vision for Atlanta problematic. In the context of this chapter, we have drawn on *Imagineering Atlanta* less for its detail and more to suggest that Atlanta, which apparently has the potential to have a vision based on being more masterful over capital, can also be interpreted as a city operating in the same opportunistic manner as those located more marginally relative to the world economy.

Glasgow

The Glasgow of the 1970s and early 1980s has been ubiquitously portrayed as a city in decline. With the decline of the traditional pillars of ship-building on the Clyde, the steel industry and associated works inland, and a demographic decline (Damer, 1990; Gibb, 1983; Lever, 1990), Glasgow was, as Cheshire et al. (1986) indicated, one of the least economically healthy cities in Europe. As such, Glasgow had many of the features of a city located towards the margins of the new global economy – a location all the more poignant, given that its former role as 'The Second City' of the Empire (Oakley, 1965) remained part of the not too distant memories of many of its citizens. Despite the lack of hard evidence, the negative image associated with this decline has been recognized by those in local government to be a significant barrier to new investment and to the path out of the spiral of decline. In an attempt to breach this perceived barrier, the Labour administration of the local council in the early 1980s adopted a new vision for the city. Under this vision, the city was to be repositioned as a leading 'European city', at the heart of the European economy and successfully competing for European inward investment (Rogerson and Boyle, 1998).

In the form of construction of new physical infrastructure and a variety of cultural projects, the concrete articulation of this vision has not always been coherent. It has included, for example, both city centre redevelopment as part of a post-industrial vision focused on office development and retail consumption, and the provision of arts and cultural infrastructure (Boyle and Hughes, 1994). The manifestations in terms of cultural provision have been mixed – short-term, high profile events such as the 1990 Year of Culture and the 1999 Year of Architecture, funded primarily from the public purse, alongside longer-term provision with private sector funding in, for example, multi-screen cinemas as the hub of area regeneration. The office and retail

elements in the city centre have been transformed
with new glass-fronted, postmodern architecture
in office complexes set alongside the cleaned
Victorian facades retained to portray the city's
heritage and vernacular styles. New shopping
mall developments such as those at St Enoch's
and the soon to be completed Buchanan Galleries
sit alongside the chainstores of the high street,
with high rates of turnover and vacancies. In
addition, the attempt to change the image –
reflected in promotional and marketing
campaigns, the most renowned of which was the
'Glasgow's Miles Better' campaign of the 1980s,
re-invented in the 1990s – has frequently accompanied and even led the re-creation of the city's
profile locally and internationally.

It is not the intention here to debate whether
such an interpretation is sustainable or to explore
further the complex articulations by many
observers of the vision and trajectory adopted by
Glasgow (Boyle, 1993; Boyle and Hughes, 1994).
As a city striving to find a new, post-industrial
form and adopting elements of place marketing,
Glasgow has been seen to be archetypal. As a city
away from the core of the global economy,
however, Glasgow would appear to lack the
power required to master the needs of this capital
and thus it could be anticipated that it would
require to have the characteristics of the short-
term approach. In some respects, the attraction of
the spectacles of the Garden Festival and the Year
of Culture, for example, has been viewed as a
form of just such short-term, opportunistic
regeneration (Keating, 1988).

Alternatively, these events can also be seen to
be part of a long-term vision, one that is built on
a re-imaging and re-aestheticizing of the city as a
cultural centre in the international scene. Thus,
for example, in writing in the early 1990s about
arts events such as Mayfest and the Garden
Festival, Boyle and Hughes (1991: 219) noted that
they were part of such a strategy and that
'Glasgow's role as the European City of Culture
1990 represents the latest stage in an intensive
campaign to regenerate the city'. Or as Booth and
Boyle (1993: 33–4) note in relation to the bidding
for the 1990 title of Year of Culture, 'the bid
emphasized the importance to the city of developing cultural tourism . . . the arts would be used
as an additional strand of economic planning
directly through the attraction of tourists to the
big event . . . the mere fact of winning the title
inevitably conferred a comparative advantage
upon Glasgow through the recognition of its
cultural specialization'. In this reading there
appears, despite the city's marginal ability to
fashion capital, a vision and trajectory which has
a greater degree of permanence than would be
suggested by a short-termist perspective.

These two examples have been, of course,
highly selective, both in terms of their representation of different types of urban areas, and in the
interpretative readings of the regeneration projects they present. As such we could be accused of
trying to make the 'facts' fit our notions. Whilst
detail is of relevance, what we wish to conclude
from these thumb-nail sketches is more general.
First, even within one overarching agenda, that of
the NUP, it is problematic to assert relatively
simple relations between cities, their trajectories
and global capital. Second, even within one city,
competing interpretations of short- versus long-
term strategies are possible, and, as in the case of
Glasgow, the rhetoric employed by key players
may confuse the issue further. Rutheiser's account
of Atlanta suggests – like Davis's account of Los
Angeles (1990) – that with meticulous trawling of
alternative sources, such as media, private papers
and council records, it is possible to explore
'behind' the public exterior to postulate different
readings of city visions and regeneration projects.
Understanding the way the volatility of capital,
city visions and trajectories impinge on each
other, therefore, is problematic, but worthy of
further exploration.

THE DISCOURSE PERSPECTIVE IN URBAN STUDIES: BEAUREGARD'S *VOICES OF DECLINE*

The above account provides a mere skeletal
outline of the manner in which the NUP agenda
is creating a context within which city authorities
are charting out development trajectories and
imagining what their potential futures might be
like. For those inspired by what might be termed
orthodox policy analysis, and by implication
arguably for most analysts both inside and
outside the academy, the NUP thesis is lived
largely as a *given*. By this we mean that it is
treated as 'the true and accurate' account of the
plight cities face, and as an objective evaluation of
the policy options that need to be considered. In
this section, we now wish to jolt the reader into a
different epistemological framework which
provides an altogether different starting point for
the interpretation of the NUP agenda. In so
doing, we are not necessarily arguing for a rejection of the NUP agenda but do wish to challenge
the apparently accepted wisdom that it is *the* only
way of viewing the city. Specifically, we wish to
reflect upon what the NUP framework might
look like if viewed as a *discourse*.

Whilst the discourse approach is now well
established in many sub-areas within the social
sciences, it seems that the boundaries around the
field of urban studies have been somewhat impermeable to its infiltration. One notable exception

to this observation has been the work of Robert Beauregard. In his ground-breaking book *Voices of Decline*, Beauregard (1993) indeed argued that he was prompted to write the work because of 'the failure or urban theory to confront in any meaningful way the issue of representation' (Beauregard, 1993: xi). He continues: "If urban theory were to be advanced, and our under-standing of cities improved, theorists would have to confront the tension between interpretive strategies and objective analyses' (p. xi). In this section, we will use *Voices of Decline* to introduce the basic tenets of the discourse approach in the sphere of urban studies. Whilst Beauregard's analysis is exceedingly rich and full of nuances, we will isolate four aspects which help to define the foundations of the discourse perspective to the study of cities.

In *Voices of Decline*, Beauregard's empirical focus is upon North American cities generally, and the larger conurbations specifically. Although the book has the sub-title 'the postwar fate of US cities', Beauregard's reflections extend back to the late nineteenth century. When reviewing a mass of writings and discussions on North American cities, including those by city officials, civil servants, newspapers and magazines, Beauregard was struck by the manner in which one could discern major sea changes in the way cities were being 'understood' through time. For the most part, the emphasis was firmly upon 'decline'. More particularly, cities were being understood in terms of the manner in which they were disinte-grating and failing to reproduce themselves successfully. However, across time, the concept of decline being deployed appeared to vary. As he investigated these various concepts of decline in more detail, Beauregard became convinced that instead of approaching writings and discussions on the fate of cities as if they reflected some underlying reality, a better approach would be to reflect upon the manner in which they were actively representing cities in certain types of ways. More forcefully, Beauregard (1993: 6) became concerned with the manner in which the mass of texts he examined were actually *creating a reality* for urban populations:

> [my] interpretive reading is premised on the notion that the discourse is not simply an objective reporting of an incontestable reality but a collection of unstable and contentious interpretations. The ways in which urban decline is represented are always prob-lematic, and although a commentator might claim privileged access to a 'self evidently solid ground of meaning', no commentator can successfully defend that exalted position.

The claim that 'language' (textbooks, political manifestos, political speeches, newspaper and magazine articles etc.) should be understood as actively *constructing cities* provides the first cornerstone notion of the discourse perspective. As Parker (1992: 4–5) notes, more generally, 'discourses do not simply describe the social world, but categorize it, they bring phenomena into sight. A strong form of the argument would be that discourses allow us to see things that are not "really" there, and that once an object has been elaborated in a discourse, it is difficult not to refer to it as if it were real.'

Second, on the basis of this point of departure, Beauregard then attempted to delineate and reflect upon the various discourses which have underpinned policy approaches to the North American city. Throughout the book, Beauregard adopts a consistent perspective on the role of discourses of urban decline in framing the policy process. According to Beauregard (1993: 5), 'the meaning of the discourse . . . can be found in the ways that it conveys practical advice about how we should respond to urban decline and mediates among the choices made available to us, the values we collectively espouse, and our ability to act.' Discourses on urban decline, therefore, both recognize different ideas about precisely what is declining and why, and produce as a result param-eters within which 'sensible' remedial measures might be undertaken. By way of illustration, Figure 26.1 attempts to summarize the main discourses recognized by Beauregard: including discourses that functioned as a prelude to the full development on the notion of decline that marked the post-war era; the turns and twists taken by the discourse since 1945; and the recent revivalist discourses which suggest that, at least for some, the fall of the largest cities in North America is being reversed. From Figure 26.1, one can begin to get a sense of the key themes associ-ated with the more important discourses, what urban problems these discourses 'identify', and what policy solutions they implicitly, and often explicitly, advocate.

Whilst the above two points try to capture the nature of discourses on the city, Beauregard explores two other key aspects of the approach concerning relationships between discourses and various power relations in the city. Third, although for the most part implicit, Beauregard's study makes use of the particular account of power outlined by Michel Foucault. For Foucault, power should be regarded as insidi-ously distributed throughout society and as such should be conceived of as a *capacity* to generate knowledge and action: 'power produces knowl-edge, and knowledge presupposes and constitutes relation of power . . . truth is already power' (Matless, 1992: 46). The value of this perspective is that it suggests that knowledge and power are two sides of the same coin and therefore have to be studied together. The very existence of a

DISCOURSE	INDICATIVE QUOTES	DEFINITION OF PROBLEMS AND FRAMING OF SOLUTIONS
Prelude to post-war decline: the progressive era 1880–1930	*The warnings were tempered by a spirit of reform and the belief that the problems increasingly concentrated in large cities could be solved. Most intellectuals, many government officials, and almost all reformers were committed to bringing order to the unruly city of industrial capitalism, yet leaving intact the basic economic relationships that created great wealth and, through exploitation, brought poverty. Civic responsibility embodied in 'scientific government' and compassionate welfare agencies was the preferred path, and optimism the guiding light. (p. 59)*	Definition of problem: Overly fast growth of cities leads to overcrowded slums. Moral corruption both within the state and civil society make cities 'dangerous' and 'sinful' places and a threat to civilization Framing of solution: Modernist faith in the capacity of the state to control and channel growth. Scientific and organizational advances of corporate capital help the engineering of better cities
Prelude to post-war decline: a temporary problem 1920–1950	*The discourse on cities that occurred prior to the Second World War then, had none of the pervasive pessimism to which people in the post-war period would become accustomed. Few commentators in the pre-war discourse doubted the potential for subsequent reinvigoration. Certainly many had been fearful of the effects of decentralization and the twin problems of slums and blight in the city's central areas, but they also, as had those in the earlier decades of the twentieth century, believed in the powers of human intervention to resist such decay. Urban growth had been interrupted, not deflected, and during the late 1940s commentators were generally confident that a return to economic prosperity would preserve the city. (p. 99)*	Definition of problem: Urban growth had been temporarily restrained by the effects of the great depression and the Second World War Framing of Solution: Post-depression and post-Second World War recovery would ensure the continuation of healthy urban growth
Escalating downward: potentially irreversible problems 1945–1960	*As the nature of urban decline changed, and its pace quickened, its perception was transformed, and the optimism of former years weakened. One commentator . . . in one of the earliest appearances of the phrase, believed that cities were 'in throes of urban crises'. The city was increasingly viewed as incapable of overcoming the endemic problems of blight, slums, and urban sprawl. This made urban decline no longer a temporary aberration, brought about by years of depression and war, but a chronic national condition. (p. 15)*	Definition of problem: City was suburbanizing itself to death. Out-migration was selective, involving mainly young middle-class families. In-migration of negroes from southern states to older industrial conurbations in the north brought new class and racial problems Framing of solution: National debate centred upon whether to rebuild existing cities, construct new suburban cities and allow existing cities to pass into a less central role, or to abandon big cities altogether
From one crisis to the next: the racial roots of urban decline 1960–mid-1970s	*The discourse shifted abruptly from the physical state of the city . . . The slum problem was transferred into the ghetto problem . . . Urban decline was no longer a list of curable ills but a fundamental contradiction. Negroes became the scapegoats for urban decline and linked to race, urban decline could not be overcome by simply adding capital and households to the city. Institutions and behaviours transcending the spatial bounds of the city had to change and change in ways that redirected decades of prejudice, discrimination, and oppression. (pp. 64–5)*	Definition of problem: Negroes made the slums and the slums could not be cured without dealing with the problems of racially generated poverty. Towards the end of the period, the negro problem was linked to the fiscal crises faced by so many cities which came more to the foreground Framing of solutions: Fundamental employment and civil rights needed to be established to overturn racial discrimination. Then, with the race problem solved, the slum or ghetto problem might also be solved
Rising from the ashes: glimmer of a future 1970–1990s	*This feverish revival produced a peculiarly 1980s representation of the city. On the one hand, revitalization displaced decline as the central theme of urban commentary. On the other hand, renewed investment and the cessation of massive population loss, not to mention a rediscovered in-migration of middle-class households, brought to the fore many long forgotten consequences of growth. Many commentators, additionally, found it difficult to overlook an increasingly visible, bothersome, and deepening bifurcation of the city into rich and poor. A fascination with ranking cities on their 'livability', however, blurred widening inequalities within cities, and thereby rescued revival from its critics. During the 1980s, critical commentary, no matter how pessimistic, simply could not dislodge the discourse from its new found hopefulness. (p. 256)*	Definition of problem: The quality of life available in a city, its leisure and cultural amenities, and its overall image, were vital in attracting business and people back to it. The timing was right to represent the city to the world as a place of life and vitality Framing of solution: City marketing exercises were to represent the city as a place in which to consume and be entertained. The city could be shown to offer new lifestyles which were attractive to those bored by life in suburbia

Figure 26.1 Beauregard's framing of 'post-war' discourses of urban decline in the United States (constructed from Beauregard, 1993)

discourse leads one to question the sources of power that have both produced and reproduced that discourse. In turn, accessing the sources of power that lurk behind certain 'ways of seeing' the world often involves asking the question: which categories of person benefit, and which lose, from the existence of these particular types of discourse? Often, the sources of power involved are institutional. Analyses of discourses, therefore, proceed by interrogating the institutional sources that benefit from the circulation of particular 'ways of seeing'.

Additionally, Beauregard draws attention to the often conscious efforts made by those institutions empowered by the discourse to promote their 'ways of thinking' and therefore to suffocate the development of alternative representations. Such institutions might indeed engage in propaganda exercises designed to promote and naturalize the discourse that they so rely upon. This act of denying contemplation of alternatives 'smothers the actual causes of our discontent. It stifles an awareness of how cities might be different. As a result, we are *unable to imagine* cities where shared prosperity, democratic engagement, and social tolerance are the norms and not the exceptions. To allow cities to be the discursive sites for society's contradictions is to be imprisoned in the cynicism of urban decline' (Beauregard, 1993: 324, emphasis added).

Despite all the twists and turns it has taken in its historical evolution, the discourse of urban decline may, at one level, be approached in terms of the manner in which it functions to legitimate capitalist development. One aspiration of the discourse, it would seem, is a search for solutions to uneven development both inside and outside cities which do not fundamentally question or thereby threaten the basic capitalist processes. As Beauregard argues (1993: 306), 'if people cannot be convinced that decline is inevitable, at least they might come to believe that it is tolerable; if not natural, at least reversible; if not curable, at least isolated and contained.' This is an intriguing argument. It suggests that even when drawing attention to capitalism's worst eyesores, the discourse should be approached as reproducing capitalist social relations, by virtue of the fact that it offers solutions 'within the system'. For Beauregard, nevertheless, discourses of urban decline are not only concerned with the power relations embedded in the class structure of capitalist cities. Beyond questions of social class, discourses have to be read in terms of their role in shaping other social relations, not least those based upon ethnicity and gender, and to these one might add age and disability, among others.

Fourthly, and finally, of course the hegemonic status of any discourse is never complete, and despite the best efforts of the institutional support apparatus which grows up around any discourse, there always exists the possibility that alternative representations will grow in persuasiveness. Indeed, when 'cities' are read from social positions at the margins of the dominant discourse, it is possible that representations will become a source of conflict. According to Jackson (1991: 200), such conflict should be thought of as constituting a *cultural politics*, defined as the 'domain in which meanings are constructed and negotiated [and] where relations of dominance and subordination are defined and contested'. The rise of a cultural politics over the way cities are represented constitutes a most fundamental and serious challenge to the established 'order'. Not only is dissent towards this order made visible but the very cultural foundations which support and legitimate its existence are questioned. It is indeed only during such a crisis of representation that the deeply naturalized discourses which buttress relations of domination become disturbed. The interrogation of fundamental concepts which then follows may be read as a moment of emancipation for the oppressed, at least at the level of consciousness.

RE-READING THE NUP AGENDA

For the most part Beauregard's focus has been upon discourses of urban decline. Perhaps because of this emphasis, we feel that his attempt to read recent revivalist agendas from the vantage of the discourse perspective remains underdeveloped. In particular, Beauregard's reading fails to foreground the central importance of the *metaphor* of *commodification* in recent and more optimistic statements on the city. In this final section, we therefore wish to turn attention to the discourse of revival and to present a fuller exposition of what a discourse reading of the NUP agenda might look like.

On the basis of the approach sketched out by Beauregard, a discourse perspective would approach the NUP agenda not as an objective analysis of the effects of globalization and the new international division of labour on the city and as an important context in relation to which cities have to formulate imagined futures. Instead, the approach would see this agenda as a *social construction*. In other words, as a discourse, the NUP represents the city in certain types of ways, therein providing an analytic logic in which problems are defined and policy measures framed. Place marketing and the aspiration to assume certain roles in the emerging global division of labour come to be seen as 'natural', 'obvious', 'self-evident' and 'legitimate'. Treated as a discourse, analysts would then seek to uncover the sources of institutional power that lie behind it, and the propaganda devices used in its production and reproduction. Finally, assuming that hegemonic status for the discourse is never total, the perspective would call upon research to search for the possibility of a cultural politics in which alternative readings of the city may succeed, however limited, in threatening the discourses' underlying assumptions and logic.

The NUP Agenda as a Discourse

What kind of social construction of the city does the NUP define? In trying to approach the NUP agenda as a discourse, it is useful to begin by locating it within a wider discursive field, often labelled somewhat haphazardly as 'Thatcherism' or 'Reaganism'. This wider discursive field involves ideological shifts within the state more generally. It has been dominated by a 'new right agenda', which champions the importance of the market system and of the necessity of freeing up private agents from state regulations. Given the importance of the local state as an ideological battleground (Duncan and Goodwin, 1988) and the political importance of the problems of inner city and urban decline, it was perhaps inevitable that new right thinking would establish an agenda to be applied to cities (Barnekov et al., 1989; Peck, 1995; Peck and Tickell, 1994).

Central to the NUP agenda is the representation of cities as *commodities*. The metaphor of 'commodification' is used to construct cities as akin to any other good which sits on a supermarket shelf. Places are represented as existing in an open competition with one another to lure private investment. They must act in subservient ways to consumers, and deploy their resources to make them more competitive in the open market. Place marketing itself is the embodiment of new right thinking. Mobile capital and tourists are the highly flexible consumers, places are the product, and local institutions and organizations are the manufacturers, marketeers and retailers. The key move in the NUP agenda, therefore, is the representation of place as a commodity in a highly competitive market place.

Given that the commodification metaphor is the central representation which underpins the NUP agenda, further strands of thought and argument about the city as a commodity therein follow. If one returns to some of the ideas outlined in section 1, once the metaphor of the commodity is revealed, it then becomes apparent that many of the ideas contained in the NUP agenda can be read as derivative of this basic notion. The following brief examples illustrate the most pertinent ideas in this context.

1 Like brands in most spheres of the market today, 'consumer' loyalty to places is never secure, and consumers are indeed becoming increasingly subject to whimsical shifts in taste. Places, therefore, are extremely vulnerable to the fast turnover times that mark the consumption preferences and habits of mobile capital, tourists and a range of other consumers. Places need to be aware that their futures are being determined by laws of the market place which can be likened to those that prevail in the case of more 'normal' commodities, and need to plan for such insecurity (Harvey, 1989a).

2 Commodities should be marketed, not merely advertised. For place marketeers, this is an important distinction. To promote a place simply on the basis of its existing attributes amounts to little more than an exercise in advertisement. 'Marketing' cities in contrast encompasses the much larger idea that commodities are *produced* to meet market demand. A survey of the market precedes the production process itself, and goods are produced in the way a priori consumer demand dictates. The notion of place marketing, therefore, suggests that cities not only should sell what they already are, but must change to become what the consumer requires. Places, thus, need to be produced to the specifications of mobile capital, tourists, etc. (Ashworth and Voogd, 1990).

3 As markets enter maturity, product specialization occurs. Firms find their own niche in the market and attempt to serve only a sub-section of the whole population of consumers. Places, likewise, are representing themselves as offering particularly suitable locations for different economic sectors, for units at different stages in the production process, and for tourists with particular tastes. As it evolves, place marketing is increasingly playing the game of the specialization of the commodity (Ashworth and Voogd, 1990).

4 Many commentators have noted that commodities are increasingly being traded not according to their functionality, but according to their image. That is, consumer taste would appear to be sensitive to the aesthetic dimension of commodities as well as their use values (Featherstone, 1988). Places would seem to be no different. Not only are they creating the hardware (infrastructure, tax arrangements, labour laws etc.) which will attract inward investment, but they are also, as a central part of their work, using image building techniques to produce soft infrastructure to create more conducive aesthetic environments. This is clearly in recognition of the fact that consumers favour not only the functional aspects of the city, but also the gift wrapping that accompanies it (Boyle, 1997).

5 Just as corporate secrecy is vital in securing market edge in the more orthodox commodity

markets, so too is corporate secrecy a central aspect in the commodification of place (Wood, 1993). Localities now jealously guard market research on the latest locational requirements of mobile capital and tourist consumption tastes, and information on projects to upgrade the local hard and soft infrastructure is treated as confidential. And, as with inter-firm rivalry, inter-city competition generates its own ways around this problem; including locating staff moles in competitors' cities, and recruiting staff from these cities.

6 Finally, whilst the essence of new right thinking champions the supremacy of market forces, like other market places, there is considerable debate between cities over the extent to which the state is subsidizing one city at the expense of others. Just as differential state intervention in areas such as coal production, farming and fisheries, creates great inter-firm rivalries, so too in the case of place marketing claims and counterclaims over the differential benefits ascribed to different localities by central government abound. The commodification of place, therefore, has not been immune from the language of subsidy and the vocabularies which have developed in other sectors of the economy over unequal state intervention in markets.

In summary, the NUP agenda can be seen to create a particular root metaphor of the city which therein informs how the city is to be understood, and the kinds of policy interventions required to 'correct' any problems it experiences (Sadler, 1993). This metaphor treats the city as a commodity. Its survival is contingent on competition with other commodities (cities) for a share of the market. To the extent that cities are rendered marginal by the wider political economy, the framing of the problem in this way leads to the conclusion that salvation will only come through the market. And, only through place marketing and the transformation of the city into something the consumer wants, will the city survive.

Institutional Power and the NUP Discourse

Beyond examining the NUP agenda as a 'way of seeing' cities, discourse analysts are interested in the sources of institutional power that both produce and reproduce the discourse. Of course, in so doing, these institutions legitimate their own activities and secure advantages for themselves. In short, the discourse perspective proceeds by asking the question: which types of organization benefit from the circulation of this 'way of thinking'? By seeking to examine the various *communities of interest* which lie behind the representation of cities as commodities, the discourse

perspective promotes the idea that far from painting a 'true' or 'real' or 'objective' picture of the world, representations are best thought of as comprising 'situated knowledge', that is, knowledge rooted in the social positions occupied by a variety of institutions and organizations. If power and knowledge are two sides of the same coin, then the key question to be asked is: what sources of power lie behind the idea that cities are fundamentally market places for consumer groups?

Clearly, if one thinks of the capitalist world in general, the democratically elected local state would appear to be the main institution that draws authority from and gives authority to the NUP discourse. It is this authority that can be found most often championing the discourse and defending its integrity and plausibility. Further, it is this institution that can be seen to benefit in both fiscal and electoral terms from the boosterism which forms a central element of the discourse. Nevertheless, depending upon which region of the capitalist world is being examined, there exist at least two other categories of institution which draw upon the discourse as a source of legitimation for their involvements in the governance of the city (Harvey, 1989a). First, in some nations, responsibility for formulating local economic development strategies falls primarily to a number of central government quasi-autonomous non-governmental organizations (quangos) which often bypass the democratic local state. Particularly in the United Kingdom (Peck, 1995; Peck and Tickell, 1994), it would appear that the local state has steadily become replaced by a network of central government bodies which have been 'parachuted' into cities without the consent of local people. These bodies have arrested many of the functions previously controlled by the democratic local state. These bodies include and are exemplified by the Urban Development Corporations, Training and Enterprise Councils and Local Enterprise Companies that now plan so much of the economic development work undertaken in British cities. Much of this work both draws upon and reproduces the NUP discourse.

A second category of interest group served by the NUP discourse is that of the variety of forms of local capital which benefit from this particular type of local accumulation strategy. Although not the only capital to benefit, North American studies, in particular, point to the disproportionate economic rewards to accrue to factions of *locally dependent capital*. Cox and Mair (1988) define local dependency to mean the degree to which certain factions of capital are wedded to the overall economic health of a locality. This weddedness may derive from the fact that substantial amounts of fixed capital have been invested or that the firm has a number of non-substitutional exchange linkages (with the local market, local labour, or local suppliers). Among the more important locally dependent capitals to benefit from the NUP discourse are property developers, banks, utility companies and local newspapers (Cox and Mair, 1988).

That such fractions of capital benefit can be observed from the manner in which they themselves actively promote 'readings' of the plight of cities which fall within the terms of reference of the NUP discourse. This would seem to be particularly important in the North American context. Logan and Molotch (1987) argue that a distinctive feature of the governance of North American cities is the manner in which locally dependent capitals form into coalitions to articulate strategic policy visions for the future of their cities. These coalitions, pressing for development which serves their interests, lobby the local state to follow policies for growth. Inevitably, to a large degree this entails an interpretation of the city consistent with the NUP discourse. Logan and Molotch (1987) characterize these coalitions as '*growth machines*'. Whilst *growth machines* do exist in other national contexts, albeit in a modified form, it would appear that their spontaneous formation and sheer power is a distinctively North American phenomenon (Rogerson and Boyle, 2000; Wood, 1996). Consequently, locally dependent fractions of capital may represent both a source of the discourse, and a beneficiary.

According to literature produced within the political economy tradition, therefore, the hegemonic status of the NUP discourse can be argued to be a reflection of the power the local state, quangos and growth coalitions have in promoting their own reading of the city. It is primarily the interest of these bodies that the representation of the city as a commodity serves. This is perhaps most evident in recent literature on the manner in which these bodies are active in propaganda exercises intent on selling their reading of events to the local population more broadly. One of the most fertile research areas within literature on the New Urban Politics has been that which seeks to examine the role of civic boosterism in the politics of local economic development (Cox and Mair, 1988). It is in this context that analysts have sought to examine the role of place marketing exercises *within* the city. In the account offered in the first section of this chapter, it was suggested that place marketing is viewed in the NUP agenda as an instrumental tool in making localities more competitive when prospecting for capital, state functions, consumer expenditure and further prestige projects themselves. At their most ostentatious, however, place marketing exercises can have effects *within* cities functioning as a form of civic boosterism or, as

Boyle (1997) calls them elsewhere, Urban Propaganda Projects. Here the manufacturing of a form of local consciousness which minimizes conflicts and galvanizes local support in line with the NUP agenda helps secure legitimacy for the agenda. As Philo and Kearns point out (1993: 3):

> There exists a social logic in that the self promotion of places may be operating as a subtle form of socialization, designed to convince local people, many of whom will be disadvantaged (by both globalization processes and local economic development projects themselves) that they are important cogs in a successful community and that all sorts of 'good things' are really being done on their behalf.

Within the political economy literature, thus, the NUP discourse is argued to be a central ideological tool used by the capitalist state, and by locally dependent capital, to secure local support and legitimacy for actions primarily designed to serve their interests. Place marketing has to be understood as playing a dual role in this context: on the one hand acting as a tool utilized in the strategy; on the other hand, in the guise of local boosterism, functioning as an ideological lubricant. Of all the discourses on the city examined by Beauregard (1993), there can surely be no other which makes such an ostentatious display to secure legitimation and self-fulfilment. Of course, this claim does not assume that every local group will be seduced by the discourse. It is towards those instances when conflict has broken out over the NUP agenda therefore, that attention now finally turns.

The NUP Discourse and Social Conflict in the City

To the extent that the NUP discourse is changing the way in which cities are being conceptualized, and therein governed, it would be fair to say that it has in turn become a central focal point within which new types of social conflicts are emerging. Given the emphasis within the discourse reading on the notion that all knowledge reflects established positions of social power, the rise of altogether new types of urban conflict not surprisingly emerges as an area of central concern within the approach. In this final section we wish to consider briefly, from the vantage of the discourse perspective, how the various conflicts that are emerging might be understood.

In so doing, a typology of potential conflicts is presented. A review of studies undertaken to date suggests that conflict may revolve around three major issues. These can be ordered in terms of how 'fundamental' they are. First, the NUP agenda has material consequences for different population groups within cities and therefore has prompted a series of conflicts over redistributive concerns. At the simplest level therefore, it can be said to have triggered a *politics of redistributive justice*. For example, there is a growing literature pointing to the existence of protests which focus upon the extent to which local economic strategies are diverting resources away from local welfare provision and towards the assistance of selective capitalist interests (Kenny, 1995; Rutheiser, 1996). Harvey calculates that this amounts to nothing less than a transition from urban managerialism to urban entrepreneurialism (Harvey, 1989a; see also Cox and Mair, 1988), and points to the macro-economic shifts across the economy from labour to capital which it brings. Inter-locality competition in this regard is seen to induce a redistribution of state expenditure away from labour and towards capital, as yet more inducements have to be offered when prospecting for business. Since all localities are involved, place marketing has the potential to become a zero-sum game, with each locality having to divert ever greater resources simply to stand still. Related to such protests, there may also arise a class based politics rooted in the nature of the economic strategy pursued. The pursuit of a post-industrial strategy may fail on a number of counts: (a) in the context of zero-sum competition for investment, the volume of new jobs created may not be sufficient to replace old ones lost: (b) the nature of jobs created may entail poor, low paid, menial, part-time service sector employment. There may be opposition, therefore, from groups for whom the trickle-down philosophy carries little weight (Imrie and Thomas, 1993; Leitner and Garner, 1993).

Second, at a more advanced level, by commodifying place, the NUP discourse fails to acknowledge the various other non-market relationships that exist between populations and cities. Particularly, sections of the population for whom the city has *use values* and not merely *exchange values*, can often prompt what we will refer here to as a *cultural politics of place*. A sophisticated account tracking the development of conflicts over how places are to be represented – as commodities with exchange values or living environments with use values – has been that provided by Logan and Molotch (Logan and Molotch, 1987). Logan and Molotch identify a variety of ways in which place performs a central role in people's efforts to make a life, ranging from concrete and practical issues like people's use of local facilities (shops, playing fields, schools etc.), to more sophisticated notions such as the ways in which social relations, situated in places, provide people with a sense of civic pride and identity. The insensitive ways in which the commodification of places threatens and undermines these all-important but fragile use values breeds tensions and leads to conflicts.

An example of the way in which the NUP discourse and place marketing have led to the development of a cultural politics of place, can be seen in Boyle and Hughes's (1991, 1994) work on Glasgow (see also Hewison's (1987) more general study). In their study of Glasgow's role as European City of Culture 1990, Boyle and Hughes sought to demonstrate how the event was hijacked by the NUP agenda and used as an exercise in place marketing for economic development. The result was that many groups in the city complained that the real, organic culture and heritage of the city was being excluded in favour of a sanitized version of consumer culture for the comfort of incoming business people and tourists. This led in turn to a heated and controversial debate on what the 'real' Glasgow was and whether its socialist past was being disrespectfully cast to one side. Whilst city authorities conceived of all aspects of the city as constituting a resource base to be used to sell their commodity (including the city's culture and heritage), for others, civic identity represented a use value that could not be disposed of so cheaply.

Finally, at the most fundamental level, the vigour with which the NUP agenda promotes the view that there is only one response to the globalization of capital – namely the commodification of place – suffocates groups with more imaginative responses. Opposition might arise therefore, challenging the view that there are no alternatives for cities. Alternative representations of what globalization means for cities might give birth, therefore, to a *cultural politics of the NUP agenda itself*. Such a politics would foreground the subversive potential of representations which create alternative responses to globalization forces. By constructing a different and more imaginative account of the plight cities face, these representations take to task ways of thinking which the NUP simply naturalizes. Given the hegemonic status of the basic thesis underpinning the NUP, it has to be said that there are few examples of conflicts at this level. Nevertheless, two examples can be cited.

Wills (1995) has argued that there exists the possibility that labour movements might in future play a more active role in combating the globalization of capital and the regressive social consequences inter-city competition generates. Her focus was upon the way in which companies with branches straddled across the world, attempt to play off workforces at each branch by threatening to transfer work to another site unless the local workforce agrees to changing labour conditions and practices. According to Wills, labour movements need to organize more effectively across national boundaries. If workforces presented a united front and were prepared to agree on a basic minimum social standard, the capacity of firms to make such threats would be greatly reduced. Pan-national coordination of the labour movement, therefore, represents a challenge to the assumption that the only solution to globalization is inter-city competition for the scraps which fall off the capitalist table.

A second example which challenges the basic representation of the world created by the NUP discourse can be found in the growth of LETS – the abbreviation Lee (1996) gives to local exchange employment and trading systems – in Australian, New Zealand, British, United States and Canadian cities. To be sure the number of people involved in LETS is miniscule. The importance of the LETS scheme, however, lies in its promotion of the idea that local communities can shield themselves from the ravages of global capitalism by formulating alternative systems of work and reward. LETS represent attempts by local communities, which might not be able to purchase work formally through the capitalist economy, to de-couple themselves from this economy and to create local forms of social and economic organization which facilitate the 'purchase' of work outside the normal concept of a currency (Pacione, 1997). Those participating in LETS give and receive all kinds of goods and services through reciprocal arrangements which do not involve a transfer of money.

Such a typology of the conflicts surrounding the NUP discourse provides a framework to guide future research. It should be noted, however, that the apparent ranking of forms of conflict from the least to the most fundamental and 'advanced' is not meant to carry any evaluative content. Each type of conflict is important in its own right. The construction of the categories in this way is meant to capture how they might be seen within a discourse reading. Conflicts over redistributive concerns do not challenge the NUP discourse directly. They point to the effects or consequences of the discourse in practice. Conceivably, redistributive concerns might be addressed within the terms of the discourse; for instance, by tinkering to ensure the proper workings of the trickle-down mechanism, or by attempting to secure a different role in the emerging global division of labour. Conflicts at this level, therefore, do not challenge the cognitive framework espoused by the discourse, but instead encourage technical fixes within its terms of reference. The second and third types of conflict, in contrast, do entail challenges to the representational qualities of the discourse: the second focusing more narrowly upon the metaphor of the commodification of place; the latter seeking to question the entire way of seeing of the discourse itself. As such, these latter alternative political challenges are arguably more subversive since they attempt to undermine the NUP at the level of representation.

CONCLUSION

In this chapter we have attempted to explore the way in which urban studies might profitably develop a new framework of analysis based upon the key concepts of power, discourse and city trajectories. We have argued that a discourse approach moves beyond orthodox policy analysis of the city in a number of key respects. Whilst such orthodoxy has tended to focus on the formulation, implementation and impact of different policies on the city, a discourse reading focuses less on policy practice and more on the cultural constructions of the city which policy naturalizes. Furthermore, it calls attention to the notion that all cultural constructions are underpinned by sources of institutional power. Instead of studying policy in action, therefore, a discourse reading foregrounds the ways of seeing cities assumed in policies, and the idea that ways of seeing always derive from an underlying position of power. Focusing primarily on the NUP agenda, we have tried to exemplify what such an approach might look like. Whilst we have raised a variety of concerns, we will conclude by drawing attention to three aspects of a discourse reading of the NUP which require development.

First, whilst Beauregard represents one of the few studies that attempt to trace the genealogy of the various discourses of the city, the whole question of writing historical geographies of the NUP discourse – and discourses of the city more generally – remains a fertile, unexplored area of research. For instance, much work still requires to be done examining the locations of origin of the NUP discourse and the geography of its diffusion. To what extent, for instance, did the discourse derive from a select number of US cities such as Baltimore, Minneapolis and Pittsburgh, diffusing, first, through the de-industrializing cities of North America and then on into the UK, through cities such as Glasgow and Sheffield? Further, it is likely that the evolution of the NUP discourse has been tortuous, encountering a variety of different local conditions and oppositions and, at least in the innovative stages, struggling to secure hegemonic status. As such, it may also be necessary to write the genealogy of the NUP discourse with reference to the historical geography of different forms of opposition in particular locations.

Secondly, in tracing out the 'key players' in particular cities who have championed the NUP agenda, and shaped its rise to fame, it would be fair to say that work to date has been biased to the political economy perspective, and as such, has approached this task within the terms of a materialist framework. That is, emphasis has been placed upon large political bodies and certain fractions of capital as the primary producers and beneficiaries of the agenda. Whilst such work has contributed greatly to our understanding of the capitalist power relations reproduced by the NUP discourse, it has done so at the expense of a thorough analysis of the many other power relations with which the discourse interacts; such as those formed around gender, ethnicity, age and disability. This is not to say that all research has ignored the importance of other power relations. It is simply to state that the origins of the NUP are taken most often to signify the rise to power of certain fractions of the state and certain types of capital. Genealogies, therefore, require more innovative ways of studying the sources of power lying beneath the NUP discourse.

Finally, given the status of the NUP discourse as *the* agenda cities have to work with, future work needs to be careful in not assuming that the discourse will lead to certain concrete outcomes in different types of cities. As noted in the first section of this chapter, some recent studies of the NUP have worked with the explicit assumption that in engaging with global capital, strategic trajectories open to cities are conditioned by their differential abilities to master global capital. For instance, there is an acceptance that some cities, such as the global cities, can construct a coherent 'blueprint' based upon a mastering of capital's volatility and flexibility, whilst for other cities strategies are, at best, short-term and opportunistic or fragmentary. The connections between the NUP discourse and the strategies pursued by different cities is seen to be more contingent than is found in this narrow reading, and is thus worthy of future research.

BIBLIOGRAPHY

Allen, F. (1996) *Atlanta Rising. The Invention of an International City 1946–1996*. Atlanta: Longstreet Press.

Allen, J. and Massey, D. (eds) (1988) *The Economy in Question*. London: Sage.

Ashworth, G. and Voogd, H. (1990) *Selling the City*. London: Belhaven.

Barnekov, T., Boyle, R. and Rich, D. (1989) *Privatism and Urban Policy in Britain and the United States*. Oxford: Oxford University Press.

Bayor, R. (1996) *Race and the Shaping of Twentieth-Century Atlanta*. Chapel Hill, NC: University of North Carolina Press.

Beauregard, R. (1993) *Voices of Decline: the Postwar Fate of US Cities*. Cambridge, MA: Blackwell.

Blair, J. (1995) *Local Economic Development: Analysis and Practice*. Thousand Oaks, CA: Sage.

Blakely, E. (1989) *Planning Local Economic Development: Theory and Practice*. Newbury Park, CA: Sage.

Booth, P. and Boyle, R. (1993) 'See Glasgow, see culture' in F. Bianchini and M. Parkinson (eds), *Cultural Policy and Urban Regeneration*. Manchester: Manchester University Press. pp. 21–47.

Boyle, M. (1997) 'Civic boosterism in the politics of local economic development – "institutional position" and "strategic orientation" in the consumption of hallmark events', *Environment and Planning A*, 29: 1975–97.

Boyle, M. and Hughes, G. (1991) 'The politics of the representation of "the real": discourses from the left on Glasgow's role as European City of Culture, 1990', *Area*, 23: 217–28.

Boyle, M. and Hughes, G. (1994) 'The politics of urban entrepreneurialism in Glasgow', *Geoforum*, 25: 453–70.

Boyle, R. (1993) 'Changing partners: the experience of urban economic policy in west central Scotland', *Urban Studies*, 30: 309–24.

Castells, M. (1989) *The Informational City*. Cambridge, MA: Blackwell.

Cheshire, P., Carbonaro, G. and Hay, D. (1986) 'Problems of urban decline and growth in EEC countries', *Urban Studies*, 23: 131–50.

Cox, K. (1993) 'The local and the global in the new urban politics: a critical review', *Environment and Planning C: Society and Space*, 11: 433–48.

Cox, K. (1995) 'Globalisation, competition and the politics of local economic development', *Urban Studies*, 32: 213–24.

Cox, K. and Mair, A. (1988) 'Locality and community in the politics of local economic development', *Annals, Association of American Geographers*, 78: 307–25.

Cox, K. and Wood, K. (1994) 'Local government and local economic development in the United States', *Regional Studies*, 28: 640–5.

Damer, S. (1990) *Glasgow: Going for a Song*. London: Lawrence and Wishart.

Davis, M (1990) *City of Quartz: Excavating the Future in Los Angeles*. Verso: London.

Duffy, H. (1995) *Competitive Cities: Succeeding in the Global Economy*. London: Spon.

Duncan, S. and Goodwin, M. (1988) *The Local State and Uneven Development*. Cambridge: Polity Press.

Featherstone, M. (1988) *Postmodernism*. London: Sage.

Friedmann, J. (1986) 'The world city hypothesis', *Development and Change*, 17, 69–84.

Friedmann, J. and Wolff, G. (1982) 'World city formation: an agenda for research and action', *International Journal of Urban and Regional Research*, 6: 309–44.

Gertler, M. (1992) 'Flexibility revisited: districts, nation-states and the forces of production', *Transactions, Institute of British Geographers*, 17, 259–78.

Gibb, A. (1983) *Glasgow – the Making of a City*. London: Croom Helm.

Harvey, D. (1989a) 'From managerialism to entrepreneurialism: the transformation in urban governance in late capitalism', *Geografiska Annaler*, 71B: 3–17.

Harvey, D. (1989b) *The Postmodern Condition*. Oxford: Blackwell.

Hewison, R. (1987) *The Heritage Industry*. London: Methuen.

Imrie, R. and Thomas, H. (1993) 'The limits of property-led development', *Environment and Planning C: Society and Space*, 11: 87–102.

Jackson, P. (1991) 'The cultural politics of masculinity: towards a social geography', *Transactions, Institute of British Geographers*, 16: 199–213.

Keating, M. (1988) *The City that Refused to Die – Glasgow: the Politics of Urban Regeneration*. Aberdeen: Aberdeen University Press.

Kenny, J. (1995) 'Making Milwaukee famous: cultural capital, urban image, and the politics of place', *Urban Geography*, 16: 440–58.

Lee, R. (1996) 'Moral money? LETS and the social construction of local economic geographies in southeast England', *Environment and Planning A*, 28: 1377–94.

Leitner, H. and Garner, M. (1993) 'The limits to local initiatives: a reassessment of urban entrepreneurialism for development', *Urban Geography*, 14: 57–77.

Lever, W.F. (1990) 'The city of Glasgow – the urban economy of a city in transformation', in S. Grundmann and W.F. Lever (eds), *Social and Economic Change in Metropolitan Areas*. Vienna: ECRDSS. pp. 133–62.

Logan, J. and Molotch, H. (1987) *Urban Fortunes: the Political Economy of Place*. Berkeley, CA: University of California Press.

Matless, D. (1992) 'An occasion for geography: landscape, representation and Foucault corpus', *Environment and Planning D*, 10: 41–56.

Oakley, C. (1965) *The Second City*. Glasgow: Blackie.

Pacione, M. (1997) 'Local exchange trading systems as a response to the globalisation of capitalism', *Urban Studies*, 34: 1179–99.

Parker, I. (1992) *Discourse, Dynamics: Critical Analysis for Social and Individual Psychology*. London: Routledge.

Peck, J. (1995) 'Moving and shaking: business elites, state localism and urban privatism', *Progress in Human Geography*, 19: 16–46.

Peck, J. and Tickell, A. (1994) 'Jungle law breaks out: neoliberalism and global-local disorder', *Area*, 26: 317–26.

Philo, C. and Kearns, G. (1993) 'Culture, history, capital: a critical introduction to the selling of cities', in G. Kearns and C. Philo (eds), *Selling the City as Cultural Capital*. Oxford: Pergamon. pp. 1–32.

Rogerson, R. and Boyle, M. (1998) 'Reluctant mavericks: towards an explanation of the limited interest of capitalist firms in shaping Glasgow's inward investment strategy', *Scottish Geographical Magazine*, 114:109–19.

Rogerson, R. and Boyle, M. (2000) 'Property, politics and the neo-liberal revolution in urban Scotland', *Progress in Planning*, 54:133–96.

Rutheiser, C. (1996) *Imagineering Atlanta*. London: Verso.

Sadler, D. (1993) 'Place marketing, competitive places and the construction of hegemony in Britain in the 1980s', in G. Kearns and C. Philo (eds), *Selling the City as Cultural Capital*. Oxford: Pergamon. pp. 175–92.

Sassen, S. (1991) *The Global City: New York, London, Tokyo*. Princeton, NJ: Princeton University Press.

Schoenberger, E. (1988) 'From Fordism to flexible accumulation: technology, competitive strategies and international location', *Environment and Planning D: Society and Space*, 6: 245–62.

Wills, J. (1995) 'Every cloud has a silver lining: globalisation and potential strategies for the labour movement'. Paper presented at workshop on Locality and Community, University of Reading, 12th July.

Wood, A. (1993) 'Organising for local economic development: local economic development networks and prospecting for industry' *Environment and Planning A*: 25: 1649–61.

Wood, A. (1996) 'Analysing the politics of local economic development: making sense of cross-national convergence', *Urban Studies*, 33: 1281–95.

Zukin, S. (1995) *The Cultures of Cities*. Oxford: Blackwell.

Part VII

CITIES IN TRANSITION

In the closing decades of the last century two changes dominated the world arena, the rapidly deepening processes of globalization and the ending of the Cold War. Both were linked to the increasing ascendancy of neo-liberalism, and both are closely linked to the different trajectories being enjoyed (or endured) by the different cities comprising what is an increasingly integrated world urban hierarchy. Subsumed within globalization processes are several major trends, including the growing significance of capital, in particular international capital, as the motor of economic development (accompanied by the relative decline in the importance of raw materials and manufacturing), rapid technological advances, and the deepening of the debt crisis among resource-exporting countries in the developing world. The ending of the Cold War signalled the demise of the command economies of the Soviet Union and other members of the Second World, and together with the economic reforms in China, the spread of marketization to virtually global coverage.

As with the spread of economic development beforehand the influence of globalization processes is uneven. Unevenness is apparent at different scales, between the regional economic blocs comprising the world economy and between cities within them. More favoured by changes in global production are the three northern regional blocs of North America, the European Union and Pacific Asia. Least favoured is Africa, particularly sub-Saharan Africa, which most accounts of globalization argue as being largely marginalized. Other regions, including Latin America and the post-Socialist economies, occupy intermediate, but far from similar, positions. Within these regions and precisely because globalization processes have undermined the significance of national borders, and economies, the trajectories of different cities vary. Declining cities characterize the affluent northern economies, just as within the more marginalized regions, individual cities, more connected within the world economy, constitute islands of relative growth.

Accompanying these trends has been the growth of the mega-city, defined as exceeding 8 million. While there were several mega-cities earlier in the twentieth century in the First World, their number had expanded rapidly by the end of it. Most are outside the First World, concentrated in Pacific and South Asia in particular, with 'outliers' in Central and South America.

While globalization is both a highly complex and differentiated set of processes, it is tempting to generalize what sorts of cities are emergent within the global arena. Ruble et al. (1996) suggest five types. First, are the post-industrial cities, dominated by the successful industrial cities of the past, particularly those of North America and Western Europe. In these early industrial nations contemporary urban renaissance is itself uneven, old industrial regions being accompanied by rapid urban growth often in greenfield locations. Rapidly emergent cities are represented by the new age boomtowns, those which have been able to lock into the fast growing clusters, notably financial services and new technology industries. They need not necessarily be large (Cambridge, England or Montpellier in France, for example), though many, such as Houston, Kuala Lumpur and Hong Kong are significant within their national urban hierarchies as well as globally. Third, the post-socialist cities form a distinct, if internally differentiated category, positioned within economies which are confronting the challenges of global competition, the dismantling of state control and spread of privatization, and, as in Russia, the socially corrosive impacts of corruption and organized crime. The remaining types of cities are concentrated in the former Third World, the 'partially marketized cities' of Latin America (some of which overlap with the new age boomtowns) and the marginalized city, found throughout the developing world but concentrated in Africa. Its assets of raw materials and cheap labour are of dwindling importance in the new global economy, many of its cities marginalized too by global investment.

Clearly, one of the drawbacks of such general-izations is that they fail to capture the diversity characterizing urban change. One of the accepted canons of globalization talk is that it is simulta-neously emphasizing the local and in turn its uniqueness. While the supranational has gained in importance this has been accompanied by the growing importance of the subnational, or the local. It is partly for these reasons that most accounts of globalization point to the weakened position of the national economy and of the nation state, arguments, which can be overstated. It is also for these reasons that cities can 'delink' from their national economies, attaining levels of economic performance above (or below) those of the national economy, reflecting the ability of the subnational agency to capitalize on the assets of the local environment to capture global invest-ment.

Globalization trends, together with the demise of the socialist economies, have universalized the processes of urban restructuring. Looked at from the macro-perspective, the world economy is in a transitional phase characterized by major changes in the global system of production. In this *Handbook* it is not plausible to capture the complexities of these overarching trends as they impinge on individual cities. Rather, the focus is on the different experience of cities within their regional context, taking those regions which have not been the focus of attention in previous chap-ters in the book. One region is ascendant in the region in the world economy, Pacific Asia, one marginalized by globalization, Africa, and the final region, from the post-socialist world, is the more obviously in transition. The accounts are illustrative rather than exhaustive, both in the sense of attempting to cover the different 'regions' comprising the world economy or in the sense of encapsulating the multiple ways, economic, cultural and political, current restruc-turing is unfolding within cities.

REFERENCE

Ruble, B.A., Tulchin, J.S. and Garland, A.M. (1996) 'Introduction: globalism and local realities – five paths to the urban future', in M. Cohen, B.A. Ruble, J.S. Tulchin and A.M. Garland (eds), *Preparing for the Urban Future: Global Pressures and Local Forces*. Washington: Woodrow Wilson Center Press.

27

Cities in Pacific Asia

DAVID W. SMITH

The images that Pacific Asian cities conjure in our minds are as numerous as the settlements themselves, each image varying further according to individuals' viewpoints – their needs, interests and status. For some, Asian cities present an image of grinding poverty, of people living in ramshackle squatter huts, eating one meal a day and queuing endlessly for work as a casual labourer; for others, the city seems a place of opportunity where endeavour and enterprise are quickly rewarded and where selling cooked food on the streets can lead to a restaurant of one's own. For many tourists the cities of Pacific Asia present exotic and absorbing scenes of gold-faced temples, floating markets, strange music or captivating dancers. For other more cynical visitors the region's cities are increasingly becoming a compilation of the familiar – international hotels, franchised fast food chains, the latest films and live sports relayed via satellite television. Above all, perhaps, Pacific Asian cities are seen as agglomerations of industrial enterprises, large and small, producing a wide range of reliable, reasonably priced products, from automobiles to silk shirts, on which our own consumer societies have come to rely. All of these interpretations are true, but their combination makes each city unique.

But where is Pacific Asia? Although the region has such a dynamic reputation, its definition and the terminology used to describe it have varied considerably over the past decade or so (Dixon and Drakakis-Smith, 1993; Hodder, 1992; Le Heron and Park, 1995). At its maximum, the region can stretch from the Russian Far East to the North to Australia in the South (increasingly, both these countries wish to identify with the region) and from Myanmar in the West to the island of New Guinea in the East. For an investigation focusing on the region's principal cities, the definition of its perimeters is less essential, for whatever boundary is chosen, the region so-defined will contain some of the most rapidly

growing economies and cities in the world. The popular assumption is that these two phenomena are closely integrated and that this symbiosis is not only mutually reinforcing but also exemplifies the growth pattern of the whole region. This is a debatable assumption. The region contains within it a wide variety of political, economic and social conditions, and the growth patterns reflect this variation. It will be the task of this chapter, therefore, to examine in general terms the economic and urban features of Pacific Asia and tease out from their variations not only the common denominators but also the way in which these differ from other parts of the world.

THE ECONOMIC IMPORTANCE OF PACIFIC ASIA

Despite a prolonged world recession, Asian economies have maintained steady growth since the mid-1980s. As Kwan (1994: 1) notes, 'this reflects not only Asian countries' higher potential growth rates compared with the world average, but also the growing ability of these countries to generate demand from within the region'. Asian economies have become much less dependent upon growth elsewhere and this independence reflects and represents a shifting balance of power in the world economy. The sustained and robust growth in the Asian economies has led to the creation of a broader-based and more substantial third axis for the world economy (see Dicken, 1992) and this is clearly reflected in regional trade structure (Figure 27.1), with Pacific Asia increasing its share of world trade from 15 per cent in 1980 to 42 per cent in 1995. Within the region some three-quarters of the total trade is accounted for by Japan and the four mature industrializing economies (MIEs) of Hong Kong, Taiwan, South Korea and Singapore all of which

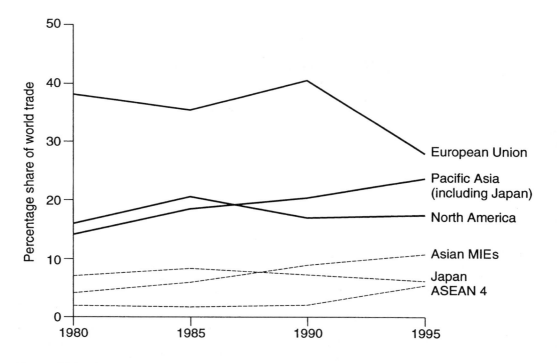

Figure 27.1   Global trade structures (Kwan, 1994; and IMF International
Financial Statistics for 1995)

rank in the world's top 20 trading countries. The
MIEs alone house half of the world's 10 leading
container ports. As a consequence of the trading
surpluses, it is not surprising that the region has
more foreign exchange reserves than the EU, with
Taiwan having the world's largest national foreign
reserves (Kwan, 1994).

The 'rise and rise' of Pacific Asian economies
has produced a plethora of economic analyses
designed to explain the nature of this growth (for
example, Shibusawa et al., 1992) and, in some
instances, to identify a model which may be
followed by other developing countries (Berger
and Hsiao, 1986). The World Bank (1993, see also
Page, 1994), in particular, likes to regard the
'miracle' of Pacific Asian development as a prime
example of a market-friendly approach that gets
the macro-economic basics right (high rates of
saving, investment, expenditure and education
etc.). As Alice Amsden (1994: 627) neatly
expresses it, 'like Narcissus, the World Bank sees
its own reflection in East Asia's success', taking
particular pride in the effect that its structural
adjustment programmes have had in guiding the
MIEs and EIEs (the Emerging Industrial
Economies of the ASEAN states) onto the
chosen path. As Amsden (1994) and many others
have shown, however, there is very little evidence
to confirm that such a model exists or, indeed,

that the World Bank has marshalled effective
evidence to indicate that getting the basics right
within a market-friendly environment can explain
much of the Pacific Asian growth. Dixon (1995a),
in contrast, has shown clearly that in the case of
Thailand, seemingly a prize example of the
success of structural adjustment, the World Bank
was complaining that the government had not put
many of its recommended measures into force at
the very time that the Thai economy was taking
off.

Part of the problem in much of this attempt to
explain and model the successes of East Asia is
the fact that the World Bank has its own political
agenda to proselytize, so that the conclusions of
its East Asian Miracle Report (World Bank,
1993) are often at odds with the 'evidence' it
assembles. Perhaps even more important is the
fact that, as Perkins (1994: 655) notes, 'there are at
least three models of East Asian development',
namely the *laissez-faire* of Hong Kong, the interventionism of Japan, South Korea and Taiwan
(with Singapore somewhere between the two),
and the resource-rich economic growth of the
ASEAN states.

The variations within the pathways to
economic development that have been, and are
being, experienced by the Asian economies are
substantial. Most obviously the process of

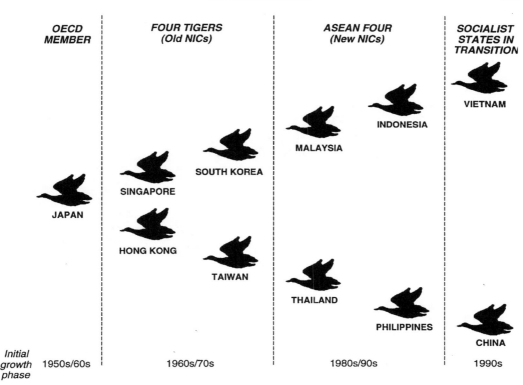

| OECD MEMBER | FOUR TIGERS (Old NICs) | ASEAN FOUR (New NICs) | SOCIALIST STATES IN TRANSITION |

Figure 27.2   The Flying Goose model of Pacific Asian development

industrialization has been spread over half a century, with the various centres following one another in the well-known 'flying geese' model (Figure 27.2) (Kwan, 1994). Over this long time period there have been many changes in the context of development – at global, regional and national levels. The regional and global economic climate of the 1960s and 1970s when the MIEs were taking off, with a buoyant world economy, abundant investment funds, the discovery of regional oil reserves and the acceleration of American involvement in Vietnam, was very conducive to industrial growth. The global and regional climate of the 1980s, dominated by world recession, growing protectionism and the emergence of a regional division of labour, was far less promising and has made the growth of the ASEAN EIEs all the more remarkable. By the 1990s the situation had altered once more as the barriers between the socialist and market economies began to dissolve in the rains of rapprochement.

These fluid contextual situations have also affected the rate at which internal structural changes have occurred in national economies,

with the shift from developing to industrialized economy being quite variable within the region (Figure 27.3). For the MIEs, the rising competition from the ASEAN countries and China, together with the resurgence of Europe and the development of large regional trading blocs, have all forced a reappraisal and a restructuring of their economies (Le Heron and Park, 1995). Labour shortages have led to increasing labour costs, fuelled in some countries by pressures from trades unions long suppressed as part of the cold war against socialism. In turn, this has resulted in much more flexible labour use in order to overcome problems of labour shortage and higher wages, such as replacing permanent workers with temporary employees or retraining. Subcontracting too has been used to tackle these problems, whilst new technologies have helped increase productivity, quality and competitiveness (Van Grunsven et al., 1995).

As part of this restructuring process, the MIEs have become increasingly active in regional investment. In a sense, there has always been regional investment, with Japan and later the MIEs seeking to develop and ensure access to regional resources,

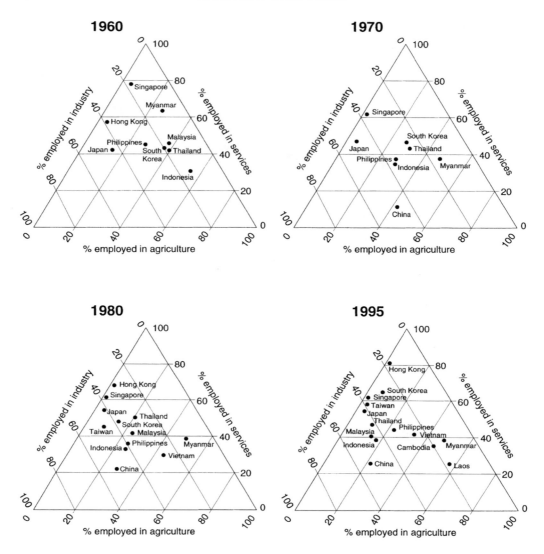

Figure 27.3   Pacific Asia: changing structures of GDP, 1960–1995
(World Bank, *World Development Reports*, various)

such as oil. Countries such as Indonesia, therefore, have long been the recipient of a regional invest- ment pattern which replicates that of the earliest International Division of Labour and with which it has always competed. Much more recent is the extent to which the investment patterns of the so- called New International Division of Labour have been challenged and to a certain extent replaced by regional investment flows, often underpinned by a desire to shift the increasingly costly labour- intensive, and more recently, environmentally damaging aspects of the production process to the industrial '*arrivistes*' of the region.

Such investment has long been a feature of the Japanese economy to the extent that growth rates in the ASEAN economies closely mirror the extent of Japanese Foreign Direct Investment (FDI) in the region. More recently, however, Japanese investment has been supplemented, challenged and even overtaken by investment from the MIEs (Figure 27.4). Within this increas- ingly complex pattern, there have been areas of special interest, with Taiwanese investment, for example, particularly favouring Malaysia and Thailand. The regional consequence of this increased investment has, of course, been an

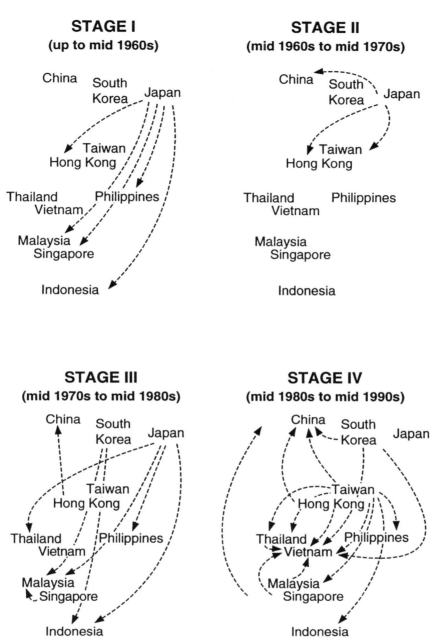

Figure 27.4 Investment flows in Pacific Asia (Alvstam, 1995)

expanding and intensifying pattern of intra-regional trade. This has risen from around one-third of total trade in the mid-1980s to 43 per cent by the mid-1990s (Table 27.1), with exports to the MIEs from all parts of Pacific Asia being a particularly important component of this.

Shibusawa et al. (1992) have argued that the increasingly inward focus of much investment and trade is not necessarily indicative of growing interdependency within the Pacific Asian region, but rather reflects an increased diversity of trade and investment away from the high degree of

Table 27.1 *Pacific subregional exports as a proportion of total exports, 1994*

	Exports to				
	Japan	MIEs	SE Asia	China	Pacific Asia
Exports from					
Japan	—	22.5	12.8	4.7	40.0
MIEs	8.7	12.3	9.3	13.4	43.7
Southeast Asia	19.4	26.2	5.2	2.3	53.1
China	34.8	29.7	3.0	—	69.5
Pacific Asia	7.4	19.7	8.1	7.3	43.0

Source: Asian Development Bank (1995) *Asian Development Outlook 1995–6*. Manila: ADB

dependency on the United States and Japan that characterized the region in the 1970s. Indeed, they point to a resurgence of links with Europe to underline this point. It is, however, different to challenge the fact that Pacific Asia has now built up a considerable internal momentum of its own, in terms of sustained economic growth. The question that needs to be asked is whether this level and form of development is sustainable.

As far as economic sustainability is concerned, the regional division of labour continues to evolve, particularly as the socialist states of the region begin to re-engage with the regional and world economies. The most recent activity in this respect has related to Vietnam where, in the wake of the removal of the US trade embargo, FDI has flooded into the country to take advantage of very low wage levels, an educated and abundant workforce and a compliant government. As Kwan (1994) indicates in Figure 27.5, however, further growth potential in the region is still considerable, particularly if one considers the interior regions of China and Russia.

Clearly, there have been substantial changes to the recent and contemporary economic growth patterns of the region. The focus of this discussion, however, is the impact this has had on the urbanization process in the region. In this context Van Grunsven et al. (1995: 155) claim that we are witnessing 'a transition from the landscape shaped by the first wave of internationalization to a mosaic of industrial complexes and regions within the Asian Pacific Rim'.

URBANIZATION IN PACIFIC ASIA

Despite Pacific Asia's economic successes and the fact that much of this manufacturing based growth is set in an urban context, the region is far from being heavily urbanized in comparison to other parts of the world (Table 27.2). With the

obvious exception of the city states of Hong Kong and Singapore, only Japan and South Korea have urban proportions of total populations that approach those of Europe or Latin America. Nor is there any straightforward correlation between the level of urbanization and the level or rate of growth of GNP, with Thailand, in particular, exhibiting very rapid economic growth within what is still predominantly a rural population structure. The other obvious characteristic of Table 27.2 is the variability of both the level of urbanization and the rate of urban population growth. Many of these rates of urban population increase are likely to persist for the next two decades or so, stabilizing only around 2020. What this means is that for the next generation population growth in most of the Pacific Asian states will be urban-centred.

In the context of the regional development process, two features stand out as worthy of attention in Pacific Asia. The first is the degree of urban primacy and the second is the extent to which the urbanization process is being concentrated into increasingly larger and complex urban-economic entities. Whilst there is considerable overlap between these two processes, they are far from being the same. Urban primacy is the extent to which urban populations are found in (usually) one or two cities, so that the urban hierarchy becomes top heavy. Usually primate cities dominate most other aspects of national development too, such as economic activities or social services. The principal example of primacy in Pacific Asia, indeed one of the most extreme examples in the world, is Bangkok (see Dixon, 1995b) which is officially some 30 times larger than the second most populous city in Thailand (incorporating the 'unofficial' population would raise this to nearer 50 times). But primacy is not necessarily correlated with overall size of settlement – it is principally a national index. There are, therefore, other regional examples of primacy, such as Phnom Penh in Cambodia or Vientiane in Laos, where the primate city is quite small even by regional standards.

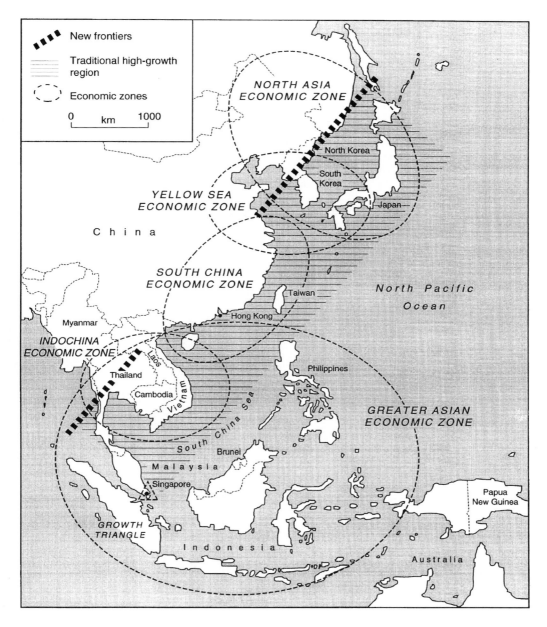

Figure 27.5   Growth potential in Pacific Asia (Kwan, 1994)

Urban primacy differs fundamentally from the second type of urban concentration within Pacific Asia, which is essentially city-centred regional growth. The phenomenon, characterized by a complex of cities, towns and urban-oriented rural population, has been known by a variety of terms over the past 10 years as it has become more pronounced and more identifiable as a new scale of development. Terry McGee

(1991) initially coined the term *desakota* (from the Indonesian words for village and city) but then moved to 'extended metropolitan regions' (Ginsberg, et al., 1991) and more recently to the even more prosaic 'mega-urban regions' (McGee and Robinson, 1995). This discussion will employ the slightly more widely used of these terms, namely Extended Metropolitan Regions (EMRs).

Table 27.2a *Pacific Asia: basic economic and urban indicators (mid-1990s)*

	GNP per capita	GNP growth per annum (1980–94)	% GDP share Ag.	% GDP share Ind.	% GDP share (Mfg)	% GDP share Serv.	Urban population % total growth (1990)	Urban population	Urban population % in largest city
Japan	31,490	3.4	2	41	(24)	57	77	0.6	19
Singapore	19,310	6.1	…	36	(27)	64	100	1.0	100
Hong Kong	17,860	5.4	…	19	(13)	81	94	1.0	100
Taiwan	10,850	–	4	37	(29)	59	57	2.1	
S. Korea	7670	8.2	7	43	(27)	50	74	2.3	35
Malaysia	3160	3.5	15	45		40	44	4.2	19
Thailand	2040	0.4	11	42	(28)	47	35	3.9	69
Philippines	380	-0.6	22	33	(23)	45	44	3.4	29
Indonesia	730	4.2	17	41	(24)	42	32	4.2	17
N. Korea	–	–	–	–	–	–	60	1.3	17
China	490	8.2	21	54	(38)	25	28	4.3	4
Laos	290	4.2[a]	57	18	(13)	25	20	6.1	52
Cambodia	200	–	45	20		35	12	4.5	90
Vietnam	170	–	28	29	(22)	43	20	2.9	24
Myanmar	–	–	63	9	(7)	28	25	3.3	32

[a] 1980–91

…, Less than 1; –, no information.

Sources: Asian Development Bank (1995) *Key Indicators of Developing Asian and Pacific Countries*. Manila: ADB. Asian Development Bank (1995) *Asian Development Outlook 1995–6*. Manila: ADB. World Bank (1995) *World Development Report*. Oxford: Oxford University Press. UNDP (1995) *Human Development Report*. Oxford: Oxford University Press

Table 27.2b *Pacific Asia: basic social and environmental indicators (mid-1990s)*

	GNP per capita	Population growth (% p.a.)	Over 65 yrs (%)	Inequality			% Urban population with access to	
				Ratio highest to lowest quintiles	Gini coeff.	% in poverty	Clean water	Sanitation
Japan	31,490	0.5	18[a]	43				
Singapore	19,310	2.2	7	9.6	0.42	10	100	99
Hong Kong	17,860	1.0	10	8.7	0.45	14	100	90
Taiwan	10,850	1.0	7					
S. Korea	7670	1.0	5	7.9	0.36	5	97	100
Malaysia	3160	2.4	4	11.7	0.48	16	96	94
Thailand	2040	1.5	5	8.3	0.47	30	89	80
Philippines	830	2.6	3	7.4	0.45	45		
Indonesia	730	1.7	4	4.9	0.31	25	68	64
N. Korea	–	2.3	–	–	–	–	–	–
China	490	1.2	6	6.5	0.31	9	87	100
Laos	290	2.6	3	–	–	–	54	97
Cambodia	200	2.7	3	–	–	–	65	81
Vietnam	170	2.4	5	–	–	54	47	34
Myanmar	–	1.9	4	5.0	0.35	35	37	39

[a] Over 60 years.

Sources: Asian Development Bank (1995) *Key Indicators of Developing Asia and Pacific Countries*. Manila: ADB. World Bank (1995) *World Development Report*. Oxford: Oxford University Press. UNDP (1995) *Human Development Report*. Oxford: Oxford University Press.

The premise of this growing literature is not only that EMRs constitute the dynamic core, driving both national and regional growth, but that they 'represent something distinctively different from urbanization in other parts of the world' (Webster, 1995: 27). This region-based rather than city-based urbanization is also 'distinctively different' for McGee (1995a: 10), both from nineteenth and twentieth centuries Europe and from the megalopolises of North America identified by Gottman (1961). And yet the blurring of urban and rural functional space has also been debated in the Western context (see Thrift, 1989). However, before assembling the characteristics of these Pacific Asian EMRs and assessing their degree of difference, it is necessary to situate their emergence in the context of the economic growth to which they contribute and by which they are shaped in a symbiotic relationship.

As Mike Douglass (1995) has clearly illustrated, EMRs are but one in a hierarchical series of changes which have been taking place in the spatial order of development. In terms of ascending scale, Douglass (1995: 48) identifies 'five emerging patterns' of political, social and economic restructuring (Figure 27.6).

- A *polarization* of national development into a limited number of core regions. Despite assurances from planners that such cores would be counterbalanced by trickle-down mechanisms, this polarization appears as strong as ever. Indeed, Douglass (1995: 50) claims that 'no country in Asia has been able to reverse spatial polarization of population and economic development during the cause of industrialization'.
- The emergence of *mega-urban regions* (EMRs) as certain core urban centres become linked through a series of technological, economic and social networks with which multinuclear development occurs. This cuts across traditional rural and urban distinctions and is not necessarily structured around a single core. Different technologies, scales of operation and labour systems operate in juxtaposition in a system that is as much the product of local as international processes.
- These EMRs constitute part of a hierarchy of *world and regional cities* that act as command centres for the accumulation and circulation of global capital within the region. In Pacific Asia, only Tokyo functions at the highest level, with most of the other major EMRs operating more at the regional level, with Singapore hovering somewhere between the two.
- A different dimension of spatial change emerging within Pacific Asia can be found in some *transborder regions*, where urban agglomerates span national boundaries, 'a functional and efficient form of integration that is new and does not fit into the conventional category of economic integration stages' (Kettunen, 1994: 1). Few, as yet, have reached the level of an EMR – Singapore's growth triangle being the more obvious exception. However, polynuclear urban centres are rapidly growing to near coalescent stage in the Pearl River area or in Bohai (encompassing China and Korea). Douglass (1995: 53–4) notes that within these transborder developments the presence of the national boundary is a prerequisite in maintaining 'the socially and politically constructed differences between the actors involved in the transitional exploitation of cheap resources and labour'.
- Finally, Douglass suggests that all of the above constitute subprocesses of an emerging international network of transportational, communicational and decision-making *growth corridors* of international development designed to integrate the region as a whole into the global system of finance, production, trade and consumption. In this sense, Douglass does not see the Pacific Asia region as an emerging independent economic force but as a closely dependent subordinate of the United States and European markets.

The processes creating EMRs within this new spatial order are clearly linked to major changes at the global level, although the nature of each EMR must also be rooted in its local history, culture and politics. McGee (1995a) has identified three broad forces operating in concert on the urbanization process in Pacific Asia which has encouraged the formation of EMRs, namely globalization, the transactional revolution and structural change. The first of these he characterizes by the so-called New International Division of Labour (NIDL), by the emergence of global EMRs (Douglass's World Cities) and by the growth of intense global and regional competition. These three are, of course, very closely interconnected, with EMRs competing with one another in the quality and efficiency of their built and transactional environments in order to attract investment. The more they offer, the more concentrated becomes this investment, and the more the EMRs acquire their world trade centres, telecommunication complexes, international schools, prestige retail stores and the like. But, of course, as earlier discussion indicated, NIDL has been supplemented, even superseded, as a formative process by a Pacific Asian Regional Division of Labour (RDL) (Dixon and Drakakis-Smith, 1993). Within the RDL, trade and investment

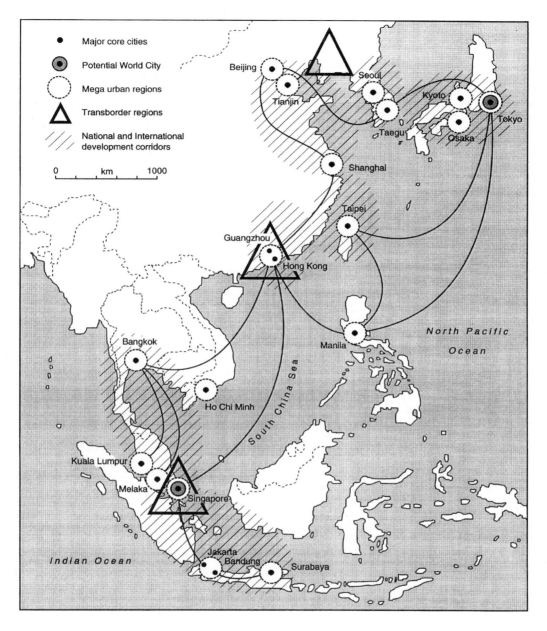

Figure 27.6   Spatial patterns of restructuring (Douglass, 1995)

flows may be expanding but the competition for regional pre-eminence is as keen as ever. Below the obvious focal point of Tokyo, Hong Kong and Singapore have emerged as the main rivals for senior status. Both form the pivot of considerable regional development, but the greater security of its independence would seem to set Singapore on a firmer base for future expansion.

In many ways the changes described above have been dependent on the second of McGee's influencing factors, namely the transformation of transactional space–time. Not all transactional changes are occurring to the same extent. Information, capital and decision-making may all be transmitted electronically, but people and commodities still have to be physically transported. Whilst electronic transactions can substitute for some physical movements, much of the interchange of increasingly complex information must be undertaken face-to-face (Rimmer, 1995) and there are indications that electronic information exchange is more of a substitute for international, rather than local, travel. Physical transportation of goods and people can be further divided into low- and high-speed categories dependent on their relationship to the production process.

The impact of these transactional changes on the formation of EMRs is often unsatisfactorily explained. The cost of new technologies often means that electronic nodes are concentrated into those EMRs that can afford them. On the other hand, cheap, personalized developments, such as mobile telephones, fax machines and personal computers, mean that within specific EMRs individuals are less place-dependent. Similarly, whilst the development of cheap, reliable minibus transportation has exacerbated greater long-distance migration or commuting, individual mobility in the city itself is facilitated by low-cost personal transportation such as cheap motor cycles. Within the various EMRs too, new media developments, particularly in television, are creating a more cohesive and identifiable image of a big-city lifestyle in the minds of its new residents or potential migrants (Webster, 1995: 32).

The consequence is what may be best described as unfocused concentration into EMRs which have multiple nodes rather than a single centre, for example, a tourism area, a business centre, several major retail districts, port zones and industrial areas. Often the key ingredient in the formation of the EMR was, or is, a major transportation link towards and along which new developments gravitated. The Shinkansen spine that linked Tokyo and Osaka is mirrored in rail and expressway developments between Taipei and Kaoshiung, as well as between Seoul and Pusan. New airports too have helped both to focus and decentralize urban development. In many ways airports have functioned like railway stations in nineteenth century urban morphological development. As their respective epitomes of urban technology and urban architecture, they have attracted new developments to their vicinity, particularly hotels. The proposed new airport for Bangkok will, therefore, play an important role in linking the existing Bangkok Metropolitan Area (BMA) with the newer, port/tourist developments along the southeast seaboard of Thailand.

The irony, of course, is that whilst the various transactional revolutions have been fundamental to the creation of EMRs in terms of both concentration and polynucleation, they have been so unevenly layered and poorly managed within the urban areas as to be responsible for some of the most pressing environmental problems, such as congestion and pollution. The Bangkok and Taipei EMRs rank first and second in the world for traffic congestion, with average 'rush hour' speeds of 10 and 11 mph respectively. Not all of this is the consequence of high-tech change; in Taipei some 160,000 people each year are fined for driving without shoes (Selya, 1995)! Perhaps even more important, however, is the fact that comparatively little research and investment has been put into essential parallel transactional movements. It is not just money, ideas, people and commodities that need to be shifted but energy, water, sewage and solid waste. The brown agenda still awaits its transactional revolution, particularly in Taipei (Selya, 1995).

The third of McGee's formative processes relates to structural change and in this context it is essential to appreciate that development planning in Pacific Asia has been constructed primarily in sectoral rather than spatial terms. Both external advisers and internal strategists have seen the principal developmental issue as a need to diversify economic growth from a reliance on agriculture and primary exports to a more manufacturing-based economy. Initially this was constructed around import-substitution industries, but since 1970 the goal has been export-based manufacturing. Development is thus equated with a sectoral shift of the economy from agriculture to industry. As almost all industrial investment has been urban-focused, so the sectoral shift in production has become reflected in a movement of population from rural to urban areas. The transactional revolution has, of course, played its part in these movements. Interestingly, the rapid change in transport technology which has kickstarted Thailand's industrial growth by enabling isolated villagers to move directly to Bangkok almost as easily and cheaply as moving to their nearest town, has also played a crucial role in taking Japan's economy to new heights through the development, production and export of the transport hardware facilitating this movement.

All of these processes have interacted to produce EMRs throughout Pacific Asia, some of which (Tokyo) have been developing for several decades, others of which are relative fledglings. But all have emerged within their own particular environmental, cultural and political contexts and, as a result, exhibit individual features. True, all can be said to be polynodal in form with the old division between rural and urban becoming very indistinct, but each has particular features which need to be taken into account not only in historical analysis but also in terms of urban management.

The most common form of EMR has emerged from high-density rural and high-density urban areas coagulating into a complex entity in which the most rapid growth is occurring outside the principal urban core(s). Bangkok, Jakarta,

Shanghai and Taipei constitute examples of this type of EMR, although the Taiwanese capital also instances a second and even more extensive type of development in which two or more major nodes are joined by a major rapid transit system to form a settlement pattern much more akin to those identified in North America many years ago. Within Pacific Asia the Tokyo–Osaka complex is perhaps the most developed form of this type and differs in scale from nascent East Asian replicas such as Seoul–Pusan and Taipei–Kiaosuing. However, in most current discussions the EMR refers primarily to the regional urban growth around the principal city, recognizing the fact that the distinction between this and the developmental corridor is increasingly becoming blurred in some countries.

Besides these two basic types of EMR – the multiple city model and the high-density model – there are other forms of mega-city evolution in Pacific Asia which underpin the need to recognize local circumstances. Kuala Lumpur, like Seoul to a certain extent, is a relatively lower-density EMR in which development has been much more controlled, largely resultant from its establishment as a Federal Territory in the mid-1970s. Although this is also a polynucleated EMR, the national state has been heavily involved in the management and planning of its development.

Finally, Singapore appears to be a unique form of EMR in that it straddles international boundaries in its composition. Over the past 10 years the neighbouring states of Malaysia (Johore) and Indonesia (Riau Islands) have effectively become part of Greater Singapore through the development of joint economic and tourism activities. These joint ventures in the SIJORI growth triangle (Van Grunsven, 1995) have exploited Malaysian and Indonesian land and labour through Singaporean management and investment and, in the process, created an integrated EMR, despite the international borders and island composition.

EMR MANAGEMENT: PROBLEMS AND RESPONSES

Rapid growth, such as that which is occurring in the Pacific Asian EMRs, does not bring universal benefits. Indeed, the gains experienced by individual firms or by national states are often at the expense of many other elements of society or the environment. This brings the discussion back to the fact that sustained growth is not the same as sustainable development and in most EMRs in Pacific Asia the former has taken priority over the latter. To be sure, most observers have identified a bundle of issues or problems related to rapid and concentrated urban-economic growth. These range from various environmental matters, such as air and water pollution or inadequate sewage facilities – the so-called 'brown agenda' (Drakakis-Smith, 1995), through to deepening social cleavages along ethnic lines, as migrants move into EMRs from increasingly wider areas and increasingly different backgrounds (see Dwyer and Drakakis-Smith, 1996), and along class lines as exploited proletariats seek to strengthen trades union movements and emerging middle classes attempt to demand and assert special privileges (Robinson, 1995).

On the other hand, apologists for EMR development as being an inevitable future growth process often seek to downplay many of these matters. Webster (1995) predicts, therefore, that currently sizeable informal sectors, including squatter settlements, will diminish and public utilities will become more widespread, contributing to even denser development of EMR interstices. McGee (1995a: 20) goes further, dismissing some of the principal concerns with macro-cephalic urbanization as myths that 'need to be defused if mega-urbanization is to be managed effectively'. In particular, he identifies what he considers to be five popular shibboleths suggesting that large cities are a drag on development:

- service provision is problematic;
- sustainability of energy supplies is difficult;
- cities are parasitic and develop at the expense of rural areas;
- poverty is excessive;
- there is disharmony and poor quality of life.

These seem to me to be a mixture of straw men, easily knocked down (such as the old assertion that there is urban bias in development), and irrefutable truths which are rather too easily dismissed. Thus, energy problems offer 'exciting prospects for recycling'; the poor quality of life is apparently countered by the presence of theatres, museums and art galleries; poverty and inequality are set against opportunities for the poor in the informal sector and so on. As for overall sustainability, it is suggested that large-scale economic collapse would have occurred if EMRs were *not* sustainable.

Such uncharacteristic dismissal of the very real problems posed by large-scale concentrations and faced daily by many of their inhabitants, fails to grasp the need to recognize the integrative nature of sustainable urbanization within which individual issues, whether related to housing, traffic congestion or poverty, must be set (Drakakis-Smith, 1995, 1996). Neglect of social issues, in particular, as noted by Anne Booth (1997), poses just as much of a threat to sustained growth than oscillations in economic factors, as events in

Indonesia recently indicated, whilst environmental deterioration can very quickly cause so many practical problems as to induce a shift of investment elsewhere. Bangkok is commonly considered to provide a clear example of this situation (Dixon, 1995b).

There is no doubt that the continued growth of EMRs is posing new challenges to urban and regional planning in Pacific Asia. Comprehensive planning is virtually impossible within rapidly changing EMRs but the focus will need to recognize the regional dimension of these entities. Webster (1995) suggests tailoring systems of governance to coincide geographically with EMRs, but given their dynamism such coincidences are unlikely to last very long. More important, national governments may need to consider carefully whether EMRs should be favoured or disfavoured in regional development policies. At this macro-level, sectoral policies will need to shift more firmly to a set of geographical bases which are fully integrated to cover the EMR – what McGee (1995a) terms metrofitting.

Webster (1995) advises setting strategic planning objectives structured around a few key priorities, but such priorities will not be realized without considerably more financial resources. This means decentralized powers in decision-making and raising revenues. At present, many EMRs coincide with capital cities so that metropolitan management has effectively been carried out by national politicians. As new or extended EMRs merge, however, such synchronization is likely to be replaced by conflict and competition for the control of the development process. Within the EMRs themselves, urban management faces increasingly intense problems, such as the transportational, environmental and social issues outlined earlier. Nor is the last of these a matter of improving human resources merely with regard to maintaining a healthy and trained labour supply. In addition to receiving a fair wage, urban populations have a range of civil and human rights, from freedom of speech to electoral enfranchisement; these too are part of sustainable urbanization. Urban management of this nature can present daunting tasks but these need not, indeed often cannot, be faced alone. Increasingly, the World Bank and similar institutions see a role for the private sector in meeting many of the pressing planning problems of EMRs – from housing provision to water supplies (UNDP, 1991; World Bank, 1991). Certainly there have been some successes in this area (Gilbert, 1994), but it must be recognized that such collaboration will not extend to those areas of activity that do not guarantee profits. The danger here is that urban management will increasingly favour user-pay strategies to make such schemes attractive to the private sector, effectively cutting out the poorest and neediest.

Such planning dilemmas have, in recent years, suggested that there may be an increasingly important role in urban management for Non-Government Organizations (NGOs) or Community Based Organizations (CBOs) (Webster, 1995). Presumably such organizations are envisaged as intermediaries between EMR populations and urban managers. Others question such assumptions and see the incorporation of NGOs and CBOs into urban management as an exercise in legitimization, in part expediting the privatization of welfare services that ought to be provided by the state, rather than as a means of improving communication between the masses and the state, thus permitting an increased involvement of the former in formulating the urban management strategies of the latter (Devas and Rakodi, 1993).

However, what ought to be and what has happened in reality are two quite different things. This chapter will proceed to describe briefly the situation in four representative EMRs in Pacific Asia with regard to their evolution, current composition and problems, and future prospects. The EMRs are selected to represent four stages in the development process within the region, namely Tokyo (advanced capitalism), Singapore (MIE), Bangkok (EIE), Hanoi (Socialist transition).

TOKYO: THE URBAN FUTURE OF ASIA?[1]

That Japan is a major economic power is axiomatic, and does not need amplification here (see Daniels, 1996). Suffice it to say that it is the world's leading nation in terms of outflowing FDI (Drennan, 1996). What is more interesting and relevant to a review of its capital city are the various attempts to explain Japan's success and replicate this elsewhere. Most analyses are sceptical of earlier claims of transferability. Peck and Miyamachi (1995: 37), for example, argue that 'the essence of Japanese competitive success is less easy to extract [and] is deeply embedded in the social institutions of Japanese capitalism'. Parker (1995) supports this position, and claims that a key factor in Japan's success has been the integrated links between Japanese finance, security, information and production systems, links which cross the private–public sector domains. This essentially constitutes a mode of Social Regulation in which considerable interlocking occurs between enterprises, labour, government and bureaucracy (Humphrys, 1995). The result is a series of structures that offer collective security through flexible production, ready access to

credit and extensive trade and technical knowledge for the pursuit of shared goals (Parker, 1995:
102).

Much of this interlocking has influenced the
distribution of economic growth in Japan, underpinned by the extent of vertical near-integration
that characterizes industrial production. However,
the physical geography of Japan has also played its
part since there are limited lowland areas for
settlement, and transportation has always been
problematic. The result has been an overwhelming
concentration in the area between Osaka and
Tokyo, with three conurbation's around Tokyo,
Nagoya and Osaka containing 60 per cent of the
industrial labour force. It would be possible to
consider the Osaka–Tokyo region as one huge
EMR in its own right – comparable to Gottman's
megalopolis in the north-eastern United states.
But for clarity of analysis the focus of the discussion will be on the capital, where a similar institutional integration to that affecting industrial
growth can be seen to operate.

Scale and Status

Although Japan was a predominantly rural
society until the early twentieth century, as early as
1750 more than 20 per cent of the population was
urban. Tokyo had a population of a million and
was probably the largest city in the world
(Humphrys, 1995), but on the eve of the Meiji
restoration it was 'unambiguously pre-modern' in
its nature. In the mid-1990s metropolitan Tokyo is
once again the largest urban agglomeration in the
world, with a population variously put at somewhere between 10 and 30 million. It is difficult to
define Tokyo – as one might expect given the
amorphous nature of the contemporary EMR
(Witherick, 1981). Cybriwsky (1991) observes that
the name can refer to the original city, as it was
prior to the creation of the contemporary administrative unit of Tokyo in 1943. The former is a
densely packed urban area of some 8 or 9 million
people; the latter comprises metropolitan Tokyo
and contains some heavily populated suburban
areas as well as less populated mountain areas.
From east to west this region covers some 80 km
and contains around 12 million people. It has an
elected governor and is equivalent to a prefecture
in its administrative composition. Within its
boundaries lie many sub-districts, mostly urban in
character. Beyond Tokyo-to and Metropolitan
Tokyo there is the continuous built-up area of
Greater Tokyo, largely coincidental with the
southern half of the National Capital Region, a
huge area of some 39,000 km^2 and 40 million
people.

Table 27.3   *Indicators of Tokyo as a primate
city*

	%
Population	15.0
Urban population	19.0
Company headquarters	57.8
Foreign companies	83.9
Foreign banks	63.8
Publishing firms	64.9
Bank deposits	46.7
College students	29.4

Source: Cybriwsky (1991: 16); World Bank Development
   Report (1995)

Tokyo EMR (conservatively taken to be Tokyo-
to) is far from being a primate city, containing
only 15 per cent of Japan's total population from
19 per cent of its urban residents. However, by all
other measures it dominates the economic,
cultural, social and political life of Japan (Table
27.3). Moreover, given Japan's global economic
importance, it would be unusual if Tokyo were
not equally prominent at that level. 'The city is the
world's principal lender and biggest creditor as
well as the number one ranking city in terms of
large corporate headquarters, total bank deposits
and numerous other economic measures' (Cybriwsky, 1991: 12). Moreover, the nature of this
activity is changing as Tokyo consolidates its
global and national roles. Thus over the past two
decades manufacturing has fallen to just 17 per
cent of all jobs whilst service employment of all
kinds has risen. Almost one-quarter of Japan's
producer services are now located in the capital
(Drennan, 1996).

The morphological product of economic
growth of this magnitude is a markedly polynuclear city (Figure 27.7), with at least 12 major
subcentres in Tokyo-to and 10 in adjacent prefectures. Of crucial importance in this multinodal
structure is the metropolitan railway system, the
inner loop of which is to be the focus of a concentrated group of seven major subcentres, several of
which are still in the developmental stage and all
of which have, on paper at least, distinctive functions, none of which are particularly well defined
so that 'it is possible to envision their eventual
coalescence into one super-giant business core'
(Cybriwsky, 1991: 209). One of these new subcentres is Teleport Town, a centre of advanced
telecommunications that will be built on
reclaimed land in Tokyo Bay and underpins
Tokyo's determination to become the world's
foremost business centre. Ironically the development of Teleport Town has been delayed by poor
transportation links with the existing city.
Although the centre will be on an axis between
the two main airports, emphasizing and facilitating

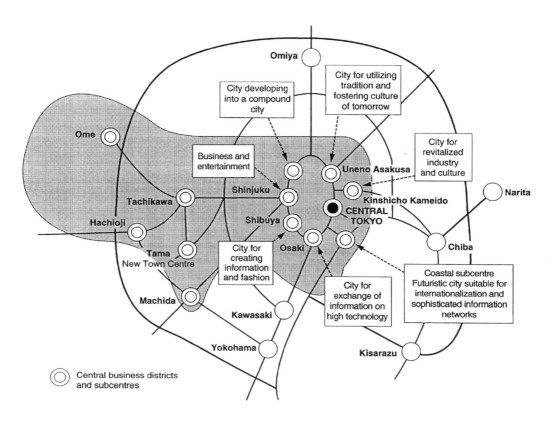

Figure 27.7   Polynuclear plans for Tokyo (Cybriwsky, 1991)

its international links, the day-to-day movement of people and commodities will be subject to the same constraints that currently make transport logistics one of the biggest planning problems facing Tokyo.

To the west of Tokyo a number of suburban developments are intended to balance this inner metropolitan growth, at least to a certain extent, but all the adjacent prefectures also have similar ambitious plans. Of these the most notable is perhaps the Chiba Industrial Triangle which is modelled on a similar high-tech development in North Carolina. Interspersed with these urban–industrial complexes are a series of small towns and more rural enclaves but their subservience to the main metropolis in terms of a wide range of functions is becoming more intensive by the day. Villages are consumed by commuter suburbs and in the interstices the leisure facilities serving 12 million people are mushrooming; Tokyo's Disneyland is halfway between Tokyo and Chiba, whilst golf courses, or at least driving ranges, replace more traditionally rural pursuits.

Not content with spreading and intensifying urban land uses outwards and upwards, Cybriwsky (1991: 233) notes that Tokyo is now seeking to expand even further downwards. In addition to the normal urban infrastructure below street level, including the world's largest subway network, Tokyo is increasingly expanding its urban functions below the surface through the additional basement levels to its new buildings. To clarify and facilitate such developments the Diet recently limited ownership of land to a depth of 50 metres, primarily to enable the construction of underground projects without the need for extensive compensation payments.

Sustainable Urbanization in Tokyo

The post-war reconstruction period in Tokyo up to the Olympics of 1964 was characterized by many of the challenges and responses that now feature prominently in most of Pacific Asia's burgeoning EMRs. Not only was there a concern with recreating a functioning city but also with re-establishing the international status of the capital, largely through the visual evidence presented to the world by its built environment. Symbolized by the Tokyo Tower, opened in 1958 'simply to stand tall and be seen' (Cybriwsky, 1991: 90), this phase culminated in a plethora of architectural prestige projects for the 1964 Olympic Games.

Whilst energies and funds were being ploughed into such projects, as they are in Kuala Lumpur, for example, in the 1990s, much less attention was being paid to the needs of the capital's citizens. At the time of the Olympics, some two-thirds of the households were without sewerage connection and still relied on night-soil trucks (Seidensticker, 1990); huge open garbage dumps also produced plagues of flies that infested the city. It is estimated that almost 30 per cent of the city's population was living in substandard accommodation, with 45 per cent sharing toilets and kitchens (Hall, 1984). Workers travelled into the city crammed inside hugely overcrowded commuter trains, often for several hours at a time, in conditions far worse than those experienced in London or New York.

These problems meant that for the majority of Metropolitan Tokyo's citizens, the sustainability of an adequate lifestyle was being threatened, despite the growing economic wealth and prestige of the capital. It proved to be politically unacceptable and those responsible for the grand projects of the pre-Olympic era were voted out of office in favour of a new socialist administration with some new priorities. In the 30 years since the administration of Metropolitan Tokyo has changed, whilst prestige projects and the internationalization (*kokusaika*) of the capital are still major motivating factors within the planning process, the city has clearly improved considerably in terms of the quality of life of its inhabitants. Making direct comparisons to metropolitan areas in Europe or North America is difficult in this context, given the different priorities and cultural characteristics of the Japanese *per se*. Overcrowding on commuter trains has, for example, improved little, but is tolerated much more than it would be in the West. Sustainability, therefore, becomes a relative rather than an absolute concept with few fixed standards.

Much of the improvement in the quality of life has undoubtedly been underpinned by continued and rapid economic growth. Considerable public and private investment has gone into Japan's capital in order to sustain its role as a world city. The structural shift in its employment pattern means that much of the medium- to large-scale manufacturing industry has moved to the peripheral areas, leaving behind a proliferation of very small enterprises. Almost half employ no more than three people and more than 80 per cent employ under 10 workers (Cybriwsky, 1991:

110). The shift to service industries not only reflects the needs of the metropolis itself, but is also linked to the goals of national economic planners who wish to see Tokyo develop as the premier producer service centre for Pacific Asia as a whole, having major foreign, as well as domestic TNCs. However, as Drennan (1996) observes, Japan has long discouraged inward investment and has failed to provide a welcoming climate for foreign firms. Given that office rents in Tokyo are also three times that of New York, it is unsurprising that Japan's economic and urban planners alike fear the competition that Singapore, Hong Kong and even Shanghai offer in this context. Hence the concern with *kokusaika*, the desire to internationalize the nature of life in Japan, partly to make it more attractive to foreigners. Cybriwsky (1991: 114) cites a serious planning meeting on the internationalization of toilet planning as one aspect of this, including a 12-country discussion on 'my experience with toilets'.

Whilst incomes may have increased as a result of the economic growth of Japan and Tokyo, the daily lives of the capital's residents may not have improved commensurately. Intensive redevelopment of the city centre has pushed many people to great distances from their workplace, whilst those left behind complain of too much traffic, high living costs, cramped living conditions, heavy air and water pollution and lack of greenery. Socially, old problems persist and new ones have arisen. The persistence and toleration of molestation of women on crowded commuter trains continues despite growing complaints by women's groups. Labour shortages have brought in increasing numbers of migrant workers with whom most Japanese have or want very little contact or sympathy, giving rise to new ethnic tensions to put alongside the more 'traditional' ones involving the Burakumin (Neary, 1997). In addition to the 600,000 Koreans working in Japan, there are estimated to be some 300,000 illegal migrants, together with almost 200,000 *nikkeijin* or returned migrants from Japanese families settled around the Pacific (Sellek, 1997). Much of this migratory movement is a recent phenomenon in a country little familiar with the ethnic tensions that Europe and the United States have been experiencing for several decades, and to which other EMRs around Asia have also been exposed.

One final social issue with which metropolitan Tokyo is having to cope and which is rapidly affecting other Asian EMRs, particularly those in the MIEs, is the ageing of the population. The proportion aged 65 or more has doubled over the past 30 years to more than 10 per cent, and is much higher in some older inner city areas. This greying of the workforce cannot be addressed here but is a major issue affecting many facets of

Japanese life – from the relatively poorly developed welfare/pension system to the nature of the labour force/dependency ratio in the coming decades. Recently the retirement age was raised and arrangements to rehire retired workers (often at much lower wages/salaries) are becoming increasingly common.

In many ways, therefore, the problems facing metropolitan Tokyo in terms of the sustainability of its urbanization process are becoming more like those of the West, although the innate cultural differences of Japanese society clearly give a particular character to the problems facing the capital. Moreover, in contrast to London or New York an underlying principle of current urban planning strategy seems to favour the continued growth of the metropolitan region as part of the drive to make Tokyo an increasingly attractive location for the international community. The fear of Pacific Asian competition is not an unimportant motive in this process. 'Thus, it seems that, in Tokyo, progress has been defined as growth . . . in all directions: up, down and out to ever more distant reaches of land and bay' (Cybriwsky, 1991: 240). Whether Tokyo provides a glimpse of what other Asian EMRs might become or aspire to become is a difficult question. Perhaps the brief reviews of Singapore, Bangkok and Hanoi which follow will shed some light on this.

SINGAPORE AND THE SIJORI GROWTH TRIANGLE

Many observers assume that Singapore is a unique city-state, so much so that they reject it as any sort of indicator as to how development may occur in the Third World. As we will see, this is not the case. Not only is Singapore envied but its current development process is providing substantial clues as to what the urban-economic future may be in many parts of Asia. Indeed, one might argue that Singapore provided such a starting point in the mid-1970s – when its early growth so impressed neighbouring states – that many governments reconstructed the administrative areas around their capital cities in order to create the semi-autonomous political units that they felt were needed to free capital cities from the problems of their rural hinterlands (Rimmer and Drakakis-Smith, 1982). It could be fairly said that such moves laid the initial bases for the subsequent expansion into EMRs.

As noted earlier, many observers consider that EMRs comprise fundamentally different forms of urbanization (McGee and Robinson, 1995); others have argued that growth triangles, particularly those that stretch across international borders, also constitute a new type of economic integration (Kettunen, 1994). As Singapore incorporates both of these innovatory development processes, it could be said to exemplify an important future scenario for the region. But to what extent is this true: how has the present situation evolved and is it sustainable?

The Development of Singapore as a City

Singapore as a city was initially created and shaped by the British during the colonial period between 1818 and the 1950s. On the eve of independence in 1965 it was a rather forlorn remnant of Empire with high unemployment, rapid population growth, a crumbling built environment and economic prospects that were far from rosy. Within 30 years this has been transformed to a near-global city with double digit growth and a per capita income which put it well into the premier league of OECD nations (for summaries see Drakakis-Smith and Graham, 1996; Rigg, 1990). This transformation occurred in marked phases. The first lasted from 1965 to 1980 and effectively encompassed the evolution of a manufacturing economy thoroughly embedded in what has become known as the New International Division of Labour (NIDL) and funded substantially by foreign investment, largely from Japan and the United States. Foreseeing the impact on Singapore of the global recession of the 1970s and 1980s, the government put in train what it called its second industrial revolution aimed at replacing 'down-market', labour-intensive manufacturing industry with high-tech industry, a range of producer services and a regional investment profile. After several initial hiccups, the economy has now been transformed along these lines (Table 27.4), but not without considerable social costs as the rather authoritarian government sought to manage its human resources in order to sustain the new directions of economic growth (Drakakis-Smith and Graham, 1996).

All of this economic and social change has had enormous impact on the built environment. A massive public housing programme, given impetus by a mixture of welfare, economic and political motives has transformed the island into a series of new towns and estates into which specific functions and groups are beginning to shift. The dwindling green areas in the periphery have in parallel shifted from low value rice production through higher value vegetable and pig farms to even higher value orchid farms and urban amenity areas. Meanwhile, the transactional revolution has also begun to alter the morphology of the city, creating an 'airpolis' around Changi airport and a series of telecommunications towers that dominated the low-lying island. The CBD

Table 27.4   *Singapore: economic transformation, 1960–1995*

Industry group	1960	1970	1980	1990	1995
Primary	4.1	3.1	1.5	0.5	0.3
Manufacturing	11.9	20.4	29.1	29.4	26.5
Utilities	2.5	2.5	2.2	2.0	2.0
Construction	3.6	6.8	6.4	5.6	6.7
Trade	35.9	29.2	21.8	17.2	17.8
Transport/communications	14.2	11.0	14.0	13.3	15.2
Finance/business	11.3	14.0	19.6	28.0	27.6
Other services	17.4	12.7	9.1	9.9	10.2

Excludes bank service charges and import duties.

Source: Department of Statistics (1995) *Monthly Digest*

has been shifted from a set of colonial buildings situated to the east of the Singapore river to a forest of towering skyscrapers on the opposite bank, mirroring most North American cities – but much cleaner. The old colonial area itself was not demolished but is now being regenerated as a 'heritage' area for both local and overseas tourists (Yeoh and Kong, 1995).

Despite all these transformations to its physical and business environments, however, Singapore appeared to be reaching limits to growth very rapidly. Some of the problems of sustainability were dealt with through a combination of good planning and economic wealth. Traffic congestion, through hugely expanded vehicle ownership, has been controlled by a wide variety of measures from massively increased car ownership and operating costs, to area licensing and the construction of a super-efficient rapid transit system (Pendakur, 1995). Environmental pollution problems have been effectively countered by stringent legislation, strongly enforced to the extent that the old warehouse areas along the sides of the formerly noxious Singapore river now constitute recreational quays that have transformed the city's nightlife.

Other aspects of sustainability have been more difficult to counter *in situ*. Most problematic have been the issues of limited human and physical resources – people and land. In combination, the rising costs of these two factors posed substantial development problems for Singapore, forcing it to encourage the outmigration of labour-intensive industries which were still highly profitable, leaving the city still short of space for the growing recreational, residential and transportational demands of its increasingly affluent population. The solution, as in most cities, was to overflow into the immediately surrounding area. The difference and the difficulty for Singapore was that these areas were part of other political entities. At this stage, in the mid-1980s, Singapore was more conventional city than incipient EMR. Its population, even in the mid-1990s was still only

around 3 million, and 10 years earlier it had not yet developed the polynodal characteristics allegedly typical of EMRs. Clear functional and ethnic zones were present, such as an Economic Planning Zone (EPZ), a Chinatown and a high-tech science park, but this was hardly comparable with Tokyo. Transformation began with the emergence of the SIJORI growth triangle.

SIJORI and the Emergence of the EMR

The SIJORI growth triangle (Figure 27.8) is often held up as an example of the economic cooperation between ASEAN nations that belatedly followed some two decades of rather less productive links in response to global recession in the 1980s. So it did, but the growth triangle can be examined through a much more complex series of filters, performing as it does a variety of functions and illustrating a range of changing developments in the region. In terms of practicalities, the Singapore link with Malaysia emerged before that between Singapore and Indonesia, and the declaration of a 'triangle of growth' was unilaterally announced by Singapore in 1990, with the two neighbouring states responding with rather different degrees of enthusiasm (Parsonage, 1992).

In essence, the triangle was driven by Singapore's initial failure to stimulate its second industrial revolution in the mid-1980s and the realization that pushing out downmarket, labour-intensive industries to neighbouring states had caused the republic considerable economic losses. These considerations were reinforced by the parallel appreciation of the future labour structure of Singapore, with its predicted downturn in overall size by the early decades of the next century (Cheung, 1990). One response to this dilemma was to encourage Singaporeans, particularly the better educated, to have more children (Drakakis-Smith et al., 1993), but in reality the economic problem which underpinned this policy was only

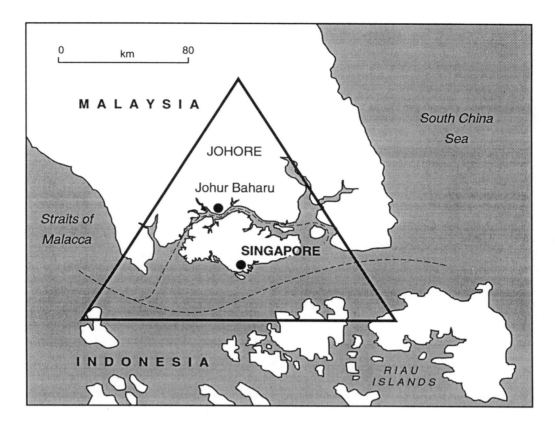

Figure 27.8   The SIJORI growth triangle

going to be resolved by the export of capital to labour. This, of course, was an option which was becoming equally attractive to the other Asian MIEs because of their own rising labour costs, and was being facilitated by the maturing nature of the economic enterprises themselves (Van Grunsven et al., 1995).

Thus, the impetus to shift investment off-shore was a function both of the development stage at which Singapore found itself, in common with the other MIEs, and of the specific features of the Singaporean labour market, principally its stagnating size and its greying nature. In addition, the increased pressures on space in Singapore meant that large-scale land users, such as industrial plant, were increasingly being threatened by rising land prices (Table 27.5). Essentially, therefore, the situation is one of vertical integration with companies headquartered in Singapore (where administrative, financial, marketing and distributive functions are located) downloading their labour-intensive operations to Johore in Malaysia or the Riau Islands, particularly Batam, in Indonesia. There are very few links between

Batam and Johore so that the arrangement is essentially hierarchical. In the Riaus especially, Singapore companies have benefited not only from the industrial integration but also from the construction and management opportunities offered in the wholesale restructuring of the island. It is worth noting in this context that most of the Indonesian labour in the Riaus comprises in-migrants from elsewhere in Indonesia.

Economic evaluation of the growth triangle is not the objective of this discussion and may be found in Lee (1995), Perry (1991) and Parsonage (1992). But in the context of the development of a Singapore EMR, it is worth noting that the growth of links between Singapore and the Riau Islands have been much more dominant in recent years than those with Johore, where Singapore is but one of many investors. In the Riaus, in contrast, a substantial proportion of Singaporean investment, amounting to over half of the current flow, is in tourism, with the islands of Batam and, more recently, Bintan being developed as major resort and leisure destinations for Singaporeans (Lim, 1992). However, it must not be assumed that

Table 27.5   *Comparative land and labour cost indices in SIJORI, 1989*

	Singapore	Johore	Batam
Land	100	96	54
Labour			
Unskilled	100	43	26
Semi-skilled	100	52	33
Skilled	100	67	33

Source: Van Grunsven (1995)

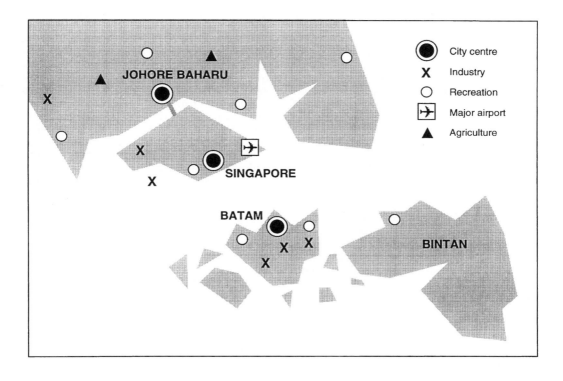

Figure 27.9   SIJORI polynuclei

Johore–Singapore links have been neglected; the planning of a second causeway bridge is well under way to cater not only for the industrial traffic, but also for the thousands of Singaporeans whose golf and recreational clubs are located in Johore state. Conversely, there is an increasing commuter flow into Singapore from Johore (Van Grunsven et al., 1995).

The physical consequence of these developments since the growth triangle was 'established' in 1990 has been the emergence of a closely integrated urban economic complex. Although it has a total population of only 4.5 million (Van Grunsven, 1995), the EMR covers an extensive area across two international boundaries and the physical barriers provided by water channels. The features of the EMR are, in spite of the physical and political barriers, comparable to those of other Pacific Asian mega-cities. Growth rates of population, economic enterprises and urban land uses are very high in the 'periphery' areas of Johore and the Riau Islands. In the former, in particular, the main transport corridor into Johore Baharu from Kuala Lumpur has itself become a magnet for subordinate industrial/residential complexes. In both of the satellite areas there has been extensive and rapid change of land use out of rural activities and into a range of EMR-oriented functions, from industry to golf resorts, and a complex patchwork of land usages has emerged.

As there are three national components to the SIJORI EMR, it is by definition polynodal (Figure 27.9), as the Malaysians and Indonesians have been eager to develop their own urban cores. Johore Baharu across the causeway in Malaysia currently has a population of over 500,000. These urban cores are being increasingly linked by surface transportation links. The new causeway between Singapore and Johore has already been noted, whilst the hydrofoil and jet-boat links to the Riau Islands offer an extremely frequent and well-used service.

Whilst these subcentre developments have been proceeding apace, so has the restructuring of Singapore itself. Although the Concept Plan came out in 1991, before the SIJORI developments began to accelerate, the built environment of Singapore was already being geared towards its role as a major regional, and possibly global, focus for development. Central to this role has been the communications revolution within and beyond Singapore. It is not only a computer literate society but almost a computer generated one. Its international communications are superb and have always featured primarily in development planning. Air links too are outstanding at the Changi Airopolis. The container port is virtually state-of-the-art (Government of Singapore, 1991).

In the wake of this structural transformation new subcentres are identifiable within Singapore itself, such as 'airopolis' and the new science parks, to set alongside the EPZ and container port of the immediate past. Meanwhile, the central city too is being reconstructed to fit Singapore's new status as a regional mega-city. The readily identifiable tourist zone, stretching from Orchard Road to the Singapore River, is being expanded through extensive reclamation projects and is also being driven upmarket by the pedestrianization and 'internationalization' of the Singapore riverside, the old China town and Little India. Not that this meets with universal approval (Yeoh and Kong, 1995), but it has created an environmentally friendly, pleasant and sophisticated side to Singapore's growth process.

A Sustainable EMR?

In a very fundamental sense the SIJORI EMR is an extremely vulnerable entity as it has no formal political existence, being based only on an informal 'understanding'. The considerable investments that have gone into this burgeoning mega-city are, therefore, at risk from any change in the political balance in the region. Admittedly all three countries are members of ASEAN, which should help, but they are also economic rivals, so that if and when Malaysia or Indonesia feels that it might be able or desirable to try to 'go it alone', then the political basis of the whole growth triangle could easily be undermined. As it is, there is no formal management structure for the EMR as a whole, although this is not unlike the situation in other mega-cities. The difference in SIJORI is that there is no overwhelming national authority to undertake any strategic planning that may be necessary in the future.

Below the political and managerial level there are other tensions observable within the EMR which could threaten future growth patterns. One of these is the ethnic unease which is inherent in predominantly Chinese capital exploiting Malay and Indonesian labour (Parsonage, 1992; Vatikiotis, 1993). Add to this the growing class differences and there is the potential for some considerable social disruption in the future, particularly in the context of regional Islamic fundamentalism which could seek to exploit this situation. Indeed, growing social cleavages in a variety of contexts, fuelled by labour migration to, from and within the various components of the EMR, could prove to be a more potent threat than economic change to the stability of the EMR (Booth, 1997).

Elsewhere, for example in terms of the 'brown agenda', the SIJORI EMR seems to be well placed compared to other urban agglomerations in the region. Singaporean urban planning sets high standards and is efficient at ensuring these are met. New developments in both Johore and the Riau Islands are maintaining these standards. Indeed, so proficient is Singapore at the management of urban-economic growth *per se* that Lee Kuan Yew has successfully sold this to the Chinese. Singaporean involvement in the development of Suzhou is to a large extent based on the Bintan Model and creates enormous profits for the Singaporean planning and construction firms involved. More recently the Vietnamese government has been impressed by Singapore's urban development achievements and is inviting its planners and administrators over in increasing numbers.

In summary, the SIJORI EMR, although limited in scale, exhibits many of the developmental features of its larger urban counterparts but perhaps fewer problems, particularly in terms of the urban environment. However, its political complexity, together with existing potential ethnic–class based tensions, hold a question mark against its future expansion. All seems well at present but the political vicissitudes of the region have long coloured the past and are likely to shape its future too. At present the international boundaries that slice across the EMR are both permeable and yet essential to the mainte-nance of production cost differentials. At some point in the future they could easily become impermeable and obstructive.

BANGKOK: THE PRICE OF SUCCESS

Thailand is often hailed by many international agencies as a new model for developing countries. Its growth rates have averaged at around 8 per cent for the past 10 years or so, and the World Bank puts this down to the structural adjust-ments which it recommended in the 1980s. However, as Dixon (1995a) has clearly shown, economic growth in Thailand was not only well under way before the World Bank recommenda-tions were made but, in addition, the structural adjustment programme was only partially enforced in Thailand. Much more influential in the economic growth process was the emergence of the regional division of labour and the flood of investment capital into Thailand from Japan and the Asian MIEs, largely into the electronics sector, taking advantage of one of the lowest levels of industrial labour costs in ASEAN. There is no doubt, however, that development in Thai-land was also much more closely aligned to the *lassez-faire* model beloved by free market capital-ists than the earlier generation of MIEs, and a free-wheeling, minimalist state interventionism has characterized growth in Thailand over the past 10 years.

Herein lie the shortcomings and future prob-lems for Thailand because the economic boom has resulted in serious social and spatial inequal-ities. Although Thailand's GNP per capita is almost three times that of Indonesia, the inci-dence of poverty in Thailand is as high, with the wealthiest 40 per cent of the population receiving well over three-quarters of national income, whilst the poorest quintile receives less than 5 per cent (Fairclough and Schwarz, 1994). What is more, this gap between rich and poor is rapidly widening (Komin, 1991). The most obvious geographical form of this inequality is the contrast between rural and urban areas, with the average household's income in Bangkok being almost four times that in the north-eastern regions (Fairclough and Tasker, 1994).

The Size and Status of Bangkok

Any debate on 'urban' issues issued in Thailand inevitably focuses, almost exclusively, on Bangkok. The Thai capital contains some 7 million people living within a sprawling metro-politan area and clearly constitutes an EMR, as

the discussion below will reveal. However, Bangkok is also one of the most primate cities in the world, containing 70 per cent of Thailand's urban population and being some 50 times larger than the second ranked urban centre in the country. In almost every economic, social and political measure Bangkok is overwhelmingly dominant (Dixon, 1995b; Wongsuphasawat, 1997) and, as a result, attracts migrants to its perceived opportunities on a colossal scale. The Thai government has, therefore, been faced with two fundamental and worsening development issues over the past four decades – the lack of development in the peripheral regions of the country and the problems of congestion and meeting basic needs in the capital.

The nature of that capital has, in fact, changed considerably over the past 10 years or so from a 'mere' primate city to a more clearly identifiable EMR, with urbanism well beyond the recognized metropolitan core. As with most EMRs, the area which falls into this conceptual categorization is somewhat fluid and variable. The core of the EMR is the Bangkok Metropolitan Area, but for the past decade or so population has spilled into the five adjacent provinces of the Inner Ring (Figure 27.10). More recently, the actual and planned expansion of economic activity into an Outer Ring has occurred apace. Seven of the twelve provinces affected are envisaged as being added to form the Extended Bangkok Metropol-itan Region (EBMR) (Dixon, 1995b; Wong-suphasawat and Parnwell, 1996).

The Outer Ring has witnessed considerable growth in recent years through relocation of industries away from central Bangkok. The cost of land and labour have been particularly impor-tant factors in this shift, with the latter being higher in Bangkok because of competition (Wongsuphasawat and Parnwell, 1996). Although the government has attempted to play a role in directing this relocation, for example through the provision of industrial estates, most decision-making appears to be the result of a combination of global/regional and local forces, in the process effectively 'selling the local to the global' (Peck and Tickell, 1994: 318). This weak-ness of the state manifests itself in a variety of ways from half-empty government industrial estates to a chaos of land uses in the peri-urban regions.

Partly because of the limitations of state control, there seems as yet to be little agreement as to the overall structure of development within the EBMR. Wongsuphasawat and Parnwell (1996) argue that with respect to industrialization, the driving force behind the spread of the EMR, there is no real pattern to recent change other than 'a seemingly random pattern of lines and dots . . . highlighting the importance of human agency in

Figure 27.10 Bangkok EBMR (Wongsuphasawat and Parnwell, 1996)

giving shape and pace to economic process' (1996: 7). Kaothien (1995), on the other hand, asserts that the northern section of the EBMR is envisaged as the principal area for industrial growth within the current development plan. However, the southeast seaboard area too will almost inevitably develop a range of industries due to the proximity to export facilities (Douglass, 1995). The planned relocation of Bangkok airport to the southeast will no doubt bring about further industrial relocation there.

In the interstices of these industrial corridors relatively little planned development is likely to take place, other than the improvement of road and rail facilities. What is actually happening is largely the consequence of local forces seeking to take advantage of the globalization of the urban/regional economy. Thus, agricultural land is being converted from rice to vegetable/fruit/ flower production for the affluent urban market and beyond, or is being taken out of agricultural use altogether to cater for other needs from the burgeoning EBMR. Golf courses, almost inevitably, form part of this scene, as do brick-works which use huge quantities of fertile local topsoil leaving behind vast areas of water-filled ponds whose fertility will take a considerable time to regenerate (Parnwell, 1994). Herein, once again, lies the real problem with *laissez-faire* urban-economic growth; the inability or unwillingness of urban management to ensure its sustainability.

Sustaining the Thai Miracle: A Worst Case Scenario

Bangkok is almost invariably cited as an example of all that is wrong with unfettered urban growth. Many of the newly emerging socialist states within the region claim unequivocally that their goal is to 'avoid another Bangkok'. Central to the many problems that currently affect Bangkok is the fact that there is no coordinated central administration. Numerous ad hoc committees have been established at one time or another but basically urban development is the responsibility of the various state sectors so that 'more than fifty agencies share the responsibility for planning, financing and managing the various departmental programmes' (Kaothien, 1995: 334). The consequence is a series of shortfalls in the provision of basic needs, eventual infrastructure and the redistribution of wealth. Social, physical and economic problems abound.

Bangkok is probably most infamous for its traffic congestion. This is not only the consequence of its rapid growth but also of poor planning. There is no integrated master plan to coordinate developments and the various suburbs have constructed their own roads, so that the overall result is a fragmented patchwork of roads. There are huge projects in the pipeline for a mass-transit system, a light rail network and super highways – estimated to cost $15 billion (Pendakur, 1995) but not scheduled for completion before the late 2000s. Meanwhile the public transportation system cannot cope and the streets are clogged with private cars, trucks and motorcycles most of which use leaded fuel. Air pollution levels are thus very high in Bangkok, well above recommended WHO levels, and over 40 per cent of its traffic policemen suffer from respiratory or related illnesses (Pendakur, 1995: 186).

Environmental pollution in general in the EBMR is severe from both industrial and domestic sources. The great majority of plant for some of Thailand's most noxious industries, such as lead smelting, is located in the region and only 2 per cent of the waste receives proper treatment. Millions of tonnes of untreated domestic waste also find their way into the major waterways of the city since only 2 per cent of the population are connected to a sewerage system. Water pollution is, therefore, a serious issue, one which is exacerbated by the rapid draining of underground aquifers. An associated and consequential problem is shrinkage and land subsidence which in turn leads to serious flooding problems during the wet season. Most of these floods are of badly contaminated water.

As might be expected, most of the worst environmental problems are experienced by the poorest households. Public low cost housing schemes have been inadequate to cope with the huge influx of migrants, in terms of quantity, location and cost. Douglass (1995: 66) reports that 70 per cent of the eligible residents sell their housing rights to higher-income families. The consequence is a proliferation of small- and large-scale squatting in and around the urban area. At one stage the Klong Toey squatter settlement on Port Authority land was estimated to house well over 100,000 people, but there are reports of up to 1,700 separate 'slum' areas housing around 1.5 million people. Many of these settlements are on private land and 'accidental' outbreaks of fire which clear the areas are not infrequent.

It may be wondered why widespread poverty and inadequate provision of basic needs does not induce a political backlash from the urban poor. But, as elsewhere in the developing world, many of the poor are recent migrants and have experienced equally bad if not worse conditions in their rural districts, and are prepared to wait for the opportunities that the city is perceived to offer to come their way. There have, of course, been spontaneous social challenges to the status quo – the most recent being in 1992. But these are largely frustrated middle-class outbursts and do little to disturb the slow political quadrille which is so stable a feature of Thailand, between the military and the landed/business elites.

The failure of the government in Thailand to grasp the nettle of sustainability for its capital city-region will, however, have a more widespread impact than the immiseration of life for so many of its residents, important though this is. The deterioration of the urban environment in its fullest sense, from widening congestion and pollution to increased poverty and crime, is already beginning to discourage international investors. Why put up with the headaches (literal and organizational) of Bangkok when a far more acceptable and efficient environment is available in Singapore, or a far cheaper labour force in Hanoi?

HANOI: EMR IN TRANSITION

Size and Status

In 1986 Vietnam began a programme of reforming known as *doi moi* which opened up the economy to international capital and introduced elements of the market economy. By the 1990s

GDP growth rate was around 9 per cent per annum, mostly from the industrial and construction sectors. It is important to remember, however, that this growth has emerged from a very low base level and the industrial sector still comprises only 20 per cent of GDP and employs only 11 per cent of the population. Recent growth has been dominated by regional investors and trading partners, particularly Japan and the MIEs.

The great majority of this reform and growth is heavily concentrated around the two main cities of Hanoi and Ho Chi Minh, where almost 90 per cent of foreign investment has occurred. It is, however, difficult to assess the impact of this activity on the urbanization process because of the unreliable and contradictory nature of the related data. According to official statistics, the urban proportion of total population in Vietnam has been around 20 per cent since the late 1980s, whilst the definition of what constitutes an urban population has changed over this period. There has also been a breakdown in the official urban registration system, and most Vietnamese planners – and almost all empirical evidence – suggest that a massive shift of population into the two main cities is taking place (Drakakis-Smith and Dixon, 1997).

In 1994 the National Institute for Urban and Regional Planning estimated that the annual rate of growth of the urban population was some 2.5 per cent, similar to the national rate of population growth. The level and rate of urbanization was not 'expected' to increase until the early decades of the next century. This suggests that the consequences of rapid urban growth are seen as problems for the future and that official statistics are being manipulated to give the present government the leeway to encourage further economic-urban growth. Official statistics indicate that Hanoi has a population of just over one million, whilst Ho Chi Minh City has just over 3 million. However, there is little doubt that rapid growth beyond the formal boundaries of both cities has taken place and more realistic estimates were in the region of 3 million and 5 million respectively (GKW-Safege, 1995). Even these totals do not reflect the transformation which is currently occurring in Vietnam, with urban land uses and economic activities moving extensively into the densely populated areas of the Mekong Delta and Red River basins around the two cities, engulfing many villages and linking them into sprawling incipient EMRs (McGee, 1995b). For Hanoi, in particular, such growth is happening at a bewildering pace. The surrounding Red River basin has rural densities of 800 per square kilometre and the provinces surrounding the capital contain some 21 million people, 29 per cent of the national population (Figure 27.11).

Sustainability and Growth in Hanoi EMR

It is the often stated aim of Vietnam's planners to avoid the emergence of another Bangkok, but this is exactly what is happening. Given the problems with data reliability discussed above, it is difficult to underpin a full review of the conflict which has appeared between sustained economic growth and sustainable urbanization. This discussion will, therefore, focus on three aspects of this debate – poverty, basic needs and the urban environment.

As with the data on urbanization *per se*, there are substantial variations in statistics on poverty in general and urban poverty in particular. Government data (GSO, 1995) suggest that only 20 per cent of the population are poor or very poor, but the most widely cited survey (the Vietnamese Living Standards Survey of 1992) suggests that over half of the population is still in poverty. These figures are used interchangeably by the main international agencies without any real challenge (see ADB, 1995; UNDP, 1995; World Bank, 1995), although all seem to agree that poverty is much less prominent in urban areas, where it is usually cited as being half the national average. Almost all reports assert that poverty has declined and greater equality has occurred since the introduction of *doi moi* reforms, but other data do not confirm this. Khanh (1994), for example, alleges that the income disparity between rich and poor had widened from three to four times in the 1970s to six to seven times in the 1980s. By the 1990s it had widened further to 20 times in rural areas and 40 times in urban areas. Moreover, despite the concentration of economic growth in the cities, the more rapid growth of population has caused urban unemployment to rise from 5.8 per cent in 1989 to 15–20 per cent in 1993 (Hiebert, 1994; Khanh, 1994).

Life in the city is made more difficult by the higher cost of living which raises the poverty datum line to some two-thirds higher than that for the country as a whole (Tiem, 1995). As a result, low-income urban households spend almost 70 per cent of their income on food and yet more than a quarter are said to be undernourished (Padmini, 1995). The introduction of structural adjustment reforms to satisfy the IMF and World Bank development funding agencies has also exacerbated this situation by reducing subsidies on a whole range of basic needs, from food to education and housing.

In environmental terms, some of the most serious problems facing Hanoi relate to water. Not only is the capital subject to serious flooding, due to the poor maintenance of its canals and bunds, but there is also a major problem of access to fresh water. Some 40 per cent of the population in the capital have no

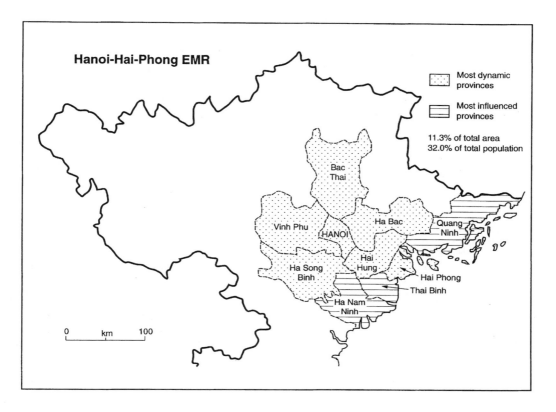

Figure 27.11   Hanoi EMR (Drakakis-Smith and Dixon, 1997)

access to the piped supply system commenced in 1985 and in the old quarters of the capital water still comes into pavement cisterns during the day and is pumped or carried up to the tenements in the evening (Minh, 1995).

Associated with these water supply and flooding problems are a range of sewerage and pollution issues. Many residents in the capital city draw their water from wells and the aquifers that supply these wells are not only becoming depleted but more polluted, and there have been serious outbreaks of cholera and dysentery amongst low-income populations. Add to this the limited treatment of both industrial and domestic water (Trouillard, 1995), and it amounts to a serious and accelerating problem of urban pollution. Growing motorized traffic and congestion is also beginning to contribute to this scenario (Hung, 1995). Hanoi has only 8.5 per cent of its surface area given over to roads, compared to 25 per cent in most Western cities; even Bangkok has 12 per cent. Congestion is likely to increase sharply in parallel with car ownership. And yet, despite these major environmental problems, remedial measures appear to

receive very low priority in recent development plans. In Hanoi's budget for 1991–4, investment in parks and zoos was twice that for roads, water supply and drainage combined. There is a National Law on Environmental Protection (1994) which sets out comprehensive responsibilities for individuals and for the public and private sectors. However, the pro-industry lobby has effectively prevented its enforcement.

Perhaps the most vivid illustration of the problems facing the sustainability of the present urban situation in Hanoi and Ho Chi Minh City relates to housing. Even before *doi moi* the state found it difficult to meet the demand for urban housing and only 30 per cent of the civil service were housed this way, often in poor quality accommodation. The state contribution to new housing construction is now down to some 16 per cent, with most of the non-state sector given over to speculative housing developments on the edge of the city designed to tap the growing domestic and foreign demand for more spacious accommodation. Often these developments flout city legislation on land use and construction standards (Khanh, 1995; Korsmoe, 1995).

The consequences of these changes have been very severe indeed for the growing low-income population of the capital. It is estimated that for Vietnam as a whole, some two thirds of the urban areas comprise slums and other dilapidated houses in need of renovation or repair. In Hanoi's ancient quarter, however, most of the renovation is for commercial redevelopment into minihotels and the pressure on low-cost accommodation has increased. In the capital, almost one-third of the population have less than 3 square metres per capita, with the average living space per capita having dropped by one-third since 1954 (Luan, 1995). The consequence is rising rents and an increasing floating population of casual labourers who come into the city each day from nearby rural areas, gathering at recognized locations and hoping to be taken on as day labourers.

These trends in the capital city of Hanoi do not bode well for the future. The optimistic claims of poverty reduction and growth of urban wealth must be treated with considerable care when the management of the capital is being planned. Wealth has risen in Hanoi but only for a few and there is considerable evidence that inequality has widened considerably. In this context, the wilful refusal to acknowledge the extent of the growth of Hanoi and the threats which this is posing not only threatens the sustainability of the present urbanization process but also the sustainability of the economic growth itself on which the future prosperity of the country as a whole is based. At present Vietnam's planners appear to be following policies that recognize the inevitability of the emergence of an EMR around Hanoi and Haiphong, complete with multinodal foci (Figure 27.12), but feel that this is under control (McGee, 1995b). This seems to be wishful thinking, and for Hanoi the worst-case scenario of becoming another Bangkok seems not too distant a prospect.

CONCLUSION

This chapter has revealed that whilst there are common demoninators characterizing the urbanization process in Pacific Asia, there are also substantial differences. On the whole economic growth can be more readily compartmentalized into a series of stages, each with identifiable geographical foci, than can the urbanization process. Notwithstanding the difference in size and scale, cities and more particularly large cities, have constituted the core of the development process in all countries no matter which rank of the flying geese they belong to. Moreover, as this chapter has indicated, the mix of economic and urban growth in Pacific Asia has given rise to EMRs or incipient EMRs across the development spectrum, a phenomenon fostered not only by rapid change within the cities themselves but also by the fact that many are surrounded by densely populated rural areas.

Within the Pacific Asian EMRs four different levels of influence can be detected – global, regional, national and local. Each one of these brings to bear a range of political, economic and cultural processes, the combination of which produces tendencies both of convergence and divergence in the nature of the eventual urban end-product. In general, the more global the influence, the more is the tendency towards convergence, whether this relates to production or lifestyle. The more local the influence, the more the EMR will retain an individual character that diverges from urban agglomerations in the West. But overlain across these four levels of influence may be placed another distinction – that between the private and public. At each level there are both institutional and individual forces operating. With regard to the former in Pacific Asia, we can clearly see, moving down the scale, the influence of the World Bank, the Asian Development Bank, ASEAN, the various national governments, municipal governments and local government. Not all have the same impact across the region and across the various levels of development – the structural adjustment programmes of the World Bank, for example, have differed in their impact in Hanoi and Bangkok. Further variation in the impact of these institutional forces is also caused by the nature and strength of the private sector involvement at each level. In general, the private sector takes considerable advantage of the gaps which exist in the administrative network, often giving rise to considerable conflict in land use patterns.

The consequence of this complex of forces as far as urbanization is concerned is an identifiable process with common features but with considerable local variation in character. In each of the EMRs discussed above, growth has been exceptionally rapid, particularly in the peripheral areas. All are multinodal in their growth patterns, with transformation of the central areas occurring as many of the earlier features of growth, such as manufacturing, move out to the boom areas of periphery, leaving the central districts to be restructured into financial centres, entertainment areas and the like. All too have overwhelmed or are overwhelming their surrounding rural populations *in situ*, with new activities being superimposed on villages which often lack the infrastructure to cope initially, resulting in major 'greenfield' development sites springing up nearby. In the interstices of these peripheral developments, often along major transport routes, land uses rapidly change to favour high value products demanded by the city, sometimes agricultural but more often recreational.

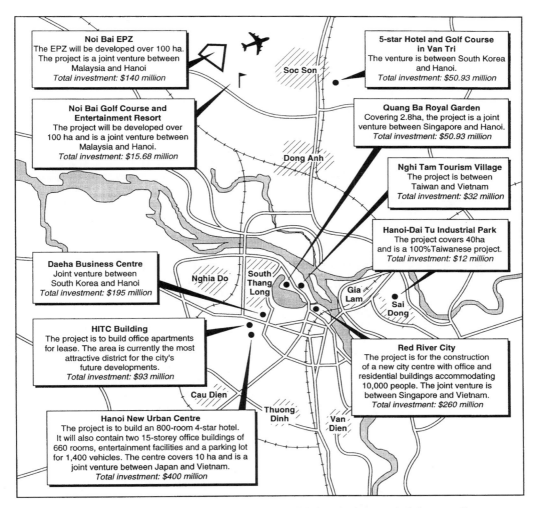

Noi Bai EPZ
The EPZ will be developed over 100 ha.
The project is a joint venture between
Malaysia and Hanoi
Total investment: $140 million

**5-star Hotel and Golf Course
in Van Tri**
The venture is between South Korea
and Hanoi.
Total investment: $50.93 million

Soc Son

**Noi Bai Golf Course and
Entertainment Resort**
The project will be developed over
100 ha and is a joint venture between
Malaysia and Hanoi.
Total investment: $15.68 million

Quang Ba Royal Garden
Covering 2.8ha, the project is a joint
venture between Singapore and Hanoi.
Total investment: $50.93 million

Dong Anh

Nghi Tam Tourism Village
The project is between
Taiwan and Vietnam
Total investment: $32 million

Hanoi-Dai Tu Industrial Park
The project covers 40ha
and is a 100%Taiwanese project.
Total investment: $12 million

Daeha Business Centre
Joint venture between
South Korea and Hanoi
Total investment: $195 million

Nghia Do

South
Thang
Long

Gia
Lam

Sai
Dong

HITC Building
The project is to build office apartments
for lease. The area is currently the most
attractive district for the city's
future developments.
Total investment: $93 million

Red River City
The project is for the construction
of a new city centre with office and
residential buildings accommodating
10,000 people. The joint venture is
between Singapore and Vietnam.
Total investment: $260 million

Cau Dien

Thuong
Dinh

Van
Dien

Hanoi New Urban Centre
The project is to build an 800-room 4-star hotel.
It will also contain two 15-storey office buildings of
660 rooms, entertainment facilities and a parking lot
for 1,400 vehicles. The centre covers 10 ha and is a
joint venture between Japan and Vietnam.
Total investment: $400 million

Soc Son: Noi Bai International Airport is located in this area. Malaysia is investing in two projects there: a golf course and an EPZ

Dong Anh: At the moment the area is one of the main vegetable supplying sources for residents in Hanoi

Nghia Do: Only five minutes from the West Lake, a property development area

South Thang Long: Attracts a lot of foreign investment. The infrastructure is good with the water supply system aided by the Finnish government

Gia Lam area: Called the 'Daewoo' area, it will soon become a satellite city of Hanoi

Cau Dien: Approved project of traditional cultural tourism villages

Thuong Dinh: Local industrial area

Van Dien: An industrial park with infrastructure not yet developed. No foreign investment at present

Figure 27.12   Hanoi polynuclear developments

Such maelstroms of development and change attract to their swirling centres inflows not only of capital but also of labour. In almost every country of Pacific Asia migration to EMRs or other major cities is extensive, not only from their own rural areas but also from other countries as the demand for both skilled and unskilled labour outstrips supply. Both of these migrational inflows can create complex ethnic and social tensions in the city – rural Malays moving into predominantly Chinese cities in Malaysia, Northeast Lao moving into Bangkok, Koreans moving into Tokyo and, soon, mainland Chinese moving into Hong Kong. The complex social tensions generated by such interaction, together with the extensive evidence of the persistence of poverty within the new urban wealth of Pacific Asia and the extensive concern with environmental deterioration, all draw attention to the fact that whilst the region's cities are undoubtedly centres for, often sustained, economic growth, they must also be evaluated as centres for sustainable development. In the latter context, many cities, including some in the more advanced economies of the region, still face serious problems. Nor will these problems be resolved until management at the urban, particularly the EMR, level is strengthened and restructured. At present many EMRs, being capital cities, are largely the creation of global and local market forces, together with some attempt at management from the national state. Urban planning at the level of the city itself, is often subsumed by national sectoral priorities, so that transport, housing or health care provision are dictated by various ministries rather than by an urban management strategy. Where such provision falls short of requirements, the prevailing institutional ethos at both global and national levels is to encourage the needy to help themselves – 'building local capacity' as this is often euphemistically termed. However, the informal sector has limited capacity to create work, raise incomes or provide improved access to water and sewerage facilities, and there is considerable evidence in Pacific Asia that unless urban management at the level of the EMR is restructured and given more resources, the growth process itself may be seriously challenged. Tokyo does not constitute a replicable model and Singapore appears, for the moment, to be the exception rather than the rule. In future, Pacific Asian EMRs may become more like Bangkok than they might wish.

NOTE

1. This account of Tokyo owes much to Roman Cybriwsky's excellent book on the city. As with his other writings, it is suffused with personal anecdotes about the city that bring the capital vividly to life.

REFERENCES

ADB (1995) *Vietnam: Country Operational Strategy Study*. Manila: Asian Development Bank.

Alvstam, C. (1995) 'Integration through trade and direct investment: Asian Pacific patterns', in R. Le Heron and S.O. Park (eds), *The Asian Pacific Rim and Globalization*. Aldershot: Avebury. pp. 107–28.

Amsden, A. (1994) 'Why isn't the whole world experimenting with the East Asian model?', *World Development*, 22 (4): 627–33.

Berger, P. and Hsiao, M. (1986) *In Search of an East Asian Development Model*. New Brunswick, NJ: Transaction Books.

Booth, A. (1997) 'Rapid economic growth and poverty declined: comparison of Indonesia and Thailand', in C.J. Dixon and D.W. Drakakis-Smith (eds), *Uneven Development in Southeast Asia*. Aldershot: Ashgate Publishing.

Cheung, P.P. (1990) 'Micro-consequences of low fertility in Singapore', *Asia-Pacific Population Journal*, 5: 35–46.

Cybriwsky, R. (1991) *Tokyo*. London: Belhaven.

Daniels, P.W. (1996) 'The lead role of the development economies', in P.W. Daniels and W.F. Lever (eds). *The Global Economy in Transition*. London: Addison Wesley Longman. pp. 193–214.

Devas, N. and Rakodi, C. (1993) *Managing Fast Growing Cities*. London: Longman.

Dicken, P. (1992) *Global Shift*. London: Chapman.

Dixon, C.J. (1995a) 'Origins, sustainability and lessons from Thailand's economic growth', *Contemporary Southeast Asia*, 17: 38–52.

Dixon, C.J. (1995b) 'The uneven development of the Thai economy'. Paper presented at the EUROSEAS Conference, Leiden.

Dixon, C.J. and Drakakis-Smith, D.W. (1993) 'The Pacific Asia region', in C.J. Dixon and D.W. Drakakis-Smith (eds), *Economic and Social Development in Pacific Asia*. London: Routledge. pp. 1–21.

Douglass, M.C. (1995) 'Global interdependence and urbanization; planning for the Bangkok mega-urban region', in T.G. McGee and I. Robinson (eds), *The Mega-Urban Regions of Southeast Asia*. Vancouver: UBC Press. pp. 45–77.

Drakakis-Smith, D.W. (1995) 'Third World cities: sustainable urban development I', *Urban Studies*, 32(4/5): 659–77.

Drakakis-Smith, D.W. (1996) 'Third World cities: sustainable urban development II – population, labour and poverty', *Urban Studies*, 33(4/5): 673–701.

Drakakis-Smith, D. and Dixon, C. (1997) 'Sustainable urbanization in Vietnam', *Geoforum*, 28 (1): 21–38.

Drakakis-Smith, D.W. and Graham, E. (1996) 'Shaping the nation state: ethnicity, class and the new population policy in Singapore', *International Journal of Population Geography*, 2 (1): 69–89.

Drakakis-Smith, D., Graham, E., Teo, P. and Ooi, G.L.

(1993) 'Singapore: reversing the demographic transition to meet labour needs', *Scottish Geographical Magazine*, 109: 152–63.

Drennan, M.P. (1996) 'The dominance of international finance by London, New York and Tokyo', in P.W. Daniels and W.F. Lever (eds), *The Global Economy in Transition*. London: Addison, Wesley Longman. pp. 352–71.

Dwyer, D.J. and Drakakis-Smith, D.W. (1996) *Ethnicity and Development*. London: Wiley.

Fairclough, G. and Schwarz, A. (1994) 'The boom continues', *Far Eastern Economic Review*, 4 August, pp. 34–5.

Fairclough, G. and Tasker, R. (1994) 'Separate and unequal', *Far Eastern Economic Review*, 14 April, pp. 22–31.

Gilbert, A. (1994) 'Third world cities: poverty, employment, gender roles and the environment during a time of restructuring', *Urban Studies*, 31 (4/5): 605–33.

Ginsberg, N., Koppel, B. and McGee, T.G. (1991) *The Extended Metropolis: Settlement Transition in Asia*. Honolulu: University of Hawaii Press.

Gottman, J. (1961) *Megalopolis: The Urbanisation of the Northeastern Seabord of the United States*. Cambridge: Cambridge University Press.

Government of Singapore (1991) *Singapore: The Next Lap*. Singapore: Times Editions.

GSO (1995) *Statistical Yearbook 1994*. Hanoi: Statistical Publishing House.

GKW-Safege (1995) 'Ho Chi Minh City water supply master plan'. Report to Asian Development Bank, Manila.

Hall, P. (1984) *World Cities*. London: Weidenfeld & Nicolson.

Hiebert, M. (1994) 'Stuck at the bottom', *Far Eastern Economic Review*, 13 January, pp. 70–1.

Hodder, R. (1992) *The West Pacific Rim*. London: Belhaven.

Humphrys, G. (1995) 'Japanese integration and the geography of industry in Japan', in R. Le Heron and S.O. Park (eds), *The Asian Pacific Rim and Globalization*, Aldershot: Avebury. pp. 129–50.

Hung, D. (1995) 'Hai Phong faces environmental damage', *Vietnam Investment Review*. 6–12 November p. 26.

Kaothien, U. (1995) 'The Bangkok metropolitan region: policies and issues in the Seventh Plan', in T.G. McGee and I. Robinson (eds), *The Mega-Urban Regions of Southeast Asia*. Vancouver: UBC Press. pp. 328–42.

Kettunen, E. (1994) 'Economic growth and integration of the ASEAN countries'. Paper presented at a NASEAS conference on 'Emerging Classes and Growing Inequalities in Southeast Asia', Aalborg.

Khanh, T. (1994) 'Social disparity in Vietnam', *Business Times*, 24–25 September, p. 4.

Komin, S. (1991) 'Social dimensions of industrialisation in Thailand', *Regional Development Dialogue*, 12 (1): 115–37.

Korsmoe, S. (1995) 'Growing pains', *Vietnam Economic Times*, September, pp. 18–19.

Kwan, C.H. (1994) *Economic Interdependence in the Asia-Pacific Region*. London: Routledge.

Lee, T.Y. (1995) 'The Johore–Singapore–Riau growth triangle; the effect of economic integration', in T.G. McGee and I. Robinson (eds), *The Mega-Urban Regions of Southeast Asia*. Vancouver: UBC Press. pp. 269–81.

Le Heron, R. and Park S.O. (eds) (1995) *The Asian Pacific Rim and Globalization*. Aldershot: Avebury.

Lim, C. (1992) 'Living investors and sun-worshippers', *(Singapore) Business Times*, 15 April, p. 9.

Luan, T.D. (1995) 'Impacts of economic reforms on urban society', in V.T. Anh (ed.), *Economic Reforms and Development in Vietnam*. Hanoi: Social Science Publishing House. pp. 134–96.

McGee, T.G. (1991) 'The emergence of *desakota* regions in Asia', in N. Ginsberg, B. Koppel and T.G. McGee (eds), *The Extended Metropolis: Settlement Transition in Asia*. Honolulu: University of Hawaii Press. pp. 3–25.

McGee, T.G. (1995a) 'Metrofitting the emerging mega-urban regions of ASEAN: an overview', in T.G. McGee and I. Robinson (eds), *The Mega-Urban Regions of Southeast Asia*. Vancouver: UBC Press. pp. 3–26.

McGee, T.G. (1995b) 'The urban future of Vietnam', *Third World Planning Review*, 17 (3): 253–77.

McGee, T.G. and Robinson, I. (eds) (1995) *The Mega-Urban Regions of Southeast Asia*. Vancouver: UBC Press.

Minh, N.V. (1995) 'Wading through H₂woes', *Vietnam Economic Times*, September, pp. 29–31.

Neary, I. (1997) 'Burakumin in contemporary Japan', in M. Weiner (ed.), *Japan's Minorities*. London: Routledge. pp. 50–78.

Padmini, R. (1995) *Report on Social Aspects, Vietnam Urban Sector Strategy Study, TA 2148-VIE*. Manila: Asian Development Bank.

Page, J.M. (1994) 'The East Asian Miracle: an introduction', *World Development*, 22 (4): 615–25.

Parker, P. (1995) 'Japanese structures: integrating state and corporate strategies', in R. Le Heron and S.O. Park (eds), *The Asian Pacific Rim and Globalization*. Aldershot: Avebury. pp. 87–106.

Parnwell, M. (1994) 'Rural industrialization and sustainable development in Thailand', *Quarterly Environment Journal*, 2: 24–9.

Parsonage, J. (1992) 'Southeast Asia's "growth triangle": a subregional response to global transformation', *International Journal of Urban and Regional Research*, 16 (2): 307–17.

Peck, J. and Miyamachi, Y. (1995) 'Reinterpreting the Japanese Miracle: regulationist perspectives on post 1945 Japanese growth and crisis' in R. Le Heron and S.O. Park (eds), *The Asian Pacific Rim and Globalization*. Aldershot: Avebury. pp. 37–60.

Peck, J. and Tickell, A. (1994) 'Searching for a new industrial fix; the after-Ford crisis and global-local disorder', in A. Amin (ed.), *Post-Fordism: A Reader*. Oxford: Blackwell.

Pendakur, V.S. (1995) 'Gridlock in slopopolis: congestion management and sustainable development', in T.G. McGee and I. Robinson (eds), *The Mega-Urban Regions of Southeast Asia*. Vancouver: UBC Press. pp. 176–93.

Perkins, D. (1994) 'There are at least three Asian models of development', *World Development*, 22 (4): 655–61.

Perry, M. (1991) 'The Singapore growth triangle: state, capital and labour at a new frontier in the world economy', *Singapore Journal of Tropical Geography*, 12 (2): 139–51.

Rigg, J. (1990) *Southeast Asia: A Region in Transition*. London: Unwin Hyman.

Rimmer, P. (1995) 'Moving goods, people and information: putting the ASEAN mega-urban regions in context', in T.G. McGee and I. Robinson (eds), *The Mega-Urban Regions of Southeast Asia*. Vancouver: UBC Press. pp. 150–75.

Rimmer, P. and Drakakis-Smith, D.W. (1982) 'Taming the wild city: managing Southeast Asia's primate cities since the 1960s', *Asian Geographer*, 1 (1): 17–34.

Robinson, R. (1995) 'The emergence of the middle class in Southeast Asia'. Paper presented to the EUROSEAS conference, Leiden.

Seidensticker, E. (1990) *Tokyo Rising*. New York: Knopf.

Selleck, Y. (1997) 'Nikkeijin: the phenomenon of return migration', in M. Weiner (ed.), *Japan's Minorities*. London: Routledge. pp. 178–210.

Selya, R. (1995) *Taipei*. London: Wiley.

Shibusawa, M., Atimod, Z. and Bridges, B. (1992) *Pacific Asia in the 1990s*. London: Routledge.

Thrift, N. (1989) 'Introduction: new models of the city', in J. Thrift and R. Peet (eds), *New Models in Geography*. London: Unwin-Hyman. pp. 43–54.

Tiem, N.V. (1995) 'Economic reforms and poverty alleviation in the rural areas', in V.T. Anh (ed.), *Economic Reforms and Development in Vietnam*. Hanoi: Social Science Publishing House. pp. 134–96.

Trouillard, P. (1995) 'HCM city choking on its soaring pollution', *Vietnam Investment Review*, 6–12 November, p. 26.

UNDP (1991) *Cities, People and Poverty*. New York: UNDP Strategy Paper.

UNDP (1995) *Poverty Elimination in Vietnam*. Hanoi: UNDP.

Van Grunsven, L. (1995) 'Industrial regionalisation and urban regional transformation in Southeast Asia: the SIJORI growth triangle reconsidered', *Malaysian Journal of Tropical Geography*, 26 (1): 47–65.

Van Grunsven, L., Wong, S.H. and Kim, W.B. (1995) 'State, investment and territory: regional economic zones and emerging industrial landscapes', in R. Le Heron and S.O. Park (eds), *The Asian Pacific Rim and Globalization*. Aldershot: Avebury. pp. 151–78.

Vatikiotis, M. (1993) 'Chip off the block: doubts plague the Singapore-centred growth triangle', *Far Eastern Economic Review*, 7 January p. 54.

Webster, D. (1995) 'Mega-urbanization in ASEAN', in T.G. McGee and I. Robinson (eds), *The Mega-Urban Regions of Southeast Asia*. Vancouver: UBC Press. pp. 27–44.

Witherick, M.E. (1981) 'Tokyo', in M. Pacione (ed.), *Urban Problems and Planning in the Developed World*. London: Croom Helm. pp. 120–56.

Wongsuphasawat, L. (1997) 'The extended Bangkok metropolitan region and uneven industrial development in Thailand', in C.J. Dixon and D. Drakakis-Smith (eds), *Uneven Development in Southeast Asia*. Aldershot: Ashgate Publishing.

Wongsuphasawat, L. and Parnwell, M. (1996) 'Between the global and the local: extended metropolitanization and industrial location decision-making in Thailand'. Paper presented at the Sixth International Conference on Thai Studies, Cheng Mai.

World Bank (1991) *Urban Policy and Economic Development: An Agenda for the 1990s*. Washington, DC: World Bank.

World Bank (1993) *East Asian Miracle*. Oxford: Oxford University Press.

World Bank (1995) *Vietnam: Poverty Assessment Strategy*. Report No 13442-VN. Washington, DC: World Bank.

Yeoh, B. and Kong, L. (eds) (1995) *Portraits of Places: History, Community and Identity in Singapore*. Singapore: Times Editions.

28

Post-Socialist Cities in Flux

GRIGORIY KOSTINSKIY

The transformation of socio-economic life which began in post-socialist countries at the end of the 1980s manifested itself first of all in the cities. Indeed, the cities were the initiators of many of the changes that were to follow. The bigger the city, the higher its administrative, financial, economic and cultural status, the more radical changes have been. Such a situation could be anticipated, where the cities, being involved both in the national and increasingly in the global economy, are particularly exposed to externally induced changes, besides being themselves the principal location through which change is mediated.

Of course, in the different countries of the former socialist bloc, whose territorial reach stretches from Central Europe to the Pacific, the depth of urban transformation is far from being identical. In this chapter the emphasis will be on the cities of Russia, whose socialist roots are much deeper than in Central and Eastern Europe. This helps explain why the extent of the transformation of the Russian and the other CIS cities lags behind that of the cities of Central and Eastern Europe. As for the rural areas of Russia (and the other CIS countries), the state-farm and the kolkhoz economy in fact still dominates, and have been practically untouched by transformation.

The current transitional processes involve a fundamental re-evaluation of the territorial economy – both cities and regions – with respect to the location, functioning and reorganization of productive activity (Hamilton, 1995). The transition has marked the trend towards the 'commodification' of places, which are exposed not only to economic, but also to social, cultural and ecological re-evaluation. Places must undergo a strict reassessment, answering the question as to how they can function as the loci for effective production within the framework of the local, regional, national and international economic systems. In such a reassessment the comparative advantages

and shortcomings of cities become clear. Market forces check the efficiency of the former functional interrelations and the division of labour formed under socialism, as well as generating new functions, and business connections.

In the socialist economy the non-economic factors of production – political, ideological, symbolic, social, military, technical – had enormous, if not over-riding importance. Their purpose was to demonstrate the superiority of communism over capitalism. This favoured the totality and 'gigantism' in the organization of space and physical planning, as well as emphasizing the significance of symbolic meaning, something in which the city, by definition, played a particularly important role. Planning was based on rigid, normative understandings, fixed for each type of settlement (French, 1995).

At first, the collapse of the planned centralized economy caused chaos. The depth of the chaos reflected the reality that the transition period was likely to be longer rather than otherwise; change, in the sense of the collapse of the 'old' and the introduction of new political institutions and of market mechanisms would not be achieved in the short-term. Indeed, a decade after the collapse the 'old', Russia remains firmly in transition.

The transition period began with liberalization of consumer prices, deregulation and the de-etatization of the economy. Following their introduction the more 'undesirable' companions of the transition period appeared: the break-up of old economic ties between enterprises which ensured the conditions of trading, unemployment, black markets and a decline in the living standards of the majority of the population. True, in parallel with this the saturation of the consumer market has occurred. If, under socialism, consumer markets suffered chronic shortages and lacked choice, now consumers are not starved of opportunities: now it is possible to get practically everything for money.

The basic stages of a transition period are:

1   a pre-transition (with a running down of the economy within the framework of the old system during which there were modest attempts at its restructuring);
2   a crisis (in which the destruction of the old economic structures outstripped the formation of new ones);
3   a post-crisis (in which processes of reconstruction and regulation begin to prevail) (Nefedova and Treivish, 1994).

The dismantling of state ownership and control together with the simultaneous development of the institutions of a market economy and the privatization of property and land in the commercial and housing sectors qualitatively strengthen the process of commodification. As the level of market exchange and commodification grew, new attitudes to urban locales and new principles of location began to emerge in the cities.

At the macro-level, that is at the level of a national settlement system, socialist countries did not differ significantly from those of the advanced capitalist countries. It was hardly possible to speak of any specifically 'socialist' hierarchy of settlements, a special rank-size order, or of any peculiarity of the leading socialist cities in terms of their primacy. The distinction between socialist and capitalist cities manifested itself rather at a local level, that is at the level of the city or the urban agglomeration. The specificity of cities in socialist countries was displayed not on a macro – and not even on a meso – scale, but at the micro-level, notably in peculiarities of intra-urban structure, in the character of their urban centres, suburban zones and in their housing formations (Musil, 1993).

CHARACTERISTICS OF THE SOCIALIST CITY

Factors Influencing Urban Development under Socialism

The specificity of cities of the former socialist countries can be better understood if we distinguish between two groups of factors. The first group includes those factors linked to the systemic properties of socialism, the centralization of power, central planning, the distributive system and the underdevelopment of a civil society. The second group unites the specific factors, connected with cultural and historical peculiarities of the separate countries, with the peculiarities of urban policy and traditions of municipal government. Looking at the present

transformation of the urban fabric, we must pay attention first of all to socio-economic factors. Thus, it is necessary to take into consideration that 'soft' elements of economy (for example, the reorientation of people towards enterpreneurial activities) change faster than do more rigid factors, such as the urban infrastructure. Adaptation to the market economy proceeds faster in those sectors of the economy that demand lower investment outlays and where the basics of entrepreneurialism have already developed (Domański, 1994).

As has been argued (Musil, 1993), several major factors influenced the development of socialist cities:

1   The non-existence of the market in land and the introduction of fixed land prices, so that for users – firms and enterprises – location within a city was almost irrelevant as an economic parameter.
2   Centralized management and the regulation of the housing sector (control of flat exchanges, purchasing of houses, sub-letting) by local authorities.
3   Nationalization of the exogeneous and endogeneous urban base, including retail trade and services; policies for their consolidation were based on the benefits of economies of scale and managerial convenience.
4   The general promotion of public interests over personal, and of interests of a higher level territorial unit (the state) over those of a lower level (sub-state) unit.

While these factors persisted throughout the socialist period, they were especially important during the initial stages of the socialist construction, characterized by the rapid progress of industrialization. Priority was given first of all to heavy industry, rather than to housing construction and the development of urban infrastructure. Only from about 1960 did a new stage in urban development begin: significantly greater attention was given to housing construction and to the development of services. The policy of mass housing construction resulted in radical changes in the shape and character of cities in the socialist countries: on the urban fringes (and in small cities, often in the centre) vast estates of standardized multi-storey houses appeared. In the socialist cities such residential districts are much more extensive than similar areas in the advanced capitalist countries. Such mass construction helped reduce the socio-spatial differentiation of the socialist cities. Yet such differentiation did exist, by virtue of distinctions in differences in income, by the development of cooperative housing (for the better-off households), and because of the occurrence of quasi-markets, and in particular the black market.

The Specificity of the Russian City

While the general concept of a 'socialist city' as applied to all socialist countries is convenient, it is necessary to take into account the specificity of cities in each of these countries. Typically 'socialist' cities were characterized by:

- State control over of urban land use.
- Complete absence of private property in land.
- State control over the housing economy (financing, realization of development, distribution of housing stock and its management).
- Wasteful land use, resulting from absence of land rent under socialism.
- Centralized organization of services and supply.
- The underdevelopment of services and locating of urban amenities quite regardless of the structure and volume of market demand.
- The domination of public over private transport.
- The exclusive importance of ideological symbols in the urban environment, including the monumental architectural style of public buildings, underlining the emphasis placed on the special importance of the urban centre.

Russian cities differ from those not only in the countries of Western Europe, but also of Central and Eastern Europe (the Czech Republic, Hungary, Poland, Slovenia, Croatia, Slovakia). A dominant feature of Russian cities is the absence of a civil society and of local participatory structures. This should not be attributed only to the notorious '70 years of socialism'. In Russian life little scope was given to any form of urban self-management before the October revolution. Thus, in the 1870s voting rights were severely restricted, being given only to one-third of home-owners.

Notions of civic duty, including participation, which in liberal democracies are an integral component of municipal political life, have not been a feature of the Russian mentality. The idea of self-management remains the exception even for the active minority of the cultural elite, who continue to place their reliance on the local politician (Glazychev, 1995).

In Russian cities any local self-organization of citizens is extremely weak. The 1960s, with the cooperative housing movement, represented the last burst of self-organization. At the present day, even in a Moscow of 9 million there are no more than two dozen local self-management structures of city-dwellers in the form of public councils (committees). Denied any rights for decades, the citizen is intimidated by the official authorities, is passive, but at the same time sceptical of change. Residents of multi-flat blocks are unable to manage them in a proper state. Neighbourhood committees, where they do exist, are only in an embryonic state, their influence on local official authorities minimal. Yet it is through the neighbourhood communities in which groups of population organized by place of residence may arise to ensure direct mass democracy. A strong civil society for Russia and other CIS countries is still a remote prospect.

New Factors Characterizing the Transition Period

The transformation of the spatial organization of cities has resulted from three main processes:

1. the spontaneous development of private business and the increase in the number of small and medium enterprises;
2. the diminishing role of the state both as a regulator of socio-political life and owner of economic enterprises;
3. the development of urban government (at both city and intra-city levels), whose purposes differ from those of previous *oblast* and state authorities.

The emergence of the market economy in the post-socialist countries has given an obvious impetus to the development of both the commercial and housing sectors of the urban economy. Typically the industrial sector has been in deep decline. In comparison with the end of the 1980s the level of industrial output and the importance of industry in the urban economy has decreased sharply; the appearance of new industrial plants is now an extraordinary event. This 'compression' of industry has been accompanied by branch restructuring. Some consumer industries – food processing for example – have, however, gained in importance.

Changes in the nature of the property market have resulted in a substantial increase in the number of new entrepreneurial agents whose activities have contributed to the apparent transformation of the spatial pattern of socio-economic life. These changes, most obvious in the central parts of cities, are connected particularly to the structure of retail trade and the development of conspicuous consumption. New office developments also characterize the centres of the bigger cities, altering in turn the city's demographic structure. The elderly and low-income households, traditionally more important in the city centre, have been displaced to more peripheral areas.

Small elegant private mansions and multi-storey rented houses, constructed in the pre-revolutionary period have been the object of considerable conversion. The former are used especially as offices (particularly as banks), the latter as luxury apartments. To live in the centre in a spacious apartment in a 'respectable' house is regarded as a status symbol of the 'new rich'. These large apartments, which in Soviet time were communal flats, have become a battleground for competing housing agencies which have made considerable profits through resettling the tenants and selling the vacated and renovated flats to the 'new rich' in the urban economy (Kostinskiy, 1994).

Commodification and rising costs of housing, diversification of housing types, widening differences in income and the expected introduction of a land market all promise further significant change in the housing sector.

THE COMMERCIAL SECTOR

The overwhelming majority of Russian cities, while traditionally 'overburdened' by industry, at the same time suffered from a serious shortage of services, retail trade outlets and infrastructure for entrepreneurial activity. Recent patterns of change in societal needs and progress along the path to a market economy urgently require the formation of such an entrepreneurial network as a necessary condition for the structural reorganization of the urban economic complex.

Changes in Retail Trade and Services

Improvements in retailing were apparent earlier and more clearly than in other spheres of economic activity. The reasons are evident: retail trade does not require large amounts of start-up capital (in comparison with manufacturing, for instance) and does not have to face the rigours of foreign competition. At the same time, under socialism this sector was recognized as being permanently under-invested, producing chronic shortages of consumer goods – consumer demand could never be satisfied. Now, with eradication of the consumer shortage problem the retail sector has developed very rapidly.

Retail trade in the socialist city was concentrated predominantly in the centre, which under the condition of the permanent commodity 'famine' drew shoppers not only from the urban fringes, but also from suburbs and surrounding rural areas. As distinct from the suburbs of Western cities, the outer areas of the socialist cities lacked big shopping centres orientated to 'motorized' customers. The spatial allocation of retail enterprises and services under socialism was determined not by free market forces, but by the decisions of urban administrators (gatekeepers, in

effect), who determined where to locate a 'trade point'. While this allocation was not completely indifferent to the needs of buyers, in practice it was buyers who needed to adjust their trip behaviour to the spatial networks of shops, and not vice versa.

The transformation has been both quantitative and qualitative, where in the retail sector first, the number of 'trade points' has grown sharply, and secondly, the quality of retail enterprises has risen remarkably. The most spectacular increase in the volume of retailing has been seen in the city centres and in a few 'outlier' establishments along the main avenues. As a result, low-order shops selling mundane goods have been quickly superseded in the city centres, where competition for space was intense, by 'up-market' shops, outlets that were simply not present in the socialist era.

The market transition began early in 1992 with the legalization of street trading. This kind of activity brought traders rapid (albeit, rather modest) returns. At first, trading was open to all, with no necessity for traders to hold a licence. Hawkers clustered outside large shops, department stores, metro, bus and railway stations or other crowded places. They sold their goods directly on the sidewalks – from boxes, and (less frequently) folding tables. The numbers engaged in street trading reached such dimensions that within a few months the urban authorities were compelled to regulate the process. First, the traders were obliged to have a licence, and secondly, trading was prohibited from certain places. Traders were forced to leave the CBD, in particular locations adjoining important official buildings. This extraordinary flowering of private sector retail trading in the post-socialist city proved short-lived.

As a second stage, traders operating along the sidewalks have been replaced by lines of booths or kiosks. Originally these kiosks were unspecialized and tended to offer the same set of consumer goods (drinks, packed foodstuffs, cigarettes, cosmetics, haberdashery). Kiosks came in all shapes and sizes, lacked electricity and refrigerators and, in general, were poorly adapted for trade. Then, as the urban authorities began to impose increasingly rigid controls, these original kiosks were replaced by others of a more attractive design and more standard format. Eventually better-equipped retail stands and trading mini-complexes began to appear. The number of kiosks is now falling, their role being taken firstly by these stands and subsequently by formal shops (Riley and Niżnik, 1994). Booths and stands were born of necessity, substitutes for normal shops which were lacking initially. The shortage of premises in the existing trade stock gave rise not to the construction of new shops (for initially businessmen simply had no money or time for new construction), but to the simple subdivision of existing premises at minimal cost to its refurbishment.

One specific development within the wholesale and retail trade in the current transition period has been the emergence of food and clothes markets outside the central parts of the cities – closer to the dormitory zones and main transport arteries. Urban authorities have allocated these markets empty sites in busy areas near railway, metro or intercity bus stations, and large stadii.

Privatizing Retail and Service Poperty

At the very beginning of the process of privatization and reform retail property was offered to potential retailers at discounted prices, much below their market value. But now in Moscow premises are sold at close to the true market price. Ninety-five per cent of potential buyers are not ready to redeem rented non-residential premises. These properties are redeemed only by those enterprises with good prospects. Emphasis has gradually shifted to competitive (auction) sales and investment contracts.

Municipal governments regulated privatization, setting auction starting prices for properties put up for sale based on their spatial location inside the city. They established a system of bid-rent pricing based on their urban location, and differentiated according to the type of shop or service enterprise (Riley and Niżnik, 1994). The aim of such a system was to retain certain socially important types of retailing and services.

Nevertheless, following privatization many of the former state-owned retail properties despite state regulation, changed their profile; and subsequent changes of (private) ownership resulted in further changes of profile. Very frequently such changes were unavoidable, as many kinds of services had simply been unavailable in socialist cities (currency exchange bureaux, tourist bureaux, casinos, etc.).

With the desperately limited supply of shopping outlets, retailers renting premises often let part of them to other enterprises. Frequently a former 'socialist' shop in the process of privatization would be partitioned into several smaller outlets. For example, the owner of a food shop might sub-let a part of its floor area to a shop selling, say, shoes or books. Only the construction of a light partition was necessary to divide the premises, but at times even this was regarded as unnecessary. The (sub-)leasing of premises to small private retailers is very widespread – this being the easiest source of profit for re-employing the use of the bigger, formerly state-owned retail outlets.

The retailing sector has undergone radical changes in its spatial pattern. Lower order retailing, such as, for example, vegetable or grocery shops, has been pushed out from the centre, while simultaneously higher order outlets are concentrating there as only they are capable of paying the high rents demanded for attractive sites. As a result, the city centre has become a more specialized trade area than before.

An even more acute shortage of premises has been experienced in the case of professional services, for whom booths and retail stands are not an option. A solution was found in the renting out of office space first in the premises of official state organizations and departments, and later in buildings belonging to the private and mixed sector. For example, 'budgetary' research institutes or the repositories of culture (museums, libraries etc.), finding themselves in straitened economic circumstances, would be compelled to let a part of their premises as offices to the new 'capitalist' sector. Even within the entertainment sphere, where state subsidy always appeared to be essential, a change of profile can be seen. For example, cinemas designed in the 1960s and in the 1970s with spacious glass halls have proved to be very adaptable as automobile and furniture showrooms.

Renting and sub-renting, evading state and municipal financial bodies, is a perfect field for financial abuse and corruption. A survey has shown that the Moscow Committee on Property officially rented out only 14 million square metres of non-residential premises of the 50 million it controlled. The remaining 36 million square metres have been the subject of illegal transactions, evading the control of the Committee.

HOUSING SECTOR

Housing Provision and Its Quality

Housing provision in Russia and the countries of Eastern Europe is poor by Western standards: the floor area per person averages 18–20 square metres, whereas in the countries of Western Europe the norm is 32 square metres (Hegedus et al., 1996). Russia also lags behind its Western neighbours in the quality of the housing stock, its operational characteristics and in the organization of the living environment outside the house. The majority of the housing, in spite of being in many cases of quite recent construction is dilapidated and badly in need of repair.

Usually, the smaller the size of Russian(/) city, the worse is the quality of its housing stock. This tendency is connected with a system of financing which persisted for much of the socialist period whereby funding prioritized the big administrative and industrial centres. As a rule, the small historical towns have amongst the worst housing conditions.

A distinctive feature of the socialist city is the standard character of houses and apartments, with a limited choice of types and styles of residential opportunities. For the overwhelming majority of the population a dwelling represents an apartment in a multi-storey (predominantly of four to nine storeys) building. Multi-storey units built of standard prefabricated concrete panels are characteristic not only in cities, but also in the small towns and suburbs, though in these areas, high-rise blocks do not dominate, their height rarely exceeding five storeys. Individual single and two-storey houses within big cities are comparatively rare and, where they do exist, are situated almost entirely on the fringes of the city (as a legacy of a time when they represented independent rural settlements). Such low-quality houses (usually made of wood and lacking indoor modern conveniences) are characteristic in small and partly for medium-sized towns, where the latter were not touched by intensive industrialization.

The majority of urban families live in two- and (less often) three-room flats in multi-storey blocks. At the starting point of the economic transformation in 1992, 42–45 per cent of the Russian housing stock comprised two-room apartments, 32–34 per cent three-room units and 15–20 per cent one-room flats. Apartments with more than three rooms accounted for only 5 per cent of the housing stock. Human needs certainly far surpass the present level of provision. For instance, a questionnaire conducted in Moscow revealed that to meet housing preferences the composition of available dwellings should be: apartments with no less than four rooms, 30 per cent; with three rooms, 33 per cent; with two rooms, 25 per cent; and with one room, 10 per cent. These figures confirm the rather modest housing claims of city-dwellers, who would be satisfied by expansion of their present dwelling by only one room, implying that the total number of rooms in the apartment should equal the number of the persons in the family plus one.

There is little differentiation of the population by housing provision as in the socialist countries the housing conditions of a family depend on income to an insignificant degree. To a much greater extent quality of housing was determined by one's place of work and ability to make use of advantageous connections and privileges.

Other things being equal, housing provision is determined by size of family. On the whole, the greater the number of family members, the less is the floor space per capita. Single people and couples without children generally enjoy a higher than average floor area per person. Particularly serious problems arise with large families (that is those that already have five members) and young families, who are frequently compelled to live together with in-laws.

A correlation between housing provision and monthly income per family member is seen only in extreme groups, i.e. the richest and poorest, but even between them the ratio in housing provision (measured in floor area per person) is only 1.5 times while the difference in income is ten-fold (Pchelintsev, 1994).

In the large cities of Russia the quality of housing corresponds to that seen in countries where household income averages $6,000 a year. Owing to a fall in living standards, however, the incomes of Russian households are now much lower (Belkina, 1994) and at current levels Russians on an average income cannot afford even the basic running and maintenance costs of housing. Consequently, the condition of urban housing stock is very poor.

Privatization of Housing

At the beginning of the economic reform process in January 1992 the urban housing stock in Russia was almost entirely in state ownership. In urban areas 79 per cent (and in large cities 90 per cent) belonged to the state – either to local authorities, or state enterprises, ministries and departments. Private ownership of dwellings was common only in villages and small towns. Before the reform a legal housing market was completely absent except for small private houses.

The transformation of the housing sector was to be based on privatization and the creation of a market that would attract citizens' savings into new construction projects. It was thought that privatization of housing would offer the population at large a sense of participation in the new system of private property ownership, as for most of them their apartment was their sole significant property.

The initial stage of the housing reform process was characterized by two main measures: the appearance of a law on mass housing privatization and the transfer of state housing stock on the balance of local authorities. All occupiers were given the opportunity to take ownership of their dwellings free of charge. This, of course, was of greatest benefit to the most well-off layers of society (French, 1995). A household privatizing its housing unit simultaneously received the right to dispose of it without impediment: to let it or to sell without any restriction. The privatization of housing was also primarily an act of legitimization, which gave citizens the additional assurance that property would not be expropriated by a new power should there be a radical change of socio-political regime.

The selling and purchasing of housing has quickly become widespread. Currently it is the existing housing stock which forms the basis of the housing market rather than new construction. The primary market is growing only very slowly.

Under current conditions the citizen does not enjoy particular advantages through possession of private property rights. The positions of a tenant and owner living in the same apartment block do not differ except that the owner of the apartment can freely sell or rent the property.

Those who prefer private ownership of their property are primarily those who intend to pass on a housing unit to successors not currently living with them. Especially interesting is the opportunity afforded by ownership to transfer housing by right of succession. Those who rent an apartment from the state can transfer it by right of succession only to those members of the family who are registered in it. Privatization, on the other hand, enables an apartment to be transferred by right of succession even to those relatives who are not registered in it and is consequently an attractive option for the elderly living in previously rented housing units.

The authorities at first tried to speed up the process of privatization, but it has become obvious that the level of privatization is approaching its 'natural' limit (50 per cent in Moscow). While in 1993 600,000 Moscow apartments were privatized, only 100,000 were transferred to private ownership in 1994 and 1995. The process tends to embrace first of all the more expensive apartments and those in the city centre. By 1996 50 per cent of the housing stock in the central administrative district of Moscow was privatized, whereas throughout the city as a whole the level of transfer was 40 per cent. Those with a clear idea what to do with their residence within the housing market proceed to privatization, while those without such confidence are more hesitant to take this step.

Municipal housing has been privatized more easily than that in the control of departments. Large firms and government departments offer resistance to the privatization of the houses they hold, having been reluctant to part with what they consider as their property. As the economic transformation continues increasing numbers of business concerns will probably be convinced that maintaining their own housing stock is too expensive, and will seek to pass it on to the local authorities.

Affordability of Better Housing Conditions

In recent years the cost of construction of housing has risen sharply such that it now approximates levels in those countries where household incomes are no less than $30,000 a year, a figure which is ten times greater than the average income of the Russian family. The prospect of purchasing new housing is thus elusive for the overwhelming majority.

The material condition for solvent demand, as the example of Western countries shows, is the appreciable excess of monthly average earnings over the average cost of construction of 1 square metre of housing and mortgage lending rates of less than 10 per cent.

If in the countries of Western Europe the average price of a new standard flat represents an equivalent of 4–6 years of average net wages, and in Eastern Europe 10 years, in the case of Russia this figure surpasses 20 years (Housing in . . . , 1996). The situation is aggravated by the virtual absence of any borrowing or tax incentive schemes to finance private home ownership.

So far the housing policy objectively works in favour of the top and bottom social layers – that is, those who are capable now of purchasing a small house or an apartment and those who, lacking the means at all, have the right under the existing rules to receive municipal housing free of charge. This policy takes little account of the interests of the urban majority with average incomes (75–80 per cent of all city dwellers). Such families have some savings, but these are obviously insufficient for purchasing a housing unit. At present, according to a survey of urban households, 80 per cent (mainly belonging to the middle class) see no opportunity to improve their housing conditions. Long-term mortgage lending would help them to solve the problem of purchasing a new residence but mortgages are still practically unknown in Russia: lending institutions do not trust the solvency of the population and, vice versa, the population does not trust the reliability of the banks. In Russia no law on mortgages can yet be devised as the question of land ownership laws has still to be resolved. These problems stem from the fact that for more than 70 years land was owned by the state. During the lifespan of several generations buildings, roads and public spaces were developed without any regard for property rights (Purgailis, 1996).

In the socialist epoch the system of granting free-of-charge municipal housing to the so-called *ocheredniks* (persons on a waiting list) dominated. Such a privilege was given to the families having less than 5 square metres of floor area per person, and among them priority was given to invalids, war veterans, large families and the inhabitants of communal flats.

Being on a waiting list for a free-of-charge flat was, and still is, a long process: for example, in Moscow in 1996 apartments were being given to families whose names had been on the list since 1982. Approximately 22–26 per cent of Russian urban families are on such waiting lists (although in Moscow, where the housing situation is in general better, the number is lower, at about 13 per cent) (Belkina, 1994). It is noteworthy that majority opinion in Russia still regards housing as something that should be provided free of charge by the state. It should perhaps be no surprise then that even now the most frequently expected way of getting housing is through a waiting list (25–33 per cent), while the second most important opportunity is thought to be an apartment exchange (20 per cent) (Abankina and Zuev, 1994). Today, however, the authorities have few possibilities for granting housing free of charge even to the most impoverished and needy families.

The construction of municipal welfare housing can only be funded by the sale of other housing on the commercial market. In order to provide a Moscow family in need of social assistance with a free-of-charge apartment, it is necessary to sell two or three apartments at the 'market' price (although until recently the ratio was almost the exact opposite).

The mechanisms of the normal, commercial way of obtaining housing, orientated to those on average incomes, are gradually developing (a housing bonded loan, savings in combination with loans on favourable terms etc.).

Those on the housing waiting-list (*ocheredniks*) can accelerate the process of receiving a dwelling if they make a partial investment from their own funds (in addition to subsidy from the urban authorities). In Moscow *ocheredniks* pay from 30 up to 95 per cent of the cost of an apartment. The size of the grant is calculated in accordance with tables that take account of duration of waiting time and per capita income in the *ocherednik's* family. The majority of subsidies are of the order of 60 to 70 per cent of the cost of the housing unit.

The authorities are ready to enter into a system of discounts, grants and other mechanisms provided that a certain part of the cost of purchasing housing will be met by the occupants themselves. Currently, the major problem in housing policy is creating an effective level of demand for housing through 'sparing' models of the housing credit. Social guarantees were formerly widespread, and thus any radical move away from them provokes a sensitive reaction.

Housing Construction

During the socialist period housing construction was based on budget assignments and plans; thus, problems with financing the work did not arise. Today, when both local and federal budgets are extremely constrained, the question of funding construction work has been pushed to the forefront.

The transition period has been characterized by a sharp reduction in the volume of housing construction. Simultaneous stagnation of the urban population, however, has prevented the problem of lack of housing from reaching crisis proportions. Moreover the standard of housing construction has been rising: new dwellings are more spacious and built to higher specifications.

Currently it is neither plans nor central authorities allocating investment to housing construction which determine its size. Construction is now largely a consequence of consumers' decisions, based primarily on household incomes (present and prospective). Demand is now the major factor, and a producer's market has been replaced by that of the investor. For the first time for decades former socialist countries faced a situation of overproduction, that is, an absence of buyers. As newly constructed dwellings failed to find buyers, it became necessary to halt many planned building projects – a phenomenon previously unknown.

The fall in demand for housing is caused by three interrelated factors:

1   General impoverishment of the population, caused by the devaluation of money savings and the decrease of real incomes.
2   Sharp growth in the cost of apartments with the withdrawal of state subsidies in the sphere of housing construction.
3   Commercialization of housing finance with the (virtual) absence of state support of investors.

As a rule of thumb, nowadays the larger the city, the more new housing is constructed in it (in relative figures per thousand inhabitants). The rate of housing construction is closely connected with the incomes of city households.

Housing construction by private investors is growing in importance and private developers have become the main builders. In Moscow in 1995 92 per cent of the new housing stock was built using non-budget means. Even during recession private companies have managed to increase the volume of construction, setting a generally high standard of new urban and suburban housing.

Large industrial and transport enterprises, ministries and organizations have sharply reduced their volume of housing construction. The 'departmental' segment, previously very strong, has now been curtailed. The industrial enterprises have no money for running and maintenance costs (central heating, water supply) and capital repairs of the houses belonging to them, and if they can afford it, local authorities are accepting responsibility for departmental housing stock and its associated social infrastructure (kindergartens, clubs, stadii etc.). Those industrial enterprises which built residential accommodation now seek new forms of providing housing for their workers (instalment selling, reducing the price of an apartment depending on length of work experience, selling a certain percentage of apartments at market prices etc.); in any case, the system of completely free-of-charge housing for employees is over.

Housing cooperatives, widespread from the 1960s to the 1980s, have reduced their building programmes in both relative and absolute terms. For instance, in Poland the volume of cooperative housing built in 1994 was one-third of the level of 1992. The reason lies in the fact that this form of construction also was significantly subsidized by the state. In the former USSR cooperative housing was only partially (30 per cent) financed out of income by the residents whereas the basic part of the charges was covered by urban and federal budgets. In Russia by 1994 federal and local subsidies had run low. As a result, support of housing cooperatives by local authorities was maintained practically only in Moscow.

To enter a housing cooperative is thus very difficult, with priority given to preferential groups of the population: first to *ocheredniks* (if they want to speed up receiving housing), then to those on the housing cooperative waiting list, where priority is given to invalids and war veterans, to the families of killed servicemen and to those born in Moscow.

In 1996 in Moscow 50 per cent of the costs of cooperative housing had to be met by the family, 40 per cent by the city budget and 10 per cent came from selling 10 per cent of apartments in each cooperative society at market prices. Intending residents had to pay 30 per cent of the cost of housing at the outset, prior to commencement of construction, while the remaining 20 per cent had to be repaid during the construction period.

SOCIAL SEGREGATION

Social disparities in socialist cities were incomparably smaller than in the cities functioning in the conditions of a market economy. An orientation towards minimization of these disparities was proclaimed as the key purpose of the state socialist policy, conducted at both central and municipal level. Nevertheless, a certain spatial differentiation of population, connected with its

social stratification, existed throughout the whole socialist epoch (French, 1995). Earlier, it is true, it had manifested itself at the level of separate houses, with differences in quality of apartments and houses, rather than in differentiation of residential areas.

The process of equalization of socio-economic disparities in the cities, if it occurred at all then, was perhaps seen only in the 1950s and 1960s. In the 1970s it was replaced by an obvious increase of distinctions in the level and quality of life of the urban population. Then an implicit tendency became apparent to the territorial redistribution of population subject to social position rather than well-being. This brought a boom in the construction of houses for privileged groups and the ruling elite, and such apartments were correspondingly strictly allocated ('distributed').

This trend correlated with the introduction to the economy in general, and housing construction in particular, first of the pseudo-market and then (with the fall of the communist regimes) of overt market elements, reflecting the growing interest of the ruling elites in providing themselves with material privileges, including housing. The idea of 'prestigious' and 'non-prestigious' districts and streets once again became popular.

The stratification of urban population by level of income should now lead to their spatial differentiation. A new suburban settlement of private single houses, surrounded by a continuous stone boundary with a projected entrance, serves as a concrete example of an overt spatial segregation by income. The growing tendency to segregation is evidenced, for example, by the fact that the 'mixing' of different income groups within an apartment block is very unstable, even 'explosive': the rich families will try to leave at the first opportunity. At the moment often one section of apartments in a multi-storey house is sold at market prices, while another is transferred to the contract organization at cost prices, with the remainder transferred to the city and allocated among those on the waiting-list free of charge. As a result one house combines families from various social and income groups. Those who paid from their own pocket are dissatisfied that their poor neighbours do not want and cannot afford additional charges to maintain the entrance and courtyard in a decent condition, are even inclined to vandalism. In their turn, the families that received accommodation free of charge feel a psychological discomfort from seeing the opportunities of a more affluent lifestyle. Not surprisingly, it is difficult to sell apartments in such 'mixed' houses, since the buyers do not want to live 'nobody knows with whom' (Kaganova and Katkhanova, 1994).

Nevertheless, despite the rapid stratification of society on a material basis, one cannot at present foresee a strong social polarization of the urban space with the formation of extensive areas of poverty and of wealth. Several circumstances support this argument:

1 The process of housing privatization in the cities is already far advanced and for the most part the apartments belong to their previous occupants, the majority of whom are not inclined to sell.

2 The state is unlikely to refuse the different forms of guardianship and social help to the citizen–owners of apartments, who cannot afford to cover the rent and running costs.

3 As the broad masses cannot afford to buy new accommodation, a widespread movement is not expected. It is solvent demand and active housing construction that have created in the Western cities not only socially more or less homogeneous residential areas of various grades, but also a mechanism that ensures a process of movement between socially stratified districts.

4 As the *nouveau riche* moving to the centre and to the prestigious suburbs comprise people originating from different social layers and from different districts, there is no corresponding obvious 'clearing' of certain residential areas.

5 Post-socialist cities lack distinct 'ethnic' regions, similar to those seen in large Western cities.

Thus, even the complete freeing of market forces will not bring radical transformation of spatial patterns and social segregation in the post-socialist cities. The structures that have formed over the decades of socialism appear resistant to widespread charge.

PROBLEMS OF THE LAND MARKET

In the socialist city no land market existed and all forms of land use were supervised by state authorities. The absence of a land market meant the absence of an effective mechanism for transition from less economic to more economic means of land use. As a result the residential density does not diminish with distance from the centre to the urban fringe and often even increases. This is the precise opposite to what is seen in Western cities, functioning under the conditions of a market economy. Moreover, the residents of multi-storey blocks located on the fringes of the cities are not rewarded for their long commuting distance by ecologically cleaner and aesthetically more pleasing environments. Plants, factories and warehouses located close to city centres have no economic incentive to withdraw outside the city limits, as land costs them practically nothing. Under the absence of a true land market the factor of land value is not strong enough to compel factories to economize in their land requirements.

Even under privatization, housing construction is still carried out in the absence of a land market. The distributive mechanism of allocating building sites and the logic of socialist city building, which continues to determine the situation in cities, completely ignores the key mechanism of urban land use in the market economy – demand (Kaganova and Katkhanova, 1994).

The major part of multi-storey construction occurs on sites allocated for mass housing construction according to former urban master plans, thus continuing the tendencies of the socialist economy. As for 'cottage' type of construction, it is conducted on a casual basis (where the developer can manage to make a bargain with the local authorities).

In the secondary housing market (purchasing existing apartments from their former owners) the location factor is distinctly reflected in price. In the primary housing market (of new apartments) on the other hand, the factor of location works poorly, as the developers are not free in the choice of building sites. Consequently, the buyers of new housing also lack choice. At first practically all new housing was bought up, but later when the most pressing demand was satisfied, developers faced the problem of non-competitiveness: the housing they offered on the urban periphery was of the same quality and price as that much better located and offered by the secondary market.

At present land in Russian cities can still not be bought or sold: it can only be leased. The zoning of urban territory and the rates of land taxation for each zone have been formulated. Nevertheless, in practice the problem of determination of land payments still exists. Land taxation tariffs are clearly understated and are many times lower than in Western cities. As a result city budgets receive much less money than they should. In Moscow in 1994 the receipts from land payments made up just 1.3 per cent of the city revenue (in the revenues of Western cities this share is equivalent to 20–40 per cent).

In the situation of the formation of market relations the authorities are afraid to lose control over urban land. They fear it will be promptly bought out and become a basis of hitherto unprecedented speculation. In addition, a psychological and xenophobic fear of foreign domination through land ownership prevails (Purgailis, 1996). Therefore it is most likely that urban land will be given the legal status of 'property for usage'. Such an approach envisages that land can be sold devised, exchanged, leased, mortgaged, included as a share in the total assets, but it must belong to the city. The 'proprietor for usage' should regularly pay land taxes to the city, unlike the proprietor who buys land in private property and repays its cost at once. In this case the object of purchasing/selling represents fixed capital, located on the site, and there is no division between cost of land and cost of the real estate. In the case of privatization of the object its owner automatically becomes the owner of the land on which the object is located. It is considered that in this case the land remains in state (city) hands and simultaneously does not impinge upon the rights of the land users, giving them opportunities for extracting entrepreneurial income.

Thus, the most urgent and dominant problem of the development of a housing market is the elaboration of a market-orientated municipal land policy. Of course the ideal would be a policy which sets up a mass market of rights in available land. But as the authorities of many cities are not prepared to pursue this kind of policy, initially it would be possible to be content with a reorientation to the 'centripetal' concept of urban development and to enter the mechanisms of the account of real territorial demand of developers and buyers of all types of real estate, housing included.

CHANGES IN THE CITY CENTRE

Especially striking transformations, as already mentioned, have taken place in the centres of post-socialist cities, and the centre of Moscow is an uncontestable leader in these changes. Moscow has not seen such a powerful renovation of its centre for several decades, as during the period of mass housing construction on the periphery (that is, since the late 1950s) the reconstruction of the historical nucleus was neglected.

The transition period has coincided with the termination of the fast population growth of Moscow and other big cities, which provoked urban sprawl. As there is practically no free space for mass construction inside the city, the Moscow authorities now pursue a course of using any inner city unused space (including former recreational space or inconvenient hill slopes) and of reconstructing the existing built-up quarters.

As the socialist cities always experienced severe shortages of the facilities necessary for retailing, and other services, in the transition period they have invaded the urban centre, at times to a spectacular degree. It is these activities that fill the formal and prim central streets with a new sense of vibrancy. In the socialist period the city centre landscape was characterized by low turnover of uses, particularly in the commercial areas. Now,

in the post-socialist city there is a high turnover of the tenants of premises in such areas, while the interiors of shops, restaurants etc. are being refurbished.

Following trade and services, the centre is now filled by numerous offices. At first existing buildings were used, but then new purpose-built office accommodation began to appear.

The metropolitan authorities have initiated the process of reconstruction of the urban centre on a wide scale, but the scope of work is restrained by lack of financial resources. Reconstruction of the city centre housing stock requires more than mere cosmetic repair. The refurbishment of seriously dilapidated buildings very often leaves nothing but a facade retained from the previous structure.

In the centre, where land is in limited supply, the built environment (for either housing or economic activities) has become increasingly dense. Single-storeyed constructions which are 'uneconomic' and which hinder development, are being demolished. Office accommodation appears in the empty courtyards, and when possible the existing buildings grow upwards: top floors or attics are added.

Any available sites, which are extremely rare in the centre, attract investors, and developers. One widespread practice which has emerged is where private investors, in return for funding construction work, receive from the local authorities the right of a long-term lease on a part of the new premises.

As the city centre is especially attractive for wealthier households, dwellers in the communal flats are being rapidly relocated. Paradoxically, the inhabitants of the centre are now moving to new apartments much more frequently even than during the years of high rates of mass construction.

The focus in commercial housing construction has also shifted from the periphery to the central quarters as those who can afford to buy new accommodation seek to settle away from the urban fringes: they are inclined to purchase more expensive accommodation, typically in prestigious central areas (the cost of an apartment in the old houses in such areas as Moscow varies from $1,000 to $2,000 per square metre). Construction in the centre is profitable for building companies: the infrastructure there is well developed and the return on capital investment is high.

SUBURBANIZATION

In the former socialist countries the processes of suburbanization did not play such an essential role as in the advanced capitalist countries. Socialist cities as compared with their Western counterparts in general were more compact and more densely populated. There was no mass rehousing of inhabitants from the nucleus of the agglomeration to its suburban zone. Under the conditions of the *propiska* restriction of settling in large cities proper, the population of their suburban zones grew due to the in-migration from outside agglomerations (that is from rural areas and small towns).

One of the main directions of present day change in the suburban zones is the construction of single- or two-storey dwellings, some with personal plots of land. The ratio between 'cottage' and multi-storey construction will change in favour of the former. High-rise construction will be retained only in big cities equipped with the engineering infrastructure needed. Cottage-style dwellings and small houses will be built on free (often reclaimed) sites within the big cities, but mostly in suburban zones. The share of such small house construction should increase even in Moscow, from 10–15 per cent up to 30 per cent.

Suburbanization in Russia is closely connected with the tradition for 'dacha' settlements, a dacha being a summer house, seasonally used by city-dwellers as a second home. Up to 20 per cent of families in big cities have such dachas; as a rule they are wooden structures, with a very limited set of conveniences (usually only electricity and running water are available) and a garden. The dacha settlements stretch along the railway networks and (to a lesser degree) along highways, radiating out from the big cities. Such dacha settlements around Moscow spread alongside certain railway radii without breaks for 30 kilometres.

The desire to have a holiday house, and attachment to the garden, is an important feature of Russian life and culture. For many households the dacha provides an opportunity to spend weekends and holidays away from their small urban apartments. In the 1970s and 1980s a large-scale planned allocation (practically free of charge) of small garden sites took place, but such sites were usually only available at some distance from the city. State policy allowed for the erection of only small houses, often in settlements which lacked basic infrastructure, including paved roads which made access difficult.

The dachas, especially those that do possess all conveniences and which are near to the city, can be used as an all-the-year-round home. Under the new economic conditions, where a car and building materials have ceased to be in short supply, these reconstructed summer houses are now becoming permanent dwellings. People now begin to think in terms of relocating to the dacha, leaving the urban apartment to the adult children.

Surveys show that 15–20 per cent of urban dwellers would like buy dachas (mainly, the summer modular houses) or, if they already possess one, then to reconstruct it in order to use it practically all the year round as a second dwelling. The demand for dacha construction, as opposed to urban construction, comes not only from well-to-do families, but also from households with average incomes.

In response to questions concerning their housing preferences, people increasingly name a small house or a single-family cottage. Previously such preferences were the exception. This reorientation of preferences of a significant part of the city population from multi-storey accommodation to houses with a garden coincides with the revision of official city planning policy placing the accent on suburban cottage construction.

A Presidential Decree in 1992 allocated sites in Moscow *oblast* for the construction of small houses and cottages. The project envisaged the construction by 2000 in the Moscow metropolitan region of 140,000 cottage-style units. Their floor area would total about half of the housing fund already existing in Moscow. However, this over-optimistic plan could not be realized. The construction, for which a few thousands hectares of suburban land were allocated, resulted in a much smaller real housing construction and in a shortage of buyers (by 1994/5 only 80 per cent of housing units had been sold). The causes are well known: the problem of discrepancy between the purchasing power of population and housing affordability, i.e. property price and the lack of mortgage opportunities. The primary demand was quickly satisfied, while the number of new buyers was insufficient.

As housing prices have risen sharply, developers have tried to reduce building costs. It is possible to reduce the cost of construction and at the same time to accelerate the building process by introducing new technology, in particular through 'modular' building. These 'modular cottages' erected using Western technology are not, to the Russian taste, which shows a preference for brick-built housing as a symbol of reliability and solidity (in 1995 they made up 78 per cent of all constructed cottages).

The developers' latest policy is to group cottages (not individual, but for several families) in separate settlements each with its own infrastructure close to the satellite cities around Moscow. The price of an apartment in such developments is about $100,000, which is several times less than its equivalent, a single-family house. The blocked form of housing of the 'town house' type has also emerged; it is characterized by vertical division, when a block-section belongs to one owner.

GENERAL STRUCTURAL SHIFTS IN THE NATIONAL SETTLEMENT SYSTEMS

The unique development pattern of cities in the former socialist countries was due not only to the absence of market mechanisms in the economy, but also to the implementation of a specific urban policy. This policy had as its long-term objective the uniform development of settlements and the liquidation of social distinctions in the conditions of life between cities and rural areas. In practice this policy was expressed in the strategy of the controlled growth of urban agglomerations and restraint of population growth of the biggest cities.

The transition period brought a partial change to these main trends which had characterized the development of the settlement systems of the socialist countries over the whole post-war period. The most prominent among recent changes has been the diminution of the overall urban population, with a great number of cities experiencing a population decline. First, the net gain of migrants from rural areas has turned into a net loss. Secondly, since the mid-1980s the natural gain of population has been diminishing and even a natural loss can now be observed. In these conditions the link between size of population and the growth of the city has become less attributable to demographic factors. The growth of a city can now be seen to be affected rather by its particular economic functions, its geographical position, the state of its infrastructure and environmental amenities, and the 'efficiency' of its local government activity.

The transition from the centrally planned 'socialist' economy to the Western market model stimulated predictions that the structure of the urban settlement would change. One such forecast predicted the amplification of the process of concentration of population in big cities and urban agglomerations accompanied by the relative decline of the small urban centres. This anticipated concentration of population has not been observed (Korcelli, 1995). As a prediction it stemmed from the assumption that barriers constraining the growth of big cities existed in the centrally planned economy and under market conditions these would cease to act. The centralized allocation of investments, a uniform country-wide system of prices and wages, an extensive use of labour force and of natural resources were all factors smoothing interregional disparities in economic development, and this in turn counteracted any spatial concentration of population. Besides, big cities operated a policy of severe restriction of inflow of non-residents by means of strict restrictions in registration (the system of *propiska*), which was necessary for getting a job as well as housing in the state or cooperative sectors. In Moscow the system of *propiska* was cancelled only in 1996 but simultaneously a system of registration was instituted. The new form also has the objective of constraining the inflow of non-residents for it requires those who move to the city to pay a considerable sum (formally for the usage of urban social and technical infrastructure).

It is of course true that in many respects the socialist economy favoured the growth of big cities. First of all, this growth was due to the development of the cities' industrial base, especially that of manufacturing. Secondly, the growth of the big cities was in many respects determined by the intensive focusing of administrative functions in them. And thirdly, the large cities were receiving the lion's share of investment in infrastructure (again at the expense of the centres of smaller size).

In the short-term the population growth of the larger cities and urban agglomerations should be favoured by the fact that in 1995–2005 the cohort born during the demographic peak of 1975–85 enters the active age (Korcelli, 1995). Changes in the agricultural sector may also result in a flow of migrants from rural areas to the big cities. However, the growth of the larger cities and agglomerations may be counteracted by strong economic factors, in particular the ongoing structural crisis. In general, although between 1995 and 2005 an acceleration of population growth in the larger cities and agglomerations is expected, the absolute population gain is not likely to be significant.

In spite of the fact that in the larger cities and agglomerations the labour markets are more balanced than outside them, it should not be forgotten that over the transition period they have already lost a large number of jobs. Secondly, a further fall in housing construction has resulted from the termination of state subsidies. Finally, the cost of living in the big cities is appreciably higher than in the smaller ones, whereas the consumer goods and services that used to be in such short supply have now become accessible in small cities as well (earlier trips to a big city and even to the capital were necessary).

CONCLUSION

The political and economic transition in the former socialist countries initiated a transformation in the territorial organization of the region's cities, in particular of its largest cities. These transformations, however, have not been accompanied by a concentration of population in big cities and urban agglomerations. The main changes have occurred within cities. The more adaptable elements of the built environment, those capable of bringing quick economic returns have been the first to undergo change under the 'marketization' and commodification of the city. These elements, often readily visable, have substantially modified cityscapes.

The overwhelming majority of changes in the urban fabric have been connected with the modification of the existing built environment rather than with new construction. Many buildings and premises in an unsatisfactory and dilapidated condition have been subject to renovation.

Marketization was very quickly able to improve the previously neglected retail sector, but could do little in the short term to alleviate an acute housing problem. This burden has now been shifted from the state building sector to the private developers. As this process will be protracted (it will take not less than several decades) and fraught with difficulty, the cities may become areas of growing social pathology. The emphasis in the city authorities' activity will gradually move from housing to environmental problems and to the improvement of the infrastructure of the city. Environmental factors will exert an increasing influence on the process of redistribution of population within the city.

In the near future the main changes in the spatial land use structure of the post-socialist cities are likely to be:

1   Gentrification of the urban centres, their territorial expansion and greater internal specialization.
2   Under-investment and, therefore, increasing dilapidation of dormitory districts (that is, the quarters characterized by socialist standard mass construction) in the intermediate and outer urban belts as an outcome of the reorientation of investment and construction activity to the centre and to the prestigious suburbs.
3   Suburbanization – the expansion of cities on to adjoining greenfield sites.

In the short term investment will be directed mainly to the emergent and expanding office and retail sectors. Substantial renovation will affect predominantly the more prestigious areas of cities that boast some distinctive historical and/or architectural 'flavour'. New housing construction will be limited predominantly to the construction of single-family houses and apartments for those in a rather narrow higher-income layer. The differentiation of function by location, affecting first of all city centres, will result from the emergence of an economic phenomenon – new to the post-socialist city differential rent.

At the inter-urban scale increasing polarization is likely. The number of cities experiencing economic decline will increase, but simultaneously the list of cities undergoing economic growth will also increase. Further polarization will be stimulated by the production cycles of the economy, with the trend towards the post-industrial types of development leading to the bankruptcy of many industrial enterprises. Polarization will be especially marked in the group of medium and small 'one company' industrial cities.

Thus, post-socialist cities in the future will need to rely to an increasing degree on their own resources, their economic base, and geographical position, where state redistributive policy has already ceased. In its place a scenario of intense competition between cities, struggling for investment, jobs and development, and prestigious projects, is emerging.

REFERENCES

Abankina, T. and Zuev, S. (1994) 'Housing strategies of urban social groups under market conditions', *Voprosy Ekonomiki*, 10: 57–67.

Belkina, T. (1994) 'Housing sector in Russia', *Voprosy Ekonomiki*, 10: 16–22.

Domański, R. (1994) 'Transformation of urban system in terms of synergetics', in R. Domański and E. Judge (eds), *Changes in the Regional Economy in the Period of System Transformation*. Warsaw: Wydawnictwo Naukowe PWN. pp. 11–23.

French, R.A. (1995) *Plans, Pragmatism and People*. London: UCL Press.

Glazychev, V.L. (1995) 'Cities of Russia a threshold of urbanization', in *City as a Sociocultural Phenomenon of Historical Process*. Moscow: Nauka. pp. 137–44.

Hamilton, I. (1995) 'Re-evaluating space: locational change and adjustment in Central and Eastern Europe', *Geographische Zeitschrift*, 83 (2): 67–86.

Hegedus, J., Mayo, S. and Tosics, I. (1996) *Transition of the Housing Sector in the East-Central European Countries*. Budapest: USAID.

Housing in the Czech Republic (1996) National Report for the United Nations Conference on Human Settlements – Habitat II. Prague.

Kaganova, O. and Katkhanova, A. (1994) 'Development of market-orientated residential construction: the case of St Petersburg and other cities', *Voprosy Ekonomiki*, 10: 131–43.

Korcelli, P. (1995) 'Aglomeracje miejskie w latach 90. Powolny wzrost, umiarkowana polaryzacja', *Biuletyn KPZK PAN*, 169: 43–58.

Kostinskiy, G. (1994) 'Moscow at a time of social upheavals and the competition for space', in C. Vandermotten (ed.), *Planification et Strategies de Developpement dans les Capitales Européennes*. Brussels: Editions de l'Universite de Bruxelles. pp. 267–72.

Musil, J. (1993) 'Changing urban systems in post-communist societies in Central Europe: analysis and prediction', *Urban Studies*, 30 (6): 899–905.

Nefedova, T. and Treivish, A. (1994) *Regions of Russia and Other European Countries in Transition in the Early 90s.* Moscow.

Pchelintsev, O. (1994) 'Housing situation and the prospects of institutional changes', *Voprosy Ekonomiki*, 10: 10–15.

Purgailis, M. (1996) 'The city of Riga, Latvia: where it is, what it is, what it wants to be', *Ambio*, 25 (2): 90–2.

Riley, R. and Niżnik, A.M. (1994) 'Retailing and urban managerialism: process and pattern in Lódź, Poland', *Geographia Polonica*, 63: 25–36.

29

The Cities of Sub-Saharan Africa: From Dependency to Marginality

RICHARD STREN AND MOHAMED HALFANI

Of all the major regions in the world today, Africa is one of the least urbanized. With a total land area of some 30 million square kilometres and a population of approximately 743 million people, only 34.7 per cent were living in urban areas in 1995. But even this relatively modest level of urbanization is recent. In 1950, for example, when most of the people of the African continent were living under some kind of colonial rule (the main exceptions at the time were Ethiopia and South Africa), the proportion in towns and cities was only 14.5 per cent. This proportion gradually increased to 18.3 per cent in 1960, 22.9 per cent in 1970 and 27.3 per cent in 1980 (United Nations, 1993: Tables A1 and A2). But by a wide margin, sub-Saharan Africa to the north of South Africa is still predominantly rural. This rural environment is one of the major conditioning factors in a continent whose cities – while often innovative and dynamic – have been shaped over time by significant external forces.

COLONIAL CITIES

Although Africa experienced major historical change and evolution before the establishment of direct European rule, it was the colonial experience – lasting for most parts of Africa from the late nineteenth century until at least the early 1960s – that had the greatest effect on urban form and function. This was the case for four main reasons. In the first place, a great number of cities that became important during the colonial and post-colonial periods, such as Nairobi, Johannesburg and Abidjan simply had not existed before colonialism. All these new towns – and many others, such as Cotonou, Libreville, and Bangui (as capital cities), and large provincial towns such as Bouaké, Tamale, Enugu, Lubumbashi and Mwanza – developed as major centres of commerce and of administrative activity. Because their major purpose was to strengthen the ties between the metropolitan country and the colonial territory, they were often located on or near the coast or a major waterway. A new colonial urban system began to emerge, displacing internal networks of trade and influence which had developed over many centuries.

A second major urban effect of colonialism in Africa was the establishment of powerful currents of rural–urban migration. Although African populations had been moving around the continent for centuries in response to commercial opportunities, variations in environmental conditions, political upheavals and the depredations resulting from the slave trade, colonial regimes and the economic activities which they promoted, raised the level of migratory activity to a much higher level. As a cash economy was introduced, and goods and services could be obtained in exchange for wages, there were incentives for African labourers to migrate to work in mines, on plantations and in urban employment.

Although most colonial migration was rural–rural, perhaps one-quarter of all migrants ended up in the towns. This movement was reinforced in some colonies by harsh administrative practices of forced labour (for example, the *corvée* in parts of Francophone Africa) and the imposition of 'hut' and 'poll' taxes in cash which individual peasants were obliged to pay. For some administrators, the imposition of non-market measures to prod peasant farmers out of their villages into the rural or urban labour markets was necessary since there was a widespread belief in the 'backward-bending supply curve for labour'. That is, it was felt that African labourers had a fixed 'target income' in mind as an objective, so that they would work only up to the point where such a target would be met, after which they would return to their rural homes.

This perspective was paralleled by the equally

fallacious – and self-fulfilling – assumption that African migrants to towns were only 'temporary visitors' whose ultimate home was in the rural 'reserves'. Since most African rural–urban migrants were single men who could not afford to bring their families to the towns, this assumption justified paying them urban wages at a scale that could not support their families, who would presumably be earning their subsistence in the rural areas. To the extent that housing and other social facilities were provided to urban African workers, their quality and extent were also limited. 'The obvious convenience of such a conception', wrote Lord Hailey in 1957, 'caused it to be cherished long after it ceased to have any relation to the actual facts of the situation' (Hailey, 1957: 565). By the mid-1950s, the 'winds of change' which nationalist movements brought to the African continent, combined with an increasing concern with African labour efficiency, led to a more positive approach to the African role in towns – at least in the regions of eastern, central and southern Africa, where government control over African urban integration was the most marked (Great Britain, 1955). By and large, the same change in attitude was occurring in both Belgian and French colonial territories at about the same time.

One of the main reasons for increasing administrative and political concern, near the end of the colonial period, for the social conditions of urban Africans was the enormous increase in population which many cities were undergoing. In Kenya, for example, where – as in so many other African countries – Africans flocked to the towns in the post-war years, the population of Nairobi, the largest city, grew from 118,976 in 1948 to 266,794 in 1962, just on the eve of independence. Although Nairobi was (and still is) a multi-racial city, almost two-thirds of this population growth was accounted for by Africans, most of whom were migrants from rural areas. During the same period, Mombasa, the second largest city in Kenya, grew rapidly as well, from 84,746 to 179,575 (Republic of Kenya, 1965: 10). Again, most of Mombasa's growth was accounted for by African in-migration rather than by a natural increase in population. While Nairobi had grown at a compound annual rate of 6.0 per cent during this period, Mombasa had grown at a rate of 5.5 per cent. To the south, Dar es Salaam grew from 69,277 in 1948 to 128,742 in 1957 (a growth rate of 7.1 per cent, and to 272,821 in 1967 (an annual increase of 7.8 per cent) (Tanzania Bureau of Statistics, reported in Kulaba, 1989a: 210). Looking at the growth of Dar es Salaam from the post-war period to the early 1970s, one study observes: 'in Dar es Salaam, after a period of virtual stagnation, the population grew rapidly, quadrupling over the period 1948–71. The annual

growth rate of nearly 7 per cent was almost three times that of the rural population and more than double what it had been during the first half of the century. Fully 78 per cent of the total population increase . . . was due to migration' (Sabot, 1979: 44–5). Similarly, across the continent in the Ivory Coast, Abidjan, with a population of 58,000 in 1948, had mushroomed to 198,000 in 1960 – for a sustained rate of growth of 10.8 per cent annually (Attahi, 1992: 37). The growth of the two largest cities in Morocco shows an aspect of this pattern even more clearly. Thus, from 1936 to 1952, over a period of 16 years, Casablanca grew from 247,000 to 682,000 at an average annual rate of 6.6 per cent; Rabat–Salé, the capital, grew during the same period from 115,000 to 203,000, at a rate of 3.6 per cent. But during the eight years (1952–1960) which straddled independence, the two cities grew much more rapidly: Casablanca attained a population of 965,000 in 1960 (for an annual growth rate of 4.4 per cent during the period), while Rabat–Salé grew to 303,000 (for an annual growth rate of 5.1 per cent) (Abu-Lughod, 1980: 248).

In general, these growth rates varied according to local political and economic circumstances. For example, most African cities grew quickly during the 1920s, slowed down during the 1930s, began to grow again during and after the Second World War, and slowed again during what were often political uncertainties during the 1950s. With independence, they began to grow rapidly, a pattern which continued through to the 1980s.

As so many Africans – most of whom were poor – moved to the cities just before Independence, colonial administrators and policy-makers became increasingly concerned about the social and economic conditions under which they lived. Poor living conditions and inadequate urban services, it was suggested, would not only reduce the productivity of African workers, but they would also create the conditions for social and perhaps even political unrest. Partly as a result, all over Africa the 1950s and 1960s (both just before and just after independence in most sub-Saharan African countries) were the decades of major efforts to plan for growing cities, and to provide social and physical infrastructure for rapidly growing populations. Decent wages and good social services, it was cogently argued, would produce a more permanent as well as a 'smaller but satisfied and efficient labour force' that would be both more productive and politically quiescent (Cooper, 1987, quotation at p. 132 from Tom Mboya).

The third major effect of colonial urbanization was the physical structure of the cities that was bequeathed to the African governments that took over the reins of power from the late 1950s (for Ghana, Egypt, Tunisia and Morocco), through the 1960s (for almost all the rest of the Anglophone and Francophone countries in middle Africa), through to the 1980s. Looking at the totality of Francophone cities in 13 West African countries at the end of the 1980s, a major study takes the position that 'today's African cities were, for the most part, established by colonizers who applied urban planning principles appropriate to their country of origin. The most striking aspect of colonial urban planning is the partition of urban space into two zones, the '"European" city and the "indigenous" city' (Poisnot et al., 1989: 11). The extreme attention to the needs of European settlers, who generally received a very high level of planned urban services and infrastructure, with, by contrast, relative indifference to the African majority was a characteristic of urban planning that was rooted in the very fabric of the colonial state.

The bifurcated nature of colonial urban space was originally conceived to 'protect' Europeans from 'disease' thought to be carried exclusively by indigenous people. In many African towns, the separation of races was promoted by town planning; one of the most important elements of this policy was the establishment of an open, neutral zone (called the *cordon sanitaire* in Francophone towns) to separate Europeans from other groups. While not all colonial cities enforced a strict separation of races – a Mombasa proposal for racial separation was rejected on grounds of cost (Stren, 1978: 116), while different groups lived at least side by side in Dar es Salaam, in Ghana and in Nigeria – the standard of infrastructure in the neighbourhoods inhabited by Europeans was many times higher than in the African areas. If Europeans tended to live in bungalows or *villas* in large plots along clean, shady streets, Africans tended to live in 'huts' or *barraques* (the French word for wooden shacks common in West Africa) in crowded 'townships' along unplanned paths, enjoying few, if any, sanitary services.

This dual system of housing and land use was justified by two major premises: first, Europeans paid both local and national taxes and contributed to the colonial economy in a way that Africans did not (in spite of their numbers); and secondly, Europeans were 'used' to living in cities with European standards of infrastructure, while Africans were essentially 'rural' and would not live permanently in the towns. Both these justifications for the dual provision of space and facilities dissolved by the 1950s as Africans moved massively and much more permanently into the towns and began to constitute an increasingly powerful urban middle class; but the form of the city was moulded by many decades of infrastructural investment and could not quickly change. The large-scale investments by colonial governments during the 1940s and 1950s were considerably enhanced by a more dynamic economy throughout Africa, and by legislative instruments in Great Britain (the Colonial Development and Welfare Act of 1940) and France (FIDES, or the Fund for Economic and Social Development, established in 1946), which supported economic planning, and the construction of significant infrastructural works.

The fourth, and perhaps most profound, impact of colonialism on Africa is the structure of its economy which was left behind. These elements have left indelible traces on urban form and function, in spite of the variation across British, French, Belgian and Portuguese colonial experiences. While it is difficult to generalize, the imposition of the colonial economic framework resulted in an intense, but very divergent spread of capitalist economic activity in mining, primary agricultural production (including plantations), and transportation and communications activities. At the same time, colonial implantation in most areas outside South Africa did not prove sufficiently robust 'to change the essential character of African societies'. Nevertheless, 'foreign capitalism, together with the fiscal demands and economic planning of the colonial state, compelled or persuaded the majority of Africans to become linked with the international economy' (Fieldhouse, 1986: 31). The interests served by this system, as variegated as it was, were very much oriented toward the needs of the metropole rather than to the needs of African society. Looking back on West Africa up to the Second World War, Michael Crowder observes that

> . . . [o]nly the railways remain as a major legacy of the economic policies of the colonial powers of that period, and they were paid for by taxes imposed on the African himself. In both British and French West Africa economic policy on the part of the newly established governments subordinated African interests to those of the needs of the Metropolis. The railways, and later the tarmac roads, tell the tale most clearly: simple feeders linking areas that produced the crops and minerals Europe needed with the ports on the coast. (Crowder, 1968: 173–4)

One of the results of this incomplete transformation that was rife with social inequalities and regional variations is that, according to most measurements, Africa has fared 'worse than almost any other world region over the last twenty or so years, becoming relatively more peripheral and in some respects, absolutely poorer. Nevertheless, small elites and bourgeoisies, overwhelmingly concentrated in major cities, have prospered by virtue of their skills, positions, control over resources and contacts' (Simon, 1992: 34–5). Thus, in spite of small islands of wealth and prosperity (the high income areas) in African cities larger and larger proportions of the population have gradually, but nevertheless dramatically, sunk into poverty and stagnation. From a dependent status, sub-Saharan African cities have become economically marginal in the new global economy.

THE URBAN CHALLENGES OF A MARGINALIZED CONTINENT: PLANNING AND POLICY INITIATIVES AFTER INDEPENDENCE

If we consider Africa to be a relatively marginal region within a more dynamic, globalizing economic system that has been energized by political and economic initiatives originating in Europe and North America (and, more recently, in Japan and Eastern Asia), it is nevertheless the case that there were a great deal of planning and policy responses to urban problems during the immediate decades after independence in the 1960s. But in spite of the fact that most African countries were nominally independent after the late 1950s and early 1960s, their urban planning policies were (a) very similar from place to place and (b) still heavily influenced by metropolitan models. Both the similarities and the local variations can be illustrated by reference to three central fields of activity relating to urban development: master plans and planning legislation; redevelopment, public housing and the regulation of squatting; and anti-poverty policies.

Master Plans and Urban Planning Legislation

After independence, many African countries attempted to respond to rapid urban growth and the contradiction between the status of independent statehood and the increasing evidence of urban poverty by major exercises in 'master planning', and by large-scale government-sponsored construction of residential dwellings. Typically, a major expatriate planning firm would produce a master plan for the future development of the capital city. The plan would contain an analysis of urban form and function, some analysis of likely future growth patterns, a number of technical maps and plans, such as a detailed land use zoning map, plans for infrastructural development and some proposals about procedures, regulations and even institutional reforms necessary to carry out the plan. These plans were the direct descendants of numerous plans produced during the colonial period, although successive documents tended to include more and more data of a sociological and economic nature. The city of Abidjan had six major planning documents from 1928 to 1990; Dar es Salaam had three plans from 1949 to 1977; Nairobi had three from 1948 to the mid-1980s. While these plans often had an important influence on the overall approach to land-use planning in the central areas of the larger African cities, they failed to capture the speed and direction of growth in the peripheral areas, and in any case were almost never supported by the level of capital expenditure necessary to implement their infrastructural projections. In the absence of major capital projects developed within the master plan framework, most urban planning decisions took place within the parameters of building and development regulations that were little more than copies of the existing legislation and bylaws in Britain, France and Portugal.

The metropolitan influence on African urban planning was reinforced by the fact that not only were virtually all urban planners working in Africa Europeans of origin, until at least the mid-1960s, but there were no schools of architecture or urban planning located on African soil (outside of South Africa) until the first decade after independence. Until at least the mid-1960s, virtually all the directors of urban planning in African countries were of European origin.

Urban planning legislation was supplemented by building codes which, in many cases, were also very closely modelled on European regulations. Thus the Kenya Building Code, applied by all local authorities in the country as a condition for legal certification, makes constant reference to a 'British Standard' or a 'British Code of Practice'. Structural steelwork, for example, 'shall be deemed to be sufficient for the purposes of . . . these By-laws, if the steel work is designed and constructed in accordance with the relevant Rules given in British Standard 449 . . . '. And on building materials in general, '[t]he use of any type of material . . . which conforms with a British Standard . . . shall, except where otherwise required in these By-laws, be deemed to be sufficient compliance with the requirements of this by-law . . . ' (Republic of Kenya, 1970: sections 47, 32).

A new factor during this early post-independence period was the establishment by international development agencies of influence over the strategic agenda for urban development. The early policy initiatives in the areas of shelter, growth-pole development and decentralization were locally designed and some were very successful. However, from 1972, when the World Bank produced its first major urban policy pronouncement (World Bank, 1972), African urban policy-makers began to defer to international development agencies with respect to their agenda for future planning and investment.

Redevelopment, Public Housing Construction and the Regulation of 'Squatting'

Within the terms of reference of the urban planners, but with a dynamic and importance of their own, were policies for the development of residential housing. Two of the hallmarks of the colonial approach to African urban housing in the 1950s were the redevelopment of decaying 'core' areas combined with the removal of 'slums' or squatter areas; and the construction of large rental (sometimes tenant-purchase) public housing estates. Once independence was accomplished, these policies were pursued by the successor governments. In a classic Nigerian study, Peter Marris focused on a large 'slum clearance' scheme covering 70 acres in central Lagos. The pursuit of this scheme illustrated the visible achievement of 'modernity' as a public goal. 'The overriding aim', he writes, 'was to rebuild the most conspicuous neighbourhood of Lagos to the standard Nigerians had set themselves, as a matter of pride' (Marris, 1961: 119). One of the main findings of this study was that the central-city 'slum' residents whose homes were originally redeveloped by the government were, on the whole, not able to take advantage of the newly built housing estate on Lagos mainland – because of either economic or sociological factors.

A number of other African countries pursued 'slum clearance' policies in their major cities during the 1960s and 1970s. Nairobi, for example, undertook a 'clean-up' campaign lasting for several months in 1970, as a result of which 10,000 squatter dwellings were demolished and 50,000 people left homeless (Stren, 1979: 186). The Tanzanian government 'redeveloped' some so-called 'slum' areas of downtown Dar es Salaam, replacing those houses destroyed with newly built units. But most of the low-income families who lost their houses could not afford to pay the rents (or tenant-purchase instalments) on the new units; others were given rights to more remote and still unserviced plots which they found difficult to develop (Kulaba, 1989a: 226–7). In Dakar, an estimated 90,000 people were displaced through the eradication of central-city bidonvilles between 1972 and 1976 (White, 1985: 512). In Lomé, government 'clearances' removed local population from major areas of land in 1968 (275 hectares were cleared for the new university), in 1974–5, in 1977 (clearing of the old zongo area near the central market), in 1979 (clearing of a marginal housing area), and in 1982 and 1983 (clearing of the population from the port and industrial area), leading to a major movement of lower income residents to the peripheral areas of the city (Adjavou et al., 1987: 20–1). And in Abidjan, between 1969 and 1973 the government displaced over 100,000 people (some 20 per cent of the total population) from low-income housing in the central areas of the city (Joshi et al., 1976: 66). '[I]n the name of "modernization" and the battle against "unsanitary conditions",' writes Alain Bonassieux, 'the authorities organized the destruction of "illegally occupied" and "unaesthetic" areas of the city' (1987: 42). While some African countries were able to build new housing for at least a proportion of those whose houses were destroyed, most of the displaced could not find alternative housing nearby to the areas where they had previously lived.

The second – and more substantial – pillar of the colonial approach to African residential development was the construction of large-scale housing estates. During the colonial period, large estate housing projects were undertaken in such countries as Southern Rhodesia (now Zimbabwe), Kenya, Senegal and the Ivory Coast, where the government (and parastatal agencies such as the railways and port authorities) needed to house their employees. After independence, this state-centred approach to housing developed more widely, with many countries establishing national housing agencies with an important mandate to improve housing for Africans. Behind this drive was the feeling of many urban Africans that their housing conditions should be upgraded in line with their changed political status. As a Lusaka city councillor put it:

> Under colonial rule, only white people lived in town, whereas Africans lived in their huts in the native quarters outside the city limits. Africans were separated from their families who remained in the villages. Jobless Africans were not allowed in the city. We are independent now. We should live in good houses with our families. (quoted in Laquian, 1983: 12)

In West Africa, large public housing projects, accompanied by systematic eradication of 'slums' and the maintenance of high standards of physical development, characterized both Senegal and Ivory Coast (Stren, 1990: 37–9). In Senegal, the Société Immobilière du Cap-Vert (SICAP) was formed as early as 1949, as a mainly public company to develop lodging in the capital. Ten years later, a fully public corporation, l'Office des Habitations à Loyer Modéré (OHLM), was formed for lower-cost housing. Both corporations depended largely on state subsidies. By 1968, the two agencies had produced close to 12,000 housing units, almost all of them in Dakar. Their production declined drastically in the 1970s, largely due to lack of resources.

But it was Ivory Coast which built the most. In a massive construction programme unparalleled in black Africa, SICOGI (Société Ivoirienne de Construction et de Gestion Immobilière) – formed out of two existing parastatals in 1965 – produced

21,897 'low cost' housing units during the next decade. At this pace it was only slightly ahead of SOGEFIHA (*Société de Gestion Financière de l'Habitat*, created in 1963), which built 17,912 units during the same period. Meanwhile, a land development agency, the SETU (*Société d'Equipement du Terrain Urbain*, established at the end of 1971) had by September 1979 prepared 1,779 hectares of land in Abidjan, of which the company itself estimated 444 hectares were for 'low-cost housing', accommodating as many as 350,000 persons (including housing provided to the public housing agencies themselves).

Leaving this massive accomplishment of the SETU aside, the two major housing parastatals produced over three times as much housing as did their Kenyan counterparts, the National Housing Corporation – the most active public housing agency in Anglophone Africa. This record is all the more remarkable given the fact that the population of Ivory Coast is less than half of that of Kenya. Like Kenya and virtually all other sub-Saharan African countries (outside South Africa) however, the role of public agencies in supplying urban housing in the Ivory Coast was drastically curtailed in the 1980s. State investment funds were at a premium, and many of the projects had strayed from their original purpose of assisting lower income groups. Thus, from 1963, when 23 per cent of Abidjan's population lived in informal, or spontaneous housing without proper services, to 1984, when 16 per cent of the population lived in spontaneous housing, the state system did its work well. After the mid-1980s, however, when the public housing and land development agencies were virtually disbanded, the proportion of people living in informal housing areas began to rise again. While the figures are not comparable with earlier estimations, it was claimed in the early 1990s that 70 per cent of the population of Abidjan lived either in 'evolving' housing (consisting of low-quality housing with limited services) or spontaneous and village housing (consisting of poor-quality housing with no services) (Attahi, 1992: 74).

The case of Abidjan is instructive, since almost all other large cities in sub-Saharan Africa had more difficulty meeting the gap between the supply of, and demand for good housing. One consequence of this difficulty – at least, outside Anglophone West Africa – is the increasing prevalence in 'informal' or 'spontaneous' housing in African cities. While there are many definitions of this type of housing, and many local variations around the main tendency, the central defining elements are usually twofold: (1) the housing in question is either illegally built (that is, without formal permission from the authorities) or (and in many cases in addition) it sits on land which has not been properly purchased through the formal system and zoned for residential develop-

ment; and (2) few, if any, services – such as water, roads, sewerage and stormwater drainage, electricity, telephones, and community facilities (such as clinics and schools) – are built in the immediate neighbourhood.

The increasing informalization of urban housing in African cities may be related to the failure of the formal sector to provide enough serviced plots and low-cost houses in the face of large-scale population growth; and it may be related to the degree to which the African state is willing to tolerate so-called 'illegal' occupation of urban land. Philip Amis has argued that, over the late 1970s and the 1980s, there was an increasing tendency among African countries to accept – at least at some level of security – the presence of informal urban housing. The underlying logic behind such an 'acceptance policy', he suggests,

> must be sought at the political level. In essence, the importance of political stability within the urban areas of the Third World seems important. In this context the acceptance of unauthorized settlements is a relatively painless, and potentially profitable, way to appease the urban poor in the Third World. The increasing mobility and ability of international capital to change locations at short notice are likely to make these political considerations more rather than less important. (Amis, 1990: 19)

Added to this argument – which is a hypothesis rather than a demonstrated relationship – is an economic one: rates of return for urban housing and land have, by all evidence, been lucrative in comparison with alternative foci for investment. To the extent that there is relative peace in the informal housing market, commercialization or 'commodification' of low-income housing is likely to grow apace. Since political elites and their business colleagues are among the most ubiquitous actors in this urban real estate market, the balance between informal housing and state regulation is likely to remain supportive of a burgeoning informal sector (Amis, 1984: 87–96). But even when large-scale commercialization has not been a major factor, informal housing has grown enormously, as is illustrated by the case of Dar es Salaam, Tanzania, where from two-thirds to three-quarters (conservatively estimated) of the demand for new housing was met by informal means by the 1990s. In spite of this growth in informal housing, however, Tanzania has never strictly enforced building regulations with regular police raids and demolitions, with the result that so-called 'squatters' feel relatively secure. Most squatter houses are constructed with at least semi-permanent materials, such as wood, cement and iron sheets; and many are built by middle- and high-income earners. Where enforcement is stricter – such as in Harare, Nairobi, or Abidjan – the quality of the informal structures (at least in central areas) is generally lower. Informal housing areas that are removed from central business districts and from high-profile main arteries such as airport highways tend to be treated more tolerantly by the authorities.

In Anglophone West Africa, the distinction between 'slums' and 'squatter' or informal housing is more important than in other parts of the continent, where the two forms of housing tend to overlap. As Margaret Peil (1976: 155–66) and Kwadwo Konadu-Agyemang (1991) have pointed out, 'squatter' settlements as such do not exist in this part of Africa, due to the fact that the owners of low-income houses 'almost invariably hold their titles from the land owners and/or their agents, or at least have permission to build, although they may not have obtained permits from the Council or Town Planning Department, and therefore may appear "illegal" in the official jargon of the authorities' (Konadu-Agyemang, 1991: 141). In spite of the fact that ownership of the land is not problematic, however, there has undoubtedly been a deterioration of housing and urban infrastructural conditions in Anglophone West African cities. Citing a large-scale housing survey of all states in Nigeria in the mid-1980s, a leading professor of planning noted that 'inadequate accommodation appears to be one of the most serious problems arising from the high rate of urbanization. A substantial proportion of urban houses . . . are either dilapidated or are in need of major repair. Overcrowding is pervasive and high rents are a major indicator of urban housing problems' (Onibokun, 1989: 78).

Growth Poles and Anti-Poverty Programmes: Statist Policies of the 1960s and 1970s

The second decade of independence was a very important phase for urban development in the post-independence period. For the first time the specificity of the 'urban question' was addressed in a coherent manner within the process of policy formulation as well as in development planning and management. The need for change was prompted by the negative externalities of urban primacy, by the failure of the modernization approach of the first decade and by the realization that a centralized state structure could not execute a development-orientated programme.

In strategic terms, this entailed devising mechanisms for containing the rapid growth of cities and balancing urban systems. The promotion of rural development and the enforcement of municipal regulations was expected to arrest rural–urban migration and haphazard settlement in urban areas. Similarly, the deconcentration of primate cities and the re-direction of urban investments to less developed areas was expected to redress the distributional imbalance of urban centres. Another strategic measure adopted was to mitigate urban poverty through the Basic Needs development strategy, which was prominently encouraged by international development agencies.

These three initiatives reflected an attempt during this period to shift the emphasis of development away from sustaining towns as mere service centres or as the loci of collective consumption into nuclei for a real economic development. The initiatives signified an attempt to create an urban system that was 'generative' and which was organically integrated with the domestic economy. The distinction between cities that were 'generative' and those that were 'parasitic' had been made originally by Bert Hoselitz in 1955 (Hoselitz, 1955), then applied by Akin Mabogunje to Nigerian cities in the 1960s (Mabogunje, 1968). While the concepts were rarely used explicitly, they continued to be important in the thinking of many African economic planners throughout the 1970s and even the 1980s. Indeed, the idea that cities in Africa were often 'parasitic' rather than 'generative' was powerfully – though indirectly – expressed in the notion of 'urban bias', a notion that still holds sway today among many planners and development agency officials. According to this notion, popularized by Michael Lipton in his book *Why Poor People Stay Poor: Urban Bias in World Development* (1977), urban groups are able to organize themselves politically to attract disproportionate investments, and other benefits, to the detriment of the more impoverished rural groups, who should be the beneficiaries of development. While this idea had considerable merit in the 1960s and 1970s, the effect of structural adjustment on African cities since the 1980s has had the consequence of significantly cutting urban real wages, while raising agricultural producer prices and abolishing the currency and price controls which, in the past, supported organized urban interests (Becker, et al., 1994).

A striking feature of the poverty alleviation programmes of the 1970s was the expansion of the public (bureaucratic) sector in the process of service delivery. State corporations were established to execute and manage the new programmes and the delivery process was supply-driven through public sector agencies. Excessive subsidies, broad standardization, and administrative rationality dominated programme management at this stage. The dependent nature of import substitution industries favoured traditional primate cities that had the advantages of not only agglomeration and economies of scale but also large markets and convenient communication for raw material imports. By the end of the 1970s, the skewed profile of urban systems was even more reinforced and the process of deconcentration had not materialized.

In the mid-1970s, there was a new wave of administrative reorganization throughout the continent. In an attempt at transforming the inherited colonial bureaucracy, particularly through the promotion of popular participation, several countries embarked on a major process of administrative decentralization. This involved the transfer of power and authority from the headquarters to the field units. For a number of countries, these reforms were limited only to deconcentrating authority in a horizontal manner to field agencies of the central government. In this respect, the decentralization exercise did not effectively devolve power so as to promote popular participation (Fisette et al., 1995; Jaglin and Dubresson, 1993; Mawhood, 1983; Wunsch and Olowu, 1990).

Despite its political shortcomings, this decentralization process led to the establishment of new administrative headquarters in smaller urban and rural settlements. But most of these administrative centres were essentially consumption driven, with little contribution to productivity change in the immediate hinterland.

While the growth pole strategy failed in its manifest aspects, the initiative in itself demonstrated an attempt to redefine the role and structure of the African city in the process of economic development. Since that time, this issue has remained part of the urban development agenda in the region. As for the Basic Needs programmes of the late 1970s, there is no question but that they opened up a flow of investment resources in the social sectors, also contributing to a change in management culture whereby the urban poor and the informal sector were acknowledged as part of mainstream urban activities. Similarly, while the decentralization exercise of the 1970s did not lead to meaningful devolution of power and authority, it introduced into the development agenda the relationship between various levels of government.

All these institutional and policy re-alignments coupled with the demographic growth of cities transformed African urban centres from mere railway stations, ports and provincial administrative headquarters into social entities which have a diversified range of socio-economic activities. None of these activities, however, created a basis for sustainable and internally driven development.

AN URBAN CRISIS?

The 1980s: Policy Shift

The global recession of the 1980s hit African countries very severely. Changes in the terms of trade and the subsequent technological restructuring completely dislocated African economies from the global trading system. The shift in the pattern of industrial accumulation (technological flexibility, computerization and prominence of the 'symbol' economy) de-valorized primary inputs and rendered superfluous much of the predominantly unskilled urban labour force. Urban centres in Africa, whose underlying function and structure had been the servicing of an import/export economy, lost their basis for sustainability. As they entered the third decade of independence, African cities were caught up in a profound crisis.

In the new economy of the 1980s and 1990s, Africa's factor advantages – cheap unskilled labour and primary inputs for industrial production – were no longer a tenable basis for economic competitiveness under emerging global conditions. In fact, industrial nations were trading much more among themselves than with the least developed countries. African urban centres did not have the requisite infrastructural and institutional endowments for attracting private investments that could promote engagement in global transactions. By the mid-1980s, African urban centres were completely marginalized. For example, from 1983 to 1991, total foreign direct investment in sub-Saharan Africa increased only marginally, seemingly stuck at a level of about 6 per cent of all direct investment in developing countries; at the same time, the region's percentage of world exports fell from 2.4 per cent in 1980 to 1.0 per cent in 1992 (Halfani, 1996: 93–4). This stagnation reflected a pattern wherein emerging technical and commercial structures of global networking completely excluded the sub-Saharan African region.

The most prominent institutional casualty of the crisis situation was the state and its agencies. The extended responsibilities it had assumed under the Basic Needs programmes of the 1970s had enlarged its own institutional size as well as its scope of operations (Sandbrook, 1982). By the early 1980s, central governments were unable to provide for even the recurrent expenditure for the provision of social services and the maintenance of law and order. In fact, even its capacity for mobilizing domestic fiscal and monetary resources was highly curtailed. At a local level, this was reflected in the form of a major municipal crisis. Deprived of central government subventions urban authorities could neither provide basic services (Stren and White, 1989), nor enforce law and order. The urban system found itself in a profound organic crisis. Its structural basis had disintegrated and the legitimacy of its superstructural agencies was impaired.

It was out of such an organic crisis that the informal sector and civil institutions found the space for a powerful resurgence. Initially evolving as survival strategies, informal activities were consolidated by the evolution of community associations and transactional procedures based on traditional norms and values. By the mid-1980s, the African city was driven predominantly through informalism (ILO, 1991; Touré, 1985). The management of land use, transportation, shelter, employment, and some of the social services took place outside state management and regulation (Bubba and Lamba, 1991; Lee-Smith and Stren, 1991).

On the one hand, the growth of informalism created a basis for a more organic and indigenous form of urbanism, once the inherent distortions are rectified. On the other hand, the spatial effect of this process was increasing sprawl, environmental damage and deepening localization. The situation was compounded by increasing rural–urban migration, amplified by the severity of the rural crisis (Baker and Aina, 1995). Some urban centres were doubling every seven years. The ensuing population growth compounded the shelter, employment and environmental crisis.

When the crisis became evident, the first reaction of both national governments and international development agencies was to suspend the social and economic investments of the 1970s that were designed for poverty alleviation. This entailed suspending shelter programmes such as sites and services (Mayo and Gross, 1985), squatter upgrading, low interest housing credit and even subsidized basic commodities. Poverty eradication ceased to be a primary priority for urban development.

In the second half of the 1980s a new programme of structural adjustment was adopted by the World Bank to deal with Africa's debt and other structural economic problems (Duruflé, 1988; Sandbrook, 1993: ch. 1). As the main accent of the new programme was to improve the balance of payment situation, primary attention was given to improving the conditions for the production of tradable goods (World Bank, 1989a). In most cases these were agricultural primary commodities. Since few urban centres had significant enterprises that produced tradable goods for export, most cities did not gain from the export drive promotion. Indeed, the wide gap between high urban incomes and low rural incomes which was such a feature of African development in the late colonial and early postcolonial years began noticeably to close (Becker et al., 1994). Meanwhile, the retrenchment of public services and reduction in government expenditures (particularly for social programmes) increased urban unemployment and accentuated urban poverty. Only the few urban elites who managed to link themselves with the export economy and had access to investment credit managed to maintain high incomes. The African city resumed its dualist morphology of the 1960s. While a very small part portrays affluence and splendour, its large bulk epitomizes abject poverty.

In 1986, with the formal establishment of the Urban Management Programme within the World Bank and the UNCHS, attention was directed to the sphere of urban management where distortions in techniques, procedures and organization of land use, financial resource mobilization and deployment, planning and programming were deemed to impinge on the efficient functioning of urban systems. Suggestions for reforms were made through the development of improved systems, training and accountability mechanisms. A global programme, financed by a consortium of donors with the UNCHS and the World Bank as the caretakers, was established (Stren, 1993). To a large extent, the management focus of this programme, with the emphasis on technical and institutional reforms as well as efficiency maximization, corresponded with the monetarist, market-orientated policies of macro-economic reforms of the structural adjustment programme.

The organizational framework for the market-driven system of urban management was underpinned by the notion of 'partnership'. It was recognized that a functional association between the public, private and civic actors was critical for the efficacy of the market urban system. A major role of government was to create an 'enabling environment' for the efficient functioning of the other two actors (UNCHS, 1987). By the late 1980s and early 1990s, these concepts were further extended through the notion of *governance* in the area of urban policy. While one approach to this concept stresses the effective role of government (Landell-Mills and Serageldin, 1991), another approach, elaborated by academics and research activists, recognizes the multiplicity of managerial regimes and promotes a more organic linkage between civil society and state (McCarney et al., 1995; Wilson and Cramer, 1996).

Anatomy of Decline: African Cities Today

During the 1980s, the increase in the level of spontaneous, or informal housing in and around African cities reflected their almost total inability to provide adequate serviced land and infrastructure to their growing populations. The 1980s and 1990s thus became, in common parlance, a period of 'urban crisis' across the continent – at least to the north of South Africa. The crisis – which itself was a reflection of declining or stagnating economies in the face of continuous rural–urban migration, had three major components: a decline in levels of formal employment, and a corresponding rapid increase in 'informal sector' activities in many key areas of the urban economy; a deterioration in both the quality and distribution of basic services; and a decline in the quality of the urban environment, both built and natural. All these changes adversely affected the quality of urban life for everyone, but particularly the poor.

An employment crisis

In the years immediately following independence in most African countries, Africanization of the public service and an expansion of parastatal agencies led to a high rate of new employment creation in urban areas, particularly in the capital cities and large regional centres. By the 1970s, these increases were paralleled by increases in both the quantity of manufacturing jobs available (most of them in the largest towns), and in average manufacturing wages. At the same time, agricultural wage employment fell, or stagnated in many countries, thus increasing the attractiveness of the larger towns to rural migrants who could not gain sufficient incomes by working in subsistence agriculture.

By the late 1960s and early 1970s, it became clear that the rate of rural–urban migration was greatly exceeding the rate of formal employment creation in Africa's cities. Total numbers of jobs created in the formal sectors (including government, the parastatal sector, manufacturing and the large-scale service sectors such as banking and tourism) were not keeping pace with the increase in the urban population, whether this was caused by rural–urban migration or by natural increase.

Figures on manufacturing employment in medium- and large-size establishments show that most countries experienced rapid growth in the 1970–75 period, followed by slower growth in formal manufacturing employment during 1975–80. While the period from 1980 to 1985 (which included the second oil shock) saw renewed or accelerated manufacturing employment growth in some countries, others – particularly those with large manufacturing sectors such as Ivory Coast, Ghana and Nigeria – reported negative employment growth rates, while such countries as Kenya, Senegal and Zambia had annual increases during this period of only 4, 3 and 1 per cent, respectively (Becker et al., 1994: ch. 5).

African countries cannot afford to maintain publicly funded social security systems such as the social assistance or unemployment insurance programmes that operate in most countries of Europe and North America. For most Africans seeking urban employment (except those who have the resources to wait), remaining in 'open unemployment' is not a realistic option. As

formal urban employment opportunities declined, the alternative of engaging in small-scale, less remunerative activities – now known collectively as the 'informal sector' – became more and more common. While there are differences of opinion over the use of this term at all,

> the African informal sector can be fairly carefully defined. In the last twenty years, urban small-scale, artisanal, residual/casual, and home production have expanded considerably in African nations. Their relative importance varies considerably across countries, in accordance with economic structure and public policy. Nonetheless, the main informal activities include carpentry and furniture production, tailoring, trade, vehicle and other repairs, metal goods' fabrication, restaurants, construction, transport, textiles and apparel manufacturing, footwear, and miscellaneous services. (Becker et al., 1994: 159–60)

Typically, enterprises with 10 or more workers tend to function within the formal sector, while smaller enterprises (down to a single person) tend to function more informally. In any case, beginning in the 1970s, and gaining strength during the economic downturn of the 1980s and early 1990s, the urban informal sector has become a powerful force for employment creation in virtually all African cities.

Although good statistics are not available for the 1980s and 1990s, the proportion of the urban labour force in informal activities has almost certainly risen. A leading Ivoirian sociologist estimated that, over the period from 1976 to 1985, the number of people 'working on the street' in a variety of informal activities had risen from 25,000 to 53,850 in Abidjan alone. During the same period, the central government complement in Abidjan rose from 31,840 to 56,940. 'Given the negligible difference between the two,' he argues, 'one could conclude that the street offers as much employment and provides a living to as many people as the public service!' (Touré, 1985: 18). One of the trends of the 1980s and 1990s has been the supplementation by informal activities on the part of a large proportion of public sector employees.

The growth in the informal economy both in terms of absolute numbers of people involved, and in terms of economic importance, has given more opportunities to certain (often previously marginal) groups in a restructured, more services-oriented economy. In particular, poor women have been prominent in food selling and food preparation trades in almost all parts of the continent. By the early 1990s, women's role in many other informal economic activities was an important focus of the research literature. As one commentator noted, 'women's activities existed in the towns well before the 1970s and 1980s. But during the aggravated crisis of recent years they have proliferated and become virtually ubiquitous' (Bonnardel, 1991: 247). Characterizing 'informal feminine enterprises', the same writer argues that they share many elements with men's informal activities, but in addition they are on average smaller than men's enterprises, they never (or rarely) employ salaried workers, the majority of them transform local primary materials (such as grains, cassava, groundnuts and local dyes), and they hardly ever use recycled material from the modern sector (Bonnardel, 1991: 264–5). The importance of women's participation in the urban informal economy is illustrated with particular force in a study of Dar es Salaam, Tanzania during the 1980s and 1990s. Based on an impressive range of personal interviews with both political actors and ordinary people working in the informal sector, Aili Mari Tripp argues that the widespread resort – especially by poor women – to 'income generating activities' (as people's personal 'projects' are often called) obliged the Tanzanian government in 1991 to revise its longstanding policy that had restricted 'leaders' and 'public servants' from engaging in economic activities outside their formal occupations (Tripp, 1997).

A crisis in urban services

As African cities continued to increase in size during the 1980s and 1990s, their declining economic situation led to a precipitous decline in the supply of basic infrastructure and urban services. In many African cities, most refuse is uncollected and piles of decaying waste are allowed to rot in streets and vacant lots; schools are becoming so overcrowded that many students have only minimal contact with their teachers; a declining proportion of urban roads are tarmacked and drained, and many that are not turn into virtual quagmires during the rainy season; common drugs – once given out freely – have disappeared from public clinics and professional medical care is extremely difficult to obtain, except for the rich; public transport systems are seriously overburdened; and more and more people are obliged to live in unserviced plots in 'informal' housing, where clean drinking water must be directly purchased from water sellers at a prohibitive cost, and where telephones and electrical connections are scarcely available. For Nigeria, research by the World Bank demonstrates 'that unreliable infrastructure services impose heavy costs on manufacturing enterprises' which, in turn, diminish their competitive advantage and raise the cost of doing business (World Bank, 1991: 38).

Nigerian cities were not alone in experiencing a serious decline in infrastructure and public services during the 1980s. For Togo, a study of urban investment levels in the four five-year plans spanning the period from 1966 to 1985 shows a steady reduction from 17.9 per cent of total investment in the first plan, to 16.7 per cent in the second plan, to 10.6 per cent in the third plan and 6 per cent in the fourth plan (Dulucq and Goerg, 1989: 149). This pattern, which appears to reflect a similar tendency in other French-speaking countries, may be connected to a decline in French overseas aid funds, both in absolute and relative magnitudes during the same period (Dulucq and Goerg, 1989). For Dar es Salaam, a Tanzanian researcher has shown an overall decline of 8.5 per cent over the period from 1978/79 to 1986/7, in terms of expenditures on services and infrastructure, measured in constant currency units (Kulaba, 1989b: 118). If the population growth of Dar es Salaam is further taken into account, the per capita decline in expenditures comes to 11 per cent per year over the period studied. While the Tanzanian economy as a whole was stagnating during much of this period, the decline of urban infrastructural fabric was occurring at a much more rapid rate. To the north, Nairobi's services were also in dramatic decline. From 1981 to 1987, another study showed the capital expenditures of the Nairobi City Commission for water and sewerage to have fallen at an annual compound rate of approximately 28 per cent. Similar calculations for expenditures on public works over the same period show an annual decrease of 19.5 per cent, compounded; and for social services an annual decrease of 20 per cent compounded (DHS and Mazingira Institute, 1988).

Two of the major urban services that have become increasingly overburdened in almost all African countries are waste management and public transport. Indeed, if housing shortages were the major urban issue of the 1970s, waste management became the major issue of the 1980s, and along with the scourge of HIV/AIDS, one of the major unresolved problems of the 1990s. In 1985, a widely circulated weekly magazine in Kenya described the situation in Nairobi in the following terms:

> It has always been proudly referred to as the Green City in the Sun. But today, Nairobi presents a picture of a city rotting from the inside. Over the past 18 months, the main city news in the local press has been about mounting piles of stinking refuse, dry water taps, gaping pot-holes and unlit streets. (*Weekly Review*, 25 January 1985: 3)

Nairobi's problems, like those of many other rapidly growing African cities, were essentially related to resources. As the city grew both in population and spatially at a rate close to 5 per cent per year, the resources available to the municipal government in order to maintain its existing fleet of refuse removal vehicles – let alone to permit it to purchase additional vehicles – were severely limited. Thus, as the city expanded, the amount of refuse collected fell from a high of 202,229 tonnes in 1977, to 159,974 tonnes in 1983, a decline of 21 per cent over six years. This meant that, over the late 1970s and early 1980s, the city was collecting, on average, almost 10 per cent less refuse per capita every year. In spite of a major internal reorganization of the Nairobi City Commission, the refuse situation was still deteriorating in August 1989, when, although officials felt 100 refuse collection trucks were needed effectively to serve the city of over a million people, only 10 trucks were functioning on a daily basis out of a total fleet of 40 vehicles (Stren, 1989: 47; 1992: 10–11). In Dar es Salaam, the situation was even more challenging, since the local government agency (the Dar es Salaam municipal council) was much less endowed with resources than its neighbour to the north. A Tanzanian study of the city's garbage collection vehicles showed that in 1985, some 20 functioning trucks (of which only six were considered to be 'in good condition') were able to collect only about 22 per cent of the estimated 1,200 tonnes of garbage produced every day. Refuse removal, as a Tanzanian researcher comments, 'is a complex financial, technical and managerial issue', but it cannot be separated from other problems of providing services in a city in which the revenue generated during the late 1980s was only $5.60 per person (Kulaba, 1989b: 133–41). While we do not have comparable figures for the 1990s, there was unlikely to have been an improvement in municipal expenditure per capita; and in any case by the mid- to late 1990s only about 10 per cent of the solid waste generated in Dar es Salaam was being collected and disposed of (Kironde, 1999: 153). Partly because of its poor record in waste management, Dar es Salaam decided in 1994 to privatize the collection of refuse in the central parts of the city (the City Council continued to collect for the outlying areas). The effort seems to have had some effect only in 10 central wards, although the private company operates under severe constraints of poor equipment and limited revenue (Kironde, 1999: 160–3). While the level of collection varies across the continent, it remains a severe problem in almost all countries.

Just as the urban poor, who tend increasingly to live in peripheral, unplanned settlements are only sporadically served by such services as water supply, refuse removal and electricity supply, their marginal location and low disposable incomes makes them more vulnerable to difficulties in public transport supply. The situation in most African cities in the 1960s and 1970s was one of monopoly supply, whereby a public (or publicly contracted) bus company organized the mass transport system throughout the whole of the municipality. Some of these companies were very large: for example, the SOTRA in Abidjan (a 'mixed' private company with both French and Ivorian shareholders) had a fleet of 1,179 vehicles, and a staff of 6,153 in 1989: the SOTRAC in Dakar (also a 'mixed' company) had a fleet of 461 vehicles and a staff of 2,871 in 1989; the KBS in Nairobi (a 'mixed' company in which the City Council held a 25 per cent share with a private British company) had 275 vehicles and a staff of 2,304 in 1989; and Harare's ZUPCO (a privatized company taken over by the government) had 666 vehicles and a staff of 2,135 in 1988 (Godard and Teurnier, 1992: 52). By the 1980s, this system was breaking down, as many city administrations could not afford to replace, let alone properly maintain their ageing bus fleets at the same time as the population was growing; and even the private companies (operating in many Francophone countries) were having difficulties maintaining an efficient service in the face of declining revenues.

As the large public companies began to falter, private transport became more and more important, especially for the poor. Often the service was more direct, faster and certainly more accessible, particularly to and from outlying areas where roads were bad and the large buses could not easily function. These smaller, privately owned vehicles had a variety of local names, derived according to a number of different principles, according to the location: for example, the Nairobi *matatus*, the Dar es Salaam *dala dalas*, the Bamako *duru-duruni* and the Brazzaville *cent-cent* are based on the coins, or fares that were charged for the service when it was first initiated; the speed of the vehicle is reflected in the *cars rapides* of Dakar, the *kimalu-malus* of Kinshasa, or the *zemidjan* of Cotonou; and the age and lack of comfort and security is indicated in the *alak-abon* of Conakry, the *congelés* of Douala and Yaoundé, the *gbakas* of Abidjan, and the *mammy-wagons* of Lagos. Typically, they are reconstituted pick-up trucks, small minibuses, or minivans, holding up to 36 passengers in various degrees of overcrowding depending on the route and the time of the day. In most cities, they have been able to carve out well over half the market for public transport. While many are concerned with their apparent recklessness (often caused by the tight financial regime under which they operate, which discounts speed and maintenance), these vehicles have widened access to many urban locations – particularly the peripheral, often informal areas.

Environmental crisis?

The decline in the built environment was sharply in evidence throughout most of urban Africa.

Thus, as more of the urban population was forced into unplanned settlements on the outskirts of large cities, or into more crowded living space in an already deteriorating housing stock in the more established 'high density' areas, as a lower proportion of the population had direct access to clean, piped water, regular garbage disposal and good health services, the quality of life for the vast majority of the population deteriorated during the 1980s and 1990s. This trend seems to have been accentuated by the effects of structural adjustment in many countries, according to which urban workers lost more than rural small-holders. Some demographers have even suggested that the decline in urban mortality rates that was clearly evident during the 1960s and 1970s may have slowed down in the 1980s (Antoine and Ba, 1993: 140).

In the 1990s, however, both traditional health problems and environmental concerns were over-whelmed by the scourge of HIV/AIDS across the continent. Of approximately 34.3 million adults and children in the world estimated to be living with HIV/AIDS at the end of the 1990s, some 24.5 million (or fully 71 per cent) live in sub-Saharan Africa (MAP Network, 2000: 3). Spread unevenly throughout Africa in a 'belt' stretching from Côte d'Ivoire in the West through to East Africa, and becoming increasingly more virulent in Southeast and Southern Africa, the pandemic affects from 12 to 25 per cent of the population in the most affected countries, hitting urban dwellers much more, proportionately, than rural dwellers. As John Caldwell has put it,

> [t]his is a disease which in a number of countries will be the cause of death of half the population. It will lower Zimbabwe's life expectancy by 2000 by 20 years to the level of half a century ago, and . . . by 2010 to 35 years. It has increased the mortality level of adults in their prime, 20–40 years, to pre-modern levels. At any one time one-third of the people one meets in cities like Harare or Blantyre are infected and have at the most only a few years to live. The situation in the cities approximates that of European cities during the Black Death or Plague. (Caldwell, 1997: 180)

While Africans are making alarming adjust-ments to the disease, and even winning some battles in the area of prevention, the social and economic effects of this affliction will almost certainly be reflected in low levels of development for generations to come (Fredland, 1998).

As public health conditions become less favourable, intra-urban differences should become more marked. While there is little data over time, it is clear, for example, that there is an increase in the number of deaths either directly or indirectly related to malaria. The peri-urban areas – where the poor have been increasingly relocating under conditions of inadequate sanitary infrastructure,

concentrate all the most ideal conditions for the reproduction of malaria vectors and the transmission of the disease: a more dispersed settlement pattern, permanent buildings which use construction mate-rials that are virtual refuges for anopheles mosqui-toes, the existence of numerous vegetable farms and finally the proximity, in some cities at least, of swampy areas, and stagnant extensions of rivers. (Antoine and Ba, 1993: 140)

Under these conditions it may not be surprising to learn that in the central districts of Brazzaville less than one infectious mosquito bite per person was recorded every three years; by contrast, the rate was more than 100 times higher in the new informal settlement areas with low population density. In Bobo-Dioulasso, the second largest city in Burkina Faso, the rate of malaria occur-rence measured in a peripheral neighbourhood was almost twice as high as that measured in a central section of the town (Antoine and Ba, 1993: 140).

One practice that some health officials have related to the increased incidence of malaria vectors is urban agriculture. Apart from a possible public health risk (which is disputable) of the increased planting of such crops as maize and sugar cane in the vicinity of cities, there is a considerable nutritional benefit, particularly to low-income families who cannot otherwise afford certain foods. A major study of six towns in Kenya showed that 29 per cent of all households grew food within the urban area where they lived, and 17 per cent kept livestock. Most of the urban crop and livestock production was consumed as subsistence by the households themselves (Lee-Smith et al., 1987). Reflecting similar findings to the Kenyan study, a survey of research on urban farming in Zambia found that most urban farmers were poor women, growing food because of the failure of incomes to keep up with prices (Rakodi, 1988). A later study of Nairobi based on a different sampling strategy (the sample was chosen from those working on the land rather than randomly from urban households in the city) found that 64 per cent of the respondents were female (very similar to the earlier Mazingira survey); that they tended to have only primary education, or no education at all; that most were born outside the city, in neighbouring districts; and that very few actually sold their crops (Freeman, 1991).

The pattern of poor women practising urban agriculture in order to provide food for them-selves and their families was also a major finding of a study of 150 urban farming households in Kampala, Uganda. This study argued that urban farming as a major activity developed in Kampala 'in response to the declining economic situation' (Maxwell and Zziwa, 1992: 29).

Across the continent in Lomé, a study of the agricultural aspects of urban farming – while not comparable to work in East Africa – observed that most of the market gardeners were men, and that very large sections of public land reserves on the periphery of the city were under cultivation, with the tacit approval of the authorities. While market gardening increased in Lomé as a direct response to the economic difficulties beginning in the late 1970s and early 1980s, the study noted that the produce was sold commercially rather than consumed by the cultivators (Schilter, 1991).

One of the effects of increasing urban agriculture in Africa is that the economic and cultural differences between city and rural areas have become blurred. This is particularly the case in peripheral, unplanned areas where formal infrastructure is scarce, and many households live in relatively large plots of land in a semi-rural environment. Where once the central business district, with its clean, wide streets and high-quality shops and offices, was the focus of urban life – in both the large capital cities and in secondary cities as well – the centre of gravity has shifted. More and more of the population is moving to the periphery of the larger cities, where land is cheaper and much more easily obtained, where shelter can be constructed economically using locally available materials, and where harassment from the police and restrictions of the formal planning system are rarely felt. In these peripheral areas, families can more easily keep farm animals such as chickens, goats, pigs and even cattle; they can cultivate subsistence crops and pursue a style of life that differs from life in nearby rural villages only in degree.

A crisis in governance

By the 1990s, as African cities continued to grow at a pace that considerably exceeded the average for most other parts of the world, it was clear that they were experiencing a crisis in governance. By 'governance' we mean both the effective functioning of the institutional structure (in this case, the structure of local government), as well as the relationship between this structure and civil society as a whole (McCarney et al., 1995). Seen in this fashion, the crisis which African cities are traversing consists of two elements: their inability to manage and coordinate a wide range of urban services at a level adequate to the needs of a poor, and growing population; and their limited ability to engage with emerging sectors of local society in finding acceptable local solutions to, and resources for, increasingly complex challenges.

Africa's historical institutional legacy is a major factor in virtually all accounts of recent problems of local governance (Jaglin and Dubresson, 1993; McCarney, 1996; World Bank,

1989b). Britain and France, as the major colonial powers on the continent from the late nineteenth century to the 1960s, provided the basic framework for two, largely parallel approaches to local government. These two approaches overlapped to some degree by the 1990s, but their essential elements could still be distinguished. The pattern most common in Francophone countries can be called the communal structure, while the pattern in Anglophone countries may be called the representative council structure. The differences between the two are largely explained by history and the accretion of many decades of legal and administrative precedent.

Decentralization in Francophone countries
Most of the Francophone countries in Africa (with the major exception of the Democratic Republic of the Congo, Rwanda and Burundi, which were Belgian colonies) are former French colonies. Since the early part of the twentieth century, urban government has been structured according to the French law dating back to 1884, which provides for communes with mayors, municipal councils and specific revenue and expenditure powers and procedures. The level of a responsibility over finances and local decisions typically depends on the size and wealth of a 'commune', although the local authority has generally been considered to be an organizational modality internal to the unitary state. In any case, by the end of the colonial period in the late 1950s, the evolution of municipal institutions was such that the municipal councils (the administrative organs of the communes) in the larger cities had become responsible for a relatively important range of local services, and were presided over by elected mayors. For example, Abidjan, the capital of Ivory Coast, was declared a 'full exercise commune' (the highest legal category) in 1955, electing a full council by universal suffrage in 1956. Its first elected mayor was Félix Houphouët-Boigny, who was also at the time Prime Minister of the government. After independence in 1960, however, and in spite of a decentralization law passed in 1961, the Ivoirian government found it necessary to suspend many of the powers and political responsibilities which the departing colonial government had extended to Abidjan. While a relatively wide range of functions was discharged by the communes and the larger 'City of Abidjan' (consisting of the central area of Abidjan and some adjoining communes), decisions on their implementation were taken by central government officials. This structure remained in place until 1980 (Attahi, 1992).

A somewhat similar situation, with much deeper historical roots, obtained in Dakar, the capital of Senegal – then the largest French-speaking city in West Africa. From 1887, when Dakar (which had earlier been designated, along with three other towns, as a 'full exercise commune' under French law) was given French-style institutions. The city had an elected council, an elected mayor and considerable influence over finance, services and the hiring of personnel. With the creation of a new administrative entity in 1924 called 'The region of Dakar and its surrounding areas' (*Circonscription de Dakar et dépendances*), however, the central government began to exert much more direct control over municipal affairs. The tendency to circumscribe the power over municipal decision-making became even more prominent in 1964 (four years after independence), when the Dakar region was put under the authority of the government-appointed governor of the region, assisted by nominated advisers. The next year, the management of the water and electricity services, previously under the aegis of the municipality, was transferred to the central government (Diop and Diouf, 1993). Discussing Dakar in the early 1980s, one knowledgeable observer concluded that the administration of the commune of Dakar was 'exclusively carried out by centrally appointed officials, which has led, as a consequence, to the setting aside of any direct participation by elected elements' (Lapeyre, 1986: 339).

By the 1980s, the balance between central and local government began to change in the former French colonies. Two of the major contextual (and external) factors explaining this shift were (1) the increasing interest of donor agencies in decentralization and democratization (World Bank, 1989b); and (2) the decentralization reforms in France, which began in 1982, and continued throughout the 1980s and early 1990s (Centre de Valorisation de la Recherche, 1989; Gilbert and Delcamp, 1993; Institut de la décentralisation, 1996; Terrazzoni, 1987). Not only was there more attention placed on the development of metropolitan government structures, but a more democratic and decentralized framework began to take shape. An exceptional case, analysed in a classic study by Sylvy Jaglin, was the decentralization of Ouagadougou's urban administration through local participation headed by a revolutionary '*Comité de défense de la Révolution*' under Thomas Sankara (until his death in a *coup d'état* in 1987) (Jaglin, 1995). While this particular initiative in Burkina Faso came to an end as a result of a larger political dynamic, the idea of decentralization did not die with the new (and current) regime (Attahi, 1996a). In any case, by the end of the 1980s, many of the major Francophone countries (such as Senegal, Ivory Coast and Bénin) were organizing regular, even multi-party elections both at the local and national level, and a number of other countries were clearly moving towards multi-party

democracy. Thus, Senegal, which began multi-party elections in the late 1970s, established in 1983 the Urban Community of Dakar (CUD), which created a working arrangement to incorporate the three newly created communes of Dakar, Pikine and Rufisque-Bargny. From 1983 to the 1990s, the Mayor of the CUD was the mayor of the commune of Dakar, a high-profile politician with previous central government ministerial experience. Indeed, the Mayor of Dakar has become, argue Momar Diop and Mamadou Diouf, 'one of the main players in the politics of national integration' (Diop and Diouf, 1993: 114).

But of all the decentralization exercises to have been initiated in Francophone Africa, the most thoroughgoing, and by many measures, effective has been that of the Ivory Coast. The initiative began in late December 1977, when a law was passed in the National Assembly confirming the establishment of the two existing 'full exercise' communes (Abidjan and Bouaké). Then, following the 7th Congress in October 1980 of the country's governing party – the PDCI – three major laws, dealing with powers and regulations, were passed. These laws still form the basis of the country's municipal legal framework. According to most observers, the decentralization exercise in the Ivory Coast has on balance been a positive experience. The role of the communes through such functions as maintaining the civil registry, public security, building and maintaining schools, maintaining urban roads, building and maintaining markets, removing household waste, and regulating abattoirs and public water taps, is clearly visible. In addition, their elected mayors have often become very proactive, using 'their networks of personal friends and supporters, as well as the bureaucracy and the party in power to mobilize support. They also attempt to obtain additional resources for their new responsibilities from foreign embassies and international NGOs' (Attahi, 1996b: 121–2).

Decentralization in Anglophone countries In English-speaking African countries, the centralist legacy of the colonial period was more ambiguous. Historically, the United Kingdom has placed more emphasis than has France on democratically elected local councils for the administration and finance of a very wide range of local services. Towards the end of the colonial period, there was a strong thrust to introduce an 'efficient and democratic system of local government' all over English-speaking Africa. But this legacy was profoundly ambivalent entering the post-independence years in the 1960s, since colonial officials often supported local councils and their locally elected leaders as a counterforce to the more broadly based nationalist, anti-colonial movements. Such support at the point of independence endowed local governments in many countries with an anti-nationalist association, an association that did not conduce to good relations with the new, centralist-oriented African governments. In any case, all over English-speaking Africa during the 1960s, local councils proved unable to cope with burgeoning demands for improved education, health, and other local services. These shortcomings were particularly acute in the large, rapidly growing cities. And their inability to raise financing, in conjunction with central government restrictions on transfers, meant that their performance fell far short of their responsibilities. Partly as a result of both political and financial factors, in most English-speaking countries the political autonomy and fiscal resource base of municipal governments was progressively restricted during the 1960s and 1970s. Important exceptions were Nigeria, where for complex political reasons, military governments were favourable to local governments; and Zimbabwe, which (after independence in 1980) opted to support local government as a major element in its development strategy.

If the decentralization reforms of the 1970s were initiated by highly centralized governments, with little involvement of local communities and other groups in civil society, the reforms of the 1980s and 1990s have involved more give and take between government and other forces in the wider society. That this relationship has involved a struggle is evident in the case of Nairobi, the capital of Kenya. With a population of 1,346,000 in 1989, Nairobi is by far the largest, and most economically important centre in the country. From the 1920s on the city was governed by an elected municipal council and mayor. In 1983, citing 'gross mismanagement of council funds and poor services to the residents', the Minister for Local Government suspended all elected officials and placed Nairobi's 17,000 municipal employees and all buildings and services under the direct control of a commission, which he himself appointed. Although the original intention of the commission had been to 'clean up' the council and re-establish elected local government, the central government passed various motions through the National Assembly extending the life of the commission until both national and local elections were held in December 1992.

Kenya's first multi-party elections ushered in a new chapter in the turbulent history of Nairobi – and of urban local governance in Kenya. One of the major new parties specifically called, in its election manifesto, for the granting of increased autonomy to local government. As has always been the case in Kenya, local and national elections were held at the same time. But whereas the governing party, KANU, had always captured both levels of seats in the urban wards and constituencies in the past, in this election the opposition parties won most of the parliamentary seats in the major urban areas, and took control of 23 of the 26 municipal councils, including Nairobi. With the end of the city commission, the new Mayor of Nairobi (elected by the sitting councillors) was himself not a member of the governing party of the country. Political differences between the central government and the newly elected municipal councils soon came to the surface, with the Minister of Local Government issuing a series of directives that curtailed the powers of the mayors. For the government, these councils were a political force to be reckoned with; but for the emerging middle class, the councils were a vehicle by which to achieve a greater measure of local autonomy (Stren et al., 1994). The elections of 1997 were another platform for conflict between a more vocal urban civil society and a government which attempted to keep the urban councils under its control. In the run-up to these elections, there was significant violence in Mombasa between indigenous coastal Africans and so-called 'up-country' Africans; the latter, nearly 100,000 of whom were displaced by riots and intimidation, might have been expected to vote for opposition parties in both national and local elections. In effect, the dislocation of many of these people probably increased the chances of the governing party of eventually winning the Coast Province elections, and the majority of seats in the Municipal Council as well. But in most other major towns of the country, as in 1992, opposition parties won a majority of local council seats. In the late 1990s a number of urban NGOs and civil rights organizations continued to press for constitutional reform, although by the end of the decade little had been formally accomplished (Lee-Smith and Lamba, 2000).

The re-emergence of South Africa During the 1990s, the emergence of South Africa as a democracy had a major impact on both research directions and policy discussions throughout Africa. The new South African Constitution, which was formally adopted in 1996, devotes a whole chapter to local government (Chapter 7, containing 14 separate articles). *Inter alia*, this chapter of the Constitution states that the objects of local government (including municipal government) are

> (a) to provide democratic and accountable government for local communities; (b) to ensure the provision of services to communities in a sustainable manner; (c) to promote social and economic development; (d) to promote a safe and healthy environment; and (e) to encourage the involvement of communities and community organizations in the matters of local government. (Section 152)

In managing its administration and planning processes, municipalities are *required* by the Constitution 'to give priority to the basic needs of the community, and to promote the social and economic development of the community' (Section 153). To strengthen the direction already taken by the Constitution and the discussions leading up to its promulgation, the South African government phased the reorganization of local government institutions from a 'pre-interim phase' (during which there were widespread discussions as to the new structure and functions of local governments), through an 'interim phase' (from municipal elections in 1995/6 through to the next elections in 1999), culminating in a 'final stage' in 1999, when a new system was finally established (White Paper on Local Government, 1998). Two subsequent pieces of legislation, the Municipal Structures Act (1998) and the Municipal Systems Bill (1999) make detailed provisions to enshrine the practical mechanisms needed to support participatory and developmental municipal governance. In addition, the government has decided that, following the municipal elections of November 2000, six major urban areas – including Cape Town, Durban and Johannesburg – will have their multiple jurisdictions amalgamated into 'unicities', rather than the existing two-tier structure. This whole process of institutional change, which began in the 1970s with the powerful movement to eradicate apartheid – in which organized workers, students, youth, women and urban residents were mobilized against both national and local state structures – led to 'local negotiating forums' in the 1990s which brought organized civil society in direct contract with government representatives. This remarkable 'bottom-up transformation of local government according to non-racial and democratic principles' (Swilling, 1996: 129) has been extremely influential in new thinking about urban governance throughout the continent.

CONCLUSIONS

The struggle of African cities to survive during the fading years of the twentieth century continues into the new millennium and has been conditioned as much by their colonial history as it has by the new global economy with which they are faced. While the history of this large and complex continent is extremely varied, both global and local dynamics require that new alternatives be found to the present situation. Clearly, attempts at structural transformation during the 1970s did not sustain real growth. In the decade that followed African cities were plunged into severe crisis – a crisis from which they have not

emerged. At the global level, sub-Saharan African cities (outside South Africa) have now been relegated from dependency to marginality; locally they are forced into attempts at institutional reform while simultaneously coping with demographic pressure, economic stagnation and social deterioration.

While their immediate economic future may not be very bright, there are some glimmers of hope for African urban centres. The example of a democratic and re-energized South Africa, the economic powerhouse of the continent, suggests new possibilities for a more effective and inclusive approach to municipal policy-making. With international assistance, decentralized urban management arrangements are being put in place that may be able more directly to respond to the needs of the low-income majority. But at a more general level, African cities are transforming both their own physical landscape and their social and political structures into new forms that better reflect their unique cultural and historical conditions. Paradoxically, any eventual integration of African cities into the world economy may first require a recognition of their real differences as a basis for institutional regeneration.

BIBLIOGRAPHY

Abu-Lughod, Janet (1980) *Rabat. Urban Apartheid in Morocco*. Princeton, NJ: Princeton University Press.

Adjavou, Akolly, Bouzy, Dominique, Chemlal, Aziz, Couchey, Alain, Dadzie, Komlan, Gonzalez, Patrice, Mura, Valérie, and Osmont, Annik (1987) *Economie de la construction à Lomé*. Paris: L'Harmattan.

Amis, Philip (1984) 'Squatters or tenants?: the commercialization of unauthorized housing in Nairobi', *World Development*, 12 (1): 87–96.

Amis, Philip (1990) 'Introduction. Key themes in contemporary African urbanization', in Philip Amis and Peter Lloyd (eds), *Housing Africa's Urban Poor*. Manchester: Manchester University Press for the International African Institute. pp. 1–31.

Antoine, Philippe and Ba, Amadou (1993) 'Mortalité et santé dans les villes africaines', *Afrique contemporaine* (special issue, *Villes d'Afrique*), No. 168 (October/December): 138–46.

Attahi, Koffi (1992) 'Planning and management in large cities. A case study of Abidjan, Cote d'Ivoire', in UNCHS (Habitat), *Metropolitan Planning and Management in the Developing World: Abidjan and Quito*. Nairobi: UNCHS. pp. 31–82.

Attahi, Koffi (1996a) 'Burkina Faso', in Patricia McCarney (ed.), *The Changing Nature of Local Government in Developing Countries*. Toronto and Ottawa: Centre for Urban and Community Studies, University of Toronto; and Federation of Canadian Municipalities. pp. 65–84.

Attahi, Koffi (1996b) 'Côte d'Ivoire', in Patricia McCarney (ed.), *The Changing Nature of Local Government in Developing Countries*. Toronto and Ottawa: Centre for Urban and Community Studies, University of Toronto; and Federation of Canadian Municipalities. pp. 109–26.

Baker, Jonathan and Aina, Tade Akin (eds) (1995) *The Migration Experience in Africa*. Uppsala: Nordiska Afrikainstitutet.

Becker, Charles M., Hamer, Andrew M. and Morrison, Andrew R. (1994) *Beyond Urban Bias in Africa. Urbanization in an Era of Structural Adjustment*. London: James Currey.

Bonassieux, Alain (1987) *L'Autre Abidjan. Chronique d'un quartier oublié*. Paris: Karthala.

Bonnardel, Régine (1991) 'Femmes, villes, informel, en Afrique au sud du Sahara', in Catherine Coquery-Vidrovitch and Serge Nedelec (eds), *Tiers-Mondes: L'informel en question?* Paris: L'Harmattan, pp. 247–69.

Bubba, Ndinda, and Lamba, Davinder (1991) 'Urban management of Kenya', *Environment and Urbanization*, 3 (1): 37–59.

Caldwell, John C. (1997) 'The impact of the African AIDS epidemic', *Health Transition Review*, supplement 2 to Volume 7, 169–88.

Centre de Valorisation de la Recherche (eds) (1989) *La décentralisation. Etudes comparées des législations ivoiriennes et françaises*. Toulouse: Université des Sciences Sociales.

Cooper, Frederick (1987) *On the African Waterfront. Urban Disorder and the Transformation of Work in Colonial Mombasa*. New Haven, CT: Yale University Press.

Crowder, Michael (1968) *West Africa under Colonial Rule*. London: Hutchinson & Co.

DHS and Mazingira Institute (1988) *Nairobi: the Urban Growth Challenge*. Nairobi: Mazingira Institute.

Diop, Momar C. and Diouf, Mamadou (1993) 'Pouvoir central et pouvoir local: la crise de l'institution municipale au Sénégal', in Sylvy Jaglin and Alain Dubresson (eds), *Pouvoirs et cités d'Afrique noire*. Paris: Karthala, pp. 101–25.

Dulucq, Sophie and Goerg, Odile (eds) (1989) *Les investissements publics dans les villes africaines, 1930–1985*. Paris: L'Harmattan.

Duruflé, Gilles (1988) *L'ajustement structurel en Afrique*. Paris: Karthala.

Fieldhouse, D.K. (1986) *Black Africa, 1945–1980. Economic Decolonization and Arrested Development*. London: Allen & Unwin.

Fisette, Jacques, Sabou, Ibrahim and Diallo, Mahamadou Aoudi (1995) 'La décentralisation au Niger: des réformes inachevées', in Mario Polèse and Jeanne M. Wolfe (eds), *L'Urbanisation des pays en développement*. Paris: Economica. pp. 307–25.

Fredland, Richard A. (1998) 'AIDS and development: an inverse correlation?', *Journal of Modern African Studies*, 36(4): 547–68.

Freeman, Donald (1991) *A City of Farmers: Informal Urban Agriculture in the Open Spaces of Nairobi, Kenya*. Montreal and Kingston: McGill-Queen's University Press.

Gilbert, Guy and Delcamp, Alain (eds) (1993) *La décentralisation dix ans après*. Paris: Librarie générale de droit et de jurisprudence.

Godard, Xavier and Teurnier, Pierre (1992) *Les transports urbains en Afrique à l'heure de l'ajustement*. Paris: Karthala.

Great Britain (1955) *East Africa Royal Commission, 1953–1955 Report*. Cmd 9475. London: Her Majesty's Stationery Office.

Hailey, Lord (1957) *An African Survey: Revised 1956*. London: Oxford University Press.

Halfani, Mohamed (1996) 'Marginality and dynamism: prospects for the Sub-Saharan African city', in Michael A. Cohen, Blair A. Ruble, Joseph S. Tulchin and Allison M. Garland (eds), *Preparing for the Future. Global Pressures and Local Forces*. Washington, DC: Woodrow Wilson Center Press. pp. 83–107.

Hoselitz, Bert F. (1955) 'Generative and parasitic cities', *Economic Development and Cultural Change*, 3 (3): 278–94.

ILO (International Labour Office) (1991) *The Urban Informal Sector in Africa in Retrospect and Prospect: an Annotated Bibliography*. Geneva: ILO.

Institut de la Décentralisation (1996) *La décentralisation en France*. Paris: La Découverte.

Jaglin, Sylvy (1995) *Gestion urbaine partagée à Ouagadougou*. Paris: Karthala.

Jaglin, Sylvy and Dubresson, Alain (eds) (1993) *Pouvoirs et cités d'Afrique noire*. Paris: Karthala.

Joshi, Heather, Lubell, Harold and Mouly, Jean (1976) *Abidjan. Urban Development and Employment in the Ivory Coast*. Geneva: ILO.

Kirond, J.M. Lusugga (1999) 'Dar es Salaam, Tanzania', in Adepoju G. Onibokun (ed.), *Managing the Monster. Urban Waste and Governance in Africa*. Ottowa: International Development Research Center. pp. 101–72.

Konadu-Agyemang, Kwadwo O. (1991) 'Reflections on the absence of squatter settlements in West African cities: the case of Kumasi, Ghana', *Urban Studies* 28: 1: 139–51.

Kulaba, Saitiel (1989a) 'Local government and the management of urban services in Tanzania', in Richard Stren and Rodney White (eds), *African Cities in Crisis. Managing Rapid Urban Growth*. Boulder, CO: Westview Press. pp. 203–45.

Kulaba, Saitiel (1989b) *Urban Management and the Delivery of Urban Services in Tanzania*. Dar es Salaam: Ardhi Institute.

Landell-Mills, Pierre and Serageldin, Ismail (1991) 'Governance and the development process', *Finance and Development*, September, pp. 14–17.

Lapeyre, Charles (1986) 'La ville de Dakar: commune et région', in Centre d'Études et de Recherches en Administration Locale (eds), *L'administration des grandes villes dans le monde.* Paris: Presses Universitaires de France. pp. 333–56.

Laquian, Aprodicio (1983) *Basic Housing Policies for Urban Sites, Services and Shelter in Developing Countries.* Ottawa: IDRC.

Lee-Smith, Diana and Lamba, Davinder (2000) 'Social transformation in a post-colonial city: the case of Nairobi', in Mario Polèse and Richard Stren (eds), *The Social Sustainability of Cities. Diversity and the Management of Change.* Toronto: University of Toronto Press. pp. 250–79.

Lee-Smith, Diana and Stren, Richard (1991) 'New perspectives on African urban management', *Environment and Urbanization*, 3 (1): 23–36.

Lee-Smith, Diana, Manundu, Mutsembi, Lamba, Davinder and Gathuru, P. Kuria (1987) *Urban Food Production and the Cooking Fuel Situation in Urban Kenya.* Nairobi: Mazingira Institute.

Lipton, Michael (1977) *Why Poor People Stay Poor: Urban Bias in World Development.* Cambridge, MA: Harvard University Press.

Mabogunje, Akin (1968) *Urbanization in Nigeria.* London: University of London Press.

MAP (Monitoring the AIDS Pandemic) Network (2000) 'The Status and Trends of the HIV/AIDS Epidemics in the World', US Census Bureau website: http://www.census.gov/ipc.

Marris, Peter (1961) *Family and Social Change in an African City: A Study of Rehousing in Lagos.* London: Routledge and Kegan Paul.

Mawhood, Philip (ed.) (1983) *Local Government for Development. The Experience of Tropical Africa.* Chichester: John Wiley.

Maxwell, Daniel and Zziwa, Samuel (1992) *Urban Farming in Africa. The Case of Kampala, Uganda.* Nairobi: ACTS Press.

Mayo, Stephen K. and Gross, David J. (1985) 'Sites and services – and subsidies: the economics of low-cost housing', in Richard M. Bird and Susan Horton (eds), *Government Policy and the Poor in Developing Countries.* Toronto: University of Toronto Press. pp. 106–43.

McCarney, Patricia (ed.) (1996) *The Changing Nature of Local Government in Developing Countries.* Toronto and Ottawa: Centre for Urban and Community Studies, University of Toronto; and Federation of Canadian Municipalities.

McCarney, Patricia, Halfani, Mohamed and Rodriguez, Alfredo (1995) 'Towards an understanding of governance', in Richard Stren and Judith Bell (eds), *Urban Research in the Developing World. Volume 4: Perspectives on the City.* Toronto: Centre for Urban and Community Studies, University of Toronto. pp. 91–141.

Onibokun, Adepoju (1989) 'Urban growth and urban management in Nigeria', in Richard Stren and Rodney White (eds), *African Cities in Crisis.* Boulder, CO: Westview Press. pp. 68–111.

Peil, Margaret (1976) 'African squatter settlements: a comparative study', *Urban Studies*, 13: 155–66.

Poisnot, Jaqueline, Sinou, Alain and Sternadel, Jaroslav

(1989) *Les villes d'Afrique noire. Politiques et opérations d'urbanisme et d'habitat entre 1650 et 1960.* Paris: Ministère de la coopération et du développement.

Rakodi, Carole (1988) 'Urban agriculture: research questions and Zambian evidence', *Journal of Modern African Studies*, 26 (3): 495–515.

Republic of Kenya (1965) *Statistical Abstract 1965.* Nairobi: Statistics Division, Ministry of Economic Planning and Development.

Republic of Kenya (1970) *Building Code.* Nairobi: Government Printer.

Sabot, Richard H. (1979) *Economic Development and Urban Migration: Tanzania 1900–1971.* Oxford: Clarendon Press.

Sandbrook, Richard (1982) *The Politics of Basic Needs. Urban Aspects of Assaulting Poverty in Africa.* Toronto: University of Toronto Press.

Sandbrook, Richard (1993) *The Politics of Africa's Economic Recovery.* Cambridge: Cambridge University Press.

Schilter, Christine (1991) *L'agriculture urbaine à Lomé: approches agronomique et socio-économique.* Paris and Geneva: Karthala and IUED.

Simon, David (1992) *Cities, Capital and Development. African Cities in the World Economy.* London: Belhaven Press.

Stren, Richard (1978) *Housing the Urban Poor in Africa. Policy, Politics and Bureaucracy in Mombasa.* Berkeley, CA: University of California.

Stren, Richard (1979) 'Urban Policy', in Joel Barkan and John J. Okumu (eds), *Politics and Public Policy in Kenya and Tanzania.* New York: Praeger. pp. 179–208.

Stren, Richard (1989) 'The administration of urban services', in Richard E. Stren and Rodney R. White (eds), *African Cities in Crisis. Managing Rapid Urban Growth.* Boulder, CO: Westview Press. pp. 37–67.

Stren, Richard (1990) 'Urban housing in Africa: the changing role of government policy', in Philip Amis and Peter Lloyd (eds), *Housing Africa's Urban Poor.* Manchester: Manchester University Press for the International African Institute. pp. 35–53.

Stren, Richard (1992) 'Large cities in the Third World', in UNCHS, *Metropolitan Planning and Management in the Developing World: Abidjan and Quito.* Nairobi: UNCHS (Habitat). pp. 1–30.

Stren, Richard (1993) 'Urban management in development assistance: an elusive concept', *Cities*, 10 (1): 125–38.

Stren, Richard and White, Rodney (eds) (1989) *African Cities in Crisis: Managing Rapid Urban Growth.* Boulder, CO: Westview Press.

Stren, Richard, Halfani, Mohamed and Malombe, Joyce (1994) 'Coping with urbanization and urban policy', in Joel D. Barkan (ed.), *Beyond Capitalism vs. Socialism in Kenya and Tanzania.* Boulder, CO: Lynne Reinner. pp. 175–200.

Swilling, Mark (1996) 'Building democratic local urban

governance in Southern Africa: a review of key trends', in Patricia McCarney (ed.), *Cities and Governance*. Toronto: Centre for Urban and Community Studies, University of Toronto. pp. 127–44.

Terrazzoni, André, (1987) *La décentralisation à l'épreuve des faits*. Paris: Librarie générale de droit et de jurisprudence.

Touré, Abdou (1985) *Les petits métiers à Abidjan. L'imagination au secours de la 'conjoncture'*. Paris: Karthala.

Tripp, Aili Mari (1997) *Changing the Rules. The Politics of Liberalization and the Urban Informal Economy in Tanzania*. Berkeley, CA and Los Angeles: University of California Press.

UNCHS (1987) *Global Report on Human Settlements, 1986*. New York: Oxford University Press for UNCHS (Habitat).

United Nations (1993) *World Urbanization Prospects: The 1992 Revision*. New York: United Nations.

White Paper on Local Government (1998) Issued by the Ministry for Provincial Affairs and Constitutional Development. Pretoria: Government of the Republic of South Africa.

White, Rodney (1985) 'The impact of policy conflict on the implementation of a government-assisted housing project in Senegal', *Canadian Journal of African Studies*, 19: 3: 505–28.

Wilson, Robert and Cramer, Reid (eds) (1996) *International Workshop on Local Governance. Second Annual Proceedings*. Austin, TX: Lyndon B. Johnson School of Public Affairs.

World Bank (1972) *Urbanization. Sector Working Paper*. Washington, DC: The World Bank.

World Bank (1989a) *Sub-Saharan Africa: From Crisis to Sustainable Growth*. Washington, DC: The World Bank.

World Bank (1989b) *Strengthening Local Government in Sub-Saharan Africa*. Washington, DC: Economic Development Institute for the World Bank.

World Bank (1991) *Urban Policy and Economic Development. An Agenda for the 1990s*. Washington, DC: The World Bank.

Wunsch, James S. and Olowu, Dele (eds) (1990) *The Failure of the Centralized State. Institutions and Self-Governance in Africa*. Boulder, CO: Westview Press.

Index

accident risk, 115, 116, 130
ACORN system (neighbourhoods), 44
administrative boundaries, 14, 27, 31, 45, 266, 341
Advisory Council on Intergovernmental Relations, 329
Advisory Regional Plans, 113
AFDC programme, 354, 355
affluence, 315
Africa (Sub-Saharan), 12, 417, 466–82
agglomeration
  economics, 26, 248–52, 276, 297, 300
  spatial, 261–2
  urban (trends), 21–5, 34–5
agriculture, 14–16, 26, 397, 478–9
allocative efficiency/inefficiency, 338–9, 343, 344, 346
annexation process, 327–8
Anti-Poverty Network, 361
ASEAN states, 420–3, 437, 446
Asian Development Bank, 125, 424, 426–7, 444, 446
Association of Metropolitan Authorities, 361
Atlanta, 331, 332, 405
atmospheric pollution, 129–30, 135
Auster, Paul, 64–5

Balsall Heath (Birmingham), 208–9, 215–16
Bangkok, 430, 432, 441–3, 448
Basic Needs programmes, 473
'beady ring' pattern, 47–9
Beauregard, R., 168–9, 299, 406–9
biologism, 36, 38–9, 43–6
biotic competition, 36, 39–42, 45, 46
Birmingham, 45, 208–9, 215–16, 290–2, 294
black box approach, 243–4, 245–6, 254
black political power, 182, 190
Booth, Charles, 2, 3, 163
Bordeaux, 329
Bretton Woods agreement, 275
Britain, 2–3
  planning system, 385–97
  race relations, 177–84
  women in, 208–13, 214–15
  see also individual cities
Brundtland Report, 127

Buchanan Report (1963), 110
built environment, 54–5, 67–8, 131–2
  see also environment, city as
burgage cycle, 73–7
business improvement districts, 345

California, 385–91, 393–7
Canada, 156, 311–13, 325, 328, 360
capital, 344, 403
  accumulation, 42, 90, 241, 273, 411
  mobility, 278, 286, 303, 404, 410
capitalism, 39, 42, 90, 310, 385
  industrial, 241, 273, 285
  NUP and, 404, 411–15
  oligopolistic, 273, 275, 395, 397
  post-welfare, 336, 341–4, 363
  social segregation and, 163, 166–7
car ownership, 107–10, 112, 114–17, 119–21
Centers for Disease Control, 233–4, 236
central business district, 41, 69–70
  transitional cities, 436–7, 454
  transport policy, 106–7, 114, 118
central city, 149–50, 327–8
centrality (space/power), 263, 264
centralization, 147, 346
Centre for Local Economic Strategies, 362
Chicago, 3, 40–1, 46, 63, 79, 107, 196, 285, 367
Chicago School, 5, 7, 36, 40–1, 43, 45, 46, 50, 194, 199
Child Support Agency, 356
chronotope, 57–8
cities
  communities in, 194–204
  crime in (US/Europe), 220–37
  definitions, 14–35
  gender issues, 206–17
  in global economy, 256–67
  housing trends, 88–100
  informal economies, 308–20
  new urban economies, 296–305
  physical form, 67–8, 69–85
  post-Fordist, 273–81
  post-industrial, 284–94

postmodern, *see* postmodern cities
race relations in, 177–91
representation of, 52–65
scale economies, 243–54
services (post-welfarism), 336–48
social polarization, 162–73
social policy and, 351–64
studying, 1–7, 406–9
sustainable environments, 124–37
transition, *see* transitional cities/economies
transport and, 102–21
urban ecology, 36–51
urban governance, 325–34
urbanization patterns, 143–59
Citizen's Charter, 339
citizenship, 304, 361–2, 373
City Action Teams, 360
city centres, 78–80, 82–3, 107, 157–8, 461
City Challenge, 202
city size, 12, 24–5, 34–5, 146, 327
city trajectories, 402–15
civic identity/society, 302, 304–5
Civil Rights Act (1964), 191
civil society, 323, 360, 453, 474, 481
Coastal Act (USA), 391, 394
Cold War, 363, 364
collective identity, 378–81
colonial cities, 466–9
commodification, 97, 304, 319, 376
  NUP agenda, 409–11, 413–14
  Russia, 452, 454–5
Common Agricultural Policy, 397
Communautés Urbaines (CUs), 328, 329
communications, 15, 31, 37, 244–5, 292
communicative planning, 369–81
community, 6, 42, 43, 235
  in the city, 142, 194–204
  citizenship and, 361–2
  forms/processes, 198–201
Community Based Organizations, 432
Community Redevelopment Law (USA), 394
commuting, 26, 30, 150, 249
competitive bidding, 210, 342, 344, 348
competitive cities, 300–2
compulsory competitive tendering, 210, 342, 344, 348
Computer Aided Manufacturing, 280
concentric rings, 5, 42, 163
conflict, planning and, 385–99
congestion (traffic), 115–16, 130–1
consensus, 371, 372–3, 376–7
Consolidated Metropolitan Statistical Area, 28
consolidation (city-county), 327–8
consumer behaviour, 7
consumer goods, 90, 273, 403, 454
consumerism, 274, 277, 339, 346, 368
consumption, 201, 274–5, 280, 301, 368, 410–11
cooperative forms, 328–31
corporatist agreements, 317
Cortese-Knox Act (USA), 394
Council for the Preservation of Rural England, 389

Councils of Government, 328, 329, 390
counter-emancipatory ideology, 374–5
counterurbanization, 141, 143, 145–6, 148, 150–3, 158
Countryside Alliance, 12
created environments, 36–8
crime, 142, 171–2, 182, 191
  race and, 221–3, 224–5
  US/Western Europe, 220–37
criminal justice model, 232–4, 236
critical planning theory, 372
CUD (in Dakar), 480
cultural imperialism, 380–1
cultural politics, 409–10, 413
culture, 40–1, 44, 280–1, 316, 405–6
'Cutteslowe Walls' (Oxford), 45

Dakar, 480
decentralization, 3, 7, 12, 278, 297
  in Africa, 479–81
  government/governance, 328, 339–42, 346
  urbanization, 145–51, 153, 158
deconcentration, 26, 29, 147, 150, 152–3, 156
defended communities, 199–200
deindustrialization, 294, 298
demand-side policy, 338, 344
demand management, 49–50, 92
democratic accountability, 346–7, 348
dependency (Sub-Saharan Africa), 466–82
deregulation, 3
  financial, 309, 319, 353
  housing, 90–1, 93, 95, 99
desakota, 12, 425
DETR, 389
Detroit, 290–2, 294
deurbanization process, 158, 341
Development Agencies, 330–1, 341
Development Land Tax, 395
development plans/control, 392–3, 395
differential urbanization, 145–6, 151
discourses, 367–8
  power and, 402–15
discrimination (racial), 178–82, 185–7
distributive efficiency, 132, 133
disurbanization stage, 147, 151, 153
division of labour, 273, 278
  international, *see* international division of labour
  sexual, 197–8
  spatial, 301, 369
DoE, 389, 392, 398
doi moi reforms, 443, 444–5
Drug Abuse Resistance Education, 235
dual city, 158, 162, 164–8, 170, 287, 304–5
Durkheim, Emile, 39, 41
dynamic externalities, 243, 251–4

e-space, 263
Earned Income Tax Credit, 355
ecology, 26, 124
  human, 3, 36, 39–44
  urban, 36–51

economic environment, 131–2
economic issues (transport), 113–14
economic opportunity, lack of, 187
Economic Planning Zone, 437, 440, 447
economic regulators, 315–16
economic restructuring, 90, 166–7, 241, 296, 370
economy, city as, 241–2
    global economy, 256–67
    new urban economies, 258–9, 296–305
    post-Fordist city, 273–81
    post-industrial city, 284–94
    urban informal economies, 308–20
    urban scale economies, 243–54
economy, society and, 129–33
edge cities, 303, 326
education, 49–50, 316, 324, 357–8
emancipatory politics, 369–81
emission trading/permits, 137
employment
    African crisis, 475–6
    informal/formal, 308, 310–20, 352
    occupational structure, 165, 167, 170, 180
    race and, 178, 180, 183–4, 187–91
    welfare to work, 319, 355–7, 362–3
    of women, 210–11
    see also full employment; labour market;
        unemployment
endogenous growth model, 251–2
energy resources, 130
Engels, F., 4, 7, 15, 38, 52, 162, 231
English house, 73–7
Enlightenment, 39, 374–5
Enterprise Zones, 360
entrepreneurial government, 340, 347, 405
environment, city as, 67–8
    housing, 88–100
    managing sustainable environments, 124–37
    physical form of cities, 69–85
    transport, 102–21
environmental awareness, 135–6
environmental crisis (Africa), 477–9
environmental equity, 132, 133
Environmental Impact Assessment, 390
environmental policy, 124–37
environmental problems, 129–31
environmental regulators, 317–18
episteme, 377, 378
Equal Employment Opportunity Commission, 184,
    191
equal opportunities, 184, 191, 207, 215, 363
equity (transport policy), 115–17
ESRC, 209
ethnic minorities, 162, 164, 199
    gender and, 208–9, 210–11
    informal work and, 313–14
    see also race; racism
Europe, 309, 390
    crime, 220, 223–30, 232–7
    new urban economies, 300, 303–4
    post-welfarist analysis, 340–1, 344

social policy, 357–60, 361, 362–3
transport, 119–21
European Charter of Local Self-Government, 340,
    341
European Healthy Cities network, 360
evolutionary theory, 38–42, 43
evolutionism, 36, 39, 46–9
exchange value, 196, 278, 413
exclusion, 351, 353–4
exports (Pacific Asia), 423–4
Extended Metropolitan Regions (EMRs), 425, 428–48
externalities, 132–3, 137
    dynamic, 243, 251–3, 254
    labour market, 244, 245–6
    nature of, 248–51
    time as positive, 135–6
extra-urban carrying capacity, 124
exurbanization, 11, 16, 149

factorial ecology techniques, 4, 5, 14
Fair Housing Act (USA), 186, 190
family, 43–4, 157, 196–7
FBI, 220, 221–2, 224, 227
fear, 212–13, 352
    of crime, 142, 222–3, 225, 235
Federal Contract Compliance Programs, 191
federations, 328
feminism, 197, 206–7, 212–14, 215, 380
FIDES, 468
financial services, 258, 260, 264–5, 287
fixed capital, 273
flâneur/flâneuse, 212
flexible specialization, 274, 275, 276
flying geese model, 421, 446
food stamps, 354, 355
Fordism, 273–8, 280–1, 297, 369–70
foreign direct investment, 422–3, 424, 432, 473
formalization (urban economies), 308–10
France, 288, 329, 358, 359
freight movement, 110–11
fringe belts, 71–3, 79
full employment, 273, 309, 318–19, 337, 338, 387
Functional Community Area, 14, 28–31
functional regions, 149–50, 152, 154–5
functional responsibilities, 341
functional specialization, 273
Further Education Corporations, 346

Gemeinschaft, 43, 195–6, 197, 379
gender, 197–8, 206–17, 311–13
General Assistance, 354
General Development Order, 393
General Planning Law (USA), 394
gentrification, 148, 156–8, 164, 167, 212, 280, 304, 464
German school, 52
Germany, 358
Gesellschaft, 43, 195–6
Giddens, Anthony, 36
Glasgow, 405–6, 413
global cities, see world cities

global economy, 256–67
global environment (transport), 114–15
global value chains, 263
globalization, 3, 141, 326
    of capital, 403, 415
    communities and, 194, 199–200, 203–4
    global economies, 241–2, 256–67, 276
    postmodern cities, 367, 402–3, 415
    see also transitional cities/economies
governance, 324, 340, 347
    crisis (Africa), 474, 479–82
    metropolitan, 325–33
    urban, 323, 325–34, 403–6
governing structures, transport and, 112–13
government, 324
    public institutions, 389–91
    services (post-welfarism), 336–48
    social policy, 351–64
    urban governance and, 323, 325–34
Gramm-Rudman law, 357
grand narratives, 375–6
grant maintained school boards, 346
Greater London Council, 330–1, 389
Green Bans (in Sydney), 200–1
growth management policies, 398–9
growth poles (Africa), 472–3

Habermas, Jurgen, 375
Hanoi, 432, 443–6, 447
Hayek, Friedrich, 49, 50
Herbert Commission, 328
heterogeneous informal labour market, 310–14
Hirschman-Herfindahl index, 250
historico-geographical approach, 69–85
HIV/AIDS, 476, 478
home ownership, 46, 78, 81, 90–1, 93–6, 98–100, 164,
    387–8
homelessness, 68, 89–90, 167, 209
homicides, 221–4, 226, 229, 232–4
Hong Kong, 429
horizontal integration, 347
House Builders' Federation, 389, 398
household structure, 211–12
houses/housing, 47, 50, 57, 137, 443
    English, 73–7
    gentrification, 148, 156–8, 164, 167, 212, 280, 304,
        464
    legislation, 329, 395
    racial segregation, 178, 180–7, 190–1, 398
    residential differentiation, 162–4
    residential growth, 69–71, 77–8, 80
    residential segregation, 4, 44–6, 141–2, 169, 180–2,
        185–7, 397–8
    Russia, 452–62
    social, 44–5, 88, 90–3, 96–8, 100, 164, 359
    Sub-Saharan Africa, 468, 470–2, 474, 476
    tenure, 46, 67–8, 78, 81, 89–91, 93–100, 164, 317,
        361, 387–8
    see also homelessness
Housing Associations, 346

Housing Discrimination Study, 187
human capital, 252–3, 273, 316
human design, 36–8
human ecology, 3, 36, 39–43, 44
human environment (novels), 55–6

identity, 198–9, 260
ILO, 474
imaginary city, 213–14
immigration, 157, 166, 177–8, 314
    Britain, 45, 179–83, 191
    USA, 184–5, 190, 191, 353
income structure, 165, 167, 170
industrial districts, 276–7, 297
industrial restructuring, 42, 326, 369
Industrial Revolution, 15, 73, 78, 84, 88, 100, 167,
    284–5, 286, 325, 385
industrial society, 286, 287, 385
industrial structure, 315, 316
industrialism, 2, 285
industrialization, 2, 15–16, 196, 241, 285
inequality, 167–8
informal economies, 308–20
informal employment, 308, 310–20, 352
information, 257, 340
    centres/cities, 292–3, 297, 302–3
    spillovers, 244–5, 251, 254
infrastructure, 67–8, 404, 405
    see also houses/housing; transport
inner city areas, 45, 89, 172, 189, 197, 208–9
    gentrification, 156–8, 164, 280
    social policy, 354–5, 359–61
    urbanization trends, 141, 152–3, 156–8
Inner London Education Authority, 331
innovations, 285, 287, 293, 302, 326
insiders, 201–2
institutional forms (governance), 327–31
institutional power, 411–12
institutional racism, 178–9, 182–4
institutional regulators, 316–17
Inter-City Network, 361
inter-generational model, 253
interactions (community), 198–9
interest group pluralism, 369–70
international division of labour, 42, 164, 241, 261, 289,
    369–70, 402
    new, 280, 403, 409, 422, 428, 436
International Monetary Fund, 276, 420, 444
interregional migration (USA), 184
intra-regional transport, 105–7
intra-urban carrying capacity, 124
intra-urban movement, 107
investment, 422–3, 424, 432, 473
irreversibility, 133–4

Japan, 287–8, 421–2, 429–36, 448
Jerusalem, 56, 60–2
Joseph Rowntree Foundation, 209, 351
'just' city, 303–4
just-in-time system, 111, 275

justice, 373, 375–7
    see also social justice

Kafka, Franz, 58, 63–4, 65
KANU party (Kenya), 481
KBS (in Nairobi), 477
Kellogg Foundation, 361–2
Kenya, 476–7, 478, 481
Keynesian economics, 49–50, 92, 98, 273, 275, 337
kinship, 43–4, 157, 196–7
knowledge, 251–2, 285–6, 292–3, 297, 376, 378, 407,
    411, 412
Kondratiev's long wave, 275
Kyoto Earth Summit, 117

labour law, 316–17
labour market, 99, 157, 171, 173, 242, 277, 279
    externalities, 244, 245–6
    gender and, 210–11
    informal/formal, 308, 310–20, 352
    local, 29–30, 151–2, 297, 300, 303, 317, 358
    public sector (reform), 343
    racial discrimination, 178, 183, 187–8
labour movement, 413–14
labour power, 42
Lakewood Plan, 329
Land Commission Act (UK), 395
land markets, 395–6, 460–1
land use, 137
    planning, 385–92, 394–6, 398–9
landscape management, 83–5
language games, 376, 377
leisure economies, 298, 304
Leontief, W., 248
LETS scheme, 414
liberalization, 3
Lille, 329
linkages (community), 198–9
literary representation (city), 52–65
Local Agency Formation Commission, 390, 394
local community (role), 194–204
local democracy, 323, 337–9, 341, 346–7, 362
local diversity, 244, 247–8
Local Enterprise Companies, 346, 411
local government, 112, 289
    governance and, 323–5
    legislation, 339, 392, 395
    planning policies, 388–92, 394–5
    post-Fordist politics, 278–80
    post-welfarist analysis, 337–48
    social policy role, 360–1
local labour market, see labour market
local planning authority, 392, 393
local politics, 278–80, 300
localism, 300
locality, 128, 301, 302
localization, 249–51, 253–4, 260, 402
location-specific knowledge, 251
locational analysis, 4, 5
locationalization economies, 276

LODE Programme, 341
London, 79, 258, 287–8, 389
    social segregation, 163–6
    transport policies, 103, 113–14
    urban governance, 328, 330–2
long-distance transport, 102–5
long-term allocative efficiency, 132–3
Los Angeles, 104, 107, 165, 288–9, 329, 367, 396
Lynch, Kevin, 52–3
Lyons, 329
Lyotard, J.-F., 375–7

macrocephaly, 16
managing sustainable environment, 124–37
Manchester, 4, 7, 103, 162–3, 285
manufacturing, 165, 475
    in global economy, 256, 262–4
    post-Fordist, 273, 276–7
    post-industrial, 285, 287
marginalized economies, 264, 310–11
    Sub-Saharan Africa, 469–73
markets/marketization, 3, 342, 395–6
Marshall-Allen-Romer externalities, 253–4
Marx, Karl, 15, 37–8, 41, 231
Marxism, 37–9, 42, 215, 375
master plans (Africa), 469
matching (firms-workers), 245–6
Medicaid, 354–5
mega-cities, 241, 327, 385, 417, 439
mega-urban regions, 425, 428–48
men, 197–8, 206–17
Mental Deficiency Committee, 232
mercantilism, 274
meta-narratives, 376, 377
metropolitan area, 184–6, 323
    definitions, 26–31
    governance, 325, 326–33
    urbanization trends, 147–8, 150–1, 153–4, 156
Metropolitan Statistical Area (MSA), 28, 29, 30
micro-foundations (scale economies), 243–8
migration, 16, 141, 145, 148, 151–3, 157, 178, 184,
    294
modal share, 108–11, 121
modern political theory, 374–5
modernism, 56–65
monopolistic capitalism, 273
morphogenetic approach, 69
morphological periods (English house), 73–7
Moscow, 453, 455–63
multinational enterprises, 274, 276, 286
Municipal Structures Act (South Africa), 482

Nairobi, 476–7, 481
narrative approach, 12–13, 53–6, 213
Nash equilibrium, 246, 248, 252
nation states, 278, 288, 300, 302, 370
National Advisory Commission on Civil Disorders
    (USA), 184–5, 188–9
National Association of Councils, 329
National Audit Office, 360

National Institute for Urban/Regional Planning (Vietnam), 444
National Law on Environmental Policy (Vietnam), 445
National League of Cities (USA), 329
national state, 264–5
nationalism, 370
natural areas, 36, 40, 41–2, 43–6
natural environment, 54, 129–30, 131–2
natural resources, 127
natural selection, 39, 44, 47
natural will, 43–6
neighbourhoods, 235–6, 328, 359, 361
neo-classical economics, 260
neo-liberalism, 194–5, 200, 300, 303, 309, 318–19, 323, 363, 370, 417
neo-Marxism, 4, 6, 7, 39
neo-Schumpeterian approach, 274, 275
networks, 276–7
    post-industrial, 288–90
    social, 261, 316, 317, 361
    spatial, 4, 5, 7
New Deal (UK), 358, 360
New Deal (USA), 352
New International Division of Labour, 280, 403, 409, 422, 428, 436
New Right, 336, 410, 411
new social movements, 274
New Towns, 37–8, 45, 90, 164, 197, 276, 387
new urban economies, 141, 258, 296–305
new urban politics, 296, 299–300, 402–15
New York, 36, 56, 63, 165–6, 169, 258, 287–8, 326, 329, 332, 393
NGOs, 358–9, 361, 432, 480, 481
noise problems, 130
North America (transport), 119–21
North American Free Trade Association, 276, 300
novels, 12–13, 54–65, 213–14

objectivist narratives, 168–9
occupational structure, 165, 167, 170, 180
OECD, 93–4, 338–43, 351, 358, 436
OHLM (in Senegal), 470
oligopolistic capitalism, 273, 275, 395, 397
Omnibus Crime Control Act, 234–5
Ontario Municipal Board, 328
ontology, 377–8
organized crime, 224, 225
otherness, 213, 216
outsider, 53, 56–7, 201–2, 213
overurbanization, 16, 22–3
Oz, Amos, 56, 61–2

Pacific Asia, 417, 418, 419–48
Pahl, Ray, 37, 38, 165–6
Paris, 288
Park, Robert, 5, 36, 40–6, 50
Parsons, Talcott, 39
partnership approach
    in Africa, 474
    social policy, 359–60, 362, 364

PDCI party (Ivory Coast), 480
peri-urban zones, 12, 478
peripheral labour, 310–11
personal movement (in cities), 108–10
PHARE programme, 341
Philadelphia, 289–90
physical environment, 67–8, 131–2
physical form (of cities), 67–8, 69–85
place (in global economy), 257–9
place marketing, 413
planning, 6
    communicative, 369–81
    power and (conflict), 385–99
    Sub-Saharan Africa, 469–72
polarization reversal, 145, 146, 151
police, 182, 235–6, 388
policy discourses, 367–8
    communicative planning, 369–81
    planning/power (conflict), 385–99
    power and city trajectories, 402–15
political decentralization, 339
political economy, 4, 6, 296, 393, 412, 415
political power, black, 182, 190
politics
    emancipatory, 369–81
    of metropolitan governance, 331–3
    new urban, 296, 298–300, 402–15
    post-Fordist city, 278–80
pollution, 68, 115, 124, 126, 129–31, 135, 137, 390, 431, 443, 445
polynuclear city, 433–4, 439–40, 447
population, 12
    deconcentration, 26, 29, 147, 150, 152–3, 156
    Pacific Asia, 424, 426
    urban agglomerations, 21–5, 34, 285
    urbanization trends, 16–21, 144–8
post-Fordism, 168, 194, 296, 301, 336
post-Fordist city, 241, 273–81
post-industrial city, 241, 284–94, 300, 353, 417
post-industrial society, 171, 376
post-socialist cities, 417–18, 451–64
post-welfarist analysis, 336–48
postmodern cities, 367–8
    communicative planning, 369–81
    planning/power (conflict), 385–99
    power and discourses, 402–15
postmodern society, 4, 6, 7, 75, 78, 326
poverty, 49, 52, 209–10
    crime and, 182, 191
    housing and, 91, 93
    Pacific Asia, 443, 444, 448
    race and, 182, 191, 294
    social polarization, 162, 171, 173
    social policy, 351–64
    spatial concentration (USA), 188–90
    Sub-Saharan Africa, 472–3, 474
power
    discourses, 367–81, 402–15
    planning and conflict, 385–99
    space and (centrality), 263

praxis, 378
Primary Metropolitan Statistical Area (PMSA), 28
private finance (services), 343–4
Private Finance Initiative, 112, 344
private individuals, 389–91
private interests, 387–9
private markets (transport), 111–17
private property, 386, 388–9, 391–5, 398–9
privatization
    housing, 90–1, 97–9, 456–7
    public spaces, 281, 304
    retail/service property, 455
    of services, 323, 339, 341–2, 345
procedural rationality, 128, 370
production (Fordist systems), 273–8, 280–1, 369
property rights, 386, 388–9, 391–5, 398–9
public choice theory, 338
public health
    in Africa, 478
    model (crime), 232–4, 236
public institutions, 389–91, 398
public interest, 338–9, 387–9, 398
public realm theory, 377–8
public regulation (of private property), 391–5, 398
public sector services, 337, 339–44
public spaces, 281, 304
public transport, 105–17, 121, 212, 476, 477
purchase-provider split, 345–6, 347

quangos, 346–7, 411
quasi-markets, 323, 339, 342–3, 452
Quebec city, 311–12, 313

race
    crime and, 221–2, 223, 224–5
    poverty and, 182, 191, 294
    relations (in cities), 177–91
    underclass and, 171–2, 187–9, 290, 353
Race Relations Acts, 183
racism, 294, 353, 356, 357
radical planning, 372
rational planning (housing), 90–1
rationality, 375–7
RCEP Report (2000), 113, 117, 120
reality, 377–8
recentralization, 278
redistributive justice, 413, 414
regime theory, 42, 299–300
Regional Boards, 113
Regional Division of Labour, 428–9
regulation approach, 274–5, 296, 317, 319, 391–5
representations of city, 53–6
research agendas, 214–15, 261–5
residential community associations, 203–4, 345
residential development, *see* houses/housing
residualization (housing market), 164
retail sector, 7, 292, 454–5
reurbanization, 141, 143, 147–8, 153–8, 278
rhetorical narratives, 168–9
Rio Earth Summit, 116, 117, 129

risk-taking, 299, 301, 303
risk aversion, 128
RMI (in France), 356, 358
Rockefeller Institute, 285
Rotterdam, 325
rural areas (Africa), 466–7, 474, 475
Rural Challenge, 344
rural population, 11–12, 14–16, 18–21, 26–7
Russian cities, 417, 451–64

SACTRA, 102–4, 117, 118, 119
Safe Cities and Crime Concern, 235–6
safety (transport policy), 114–16
Sassen, Saskia, 165–7, 170, 287
Save Our Sons and Daughters, 235
scale economies, 260, 273, 275–6, 341
    urban, 243–54
scientific knowledge, 374, 376
scientific management, 273
search-matching externalities, 245–6
segregation
    racial, 180–2, 185–7
    residential, 4, 44–6, 141–2, 169, 180–2, 185–7, 397, 398
    social, 141–2, 162–73, 459–60
self-employment, 310–11, 313, 315, 357
SERPLAN, 390
service economy, 256, 263–4
service exchange model, 339
services sector, 277, 280, 323
    Africa, 476–7
    post-welfarist analysis, 336–48
    Russia, 454–5
settlement
    patterns, 28–9, 31, 47–9, 145–6, 152
    size, 145–6, 317
    systems, 144–6, 463
SETU (Ivory Coast), 471
sexual division of labour, 197–8
SICAP (in Senegal), 470
SICOGI (Ivory Coast), 470–1
SIJORI growth triangle, 431, 436–41
Simmel, G., 52, 195, 196
simulation effect, 378
Singapore, 428–9, 431–2, 436–41, 448
Singer, Isaac Bashevis, 54, 56–61
single-tier government, 325, 327–8
Single European Market, 276
single mothers, 164, 211–12
Single Regeneration Budget, 202, 280, 344, 360
Smith, Adam, 244, 247–8
social capital, 202
social character (urban concept), 26–7
social class, 15, 42, 52, 141, 196–7, 409
    gender and, 206–7, 210
    housing and, 44, 73, 75, 77–8, 81, 106, 157
    segregation/polarization, 162–73
    see also underclass
social conflict (NUP discourse), 412–14
social contributions, 315–16
social Darwinism, 39

social democracy, 363
social engineering, 50, 90, 375
social environment, 130–2
social evolution, 38–9
social exclusion, 351, 353–4
social housing, 44–5, 88, 90–3, 96–8, 100, 164, 359
social justice, 163, 207, 209, 215–17, 364, 370, 374, 377–81
social networks, 261, 316, 317, 361
social organization of cities, 36–51
social polarization, 89, 141–2, 162–73, 260, 264, 280–1, 367
social policy/problems, 351–64
social segregation, 141–2, 162–73, 459–60
socialism, 37–8, 50, 90, 141
    post-socialist cities, 451–64
society
    changes affecting planning, 370–1
    economy/environment relation, 129–33
    see also civil society
socio-cultural approach, 4, 5
socio-economic mix, 316
socio-political approach, 4, 6
sociological approach, 4, 5–7
SOGEFIHA (Ivory Coast), 471
solid waste disposal, 130
SOTRA (in Abidjan), 477
SOTRAC (in Dakar), 477
South Africa, 333, 481–2
Southall Black Sisters, 208–9
space, power and, 263
space syntax, 48–9, 50
spatial agglomeration, 261–2
spatial approach, 4–5, 6–7, 11–12
spatial forms, social logic of, 46–9
spatial networks, 4, 5, 7
specialization, 15, 244, 247–50, 254, 258, 273
Spencer, Herbert, 39
sports, 301–2, 402, 405
squatting (regulation), 470–2
Standard Consolidated Area, 28
Standard Consolidated Statistical Areas, 28
Standard Metropolitan Area, 28, 149
Standard Metropolitan Labour Area, 149
Standard Metropolitan Statistical Area, 28, 325–6
statist policies (Africa), 472–3
structural adjustment programmes, 3, 323–4, 363, 420, 441, 444, 446, 474, 478
structural functionalism, 39, 42
Students Against Violence Everywhere, 235
Stuttgart, 325, 330
sub-contracting, 315, 421
sub-national units, 257, 265
Sub-Saharan Africa, 12, 417, 466–82
Subdivision Map Act, 394
subsidiarity, 340–1
suburbanization, 15–16, 89–90, 106, 124, 185, 275, 278, 325
    developments, 141, 143, 147–51, 153, 156, 158, 163
    Russia, 461–2, 464

Supplemental Security Income, 355
supply-side policy, 338–9, 344, 348
sustainable development, 124–37, 374
Sydney, 200–1, 402

Take Back the Night, 235
taxation, 112–13, 116, 117, 314–16, 395, 396
Taylorism, 273
technocracy, 274, 286, 300, 370
technology, 370–1
technopoles, 297–8, 301, 326
telecommunications, 266, 292, 297
telematics, 256, 261–4, 297, 302, 303
Thailand, 430, 432, 441–3, 448
time, role of, 133–6
Tokyo, 287–8, 429–36, 448
Tönnies, Ferdinand, 2, 3, 43–5, 195–6
Toronto, 156, 325, 328
total quality management, 339
Town and Country Planning Acts, 49, 81, 83, 291, 386, 391–2, 394, 395
trade (Pacific Asia), 419–21, 423–4
training, 357–8
    TECs, 330, 346, 411
Trans-European Network strategies, 104
transborder regions, 428–9
transition process, 252
transitional cities/economies, 97, 296, 417–18
    Pacific Asia, 419–48
    post-socialist, 451–64
    Sub-Saharan Africa, 466–82
transnational corporations, 403
transnationality, 257–8, 263, 265, 286–8
transport policies, 67–8, 102–21, 134
truth (in planning), 372–4
Tudor Walters Report (1918), 70, 75
two-tier government, 325, 328
Tyne and Wear Development Corporation, 202–3

underclass, 158, 210, 277, 290, 304–5, 370, 397
    crime and, 220, 230–2
    emerging (USA), 187–9
    social polarization, 162–4, 171–3
    social policy, 353–5
unemployment, 49, 210, 273, 290–1, 298, 301
    Africa, 475
    crime and, 224–5, 229, 234
    housing and, 93, 100
    informal employment and, 310–14, 316–19
    race and, 180, 187
    social polarization and, 170–3
    social policy, 351–3, 355, 357–8, 363–4
unitary development plans, 389, 392
United Nations, 144, 227–9, 304
    Centre for Human Settlements, 16, 89, 341, 474
    Conference on Human Environment, 124
    UNICEF, 351
urban agglomerations, 14, 21–5
urban area (defining), 25–9

urban development, 452
   stages, 145–8, 150–1, 153–6
Urban Development Corporations, 202–3, 411
urban ecology, 36–51
urban economies
   informal, 308–20
   new, 296–305
urban environment, sustainable, 124–37
urban governance, 323, 325–34, 403–6
urban growth (trends), 14–25
urban hierarchy, 276–8
urban landscape, 69, 80, 83–5
urban managerialism, 6, 37–8, 42, 163
urban networks, *see* networks
urban planning, *see* planning
urban politics, *see* politics
Urban Programme partnerships, 360
urban scale economies, 243–54
urban studies, 1–7, 406–9
urban sustainability, 124–37
urban transport, 67, 68, 102–21, 134
urbanism, 27, 52, 53, 56, 225–6
urbanization, 2–3, 12, 14–15, 196, 225–6, 249
   definition, 25–9
   developments, 141, 143–59
   externalities, 253–4
   growth (trends), 16–21
   housing and, 88, 92, 99
   movement and, 105–11
   Pacific Asia, 424–31, 446
   post-welfarist analysis, 336–7
   Tokyo, 434–6
USA
   crime, 220–3, 225–7, 228–30, 232–7
   race relations, 177–8, 184–91
   social policy, 354–7, 361–3
   urban areas, 27–9
   see also individual cities
Use Classes Order, 393
use value, 413
user-charges (public services), 343–4
Uthwatt (1942) Final Report, 395

VASE model, 135–6
verbal environment (in novels), 56
vertical integration, 347, 438

Vietnam, 432, 443–6, 447
vigilantism, 208–9
violence, 190, 212
violent crime, 189, 220–1, 223–30, 232–6, 362
voluntarism, 347, 363
voluntary sector, 348

wages, 353, 363, 466–7, 475
   informal/formal work, 311–13, 315
Warsaw, 54, 56, 57, 58–61
water pollution, 130
Weber, Max, 40, 52
welfare benefit regulations, 317
welfare regimes, 354–60
welfare state, 49–50, 167, 170, 230, 280, 303, 369, 387
   informal economies, 309, 318–19
   post-welfarist analysis, 336–48
   social policy, 351–64
welfare to work policy, 319, 355–6, 357, 362–3
welfarism, 98
   post-welfarist analysis, 336–48
Western Europe (crime), 220, 223–30, 232–7
women, 197–8, 206–17, 311–13, 380, 475
Women's Design Service, 212
Woolf, Virginia, 56, 58, 206–7
Working Families Tax Credit, 355
world-view (of planners), 372–4, 376
World Bank, 276, 341, 420, 427, 432–3, 441, 444, 446, 469, 474, 476
world cities, 3, 15, 241, 286, 288, 297, 326, 332, 402–3, 428–9
   duality, 158, 162, 164–8, 170, 287, 304–5
   in global economy, 256–67
World Health Organization, 234
World Teleport Association, 302

X-efficiency, 338–9, 343, 346

Young, Iris Marion, 216–17
young people, crime and, 222, 225–6
Youth Violence Prevention Programmes, 235, 236

zones/zoning, 31, 69–70, 73, 77–8, 106–7, 388, 390, 393–6, 398–9
ZUPCO (in Harare), 477